The Physiological Ecology of Woody Plants

Physiological Ecology

A Series of Monographs, Texts, and Treatises

KOZLOWSKI, T. T. Growth and Development of Trees, Volumes I and II, 1971

HILLEL, D. Soil and Water: Physical Principles and Processes, 1971

YOUNGER, V. B., and MCKELL, C. M. (Eds.) The Biology and Utilization of Grasses, 1972

KOZLOWSKI, T. T., and AHLGREN, C. E. (Eds.) Fire and Ecosystems, 1974

MUDD, J. B., and KOZLOWSKI, T. T. (Eds.) Responses of Plants to Air Pollution, 1975

DAUBENMIRE, R. Plant Geography, 1978

SCANDALIOS, J. G. (Ed.) Physiological Genetics, 1979

MURRAY, B. G., JR. Population Dynamics: Alternative Models, 1979

LEVITT, J. Responses of Plants to Environmental Stresses, Second Edition. Volume I: Chilling, Freezing, and High Temperature Stresses, 1980. Volume II: Water, Radiation, Salt, and Other Stresses, 1980

LARSEN, J. A. The Boreal Ecosystem, 1980

GAUTHREAUX, S. A. JR. (Ed.) Animal Migration, Orientation, and Navigation, 1981

VERNBERG, F. J., and VERNBERG, W. B. (Eds.) Functional Adaptations of Marine Organisms, 1981

DURBIN, R. D. (Ed.) Toxins in Plant Disease, 1981

LYMAN, C. P., WILLIS, J. S., MALAN, A., and WANG, L. C. H. Hibernation and Torpor in Mammals and Birds, 1982

KOZLOWSKI, T. T. (Ed.) Flooding and Plant Growth, 1984

RICE, E. L. Allelopathy, Second Edition, 1984

CODY, M. L. Habitat Selection in Birds, 1985

HAYNES, R. J., CAMERON, K. C., GOH, K. M., and SHERLOCK, R. R. Mineral Nitrogen in the Plant–Soil System, 1986

The Physiological Ecology of Woody Plants

THEODORE T. KOZLOWSKI

Environmental Studies Program and
Department of Biological Sciences
University of California
Santa Barbara, California

PAUL J. KRAMER

Department of Botany
Duke University
Durham, North Carolina

STEPHEN G. PALLARDY

School of Natural Resources
University of Missouri
Columbia, Missouri

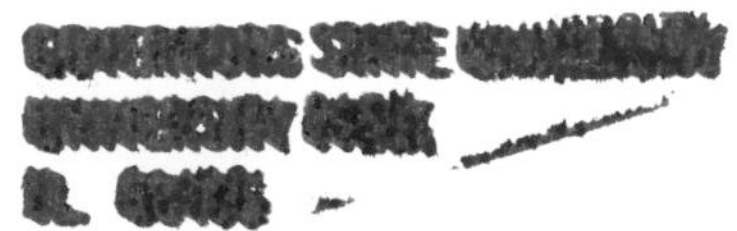

ACADEMIC PRESS, INC.
Harcourt Brace Jovanovich, Publishers
San Diego New York Boston
London Sydney Tokyo Toronto

Cover photograph: Englemann spruce–subalpine fir stand with several age classes of competing trees in the Rocky Mountains. Courtesy of the United States Forest Service.

Corrections for reprint.
Kozlowski: The Physiological Ecology of Woody Plants

This book is printed on acid-free paper. (∞)

ACADEMIC PRESS, INC.
1250 Sixth Avenue
San Diego, California 92101

United Kingdom Edition published by
Academic Press Limited
24–28 Oval Road, London NW1 7DX

Library of Congress Cataloging-in-Publication Data

Kozlowski, T. T. (Theodore Thomas). Date.
The physiological ecology of woody plants / Theodore T. Kozlowski, Paul J. Kramer, Stephen G. Pallardy.
p. cm. -- (Physiological ecology.)
Includes bibliographical references and index.
ISBN 0-12-424160-3 (alk. paper)
1. Woody plants--Physiological ecology. 2. Trees--Physiological ecology. I. Kramer, Paul Jackson, Date. II. Pallardy, Stephen G. III. Title. IV. Series.
QK905.K69 1990
582.1'5045--dc20 90-795
CIP

PRINTED IN THE UNITED STATES OF AMERICA
92 93 94 95 96 EB 9 8 7 6 5 4 3 2

Contents

Chapter 7

Water Stress

Chapter 9

Air Pollution

Chapter 10

Carbon Dioxide

Chapter 12

Wind

Chapter 13

Cultural Practices

Preface

This book was written for use as a text by students and teachers and as a reference for investigators and growers who desire a better understanding of how woody plants grow and communities of woody plants are established and develop. Because of its interdisciplinary scope, the book will be useful to a wide range of plant scientists including agronomists, arborists, plant ecologists, foresters, horticulturists, geneticists, plant breeders, plant physiologists, soil scientists, and landscape architects.

The book is based on the premise that efficient management of growth of shade, orchard, and forest trees and other woody plants depends on our understanding of the physiological processes that control growth, the environmental complex that controls those processes, and our ability to modify the environment to maintain conditions that will be conducive to favorable rates of physiological processes. Accordingly this book:

1. emphasizes the interactions of heredity and environment in influencing growth of woody plants, and points out the importance of various environmental factors and the interactions among them on growth;
2. outlines differences in responses of individual trees and of communities of trees to environmental stresses. The primary emphasis is on the impact of environmental stress factors (alone and in combination) on growth and development of communities of woody plants; and
3. provides information about various cultural practices useful for efficient management of shade, forest, and fruit trees as well as shrubs and woody vines.

The first chapter describes growth characteristics of woody plants, emphasizes the importance of the physiological processes that are the critical intermediaries through which heredity and environment interact to influence growth, and introduces the complexity of environmental control of woody plants. The second chapter describes the compounds essential for plant growth, including foods (carbohydrates, proteins, lipids), water, mineral nutrients, and hormonal growth regulators. Separate sections are devoted to sources and functions of these compounds.

The third chapter deals with stand regeneration, seedling establishment, and subsequent growth and development of stands of trees. Considerable attention is given to competition among canopy trees and between canopy trees and subordinate vegetation for resources (light, water, mineral nutrients). Also included are sections on plant succession and accumulation and partitioning of biomass.

Chapters 4 to 12 deal with effects of important environmental factors on physiological processes and on vegetative and reproductive growth of communities of woody plants. Separate chapters are devoted to radiation (light intensity, light quality, and duration of exposure); temperature; soil physical properties and mineral nutrition; water deficits; soil aeration, compaction, and flooding; air pollution; carbon dioxide; fire; and wind. Special attention is given to environmental preconditioning and interactive effects of various environmental stress factors on growth.

The final chapter describes a variety of cultural practices that can be employed to improve propagation and production of planting stock in the nursery and growth of established landscape trees, fruit trees, and forest trees in plantations and natural forests. The cultural practices discussed include site preparation treatments (slash disposal, prescribed burning, harrowing, disking, bedding, use of herbicides, and drainage), fertilizer application, irrigation, thinning of forest stands, pruning of branches and roots, use of chemical growth retardants and pesticides, and integrated pest management. A final section on planting for high yield discusses short-rotation forestry, agroforestry, and high density fruit and nut orchards.

A summary list of general references has been included at the end of each chapter and papers cited in the text are listed in the Bibliography at the end of the book. We have selected significant references from a voluminous world literature in order to make the work authoritative, well-documented, and up-to-date. Where appropriate, we have presented contrasting views and often have given our personal conclusions, on the basis of the weight of evidence, on controversial issues. As new research data become available, some conclusions will require revision, and we hope that readers will join us in modifying their views when changes are appropriate.

In the text we have used common names for most well-known species of plants. Scientific names are used for a few unusual plants that do not have widely used names. Separate lists of common and scientific names, and scientific and common names, are given following the text. Names of North American forest trees are largely based on E. L. Little's *Check List of Native and Naturalized Trees of the United States* (1979), Agriculture Handbook No. 41, U.S. Forest Service, Washington, D.C. However, to facilitate use of the common name index for a wide audience, we chose not to employ the rules of compounding and hyphenating recommended by Little (1979). These rules, while reducing taxonomic ambiguity in common names, often result in awkward construction and unusual placement within an alphabetical index. Names of other species are from various sources.

We express our appreciation for the contributions of the many people who assisted directly and indirectly in the preparation of this book. Much information and stimulation came from our graduate students, research collaborators, and from arborists, foresters, plant ecologists, horticulturists, and plant physiologists all over the world with whom we have worked and discussed problems.

Various chapters were read by C. E. Ahlgren, N. L. Christensen, R. Oren, P. B. Reich, W. H. Schlesinger, and W. E. Winner. However, the text has been revised since they read and commented on individual chapters, and they should not be held responsible for errors that may occur. W. Ferren, R. Haller, R. Miller, C. H. Muller,

J. L. Rhoads, and R. M. Wilbur assisted with preparation of the list of scientific names of species.

T. T. Kozlowski
P. J. Kramer
S. G. Pallardy

Chapter 1

How Woody Plants Grow

Introduction

The objective of this book is to explain how individual trees and shrubs grow, how they develop into communities, and how environmental factors and cultural practices affect the quantity and quality of growth. Ecologists, arborists, foresters, horticulturists, and gardeners know that some kinds of trees and shrubs grow better than others on dry sites; that only a few thrive in wet soil; and that cultural practices such as thinning, pruning, fertilization, irrigation, and drainage, used properly, often improve growth. However, they seldom know why plants respond as they do to environmental stresses and cultural treatments. Over a century ago Johnson (1868) wrote that in order to grow plants efficiently it is necessary to understand how plants grow. Such an understanding is even more important today because of the increasing complexity of the environment caused by industrial air pollution and by the increasing concentration of atmospheric CO_2 and associated changes in climatic conditions. Also, forest trees are being treated more as crop plants (Cannell and Jackson, 1985). Furthermore, modern biotechnology is increasing the possibility of producing superior genotypes if the desirable attributes can be linked to specific genes. Thus there are increasing pressures from several directions for a better understanding of the interaction between the physiological processes of woody plants and their environment (Osmond *et al.*, 1987).

Particularly important to ecologists and foresters is an understanding of the factors involved in successful competition. For example, why are plants such as birch and white or loblolly pine, which are successful pioneers on recently disturbed land, unable to perpetuate themselves and therefore are succeeded by other species? Will the increasing concentration of atmospheric CO_2 and associated climatic changes affect the competitive capacity of competing species differently and change natural succession or the choice of species for planting? Will shorter rotations increase the severity of mineral deficiencies and the need for fertilization? Why do some tree seedlings and shrubs thrive in the shade where others fail? Answers to such questions require an understanding of how plant growth is affected by environmental factors, and in this book we attempt to provide physiological explanations for those effects.

Heredity and Environment

The growth of woody plants, like that of all other organisms, is controlled by their heredity and environment, operating through their physiological processes as shown in Fig. 1.1. Although all trees exhibit some common growth characteristics, such as increase in height and stem diameter, they also show considerable variability in growth because of their hereditary differences. Genetic variations in size, crown form, straightness of stems, wood density, leaf retention, and longevity are particularly well known. Less obvious but often equally important are hereditary differences in rates of growth, winterhardiness, drought tolerance, flooding tolerance, and disease and insect resistance. Genetic variations in physiological and growth characteristics are responsible for the differences between tropical and temperate zone trees and evergreen and deciduous trees, as well as differences among and within species (e.g., differences in growth among clones, ecotypes, and seed sources). For example, Arizona–New Mexico sources of white fir grew 50 to 100% faster than Utah–Colorado sources when planted in the eastern United States (Wright, 1976). Such differences are important in establishing plantations of Christmas trees or trees for pulp and timber because trees from some sources can be harvested at half the age of trees from other seed sources. Important differences have been found in growth, cold and drought tolerance, length of growing period, and other characteristics of seedlings grown from seed of the same species obtained from various geographic regions or provenances (e.g., Lester, 1970; Scholz and Stephan, 1982).

Experience also shows that trees and other plants sometimes grow much better in a new and different environment than in their native environment (Jones, in Cannell and Jackson, 1985, p. 69). Monterey pine from California and several species of

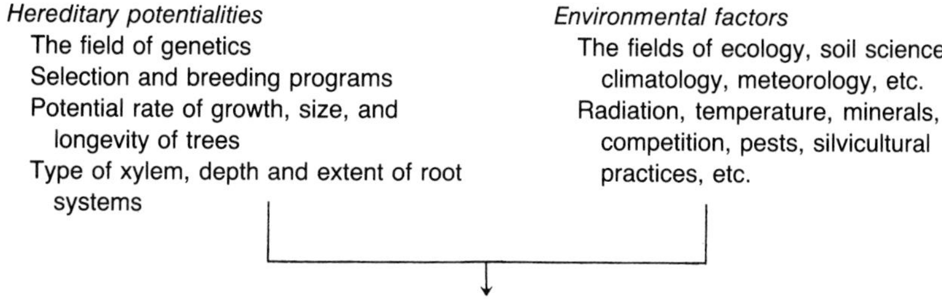

Physiological processes and conditions
The field of plant physiology
Photosynthesis, carbohydrate and nitrogen metabolism
Respiration, translocation
Plant water balance and its effects on growth and metabolism
Growth regulators, etc.

Quantity and quality of growth
The field of arboriculture, forestry, and horticulture
Amount and quality of wood, fruit, or seed produced
Vegetative versus reproductive growth
Root versus shoot growth

Figure 1.1 Diagram showing how the hereditary potentialities of plants and the environment operate through physiological processes and conditions to determine the quantity and quality of plant growth. (From Kramer and Kozlowski, 1979, by permission of Academic Press.)

pine from the southeastern United States grow better in New Zealand than in their native habitats (Jackson, 1965). In contrast, Rehfeldt (1986) claimed that in central Idaho, ponderosa pine should not be planted more than 290 km from the seed source, and Lowe *et al.* (1977) reported that balsam fir from local seed grows best in northern New England. However, several North American tree species grow well in other parts of the world, such as Sitka spruce in Scotland and Germany and bald cypress in China. The early history of successes and failures in the introduction of exotic trees was reviewed by Moulds (1957), and Zobel *et al.* (1987) published an interesting book on exotic trees.

The environment determines the extent to which the hereditary potentialities of plants are attained, as shown by the differences in size of trees of the same species growing in moist fertile soil and in dry infertile soil. An extreme effect of environmental and cultural conditions on growth is seen in the dwarf trees (bonsai) of China and Japan, which have been so dwarfed by cultural treatments that they are less than a meter high after a century or more, although their normal height may be 15 or 20 m (Fig. 1.2). Another example is the short life expectancy for urban trees (10 years

Figure 1.2 A bonsai of eastern red cedar dwarfed by a restricted root system. In nature, eastern red cedar grows to heights of 50 to 75 ft or more (15–23 m). (Photo courtesy of Arnold H. Webster.)

for streetside trees) compared to that of the same kinds of trees growing in the forest (Foster and Blaine, 1978).

The physiological processes of plants comprise the machinery through which cultural treatments—such as thinning, fertilization, and irrigation—and environmental stresses—such as drought, flooding, abnormally high or low temperatures, air pollution, insect pests, and diseases—influence growth. If a tree breeding program or a cultural treatment increases growth, it does so by improving the functioning of the physiological machinery. The environmental changes that alter growth do so through their influence on rates and balances among such processes as photosynthesis, respiration, hormone synthesis, absorption of water and minerals, and translocation of substances needed for growth (including carbohydrates, nitrogen compounds, hormones, water, and mineral nutrients). Furthermore, reduction of growth or death of trees is preceded by a series of abnormal physiological events. Hence, some understanding of the physiological processes involved in growth and the manner in which such processes respond to variations in environmental factors is essential to formulation of management programs for growing trees and shrubs efficiently.

The importance of the role of physiological processes in the response of plants to changes in environmental factors and to cultural practices can be illustrated by a few examples:

1. Defoliating insects damage plants by reducing the amount of foliage available for photosynthesis. See Chapter 2.
2. Fungal pathogens damage leaves and cause their premature shedding (abscission), sometimes block the movement of water in the xylem or carbohydrates in the phloem, and produce toxins that adversely affect physiological processes. See Chapter 2.
3. Abnormally short nights caused by artificial lighting retard development of dormancy in some trees and shrubs, resulting in winter injury. See Chapter 4.
4. Fertilization supplies the mineral nutrients associated with enzymes and the formation of new protoplasm and cell walls. Such nutrients also have an important role as buffers and as solutes that maintain osmotic pressure in cells. See Chapter 6.
5. Irrigation prevents water deficits that inhibit cell expansion, induce stomatal closure, and reduce photosynthesis. See Chapter 7.
6. Compaction or flooding of soil reduces availability of soil oxygen and inhibits root respiration, thus decreasing root growth and absorption of water and mineral nutrients and affecting the production of certain hormones. Anaerobic soil conditions also favor activity of fungi that cause death of roots and suppress development of mycorrhizae, further decreasing uptake of mineral nutrients and water. See Chapter 8.
7. Thinning of tree stands reduces competition for water and minerals among the remaining trees and postpones the reduction in growth that is characteristic of aging overstocked stands. See Chapters 3 and 13.

Growth Characteristics

Before going into a detailed discussion of the physiological processes and environmental conditions controlling the growth of woody plants, we will discuss some characteristics of growth and yield itself. Foresters are interested primarily in yield of wood, whereas horticulturists and gardeners are interested in fruit, flowers, or aesthetically pleasing plants. In any event, yield is the result of growth, which involves a complex series of processes beginning with the formation of new cells and progressing through the differentiation of cells into tissues, and finally integration of the tissues into the organs of a plant. Growth usually is measured as increase in height, stem diameter, or dry or fresh weight, and yield in terms of the commercially valuable product such as volume of wood, fruit, or nuts, number and quality of flowers, or the aesthetic success of ornamental shrubs and trees. Growth and high-quality yield are not necessarily closely correlated. More rapid tree growth can be obtained by wider spacing, but at the cost of reduction in yield per acre and possibly in quality of wood.

Growth of a shrub or tree can be considered analogous to a complex biochemical factory that builds itself from its own products, following a pattern provided by its heredity. The chloroplasts represent the machines that produce the primary carbohydrates in photosynthesis. Carbon dioxide and water are the raw materials, and sunlight supplies the energy. The products of photosynthesis go through intermediate processes, including fat and protein synthesis, and eventually some are transformed into new tissues.

At the cellular level, growth involves cell division, enlargement, and maturation into the various tissues constituting a plant. At the whole-plant level, it involves integration of tissues into the various organs of the plant; the roots, stems, and leaves, and eventually flowers, fruits, and seeds. Cell differentiation and maturation seem to be regulated by organizing factors that influence form and function by controlling relative rates of growth and degrees of differentiation. As all cells in a tree have the same genome, new cells formed in a meristematic region might develop into any kind of tissue. However, cell differentiation is so controlled that cells in various parts of plants develop quite differently, resulting in the diverse tissues of roots, stems, and leaves. The formless masses of cells often found in tissue cultures show what occurs in the absence of control by the proper hormonal growth regulators. Usually such masses of undifferentiated cells can be induced to differentiate into plantlets only by applying the proper combination of growth regulators in the appropriate sequence (Henke *et al.*, 1985; Durzan, in Kossuth and Ross, 1987). This situation supports de Bary's aphorism: "The plant forms cells, not cells the plant" (Barlow, 1982). At the physiological level, growth processes are dependent on a supply of food provided by photosynthesis, followed by a complex series of biochemical processes ending in the production of new protoplasm and its products. As stated earlier, all of this is under control of growth regulators and biochemical and physical equilibria that tend to keep various processes and structures in balance. This will be discussed in more detail in Chapter 2, but the complex constellation of processes involved in growth and development presents many puzzling problems.

Location of Growing Regions

In plants, unlike in animals, growth is localized in apical and lateral meristems, tissues that contain cells with a capacity to divide. In woody plants, apical meristems are localized in stem and root tips. Height growth of trees and elongation of branches result from activity of apical meristems contained in buds. Growth in diameter of stems, branches, and major roots results from activity of the vascular cambium, a thin sheath of meristematic tissue located between the wood and bark. Cambial cells divide and the daughter cells on the inside become xylem (wood), whereas those on the outside become phloem (bark). In this manner new layers of wood are inserted each year between the previous year's layers of wood and bark, causing an increase in diameter of stems and branches. Because of this mode of growth, a tree stem consists of annual increments of wood, the annual rings, each new one being added outside of that of the preceding year. A longitudinal section of a tree stem shows annual increments of wood as a series of overlapping cones (Fig. 1.3). A small amount of increase in stem diameter is traceable to the activity of a cork cambium (phellogen), but the outer bark tends to be shed so, with a few exceptions such as in redwood, the bark usually remains relatively thin.

The annual rings of wood in a stem cross section can be distinguished because of differences in size of the cells produced early and late in the growing season. The wood formed early in the season (earlywood or springwood) has larger cells with thinner walls and a lower density than the wood formed later in the season (latewood or summerwood). Thus annual rings often are prominent in stem cross sections because of differences in cell diameter and density of the earlywood of one year, which is next to the smaller, last-formed latewood cells of the previous year (Fig. 1.4). Although trees of the temperate zone generally produce one ring of wood each year, they sometimes produce more than one. In the tropics, however, trees may grow continuously throughout the year and may not produce distinct annual rings of wood (Fig. 1.5).

Growth follows the same sequence of processes in apical and lateral regions, beginning with cell division and followed by cell enlargement and differentiation into the tissues characteristic of each organ. An important difference between apical and lateral growth is that cell division in apical meristems occurs chiefly in a plane perpendicular to the long axis of roots and stems, thus adding to their length, whereas division in lateral meristems occurs primarily in a plane parallel to their length, which adds to stem diameter. Other meristems develop at various locations in roots and stems, giving rise to branches. The meristems that produce branch roots originate in the pericycle, a ring of cells lying just outside the vascular tissue. In seed plants, stem branches usually arise from meristematic regions in axillary buds that originate in the axils of leaves. In addition, meristems arise in various regions

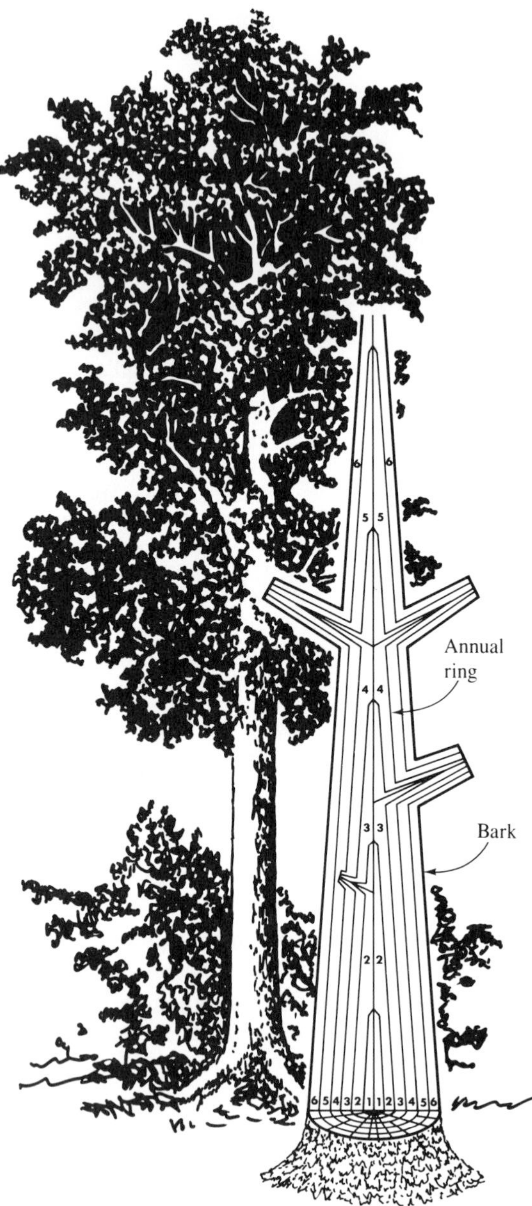

Figure 1.3 Diagrammatic median longitudinal section of a tree showing the pattern of annual xylem increments or "rings" in the stem and major branches. (From Kramer and Kozlowski, 1979, by permission of Academic Press.)

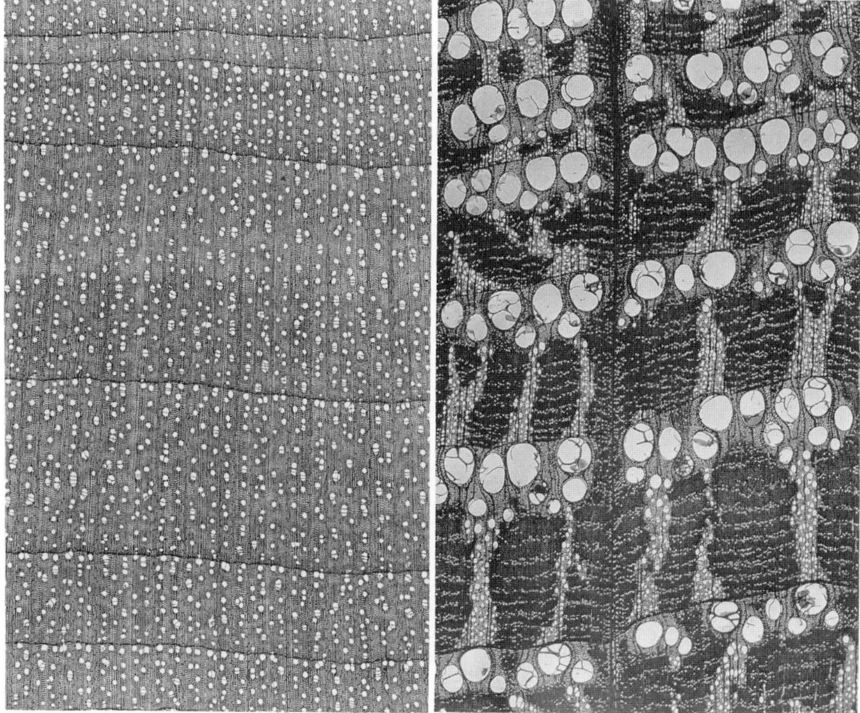

Figure 1.4 Cross sections of stems of a diffuse porous species, silver maple (*left*), and a ring porous species, white oak (*right*), showing differences in diameter and distribution of vessels in springwood and summerwood. (U.S. Forest Service photograph from Kramer and Kozlowski, 1979, by permission of Academic Press.)

of trees and shrubs as a result of wounding, flooding of soil, and other stimuli that produce adventitious roots and stems. The rooting of cuttings is an important example of adventitious root development. Most stump sprouts and epicormic branches (water sprouts) develop from dormant buds that were produced during primary growth but did not develop into visible branches until subjected to a special stimulus such as severe pruning, removal of the treetop, or increased exposure of the stem to light by thinning a stand. Tree growth is discussed in detail in the two-volume monograph by Kozlowski (1971a,b).

Variations in Rates of Growth and Apical Dominance

The relative rates of growth of the various stem tips of a tree or shrub control its shape. For example, in most conifers the apical meristems (buds) of terminal shoots grow more rapidly than those of lateral branches, resulting in their typically conical shape. Occasionally, apical dominance is lost with increasing tree age, resulting in

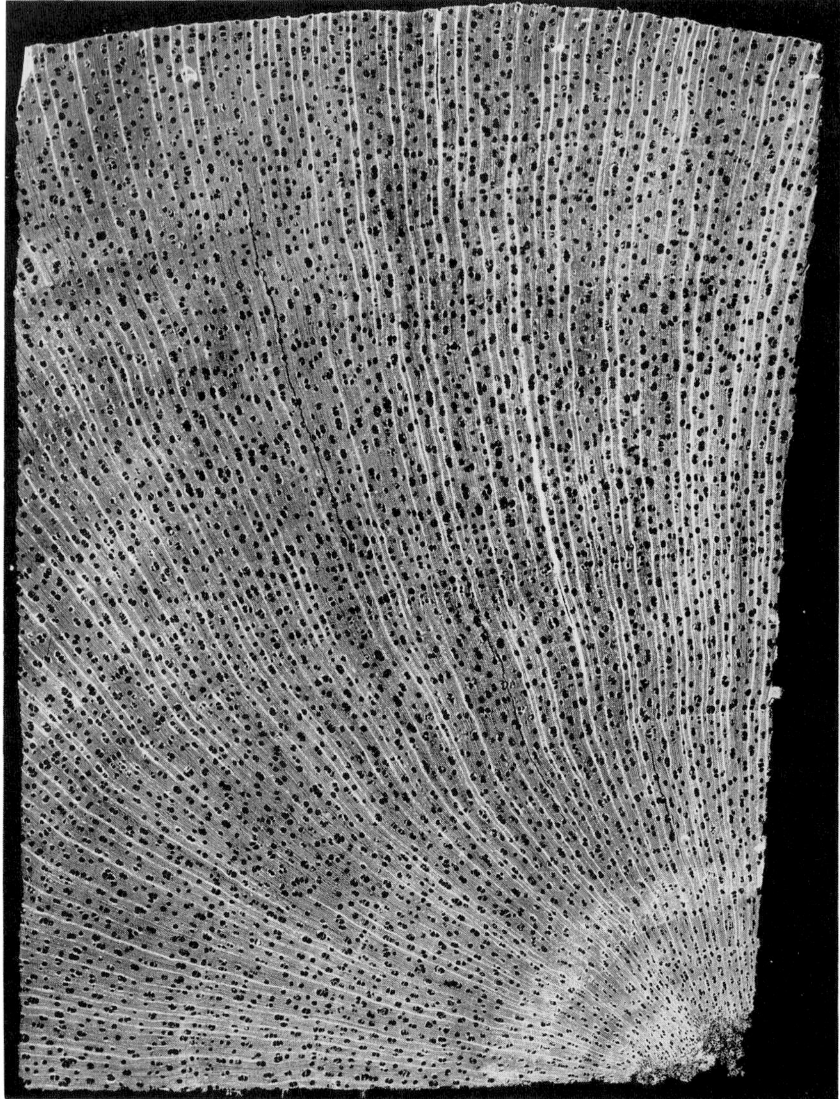

Figure 1.5 Cross section of stem of *Boswellia serrata,* a tropical tree, showing indistinct growth rings. (From Kozlowski, 1971b, p. 75, by permission of Academic Press.)

flat-topped trees such as stone pine. In most angiosperms there is less difference in the rate of growth among various shoot meristems, resulting in more rounded treetops. American elm, for example, is notable for its deliquescent form, characterized by a broad, spreading top. Its characteristic crown form develops because the terminal buds of elm die and numerous axillary buds develop into branches. Occasionally, upright or conical forms of angiosperms are found, such as Lombardy poplar and some selections of English oak and linden, which are useful in ornamental plantings. Some examples of tree form are shown in Fig. 1.6, and apical regulation of shoot growth is discussed by Kozlowski (1971a, pp. 282–295).

Differences in apical dominance and growth form are important to arborists and landscape designers because they enable them to create various visual effects. Differences in tree form must also be taken into account by horticulturists with regard to pruning, spraying, exposure of fruit to sun, and fruit picking. Apical dominance is very important to foresters, who desire strong, upright stems, because loss of the terminal bud of the main stem by insect or other injury results in undesirable stem forking. In contrast, growers of Christmas trees shear young trees

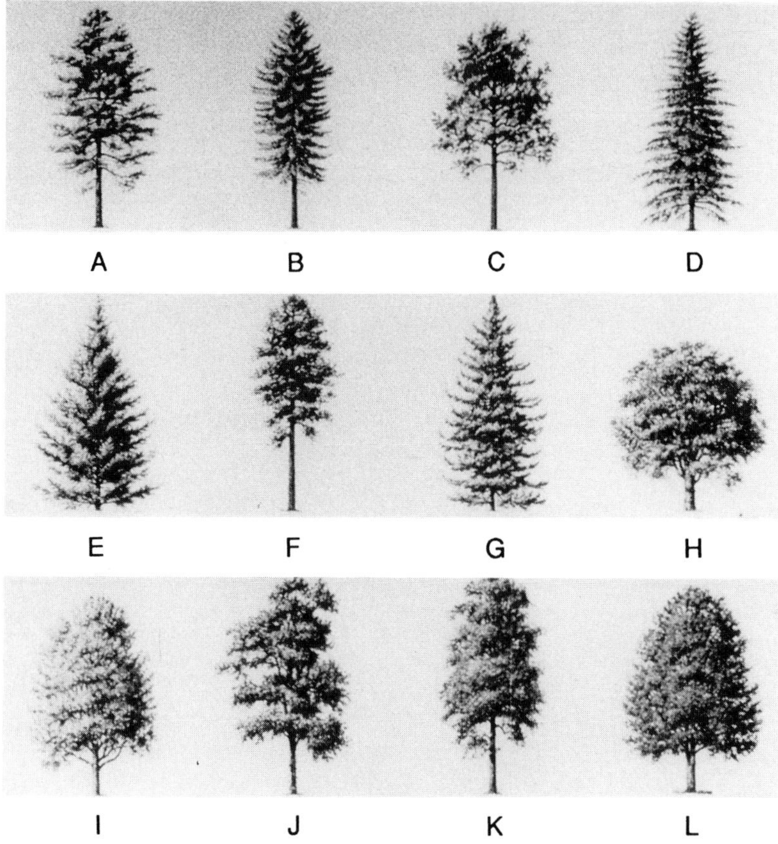

Figure 1.6 Variation in form among 12 species of open grown trees. (A) Eastern white pine; (B) Douglas-fir; (C) longleaf pine; (D) eastern hemlock; (E) balsam fir; (F) ponderosa pine; (G) white spruce; (H) white oak; (I) sweet gum; (J) shagbark hickory; (K) tulip tree; (L) sugar maple. (Courtesy of St. Regis Paper Co.)

to discourage apical dominance and encourage development of lateral buds and dense branching (Fig. 13.12). Heavy pruning of apical branches is also used to stimulate dense branching of ornamental shrubs, especially those used for hedges. Growth of trees in dense stands tends to suppress lateral branches by shading and encourages apical dominance. However, by careful pruning, a row of closely planted trees can be trained into a dense hedge that retains its lower branches (Fig. 1.7). There is currently a trend toward growing fruit trees in rows and pruning them into hedgelike arrangements that can be shaped and sprayed more efficiently than trees standing alone (Jackson, 1985) (see Chapter 13).

Duration of Growth

The duration of growth refers both to the length of the annual growing season and to the length of life of a tree or shrub. While plants are classified as annuals, biennials, or perennials on the basis of length of life, in this book we are interested only in woody perennials.

Life Span

There is a wide difference in the life span of various species of woody plants, ranging from a few decades to a few thousand years. Peach trees are old at 20 to 30 years, oaks often live to be several hundred years old, bald cypress may reach 1600,

Figure 1.7 An apple orchard with trees trained as a hedge. (From Westwood *et al.*, 1976.)

coast redwoods 3000, and bristlecone pine at least 5000 years. Some data on tree life spans are given by Harcombe (1987). Aging forest trees have some common characteristics such as reduced rates of growth, slow wound healing, decreased ratio of photosynthetic to nonphotosynthetic tissue, and increase in top dieback. However, it is not clear why apple trees typically live longer than peach trees and oak trees live longer than apple trees.

Good opportunities exist for some interesting and useful research on plant aging. Among the interesting problems is the fact that individual trees such as Lombardy poplar and peach are shortlived although the cultivars have been propagated vegetatively for many decades or even centuries. Lombardy poplar is said to have originated as a mutation in northern Italy between 1700 and 1720, and the genotype has been propagated successfully by cuttings for nearly three centuries even though individual trees are rather short-lived. Perhaps short-lived trees are more susceptible to injury from environmental and biotic stresses than long-lived trees. Loehle (1988a) suggested that slowly growing trees probably live longer than rapidly growing trees because they allocate more photosynthate to protective characteristics such as thick bark and decay-inhibiting chemicals. Some of the changes associated with aging were discussed by Kozlowski (1971a, Chapter 4) and plant senescence was discussed in detail by Thimann (1980).

Seasonal Span of Growth

Although the length of the growing season of woody plants in the temperate zones is controlled in general by temperature, photoperiod, and rainfall patterns, other factors seem to be involved. Shoot growth of most woody plants begins in the spring before the danger of frost is past, and this results in occasional damage to leaves and flowers from late frost. However, there are wide differences among species in the proportion of the potential growing season actually used and shoot growth ceases in some species long before temperatures are low enough to be the cause (Fig. 1.8). This suggests that length of growing season is also affected by internal factors. In North Carolina, red pine and eastern white pine seedlings imported from New York State and planted out-of-doors made most of their height growth in one flush in April and ceased growth by midsummer, whereas loblolly, slash, and shortleaf pine seedlings made approximately the same amount of shoot growth in several flushes over a period of more than 4 months. Likewise, white ash and various species of oaks make most of their height growth in one flush early in the season, whereas tulip poplar grows most of the summer. In general, shoot growth is said to occur over a longer season in young trees than in older trees (Kozlowski, 1971a, pp. 117–163), but Reich *et al.* (1980) observed a shorter growing season for white oak seedlings than for canopy trees in Missouri.

In the lowland tropics, where temperature is never limiting, the length of the growing season often is restricted by the occurrence of dry seasons during which water becomes limiting. Even where growth is not limited by a dry season, tropical trees and shrubs usually show marked periodicity in alternating between growth and dormancy in a manner that is not easily explained (Kozlowski, 1971a,

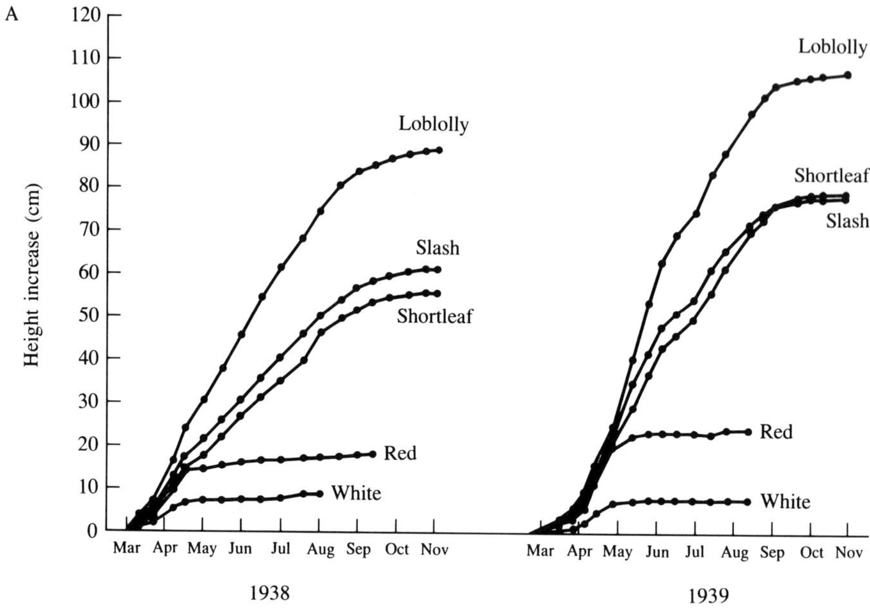

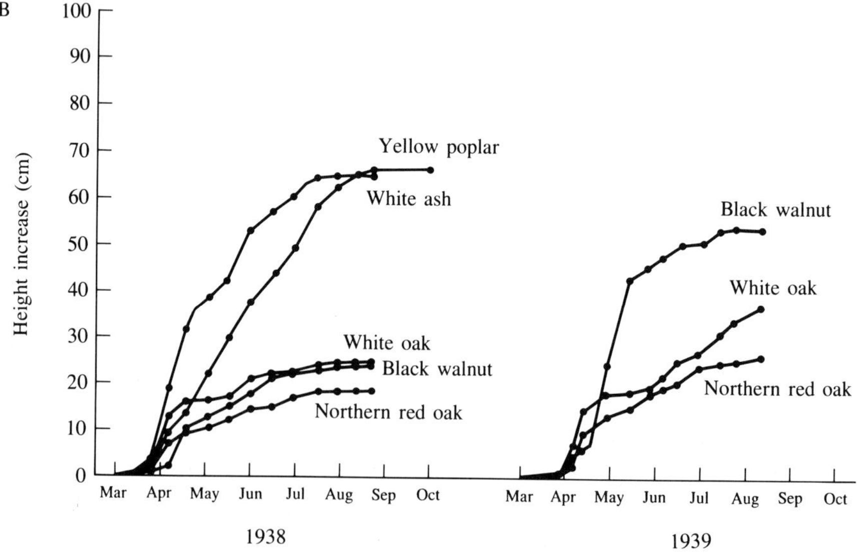

Figure 1.8 (A) Height growth of pine seedlings growing out-of-doors at Durham, North Carolina. The red and white pine seedlings, brought from upper New York State, grew for only a few weeks early in the season, whereas the southern pines grew all summer. (B) Growing period of hardwood seedlings during second and third seasons after planting. White ash and yellow poplar were removed at the end of the second season because they were overtopping the other species. (From Kramer, 1943.)

pp. 226–228). Borchert (1973) suggested that rhythmic shoot growth in a constant environment may result from feedback between root and shoot growth, tending to maintain a constant root–shoot ratio, but this needs more research.

As pointed out by Kozlowski (1971a, p. 37), smooth growth curves such as those of Fig. 1.8 may present a misleading idea of continuous growth, because it usually is discontinuous. In one experiment with eight loblolly pine seedlings in a controlled environment, shoot growth occurred over a period of 34 weeks, but the average seedling grew only 20 of the 34 weeks and in only 3 of the weeks were all eight seedlings growing (Kramer, 1957). It also should be emphasized that duration of growth varies among the various organs of the same tree, as shown in Fig. 1.9. In white pine in New Hampshire, the roots grew from early April to November but shoot growth ceased by midsummer, while cambium growth continued until early October. In the North Carolina Piedmont, roots of loblolly pine grow every month in the year and shoot growth can occur from April to the end of October, and in Missouri the roots of white oak trees grow over 60% of the time between leaf fall and spring bud opening (Reich *et al.*, 1980). Other examples of differences in duration of root and shoot growth are given by Kramer and Kozlowski (1979, p. 103).

Variations in Cambial Growth

Increase in stem diameter is highly variable both in space and in time. In the spring, cambial activity that produces new xylem starts in the top of the tree stem and

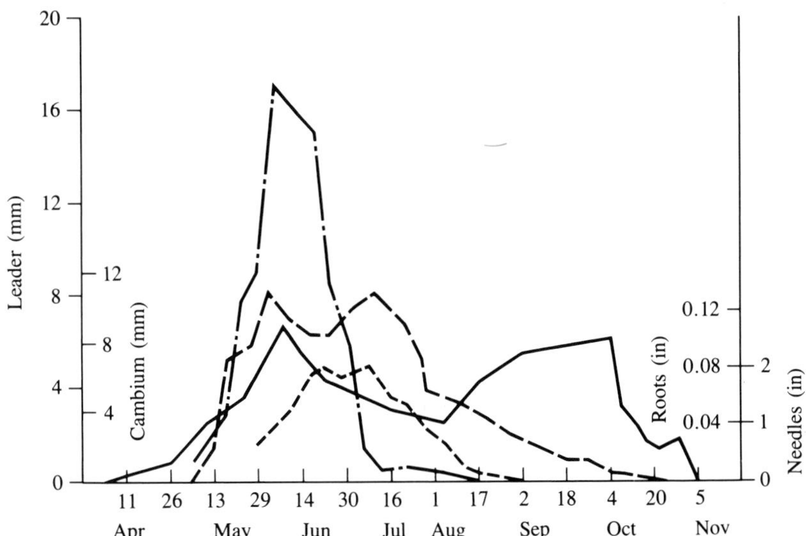

Figure 1.9 Seasonal duration of height (leader) growth (–·–·), cambial growth (– –), root elongation (——), and needle elongation (-----) of a 10-year-old white pine tree in southern New Hampshire (After Kienholz, 1934; from Kramer and Kozlowski, 1960.)

branch tips, apparently activated by auxins produced in the physiologically active buds, and progresses downward (see Little and Savidge, in Kossuth and Ross, 1987). However, the amount of xylem added often varies on different sides of the stem and also varies considerably with stem height. The width of the annual sheath of xylem laid down by the cambium usually is thickest at the height of greatest leaf volume. Below the crown the thickness of the annual ring varies with tree vigor. In large-crowned vigorous trees the annual ring become narrower down the stem, but near the base it usually thickens again, sometimes producing conspicuous buttresses. Knees apparently develop on bald cypress as a result of localized cambial activity on the better aerated upper sides of roots (Whitford, 1956). In slow-growing suppressed trees with small crowns, the ring is thinner than in vigorous trees, and little xylem, or sometimes none, is laid down in the lower part of the stem. This growth pattern is of considerable importance to forest managers because stems of trees too widely spaced in stands will be strongly tapered and may be less desirable for sawlogs (see Chapter 13). The amount of cambial activity is sensitive to environmental stresses and diameter growth often is interrupted by drought, sometimes resuming after rains occur. Cambial activity is so strongly affected by water supply that variation in the width of annual rings of old trees and timbers in dwellings has been used to establish the age of ancient buildings and rainfall patterns in prehistoric times (Fritts, 1976; Stahle *et al.*, 1988). Variation in annual rings in relation to water supply is discussed in more detail in Chapter 7, and by Kozlowski (1971b, pp. 126–131), and the effects on wood quality are summarized in Zobel and van Buijtenen (1989, pp. 197–204).

Another variable in cambial activity is the relative proportion of earlywood and latewood in an annual ring. In wet summers the transition occurs later than in dry summers, resulting both in wider rings and in a higher proportion of earlywood in wet summers (Fig. 7.8). The proportion of earlywood to latewood has important effects on utilization of wood because it affects wood density, strength, and durability (Zahner, 1968).

Special Interests of Arborists, Foresters, and Horticulturists

Having dealt with the growth of woody plants, we will consider briefly their uses and the significance of this to the objectives of this book. In general, arborists, foresters, and horticulturists use shrubs and trees in quite different ways, while ecologists are concerned chiefly with their role as competing components of plant communities rather than in their uses.

Foresters grow trees in stands, primarily for saw timber, pulp, and other wood products. They are, therefore, concerned principally with the average behavior of large numbers of trees that often grow on a variety of contiguous sites of varying quality. It seems likely that in the future there will be more selection and breeding of trees that are best adapted to particular sites and for specific purposes. For example, some families of loblolly pine trees probably can be found that are best suited to dry sites and others that are better suited to wetter sites. Likewise, one might expect that the tree families best adapted to use in short-rotation pulpwood production would not necessarily be best adapted to production of structural timber, where strength of wood is a primary consideration. This search will require extensive exploration of the ecology, physiology, and wood properties of various families of the principal species being planted and selection for specific purposes.

Sometimes trees are grown chiefly to protect watersheds and stabilize runoff water, and in national parks and certain national forests they are grown primarily for their aesthetic value. The ability of pines to establish themselves on abandoned, eroded farmland in the southeastern United States has been an important factor in controlling erosion and rebuilding the soil of that region. In many less developed countries trees are the primary source of fuel. In fact it is estimated that about one-half of the total wood consumption of the world is for fuel (Burley and Plumptre, in Cannell and Jackson, 1985). In some areas, leaves and twigs of trees are used as feed for livestock (Robinson, in Cannell and Jackson, 1985), and some kinds of trees are important sources of such diverse compounds as rubber, turpentine, essential oils, quinine, and maple syrup (Raven, in Cannell and Jackson, 1985). Selection and management of species may be quite different for these different purposes, but all plant growth depends on the same basic principles.

Arborists and ornamental horticulturists, in contrast to foresters, select and plant individual trees and shrubs largely on the basis of aesthetic considerations, often with little consideration of the ability of such plants to thrive in the prevailing environmental regimes. This greatly complicates the problems of growing shade trees and ornamental shrubs because many urban sites that would benefit aesthetically from their presence are unsuitable for their growth. Soil compaction, inadequate drainage and poor soil aeration, and limited space for root growth characterize many urban planting sites (Craul, 1985; Kozlowski, 1985a: Krizek and Dubik, 1987). The soil around homes and in golf courses, parks, and campsites often becomes compacted by pedestrian and vehicular traffic and changes in grade in housing developments often cause slow decline of existing trees. Furthermore, toxic air pollutants are heavily concentrated in urban areas and the amounts of some toxic air pollutants are likely to increase in future years (see Chapter 9). Additionally, when incorrectly applied, pesticides, herbicides, and deicing salts adversely affect physiological activity and injure shade trees (Kozlowski, 1986a–c). Hence, urban trees often are less vigorous and shorter-lived than the same species of trees growing in a forest. These considerations emphasize that in selecting plant materials for unfavorable sites, arborists should give as high a priority to the tolerance of species and cultivars for prevailing environmental stresses as to their aesthetic attributes. Some of the ecological problems encountered in ornamental plantings are discussed by Bradshaw *et al.* (1986).

Horticulturists grow trees, shrubs, and vines for their fruit, seed, or flowers. Usually the plants are fertilized, irrigated, and intensively managed to produce the best possible crops. The primary objective often is reproductive growth, and the emphasis is on flowering and fruiting. The vegetative structure simply provides physical and physiological support for the crop. Thus the physiological factors controlling flowering and fruiting are of primary importance to horticulturists.

In all of these situations, growth of trees is controlled by the same basic principles and the amount and quality of growth are determined by heredity and environment, operating through physiological processes. Thus an understanding of the basic ecological and physiological principles is equally important to arborists, foresters, and horticulturists. However, the most desirable allocation of photosynthate is quite different among the different uses. Foresters desire the maximum possible growth of wood, arborists desire well-rounded, heavily branched trees and shrubs, while horticulturists desire the maximum possible allocation of photosynthate into fruits. The partitioning of photosynthate is discussed in Chapter 2 and in more detail by Cannell in Cannell and Jackson (1985).

Complexity of Environmental Control of Growth

The environmental requirements for growth are relatively few and simple: carbon dioxide, water, and minerals for raw materials, light as a source of energy and for certain morphogenetic or formative effects, enough water to maintain cell turgor, oxygen, and temperatures favorable for the numerous processes involved in growth. Nevertheless, growth of trees is limited more often by unfavorable environmental conditions than by limitations in the capacity of the physiological processes. For example, most plants, including trees, have a photosynthetic capacity to produce large amounts of carbohydrates, but their full capacity seldom is utilized because of limitations imposed by deficiencies in light, water, minerals, and other environmental factors. As a result, the production of dry matter by trees usually is far lower than the potential maximum production (Farnum *et al.*, 1983). Thus it becomes important to examine the role of various environmental factors in limiting growth and how the effects of environmental stresses can be minimized. This will be done in detail in later chapters, but a few general principles will be discussed here.

Growth of woody plants is influenced by a wide variety of abiotic and biotic environmental stresses, operating through their effects on physiological processes. The major abiotic stresses include drought, flooding, low soil fertility, poor soil structure, temperature extremes, pollution, fire, and wind. Important biotic stresses include plant competition, insect attacks, plant diseases, activities of humans, and occasionally feeding by animals such as deer and beaver.

Some environmental stresses act persistently or chronically on woody plants while others are randomly imposed for short periods. Whereas shading affects understory trees more or less continually, droughts, unusually early or late frosts, and outbreaks of insect attacks and diseases are unpredictable and usually of short duration, and their inhibitory influences on growth depend on their intensity and duration as well as on the stage of plant or ecosystem development at which they occur.

A voluminous literature emphasizes the importance of various environmental stresses in inhibiting growth of woody plants. However, environmental control of growth is complex and it is difficult to quantify precisely the impact of single environmental factors because growth of plants is the integrated response to the influences of numerous continuous and periodic stresses.

The relative importance of various environmental stresses changes over time. Abrupt imposition of a severe stress, such as an insect attack or pollution episode, may suddenly dominate over other milder chronic stresses that previously were the chief inhibitors of growth. A given amount of precipitation early in the growing season, when the soil is fully charged with water, will have little effect on growth, whereas one-half that amount later in the season during a drought will greatly influence growth. In Wisconsin the correlation of cambial growth of northern pin oak trees with temperature decreases in late summer as soil moisture is progressively depleted and growth is increasingly limited by water deficit (Kozlowski *et al.*, 1962). Another problem is that correlation between the intensity of a certain environmental stress and degree of inhibition of growth may suggest that the stress was actually controlling growth, when it may only have been coincidental with some other stress factor that was equally or more inhibitory but was not measured in the analysis.

Quantitative assessment of environmental effects on woody plants is further complicated by differences in sensitivity of various physiological processes to stress, lag in responses, interactive influences of environmental factors, environmental preconditioning of plants, and genotypic variations in response to stress. Some of these difficulties can be minimized or eliminated by conducting experiments in controlled environments (Downs and Hellmers, 1975; Kramer, 1978).

Responses to Stresses

In nature, plants are almost universally subjected to a variety of stresses. Their responses can usually be grouped into one of three categories: visible injury, retarded growth and development, or interference with physiological processes, which is responsible for the other two categories of injuries. The most familiar category is visible injury as manifested by leaf death and abscission, obvious reduction in growth, and changes in ecosystems. Growth responses include reductions in height and diameter growth, leaf area increment, root growth, flowering, and production of flowers and seeds. Because large plants usually show larger growth increments than small plants, growth often is reported as relative growth rate (RGR) or increase in dry weight per unit of plant weight. This permits the comparison of the effects of environmental factors on the rate of growth independently of the size of the plants that are being compared. However, it should be remembered that older trees generally have a lower RGR than young trees under the same conditions. Among the physiological processes and conditions often measured for plants under stress are water status, stomatal conductance, rates of photosynthesis and transpiration, membrane permeability, enzymatic activity, and metabolite pools. A book edited by Jones *et al.* (1989) discusses stress effects at the physiological and molecular level in detail and considers the problems involved in developing stress-tolerant strains of plants.

Some plant processes are much more sensitive than others to environmental stresses. For example, water stress usually reduces cell enlargement before cell division, shoot growth before root growth, and cambial growth before height growth. Pollution with SO_2 decreased leaf growth, dry weight increment, and RGR of bald cypress seedlings within 6 weeks, but height and diameter growth were not affected (Shanklin and Kozlowski, 1985a). Tsukahara *et al.* (1985) noted that the rating of tolerance of three species of conifers to SO_2 differed with the growth parameter measured.

As mentioned earlier, various parts of trees grow at different times of the year. The seasonal duration of shoot growth usually is shorter than that of cambial growth of the same tree, and growth in length of roots lasts longer than either of those (Fig. 1.9). In addition, the amount and duration of cambial growth are different in the upper and lower parts of the stem. Flower bud formation, flowering, and seed development may cover two growing seasons in pines and red oaks, hence weather, fertilization, and size of seed crop in one year can affect seed production the next year (Shoulders, 1967, and others). All of this complicates the establishment of clear relationships between environmental stresses, cultural treatments, and plant growth.

Because of these variables, assessment of environmental control of growth will depend on what is measured and in which part of the plant it is measured. Hence great care is advised when developing models of responses of woody plants to environmental stresses and using data from different investigators who did not use the same response parameters. Growth analysis involving measurement of the growth of various organs often helps to explain how various factors affect overall growth. Chiarello *et al.* (1989) discussed some problems encountered in measuring plant growth.

Lag in Response to Environmental Stresses

The effects of environmental stresses on growth of some woody plants may not be obvious for a long time, as shown by the lag in response of shoot growth to environmental regimes. Length of the new shoots of some species such as northern pines and spruces, and some broad-leaved trees, is predetermined during bud formation. Buds form during one year and the preformed parts within them (internodes and leaves) expand into shoots the next year, a pattern called "fixed" growth. No matter how ideal the environment is during the year of bud expansion, the shoots expand for only a few weeks during the early part of the summer. A favorable environment during the year of bud formation results in formation of large buds that produce long shoots with many leaves the next year (Table 1.1). Thus the environmental conditions during the year of bud formation affect shoot growth more than those during the year of bud expansion into a shoot. When late-summer temperatures were low, small buds formed in Norway spruce trees and expanded into relatively short shoots the following year (Heide, 1974). In Finland, height growth of Scotch pine depends more on air temperature of the preceding growing season than on that of the current season (Mikola, 1962).

In contrast to species that exhibit fixed growth, some species exhibit "free" growth, which involves elongation of shoots by simultaneous initiation and elongation of new shoot components as well as expansion of preformed parts. Such plants, which include poplars, birches, apple, eucalyptus, and larch, continue to expand their shoots late into the summer (Kozlowski and Clausen, 1966). In still other species, such as the southern pines (e.g., loblolly, shortleaf, slash, and longleaf) and many tropical broad-leaved trees, shoot elongation consists of recurrent growth flushes from the opening of a series of buds produced during the same growing season. In some tropical pines, shoots even grow continuously throughout the year rather than in periodic flushes (Kozlowski and Greathouse, 1970). Sometimes this results in "foxtailing," the production of long terminal shoots lacking lateral branches (Kramer and Kozlowski, 1979, p. 80). Shoot growth of species exhibiting free, recurrently flushing, or continuous growth is affected much more by the environmental regime during the year of shoot expansion than is shoot growth of species exhibiting fixed growth. Variations in seasonal duration of shoot growth of several species are shown in Fig. 1.8. Shoot growth is discussed in more detail by Kozlowski (1971a, pp. 164–206).

Long lag times in response to environmental changes also are shown by changes in cambial growth of trees in thinned stands. After a stand of trees is thinned, the gradually increasing crown size of the residual trees results in increased production

Table 1.1
Effect of Bud Size on Shoot Growth of 8-Year-Old Red Pine Trees[a,b]

	Bud diameter (mm)	Bud length (mm)	Shoot length (mm)
Terminal leader	8.2 ± 0.7	38.0 ± 2.8	742.0 ± 26.7
Whorl 1 shoots	5.9 ± 0.1	27.3 ± 0.7	484.8 ± 11.0
Whorl 2 shoots	5.5 ± 0.1	22.9 ± 0.8	403.2 ± 13.0
Whorl 3 shoots	4.5 ± 0.2	16.6 ± 0.9	271.4 ± 19.1
Whorl 4 shoots	3.8 ± 0.3	12.5 ± 1.0	132.1 ± 20.6
Whorl 5 shoots	3.7 ± 0.3	9.9 ± 0.8	65.2 ± 16.0
Whorl 6 shoots	3.3 ± 0.4	8.6 ± 1.4	74.4 ± 31.5

[a] From Kozlowski *et al.* (1973).

[b] Data are means and standard errors of bud diameters and lengths before initiation of shoot expansion (March 20, 1970) and final shoot lengths (August 19, 1970) at different stem locations.

and downward translocation of carbohydrates and hormonal growth regulators needed for cambial growth. Eventually wood production in the lower stem is increased, but this may not be apparent for a few years, as discussed in Chapter 13.

Interactions of Environmental Factors

Interactive effects of environmental factors on physiological processes and growth of woody plants have been well documented. For example, injury to the photosynthetic mechanism by bright light (photoinhibition) is increased when exposure to high light intensity is associated with drought or very high or low temperature (Powles, 1984). An interactive effect of light intensity and relative humidity on stomatal aperture is shown in Table 1.2. The stomata of white ash opened when the relative humidity was high and closed when it was low. However, both opening and closing of stomata occurred faster when the light intensity was high than when it was low. Interaction of temperature and light intensity on stomatal aperture of sugar maple is shown in Fig. 1.10. Addicott (1982) showed that photoperiod and drought interacted in inducing leaf abscission. The complex interactions caused by increasing CO_2 concentration are shown in Fig. 10.7.

Other abiotic stresses often interact with environmental pollutants. Prevailing light intensity, humidity, temperature, and soil moisture regimes influence stomatal aperture and thus affect the amount of pollutant absorbed (Kozlowski and Constantinidou, 1986b). Because air pollutants seldom exist singly, much interest has been shown in interactive effects of combined pollutants on woody plants. Synergistic effects of SO_2 and O_3 mixtures have been shown on growth and leaf injury and on photosynthesis (Constantinidou and Kozlowski, 1979a,b; Jensen, 1983). Interactive effects of air pollutants on woody plants are discussed in more detail in Chapter 9.

Environmental Preconditioning

Physiological and growth responses of woody plants to abiotic and/or biotic stresses often vary appreciably with the environmental regime in which the plants were previously grown. For example, stomatal aperture usually is less sensitive to water stress in plants grown in the field than in plants grown in greenhouses or controlled-

Table 1.2
Time for Stomata of Normal and Mineral-Deficient Plants of Sugar Maple and White Ash to Open or Close after a Change in Light Intensity from 0 to 32,000 Lux and the Reverse[a]

	Sugar maple		White ash	
	Open	Close	Open	Close
Green	18.2 ± 1.02[b]	19.0 ± 0.89	26.3 ± 0.83	31.3 ± 1.29
Yellow	27.0 ± 1.35	29.2 ± 0.76	30.6 ± 0.77	36.8 ± 0.81

[a] From Davies and Kozlowski (1974).
[b] Standard error of mean. Time given in minutes.

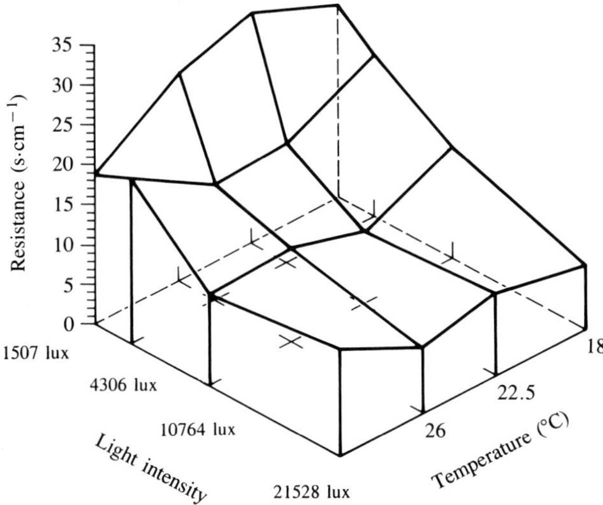

Figure 1.10 Interaction of light and temperature on stomatal aperture of sugar maple leaves, measured as leaf resistance. (From Pereira and Kozlowski, 1977b.)

environment growth chambers and never subjected to water stress (Davies, 1977). The soil fertility regime in which plants are grown also influences stomatal sensitivity. The stomata of vigorous ash and sugar maple seedlings responded faster to changes in light intensity and humidity (Table 1.3) than the stomata of slow-growing, mineral-deficient plants. Flooding of soil generally induces rapid stomatal closure. Exposure to SO_2 reduced growth more in unflooded bald cypress seedlings than in those previously flooded for 8 weeks, because unflooded seedlings had more open stomata and greater SO_2 uptake (Shanklin and Kozlowski, 1985a).

Environmental preconditioning also influences the rates of photosynthesis. Because thick, sun-grown leaves have lower stomatal and mesophyll resistances to CO_2 diffusion, they often have higher rates of photosynthesis per unit of leaf area

Table 1.3
Interaction of Light and Relative Humidity on Opening and Closing of Stomata of White Ash and Sugar Maple Seedlings[a,b]

	Light intensity, 6500 lux		Light intensity, 32,000 lux	
	RH 80% to 20%, closing	RH 20% to 80%, opening	RH 80% to 20%, closing	RH 20% to 80%, opening
White ash	19.73 ± 0.79[c]	18.45 ± 0.64	14.80 ± 0.52	13.80 ± 0.77
Sugar maple	8.30 ± 0.88	4.40 ± 0.51	5.00 ± 0.56	2.60 ± 0.36

[a] From Davies and Kozlowski (1974).
[b] Data are time in minutes for stomata to reach equilibrium after a change in relative humidity (RH).
[c] Standard error of mean.

and become light saturated at a higher light intensity than thinner shade leaves (Nobel, 1976).

Exposing plants to either a high or low temperature affects the subsequent rate of photosynthesis at another temperature (Pearcy, 1977; Pharis *et al.*, 1970). Although subfreezing temperatures injure the photosynthetic mechanism, the injury often can be reversed when the temperature rises above freezing. However, the time required for recovery varies with the severity and duration of the subfreezing treatment. High-temperature injury to the photosynthetic apparatus also occurs, but it may be repaired when the temperature falls if the injury is not too severe.

Predisposition to Disease

Abiotic stresses often predispose woody plants to biotic stress. Outbreaks of stem cankers, diebacks, declines, and some root rots follow loss of tree vigor as a result of drought, flooding, mineral deficiency, or air pollution, as well as combinations of these (Kozlowski, 1985b). Such stresses render the host plant more susceptible to facultative parasites, especially weak and unaggressive ones that would not be a threat to vigorous plants. In many instances woody plants are invaded regardless of prevailing stresses, but the invading pathogens usually are not pathogenic until the host has become physiologically predisposed under stress (Schoeneweiss, 1978a,b).

Development of disease varies with availability of nutrients at the infection site and presence of inhibitors and toxic substances produced by the host. Examples of compounds that increase resistance to fungi include phenols, tannins, chlorogenic acid, and enzyme inhibitors. In tropical trees, alkaloids, rotenoids, and saponins may contribute to disease resistance. Some compounds that are not very effective alone may act synergistically with other compounds and together provide resistance (Kozlowski, 1969). Abiotic stresses may alter activities of the host in such a way as to either reduce growth of the pathogen or increase tolerance of its toxic products, thereby altering the biotic stress. Or the effect of the concurrent abiotic and biotic stresses may be additive, with their sum being sufficient to induce injury where none existed before, or to increase existing injury and reduce the possibility of recovery (Ayres, 1984).

There are many examples of environmental predisposition of woody plants to disease and only a few will be given. Either soil water deficit or excess predisposes woody plants to certain diseases. A variety of dieback diseases of forest trees were induced by the drought years of the 1930s in the United States (Schoeneweiss, 1978a). Dieback of ash was induced by drought, followed by production of cankers caused by *Cytophoma pruinosa* and *Fusicoccum* sp. Although the fungi were present in the bark of unstressed trees, they induced cankers only after the trees were predisposed by drought (Silverborg and Ross, 1968). Cankers on sweet gum trees caused by *Botrysophaeria dothidea* are associated with dry summers (Neely, 1968). Susceptibility to disease may involve both the degree and duration of drought. When paper birch seedlings were exposed to drought their susceptibility to *Botrysophaeria dothidea* did not increase until the plant water potential fell below

− 12 bars, when susceptibility increased greatly and cankers formed within 4 days (Schoeneweiss, 1978b). The changes in chemical composition, such as the accumulation of proline that occurs in water-stressed plants, may be involved in increased susceptibility to infection (Griffin *et al.*, 1986). Flooding of soil often increases root diseases because several pathogenic fungi grow vigorously in poorly aerated soil. This is discussed in Chapter 8.

Temperature extremes also predispose trees to disease. Frost injury increased susceptibility of birch and spruce to *Nectria* (Gäumann, 1950), of sweet cherry to *Cytospora* (Kable *et al.*, 1967), of European white birch to *Botryosphaeria dothidea* (Crist and Shoeneweiss, 1975), and of citrus to *Botrytis, Sclerotinia, Alternaria, Phomopsis, Diplodia,* and *Dothionella* (Fawcett, 1936).

Warm temperatures before inoculation may predispose plants to disease at temperatures that normally allow resistance. Woody plants adapted to cool climates often show a decrease in disease resistance as the temperature is increased. For example, elms progressively lost resistance to Dutch elm disease as the temperature was raised from 16 to 26°C (Birkholz-Lambrecht *et al.*, 1977). Resistance of many woody plants to certain pathogens commonly breaks down at very high temperatures (30 to 45°C) because of adaptation of the pathogen to such high temperatures (Bell, 1981). In contrast, some plants can be freed of viruses by growing them at high temperatures. The branch wilt-fungus, *Hendersonula toruloidea,* invaded black walnut trees through cracks caused by sun scald. The fungus was saprophytic in dead tissue but became parasitic when the branches were exposed to further sun scald (Sommer, 1955).

Fungus infections sometimes predispose plant tissues to further infection by the same or different pathogenic organisms. For example, *Monilia* and *Sclerotinia* release pectic enzymes that predispose plant tissues to further attack by the same fungi, and *Dothidia* may predispose plant tissue to attack by releasing toxins (Yarwood, 1959). Examples of predisposition by one pathogen to attack by another one include infection of grape with *Peronospora,* which predisposes tissue to *Botrytis* (Boubals *et al.*, 1955), and infection of raspberry with *Thomasiana,* which predisposes the tissue to *Leptosphaeria, Fusarium,* and *Didymella* (Pitcher and Webb, 1949). The biochemistry of disease resistance was reviewed by Bell (1981) and Hogue (1982).

Predisposition to Insect Attack

Physiological changes in woody plants that are induced by environmental stresses commonly are prerequisites for attacks by certain insects. Survival and growth of insects in stressed trees have been attributed to lowered chemical defenses, improved nutrition for the insect, and a better physical environment for insect growth.

Successful attack of conifers by bark beetles depends to a large degree on lowered defensive capacity of trees, which in turn depends on the amount of oleoresin available at the time of attack and the capacity of trees to mobilize it near the point of attack. Vité (1961) demonstrated that abundant exudation of oleoresins was associated with resistance of ponderosa pine to attacks by bark beetles. Copious production of oleoresins by resin ducts repelled or "pitched out" the invading bark beetles. Variations in susceptibility to attack on different sites were related to rates of flow of oleoresins, which decreased during droughts, thus increasing susceptibility of trees to beetle attacks. Variations in composition of monoterpenes also are important in resistance to bark beetles, as shown by three lines of evidence: (1) monoterpenes of ponderosa pine vary in their toxicity to *Dendroctonus* beetles; (2) monoterpene composition of attacked and unattacked trees differs appreciably; and (3) the composition of monoterpenes of ponderosa pine varies greatly (Smith, R. H., 1966a,b).

Berryman (1986) described two types of plant defenses against insects: static defenses, characterized by normal production of toxic or inhibitory chemicals (e.g., phenols, terpenes, and tannins) that make plants less palatable; and dynamic defenses developed in reaction to attack, which repel insects or inhibit their development.

Production of defensive chemicals requires large amounts of carbohydrates. Christiansen *et al.* (1987) speculated that the reduced capacity of stressed trees to withstand bark beetle attacks and their associated fungi is correlated with low availability of carbohydrates for use in defensive wound reactions. Hence, a number of environmental stresses that reduce crown size or photosynthetic efficiency might lower resistance of trees to bark beetle attack. Lorio and Sommers (1986) reported that mild water stress that limits growth but not photosynthesis of loblolly pine is favorable for synthesis of oleoresin, the primary defense against southern pine beetles (*Dendroctonus frontinalis*). The relationship between plant chemistry and insect attacks is discussed further in Chapter 2.

Mattson and Haack (1987a,b) emphasized that environmental stresses have important effects on the attacking insects as well as on the host plants. They attributed the increased susceptibility of water-stressed plants to insects to: (1) a more favorable thermal environment for insect development; (2) greater attractiveness of plants as a result of leaf yellowing, higher temperature, and greater reflection of infrared radiation; (3) improved physiological suitability of the host (more concentrated and better balanced nutrients and weakening of defense mechanisms such as reduced production of oleoresins); (4) increase in mutualistic microorganisms but not in natural insect enemies; and (5) genetic changes in plants. Some of the sequential events that follow environmental stress (e.g., drought), involving host plants, insects, and insect enemies, and that may lead to insect outbreaks are summarized in Fig. 1.11.

Attacks by some insects may physiologically predispose woody plants to attacks by other insects. For example, defoliation by the Douglas-fir tussock moth (*Orygia pseudotsuga*) increased susceptibility of grand fir to the fir engraver beetle (*Scolytus ventralis*) (Wright *et al.*, 1979, 1984). During and one year after defoliation, the beetles generally infested trees that had been more than 90% defoliated. Defoliation reduced the amounts of carbohydrates and monoterpenes, with trees producing the lowest amounts of monoterpenes being most susceptible to beetle attacks.

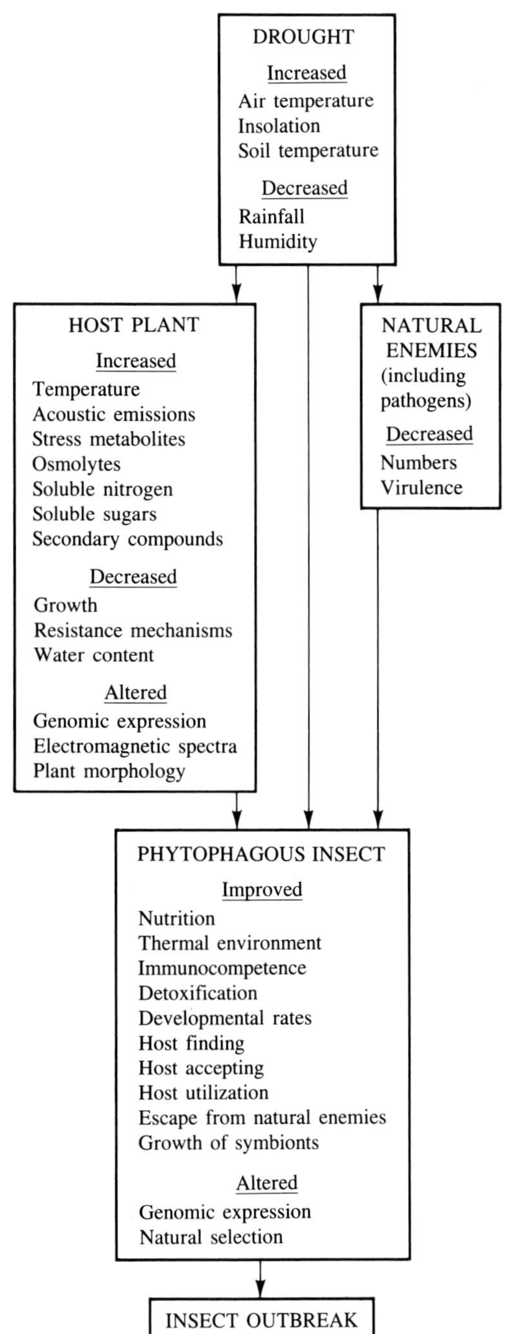

Figure 1.11 Interactions of drought on host plants, attacking insects, and natural enemies of phytophagous insects, resulting in insect outbreaks during dry seasons. (From Mattson and Haack, 1987b. Copyright © 1987 by the American Institute of Biological Sciences.)

Plant diseases may also predispose trees to attack by bark beetles. For instance, following fires, outbreaks of the mountain pine beetle (*Dendroctonus ponderosae*) in lodgepole pine forests occur mainly in trees 80 to 150 years old that have been weakened by the fungus *Phaeolus schweinitzii*. The surviving fire-scarred trees are infected by the slowly spreading fungus, which lowers their vigor and predisposes them to another attack by bark beetles (Fig. 1.12). The outbreaks subside after most of the larger trees are killed. Waring and Pitman (1985) reported that reduction of crown density in stands of lodgepole pine by manual thinning or insect attack increased the resistance of the remaining trees to pine beetle attack for at least the next 3 years. However, the real reason may be a less favorable microclimate for beetles in thinned stands (Amman *et al.*, 1988).

Genetic Variation in Stress Tolerance

Another complexity in evaluating effects of environmental stresses on different species of woody plants is the considerable variation among individual trees within a species in their response to stress. Differences in drought tolerance of poplar clones were associated with variations in stomatal frequency, stomatal size, and

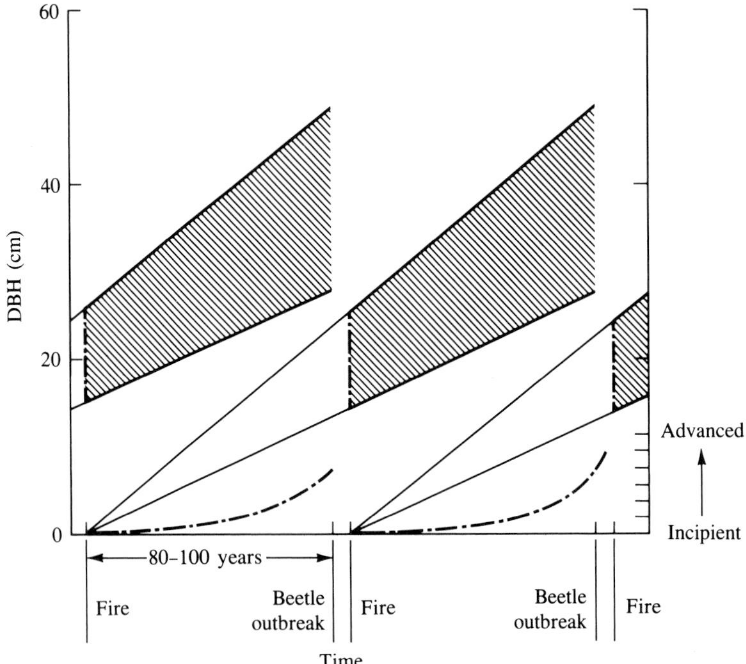

Figure 1.12 A conceptual model of the interaction between fire, fungus (---), and the mountain pine beetle in a lodgepole pine forest. Trees of similar age are represented by wedges. The shaded portions indicate fungal infection, and the dashed lines indicate the degree of decay in living trees. (From Geiszler *et al.*, 1980.)

speed of stomatal closure during drought (Siwecki and Kozlowski, 1973; Pallardy and Kozlowski, 1979a; Mazzoleni and Dickmann, 1988, and others). On a day when the vapor pressure deficit (VPD) was low, the stomatal aperture of two poplar clones varied little. However, when the VPD was high, the adaxial stomata of one clone closed more than those of the other clone (see Fig. 7.22). Another study showed that the stomata of *Populus candicans* × *Populus berolinensis* were more responsive to a change in VPD and less responsive to a change in light intensity than were the stomata of a clone of *Populus deltoides* × *P. caudina* (Pallardy and Kozlowski, 1979e). Wide intraspecific variation in frost tolerance (Sakai and Larcher, 1987) and in tolerance of air pollution (Kozlowski and Constantinidou, 1986a) have also been shown. The existence of this variability provides opportunities for tree improvement programs. Such variations in tolerance of various stresses are discussed further in several other chapters.

Summary

Growth of woody plants, like that of other plants and of animals, is controlled by their hereditary potentialities, which operate through their physiological processes, and by the environment in which they grow. The environment, which includes light, water, temperature, minerals, atmospheric gases, wind, other plants and animals, and cultural treatments imposed by humans, determines the manner in which the hereditary potentialities are expressed. For example, a tree that grows to a height of 20 m in nature can be restricted to a height of less than a meter at the same age by the dwarfing treatment used to produce bonsai trees. There are wide differences among trees with respect to life span, ranging from peach trees that are old at 30 years to giant redwoods and bristlecone pines that live several thousand years. However, the manner in which growth occurs and the basic requirements for growth are similar for all kinds of trees, large or small.

Environmental limitations generally prevent plants from attaining their maximum potential for growth. However, the interactions between growth and environmental limitations are so complex that it is difficult to quantify the impact of any single factor. Growth represents the integrated response to numerous continuous and intermittent stresses to which a plant is subjected. For example, water stress not only reduces growth, but may also render a plant more susceptible to attack by insects and pathogens. It seems likely that there are significant differences in tolerance of various stresses among individuals of a species that should be exploited by arborists, foresters, and horticulturists.

General References

Ayres, P. G. (1984). The interactions between environmental stress injury and biotic disease physiology. *Ann. Rev. Phytopath.* **22**,53–75.

Bell, A. A. (1982). Plant interactions with environmental stress and breeding for pest resistance. *In* M.

N. Christiansen and C. F. Lewis, eds., "Breeding Plants for Less Favorable Environments," pp. 335–363. Wiley, New York.

Bradshaw, A. D., Goode, D. A., and Thorpe, E. H. P., eds. (1986). "Ecology and Design in Landscape." Blackwell Scientific, Oxford.

Cannell, M. G. R., and Jackson, J. E., eds. (1985). "Attributes of Trees as Crop Plants." Institute of Terrestrial Ecology, Huntingdon, England.

Cannell, M. G. R., and Last, F. T., eds. (1976). "Tree Physiology and Yield Improvement." Academic Press, London.

Harris, R. W. (1983). "Arboriculture: Care of Trees, Shrubs, and Vines in the Landscape." Prentice-Hall, Englewood Cliffs, New Jersey.

Jones, H. G., Flowers, T. J., and Jones, M. B. (1989). "Plants under Stress." Cambridge University Press, Cambridge.

Kossuth, S. V., and Ross, S. D., eds. (1987). Hormonal control of tree growth. *Plant Growth Regulation* **6**(1 & 2), 1–215.

Kozlowski, T. T. (1971). "Growth and Development of Trees." Academic Press, New York.

Kozlowski, T. T. (1985). Tree growth in response to environmental stress. *J. Arboric.* **11,** 97–111.

Kramer, P. J., and Kozlowski, T. T. (1979). "Physiology of Woody Plants." Academic Press, New York.

Landsberg, J. J. (1986). "Physiological Ecology of Forest Production." Academic Press, New York.

Larcher, W. (1983). "Physiological Plant Ecology." Springer-Verlag, Berlin and New York.

Mooney, H. A., Pell, E., and Winner, W. E., eds. (1991). "The Integrated Response of Plants to Stress." Academic Press, San Diego.

Schoenweiss, D. F. (1981). The role of environmental stress in diseases of woody plants. *Plant Dis.* **65,** 308–314.

Waring, R. H., and Schlesinger, W. H. (1985). "Forest Ecosystems: Concepts and Management." Academic Press, Orlando, FL.

Wright, J. W. (1976). "Introduction to Forest Genetics." Academic Press, New York.

Zobel, B. J., van Wyk, G., and Stahl, P. (1987). "Growing Exotic Trees." Wiley, New York.

Zobel, B. J., and van Buijtenen, J. P. (1989). "Wood Variation: Its Causes and Control." Springer-Verlag, Berlin.

Chapter 2

Physiological and Environmental Requirements for Tree Growth

Introduction

Chapter 1 discussed the interaction of the hereditary and environmental factors that control growth and described how trees grow in height and diameter. This chapter continues by describing briefly some of the physiological processes that control growth and yield and the manner in which they are affected by environmental stresses. Tree breeding and selection programs are providing foresters with trees having good form, rapid growth, and increased resistance to disease and insects. Likewise, horticulturists have developed trees and shrubs with the potential for producing high yields of fruit and flowers. However, actual yields seldom approach the maximum possible yields. For example, over a 10-year period, the highest fruit yield in a certain British apple orchard was 932 bu per acre, the lowest 410, and the average 532 bu per acre (Hudson, 1977). According to Farnum *et al.* (1983), the average yields of well-managed loblolly pine and Douglas-fir plantations in the United States are only about half of their potential yields. This failure to attain maximum yields is caused chiefly by environmental limitations on the physiological processes that determine the amount of growth. It therefore is important to understand how environmental factors limit physiological processes essential for growth and how these limitations can be reduced by management and tree improvement programs.

Before discussing growth requirements, readers are reminded that successful growth of managed forests and cultivated crops is measured by criteria that are somewhat different from those that apply to unmanaged natural ecosystems. In natural systems, success of a species depends first of all on reproduction, but in managed systems, reproduction and establishment usually are a secondary problem because seedlings usually can be protected from competition and natural enemies. In managed systems, success is measured by the efficiency of production of the economic product, such as wood, flowers, fruit, or seed.

In this chapter we will discuss the major internal processes and environmental factors involved in growth as preparation for a more detailed discussion of the effects of environmental factors in later chapters. The requirements for the growth of plants are relatively simple: a supply of carbohydrates produced by photosynthesis, plant hormones that aid in integrating various physiological processes, nitrogen and about a dozen other mineral elements, enough water to maintain cell turgor, and environmental conditions favorable for the biochemical and biophysical processes involved in growth.

The first requirement for growth and reproduction of plants is a supply of food. In spite of the advertising of fertilizers as "plant foods," plants really use the same kinds of foods as humans, namely, carbohydrates, fats, and proteins. The major difference is that plants manufacture their basic food materials whereas humans cannot do this. In this chapter we will discuss how foods are manufactured and used by plants, and discuss briefly the role of water, mineral nutrients, and growth regulators in plant growth. More detailed discussions of some of these topics are found in later chapters.

Photosynthesis

Photosynthesis sometimes is called the most important chemical process in the world because it supplies the energy used by all living organisms, except for a few chemosynthetic bacteria. It also originally supplied the energy now stored in the fossil fuels that provide industry with most of its energy and the heat and light for our homes. Lieth (1975) estimated that land plants produce 100 to 125 $\times$ 10^9 metric tons of dry matter per year and two-thirds of this is produced by trees. The dry matter production of various tree species and forest types was discussed in detail by Jarvis and Leverenz (1983). It can be argued that the major objective of most cultural practices really is to ensure a high rate of photosynthesis.

Photosynthesis can be described as the process by which light energy is trapped by chlorophyll in green plants and used to produce sugar from carbon dioxide and water. More specifically it consists of the reduction of carbon dioxide and recombination of the carbon into carbohydrates by the use of light energy, by the splitting of water, and hydrogen obtained as shown in the following simplified equation:

$$CO_2 + H_2O \xrightarrow[\text{chloroplasts}]{\text{light}} (CH_2O) + O_2 + H_2O$$

The photosynthetic process actually is much more complex than indicated by this equation and can be better described as a sequential series of events, including (1) trapping of light energy by chloroplast pigments, (2) splitting of water and release of O_2 and high-energy electrons used to reduce $NADP^+$, (3) generation of chemical energy in ATP and reducing power in $NADPH_2$, and (4) use of the reducing power of the $NADPH_2$ and energy stored in ATP to fix CO_2 in phosphoglyceric acid and reduce it to triose phosphate, from which glucose and other carbohydrates are synthesized.

This series of processes is known as the C_3 carbon fixation pathway because a three-carbon compound, phosphoglyceric acid, is the first detectable product. The carboxylating enzyme is ribulose bisphosphate carboxylase (RuBP carboxylase or Rubisco). Unfortunately, this enzyme also functions as an oxygenase, hence oxygen is a competitive inhibitor of CO_2 fixation. In some plants, including sugarcane and maize, but in very few woody plants, there is an additional carbon fixation pathway called the C_4 pathway because the first products are the C_4 acids, malic and aspartic, formed in the mesophyll cells. Those acids are transferred to the bundle sheath cells, where they are decarboxylated, releasing CO_2, which then goes through the C_3 pathway. In the C_4 pathway the initial carboxylating enzyme is phosphoenolpyruvate carboxylase (PEP carboxylase), which has a high affinity for CO_2 and is not inhibited by oxygen. Plants with the C_4 pathway have a very low CO_2 compensation point, are not light saturated in full sun, and show no detectable photorespiration. The CO_2 compensation point refers to the concentration of CO_2 at which the fixing of CO_2 by photosynthesis just balances its release by respiration. In spite of these seeming advantages of the C_4 carbon pathway, the only woody plants known to possess it are a few woody euphorbias (Pearcy and Troughton, 1975; Pearcy and Calkins, 1983) and some shrubs in the Middle East and USSR (Winter, 1981).

Another type of carbon fixation is crassulacean acid metabolism (CAM) found in many succulents and a few woody cacti and euphorbias (Ting, 1985). CAM plants have stomata that are open at night and closed during the day. CO_2 is absorbed at night and fixed in organic acids, chiefly malic, resulting in a large increase in acidity at night. In the light, malic acid is decarboxylated, yielding pyruvic acid and CO_2 that is fixed by the C_3 photosynthetic cycles. Daytime closure of stomata in CAM plants reduces transpiration and most such plants are found in arid regions. However, Popp *et al.* (1987) found CAM trees of the genus *Clusia* in both moist and dry habitats in Venezuela.

The actual reduction of CO_2 can occur in darkness, but it requires $NADPH_2$ and ATP, which are generated in the light. The dark reactions are relatively slow and are temperature sensitive, whereas the light reaction is rapid, and like other photochemical reactions is little affected by temperature. Readers who desire a more detailed discussion of the mechanism of photosynthesis are referred to textbooks of plant physiology such as Salisbury and Ross (1978), Ting (1982), and Kramer and Kozlowski (1979). More specialized treatments include those described by Bidwell (1983), Foyer, (1984) several chapters in Baker and Long (1986), and Volumes 5 and 6 of the "Encyclopedia of Plant Physiology, New Series." Osmond (1987) presents an interesting review of photosynthesis in an ecological context.

Tissues Involved in Photosynthesis

Although most photosynthesis occurs in foliage leaves, some occurs in cotyledons, buds, stems, flowers, and fruits. Photosynthesis of epigeous cotyledons is very important for establishment and early growth of tree seedlings (Kozlowski, 1979). Even mild environmental stresses inhibit photosynthesis of cotyledons, leading to growth inhibition and often death of seedlings of angiosperms (Marshall and Kozlowski, 1974a, 1976) and gymnosperms (Sasaki and Kozlowski, 1968d, 1970). Photosynthesis in most fruits (apple, orange, grape, locust) is too limited to contribute significant amounts of carbohydrate (Kozlowski and Keller, 1966), but it is said to contribute up to a fourth or more of the dry weight of fruits of *Acer, Cercis, Liquidambar, Magnolia,* and *Tilia* (Cannell, 1975; Bazzaz *et al.*, 1979). Some photosynthesis also occurs in pine cones (Linder and Troeng, 1981), but it is insufficient to supply all the carbohydrate required (Dickmann and Kozlowski, 1968). According to Blanke and Lenz (1989), in some fruits photosynthesis differs somewhat in its biochemistry from that in leaves.

Measurable photosynthesis occurs in the bark (Schaedle, 1975) but it usually is only sufficient to reduce the loss from respiration (Foote and Schaedle, 1976). However, Schaedle and Brayman (1986) reported that the Rubisco activity in twigs of quaking aspen is comparable to that in leaves. Some photosynthesis also occurs in wood, usually in xylem rays (Kriedemann and Buttrose, 1971; Wiebe, 1975). Although winter photosynthesis has been demonstrated in twigs of a number of species of deciduous woody plants, the rate seldom exceeds the loss by respiration (Keller, 1973a,b; Coe and McLaughlin, 1980). Stem and twig photosynthesis is important in some desert shrubs such as palo verde that are leafless most of the year (Adams and Strain, 1969) and in woody species of cacti and euphorbias.

Rates of Photosynthesis

The rate of photosynthesis varies widely among tree species and provenances, between sun and shade leaves, during the course of a day, and during the growing season. These variations result from interactions among plant factors such as leaf age, structure and exposure, canopy development, stomatal behavior, and amount and activity of Rubisco, and environmental factors such as light intensity, temperature, water supply, concentration of atmospheric CO_2 and air pollutants, and soil conditions. The effects of environmental factors will be discussed briefly in this chapter and in detail in later chapters.

The methods used to measure and express the rate of photosynthesis also affect the values reported. For example, Tranquillini (1962) reported that the apparent photosynthesis of European larch was twice that of Swiss stone pine on a dry weight basis, but similar on a leaf area basis. Kozlowski (1949) found that the rate of CO_2 uptake of oak seedlings exceeded that of pine seedlings both per unit of leaf dry weight and per unit of leaf surface, but the pine seedlings bore more leaves, which tended to compensate for the lower rate per unit of leaf surface. Capacities estimated from increase in dry weight are especially valuable because they average the production of dry matter over long periods of time and integrate the effects of varying environmental factors. Unfortunately, increase in dry weight is difficult to measure accurately for large organisms such as trees, particularly because of the large amount hidden underground in roots. From a practical standpoint the most useful indicator of long-term photosynthetic efficiency is the yield of merchantable products such as pulpwood, sawlogs, or fruit. Perhaps the least valuable indicators are short-term measurements of CO_2 uptake by single leaves or branches, yet such measurements provide most of the information available concerning rates of photosynthesis.

Differences in Rates among Species and Families

Data collected by Larcher (1980, p. 94) indicated that woody plants generally have lower rates of photosynthesis than herbaceous plants and evergreen trees have lower rates than deciduous trees. However, Nelson (1984) summarized considerable data and concluded that the rates are too similar to justify such a generalization (Table 2.1). Kramer and Decker (1944) found that the CO_2 uptake per unit of leaf surface was greater for northern red and white oak than for loblolly pine seedlings and Avery (1977) reported that it was twice as high for pear and apple leaves as for citrus leaves. The rate also differs among species within a genus, being higher for blue gum than for jarrah. There also are differences among clones of the same species, among geographic races, and even between sun and shade leaves. For example, Luukkanen and Kozlowski (1972) observed large differences among poplar clones with respect to rate of CO_2 uptake and light compensation point. Krueger and Ferrell (1965) found that at certain temperatures Douglas-fir seedlings from Vancouver Island generally had higher rates than those from Montana, but individual

Table 2.1
Net Photosynthesis Rates of Various Agricultural, Herbaceous, and Woody Plants under Favorable Conditions[a]

Type of plant	Species or hybrid	CO_2 uptake	
		mg CO_2 m^{-2} sec^{-1}	μg CO_2 g^{-1} sec^{-1}
Agricultural plants		0.56–1.25	8.34–16.68
Herbaceous sun plants		0.56–1.39	8.34–22.24
Exceptional herbaceous plants	*Typha latifolia*	1.20–1.90	
	Helianthus annuus	1.29–1.60	
Deciduous nonconiferous trees			
	Populus nigra × *P. trichocarpa*	1.26	
	P. deltoides × *P. nigra*	1.05	12.30
	P. deltoides	0.92	
	P. tremuloides	0.85–0.94	
	Salix (several spp.)		13.51–17.24
	Malus domestica	0.97–1.19	
Conifers			
	Pinus silvestris[b]	0.88	9.73–10.52
	P. rigida		14.46
	P. taeda		11.11
	Pseudotsuga menziesii		9.73

[a] After Nelson (1984).
[b] Rates for pines probably are below possible maximum because of self-shading by needles, as reported by Kramer and Clark (1947).

differences among seedlings from one source sometimes exceeded differences among sources.

It is probable that many additional differences in rates of photosynthesis would be found among families and geographic races if a systematic search were made. However, the importance of such differences with respect to growth and productivity is uncertain, because, as we will point out later, the rate of photosynthesis is not necessarily closely correlated with the rate of growth. Dry matter production is related to leaf area, leaf exposure, leaf retention, changes in photosynthetic capacity, rate of respiration, sink strength, and partitioning, in addition to the rate of CO_2 uptake per unit of leaf surface.

Diurnal Variations in Photosynthesis

Photosynthesis is limited to the daytime, except in CAM plants, because light energy is required for generation of the reducing power required to fix CO_2 into carbohydrates and to keep the stomata open to permit the entrance of CO_2. Even on a clear summer day the rate is low early in the morning because of low light intensity, low temperature, and closed stomata. As light intensity and temperature increase and stomata open, the rate increases toward a maximum at midday then decreases late in the day, as shown in Fig. 2.1. On hot, sunny days there often is a temporary midday decrease in rate of photosynthesis, as seen in Fig. 2.1A. This has been attributed to stomatal closure caused by water stress, but some investigators claim that it is caused by inhibition of the photosynthetic mechanism by water stress and stomatal closure is merely coincidental (e.g., Farquhar and Sharkey, 1982). This is questioned by other researchers (Downtown *et al.*, 1988). Certainly in many cases, stomatal closure and reduction in rates of photosynthesis and transpiration occur simultaneously with leaf dehydration (Regehr *et al.*, 1975; Bacone *et al.*, 1976; Teskey *et al.*, 1986) (Figs. 2.2 and 7.15), but in other cases they apparently do not (Scarascia-Magnozza *et al.*, 1986). The relative importance of stomatal and nonstomatal inhibition of photosynthesis is discussed further in the section on photosynthesis in water-stressed plants in Chapter 7.

Seasonal Variations in Photosynthesis

The seasonal course of photosynthesis in deciduous species is affected by changing leaf area and photosynthetic capacity per unit of leaf area, and by environmental factors such as light intensity, temperature, photoperiod, and water supply. Leaf area is more constant in evergreen than in deciduous trees, but changes in photosynthetic capacity of leaves and seasonal environmental changes are important in both. The seasonal course of photosynthesis and important environmental factors are shown in Fig. 2.3 for an apple tree growing in an orchard at Ithaca, New York. Respiration exceeded photosynthesis until mid-May, but as the leaf area increased, photosynthesis increased rapidly and remained high until September. It then began to decrease along with declining temperature and daylength and decrease in leaf area. Leaves began to fall in mid-October, but the few leaves remaining on the tree carried on a surprising amount of photosynthesis until early November. According to Friedrich and Schmidt (1959), apple trees have a higher rate of photosynthesis than pear, cherry, or plum trees and continue the process later in the autumn. Nelson and Isebrands (1983) reported that a poplar hybrid that retained its leaves after several frosts exported considerable photosynthate to the lower stem and roots late in the season. This suggests that late leaf retention should be considered in tree improvement programs for deciduous species.

Winter photosynthesis is important for evergreen trees, especially those growing in regions with mild winters and dry summers. Emmingham and Waring (1977) and Waring and Franklin (1979) reported that conifers in the Pacific Northwest fix 30 to 65% of their annual total carbon during the dormant season. In contrast, in central Sweden only 5% of the carbon fixation of Scotch pine occurs during the winter. However, some winter photosynthesis occurs in the coastal region of Norway and in northern Italy (Larcher, 1961), and seedlings of Sitka spruce doubled in dry weight from late September to mid-May in southern Scotland (Bradbury and Malcolm, 1978). In the southern United States, pine seedlings in nurseries usually make

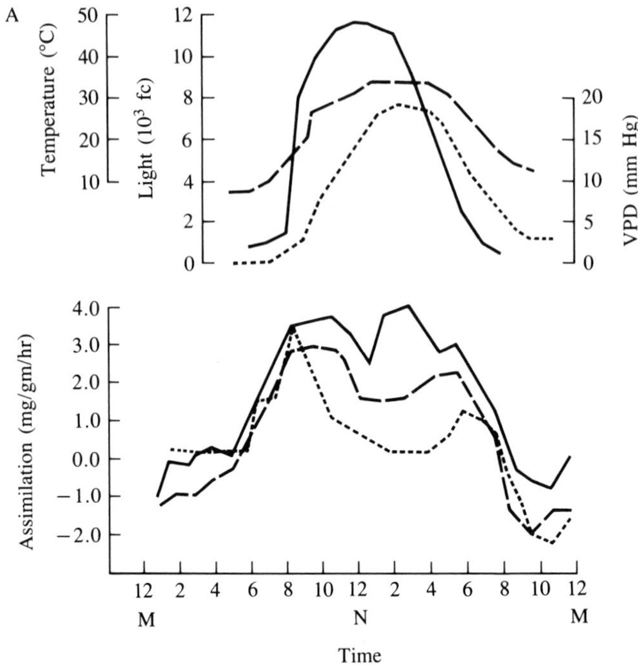
A
Temperature (°C)
50
40
30
20
10
Light (10³ fc)
12
10
8
6
4
2
0
VPD (mm Hg)
20
15
10
5
0
Assimilation (mg/gm/hr)
4.0
3.0
2.0
1.0
0.0
−1.0
−2.0
12 2 4 6 8 10 12 2 4 6 8 10 12
M
N
M
Time

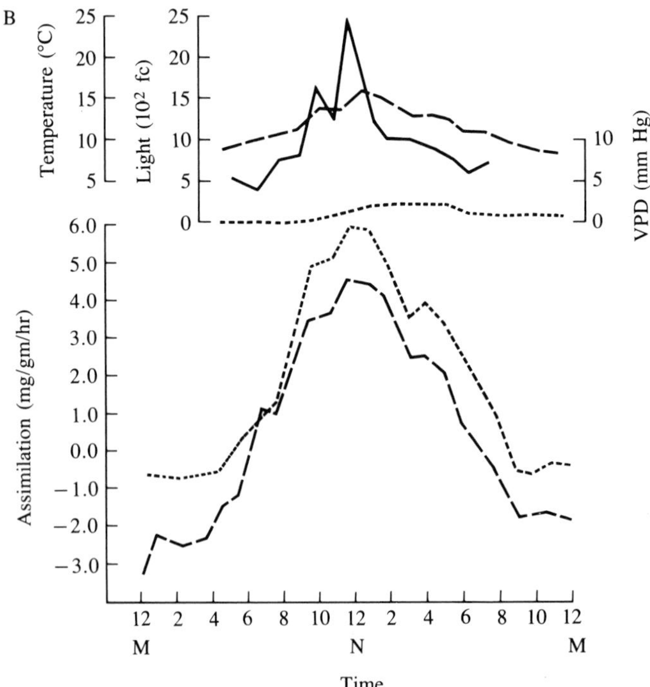
B
Temperature (°C)
25
20
15
10
5
Light (10² fc)
25
20
15
10
5
0
VPD (mm Hg)
10
5
0
Assimilation (mg/gm/hr)
6.0
5.0
4.0
3.0
2.0
1.0
0.0
−1.0
−2.0
−3.0
12 2 4 6 8 10 12 2 4 6 8 10 12
M
N
M
Time

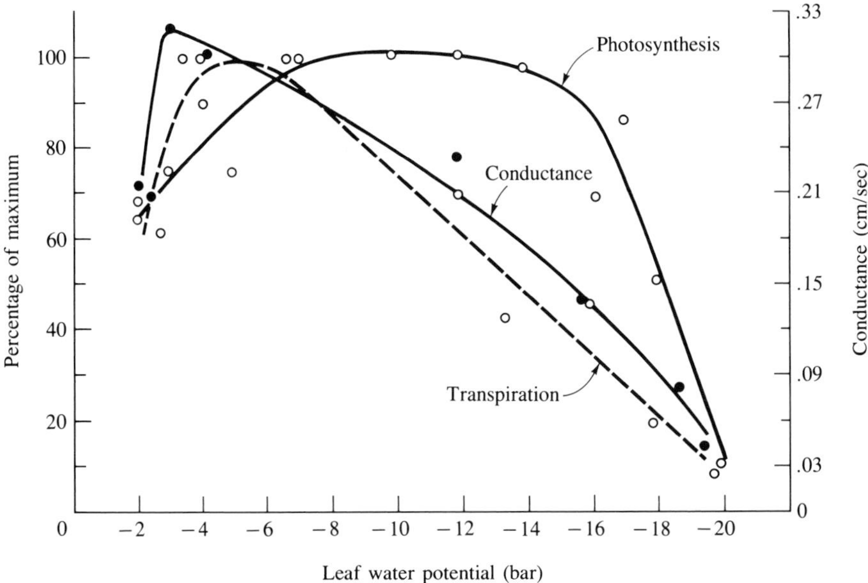

Figure 2.2 Changes in photosynthesis (○——○), transpiration (○−−○), and stomatal conductance (●——●) of winged elm with decreasing leaf water potential. (From Bacone *et al.*, 1976.)

significant increases in dry weight over the winter. Figure 2.4 shows the effects of summer drought and winter freezes on potential photosynthesis of Douglas-fir in Oregon. Winter photosynthesis is discussed further in Chapter 5.

A marked seasonal change in capacity to carry on photosynthesis was observed in a group of loblolly and eastern white pine seedlings kept out-of-doors at Durham, North Carolina, and brought in at intervals to measure photosynthesis under standard conditions. Figure 2.5 shows a steady increase in rate from February to June, a maximum rate in July, August, and early September, and a gradual decrease from October to January. The increase from February to April and the decrease in the autumn must have resulted from changes in photosynthetic capacity. The fact that loblolly pine had three flushes of shoot growth and eastern white pine only one did not affect the average summer rates. Strain *et al.* (1976) also observed seasonal changes in photosynthetic capacity of loblolly pine needles in the field. Winter chlorosis is common in pines of the southern United States and Perry and Baldwin (1966) observed considerable disorganization of chloroplasts in loblolly pine needles during the winter, which probably limits photosynthesis.

Figure 2.1 The daily course of photosynthesis of three species of well-watered conifers (——, Scotch pine; −−, noble fir; ---, grand fir) on a (A) sunny and a (B) cloudy day. On sunny days there often is a midday decrease in photosynthesis caused by leaf water deficit. VPD (Vapor pressure deficit), ---; light, ——; temperature, −−. (After Hodges, 1967; from Kramer and Kozlowski, 1979, by permission of Academic Press.)

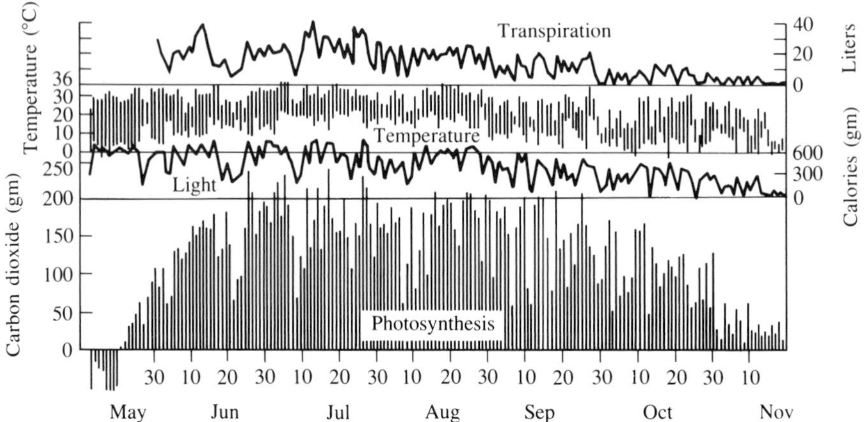

Figure 2.3 Seasonal course of photosynthesis and transpiration of an 8-year-old McIntosh apple tree enclosed in an air-conditioned greenhouse in an orchard at Ithaca, New York. Variations in rate of CO_2 uptake of the tree were correlated with variations in light intensity over most of the season. When growth started in early May, respiration released more CO_2 than was used in photosynthesis. (After Heinicke and Childers, 1937; from Kramer and Kozlowski, 1979, by permission of Academic Press.)

Photosynthesis in Relation to Leaf and Tree Age

The use of food in assimilation and respiration by very young leaves usually exceeds production by photosynthesis, hence growth is dependent on food imported from the woody structure or on photosynthate from nearby older leaves (Dickmann and Kozlowski, 1968; Dougherty *et al.*, 1979; Hanson *et al.*, 1988b). In general, the rate of photosynthesis increases until the leaves are fully expanded, then gradually declines. The transition from dependence on an outside source of food to independence is discussed by Turgeon (1989). The rate of photosynthesis of leaves living more than one year usually decreases after the first year (Fig. 2.6). Such changes probably are related to alterations in leaf structure, stomatal behavior, and activity of enzymes such as RuBP carboxylase, and shifts in photorespiration and dark respiration also occur (Dickmann and Kozlowski, 1968). Miller (1986) cited data indicating that growth of evergreens is much better correlated with the amount of current year foliage than with total foliage. This suggests that only the foliage of the current year contributes a significant amount of photosynthate to growth and raises questions concerning the value of the longer life of leaves on evergreens. Matyssek (1986) argued that the evergreen habit conserves nitrogen and water, but Gower *et al.* (1989) found that deciduous western larch makes more efficient use of foliage nitrogen than evergreen loblolly pine. More research is needed on the contribution of older leaves.

There also are variations in rate related to changes in relative size of sources and sinks for photosynthate. According to Maggs (1965), removal of part of the leaves of an apple tree results in an increase in rate of photosynthesis of the remaining

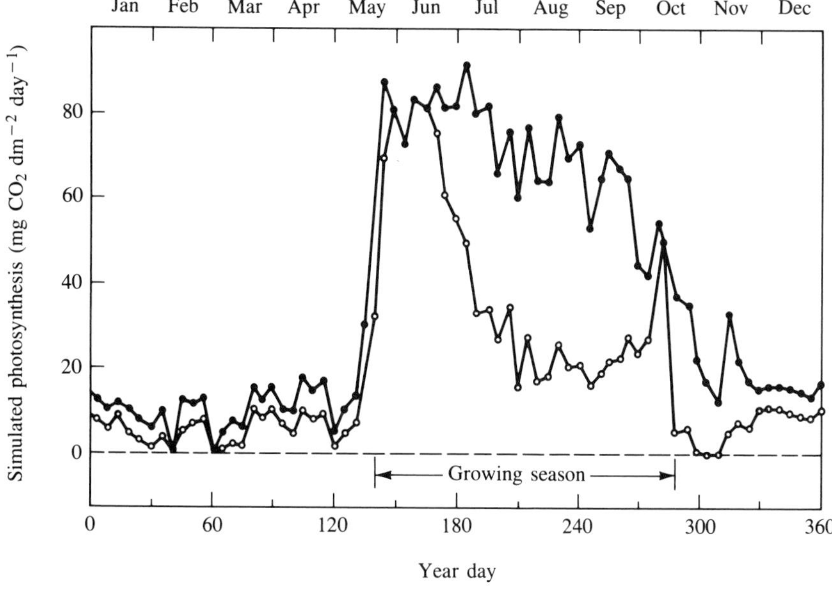

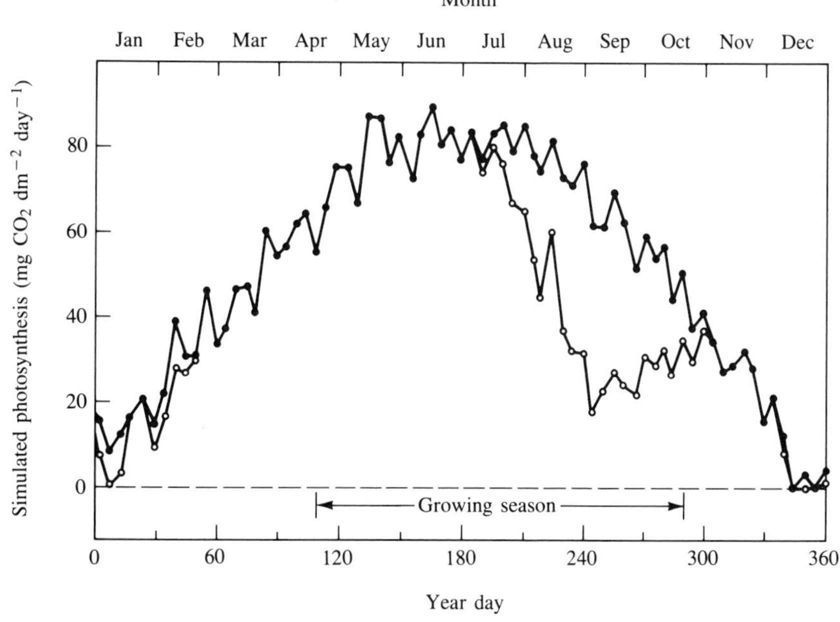

Figure 2.4 Potential rate of photosynthesis of Douglas-fir in western Oregon adjusted for seasonal changes in average temperature and irradiation, but assuming no drought or severe winter freezes (●), and predicted rate when adjusted for effects of the usual summer drought and winter freezes (○) that decrease the actual rate below its potential rate. Sites were in (A) the Cascade Mountains, where long, dry summers and winter frosts reduce predicted photosynthesis to 40% of the potential, and (B) the coastal region, where the summer drought is shorter and mild winters are the rule and photosynthesis is predicted to be 74% of the estimated potential. (From Emmingham and Waring, 1977.)

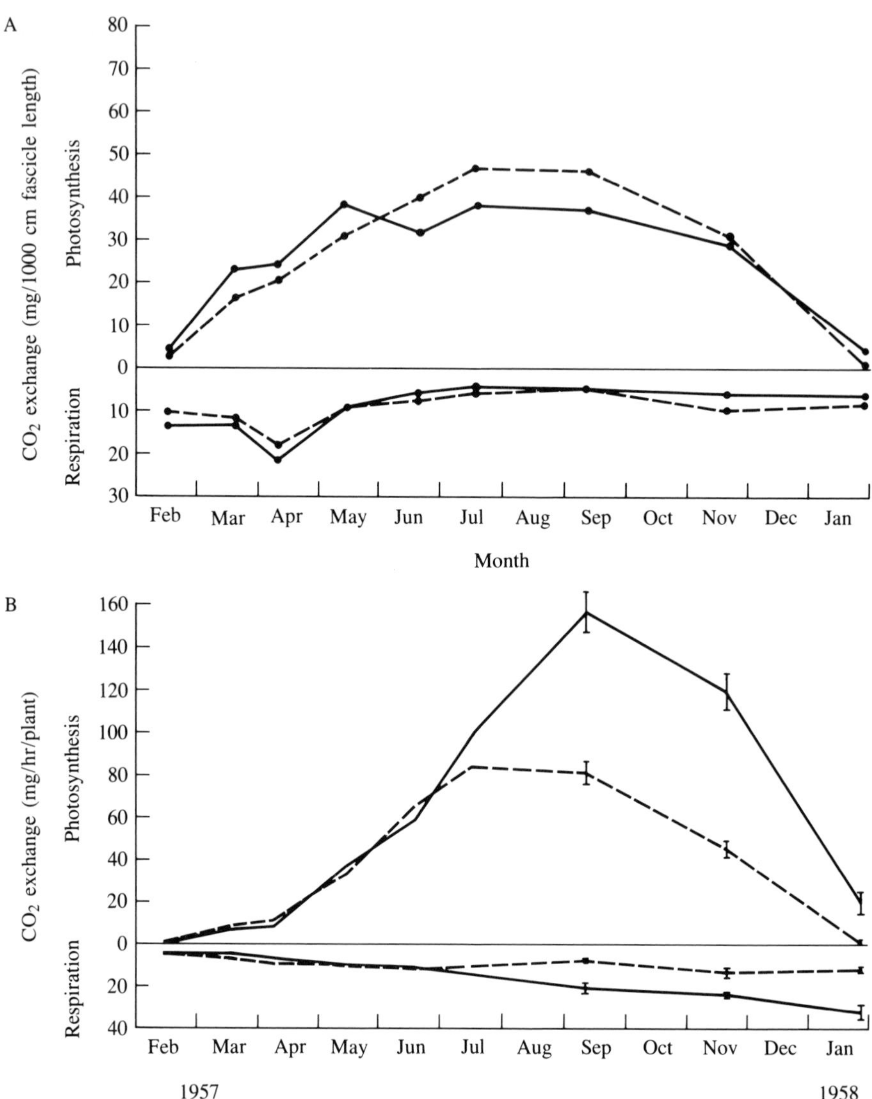

Figure 2.5 (A) Seasonal course of photosynthetic capacity and respiration per 1000 linear cm of needle fascicles of loblolly (——) and white pine (– –) seedlings brought indoors at intervals for measurement of CO_2 exchange at 25°C and 4000 fc and in darkness. (B) Seasonal changes in net photosynthesis and respiration per seedling. (After McGregor and Kramer, 1963; from Kramer and Kozlowski, 1979, by permission of Academic Press.)

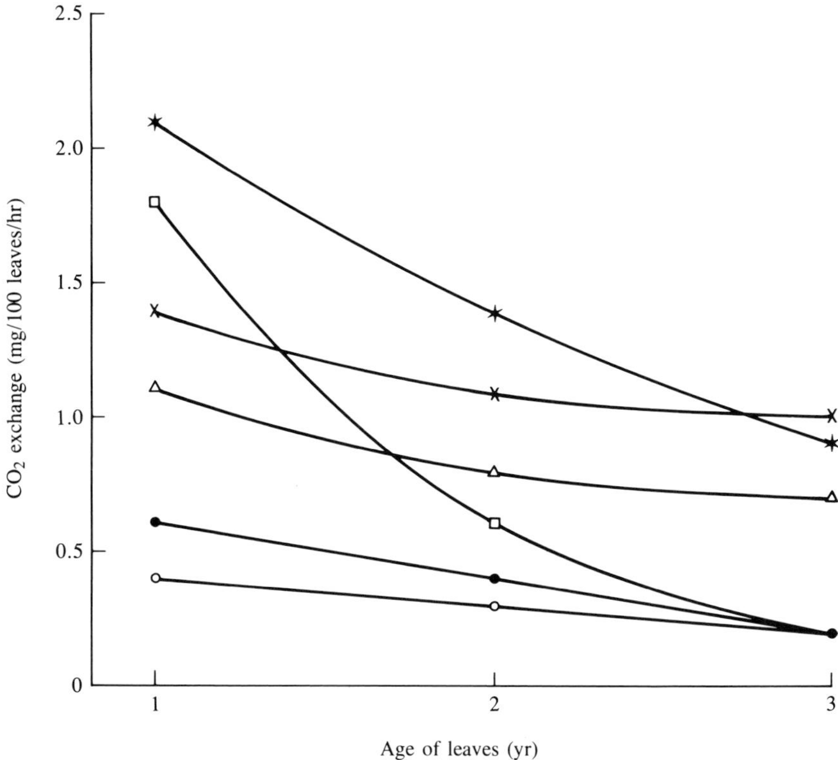

Figure 2.6 Effect of increasing age of conifer needles on net photosynthesis measured under standard conditions. Austrian pine, ★; white pine, △; Scotch pine, X; Western yellow pine, □; white fir, ●; Colorado spruce, ○. (After Freeland, 1952; from Kramer and Kozlowski, 1979, by permission of Academic Press.)

leaves and removal of part of the root system also causes an increase in photosynthesis. Development of a crop of fruits on an apple tree also is said to cause an increase in rate of photosynthesis (Avery, 1977). For example, the rate of photosynthesis was much higher for flowering and fruiting branches of apple than for nonfruiting branches (Hansen, 1970; Fujii and Kennedy, 1985). Hanson *et al.* (1988b) reported that the rate of photosynthesis of second flush leaves of northern red oak increases when third flush leaves begin to develop and act as a sink. Thus there appears to be a strong interaction between size of sinks and rate of photosynthesis. There also is debate concerning the role of carbohydrate sink size versus hormone relations in the stimulation of photosynthesis when sink size is increased relative to source size (Neales and Incoll, 1968; Wareing *et al.*, 1986; Hanson *et al.*, 1988b). Probably both are involved. Perhaps some day the allocation of photosynthesis can be controlled and more efficient utilization of the products of photosynthesis will become possible.

Photosynthesis in Relation to Growth

Because growth is dependent on the carbohydrates produced by photosynthesis, there has been hope that if the rate of photosynthesis could be increased there would be an increase in growth. In broad terms, there must be some correlation because the dry matter is supplied by photosynthesis. However, poor correlation often exists between short-term measurements of the rate of photosynthesis per unit of leaf surface and plant dry matter production. Briggs *et al.* (1986) reported that aboveground dry matter increase of bigtooth aspen was twice as great on a good as on a poor site, but the rate of photosynthesis was only 50% greater. Ledig and Perry (1967) found a slightly negative correlation between rate of growth and rate of photosynthesis in 18 families of loblolly pine seedlings. However, Ceulemans and Impens (1983) found that differences in rate of photosynthesis with light saturation were significantly correlated with differences in growth among young poplar clones. Carter (1972), Helms (1976), and Ledig (1976) discussed the difficulty in relating tree growth to short-term measurements of the rate of photosynthesis, and Evans (1975) found the same difficulty in relating rate of photosynthesis and yield of crop plants. This problem also is discussed in Chapter 4.

The poor correlation probably results partly from the fact that dry matter production depends on leaf area, leaf duration, and leaf exposure, and all of these vary during a growing season and during the life of trees. Also, photosynthesis usually is measured on only a few leaves or branches a few times during a growing season, but as Woodman (1971) found, the rate of photosynthesis varies over time and in various parts of the tree crown (Fig. 2.7). Therefore, it is not surprising that such limited sampling fails to provide a reliable estimate of a process that occurs over many growing seasons with wide variations in plant and environmental factors. One of the best correlations between photosynthesis and tree growth was that obtained for an apple tree enclosed in an air-conditioned greenhouse for an entire growing season (see Fig. 2.3) (Heinicke and Childers, 1937).

Another major problem arises from the fact that not all the photosynthate is used in producing new tissue, but a large and somewhat variable fraction is used in respiration and some is used in the production of various secondary substances found in trees and other plants. The various factors affecting dry matter production were discussed by Jarvis and Leverenz (1983). We will next turn to a consideration of the various uses made of the products of photosynthesis by plants.

Uses of the Products of Photosynthesis

Photosynthesis is only the first step in the complex series of biochemical processes that produce a plant. These include conversion of the primary products of photosynthesis into other carbohydrates and into fats, proteins, and a variety of secondary compounds such as alkaloids, tannins, and rubber; assimilation into new plant tissue; and oxidation in the process of respiration. The surplus food left over from

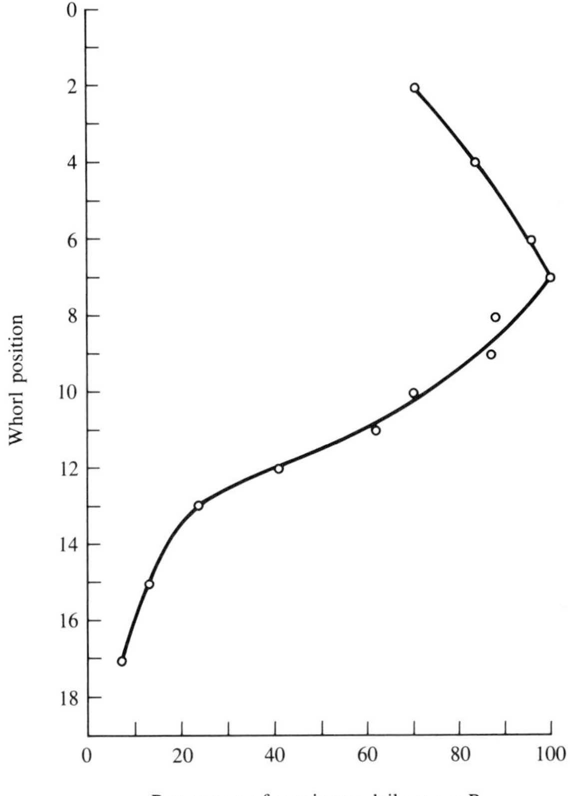

Figure 2.7 Variation in rate of photosynthesis at various distances below the treetop of a 38-year-old Douglas-fir tree growing in a closed canopy. (After Woodman, 1971; from Kramer and Kozlowski, 1979, by permission of Academic Press.)

these biochemical activities is accumulated in seeds, fruits, and vegetative tissues. Many of the interrelationships among important compounds are shown in Fig. 2.8. The complex biochemistry of these transformations lies outside the scope of this book. It is discussed briefly in Kramer and Kozlowski (1979) and readers are referred to plant biochemistry texts such as Stumpf and Conn (1980–) for more detailed discussions.

Carbohydrates

Carbohydrates are not only the first products of photosynthesis but also the chief constituents of plant dry matter and the substrate for respiration. They are composed of carbon, hydrogen, and oxygen in the proportions of $(CH_2O)n$, and some contain nitrogen or phosphorus. They are classified into three groups: mono-, oligo-, and polysaccharides. Monosaccharides are simple sugars such as glucose ($C_6H_{12}O_6$) and fructose with six carbon atoms and pentose with five carbons ($C_5H_{10}O_5$). As indicated in Fig. 2.8, glucose phosphate is the starting point for a number of carbohydrate transformations. The process of phosphorylation can be illustrated by the following equations:

$$\text{glucose} + \text{ATP} \xrightarrow{\text{hexokinase}} \text{glucose 6-phosphate} + \text{ADP}$$

$$\text{glucose 6-phosphate} \xrightarrow{\text{phosphoglucoisomerase}} \text{fructose 6-phosphate}$$

The oligosaccharides consist of two, three, or four monosaccharide units linked together. They include the common disaccharides ($C_{12}H_{22}O_{11}$) sucrose and maltose, the trisaccharides ($C_{18}H_{32}O_{11}$) raffinose and melezitose, and the tetrasaccharide stachyose ($C_{24}H_{42}O_{11}$). Sucrose is most important because it is abundant, it is the carbohydrate most commonly translocated, and it is an important reserve carbohydrate. Maltose is common, but in lower concentration, and the other oligosaccharides are found only in certain plant families or species.

The polysaccharides consist of large numbers of monosaccharide units connected by ether linkages. Cellulose is a polymer that consists of about 3000 glucose residues linked together by β-1,4 glycosidic bonds in straight chains. It is the chief constituent of cell walls, contains about one-third of all the carbon fixed by plants, and probably is the most abundant biochemical compound in the world. The role of cellulose in the evolution of land plants was discussed by Duchesne and Larson (1989). Another polysaccharide, starch, an abundant reserve carbohydrate in plants, is formed by the condensation of hundreds of glucose molecules linked together by α-1,4 glycosidic bonds into long, spiral, branched chains. Starch is easily hydrolyzed enzymatically through maltose to glucose, which can be recycled.

The hemicelluloses are another group of polysaccharides found in the cell walls of all woody and some herbaceous plants. They also are found as reserve foods in some seeds, including palm and persimmon. Pectic compounds are hydrophilic substances found in the middle lamella and primary wall of cells and are abundant in some fruits. Such fruits usually make good jelly. Gums and mucilages are polysaccharides found in seeds and stems of many woody plants. Gum arabic or gum acacia comes from an African acacia (*Acacia senegal*), and gum exudes from wounds in stems of cherry, peach, plum, and spruce. Gums should not be confused with the resins released by injured pines, which are terpenes belonging with the lipids, nor with latex substances such as rubber and chicle, which are obtained from certain trees.

Nitrogen-Containing Compounds

Although nitrogen forms less than 1% of the dry weight of a tree, nitrogen-containing compounds are extremely important physiologically. The most abundant

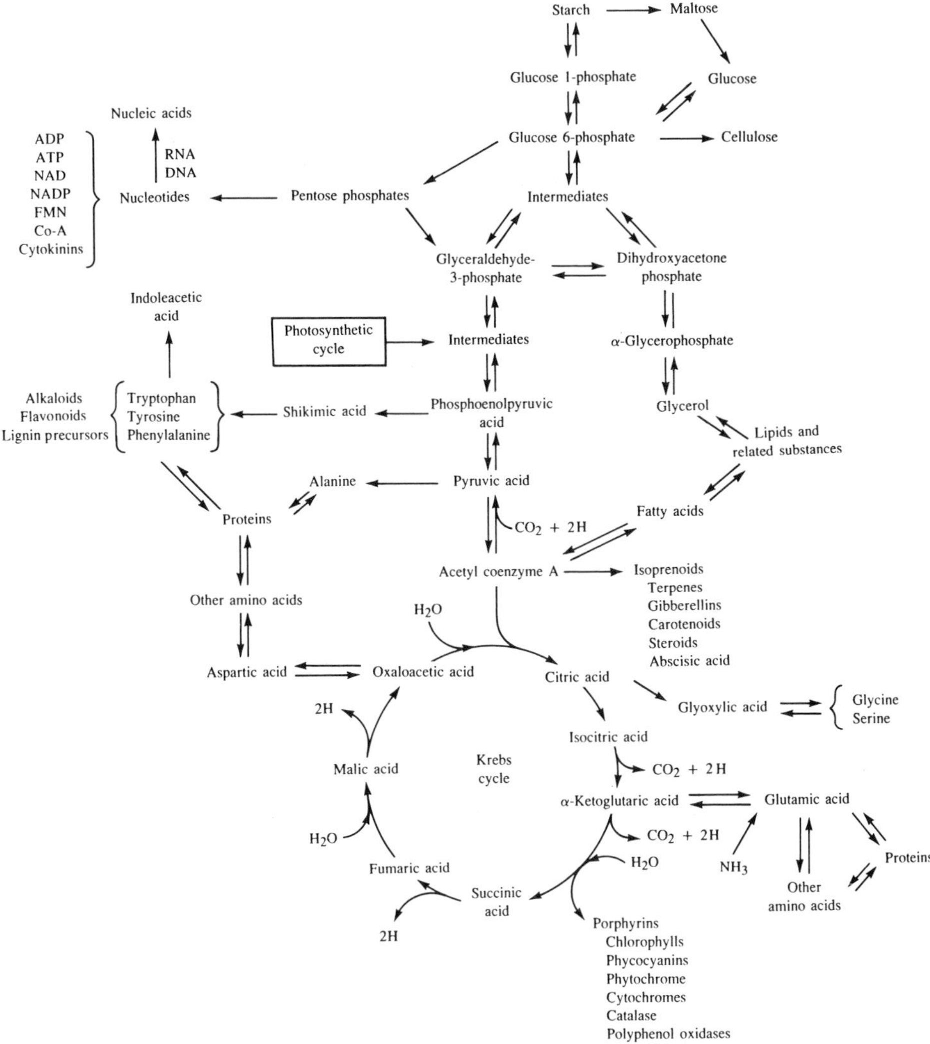

Figure 2.8 Diagram showing some metabolic pathways and the origin of important compounds in plants. (From Kramer and Kozlowski, 1979, by permission of Academic Press.)

compounds are the proteins in protoplasm and in the hundreds of enzymes that catalyze biochemical reactions. Proteins also are important storage compounds in seeds, particularly those of leguminous plants. Other important nitrogen-containing compounds include amides, amino acids, nucleic acids and nucleotides, chlorophyll, and alkaloids. Most of these compounds play essential roles in plants, but some, such as many of the alkaloids, seem merely to be by-products of plant metabolism. This question will be discussed later in the section on Secondary Compounds.

The highest concentration of nitrogen occurs in meristems and other physiologically active regions. Twenty to twenty-five percent of the nitrogen in apple and pine trees occurs in the leaves, perhaps half of it in the photosynthetic enzyme RuBP carboxylase. Twigs, small roots, and the phloem also contain considerable amounts of nitrogen. The amount of nitrogen present varies with the age, stage of development, and physiological activity of the tissue. In general, the nitrogen content of branches increases during the dormant season, decreases abruptly in the spring when growth starts, and remains low during the growing season. The heavy use of nitrogen during periods of rapid growth suggests the desirability of coordinating application of fertilizer with growth and bud formation (see Chapter 13). The nitrogen content of young leaves as a percentage of dry weight is high, but tends to decrease with increasing age because of loss by leaching and dilution caused by increase in cell wall components of leaf dry weight. As leaves begin to senesce, much of the nitrogen is translocated back into the twigs, where it is stored until growth is resumed the next spring. There is some evidence that the nitrogen stored in the bark (phloem) of apples, peaches, and grapes is more available when growth is resumed than is new nitrogen applied as fertilizer (Taylor and May, 1967; Possingham, 1970; Tromp, 1970).

The Nitrogen Supply

The most frequent limitation on plant growth after water stress probably has been nitrogen deficiency. However, in recent decades there has been a dramatic increase in deposition of airborne nitrogen compounds that seems to be saturating some forest soils, causing unbalance in mineral nutrition that is reducing tree growth (Aber *et al.*, 1989; Schulze, 1989). Because of their longer growing season and slower growth, woody plants suffer less than annual crops from nitrogen deficiency, but most ornamental trees and shrubs and many stands of forest trees benefit from fertilization with nitrogen. Much of the benefit, at least in conifers, seems to result from an increase in leaf area (Brix, 1983), although there sometimes is a significant increase in rate of photosynthesis. The role of nitrogen is discussed in Chapter 6 and nitrogen fertilization in Chapter 13.

Lipids and Related Compounds

Fats and oils belong to a heterogeneous group of compounds known as lipids, which have the common characteristics of being soluble in organic solvents such as benzene and ether and relatively insoluble in water. Other compounds belonging to this group include cutin, suberin, waxes, phospholipids, and the group of compounds called isoprenoids, which include terpenes, oleoresins, abscisic acid, the gibberellins, carotenoids, and rubber. True fats and oils are esters of glycerol and fatty acids, while waxes, cutin, and suberin contain long-chain fatty acids and alcohols other than glycerol. The simple lipids occur in low concentrations in all cells as droplets and as constituents of cell membranes. They also are important forms of stored food in fruits such as olive and avocado and in many kinds of nuts and seeds.

Lipids form protective coatings on the outer surfaces of leaves, stems, and fruits in the form of wax, cutin, and suberin. The cuticle covering the epidermal cells of plants often is rendered relatively impermeable to water by the wax embedded in it and extruded onto its surface (Kolattukudy, 1975; Schönherr, 1976). The wax on leaf surfaces varies from traces to as much as 15% of the dry weight (Eglinton and Hamilton, 1967), and where a large amount is present it makes leaf surfaces so difficult to wet that wetting agents must be added to spray materials. A layer of wax creates "bloom" on some leaves, adding to their attractiveness. Wax sometimes accumulates in stomatal pores of conifers and reduces stomatal conductance. Electron micrographs of wax deposition on leaf surfaces are shown in Fig. 2.9. The nature, amount, and arrangement of wax particles probably are more important than the thickness of the cuticle in controlling cuticular transpiration. Cutin also is a barrier to penetration by germinating fungal spores unless they are induced by contact with cutinized surfaces to produce cutinase, which digests cutin to form pathways for germ tube penetration (Padilla *et al.*, 1988).

Many lipids and related compounds are of commercial importance. Olive oil and palm oil are widely used, the oil from tung seeds is an excellent drying oil for varnishes, and the wax from the leaves of a palm (*Copernicia cerifera*) supplies the carnauba wax used in polishes. The liquid wax found in the seeds of the desert shrub jojoba is of commercial interest because it resembles sperm oil.

Suberin is important as the waterproof material in bark. In older woody stems and roots, the outer layers of cells of the phloem become impregnated with suberin, rendering them relatively impermeable to water. Internal suberization also occurs, that in the radial walls of the endodermis being of particular interest. Internal layers of lipid materials are fairly well known in plants. Scott (1964) reported that a hydrophobic layer develops wherever cell walls are exposed to air in the intercellular spaces, and Norris and Bukovac (1968) found that the inner, exposed surfaces of the cells of the lower epidermis of pear leaves are cutinized. The amount and importance of cutinization of internal surfaces in leaves deserve more study.

Secondary Compounds

In addition to the groups of primary compounds just discussed, many other compounds are produced by secondary reactions from primary carbohydrates, amino acids, and lipids. These secondary compounds are responsible for many of the differences among plants with respect to color, odor, taste, and resistance to pathogen and insect attacks. They usually are products of metabolic bypaths and most of them can be classified into one of three groups: alkaloids, phenolics, or terpenes. The alkaloids are a large and heterogeneous group of compounds that have few known functions in plants. However, they include numerous compounds of commercial and pharmaceutical importance, such as atropine, caffeine, cocaine, nicotine, opium, quinine, and many others. The most common phenolic compounds are the lignins, which stiffen the walls of wood cells and are second only to cellulose in abundance. Tannins are phenolic compounds of commercial importance in tanning

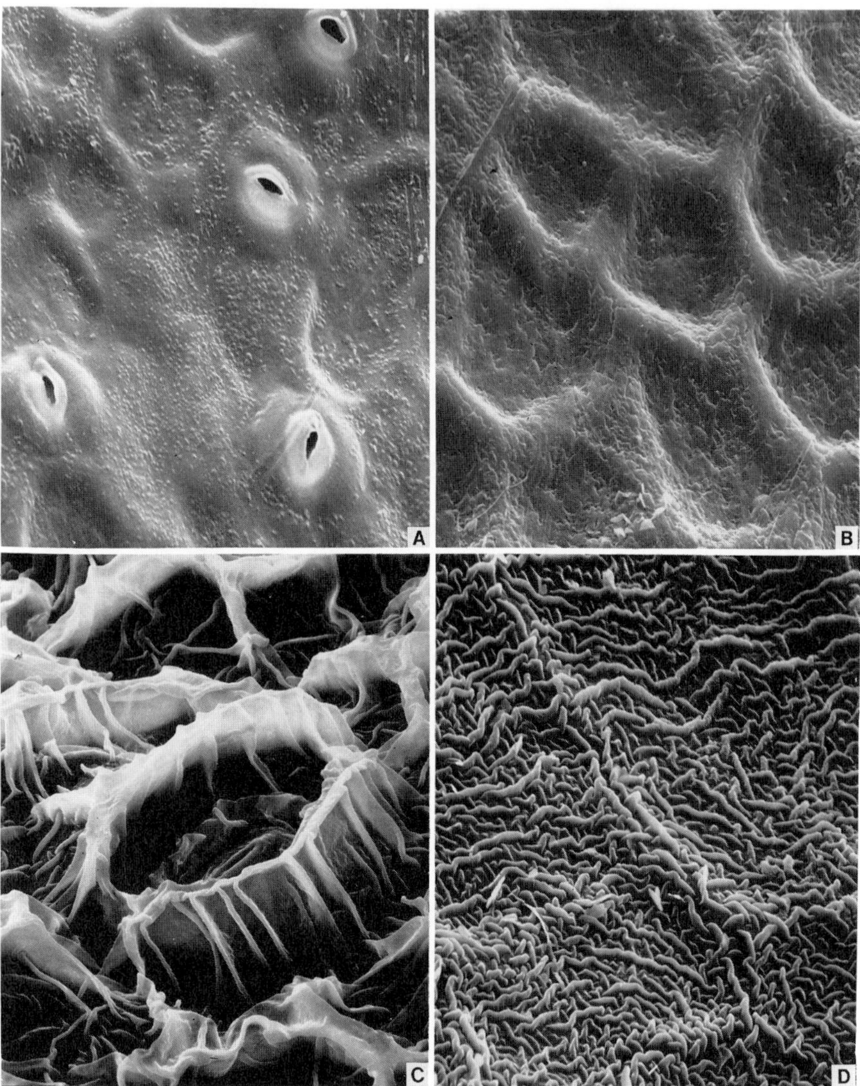

Figure 2.9 Scanning electron micrographs of variations in the appearance of wax deposits on leaves of (A) cacao, lower surface; (B) cacao, upper surface; (C) rubber, lower surface; and (D) rubber, upper surface (1500×). (From Sena Gomes and Kozlowski, 1988a.)

leather and as deterrents to insect feeding, but are a great nuisance to the plant chemist. The flavonoids include anthocyanin and other water-soluble pigments.

The isoprenoids or terpenes are another large group of secondary compounds of much biological and commercial importance. They include essential oils, which provide most of the odors produced by plants and are used in perfumes. Examples are attar of roses and the oils of eucalyptus and mint. Terpenes also include resin gums such as copal, dammar, and kauri, carotenoids, rubber, and the oleoresins of pines used in the form of turpentine and rosin. The plant hormones abscisic acid and gibberellin also are terpenes. The chemistry and economic uses of some of these compounds were discussed by Raven in Cannell and Jackson (1985). According to Monson and Fall (1989), 2.5 to 8% of the carbon fixed in photosynthesis is lost from quaking aspen leaves as volatile isoprenoids. The biochemistry of isoprenoid synthesis was reviewed by Kleinig (1989).

Some secondary compounds such as carotenoids and hormones clearly have important functions in plant metabolism and growth. Others, such as most of the alkaloids and tannins, and terpenes such as rubber, are of economic importance but have few or no obvious roles in plant metabolism and any indirect benefits to plants are largely fortuitous. In recent decades there has been increasing interest and even controversy concerning the role of secondary chemical compounds in protecting plants from predators, ranging from bacteria and fungi to insects and grazing animals. Long ago it was observed that some insects feed chiefly or exclusively on one kind of plant and grazing animals often prefer one kind to another. This led to consideration of the probability that plants may produce chemicals that are either attractive or repellent to herbivores. There has been extensive discussion of the possible coevolution of insects and plant chemical defenses, some of which is reviewed in the journal *Ecology* (Vol. 69, No. 4, 1989). The complex interactions between plants and their predators have been discussed in many papers and in books edited by Rosenthal and Janzen (1979), Heinrichs (1988), and Mattson *et al.* (1988), and in Vol. 19 of *Recent Advances in Phytochemistry.* A paper in the latter by Rhoades advances the interesting idea that trees attacked by leaf-eating insects may produce a volatile substance, probably ethylene, that causes adjacent trees to produce substances that render their foliage less palatable to insects. There is a need for more research in this area.

Of special interest are the occasional explosions of populations of defoliating insects and borers that often persist for several years and cause damage over large areas. Examples are the bronze birch borer infestation that occurred in New England and Canada in the 1930s, the gypsy moth infestation in the northeastern United States in the 1980s, and the periodic outbreaks of southern pine bark beetles in the pine forests of the southern United States. Such outbreaks occur all over the world and probably have been occurring ever since plants and insects came into existence. For example, Duncan and Hodson (1958) stated that severe defoliation of aspen forests in Minnesota by tent caterpillars has occurred about every ten years since observations began in 1870.

There has been much discussion concerning the causes of these cyclical infestations, but they are most often ascribed to loss of tree vigor caused by unfavorable environmental conditions, senescence, or, more recently, air pollution. There probably are multiple causes. Mattson and Haack (1987a,b) suggested that moderate drought often is favorable for insect colonization and reproduction. However, severe drought can result in physical changes in leaves such as thickened cell walls and increased toughness, and chemical changes such as increase in phenolics, terpenes, and proline and decrease in osmotic potential and starch reserves, any or all of which might decrease their palatability.

Wagner and Evans (1985) reported that the new foliage formed on ponderosa pine after partial defoliation is more palatable than the old foliage even though it is high in phenols. Foliage formed after release from water stress may also be more palatable. It seems possible that the physical condition of foliage is just as important for insect feeding as the chemical composition. Coley (1987) found that over 70% of the wide differences in extent of insect feeding among species in a tropical rain forest could be attributed to differences in leaf toughness or nutritional value.

Not only do plants provide nutrients, but they occasionally provide substances essential for insect development, and it has been suggested that certain kinds of plants could be protected from certain insects if synthesis of these substances could be prevented. A few kinds of plants such as the Podocarpaceae produce substances that are inhibitory to insect development. The interrelationships among plant nutrition, rate of plant growth, and the production of secondary compounds that discourage herbivory are a topic of lively discussion (Coley *et al.*, 1985, and papers cited earlier). In spite of the strong interest in protective compounds in specific instances, it is difficult to prove that overall they are essential to survival because many plants succeed without them.

Some of the same compounds that discourage herbivory also discourage attacks by pathogenic fungi and decay of wood by saprophytic fungi. Loehle (1988a), following Shigo (1984b), classifies the defenses against pathogens and decay into three groups: increase in wood density, accumulation of chemicals such as phenolics and tannins in the wood, and compartmentation or sealing off of infected regions. Shigo (1984b) emphasizes the importance of the formation of barriers around wounds and dying branches (compartmentalization) that limit the spread of organisms. Unfortunately, according to Shigo these barriers also cause mechanical weakness, resulting in cracks that reduce the value of wood and increase the danger of wind damage.

Some secondary compounds are being used as a means of differentiating among cultivars and geographic races. For example, Mirov (1961), Thor and Barnett (1973), and Squillace (1971) used differences in terpene composition to establish taxonomic relationships among pines. Flake *et al.* (1969) used terpene differences to distinguish among populations of eastern red cedar and Fretz (1977) used similar criteria to distinguish among the numerous cultivars of creeping juniper. Differences in flavonoids (Asen, 1977) and polyphenols (Thielges, 1972) also have been used to identify cultivars. Alston and Turner (1963) and Gibbs (1974) have written books on chemical taxonomy.

Allelochems and Allelopathy

Substances produced by one organism that are harmful to other organisms are called allelochems or allelochemics and production of allelochems by plants is termed allelopathy. Theophrastus, Pliny, and other observers of ancient times noted that walnut trees and some herbaceous plants such as chick-peas appeared to injure neighboring plants. It is now known that plants of the genus *Juglans* produce a toxic substance, juglone, that inhibits growth of some plants. A variety of substances released from living plants and during the decay of plant residues cause allelopathic effects (Kramer and Kozlowski, 1979, p. 620). Rice (1984) reviewed allelopathy in detail and its ecological importance is discussed in Chapter 3.

Assimilation

The term "assimilation" is used here with reference to the conversion of carbohydrates, lipids, and amino acids into new tissue. Assimilation naturally is most important in meristematic regions such as root and stem tips and cambia where new tissue is being produced. However, it also occurs in fully expanded tissue, where processes such as cell wall thickening continue. The sugars translocated to growing regions are converted into cellulose, lignin, and pectic compounds, or used in respiration, while the amino acids and amides are incorporated into enzymes and other protein of new cells. Small quantities of lipids go into cells and cell membranes and larger quantities into the suberin, cutin, and waxes of the protective coatings on leaves, stems, and fruits. The orderly synthesis and incorporation of these numerous compounds into the structure of new cells require a degree of control at the molecular level that taxes the imagination.

Assimilation requires large amounts of chemical energy, especially for the synthesis of highly reduced compounds such as lignin and lipids, as shown in Table 2.2 from a study by Chung and Barnes (1977). For example, about 1.2 g of glucose is used to synthesize a gram of carbohydrate, but 1.9 g is required per gram of lignin and 3 g per gram of lipid. The energy required for these syntheses is supplied by respiration, which will be discussed in the next section.

Table 2.2
Amounts of Substrate Used and CO_2 Produced during Synthesis of the Principal Constituents of the Shoots of Loblolly Pine[a,b]

	Substrate used		By-products	
Constituents	Glucose	O_2	CO_2	H_2O
Nitrogenous compounds	1.58	0.28	0.40	0.65
Carbohydrates	1.18	0.11	0.13	0.16
Lipids	3.02	0.30	1.50	0.82
Lignin	1.90	0.04	0.27	0.66
Organic acids	1.48	0.35	0.48	0.35
Phenolics	1.92	0.37	0.56	0.73

[a] From Chung and Barnes (1977).
[b] Amounts are in grams per gram of constituent.

Respiration

Respiration refers to the process by which the energy stored in reduced carbon compounds during photosynthesis is released by oxidation in a form that can be used in assimilation and growth and in maintenance of cell structure and function. It occurs in all living cells, but more rapidly in growing regions than in mature tissue, and is very slow in dormant tissue. Respiration is sensitive to temperature, having a Q_{10} of about two, meaning that the rate doubles for every increase of 10°C. Most fruits, vegetables, and nursery stock therefore keep better if stored at low temperature. Exceptions are some tropical fruits that are injured at temperatures below 10 or 12°C and photosynthetic tissues that suffer chilling injury at low temperatures, especially if exposed to the sun (see Chapter 5).

The oxidation process can be summarized as follows:

$$C_6H_{12}O_6 + 6O_2 \rightarrow 6CO_2 + 6H_2O + \text{energy}$$

The carbon in glucose is split off in a series of steps, releasing hydrogen, which eventually combines with oxygen to form water. Energy is released during the accompanying rearrangements of atomic bonds. If the glucose were burned the energy would be released as heat, but during oxidation in cells about two-thirds of the energy is stored in ATP (adenosine triphosphate). Energy can be released from ATP to do the chemical work of assimilation, transport of materials, and maintenance of tissue structure. The respiration process requires about 25 enzymes and numerous intermediate steps (see Kramer and Kozlowski, 1979, Chapter 6). ATP is the principal form in which energy is stored and transferred in living organisms. It is particularly effective because when the terminal phosphate group is split off a large amount of energy is released.

In the absence of an adequate supply of oxygen to completely oxidize the substrate to carbon dioxide and water, intermediate compounds accumulate and much less energy is released than if oxidation were complete. Thus not only is anaerobic respiration inefficient, but the accumulation of intermediate compounds such as alcohols and organic acids is generally regarded as injurious. However, Jackson *et al.* (1982) claim that alcohol does not accumulate to a toxic concentration in roots of flooded plants. The relationship of anaerobic respiration to root injury in flooded soil is discussed in Chapter 8. Anaerobic respiration sometimes causes injury to fruit, vegetables, and tightly bundled tree seedlings kept too long in poorly ventilated storage chambers. Kimmerer and Stringer (1988) found ethanol and alcohol dehydrogenase activity in the cambium of trees of several species, suggesting but not proving that respiration in the cambium is oxygen limited.

Because the rate of respiration approximately doubles with each increase in temperature of 10°C, warm nights cause excessive use of food in respiration and a corresponding decrease in that available for growth. Decker (1944) found that although the respiration rate of loblolly and red pine seedlings increased up to 35°C, photosynthesis decreased at higher temperatures (Fig. 2.10), and Kramer (1957)

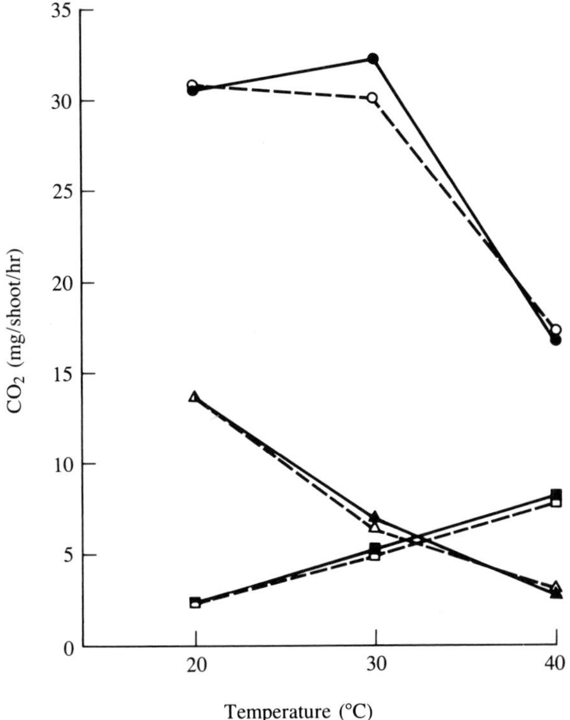

Figure 2.10 Rates of net photosynthesis (circles) and dark respiration (squares), and ratio (triangles), of loblolly (open) and red pine (filled) seedlings measured at 20, 30, and 40°C and 4500 fc. (From Decker, 1944.)

observed that decreasing the night temperature from 23 to 17°C resulted in increased height growth of loblolly pine seedlings. Hellmers and Rook (1973) found cool nights to be favorable for increase in dry weight and stem diameter of Monterey pine seedlings.

The rate of respiration is not tightly coupled to energy requirements and much food can be wasted by high night temperatures. It seems probable that reduction in rate of dark respiration might allow more food to be converted into new tissue. For example, pruning of lower branches as soon as their photosynthesis approaches the compensation point may be advantageous. In general, it seems probable that a decrease in the fraction of photosynthate used in respiration often will be accompanied by an increase in vegetative growth. It also appears that the extensive annual replacement of roots under forest stands may exceed the probable benefits, although this is debatable (Bowen, 1985; Teskey *et al.*, 1985; Carlson *et al.*, 1988). Writers from Lundegärdh (1931) to Lambers (1985) and other investigators have discussed the importance of low rates of respiration to the success of plants, and Loach (1967) suggested that the low rate of respiration of shade-tolerant species is a major factor in their success in deep shade. The ecological aspects of respiration are discussed further in Chapter 3, and Lambers (1985) reviewed factors affecting the rate of respiration and its relationship to growth. Jarvis and Leverenz (1983) discussed maintenance and growth respiration of trees in relation to forest productivity.

Storage of Food

The food budget of a plant can be expressed much like a bank balance:

Income	=	carbohydrate manufactured by photosynthesis
Expenditures	=	food used in growth (assimilation) and respiration
Balance	=	food accumulated

There is little accumulation of food in rapidly growing plants and starch accumulation often is at its minimum in spring and early summer when growth is most rapid. In the late summer and autumn, after vegetative growth has slowed or ceased, carbohydrates accumulate in roots, stems, and twigs and the next spring they are used when growth is resumed. Adams *et al.* (1986) reported that the starch content of roots of 8-year-old trees of loblolly pine was highest in the spring before buds opened and lowest in the autumn. As might be expected, seasonal cycles in accumulation of food are more pronounced in deciduous than in evergreen plants (Fig. 2.11). However, there are detectable cycles in carbohydrate accumulation even in tropical plants, with decreases occurring each time a new flush of growth occurs (Alvim and Kozlowski, 1977).

The situation may be somewhat different in individual branches such as those studied by Chung and Barnes (1980a,b). Figure 2.12 shows seasonal changes in photosynthate production, consumption, and the surplus available for export in a first year branch of loblolly pine, including a second flush of growth. Much of the annual accumulation probably occurred after the termination of this experiment. Accumulated carbohydrates are used to maintain respiration during periods when photosynthesis is reduced or ceases, and for the resumption of growth. The accumulation and storage of food are particularly important in deciduous plants because considerable root, shoot, and cambial growth usually occurs before leaves have fully expanded or attained their full photosynthetic capacity (Dougherty *et al.*, 1979). This suggests that seedlings and nursery stock should be left in the field in the autumn until they have accumulated enough food to carry them through a period of storage and permit resumption of growth when outplanted. Food accumulation also affects stump sprouting, which usually is less vigorous on stumps of trees cut when reserves are low than on those of trees cut or girdled during the dormant season when food reserves are high (see Chapter 3). Of course hormones may also play a role in stump sprouting.

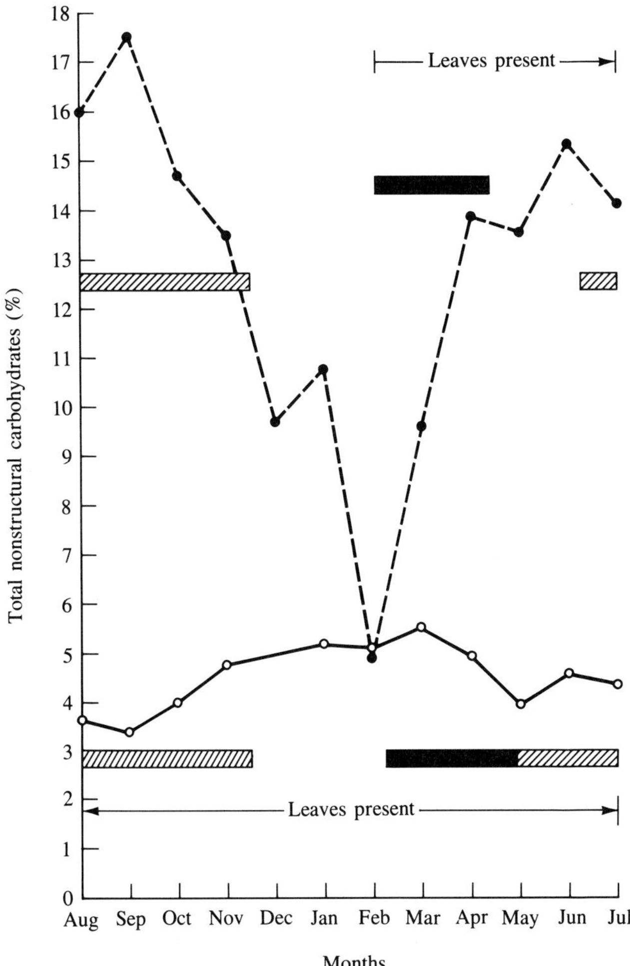

Figure 2.11 Seasonal change in total nonstructural carbohydrates (TNC) in twigs of drought deciduous California buckeye (– –) and evergreen California live oak (——). The deciduous buckeye shows a large decrease in carbohydrate content while leafless, and the evergreen oak shows only small seasonal variations in carbohydrate content. Hatched region shows the fruiting season; filled region shows stem growth. (After Mooney and Hays, 1973; from Kramer and Kozlowski, 1979, by permission of Academic Press.)

Translocation

Translocation refers to the long-distance transport of water, minerals, and organic substances in plants, but this section will deal only with organic substances. The use of food in assimilation and respiration in growing regions requires that large

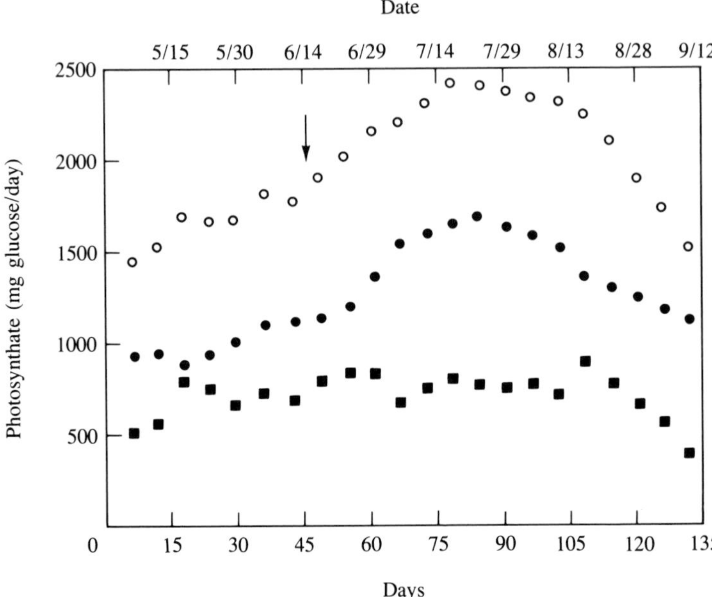

Figure 2.12 Seasonal change in photosynthate production (○), consumption (●), and the surplus available for export (■) in a first year branch of loblolly pine, including a second flush of growth (↓). (From Chung and Barnes, 1980b.)

amounts be translocated from the point of origin in photosynthetic tissue to distant parts of the plant where they are used or stored. Translocation is particularly important in trees and large vines in which organic compounds must move over distances of many meters to maintain root systems. In addition to carbohydrates, nitrogen-containing compounds, and hormones of endogenous origin, externally supplied growth regulators, viruses, herbicides, fungicides, and systemic insecticides are translocated considerable distances in the phloem, often along with carbohydrates. Although inorganic nitrogen supplied to the soil usually is reduced to organic compounds in the roots of trees and moved up in the xylem sap as amides and amino acids, most translocation of organic compounds occurs in the phloem. This has been demonstrated by various kinds of girdling experiments, which are relatively easy to perform on woody plants.

Carbohydrates are translocated chiefly as sucrose, but sorbitol moves in the phloem of apple and cherry, mannitol (from which "manna" is obtained) in ash, and raffinose, stachyose, and verbascose are constituents of the phloem sap of other woody plants. For the most part, translocation of hormones and herbicides seems to be correlated with that of carbohydrates. Nitrogen-containing compounds move chiefly as amides and amino acids, but large molecules such as proteins can be translocated as demonstrated by translocation of viruses through the phloem (Bennett, 1956).

Although there is some uncertainty concerning the mechanism of phloem transport, the mass flow hypothesis seems to be the most acceptable explanation. It assumes that movement occurs along a pressure gradient produced by differences in the osmotic potential in source and sink regions. Flow is a passive pressure-driven process, but phloem loading, the transport of solutes into the phloem from surrounding cells, is an active transport process. Although actual movement of solutes is little affected by temperature, phloem loading and perhaps unloading is sensitive to temperature and respiration inhibitors. Measurement of velocities gives translocation rates ranging from 10 to 250 cm hr^{-1}, and up to 300 cm hr^{-1} in cucurbits. Readers are referred to plant physiology texts and reviews for the details of the mass flow theory and more detailed discussions of phloem transport (Zimmermann and Milburn, 1975; Dale and Sutcliffe, 1986). The important fact is that translocation is essential for the growth of plant tissues that cannot carry on photosynthesis. It is particularly important with respect to the development of root systems, fruits, and seeds. Girdling of stems kills trees by interrupting the food supply to the roots, thus stopping their growth and finally killing them, thereby reducing the absorption of minerals and water. Eventually the tops of girdled trees usually are killed by water stress, although lack of hormones normally supplied by the roots may be a contributing factor.

Internal Competition and the Partitioning of Food

As suggested earlier in this chapter, there is competition for food between reproductive and vegetative growth. There also is competition between roots and shoots, and shading often results in reduced root growth. In fact a growing plant can be regarded as a system of potentially competing organs and its success depends on the manner in which food is partitioned among its competing sinks. The history of this viewpoint was reviewed by White (1979). The competition in terms of processes and structures tends to maintain a fairly constant relationship regarding relative sizes among the various organs. If the leaf surface becomes too large in proportion to the root system, water stress will reduce shoot growth. Moderate water stress often reduces shoot growth more than it reduces photosynthesis, resulting in an increased supply of photosynthate for increased root growth and an increase in the root–shoot ratio. Conversely, if the photosynthetic surface is much reduced by defoliation, root growth is decreased temporarily by the reduced supply of photosynthate, but leaf growth is stimulated and leaf area tends to increase. As a result of the internal control exerted by water, carbohydrates, nitrogen, and hormonal balances, relatively stable relationships usually are maintained between leaf area, stem diameter, sapwood cross-sectional area, and root biomass (Waring and Schlesinger, 1985, pp. 29–37; Carlson and Harrington, 1987). The tendency of living organisms to maintain a dynamic equilibrium in structures and processes sometimes is termed homeostasis.

The proportion of photosynthate allocated to various uses and into various compounds change with the stage of growth and sometimes with the environment. The proportions of photosynthate allocated to various uses in loblolly pine were studied in detail by Chung and Barnes (1980a). The total budget for a 1-year-old loblolly pine branch is shown in Fig. 2.12. Protein constitutes only a small percentage of the total and most of that is produced early in the season (upper part of Fig. 2.13),

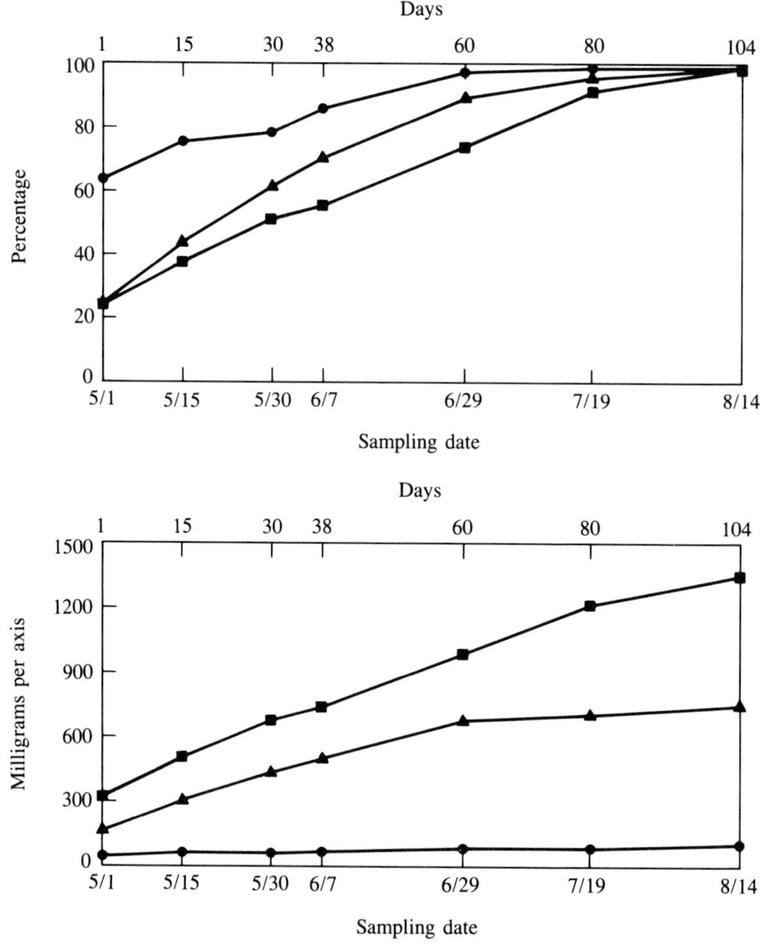

Figure 2.13 Seasonal changes in protein, lignin, and polysaccharide content of a first flush loblolly pine branch. *Top,* Percentages of total constituents present at each date. *Bottom,* The total amounts of constituents present at each sampling date. Protein (●), lignin (▲), and polysaccharide (■). (From Chung and Barnes, 1980a.)

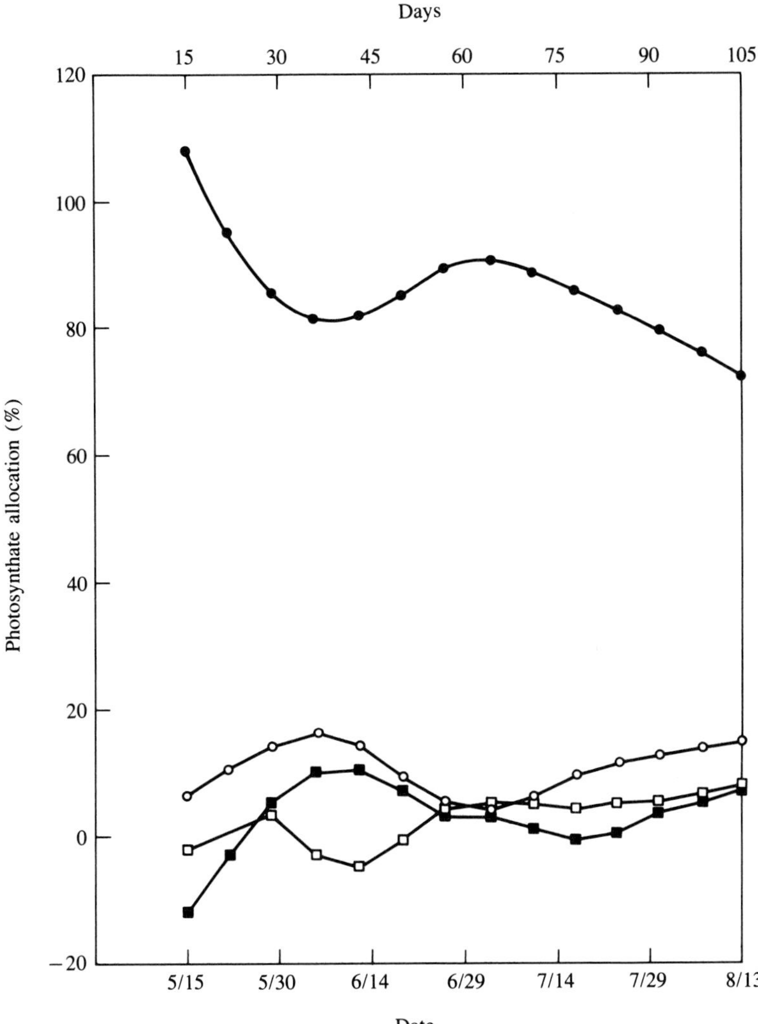

Figure 2.14 Seasonal changes in percentages of current photosynthate allocated to various functions in first flush loblolly pine branch. Structure (●), protection (○), metabolism (■), and storage (□). (From Chung and Barnes, 1980a.)

whereas most of the lignin and polysaccharides are formed later in the season. Figure 2.14 shows seasonal changes in percentage of current photosynthate allocated to various functions, classified roughly as structure (lignin and cellulose), protection (phenolics), metabolism (nitrogen compounds, organic acids, and lipids), and storage (starch and soluble carbohydrates). Readers are referred to the original papers by Chung and Barnes (1980a,b) for details.

Differences in sink strength result in changes in the ability to compete for the available food. For example, early in the season new roots, and later new leaves, are strong sinks in loblolly pine (De Wald and Feret, 1987), and this probably is true in many other plants. Later in the season cambial activity and diameter growth dominate, and if seeds begin to develop they usually are very strong sinks with priority over vegetative growth. During periods of rapid twig and leaf expansion, root and cambial growth of white oak cease (Dougherty *et al.*, 1979) and an increase in shoot growth reduces root growth in apple (Head, 1967). Fruit development competes with vegetative growth in peaches, in which fruit sink strength seems to be controlled endogenously (Chalmers and Van den Ende, 1975a). Tuomi *et al.* (1982) found that leaves were smaller on dwarf shoots of birch bearing catkins than on shoots without catkins, and there are reports that heavy seed crops reduce diameter growth and even cause dieback (Gross, 1972). Nutman (1933) reported that heavy crops of coffee berries sometimes cause severe loss of roots, and other examples of reduction in root growth on fruiting plants were given by Kramer (1983a, p. 162).

If browsing, insect feeding, or pruning materially decreases the leaf area, growth of new leaves is likely to regain a high priority, and loss of terminal buds often results in increased allocation of food to lateral branches. Removal of lower branches is likely to stimulate cambial activity in the upper stem at the expense of diameter growth near the base. Environmental factors such as soil fertility, weather, fertilization, and irrigation also affect the partitioning of food. High night temperatures increase the proportion of food used in respiration (see Fig. 2.10) and may result in a lower ratio of roots to shoots than lower temperatures (Hellmers and Rook, 1973). Axelsson and Axelsson (1986) reported that in long-term experiments in Sweden, irrigation and fertilization of conifers increased the ratio of above- to belowground biomass and suggested that differences in root–shoot partitioning of food may be important in competition among species. Other experiments indicate that a larger proportion of the total carbohydrate pool is allocated to growth of fine roots of trees on infertile soil than on fertile soil (Waring and Schlesinger, 1985, pp. 32–34).

The Harvest Index

In agriculture much of the increase in crop yield has resulted from an increase in the proportion of photosynthate directed into the desired product, or the "harvest index" (Evans, 1980; Gifford *et al.*, 1984). For example, the proportion of food partitioned into fruits of modern apple, pear, and peach trees is far greater than in their wild ancestors and more nuts are produced by cultivated pecan trees than by wild pecan trees. Likewise, more food goes into flowers of modern cultivars of plants such as roses and ornamental crab apples and cherries than into flowers of their ancestors.

Increasing the partitioning of photosynthate into economic yield has been successful in horticulture, especially in apples (Jackson, in Cannell and Jackson,

1985). Apples typically allocate a larger proportion of their total photosynthate to fruit than do plums or cherries. When grown on a dwarfing rootstock such as M9, they flower at an earlier age and allocate a much larger fraction of photosynthate into fruit than when grown on a more vigorous rootstock. Denser spacing of trees in hedgerows has further increased yields per unit of land area and decreased labor costs. The experience with apples suggests the desirability of better dwarfing rootstocks for other fruits. Development of rootstocks with a strong capacity to resprout might be worthwhile for poplar and other trees grown for firewood or fiber with the expectation of regenerating stands from stump sprouts.

Perhaps more attention should be given to the possibility of increasing the harvest index or proportion of total photosynthate incorporated into the stem wood of forest trees (Cannell, 1985; Pulkkinen *et al.*, 1989). Over half of the photosynthate goes into root replacement in some forests (Harris *et al.*, 1977; Kramer, 1983a, pp. 153–156; Bowen, 1985; Cannell, 1985). While root extension and complete occupation of the soil under a forest stand are important, it seems possible that an excessive amount of food goes into root growth, at least for trees growing on deep, well-watered soil. Perhaps trees also produce, or at least retain, more branches than are necessary. It seems probably that research on control of partitioning of food and selection for allocation of a larger percentage of food into stem growth, especially after the canopy has closed, might lead to ways to increase wood production (Cannell, 1985).

In young stands, rapid increase of leaf surface is very important in increasing the efficiency with which incident radiation is used and, according to Vose and Allen (1988), wood volume production is linearly correlated with leaf area index in young loblolly pine stands. However, after canopy closure occurs, only leaves in the upper part of the canopy are effective in photosynthesis (see Fig. 2.7), and the lower shaded branches can even become a physiological and economic liability. In general, only the branches on the upper third of a tree trunk in a closed stand contribute significantly to trunk growth (Kramer and Kozlowski, 1979, pp. 674–676). The role of the leaf area index in forest productivity is discussed in Chapter 3 of Waring and Schlesinger (1985) and in Chapter 13 of this book. Some problems involved in selection for yield and other desirable characteristics of forest trees were discussed in Volume 33 of *Forest Science*.

Cannell (1985) discussed partitioning and competition among organs of trees in detail and suggested the possibility of increasing yield by increasing the ratio of stem wood to branch and root wood. Pulkkinen *et al.* (1989) discussed the concept of harvest index and, like Cannell (1985), suggested that it might be increased by characteristics such as a large leaf area index on tall slender stems, high rate of photosynthesis in shaded foliage, low maintenance respiration, and efficient mineral cycling. They summarized a mass of data from various sources indicating wide variations in harvest index, ranging from 30 to 80%. This suggests opportunities for selection and improvement. Modern techniques such as cell fusion and propagation from tissue cultures probably will be useful in such programs.

Hormones and the Regulation of Growth

It is believed that there is some internal mechanism or mechanisms for balancing such phenomena as root–shoot ratio, vegetative versus reproductive growth, maturation and senescence, and polarity and tree form. This correlation takes two forms: regulation exerted by the partitioning of food, minerals, and water, and regulation brought about by plant hormones or growth regulators. To some extent the two kinds of regulation overlap, although the relative importance of source–sink versus hormonal regulation is debatable (Neales and Incoll, 1968; Peet and Kramer, 1980). We will review hormonal regulation briefly in this section.

Hormones usually are defined as substances synthesized in one part of an organism that in very low concentrations have important physiological effects in some distant part. For example, cytokinins and gibberellic acid synthesized in the roots have important effects on shoot growth (Davies *et al.*, 1986; Zhang and Davies, 1989b), and auxin produced in opening buds stimulates cambial activity far down the stems (Evert and Kozlowski, 1967; Evert *et al.*, 1972; Little and Savidge, in Kossuth and Ross, 1987). Five important groups of plant hormones are recognized: auxins, gibberellins, cytokinins, abscisic acid, and ethylene gas, plus the possible existence of a florigen or flowering hormone. Additional groups of exogenous compounds such as synthetic auxins and maleic hydrazide also act like hormones, and other endogenous growth regulators are likely to be identified.

Auxins increase cell wall extensibility and stimulate stem elongation, gibberellins promote cell and stem elongation and flowering, cytokinins promote cell division and delay senescence in fruits and leaves, and abscisic acid (ABA) is a growth inhibitor, promotes stomatal closure, and may be involved in plant dormancy. Ethylene, which is exceptional in being a gas, is produced by injured tissue and is involved in epinasty, leaf abscission, flower induction, and fruit ripening. Various phenolic compounds also are regarded as hormones because of their inhibitory effects; *trans*-cinnamic acid is an antiauxin, whereas its isomer, *cis*-cinnamic acid, is an auxin. Polyamines and their precursors also are said to have regulatory effects on membranes and plant growth (Smith, 1985), although this is debatable (Evans and Malmberg, 1989). There has been an intensive but thus far unsuccessful search for a florigen. One reason for supposing that it exists is the fact that in photoperiod-sensitive plants something must transmit a flowering stimulus from the leaves exposed to a suitable photoperiod to the stem tips, where initiation of flower primordia occurs. However, it seems probable that several factors interact to cause flowering.

The mechanism of action of plant hormones is not well understood because, unlike animal hormones, they do not have very specific targets. For example, auxin promotes coleoptile elongation and root formation on cuttings and initiates cell division in the vascular cambium. Among the possible targets of hormones are enzyme activity, membrane permeability, and cell wall relaxation. It now seems probable that growth and reproduction are controlled by interactions of groups of hormones and their relative concentrations may be more important than the concentration of any one hormone. Present explanations of hormone action at the molecular level are largely speculative, as indicated by Hanke's (1989) review of the survey of the field in the book edited by Boss and Morré (1989). There is an increasing tendency to involve changes in relative amounts of various hormones to explain the effects of environmental stresses such as water deficits and excesses, temperature extremes, mineral deficiencies, and mechanical stress (Johnson, in Kossuth and Ross, 1987).

The biochemistry of stress reactions requires more study and at present it is difficult to decide the extent to which changes in hormone concentration are merely the result of disturbed metabolism and the extent to which they are responsible for symptoms of stress. Readers are referred to plant physiology texts and the review by Kossuth and Ross (1987) for more detailed discussions of the role of growth regulators in woody plants. Pharis (1977) briefly reviewed the role of hormones in growth of coniferous trees. Trewavas (1986) argued that simple "cause and effect" relationships are inadequate to explain the role that any single hormone plays in the complex network of processes involved in plant growth and development.

Environmental Limitations on Growth

Having discussed some of the internal factors involved in plant growth, we will now turn to the environmental factors. There is general agreement that light, temperature, water, and nitrogen supply most often limit growth of field crops (Sinclair *et al.*, 1981), and this seems to be equally true for woody plants (Larcher, 1980). The effects of environmental stresses on cultivated crops, including some woody plants, were discussed in detail in Christiansen and Lewis (1982). We shall discuss environmental factors only briefly as they will be considered in more detail in later chapters.

Light

The importance of light for photosynthesis is well known and differences among plants with respect to tolerance of shading is important to ecology, forestry, and gardening. However, light, or more broadly radiation, has important effects on plant growth in addition to its role as a source of energy for photosynthesis. As these effects will be discussed in Chapter 4, here we merely remind readers that light is an important environmental factor and that differences in shade tolerance are important to everyone who grows plants.

Temperature

The kinds and amounts of plant growth on the earth's surface are strongly limited by temperature. The fact that water is frozen in the polar regions and at high altitudes prevents plant growth in those regions, and the distribution of most tropical plants north and south of the equator is limited by the lowest temperature that they can tolerate. Obvious examples are the limitations on growth of citrus and other tropical fruits and eucalyptus in temperate zones. Also, the growth of many tropical plants is limited northward by injury from "chilling" temperatures in the range of 5 to 10°C above freezing (Levitt, 1980a, Chapter 2; McWilliam, 1983). Other plants survive several days of freezing temperatures, while plants of high latitudes and high altitudes can survive months of extremely low temperatures. Temperature limitations are obvious in cold regions where delicate plants are killed, but sometimes the effects are indirect. For example, loblolly pine survives in the Ohio Valley, far beyond its normal range, but it does not reproduce because spring frosts limit seed production (Wenger, 1953). Likewise, high night temperatures may be inhibitory to growth (Kramer, 1957; Hellmers and Rook, 1973), probably because of their effect on dark respiration. In contrast, the extension of temperate zone species into warmer regions often is limited by lack of sufficient low temperature to break dormancy, resulting in poor flowering and fruit set. According to Hill and Campbell (1949), this problem has been observed in Africa, Australia, India, and Palestine, as well as in the warmer regions of the Americas. The ecological and physiological effects of temperature are discussed in detail in Chapter 5.

Water

The physiological importance of water also makes it of great ecological importance, with an excess resulting in bogs and swamps and a deficit in savannas, grasslands, or deserts. Water deficit caused by inadequate rainfall is the most common and widespread limitation on plant growth within the geographic limitations imposed by temperature. The quantity and quality of plant cover decrease with decreasing rainfall, with cover ranging from dense rain forests through savannas and grasslands to sparse desert scrub. Even in humid climates most of the year-to-year variation in width of annual rings of trees is related to variation in rainfall (Zahner, 1968). This relationship has given rise to the field of dendrochronology (Fritts, 1976), in which width of rings in old trees and timbers in old buildings is used to determine past rainfall conditions and the age of ancient structures (Giddings, 1962). It is now known, by examining a reduction in tree ring widths, that a serious drought occurred in the southwestern United States in the latter part of the thirteenth century. Thus lack of water may have been a major reason for the general abandonment of the cliff dwellings of that region near the end of the thirteenth century. Stahle *et al.* (1988) concluded from study of variation in ring width of 1600-year-old bald cypress growing in the coastal plain of North Carolina that wet and dry periods with an average duration of 30 years have occurred in that region over the past 1600 years. In addition, the narrow tree rings caused by water stress resulting from root injury to trees growing on fault lines are being used to date earthquakes (Jacoby *et al.*, 1988).

Even the effects of temperature are partly exerted through effects on water relations, because an amount of rainfall adequate for forests in a cool northern climate, where the rate of evaporation is low, is inadequate farther south, where higher temperatures result in more rapid evaporation and transpiration. Climates with cool rainy winters and hot dry summers have resulted in the shrubby vegetation characteristic of the Mediterranean and southern California coasts and part of the Pacific coast of South America.

The strong relationship between temperature and water in controlling the development of vegetation is indicated by observations that plant distribution in eastern North America is better correlated with the precipitation–evaporation ratio than with precipitation alone (Transeau, 1905) (see Fig. 7.1). Rosenzweig (1968) surveyed a mass of data and concluded that net annual aboveground productivity is well correlated with the rate of evapotranspiration calculated by the method of Thornthwaite and Mather (1957). Currie and Paquin (1987) reported that three-fourths of the variation in richness of tree species in North America, Great Britain, and Ireland can be explained by differences in annual evapotranspiration. These relationships exist because evapotranspiration integrates the effects of precipitation, irradiation, and temperature, three basic factors affecting plant growth.

Summary

Although breeding and selection programs are providing fruit and forest trees that have high potential yields, actual yields often are less than 50% of potential yields because of environmental limitations. Therefore it is important to understand how environmental factors limit tree growth and what can be done to minimize these limitations. The first requirement for growth is a supply of carbohydrates manufactured by photosynthesis, which requires CO_2 and H_2O, and light as a source of energy. Other requirements are nitrogen and a dozen other mineral elements, water for maintenance of turgor, oxygen, and suitable temperatures. The primary products of photosynthesis are used in assimilation to produce new plant tissue, oxidized in respiration to provide energy for various other reactions, or stored in fruits, seeds, and vegetative structures. Some products are converted into a wide variety of secondary compounds that provide the distinctive odors, tastes, and colors of plants, commercially important products such as terpenes, tannins, rubber, caffeine and other alkaloids, and plant hormones such as abscisic acid and the gibberellins.

The environmental factors that most often limit plant growth are light, water, temperature, and nitrogen supply. Other factors that become important under some conditions include plant competition, air pollution, and attacks by insects and pathogens. Most of these factors are discussed in more detail in later chapters.

General References

Binkley, D. (1986). "Forest Nutrition Management." John Wiley and Sons, New York.

Boss, W. F., and D. J. Morré, eds. (1989). "Second Messengers in Plant Growth and Development." Liss, New York.

Cannell, M. G. R., and Last, F. T., eds. (1976). "Tree Physiology and Yield Improvement." Academic Press, London.

Cannell, M. G. R., and Jackson, J. E., eds. (1985). "Attributes of Trees as Crop Plants." Institute of Terrestrial Ecology, Huntingdon, England.

Cherry, J. H., ed. (1989). "Environmental Stress in Plants." Springer-Verlag, Berlin.

Christiansen, M. N., and Lewis, C. F., eds. (1982). "Breeding Plants for Less Favorable Environments." Wiley, New York.

Crawford, R. M. M. (1989). "Studies in Plant Survival." Blackwell Scientific, Oxford.

Gifford, R. M., and Evans, L. T. (1981). Photosynthesis, carbon partitioning, and yield. *Ann. Rev. Plant Physiol.* **32,** 485–509.

Jennings, H., (1986). Salt relations of cells, tissues, and roots. *In* F. C. Steward, J. F. Sutcliffe, and J. E. Dale, eds., "Plant Physiology," Vol. IX, pp. 225–359. Academic Press, Orlando, FL.

Kossuth, S. V., and Ross, S. D., eds. (1987). Hormonal control of tree growth. *Plant Growth Regulation* **6,**(1 & 2), 1–215.

Kozlowski, T. T. (1971). "Growth and Development of Trees," Vols. I, II. Academic Press, New York.

Kramer, P. J. (1983). "Water Relations of Plants." Academic Press. New York.

Landsberg, J. J. (1986). "Physiological Ecology of Forest Production." Academic Press, London.

Luxmoore, R. J., Landsberg, J. J., and Kaufmann, M. R., eds. (1986). "Coupling of Carbon, Water and Nutrient Interactions in Woody Plant Soil Systems." Published as Tree Physiology, Vol. 2.

Marcelle, R., ed. (1975). "Environmental and Biological Control of Photosynthesis." Elsevier, New York.

Mattson, W. J., Levieux, J., and Bernard-Dagan, C., eds. (1988). "Mechanisms of Woody Plant Defenses against Insects. Search for Pattern." Springer-Verlag, New York.

Osmond, C. B., Björkman, O., and Anderson, D. J. (1980). "Physiological Processes in Plant Ecology." Springer-Verlag, Berlin.

Pearcy, R. W. Ehleringer, R. W., Mooney, H. A., and Rundel, P. W., eds. (1989). "Plant Physiological Ecology." Chapman and Hall, London and New York.

Stumpf, P. K., and Conn, E. E., eds. (1980–) "The Biochemistry of Plants," Vols. 1–12. Academic Press, New York.

Wardlaw, I. F., and Passioura, J. B., (1976). "Transport and Transfer Processes in Plants." Academic Press, New York.

Waring, R. H., and Schlesinger, W. H. (1985). "Forest Ecosystems." Academic Press, Orlando, FL.

Chapter 3

Establishment and Growth of Tree Stands

Introduction

During their development, forest stands exhibit both differences and similarities. The structure of forests varies from pure, even-aged stands consisting of a single species to mixed stands of several species and different age classes. The greater complexity of mixed over pure forests reflects variations among species in crown form, phenology, growth rate, size, and longevity. Differences between the growth and structure of tropical forests and temperate zone forests and between evergreen and deciduous forests are well known (Kozlowski, 1982c).

Both natural forests and plantations either are or eventually become multilayered (Fig. 3.1). Vertical stratification of forests may result from separate, but closely

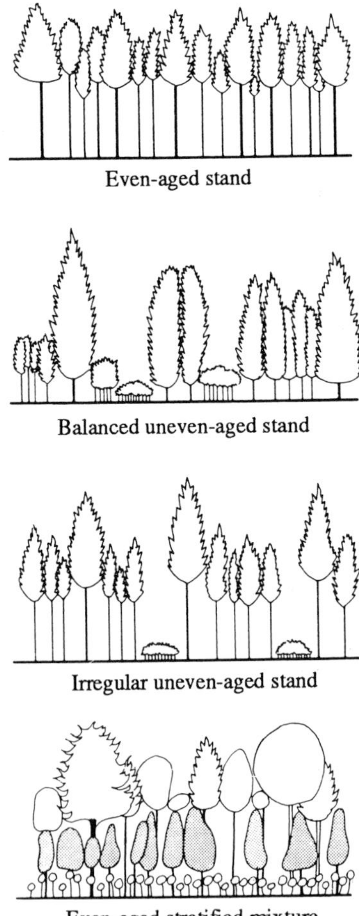

Figure 3.1 Some variations in structure of forest stands. The trees of the first three stands are composed of the same species. The fourth stand consists of several species of the same age. (From Smith, 1986. Copyright © 1986 by John Wiley & Sons.)

related, phenomena such as stratification of leaf mass, individuals, or species (Smith, 1973). The stratification of mixed forests results largely from variations in time of establishment and growth rates of the component species. In addition, differences in height of individuals of the same species occur because of variations in the time of germination of seeds within a seed lot. Single-species plantations, initially of rather uniform height and crown shape, eventually become stratified into crown classes and trees of varying vigor, as will be discussed later.

Tropical forests are well known for their layered structure (Whitmore, 1984). The shapes of tree crowns vary greatly among species occupying the different layers of tropical forests, with the tallest trees generally having the widest and flattest crowns (Fig. 3.2). The young trees of species that eventually occupy the upper levels of tropical forests and the shrub layers have diverse forms. Whereas most shrubs have a main stem and resemble dwarf trees, other shrubs lack a main stem and branch profusely near ground level.

Crown forms of tropical trees of the upper canopy change progressively during their development. When young they have long tapering crowns, characteristic of trees of lower strata; when nearly adult their crowns assume a more globose form; and when fully mature their crowns become flattened and wide (Richards, 1966).

As mentioned, forest stands exhibit some similarities. For example, during their development they change in a predictable way. Oliver (1981) and Peet and Christensen (1987) described the progression of forest stands through four sequential stages (Fig. 3.3).

1. *Stand Initiation or Regeneration Stage* Tree seedlings that will eventually form a new stand invade the space created by harvesting or disturbance.
2. *Thinning or Stem Exclusion Stage* When the tree canopy closes, plant competition intensifies and slow-growing trees die. The number of trees in the stand steadily decreases. Crowns of trees of one species may overlap those of another species, resulting in a stratified stand.

Figure 3.2 Stratification of trees in tropical forests. (After Beard, 1946; from Kramer and Kozlowski, 1979, by permission of Academic Press.)

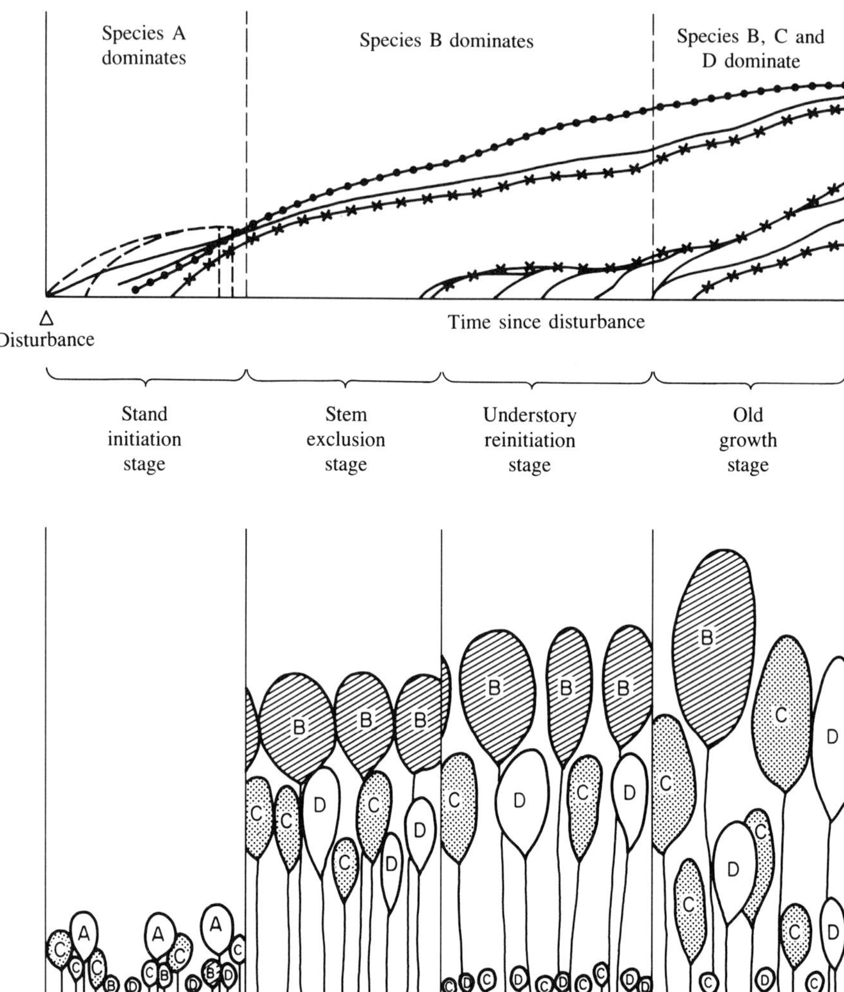

Figure 3.3 Stages of development of a forest stand following a major disturbance. All trees comprising the forest start shortly after the disturbance. As competition among trees intensifies, the number of stems decreases, and the dominant tree type changes. Species A and C are pioneers with relatively fast early growth. Species B and D, although present early, do not assert themselves in early competition. Species D eventually becomes the dominant species. (From Oliver, 1981.)

3. *Transition or Understory Initiation Stage* Gaps form in the canopy and advanced regeneration reinvades the understory.
4. *Steady State or Old Growth Stage* As the overstory trees gradually die, the understory trees slowly fill in the released space and replace the overstory trees. The forest now consists of a mosaic of patches of various sizes and ages. Within each patch the processes of establishment, thinning, and gap

formation are repeated. Actually, the steady-state stage is seldom achieved because of the long time required for a forest to develop to this stage, and the likelihood of additional major disturbances before this stage can be reached.

With the foregoing considerations in mind we will now present an overview of initiation and development of forest stands.

Stand Regeneration

Harvested or disturbed forests can be regenerated by growth of suppressed seedlings, buried or dispersed seeds, stump or root sprouts, and rooted ends of branches (layering). The severity of disturbance often determines the method of regeneration. After mild disturbances involving loss of one or a few canopy trees, a forest stand usually regenerates by growth of suppressed seedlings and saplings. However, when both dominant and suppressed trees are eliminated by severe disturbances, regeneration generally occurs from germination of seeds in the seed pool (Marks, 1975). Exceptions are species that reproduce by sprouts (e.g., oaks) and root suckers (e.g., trembling aspens).

Regeneration from Seeds

Important prerequisites for regeneration of many forest stands are an adequate supply of viable seeds and environmental conditions conducive to seed germination and seedling establishment.

Unfortunately, seed production of many forest trees is irregular and unpredictable. Some species (e.g., tupelo gum, southern magnolia) tend to produce good seed crops each year; others (American beech) have good seed crops at several-year intervals. Some species (white ash, sugar maple) have good seed crops at fairly regular intervals; others (black walnut, yellow poplar) at irregular intervals. Seed production by different species in the same forest stand often differs greatly. For example, western hemlock tends to produce large seed crops, while such associated species as western white pine and grand fir usually produce relatively small seed crops (Kramer and Kozlowski, 1979).

To ensure regeneration, foresters often leave a few seed trees when harvesting a stand of trees. Once regeneration is assured, the seed trees can be harvested or left indefinitely. Trees selected for seed trees usually have wide, deep crowns, a high live-crown ratio (percentage of length of stem with live branches), and tapering stems. Such trees are not easily toppled by wind and are likely to have large seed crops. The number of seed trees retained will depend on the amount of seed produced per tree and the number of seeds needed to satisfactorily repopulate a site (Smith, 1986). Unfortunately, good seed crops of many species can be predicted only a few months in advance when fruits or cones have already formed. When development of cones or fruits requires 2 years, as in many pines and species of the red oak subgroup, seed crops can be predicted earlier. If seed trees are uniformly distributed over an area, the amount of seed produced is more important for stand regeneration than is the distance of seed dispersal (Smith, 1986). Dispersal of seeds is discussed further in Chapter 12.

Seed Germination

Germination of seeds can be treated as the resumption of growth by embryos that are activated by exposure to favorable conditions. It involves the following steps: (1) imbibition of water, swelling of existing tissue, and rupture of the seed coat and other enclosing structures, (2) cell division in the embryo and elongation of the radicle, thus pushing the cotyledons (epigeous germination) or the epicotyl (hypogeous germination) above ground; and (3) hormones formed in the meristematic regions and perhaps elsewhere start enzymatic activity, hydrolysis of stored food, and translocation to growing regions. The process is complete when enough photosynthetic tissue is exposed to support seedling growth. Of course this does not really explain what causes cell division to start as soon as the seed begins to imbibe water or why certain hormones activate enzyme activity, but it is a systematic way to look at the process.

In early stages of germination, seedlings depend entirely on stored foods. After seed reserves are depleted, continued growth of seedlings depends on carbohydrates synthesized by cotyledons as they emerge above ground. Such cotyledons are relatively short-lived, however (about 1 to 3 months depending on species)(Marshall and Kozlowski, 1974a,b). Continued seedling development depends on carbohydrates produced by the first true foliage leaves. When the cotyledons stay below ground, as in oaks, walnut, and buckeye, the young seedling at first depends on carbohydrate reserves in the seed and, when these are exhausted, on carbohydrates synthesized by the first true leaves.

Seed Dormancy

Seeds of most species of woody plants of the temperate zone exhibit some degree of dormancy. Not all seeds of a given lot are equally dormant, hence a few may germinate promptly and others only after a long time, or never. Among the several causes of seed dormancy are (1) immaturity of the embryo, (2) impermeability of the seed coat to water, (3) mechanical resistance of the seed coat to growth of the embryo, (4) low permeability of the seed coat to gases, and (5) endogenous dormancy of the embryo (Villiers, 1972).

Seed dormancy may be advantageous or disadvantageous. The prolonged chilling requirement for breaking of embryo dormancy of some kinds of seeds (Chapter 5) prevents germination until spring. This is an advantage because earlier germination would expose young seedlings to injury by freezing. Seeds of some plants may be dormant for several years, causing seed germination to be spread out over a long time. This characteristic provides for survival of some individuals of a species, even though the seedlings that emerged first may succumb to environmental stresses. However, seed dormancy often plagues nursery operators who would like all seeds to germinate promptly so as to produce large and uniform crops of seedlings.

The seeds of the majority of tropical trees have little or no dormancy and germinate rapidly, but some remain dormant for years. For example, in Malaysia all the seeds of two-thirds of the woody flora germinated within 3 months after the seeds were shed (Ng, 1980). In the Ivory Coast, 79% of 61 species showed very rapid seed germination (Alexandre, 1980). Unusually fast germination occurs in mangroves, with some seeds germinating while the fruits are still attached to the tree. This type of germination (vivipary) also occurs in *Pithecolobium racemosum*, *Pouteria ramiflora*, *Magonia pubescens*, and *Dryobalanopsis aromatica* (Sasaki *et al.*, 1979; Vazquez-Yanes and Segovia, 1984).

The seeds of tropical species that do not germinate promptly often have impermeable and hard seed coats, including *Podocarpus* spp., *Intsia palembanica*, *Parkia javanica*, *Sindora coriacea*, and *Dialinum maingayi* (Sasaki, 1980a,b).

Environmental Control of Seed Germination

Viable seeds often are prevented from germinating by unfavorable environmental conditions such as drought, low temperature, oxygen supply, and, in the case of some seeds, light quality.

The first step in germination is imbibition of water followed by resumption of physiological activity. For example, seed respiration increases greatly with an increase in seed hydration (see Fig. 7.16) (Kozlowski and Gentile, 1959). Following germination, a continuous supply of water is needed, and even very temporary droughts may be catastrophic. Practically all nondormant seeds (except those with impermeable seed coats) can absorb enough water for germination from soil at field capacity. In progressively drier soils the rate of germination decreases (Kramer and Kozlowski, 1979, p. 503).

Exposure to near-freezing temperature is necessary to break dormancy of some seeds. However, once dormancy is broken by prolonged chilling, much higher temperatures are necessary to set germination processes in motion (Chapter 5). Optimal temperatures for seed germination vary widely for seeds of different species and are lower for temperate zone species than for tropical species. They also vary for different seed sources of the same species.

Seed germination often is prevented by oxygen deficiency because oxygen is required for initiating respiration during early phases of germination. Hence, seeds of many species fail to germinate in poorly aerated, compacted, or flooded soils (Chapter 8). Oxygen requirements are higher for germination than for subsequent growth, because seed coats often impede diffusion of oxygen into seeds (Kozlowski and Gentile, 1959).

Although most seeds germinate as well in the dark as in the light, those of some species require low light intensities for germination. The seeds of most species germinate best in photoperiods of 8 to 12 hr. An exception is Douglas-fir, whose seeds germinate better in long days or continuous light than in short days (Jones, 1961).

Germination of seeds of some species is promoted by red light and inhibited by far-red light. This response reflects stimulation of both metabolic activity and mitosis in embryos by red light (Nyman, 1961) and suggests involvement of a phytochrome system. This is discussed further in Chapter 4.

Seedbeds

Regeneration of trees varies among seedbeds because of differences in water supply, temperature, nutrient availability, and capacity of rootlets to penetrate a seedbed. Mineral soils generally are good seedbeds because of their high infiltration capacity, suitable aeration, and close contact between soil particles and imbibing seeds. Litter and duff are less suitable media than mineral soil because they prevent seeds from contacting mineral soil, dry rapidly, inhibit root penetration, and shade small seedlings. Natural establishment of some conifers before fire is particularly difficult because, although the seeds may germinate, the rootlets cannot penetrate the leaf litter and reach mineral soil before they dehydrate.

Rootlets of yellow birch germinating in humus or sandy loam readily penetrated the rooting medium and the seedlings rapidly assumed an upright position. In contrast, rootlets of seedlings germinating on leaf litter tended to grow horizontally over the surface of the leaf mat. The main rootlets commonly remained above the leaf mat and only very fine secondary rootlets penetrated the leaves. Sometimes the levering action of a rootlet growing against the leaf surface of fallen leaves overturned the seedling. The root tip, then pointing upward, turned downward again, but seldom rooted successfully (Winget and Kozlowski, 1965). Litter is more of a barrier to germination of seeds of some species than those of others. Whereas seeds of pines require a mineral seedbed and a thin litter layer for germination and survival, some oak seedlings can survive when the litter layer is several centimeters deep. Also oak and hickory seedling taproots penetrate more deeply than roots of pine seedlings, making them more drought tolerant (Coile, 1937)

Vegetative Propagation

Some species of woody plants are propagated vegetatively by sprouting. The sprouts that originate from root collars and the lower part of the stem arise from dormant buds. Such "stump sprouts" represent the most important type of vegetative propagation of broad-leaved forest stands. For example, much of the New England hardwood forests and as much as three-fourths of the harvested oak forests of the United States originate from stump sprouts. Capacity for sprouting is common in broad-leaved trees but uncommon in conifers. Nevertheless, a few species of conifers sprout prolifically, especially after fire, including redwood, pitch pine, shortleaf pine, and pond pine.

Following disturbance of forest stands, some species regenerate largely by root suckers (shoots that originate from roots). Reproduction by root suckers is well known in trembling aspen and bigtooth aspen, but it can also occur in sweet gum, American beech, and black locust (Brown and Kormanik, 1967). In heavily cut aspen stands, over 2800 root suckers per acre developed in comparison with only 377 per acre in undisturbed stands (Stoeckeler and Mason, 1956). Seed production and seedling establishment of the shade-tolerant American beech are heavy in some years and light in others. Successful regeneration depends on survival of seedlings and root sprouts under closed canopies (Jones and Raynal, 1987).

Sprouting is least abundant from stumps of trees cut in early summer when trees have just leafed out and have low carbohydrate reserves, and greatest from stumps of trees that were cut during the dormant season (Fig. 3.4).

In some species, roots form on attached branches where they contact a moist medium. For example, black spruce and northern white cedar in peat swamps may reproduce by such "layering." Layering also occurs in hemlock, spruce, fir, yew, and cedar (*Chamaecyparis*) but only rarely in pines. Natural layering also occurs commonly in raspberries, blackberries, currants, and gooseberries when the lowest branches contact the soil and in some ornamental shrubs such as forsythia.

Some plants that are not easily propagated by rooting of cuttings can be propagated by layering. For example, layering is often used to propagate apple rootstocks, filbert, and muscadine grape. Hartmann and Kester (1982) present a good review of the use of air layering in plant propagation.

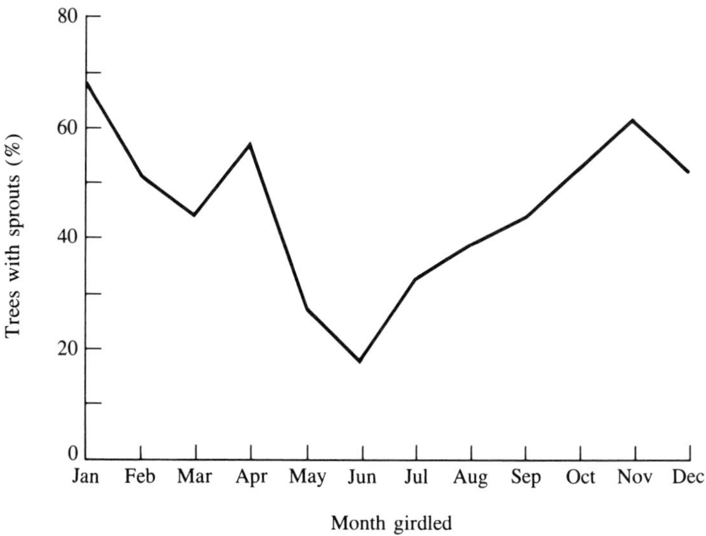

Figure 3.4 Effect of season of stem girdling on sprouting by blackjack oak trees. (After Clark and Liming, 1953; from Kramer and Kozlowski, 1979, by permission of Academic Press.)

Colonization of Gaps

In both temperate zone and tropical forests, gaps form at average rates of approximately 1% per year of the total land area (Waring and Schlesinger, 1985). Thus gap colonization is important in the regeneration of forests (Spies and Franklin, 1989; Lorimer, 1989).

In several old growth forests of the eastern United States, gaps averaged 65 m^2 and varied from 1 to 1490 m^2 in size (Runkle 1981, 1982). Small gaps are formed frequently and largely randomly as a result of death or windthrow of a few trees; large gaps covering a few to thousands of hectares are caused occasionally by fires, hurricanes, or attacks by insects.

Both the intensity and duration of light in gaps depend on their size, shape, slope, height of surrounding trees, and characteristics of the surviving vegetation. Within gaps, the light intensity, duration, and fluctuation, the soil and air temperatures, the soil moisture supply (except in the few surface centimeters), and evaporation from the soil surface are higher, but relative humidity is lower than in the surrounding forest. The amount of water in the soil surface profile in gaps may be two to eight times that in adjacent wooded areas (Minckler *et al.*, 1973). However, some environmental conditions within gaps vary irregularly. Whereas some portions of gaps are dark and cool, others are hot and dry, depending on the arrangement of plants and debris (Brokaw, 1985).

Gap Colonization in Temperate Zone Forests

Plants that colonize gaps usually were already present at the time of disturbance as suppressed trees or overstory trees (for sprouting and suckering species), or as seeds buried in the soil or newly dispersed into the area. Closure of very small gaps occurs by extension of branches of surrounding trees, and closure of larger gaps by height growth of saplings and sprouts within the gap. Tolerant species (e.g., sugar maple, American beech, eastern hemlock), with a long-suppressed seedling stage, grow vigorously after small recurrent disturbances but can also grow rapidly in larger gaps. By comparison, intolerant species such as tulip poplar usually grow rapidly in large gaps but not in small ones. Generally they are confined to gaps that are large enough so they are not closed by lateral growth of branches of adjacent trees or by previously suppressed seedlings (Runkle, 1985). Sometimes, however, such intolerant species become established in gaps that vary in size from 50 to 250 m^2 (Williamson, 1975).

Large gaps may be first occupied by early successional species and later replaced by shade-tolerant species. Alternatively, species of both these groups may arrive at the same time, with the faster-growing, short-lived pioneer species eventually being eliminated after a period of dominance. In a northern hardwood forest most of the species found early in gaps were those that were present in the forest at the time it was clear-cut. As gaps developed, different species sequentially dominated the plant community in accordance with their growth characteristics. The successional pattern varied appreciably from that on land that had been cultivated, with most species appearing gradually in a sequence that was based on frequency of seed years and capacity for seed dispersal (Bormann and Likens, 1979).

Mixed stands of *Abies delavayi* and *Betula utilis* occur in subalpine forests in China. In some stands a dense bamboo understory is present. After canopy gaps form, rapid growth of bamboo inhibits establishment and growth of other species (Taylor and Zisheng, 1988). In western Japan, dwarf bamboo (*Sasa borealis*), which dominates the forest floor of old-growth *Fagus crenata–Abies homolepis* stands, interferes with regeneration of trees in gaps (Nakashizuka, 1989). Rapid establishment of bamboos in gaps of Chilean *Nothofagus* forests also prevents colonization by other woody species (Veblen, 1985).

Many successional species depend on continuous dispersal of seeds from outside the site ("fugitive species") or storage of dormant seeds in the soil. Among the requirements for colonization of disturbed sites are (1) a good seed year, (2) migration of seeds from another site, (3) arrival of seeds near the time of disturbance, and (4) a favorable microclimate for seed germination and seedling establishment.

Gap Colonization in Tropical Forests

Recurrent creation and filling of gaps are dominant features in the regeneration of tropical forests (Lieberman *et al.*, 1985). Small gaps are created by falling tree branches, and larger gaps by falling single trees or groups of trees due to old age, disease, wind, fire, earthquakes, or lightning. The majority of the gaps are small but most of the total gap area is in a few large gaps. Very small gaps usually are filled by growth of branches of adjacent trees, and somewhat larger gaps by accelerated growth of suppressed seedlings that were present prior to opening of the canopy, and by resprouting. Still larger gaps are colonized by plants from seeds present in the soil. Very large gaps are filled primarily by plants growing from seeds that were brought in largely by animals and wind after the gap was formed (Fig. 3.5).

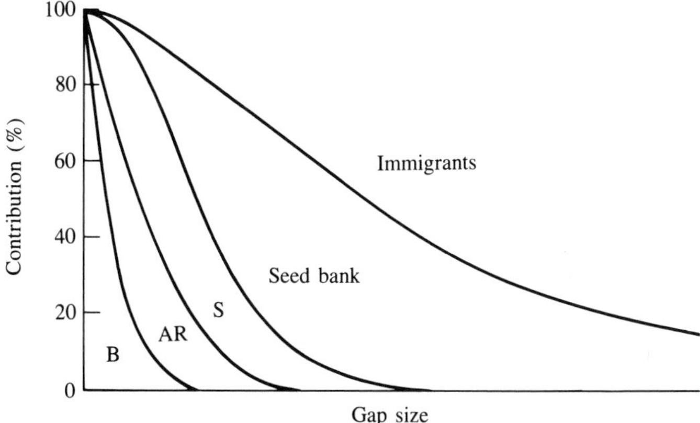

Figure 3.5 Relation between gap size and contribution of various sources of plants to gap filling. Increased severity of disturbance during creation of gaps moves the time axis to the right. B, branches; AR, advance regeneration; S, sprouts. (From Bazzaz, 1984.)

Whitmore (1989) emphasized that small and large gaps of tropical forests generally are colonized by different species of trees. In small gaps, seedlings that became established in the shade of the closed forest are released. Seeds of species that colonize large gaps germinate only in the open, hence seedlings occur only after gaps are formed. Swaine and Whitmore (1988) called these two groups climax (nonpioneer) species and pioneer species, respectively, the latter regenerating only in large gaps. Because of these differences, Whitmore (1989) considered the gap phase to be the most important part of the growth cycle for determining species composition of tropical forests. Competition among tree species and their different light requirements at the understory initiation and old-growth stages of stand development were considered less important. However, as emphasized by Martinez-Ramos *et al.* (1989), Whitmore's (1989) pioneer/climax framework distinguished among species with different average life histories. When variability within populations is considered, distinctions between these two groups of species tend to become somewhat diffuse. Schupp *et al.* (1989) pointed out that the probability that a tree of a given species will enter the forest canopy is a function of the joint probabilities of its arrival and subsequent survival in a particular habitat. The probability of arrival is determined by the mode of dispersal; the probability of survival by both physiological attributes and biotic interactions (e.g., disease and seed and seedling predation).

The early colonizers, whatever their source, grow rapidly at first but soon competition for resources becomes severe and many die. During their development in gaps, the ultimate canopy trees are exposed to a variety of changing environmental conditions and hence must exhibit wide tolerance to environmental stresses. Both a decrease in growth and recruitment of new plants characterize transition from the gap phase to the building and mature phases of the growth cycle of tropical forests (Brokaw, 1985).

Rapid seed germination follows altered light, temperature, and moisture regimes of the soil surface in gaps. Changes in radiation appear to be more important than changes in temperature. Seeds of many species are sensitive to changes in both light intensity and quality. For example, seeds of *Cecropia obtusifolia*, *C. peltata*, *Trema micrantera*, and *T. orientalis* require light for germination (Whitmore, 1983). In Ghana, seeds of 96 species germinated readily in full light and seeds of 25 species in the shade (Hall and Swaine, 1980). Light quality also affects germination of some species, with red light essential for germination (Whitmore, 1983). This is discussed further in Chapter 4.

Delay in seed germination often is linked to several environmental changes. For example, where well-defined dry and wet seasons occur, seeds often remain viable during the dry season but germinate readily when the rainy season begins.

Germination of seeds of some species is favored by temperature fluctuations in gaps. For example, dormancy of seeds of *Didymopanax*, *Ochroma*, and *Heliocarpus* is associated with hard and impermeable seed coats. These become permeable when exposed to the high temperatures of the soil surface in gaps (Vazquez-Yanes and Segovia, 1984).

Competition

Dense populations of trees inevitably compete with each other and with understory shrubs and herbaceous plants for available resources. As a result, trees and other woody plants become involved in a dynamic struggle for existence in which many are eliminated. Young forest stands with several thousand trees per acre may be reduced to less than two hundred trees when mature. Competition may be one-sided, with large plants decreasing the growth of small neighbors but not vice versa (Cannell *et al.*, 1984), or it may result in mutual inhibition of growth of competing plants. The amount of competitive stress on individual trees in forest stands has been estimated by several indices, including (1) the amount of overlap of "zones of influence" among adjacent trees (Bella, 1971), (2) growing space polygons that measure the area potentially available to each tree (Moore *et al.*, 1973), (3) indices that incorporate relative stem diameters and distances between individual trees and their competitors (Pielou, 1959, 1960; Pukkala and Kolström, 1987), and (4) pattern analysis (Laessle, 1965; Kent and Dress, 1979). Methods of quantifying competition among trees are discussed by Lorimer (1983).

In young plantations the even-aged trees are relatively evenly spaced and have crowns that are more or less similar in size and shape. As the trees grow larger, the crowns of the slower-growing trees are shaded. The competitive capacity of shaded branches for water and perhaps nitrogen is reduced, as a prelude to their death. The resulting inhibition of photosynthesis of the suppressed trees reduces their supply of carbohydrates and possibly of growth hormones, leading to a reduction in cambial growth and root growth. The inhibition of root growth decreases absorption of water and minerals, which further reduces growth. Eventually the slower-growing suppressed trees are likely to die.

There are many studies of plant competition involving woody plants and only a few examples will be given here. As the density of seedlings in forest nurseries is increased, competition for available water and mineral nutrients intensifies, resulting in small seedlings (Duryea and McClain, 1984). In chaparral shrublands, *Ceanothus megacarpus* does not resprout following fire, and dense, nearly pure, even-aged stands often grow from buried seeds. Self-thinning through intraspecific competition begins about 5 years after fire and continues for the next 10 to 15 years, depending on the density of shrubs. Thinning often reduces stand density by 50% or more (Schlesinger and Gill, 1978). Competition for resources in agroforestry systems also is well documented. Availability of water and mineral nutrients to the various plants within an agroforestry system is determined by the extent of each root system and its depth, density, and capacity to absorb water and mineral nutrients (Connor, 1983).

Grasses often compete with trees and restrict their growth. Perennial ryegrass inhibited root growth and decreased seasonal duration of growth of sycamore maple trees (Richardson, 1953). Elimination of turf grass around shade trees greatly increased growth of fine roots of 20-year-old green ash, littleleaf linden, sugar maple, red maple, and Norway maple trees (Watson, 1988). Bermuda grass (*Cynodon dactylon*) is one of the most severe competitors among the lawn grasses. Keeping an area free of Bermuda grass around newly planted trees and shrubs greatly increases their growth.

Competition for Resources

An extensive literature shows that woody plants compete with other woody or herbaceous plants for light, water, and mineral nutrients.

Light

The amount of light available to the lower part of the canopy of trees in dense stands is greatly restricted because of shading by neighboring trees. Because of mutual shading by leaves, the light intensity in different parts of a tree crown also varies greatly. Light intensity that penetrates a forest canopy may approximate 50 to 80% of full sunlight in leafless deciduous forests, 10 to 15% in open, even-aged pine stands, and less than 1% in some tropical rain forests (Spurr and Barnes, 1980). The light intensity decreases rapidly through the canopy and then more slowly below the level of the tree crowns (Fig. 3.6), resulting in the decrease in photosynthesis shown in Fig. 2.7.

About 90% of the light intensity was intercepted on a sunny day by tree crowns of a 150-year-old American beech stand, and 80% on an overcast day. Less than 5% was transmitted to the forest floor (Trapp, 1938). The several layers of leaves in uneven-aged stands of trees transmit less light to the forest floor than do the crowns of even-aged stands. In multilayered canopies the light reaching the forest floor is controlled by the thickest layer. For example, a dense second story of trees at a height of 45 to 65 ft regulated the amount of sunlight reaching the forest floor of a tropical rain forest (Ashton, 1958).

Some idea of seasonal and diurnal changes in light intensity above and within a deciduous forest may be gained from the data of Hutchinson and Matt (1977) for a 50-year-old tulip poplar forest in Tennessee. The total amount of radiation, as well as the direct and diffuse components, at any time in the forest varied directly with the amount of incident radiation at that time. However, the proportions of direct and diffuse radiation changed over time with variations in amounts of cloud cover, solar elevation, and phenological changes in forest structure (Figs. 3.7 and 3.8).

The greatest amount of radiation was received on the forest floor in the spring before leaf expansion began, and the lowest amount when the sun was low and the days were short early in the autumn while the forest was still in full leaf. After the leaves were shed later in the autumn, the light intensity in the forest decreased again with the winter decline in intensity. Throughout the year the amount of radiation decreased progressively from above the canopy to the forest floor (Table 3.1).

The amount of light transmitted to the forest floor decreases as the number of trees per unit area increases (Fig. 3.9 and Table 3.2). Light transmission by tree stands has been related to several different measures of stand density, such as leaf area index, spacing of trees, number of trees per unit of land area, crown closure,

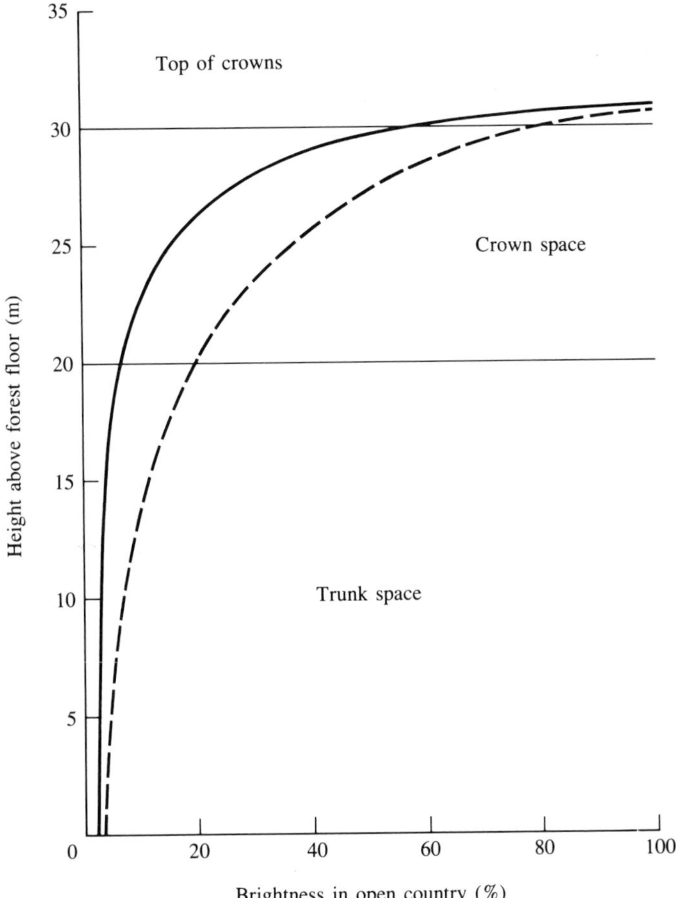

Figure 3.6 Penetration of solar radiation in a mature beech stand on sunny (——) or overcast (– – –) days. (After Trapp, 1938; from Reifsnyder and Lull, 1965, U.S. Forest Service photo.)

basal area of stems, and stem density (Reifsnyder and Lull, 1965). Many investigators consider leaf area index as the best indicator of light transmission. Basal area does not give much weight to effects of small trees and continues to increase as trees age. By comparison, stem density, which decreases as suppressed trees die, is better correlated with the resulting increase in transmission of light. Crown closure is not as well correlated as stem density with light intensity (Fig. 3.9).

Much interest has been shown in the interception of light by fruit trees. Particular attention has been given to (1) maximizing interception of light by orchard trees because the light that falls in the space between trees does not increase fruit yield and (2) optimizing light distribution within the canopy to increase the efficiency of light in photosynthesis, formation of fruit buds, and fruit coloring (Jackson, 1980).

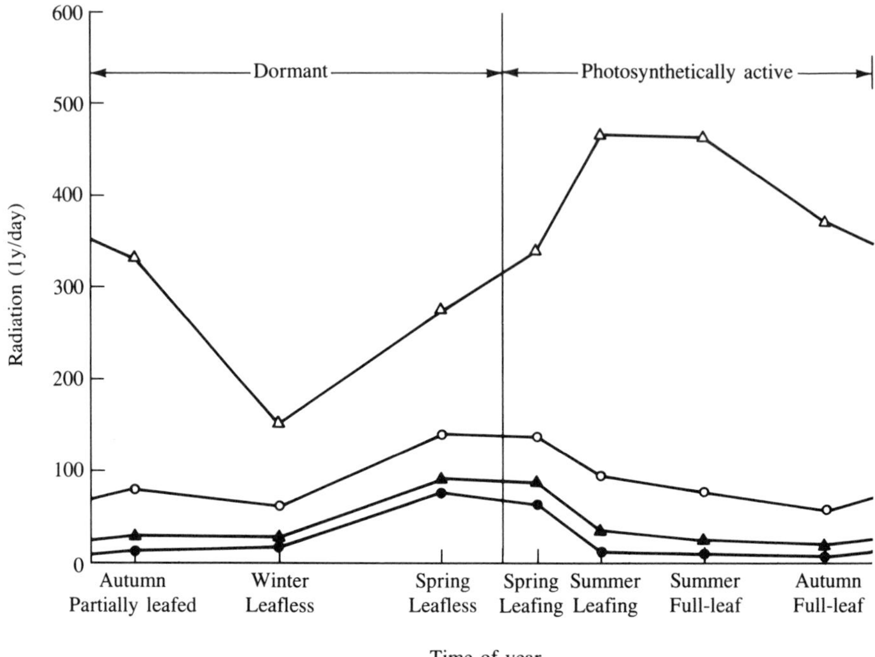

Figure 3.7 Annual regimes of total radiation in and above a yellow poplar forest. Above canopy, 32 m (△); upper canopy, 16 m (○); midcanopy, 3m (▲); forest floor, 0 m (●). (From Hutchinson and Matt, 1977.)

The canopies of trees in fruit orchards prevent transmission of large proportions of the incident light. For example, the amount of shortwave radiation transmitted through the canopy of a peach orchard to the soil surface decreased from near 80% during early leaf expansion to 15% when the trees were in full leaf (Chalmers, 1983). This compares with minimum values of 13% for a mature citrus plantation (Kalma and Stanhill, 1969), 10 to 15% in open pine stands, and 1% in tropical forests.

The efficiency of light capture by apple trees is greatly influenced by arrangement and spacing of trees as well as density of the canopy and its leaf characteristics (Jackson, 1985). The amount of light intercepted by hedgerows depends on the proportion of the ground that is covered and the height of the hedge. Figure 3.10 shows some typical values of light interception by hedgerow orchards of different geometries. Tall, triangular section hedgerows intercept a higher proportion of the available light than low hedgerows do. The percentage of available direct light that is intercepted is controlled by the relation of the hedgerow to the sun and distance between trees. Hence the influence of row orientation or interception varies with time of day and season, latitude, and orchard geometry.

Apple trees attain their maximum leaf area by midsummer but usually do not

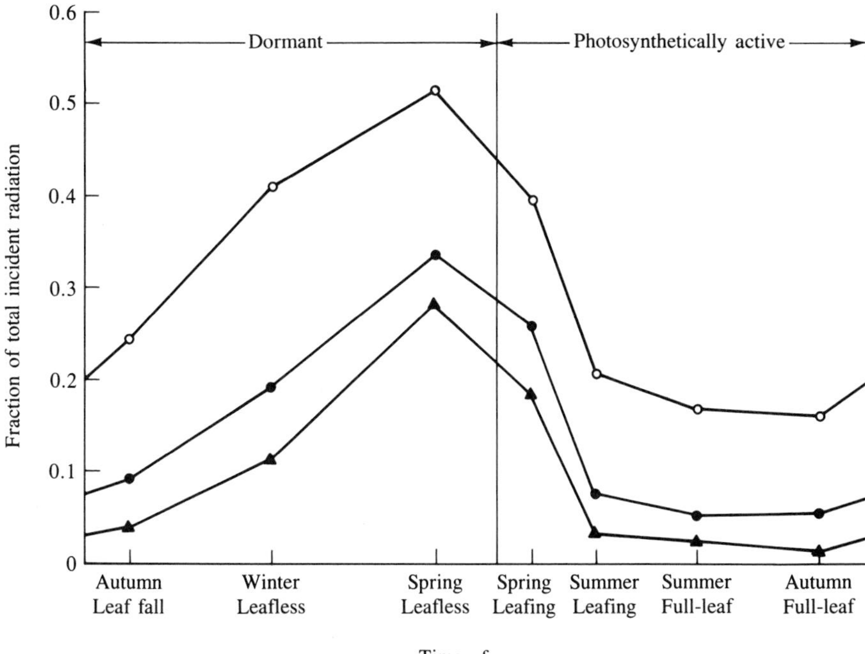

Figure 3.8 Annual course of daily fractional penetration of solar radiation in a yellow poplar forest. 16 m elevation (○); 3 m elevation (●); 0 m elevation (▲). (From Hutchinson and Matt, 1977.)

intercept more than 65 to 70% of the available light at full canopy, and several years are required to achieve this level (Table 3.3). Much can be done to increase light interception in young orchards by (1) planting trees that are tall with many lateral branches at the time of planting and (2) planting trees at high densities, in either hedgerows or multirow systems, to minimize the amount of space in alleyways between rows. Jackson (1980,1981,1985) published much useful information on methods of maximizing light interception by fruit trees.

Water

Competition for water in dense forest stands is emphasized by the greater availability of soil moisture to residual trees following thinning. For example, thinning of Douglas-fir stands led to a substantial increase in the soil water content, especially during the first 3 years after thinning (Fig. 3.11). However, in another study, Brix and Mitchell (1986) reported that removal of understory plants in a thinned and fertilized Douglas-fir stand in British Columbia did not improve soil or shoot water potential. Thinning increased the soil water supply, but improved only the predawn and early morning water potential. Both soil moisture content and leaf hydration were higher in a thinned red pine stand than in an unthinned stand (Sucoff and

Table 3.1
Amount of Radiation Received within a Yellow Poplar Forest throughout a Year[a]

		Total radiation received [langleys (percentage of yearly total)]			
Phenoseason	Duration (days)	Above canopy (32 m) (%)	Upper canopy (16 m) (%)	Midcanopy (3 m) (%)	Forest floor (0 m) (%)
Winter leafless	91	13,300 (11.5%)	5,400 (17.5%)	2,400 (16.9%)	1,500 (16.5%)
Spring leafless	55	15,000 (13.0%)	7,600 (24.6%)	4,900 (34.5%)	4,100 (45.0%)
Spring leafing	30	10,200 (8.8%)	4,000 (12.9%)	2,500 (17.6%)	1,800 (19.8%)
Summer leafing	26	11,700 (10.1%)	2,400 (7.8%)	800 (5.6%)	300 (3.3%)
Summer full-leaf	67	31,500 (27.2%)	5,100 (16.5%)	1,500 (10.6%)	700 (7.7%)
Autumn full-leaf	57	21,100 (18.2%)	3,300 (10.7%)	1,000 (7.0%)	200 (2.2%)
Autumn partial-leaf	39	12,900 (11.2%)	3,100 (10.0%)	1,100 (7.8%)	500 (5.5%)
Photosynthetically active period total		74,500 (64.4%)	14,800 (47.9%)	5,800 (40.8%)	3,000 (33.0%)
Dormant period total		41,200 (35.6%)	16,100 (52.1%)	8,400 (59.2%)	6,100 (67.0%)
Yearly total		115,700	30,900	14,200	9,100

[a] From Hutchinson and Matt (1977).

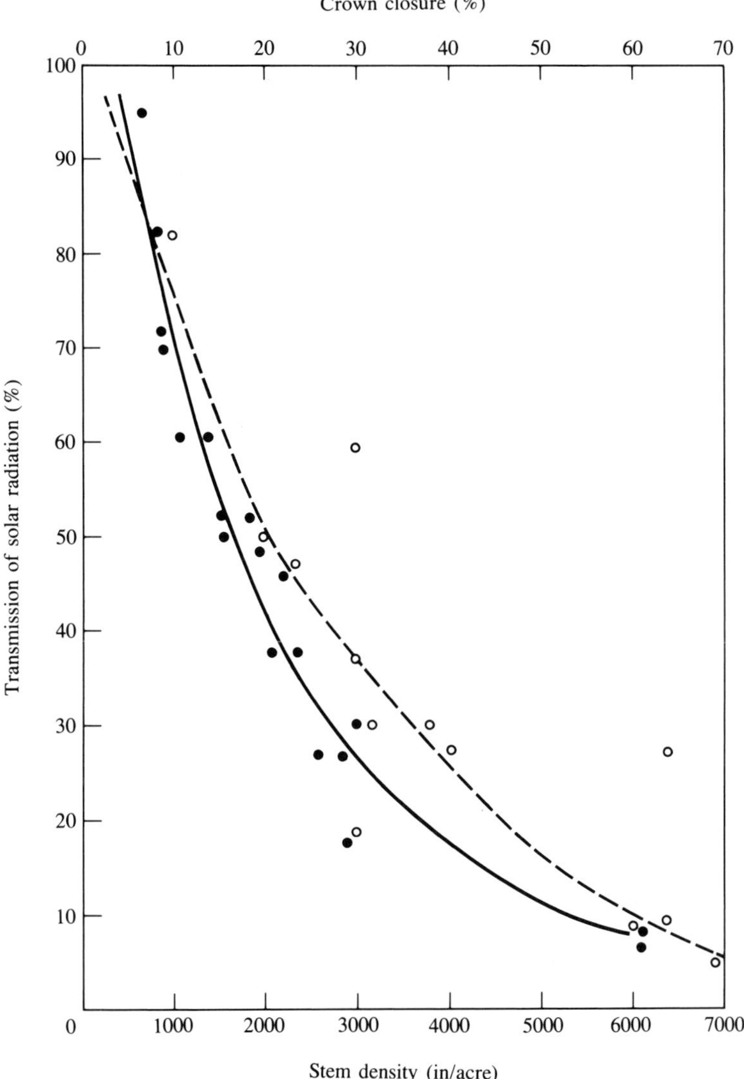

Figure 3.9 Effects of stem density (——) and associated crown closure (– – –) on transmission of solar radiation in pine forests. (After Miller, 1959; from Reifsnyder and Lull, 1965, U.S. Forest Service photo.)

Hong, 1974). Late-summer leaf hydration also was higher in thinned than in unthinned lodgepole pine stands (Donner and Running, 1986). Cacao and rubber trees, which sometimes are grown together in agroforestry systems, compete for available soil moisture (Sena Gomes and Kozlowski, 1988a).

There is much evidence of competition for water between trees and shrubs in

Table 3.2
Effect of Stem Density on Light Penetration as Influenced by Distribution of Crown Size in Conifer Forests[a]

Stem density (in $acre^{-1}$)	Basal area (ft^2 $acre^{-1}$)	Percentage of small-crowned trees		
		0–33	34–67	68–100
		(% light intensity)		
200	20	87	90	94
700	60	57	70	78
1200	100	34	50	63
1900	180	13	30	43
3700	400	7	10	12

[a] From Wellner (1948). Reprinted from *Journal of Forestry*, published by the Society of American Foresters, 5400 Grosvenor Lane, Bethesda, MD 20814-2198.

Table 3.3
Light Interception by Golden Delicious Apple Orchards of Different Ages[a]

Age of orchard (years)	Spacing (m)	Month	Light interception (%)
1	2.7 × 0.9	October	11
5	2.7 × 0.9	September	50
7	3.0 × 1.0	September	67
9	3.0 × 1.0	September	70

[a] From Jackson (1980).

forest stands. In California, white fir saplings often become established under chaparral shrubs. Competition is so intense that white fir trees 25 to 30 years old may be less than 1 m high and still below the shrub canopy. Where artificial shade was provided and competing chaparral shrubs were removed, white fir growth was doubled within 4 years (Conard and Radosevich, 1982). Growth increase was less when shade was not provided but shrub control was greater than 80%. The data emphasized the paramount importance of soil moisture for release of white fir. The added benefit of shade appeared to operate largely by improving the water balance of shaded trees.

Severe competition for soil water between Douglas-fir trees and snowbrush shrubs and forbs was shown by Petersen *et al.* (1988). Where shrubs and forbs were present, the soil water potential during late summer was less than -1.5 MPa at 100 cm depth. However, when shrubs and forbs were eliminated by herbicides, the water potential was near field capacity. Predawn stem water potential of trees competing with shrubs and forbs was lower than that for trees without competitors.

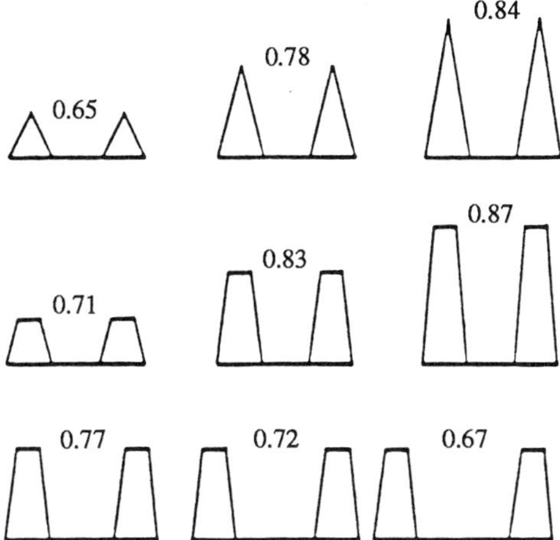

Figure 3.10 Effect of hedgerow geometry on the fraction of light intercepted annually by apple trees in England. (From Jackson, 1985.)

Shainsky and Radosevich (1986) showed that growth of ponderosa pine seedlings progressively increased as shrub competition was reduced. Soil moisture was depleted more in mixed stands than in pine monocultures but there was no significant change in the light environment. The pine seedlings were more dehydrated in mixed stands than in monoculture, emphasizing that competition with greenleaf manzanita for soil moisture reduced growth of ponderosa pine. Removal of salal (*Gaultheria shallon*), an understory shrub, increased not only the soil moisture content but also the rate of photosynthesis of Douglas-fir trees (Price *et al.*, 1986). Grasses also competed with ponderosa pine seedlings for water. *Agropyron desertorum* and *Dactylis glomerata* were more serious competitors than *Bouteloua gracilis*, because of the shallow root system of *Bouteloua*. In addition, *Bouteloua* puts on growth later in the growing season and does not compete excessively for soil moisture during early-season droughts (Elliott and White, 1987).

Mineral Nutrients

Competition for mineral nutrients has been shown for both forest and fruit trees. In agroforestry systems, competition for nutrients occurs between woody perennials and the herbaceous components as well as among individuals of each of these groups (Connor, 1983).

In forest nurseries, competition for mineral nutrients among the closely grown plants is particularly intense. As the number of seedlings per unit area is decreased, the amount of nutrients available to each seedling increases. Nitrogen concentrations of spruce needles were 1.46% for seedlings planted at a density of 788 /m^{-2} and 1.99% when seedling density was 98 /m^{-2} (Bell, 1968). Increasing the spacing among plants in seedbeds also increased the 2–0 needle N content of Douglas-fir

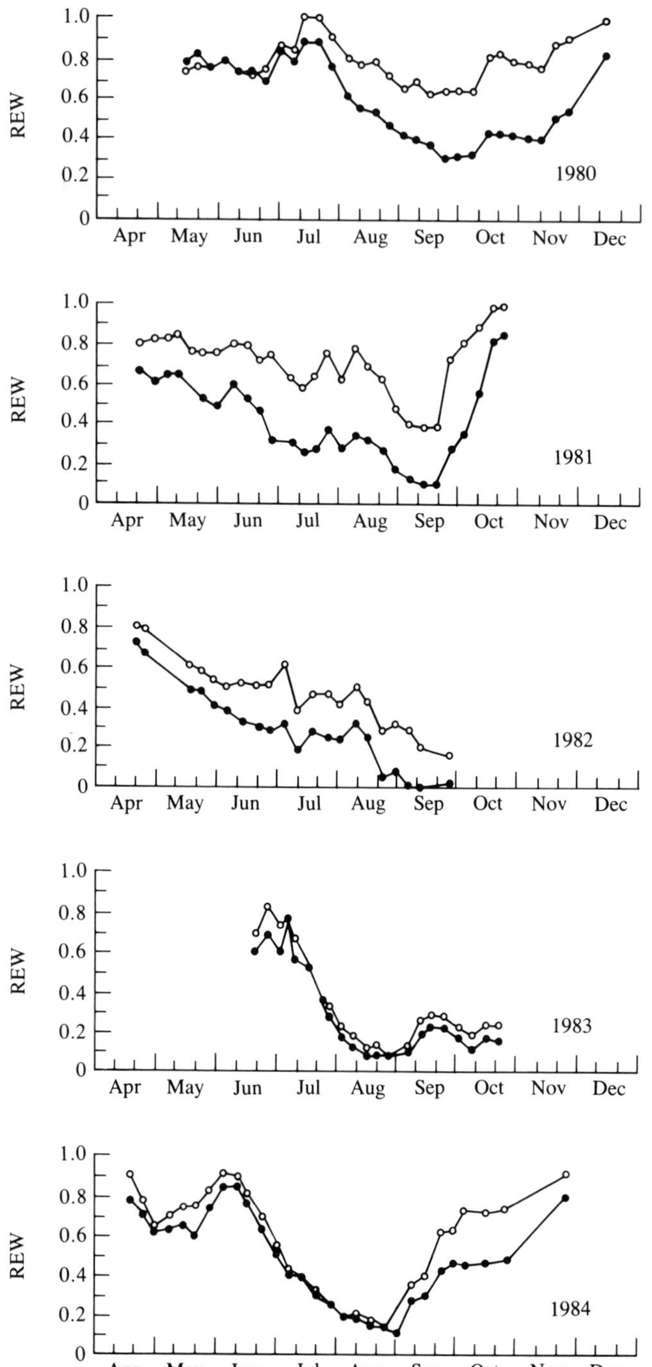

Figure 3.11 Effect of thinning a Douglas-fir stand on extractable soil water reserves (REW) in control (●) and thinned (○) stands. (From Aussenac and Granier, 1988.)

seedlings (van den Driessche, 1984). In 5-year-old Douglas-fir plantations both foliar N and P concentrations were lower in closely spaced than in widely spaced trees (Cole and Newton, 1986).

Grass interplantings are routinely used in orchards to reduce erosion, increase infiltration of water into the soil, improve soil structure, hinder weed invasion, and moderate soil temperature (Butler, 1986). However, grass swards compete with fruit trees for mineral nutrients. Atkinson and Johnson (1979), Haynes and Goh, (1980), and Haynes (1981) reported decreases in N and increases in K concentrations of apple trees as a result of competition with grass swards. Whereas the N decrease was the result of competition, the increase in K was associated with inhibition of tree growth and thus a concentration effect. Greenham (1976) reported that 24 years of fertilizer application increased the N content of apple leaves by only 0.15%, but a grass sward reduced it by 0.45% within 2 years. In Western Europe, fruit trees often are planted in weed-free, herbicide-treated strips that are separated by grassed alleys. This is done so the reduction in competition from grass will increase availability to the trees of mineral nutrients in the surface soil (Atkinson and White, 1980).

On calcareous soils in Denmark, severe competition for mineral nutrients occurs between woody and herbaceous plants. In European beech forests, small glades were dominated by a rich grass flora and other low-growing herbs. The absence of trees and shrubs in those glades was attributed to their inability to compete with the herbaceous plants for Fe and N and, to a lesser degree, for K and P (Olsen, 1961).

Effects of Competition on Form and Growth of Trees

Crown Form

As mentioned earlier, in a young, even-aged stand all trees tend to have more or less similar crown shapes. However, as the crowns begin to close some trees become suppressed and occupy low positions in the canopy whereas others express dominance and become the largest and most vigorous trees (Fig. 3.12). Foresters use the following classification for crowns of trees in even-aged stands.

1. *Dominant Trees* Crowns are above the average level of the canopy and receive full light from above and partly from the sides. Dominant trees have larger stem diameters and are more vigorous than average trees in the stand. The crowns are well developed but they may be crowded on the sides.
2. *Codominant Trees* Crowns form the level of the canopy and receive full light from above but little from the sides. These medium-sized crowns are somewhat crowded from the sides.
3. *Intermediate Trees* The trees are shorter than either dominant or codominant trees. The crowns, which extend into the crown cover formed by codominant and dominant trees, receive little light from above and none from the sides. The crowns are small and usually crowded on the sides.

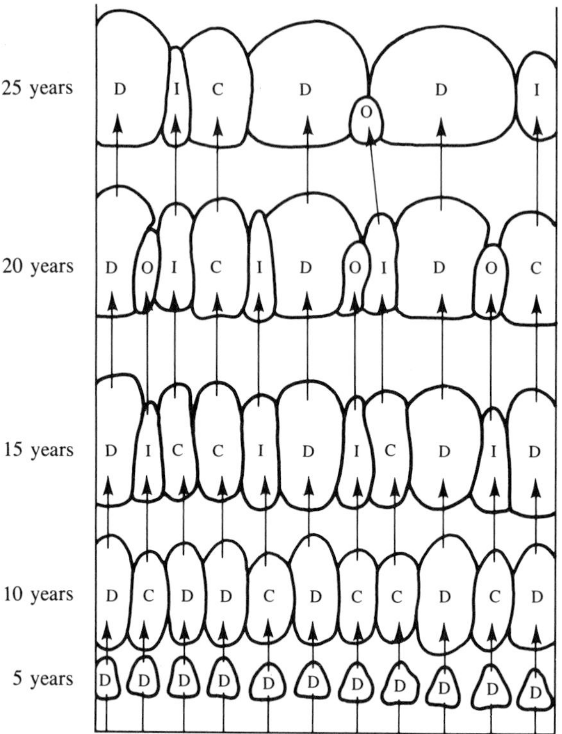

Figure 3.12 Differentiation of trees of a pure even-aged stand into crown classes. D, dominant; C, codominant; I, intermediate; O, overtopped (suppressed). (From Smith, 1986. Copyright © 1986 by John Wiley & Sons.

4. *Overtopped (Suppressed) Trees* Crowns, which are below the level of the canopy, do not receive direct light from either above or below.

Much interest has been shown in crown sizes of forest trees because of their influence on wood production. In open-grown trees the crown covers a large proportion of the main stem, whereas in mature stands the tree crowns are restricted to the upper parts of the stems. After the crowns of forest trees close, the lower branches begin to die, causing a progressive decrease in the live-crown ratio. When the live-crown ratio decreases to a critical value the rate of wood production decreases greatly. Ideally, competition among forest trees should be decreased just before this stage is reached by thinning the stand to remove undesirable trees. This generally results in increased wood production in the residual crop trees. Thereafter, thinning of the stand should be repeated often enough to maintain a relatively constant live-crown ratio. If the live-crown ratio is allowed to drop below 30 to 40% before a stand is thinned, the residual trees are unlikely to respond favorably in terms of increased diameter growth and often some die (Smith, 1986). The effects of stand thinning are discussed further in Chapter 13.

Cambial Growth

Increasing competition greatly reduces the rate of cambial growth and wood production, hence the entire annual ring of suppressed trees is thinner than in dominant trees (Fig. 3.13). Often, very suppressed trees do not show any xylem increment at the base of the stem. In competitive situations, the reduction in diameter growth is caused by late initiation and early cessation of cambial activity, as well as by a decrease in the growth rate (Winget and Kozlowski, 1965).

Cambial growth is much more sensitive than height growth to competition and especially to the relation of crown size to the length of stem that is without branches. Whereas dominant and codominant 30-year-old loblolly pine trees continued to increase in diameter into October and a few into November, suppressed trees stopped growing by late July (Bassett, 1966). On the average, diameter increase of dominant and codominant trees occurred during 80 and 70%, respectively, of each of five growing seasons. Diameter increase of large-crowned trees occurred continuously from early March through June, and then recurrently during two-thirds of the rest of the growing season. Intermediate and suppressed trees increased in diameter during only 53 and 28%, respectively, of the growing period.

Height Growth

Unlike diameter growth, height growth of canopy trees is rather insensitive to initial spacing of trees and to changes in spacing that follow the thinning of stands.

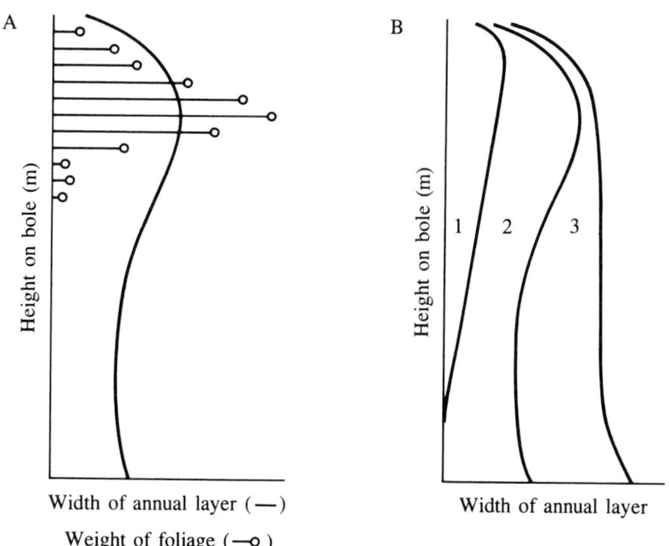

Figure 3.13 Variations in thickness of the annual ring at various stem heights in (A) dominant tree in relation to distribution of foliage and (B) suppressed (1), dominant (2), and open-grown (3) trees. (After Farrar, 1961; from Kramer and Kozlowski, 1979, by permission of Academic Press.)

According to Lanner (1985), these differences are associated with variations in sink strength and phenology of apical and lateral meristems, as discussed in Chapter 13.

Root Growth

Root biomass in forest stands generally increases with stand age as does the biomass of aboveground parts. For example, root biomass of aspen stands in Wisconsin increased from 13.8 Mg ha^{-1} in an 8-year-old stand to 23.3 Mg ha^{-1} in a 63-year-old stand (Table 3.4). Despite the increase in root biomass the root–shoot ratio declines progressively during stand development. It is sometimes claimed that the soil is fully occupied by roots at about the time of canopy closure, and most root growth thereafter involves replacement of dying roots.

Competition tends to reduce the extent, depth, configuration, and density of roots. Root systems of 10-year-old Douglas-fir trees consisted of a taproot and a few main laterals (Table 3.5). As the trees aged, the root system increased in size in proportion to increase in crown size. Much of the increase in root growth was traceable to secondary, tertiary, and quaternary branching of roots (McMinn, 1963). Within each age class the amount of root growth varied with crown class in the following order: dominant > intermediate > suppressed crown class. Often the total length of roots of dominant trees was greater than that in much older intermediate trees.

Much interest has been shown in the production of small (fine) roots because of their importance in absorption of water and mineral nutrients. The fine roots represent a large portion of the belowground biomass and a significant part of net primary

Table 3.4
Root Biomass for an Age Sequence of Trembling Aspen in Wisconsin[a]

Age (years)	Small (<0.3 cm)	Large (0.3–3.0 cm)	Stump	Total	Root–shoot ratio
		Overstory			
8	2.2 (16)[b]	7.5 (54)	4.1 (30)	13.8	0.58
14	1.8 (12)	8.3 (54)	5.2 (34)	15.3	0.38
18	1.7 (11)	8.4 (53)	5.6 (36)	15.7	0.34
32	2.1 (10)	8.0 (37)	11.3 (53)	21.4	0.21
63	2.6 (11)	8.1 (35)	12.6 (54)	23.3	0.18
		Understory			
8	7.8 (68)	3.7 (32)		11.5	5.0
14	6.8 (65)	3.6 (35)		10.4	5.3
18	6.9 (67)	3.4 (33)		10.3	3.8
32	7.0 (60)	4.7 (40)		11.7	4.1
63	6.5 (64)	3.6 (36)		10.1	3.8

[a] From Ruark and Bockheim (1988).
[b] Percentage of ecosystem total shown in parentheses. Biomass given in Mg ha^{-1}.

Table 3.5
Root Development in Three Age Classes and Crown Classes of Douglas-fir[a]

Stand age and crown class	Average height (m)	Average diameter at 1.5 m (cm)	Average length of roots 1 cm in diameter (cm)				Total and range
			Primary	Secondary	Tertiary	Quaternary	
10-Year-old stand							
Dominant	2.5	2.5	99	38	—	—	137 (65–187)
Intermediate	1.8	1.0	31	—	—	—	31 (20–44)
Suppressed	0.9	—	7	—	—	—	7 (2–17)
25-Year-old stand							
Dominant	20.0	15.7	1,129	887	156	—	2,172 (1,747–2,561)
Intermediate	10.0	8.1	484	257	10	—	751 (530–971)
Suppressed	5.0	5.1	55	51	—	—	106
55-Year-old stand							
Dominant	39.0	45.7	2,015	6,400	3,945	1,555	13,915
Intermediate	26.0	26.0	1,270	2,205	1,245	350	5,070

[a] After McMinn (1963); from Kozlowski (1971b), by permission of Academic Press.

production. Although fine roots may comprise less than 1% of the total biomass, they may account for as much as two-thirds of annual biomass production (Marshall and Waring, 1985) because of their short life and rapid turnover.

Death and replacement of fine roots occur simultaneously. Shedding of fine roots is ecologically important because it often contributes more N to the ecosystem than does aboveground litter fall (Vogt *et al.*, 1986). Fine root litter often approximates foliar litter production. For example, production of fine roots was slightly higher than leaf production in an 80-year-old stand of mixed hardwoods and slightly less than needle production in a 53-year-old red pine plantation (McClaugherty *et al.*, 1982). More fine roots often are produced by understory plants than by overstory trees (Ruark *et al.*, 1982).

Rapid occupation of a forest site by fine roots occurs early in stand development, peaks, and then levels off, independently of large-root and aboveground biomass (Santantonio *et al.*, 1977). This steady-state fine root biomass apparently coincides with development of steady-state leaf biomass (Vogt *et al.*, 1981). The growth of fine roots is very sensitive to environmental stresses, as discussed later in this chapter and in Chapter 7.

Reproductive Growth

Competition results in inhibition of reproductive growth, particularly in the more suppressed trees. For example, in forest stands most seeds are produced by dominant trees; intermediate and suppressed trees produce negligible amounts. Over a 16-year period practically all the cones borne by ponderosa pine and sugar pine were on dominant trees; only 1 to 1.5% were on codominant trees; and negligible amounts were on suppressed trees (Fowells and Schubert, 1956). In red pine, smaller cones were produced on the shaded lower branches than on the vigorous branches in the upper crown, which were exposed to a high light intensity. The larger cones in the upper crown contained more seeds than the small or lower cones (Dickmann and Kozlowski, 1971). Thinning of stands stimulates reproductive growth (Chapter 13).

Mortality

Mortality rates of trees vary during stand development, with risk of death very high during the stand initiation stage. High losses of seeds are associated with lack of seed viability, seed dormancy, injury by fungi, or consumption of seeds by insects and higher animals. If a seed germinates the young plant is subject to attacks by various organisms and to environmental stresses. Seedlings in the cotyledon stage of development are in danger because of their low reserves of carbohydrates and mineral nutrients. Even temporary mild environmental stresses at this stage can lead to death (Kozlowski, 1976a, 1979). From the beginning of shedding of Douglas-fir seeds until the end of germination, only 12% of the seeds survived (Gashwiler, 1967). Germination of eucalyptus seeds in Australia under favorable conditions exceeds 80%, but survival of seedlings often is less than 1% (Jacobs, 1955). On sandy soils in Japan, seedlings of Japanese black pine emerged in April and grew vigorously, but almost all died during the dry summer (Tazaki *et al.*, 1980). Very high mortality of seedlings during the first year after germination also has been reported by Hett and Loucks (1971), Tappeiner and Helms (1971), and Good and Good (1972).

In established forest stands undergoing competition, mortality is high in young stands and usually declines as trees age (Fig. 3.14). The actual number of trees decreases exponentially but the annual rate of death is reasonably consistent, between 1 and 2% in both temperate zone and tropical forests (Waring and Schlesinger, 1985; Swaine *et al.*, 1987). However, when unusually severe environmental stresses (e.g., sudden windstorms, fire, disease, or insect attacks) are superimposed on the stresses associated with competition, tree mortality may be appreciably increased. For example, a wavelike pattern of mortality of old mountain hemlock trees in Oregon was attributed to laminated root rot caused by *Phellinus weirii* (McCauley and Cook, 1980).

Because relatively few trees can survive the intense competition during the thinning phase of stand development, the trees that start with a competitive disadvantage (e.g., small trees, trees of low vigor, and trees close to other trees) almost certainly will die. Because the capacity to extract water and mineral nutrients is greater in large than in small trees, mortality is concentrated in the small trees. Over a 50-year period mortality in an even-aged loblolly pine stand was largely confined to the small trees. The trees with a stem diameter of 10–20 cm at first were the largest trees, and all survived for many years. Eventually, however, when some of these trees became the smallest trees, their death rate increased, emphasizing that

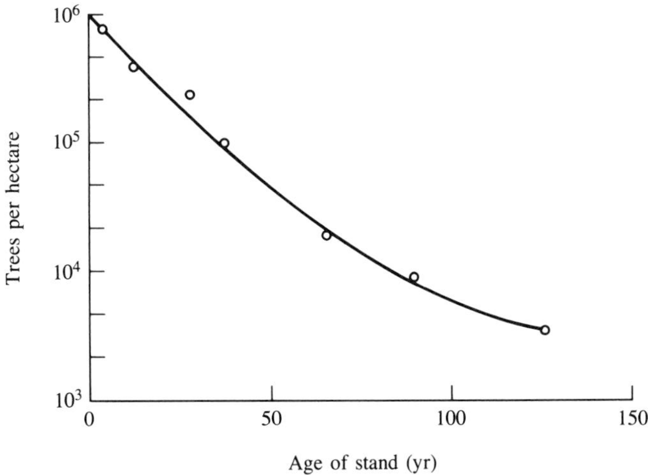

Figure 3.14 Self-thinning in *Abies* stands in Japan results in a relatively constant rate of mortality. (After Tadaki *et al.*, 1977; from Waring and Schlesinger, 1985, by permission of Academic Press.)

mortality was related to the relative size of trees in the stand rather than to their absolute size (Weiner and Thomas, 1986). In loblolly pine stands in North Carolina, practically all the dominant trees became established in the first 5 years or less. Those that became established later usually died during the self-thinning phase (Peet and Christensen, 1987). During early years of chaparral development after fire, competition for water among shrubs results in mortality of up to 50% of the initial population. Mortality is highest among small plants, which are more severely stressed than large ones (Schlesinger *et al.,* 1982).

In contrast to the thinning phase of stand development, the death of canopy trees during the transition and steady-state phases usually is not associated with deficient resources, but rather with catastrophic events such as windthrow, fire, disease, and insect attack (Peet and Christensen, 1987). For example, 21% of the mortality in a mature American beech–southern magnolia forest was attributed to wind (Harcombe and Marks, 1983). According to a citation in Loehle (1988a), Reiners found that over a 15-year period twice as many trees were killed by wind snap in a New Jersey forest as by all other causes. However, Moore (1988) considered upturning of trees more important than snapping.

Healthy trees possess carbohydrate and mineral nutrient reserves that can be used during periods of stress. By comparison, less vigorous trees lack sufficient resources to heal injuries or maintain physiological processes at levels necessary to sustain life, hence they tend to die when subjected to stress (Waring, 1987). Prolonged droughts result in lowered rates of photosynthesis, leading to depletion of carbohydrate reserves and defensive compounds, eventually causing reduction in the amount of canopy and allocation of an abnormally large fraction of photosynthate to the roots (Waring, 1983). Synchronous death of large numbers of senescent balsam fir trees in the harsh subalpine environment of New Hampshire was preceded by depletion of carbohydrate reserves associated with susceptibility to stresses, particularly pathogens (Sprugel, 1976). Mineral deficiency, like drought, leads to an increase in the proportion of the carbohydrate pool that is allocated to growth of small roots, resulting in reduction in leaf and stem growth. The low vigor of mineral-deficient trees often is associated with outbreaks of insects or disease (Kozlowski, 1985b). McCauley and Cook (1980) attributed the high mortality of mountain hemlock trees on nitrogen-deficient soils to infection by root rot. Matson and Waring (1984) noted that fertilizing increased carbohydrate and nitrogen reserves of mountain hemlock seedlings and simultaneously reduced infection by root rot.

Much concern has been shown about causes of synchronous death of large groups of neighboring forest trees, a phenomenon called "stand-level dieback." For many years such diebacks were attributed to insect pests, disease, and long-term climatic changes. More recently attention has shifted to air pollution as a possible major cause (Schütt and Cowling, 1985). Considerable evidence shows, however, that it is normal for large groups of trees to die in response to aging as well as to environmental stresses (e.g., fire or pollution) induced by humans (Jacobi *et al.,* 1983; Grodzinski *et al.,* 1984; Schütt, 1985; Mueller-Dombois, 1986, 1987).

Mueller-Dombois (1987) reviewed causes of breakdown of forest stands in different parts of the world. In England, large, even-aged stands of common juniper senesced and died at about the same time. In North America, examples of stand-level dieback include pole blight of western white pine, little-leaf disease of short-leaf pine, dieback of yellow birch and paper birch, and declines of scarlet oak, northern red oak, and red spruce. These diebacks often have been linked to various natural causes such as aging rather than to a common biotic or abiotic stress factor such as air pollution. Dieback of yellow birch was attributed to climatic change; maple decline to senescence and drought; oak decline to drought and unseasonable freezing; little-leaf disease of shortleaf pine to extension of its range to nitrogen-deficient soils interacting with attacks by the root pathogen *Phytophthora cinnamomi*; and pole blight of western white pine to drought (Mueller-Dombois, 1987). Other examples of stand-level mortality from natural causes include diebacks of balsam fir in the northeastern United States, of *Abies veitchii* and *A. mariesii* in Japan, of nothofagus (*Nothofagus solandri*) in New Zealand, and of rain forests dominated by *Metrosideros polymorpha* in the Hawaiian Islands.

According to Wardle and Allen (1983), stand diebacks may be expected when three conditions occur sequentially: (1) a forest stand is in a susceptible developmental stage (usually an old, even-aged stand, or a pole-sized stand composed of closely grown trees undergoing severe competition); (2) a stress is imposed on the stand; and (3) pests or pathogens are attracted in response to the stressed condition of the stand. As emphasized by Mueller-Dombois (1987), as long as a stand of trees is vigorous it can recover from reasonably severe environmental perturbations. However, when it becomes senescent and loses vigor, a climatic instability is more likely to trigger dieback.

Some investigators called attention to the importance of chemical and structural defenses to resist decay, herbivory, wind, and fire and thereby prevent early mortality of forest stands. Many species are susceptible to windthrow after their root systems have been invaded by decay fungi. Furthermore, extreme longevity of trees can be achieved only on sites that are free of fire or those with low-intensity fires, or in species such as redwood that have structural defenses (Chapter 11). Some species may resist pathogens and insects through increased wood density and deposits of defensive chemicals such as phenolics and resins in the wood, as well as compartmentalization at wound sites. Loehle (1988a,b) classified some forest trees (e.g., birch, ash, maple) as "low defense" types with wood that is not very resistant to decay. By comparison, such "high defense" types as bald cypress and redwood produce wood that is highly resistant to decay. Environmental stress may reduce vigor of the latter group but the wood and roots are decay resistant and the foliage is not very edible, so opportunistic pathogens are unlikely to wound these trees. According to Loehle (1988b), synchronous decline is more likely to occur in stands of species with low defensive investments than in stands composed of "high defense" species.

Succession

Plant succession has been characterized as a process by which disturbed plant communities regenerate to their previous condition if not exposed to additional disturbance. Competition among trees of the same species does not affect species composition of stands and does not influence species succession. In contrast, competition among species results in succession, which progresses toward a community of trees with species that have a high capacity to tolerate the stresses of competition.

During uninterrupted succession several characteristics of plant communities change progressively. Increases occur in the number of species, productivity, biomass, community height, and structural complexity. Finally, a climax plant community may evolve with relatively stable species composition (Whittaker, 1975). However, reestablishment of climax forests after disturbance seldom is achieved because of frequent recurrences of major and minor disturbances. Most forests are commonly subjected to periodic major disturbances such as crown fires, hurricanes, tornadoes, snow avalanches, landslides, mudflows, soil erosion and deposition, and harvesting. In addition, forests frequently undergo minor perturbations such as surface fires, windthrow of some canopy trees, lightning strikes, insect and disease attacks that kill only canopy trees, and partial cuttings and thinnings. Hence, even mature forest ecosystems are never perfectly stable, rather they are maintained in an oscillating steady state characterized by continuous elimination of suppressed and old trees and addition of new ones.

The kind, frequency, and magnitude of disturbance determine the composition and successional stage of stands of trees. Whereas catastrophic disturbances in mature forests induce reversion to pioneer stages of succession, very mild disturbances tend to keep a mature forest in a relatively steady state. In general, tropical forests are much more fragile than temperate forests (Kozlowski, 1979). Most tropical forests are closed ecosystems that exist on small mineral nutrient budgets that are tightly recycled. In contrast to the situation in temperate forests, most of the cycling nutrient capital of tropical forests is in the trees, particularly in leaves and twigs. Once dead plant tissues are broken down, the mineral nutrients are rapidly absorbed by roots of trees and other plants. Hence, although the soils are relatively infertile, recycling is rapid and efficient and there is little nutrient loss from the ecosystem (Vitousek, 1984).

Successional Patterns

Succession may be fast or slow and may lead to different types of plant communities. For example, succession in severely disturbed tropical forests to reestablishment of the original forest is particularly slow. Clearings in tropical forests are soon covered with a dense growth of weeds, shrubs, vines, and young trees. Fast-growing, short-lived trees with a life span of less than 20 years soon dominate. These are succeeded by intermediate, slower-growing, and longer-lived species. Succession then proceeds slowly until finally the climax species are reestablished. The time from clearing to reestablishment of the original forest may be hundreds of years. Often such succession is further interrupted by harvesting of firewood and clearing for farming. Such additional disturbance results in loss of mineral nutrients from the soil so the land cannot support even early successional forest trees. It is then occupied by savanna, grasses, bamboo, and ferns. The result is indefinite postponement of establishment of primary forest (Kozlowski, 1979).

Primary succession, which is initiated on bare areas where no plants grew before, may occur on glacial moraines exposed by recession of ice, newly created islands, areas of severe erosion, volcanic ash, or rock. Secondary succession results when an existing plant community is disrupted by factors such as fire, cultivation, harvesting of timber, windthrow, or other disturbances that destroy the major species of an established community. Secondary succession sometimes is very slow, for example, when a fire destroys all the organic matter on a site (Oosting, 1956).

Attributes of Early and Late Successional Species

Species replacement during succession involves interaction of competition with tolerance to environmental stress factors. Both the seeds and seedlings of early and late successional species exhibit differences in physiological characteristics that account for establishment and subsequent success or failure as competition intensifies (Table 3.6). Seeds of early successional plants usually are longer-lived and require more light for germination than do seeds of late successional plants. Germination of early colonizers is suppressed by the high far-red/red light ratios in the deep shade of forests. Because seedlings of early successional plants of both temperate and tropical forests generally are competitively inferior to those of late successional plants, they must grow and use available resources rapidly. The high relative growth rates of early successional species are particularly important. For example, sun-adapted species such as smooth sumac and white ash grow faster than the shade-adapted climax species sugar maple (Grime, 1966). In gaps of southern Appalachian cove forests, height growth of saplings of component species averaged about 30 cm year^{-1}, with species averages ranging from 18 to 49 cm year^{-1}. Both American basswood and yellow poplar grew faster than the more shade-tolerant American beech and sugar maple. Tall saplings grew faster than short ones of the same age and saplings in large gaps grew faster than those in small gaps (Yetter and Runkle, 1986; Runkle and Yetter, 1987).

Growth of early successional trees is especially rapid in the tropics, with annual rates of height growth of 5 m often reported. Fast growth of early successional species is associated with rapid accumulation of mineral nutrients (Bazzaz and Pickett, 1980).

Differences in photosynthetic rates are important factors in succession (Table 3.7). In general, early successional plants have high maximum rates of photosynthesis at high light intensities. In deep shade, rapid respiration, water stress, disease and predation lead to death of such species. The rate of photosynthesis of early successional species does not change much at light intensities above saturation, whereas the rate of late successional species may decline. As succession proceeds, the rate of decrease of respiration is lower than the rate of photosynthesis, resulting in a progressive decline in the ratio of photosynthesis to respiration (Fig. 3.15).

Table 3.6
Characteristics of Early and Late-Successional Plants[a]

Attribute	Early successional plants	Late successional plants
Seeds		
Dispersal in time	Long	Short
Secondary (induced) dormancy	Common	Uncommon?
Seed germination		
Enhanced by:		
Light	Yes	No
Fluctuating temperatures	Yes	No
High NO_3^- concentrations	Yes	No?
Inhibited by:		
Far-red light	Yes	No
High CO_2 concentrations	Yes	No?
Light saturation intensity	High	Low
Light compensation point	High	Low
Efficiency at low light	Low	High
Photosynthetic rates	High	Low
Respiration rates	High	Low
Transpiration rates	High	Low
Stomatal and mesophyll resistances	Low	High
Resistance to water transport	Low	High
Acclimation potential	High	Low
Recovery from resource limitation	Fast	Slow
Ability to compress environmental extremes	High	Low?
Physiological response breadth	Broad	Narrow
Resource acquisition rates	Fast	Slow?
Material allocation flexibility	High	Low?

[a] From Bazzaz (1979). Reproduced, with permission, from the Annual Review of Ecology and Systematics, Volume 10, © 1979 by Annual Reviews, Inc.

Allelopathy

Considerable interest has been shown in the effects of plant extracts on inhibition of seed germination and growth of neighboring plants. Potentially allelopathic compounds may be produced by any part of the plant, but the highest concentrations are found in leaves and fruits (Rice, 1984).

Various investigators claimed that allelopathy may play a role in regeneration and development of forest stands. Several genera of plants common in red pine stands include species that have been claimed to be potentially allelopathic. These include *Prunus* (Brown, 1967), *Aster*, and *Solidago* (Fisher *et al.*, 1979). Water extracts of the leaves of these plants did not affect seed germination but inhibited root growth, height growth, formation of secondary needles, and dry weight increment of red pine seedlings (Table 3.8). Total dry weight of seedlings treated with *Lonicera* extract was only 46% of that of control plants. Extreme caution is advised in

Table 3.7
Variations in Rates of Photosynthesis and Respiration of Tropical Plants[a]

	Early successional trees	Canopy trees	Understory plants
Photosynthesis	14.1[b]	6.9	2.9
Respiration	2.0	1.0	0.3

[a] Reproduced, with permission, from the Annual Review of Ecology and Systematics, Volume 11, © 1980 by Annual Reviews, Inc.
[b] Rates are given in mg CO_2 dm^{-2} hr^{-1}.

speculating, on the basis of laboratory experiments, on the ecological significance of allelopathy in plant succession. It is relatively easy to demonstrate allelopathic effects in laboratory experiments in which seeds or seedlings are treated with high concentrations of allelochems. However, as emphasized by Lerner and Evenari (1961), such concentrations often are higher than those to which plants in the field are subjected. Compounds extracted from leaves of *Eucalyptus rostrata* inhibited seed germination, but allelochems in soil beneath *Eucalyptus* trees did not accumulate to toxic concentrations. Furthermore, accumulation of allelochems in the field is modified by soil moisture and soil type, and allelochems often are destroyed by soil microflora.

Horsley (1977a,b) attributed the failure of regeneration and development of black cherry trees in old fields and forests to allelopathic interference by herbaceous

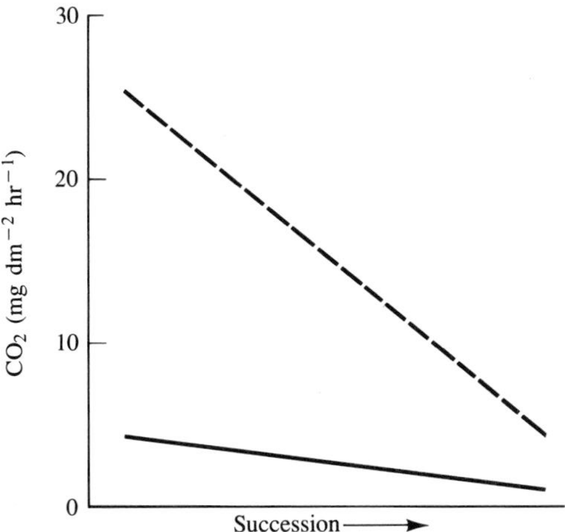

Figure 3.15 General trends of photosynthesis (– – –) and dark respiration (——) in relation to successional pattern. (From Bazzaz, 1979. Reproduced, with permission, from the Annual Review of Ecology and Systematics, Volume 10, © 1979 by Annual Reviews, Inc.)

Table 3.8
Effect of Water Extracts of Leaves from Six Plant Species and Distilled Water Control on Dry Weights of Roots and Shoots and Root–Shoot Ratio of Red Pine Seedlings after 7 Weeks of Treatment[a]

	Root		Shoot		Root–shoot ratio	
Extract	Weight[b] (mg)	Percentage of control	Weight[b] (mg)	Percentage of control	Ratio[b]	Percentage of control
Control	185.3 ± 18.6	100.0	412.9 ± 43.5	100.0	0.46 ± 0.03	100.0
Aster	139.1 ± 8.6[c]	75.1	246.4 ± 24.1[d]	59.7	0.62 ± 0.08	135.2
Lonicera	109.2 ± 10.2[d]	58.9	163.5 ± 18.4[e]	39.6	0.68 ± 0.03[e]	149.2
Prunus	143.1 ± 14.9	77.2	240.5 ± 29.0[d]	58.2	0.63 ± 0.05[c]	138.1
Rubus	126.7 ± 8.3[d]	68.4	207.6 ± 12.9[e]	50.3	0.62 ± 0.03[e]	134.5
Solanum	142.8 ± 14.2	77.1	274.5 ± 34.3[c]	66.5	0.55 ± 0.04	120.5
Solidago	124.7 ± 12.6[c]	67.3	204.2 ± 27.0[e]	49.4	0.65 ± 0.06[c]	143.1

[a] From Norby and Kozlowski (1980).
[b] Mean of 10 seedlings ± standard error.
[c] Significantly different from control at $P < 0.05$.
[d] Significantly different from control at $P < 0.01$.
[e] Significantly different from control at $P < 0.001$.

vegetation, including hayscented fern. However, when he grew black cherry seedlings in soil cores removed from areas with and without hayscented fern, he did not find differences in height growth, indicating that allelochems from hayscented fern were not accumulating in the soil. Many of the sites on which allelopathy has been indicated have poor drainage, which may prevent leaching away of allelochems or create anaerobic conditions that are unfavorable for microbial decomposition of allelochems.

Successional Pathways and Mechanisms

Following Clements' (1936) characterization of the ecosystem as a form of ontogeny, various investigators described replacement sequences in disturbed forests and suggested possible mechanisms of plant succession (Table 3.9). Some of the more recent succession models will be discussed briefly.

Connell and Slatyer (1977) described three models of mechanisms that could account for successional changes after disturbance, assuming that no further change occurs in the biotic environment:

1. *Facilitation Model* (model 1 in Fig. 3.16) Only certain early successional species can invade a site soon after stand disturbance. These species modify the site, making it more suitable for invasion and growth of later successional species. This concept is similar to the classic relay floristic pathway of Egler (1954) and mostly applies to primary successions.
2. *Tolerance Model* (model 2 in Fig. 3.16) Changes to the environment by early-colonizing plants neither increase nor decrease rates of plant recruitment and

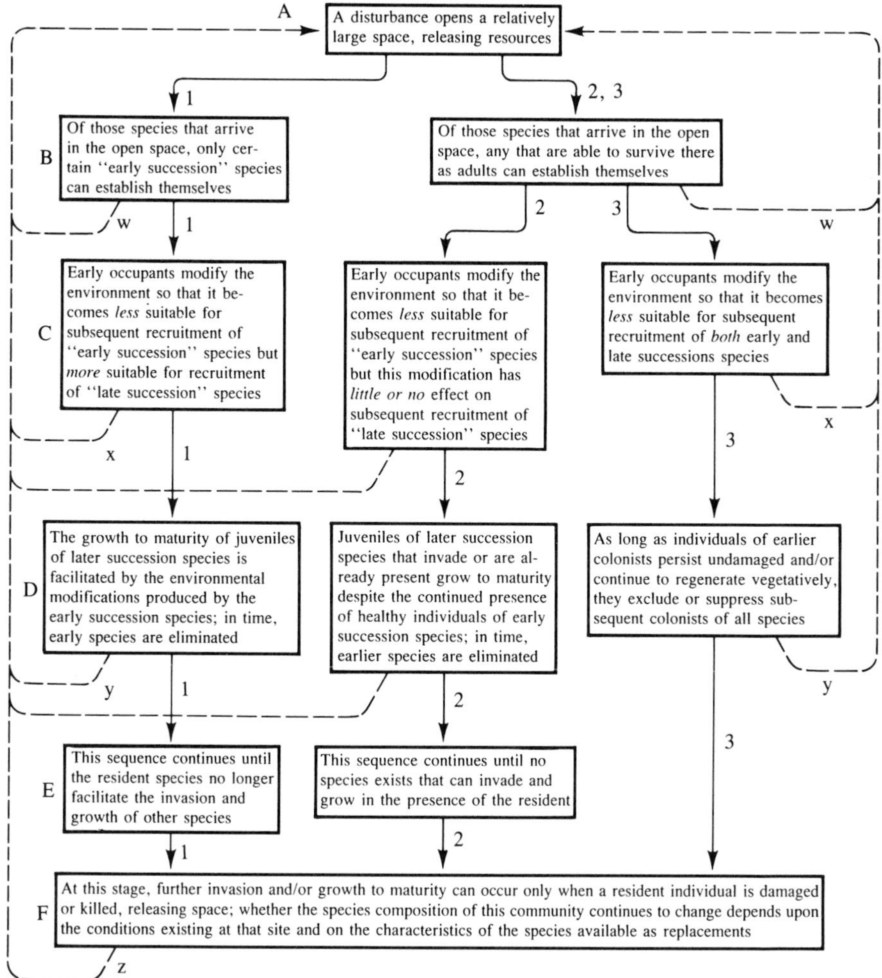

Figure 3.16 Three models of mechanisms producing the sequence of species in succession. The dashed lines represent interruptions of the process, in decreasing frequency in the order w,x,y, and z. (From Connell and Slatyer, 1977.)

growth to maturity by later-colonizing plants. The species that appear later arrive either early or late and grow slowly. The sequence of various species is determined entirely by their life history characteristics. Because they have low requirements for resources, the later-arriving species can become established and grow to maturity in the presence of other species.

3. *Inhibition Model* (model 3 in Fig. 3.16) Once early-colonizing plants secure space and available resources, they either inhibit invasion of subsequent colonists or suppress those already present.

Table 3.9
Successional Replacement Sequences as Proposed by Various Investigators[a]

	Theory				
Model[b]	Clements	Connell and Slatyer	Whittaker and Levin	Horn	Egler
 → A → B → C → D ↺	Classical	Facilitation	Replacement	Obligate	Relay floristics
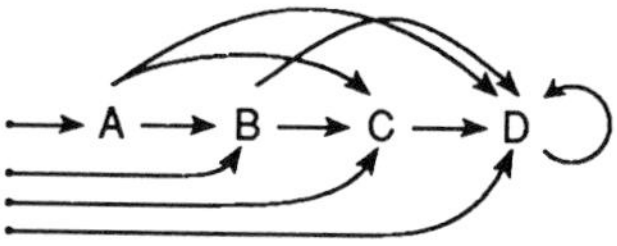 → A → B → C → D ↺		Tolerance		Competitive hierarchy	
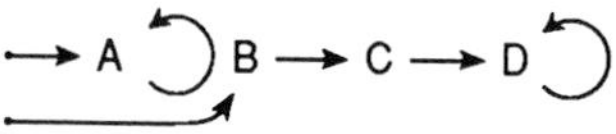 → A ↺ B → C → D ↺		Inhibition	Plateau		
 → A ↺			Direct		
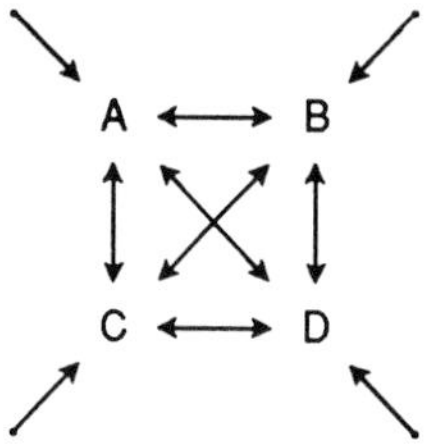 A ↔ B; C ↔ D; A ↔ C; B ↔ D; A ↔ D; B ↔ C			Chronic disturbance		
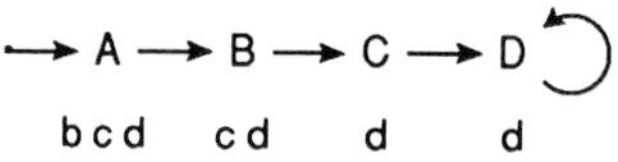 → A → B → C → D ↺ bcd cd d d					Initial floristic composition

[a] From Noble, 1981.

[b] The letters represent hypothetical dominant species; uppercase letters indicate dominance and lowercase letters indicate subdominance. The arrows marked with a dot show alternative starting points for the replacement sequence after a disturbance.

Connell and Slatyer (1977) predicted that the facilitation model would apply to many primary successions, and the tolerance and inhibition models to most secondary successions. Although evidence for the facilitation and inhibition models is strong, evidence is weak for the tolerance model.

Whittaker and Levin (1977) described four types of succession: (1) replacement succession, which strongly resembles the facilitation pathway of Connell and Slatyer (1977); (2) direct succession, with species becoming reestablished directly after a disturbance but without intermediate stages; (3) cyclic succession, involving cycles such as those associated with recurrent fires in chaparral; and (4) mosaic succession, involving interactions of recurrent disturbances with the basic set of pathways.

Noble and Slatyer (1980) focused on a small number of "vital attributes" needed to describe the behavior of a species at recurrently disturbed sites. Their approach did not emphasize mechanisms accounting for behavior patterns in disturbed communities, but rather the result of the mechanisms involved. As a first step they recognized four life stages in the population of a species: (1) juvenile stage, (2) mature stage (majority of plants mature but some juveniles may be present), (3) propagule stage (species present as a store of seeds, bulbs, dormant rootstocks, etc.), and (4) locally extinct stage (all traces of the species absent). These stages provided a basis for derivation of vital attributes of a species related to (1) method of arrival or persistence of a species at a site during and after a disturbance, (2) capacity to establish and grow to maturity in a developing community, and (3) time required for the species to reach initial stages in its life history. Important life stage parameters were the time for a species to reach reproductive maturity, its longevity, and longevity of the propagule store.

Noble and Slatyer placed particular emphasis on the method of persistence and initiation of regeneration after disturbance and less on subsequent growth and competitive interactions. This was done primarily because, on sites subjected to recurrent disturbances, the events that require persistence and regrowth occur frequently and competitive interactions of adult plants do not have time to have an impact on species composition before another disturbance occurs. According to Noble (1981), emphasis on succession has changed from community processes to species properties and especially to the regeneration phase of disturbed stands. Connell (1975), for example, emphasized that many species rarely reach population densities high enough to compete for resources. Rather the effects of environmental stresses and predation acting on young plants often reduce populations to such an extent that competition is not important.

Our knowledge of succession is incomplete because of the lack of a relation of the patterns of succession to mechanisms, poor articulation of concepts of succession, and the limited scope of available succession models. For these reasons Pickett *et al.* (1987) expanded on the Connell and Slatyer (1977) models, which focused on specific aspects of the successional process, and proposed a comprehensive framework for investigating mechanisms of succession. Their proposal is useful in providing a context for studies of succession at specific sites and a scheme for formulating general and testable hypotheses about the nature of succession.

In the past, complex phenomena like competition and succession have been traditionally studied by measuring such variables as plant dimensions, plant weight, or seed number, and disturbance has been characterized in terms of biomass destruction and available space rather than changing resource patterns. Bazzaz and Sipe (1987) suggested that a more useful approach may be to focus attention on interactions among plants through their mutual influences on the physical and chemical fluxes (resources and modifiers) that comprise their shared microenvironment. Equally important is an emphasis on differences in physiological processes of competing species that lead to succession.

Biomass Accumulation and Partitioning

Growth of individual trees and stands of trees can be quantified by various methods. Most important to foresters is measurement of increment of wood produced by tree stands. This can be determined from measurements of stem diameter and height and is modified by the amount of stem taper. Wood production usually is expressed as volume increment per unit of land area, but for some purposes it is better expressed as weight increase per unit of land area. Ecologists are interested in dry weight increases and losses of organic matter per unit of land area and the partitioning of dry matter into different organs and tissues. Yields of fruits, cones, and seeds usually are assessed by numbers produced, size, and weight. Detailed descriptions of methods of measuring plant growth are beyond the scope of this book. For such descriptions the reader is referred to Whittaker and Marks (1975), Avery and Burkhart (1983), Kozlowski (1985d), and Lassoie and Hinckley (1988).

Biomass Changes

The rate of biomass accumulation is much lower than the potential rate. This difference results primarily because of low photosynthetic rates of trees (discussed in Chapters 2 and 4), losses of carbohydrates by respiration, shedding of plant parts, and consumption of plant tissues by animals, fungi, and microbial decomposers. In addition, some carbohydrates are diverted to production of defensive chemicals and are not available for growth, and small amounts of carbohydrates are lost by leaching from tree crowns, exudation from roots, and loss to parasites, such as mistletoes. In many forests the consumption of plant tissues by animals is not very significant, except where outbreaks of defoliating insects occur (Waring and Schlesinger, 1985)

Respiration

A large portion of the food produced by photosynthesis is used in respiration and is unavailable for incorporation into plant tissues. Respiratory losses of carbohydrates have been estimated to be 58% of gross primary production (GPP) in 14-year-old loblolly pine trees (Kinerson 1975) and 65% in tropical rain forests (Kira, 1975). As trees grow older the ratio of photosynthetic tissue to respiring tissue decreases, therefore, the proportion of the photosynthate used in respiration increases. For example, respiration consumed about 40% of the carbohydrate pool in 25-year-old European beech trees and 50% in 85-year-old trees (Fig. 3.17).

Shedding of Plant Parts

Dry weight increment of woody plants is reduced because a variety of tissues and organs are shed periodically. Abscising vegetative parts include buds, branches, prickles, cotyledons, leaves, leaflets, leaf stalks, bark, and roots. Reproductive parts that abscise include inflorescences, pedicels, flowers, buds, calyxes, corollas, stamens, styles, ovaries, fruits, fruit hulls, strobili (cones), scales of strobili, and seeds.

Some idea of the extent of losses of organic matter by abscission of leaves may be gained from studies of annual litter production (Table 3.10) because leaves

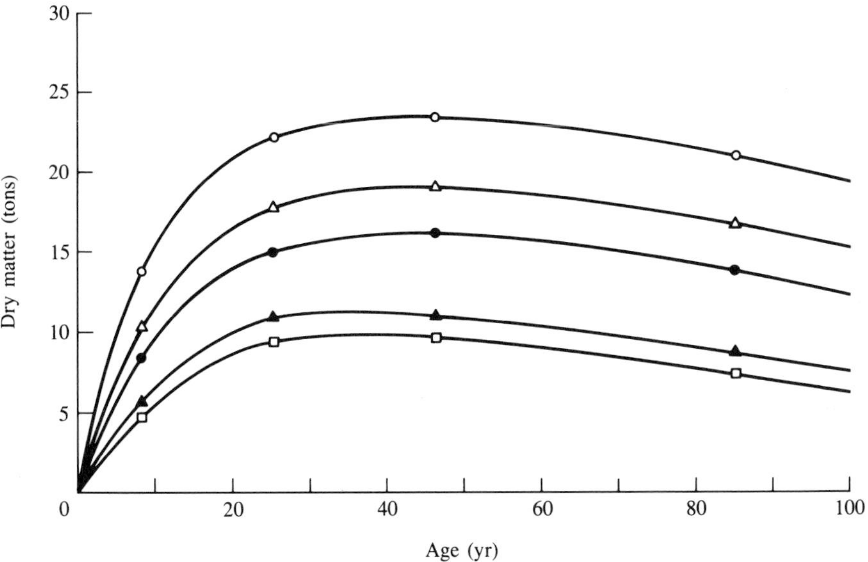

Figure 3.17 The amounts of dry matter used in various processes by beech stands of various ages growing on good sites in Denmark. The ordinate represents metric tons of dry matter ha^{-1} yr^{-1}. The upper curve (○) indicates total or gross production by photosynthesis. The middle curve (●) indicates net photosynthesis, and the lowest curve (□) indicates the amount of dry matter added in growth each year. The area between the top two curves (○ and △) represents loss of dry matter by leaf respiration, and that between the next two curves (△ and ●) loss by leaf fall. The area between the next curves (● and ▲) represents loss of dry matter by respiration in roots, stems, and branches, and the area between the bottom curves (▲ and □) represents dry matter lost by death of roots and branches. (After Möller *et al.*, 1954; from Kramer and Kozlowski, 1979, by permission of Academic Press.)

Table 3.10
Litter Production by Various Forest Types[a]

Forest type	Forest floor mass (kg ha^{-1})	Litter fall (kg ha^{-1} $year^{-1}$)
Tropical broadleaf		
deciduous	8,789	9,438
evergreen	22,547	9,369
Warm temperate broadleaf		
deciduous	11,480	4,236
evergreen	19,148	6,484
Cold temperate broadleaf		
deciduous	32,207	3,854
Cold temperature needleleaf		
deciduous	13,900	3,590
evergreen	44,574	3,144
Boreal needleleaf		
evergreen	44,693	2,428

[a] From Vogt *et al.* (1986).

comprise the major litter component. In a study of eight evergreen and deciduous forests, leaves comprised 60 to 76% of the litter, branches 12 to 15%, bark 1 to 14%, and fruits 1 to 17% (Bray and Gorham, 1964).

The extent and pattern of abscission of bark vary enormously among species and may involve shedding of thin flakes or large, thick pieces. In Australia the bark shed by *Eucalyptus viminalis* trees may accumulate to 110 tons ha^{-1} (Jacobs, 1955). Under ponderosa pine trees the pile of abscised bark may be 2 ft high. By comparison, redwoods shed little bark. The phloem of redwood contains large amounts of fibers that interlock tissues of the dead outer bark (rhytidome), and cells of the periderm adhere tightly to other cells of the rhytidome. This results in a tough adherent bark that may be 2 ft thick on older trees and provides an unusual amount of insulation from fire injury.

Loss of portions of large perennial roots and entire small rootlets is well documented (Kozlowski, 1971b). By far the greatest natural loss of roots occurs among the rootlets that grow to only a few millimeters in length. The turnover of fine roots has been estimated to range from 3.5 to more than 11.0 tons ha^{-1} $year^{-1}$ (Cannell, 1985).

Individual plants produce far more flowers than can develop into mature fruits. Excess fruits or potential fruits are abscised at various stages of their development, most commonly in the early flower bud stage, before flower opening, as young fruits are developing, or at fruit maturity (Addicott,1982). Premature abscission of flowers also is common. For example, about 90% of white oak flowers may be shed

Table 3.11
Percentage of Abscission of White Oak Acorns at Various Development Stages[a]

	1962		1963	
Development stage	Starting date	Percentage of abscissions	Starting date	Percentage of abscissions
Pollination	April 30	55.6	April 24	28.4
Ovule development	May 17	10.6	May 20	37.9
Fertilization	June 4	15.8	June 6	18.1
Embryo development	July 6	16.7	July 3	10.5
Maturation	Sept. 19	1.3	Sept. 20	5.1
Total		100.0		100.0

[a] After Williamson (1966); from Kramer and Kozlowski (1979), by permission of Academic Press.

prematurely (Table 3.11). Mortality of conelets of shortleaf pine ranged from 35 to 97% (Bramlett,1972). In addition to normal abscission of uninjured reproductive structures, premature abscission is associated with injury by frost, drought, high temperature, and attacks by insects, fungi, birds, and mammals (Sweet, 1975).

Productivity

Changes in biomass or weight of forest trees are used to study growth, nutrient cycling, and energy flow in forest stands. Growth usually is expressed in terms of productivity or weight increase in metric tons ha^{-1} $year^{-1}$ or g m^{-2} $year^{-1}$. The important components of productivity of interest to ecologists include:

1. *Gross Primary Production (GPP)* The weight of carbon compounds produced by photosynthesis that are retained in the plant plus the weight of dry matter lost by plant respiration.
2. *Net Primary Production (NPP)* The sum of (1) increase in biomass, including leaves, stems, roots, and reproductive structures, (2) litter production, and (3) the amount of biomass consumed by animals and microbial decomposers.
3. *Net Ecosystem Production (NEP)* GPP minus losses of dry matter due to both heterotrophic and autotrophic respiration.

The NPP of forest stands varies widely with site, forest type, age of stands, species and genotype, and climate. On a global basis, NPP of forests ranges from less than 1 to nearly 40 tons ha^{-1} $year^{-1}$, with a mean of about 15 tons ha^{-1} $year^{-1}$ (Waring and Schlesinger, 1985). NPP increases from arctic to tropical regions (Table 3.12), with increasing water supply and soil fertility, and with decreasing elevation. Productivity of stands of the same age often varies appreciably on different sites (Table 3.13).

Table 3.12
Variations in Average Biomass and Net Primary Production of Various Forest Types

Forest type	Biomass (g m^{-2})	Net primary production (g m^{-2} year^{-1})	Reference
Boreal forest	20,000	800	Whittaker (1970)
Temperate forest	30,000	1,300	Whittaker (1970)
Subtropical deciduous forest	41,000	2,450	Rodin and Bazilevich (1967)
Tropical forest	45,000	2,000	Whittaker (1970)
Tropical rain forest	27,090	—	Ovington (1965)
Tropical rain forest	—	5,000[a]	Westlake (1963)

[a] Probable maximum.

Across the United States, productivity of forest stands increases southward from the Canadian border to the Gulf Coastal Plain in the eastern United States, decreases westward from the Mississippi River into the Great Plains, and increases again because of the maritime influence of the Pacific Ocean. Overall, productivity is greatly affected by north–south-oriented mountain ranges (Fig. 3.18). As mentioned in Chapter 2, evapotranspiration is a good predictor of productivity in mature

Table 3.13
Net Primary Production of Plantations and Natural Stands of the Same Age on Different Sites

Species	Location	Age (years)	Net primary production (g m^{-2} year^{-1})	Reference
Plantations				
Pinus densiflora	Japan	13	1222	Satoo (1966a)
Pinus densiflora		13	1101	
Pinus densiflora		13	1254	
Pinus densiflora		13	1503	
Pinus nigra	England	22	587	Ovington (1956)
Pinus nigra		22	686	
Natural stands				
Abies balsamea	Canada	40–50	935	Baskerville (1965, 1966)
Abies balsamea		40–50	962	
Abies balsamea		40–50	1058	
Abies balsamea		40–50	1164	
Abies balsamea		40–50	1258	
Castanopsis cuspidata	Japan	40	3100	Kan *et al.* (1965)
Castanopsis cuspidata		40	2600	
Castanopsis cuspidata		12	1800	
Castanopsis cuspidata		12	2600	

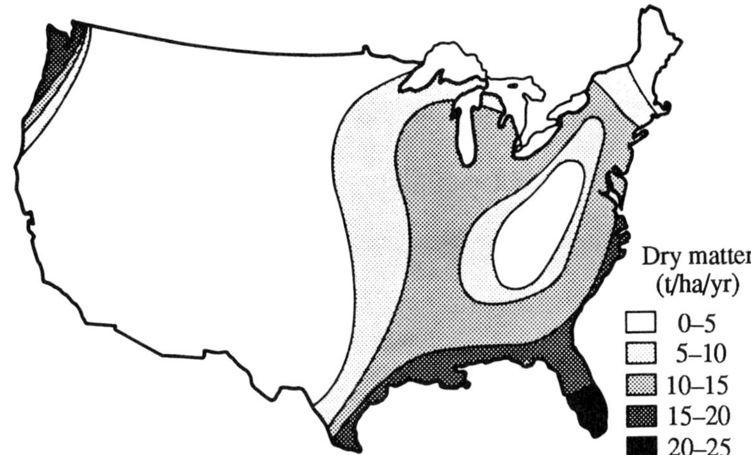

Figure 3.18 Estimated maximum woody biomass production in the United States. (After Ranney and Cushman, 1982; from Waring and Schlesinger, 1985, by permission of Academic Press.)

terrestrial ecosystems, because actual evapotranspiration measures the simultaneous availability of water and solar energy, the most important rate-limiting resources in photosynthesis (Rosenzweig, 1968). However, some investigators reported that productivity was correlated with other environmental factors. In the eastern United States, where maximum canopy leaf areas are rather similar, mean annual temperatures and length of the growing season are correlated with gross production (Lieth and Whittaker, 1975). By comparison, in the western United States, productivity is strongly related to water supply during the growing season and mean minimum air temperature in January (Gholz, 1982).

In England, growth of forest stands varies appreciably because of the very wide range of soils and elevations at which trees are grown (Jarvis, 1981a). Aboveground net productivity (ANPP) of five major forest types in Japan varied in the following order: evergreen broad-leaved forests of the warm temperate zone (central Honshu and southeastward) > pines and other temperate conifers > conifer forests of the boreal zone (northeastern half of Hokkaido and the subalpine zone of Honshu) deciduous broad-leaved forests of the cool temperate zone (Kira, 1975).

Age of Stands

Biomass changes during stand development, with the time required to achieve a maximum varying for different sites and stands. The range for natural stands of conifers is from approximately 15 to 80 years.

In plantations and even-aged forests, NPP increases with stand age to a maximum near the time of full canopy closure and usually declines thereafter, reflecting reduced photosynthetic efficiency (Fig. 3.19). However, the length of time required

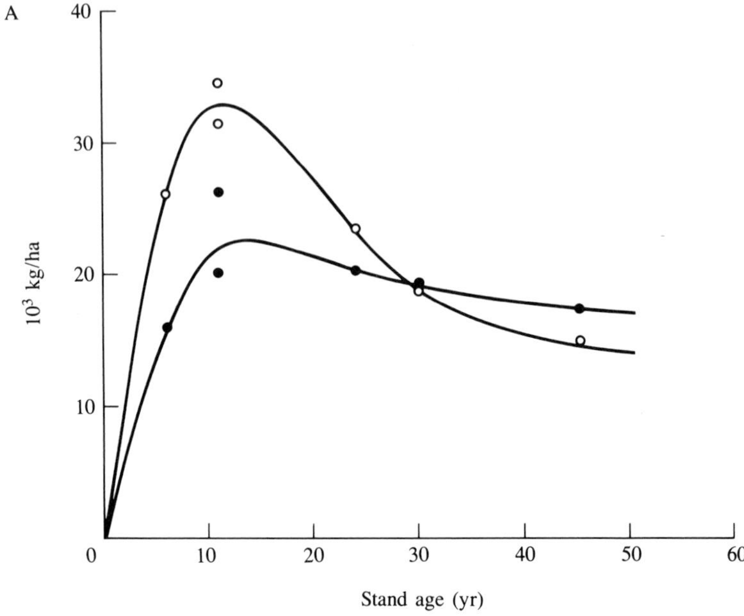

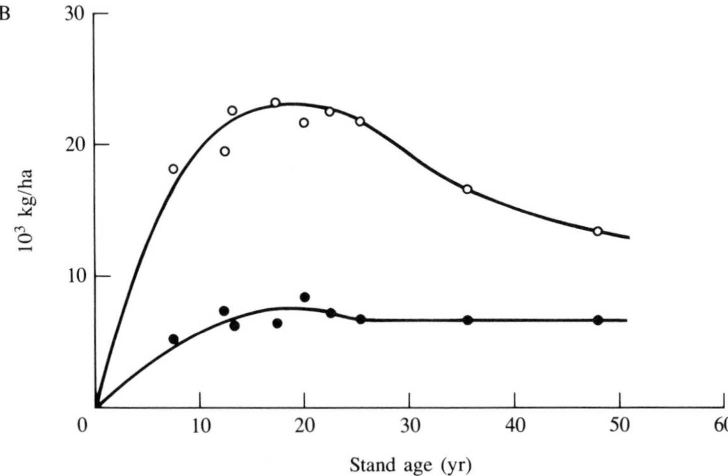

Figure 3.19 Effect of age of cryptomeria plantations (A) and even-aged Japanese red pine forests (B) on net productivity (○) and leaf biomass (●). (After Kira, 1976; from Kira and Kumara, 1983, by permission of the International Council for Research in Agroforestry.)

to reach maximum NPP is modified by spacing of trees. When trees are planted close together, NPP may reach a maximum in a few years. In the temperate zone,

Table 3.14
Dry Weight Production and Photosynthetic Efficiency of Even-Aged Stands of American Beech and Scotch Pine[a,b]

Species and age (years)	Current annual dry weight production		Photosynthetic efficiency (%)
	g acre^{-1}	cal acre^{-1}	
Beech			
8	3.0×10^6	12.1×10^6	1.4
25	5.4×10^6	21.8×10^6	2.5
46	5.4×10^6	21.8×10^6	2.5
85	4.6×10^6	18.5×10^6	2.1
Scotch pine			
12	4.0×10^6	16.0×10^6	1.0
22	8.1×10^6	32.4×10^6	2.0
28	8.8×10^6	35.2×10^6	2.2
33	8.1×10^6	32.4×10^6	2.0
41	6.1×10^6	24.4×10^6	1.5
50	4.0×10^6	16.0×10^6	1.0

[a] After Hellmers and Bonner (1960); from Kramer and Kozlowski (1979), by permission of Academic Press.

[b] Data are based on yield data of Möller *et al.* (1954) and Ovington (1957).

however, many unthinned stands, reach the stage of maximum production in 10 to 30 years (Kira and Kumura, 1983). In Scotch pine stands, productivity increased to a maximum in about 35 years and then decreased slightly (Ovington,1962). By comparison, in slash pine stands, NPP increased to a maximum at 26 years (Gholz and Fisher, 1982).

Some of the reduction in the rate of dry weight increase of old stands may be associated with inhibitory effects of aging of trees. The underlying mechanism of aging stress is an increase in respiring tissues in proportion to photosynthetic tissues. Cambial growth accelerates annually for a number of years, and then declines, at first rapidly and then more gradually. Each successive xylem ring is narrower than the previous one, unless the pattern is altered appreciably by environmental changes such as those associated with thinning. After maximum ring width is attained, the narrowing of annual rings as an aging phenomenon often amounts to less than 1% annually, but sometimes it is much more. For example, in Finland, xylem rings narrowed annually by 4 to 5% as trees aged (Mikola, 1950).

As may be seen in Table 3.14, dry weight production of 8-year-old stands of American beech was low. It was higher and relatively constant in stands 25 to 46 years old. Thereafter a decline occurred until 85-year-old trees produced dry matter at about 85% of the maximum value. By comparison, dry weight production of young Scotch pine stands increased with age up to a maximum of about 20 years. At approximately age 30, a slow decline in dry weight increment was evident. The decrease accelerated after age 40, with 50-year-old trees producing only half as much dry matter as 33-year-old trees.

Leaf Area

Production of dry matter is related to the capacity of trees to synthesize carbohydrates, hence growth of individual trees often is related to crown size. Dominant trees with large, fully exposed crowns produce more dry matter than suppressed trees with small, shaded crowns.

Many studies have been conducted on the relationship between leaf biomass or leaf area index (LAI) and productivity of forest stands. The LAI of forest stands of the temperate zone usually varies from 1 to 20 depending on the type of stand, its age, and site conditions, especially water supply and soil fertility. The LAI of deciduous forest stands generally is lower (3 to 6) than that of conifer stands (up to 20)(Waring and Schlesinger, 1985). Communities of dry sclerophylls have still lower values (1 to 3) and some fast-growing deciduous trees have much higher values than those given for the preceding deciduous stands. For example, hybrid poplars grown under intensive culture developed LAIs of 16 to 45, depending on spacing of trees (Isebrands *et al.*, 1977).

As a forest stand grows, LAI increases to a maximum and subsequently stabilizes or declines slightly. The maximum sustained LAI of even-aged stands may be up to 20% lower than the peak LAI (Ford, 1982). The LAI of Sitka spruce stands in England peaked at age 16 (Ford, 1982); of Scotch pine in Sweden at age 16 (Albrektson, 1980); of aspen in Wisconsin at age 32 (Ruark and Bockheim, 1988); and in a fir forest in Japan at age 50 (Tadaki *et al.*, 1977). In northwestern United States, the LAI of conifer stands increased for 40 years. The LAI of some conifer stands reaches an equilibrium value that does not change much thereafter, as in loblolly pine (Switzer *et al.*, 1968) and Japanese red pine stands (Kira and Shidei, 1967).

The NPP of forest stands usually is positively correlated with LAI up to some value; at higher values NPP often decreases, largely because of tree mortality and inhibitory effects of competition on photosynthetic efficiency of the lower leaves. The NPP of a Douglas-fir forest peaked at about half the maximum LAI and decreased at higher values as growth losses from death of trees exceeded the small growth increments of the remaining trees (Waring, 1983). In slash pine stands in Florida, ANPP increased rapidly for the first 10 to 15 years, following development of the canopy, and decreased rapidly thereafter (Fig. 3.20). The ANPP was closely correlated with LAI until late in the rotation. After 18 years, ANPP declined as accumulation of biomass slowed to a negligible value, while LAI was maintained. Between 26 and 34 years (after the normal rotation age), both ANPP and LAI decreased as stand structure degenerated (Gholz, 1986).

In young stands the importance of LAI to growth is modified by the length of time during the year that the foliage is photosynthetically active. Hence productiv-

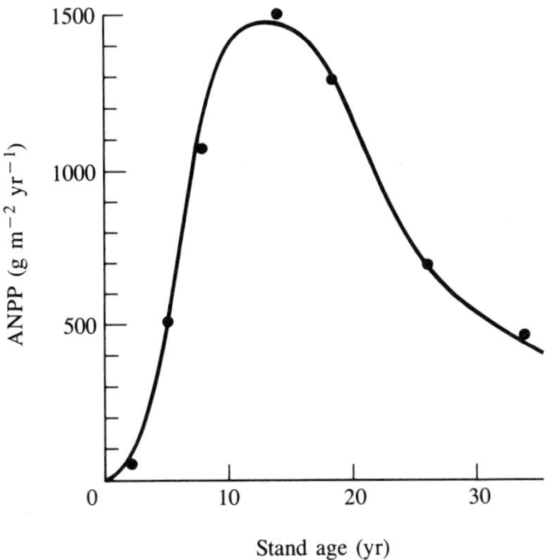

Figure 3.20 Effect of age of slash pine stands on aboveground net primary productivity. (From Gholz, 1986.)

ity of evergreen stands is favored over that of deciduous stands. Kira (1975) reported a fair correlation between gross production of both deciduous and evergreen forests and leaf area duration (defined as LAI x length of the growing season in months).

Although studies limited to specific regions show positive correlations between productivity and lower-end values of LAI, there appears to be less correlation between dry weight increase of stands on different sites and leaf biomass or LAI. Large variations in correlation between ANPP and LAI of widely separated stands are associated with differences in net photosynthesis, respiration of nonphotosynthetic tissues, and the amount of carbon allocated to roots (Gholz, 1986). Waring (1983) emphasized that correlation between productivity and leaf area breaks down in regions where drought, infertile soils, or mild winter climates predominate. Miller (1986) claimed that in evergreen trees there was little correlation between tree growth and total foliage, but good correlation with the amount of current-year foliage, suggesting that the contribution of older leaves was unimportant. Some investigators found that cambial growth at a particular stem height is related to leaf area above that point (Isebrands and Nelson, 1982; Remphrey *et al.*, 1987). Although canopy leaf area of Douglas-fir stands on four sites with different environments varied little, productivity ranged from approximately 250 to as much as 500 tons ha^{-1} year^{-1} (Emmingham and Waring, 1977). Hence, general models of productivity of forest stands should incorporate effects of the major environmental

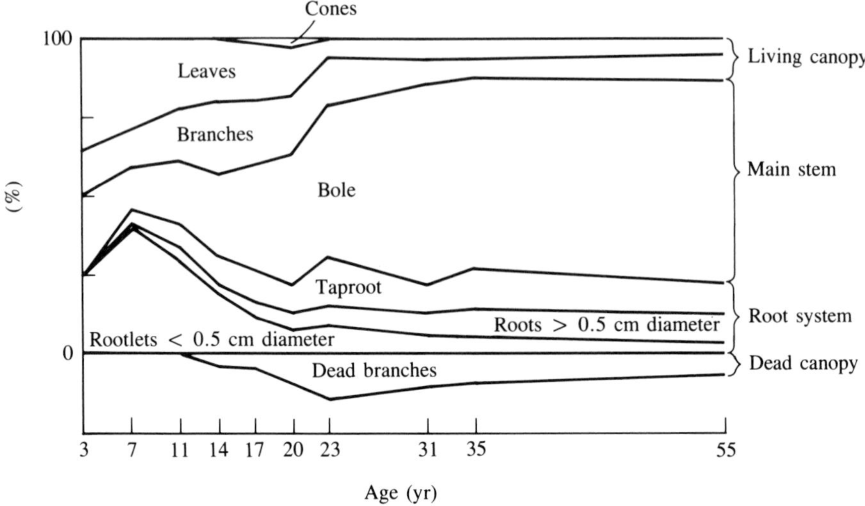

Figure 3.21 Changes with age in distribution of dry weight of Scotch pine trees. (After Ovington, 1957; from Kramer and Kozlowski, 1979, by permission of Academic Press.

factors that influence photosynthesis, growth, and carbohydrate reserves (Specht, 1981).

Partitioning of Dry Matter

The relative weights of crown, stem, and root systems vary predictably with age of trees. In young trees a very high proportion of the photosynthate is used in leaf production. In old trees more of the dry weight is in the main stem and proportionally less in the crown and root system (Fig. 3.21). Whereas roots of young Scotch pine trees accounted for almost half the weight of the plants, in old trees the proportion was much lower (Ovington, 1957). Partitioning of photosynthate also was discussed in Chapter 2.

Relative Growth Rates

For a long time investigators have been concerned about methods of comparing growth of plants of different sizes. Classical growth analysis, which involves collecting primary data on leaf surface areas and component dry weights from periodic harvests of sample plants, has been used widely in agriculture. Variables with physiological meaning are derived from the basic data and analyzed as mean stand values from evenly spaced harvests (classical growth analysis). Growth analysis has been used less widely to study growth of woody plants than herbaceous plants, because of the problems of size and longevity of the former. However, it has been used to study responses of forest trees to fertilizers and to thinning (Brix, 1983) and development of stand structure (Ford, 1984).

It was established long ago that plant growth follows the compound interest law, in which the amount of growth made in a unit of time depends on the amount of material or size of the plant at the beginning of the period. Thus seedlings developing from large seeds are likely to increase in size more rapidly than those of the same species developing from small seeds and large seedlings would be expected to grow faster than small ones. This fact led to development of the concept of relative growth rate (RGR) or measurement of increase in dry weight per unit of time per unit of growing material, often in grams per gram dry weight per week. This permits comparison of the effects of various factors in the environment on rate of growth independently of the size of the plants being compared.

The equation for calculating mean RGR is

$$\mathrm{RGR} = \frac{\ln W_2 - \ln W_1}{t_2 - t_1}$$

W_1 and W_2 are dry weights at the beginning and end of the sampling period, t_1 and t_2 are the dates of sampling, and 1n is the natural logarithm of the numbers.

Because growth depends on the rate of photosynthesis and the area or weight of photosynthetic tissue, the relationship can be shown as

$$\mathrm{RGR} = \mathrm{NAR} \times \mathrm{LAR}$$

In this expression NAR (net assimilation rate) is dry weight increment per unit of leaf area and LAR (leaf area ratio) is the ratio of leaf area to total dry weight, all for the specified time interval t_2 - t_1, which usually is 1 or 2 weeks.

There are several problems in directly applying classical growth analysis to trees. Because of cumulative production of physiologically inactive tissues, the annual increment of growth of a tree decreases progressively as a proportion of tree size or weight. Hence, RGR of mature trees is very small, relatively insensitive to environmental stress, and inversely related to accumulated biomass. For these reasons Ledig (1974) suggested using RGR of only the physiologically active tissues such as leaves. In studying the effects of fertilizers on tree growth, Ballard (1984) used the basal area increment after treatment in relation to increment before treatment. Blanche *et al.* (1985) assessed stem growth per unit of leaf area. Unfortunately data derived by these methods are influenced by differences in leaf retention or they have not been developed to provide the same insight to investigators who use RGR and its components.

Brand *et al.* (1987) derived a useful formulation, called the relative production rate (RPR), for studying tree growth. This measure and its yield components are similar to classical growth analysis but use annual increments of growth rather than total accumulated growth. The RPR eliminates accumulated past growth, a major determinant of RGR, and operates independently of tree size as a measure of growth.

In modeling growth responses of trees to environmental stresses and cultural practices, extreme caution is advised in comparing data of investigators who used different criteria for assessing plant growth. On the basis of dry weight increment, diameter growth, and RGR, the tolerance to SO_2 of three species varied in the following order: Japanese black pine, Japanese red pine, and Japanese larch. However, when height growth was used as the response criterion, the order of SO_2 tolerance was: Japanese red pine, Japanese black pine, and Japanese larch (Tsukahara *et al.*, 1985). Effects of flooding of soil and SO_2 on growth of bald cypress seedlings also differed somewhat when based on analysis of dry weight increment or RGR. SO_2 reduced the dry weight increment but not the RGR of leaves. Flooding inhibited the dry weight increase of seedlings, leaves, stems, and roots but only the RGR of stems was reduced (Shanklin and Kozlowski, 1985a). The development and application of the concepts and methodology of studying plant growth are discussed in more detail by Ledig (1974) and in books by Hunt (1982) and Charles-Edwards (1982).

We will turn our attention in the next chapters to a more detailed consideration of some of the individual environmental stress factors that influence growth of woody plants.

Summary

Forests vary from pure even-aged stands of one species to stands of several species and different age classes.

Harvested or severely disturbed stands progress sequentially through (1) a stand regeneration stage, (2) a thinning stage, (3) a transition (understory initiation) stage, and (4) a steady state or old-growth stage.

Harvested or disturbed forests are regenerated by growth of suppressed seedlings, buried or dispersed seeds, stump or root sprouts, and rooted ends of branches. Seed germination often is impeded by drought, low temperature, oxygen deficiency, and (rarely) lack of light.

Trees compete with each other, with understory shrubs, and with herbaceous plants for light, water, and mineral nutrients. As a result of competition some trees become suppressed while others become dominant.

Mortality rates of trees vary during stand development, with a very high rate during the stand initiation stage. The rate is high in young stands and declines as trees age. In contrast to the thinning phase of stand development, when mortality is associated with deficiency of resources, death of canopy trees during the transition and steady-state phases is associated more with catastrophic events.

During succession in mixed forest stands, increases occur in the rate of formation of biomass, height and stratification of trees, mineral nutrient pools, and species diversity. Although succession proceeds toward a climax plant community, this is seldom achieved because of recurrent minor and major disturbances.

Large differences between potential production and biomass accumulation of forest stands result mostly from losses by respiration, shedding of plant parts, and consumption of plants by animals, fungi, and microbial decomposers.

Productivity of forests varies with forest type, age of stands, species and genotype, and climate. Net primary production increases from arctic to tropical regions, with increasing water supply and soil fertility, and with decreasing elevation. Partitioning of dry matter into various organs and tissues varies among species and genotype, age of stands, and environmental conditions.

Relative growth rate, the measurement of increase in dry weight per unit of time per unit of growing material, permits comparison of the effects of environmental factors on the rate of growth independently of the size of plants being compared.

General References

Bazzaz, F. A. (1979). The physiological ecology of plant succession. *Ann. Rev. Ecol. Syst.* **10,** 351–371.

Bazzaz, F. A., and Pickett, S. T. A. (1980). Physiological ecology of tropical succession: A comparative review. *Ann. Rev. Ecol. Syst.* **11,** 287–310.

Cairns, J., ed. (1980). "The Recovery Process in Damaged Ecosystems." Ann Arbor Sciences, Ann Arbor, MI.

Cannell, M. G. R., and Jackson, J. E. (1985). "Attributes of Trees as Crop Plants." Inst. of Terrestrial Ecology, Midlothian, Scotland.

Cannell, M. G. R., and Last, F. T. (1976). "Tree Physiology and Yield Improvement." Academic Press, London.

Charles-Edwards, D. A. (1982). "Physiological Determinants of Crop Growth." Academic Press, Sydney.

Connell, J. H., and Slatyer, R. O. (1977). Mechanisms of succession in natural communities and their role in community stability and organization. *Am. Naturalist* **111,** 1119–1141.

Fowler, N. (1986). The role of competition in plant communities in arid and semiarid regions. *Ann. Rev. Ecol. Syst.* **17,** 89–110.

Gorham, E., Vitousek, P. M., and Reiners, W. A. (1979). The regulation of chemical budgets over the course of terrestrial ecosystem succession. *Ann. Rev. Ecol. Syst.* **10,** 53–84.

Hunt, R. (1982). "Plant Growth Curves. The Functional Approach to Plant Growth". Edward Arnold, London.

Jarvis, P. G. (1981). Production efficiency of coniferous forest in the UK. *In* "Physiological Processes Limiting Crop Growth" (C. B. Johnson, ed.), pp. 81–107. Butterworth, London.

Kimmins, J. P. (1987). "Forest Ecology." MacMillan, New York.

Kozlowski, T. T., ed. (1972). "Seed Biology," Vols. I–III. Academic Press, New York.

Kozlowski, T. T., ed. (1973). "Shedding of Plant Parts." Academic Press, New York.

Kramer, P. J., and Kozlowski, T. T. (1979). "Physiology of Woody Plants." Academic Press, New York.

Larcher, W. (1983). "Physiological Plant Ecology." Springer-Verlag, Berlin.

Mooney, H. A., Pell, E., and Winner, W. E., eds. (1991). "The Integrated Response of Plants to Stress." Academic Press, San Diego.

Odum, E. P. (1983). "Basic Ecology." Saunders, Philadelphia.

Pickett, S. T. A., Collins, S. L., and Armesto, J. J. (1987). Models, mechanisms, and pathways of succession. *Bot. Rev.* **52,** 335–371.

Pickett, S. T. A., and White, P. S., eds. (1985). "The Ecology of Natural Disturbance and Patch Dynamics." Academic Press, Orlando, FL.

Russell, G., Marshall, B., and Jarvis, P. G. (1989). "Plant Canopies: Their Growth, Form, and Function." Cambridge University Press, New York.

Sousa, W. P. (1984). The role of disturbance in natural communities. *Ann. Rev. Ecol. Syst.* **15,** 353–391.

Waring, R. H. (1983). Estimating forest growth and efficiency in relation to canopy leaf area. *Adv. Ecol. Res.* **13,** 327–354.

Waring, R. H., and Schlesinger, W. H. (1985). "Forest Ecosystems: Concepts and Management." Academic Press, Orlando, FL.

Chapter 4

Radiation

Introduction

The importance of light and its necessity for the growth of plants is self-evident. If a box or other object is left on the lawn for few days, the grass yellows and eventually dies. If seeds are germinated in darkness or deep shade, the seedlings are tall and spindly. Foresters know that pine seedlings grow poorly in the shade, where hardwood seedlings flourish, and hence that pine forests will eventually be replaced by hardwood stands if left alone. Many plants flower in the spring or early summer, but others such as chrysanthemums and witch hazel flower in late autumn or early winter. All of these phenomena represent reactions to light.

Light controls the temperature that influences the rates of most physiological processes, it supplies the energy that drives photosynthesis and transpiration, and it produces formative or morphogenetic effects that are seen in variations in leaf size, stem growth and sturdiness, root–shoot ratio, and photoperiodic control of flowering. In fact the radiation environment has a wider variety of effects on plant physiology and morphology than any other abiotic environmental factor to which plants are exposed. These effects are accentuated ecologically by the systematic variations in amount and quality of radiation related to latitude, season, and time of day, in addition to random variations related to weather and plant cover.

This chapter will discuss the nature of radiation reaching the earth's surface, the effects of various wavelengths of light on plants, how wavelength varies with latitude, season, and plant cover, and finally how these variations affect physiological processes and plant growth.

The Radiation Environment and Energy Balance of Leaves

Solar Radiation

Electromagnetic radiation produced by the nuclear reactions carried on in the photosphere region of the sun provides the energy we perceive as sunlight and heat. The radiation emitted covers a broad spectrum of wavelengths, from very long radio waves to relatively short wavelength ultraviolet radiation (Figs. 4.1 and 4.2). Of the vast amount of solar energy produced, the earth receives an exceedingly small fraction because of its small size and great distance from the sun.

Outside the earth's atmosphere the irradiance (energy striking a unit area per unit time) to a surface perpendicular to the sun's rays is about 1360 W m^{-2} or 1.95 cal cm^{-2} min^{-1}. Direct solar radiation striking the earth's surface is reduced below this value by deviation of the surface from the perpendicular and by reflection, absorption, and scattering of light in the atmosphere (Fig. 4.3). As a result, the solar radiation flux varies with zenith angle, latitude, time of year, slope aspect and percent, haziness of the atmosphere, and cloud cover. To fully characterize the radiation environment at ground level, a diffuse radiation component, which arises from interactions of sunlight with gas molecules of the atmosphere and with dust particles, must be added to direct sunlight. Molecular scattering produces the blue color of clear skies because light of blue wavelengths is scattered much more than that of red wavelengths. Light scattering by dust, which depends much less on wavelength, produces the white-colored sky associated with hazy and polluted atmospheric conditions. The combination of direct and indirect radiation (called *global* solar radiation) amounts to between 50 and 1250 W m^{-2} on a horizontal surface, depending on sky conditions (Gates, 1980).

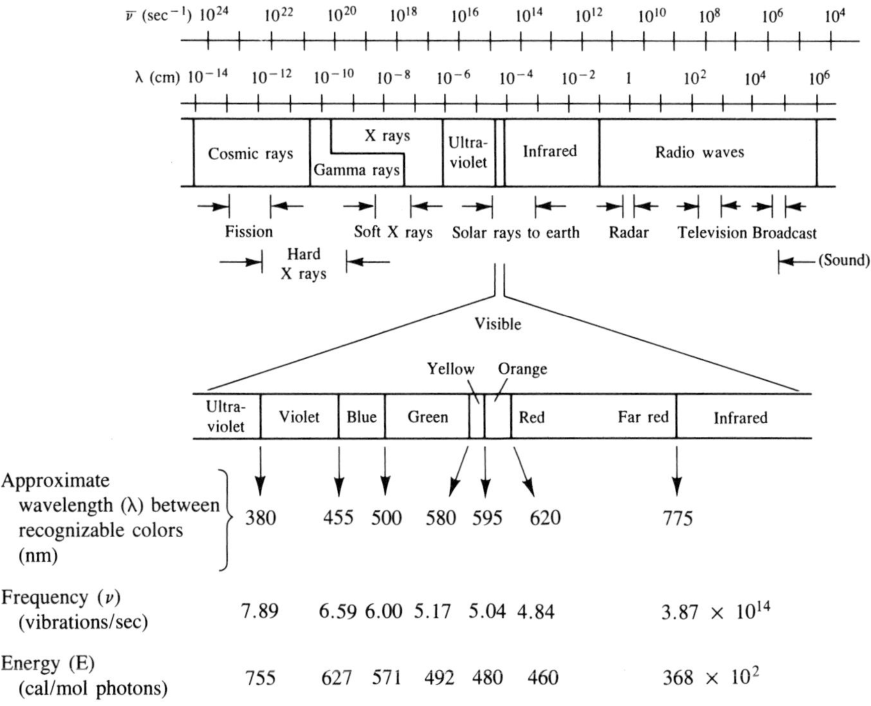

Figure 4.1 The electromagnetic spectrum using both wave number (ν) and wavelength (λ) in centimeters. The visible portion of the spectrum is expanded to show the relationships between wavelength, wave number, and color. (From *Plant Physiology,* 2E, by Frank B. Salisbury and Cleon W. Ross. © 1978 by Wadsworth Publishing Company, Inc. Reprinted by permission of the publisher.)

The diurnal pattern of global solar radiation on level land on clear days is described by a bell-shaped curve that peaks at solar noon. Slopes receive different radiation patterns depending on season, slope angle, and aspect (Fig. 4.4). During the growing season in the Northern Hemisphere, south-facing slopes receive substantially more radiation than those that face north, causing the former to be considerably warmer and drier. In extreme cases there may be a 20-fold difference in light intensity between north- and south-facing slopes (Hart, 1988). The resulting microclimate differences influence species composition, and adjacent north- and south-facing slopes can have quite dissimilar forest communities. For example, in the Cumberland Mountains of West Virginia, American beech, sugar maple, and eastern redbud were found only on north-facing slopes, whereas pine and black, chestnut, and scarlet oaks were restricted to south-facing slopes (Fig. 4.5). At the arid boundary of the geographic range of many species, individuals may only be found on north-facing slopes or in other unusually moist habitats. In mountains that have well-developed vegetation zones, similar plant communities are found at higher altitudes on south-facing slopes (Whittaker, 1970), emphasizing the more xeric

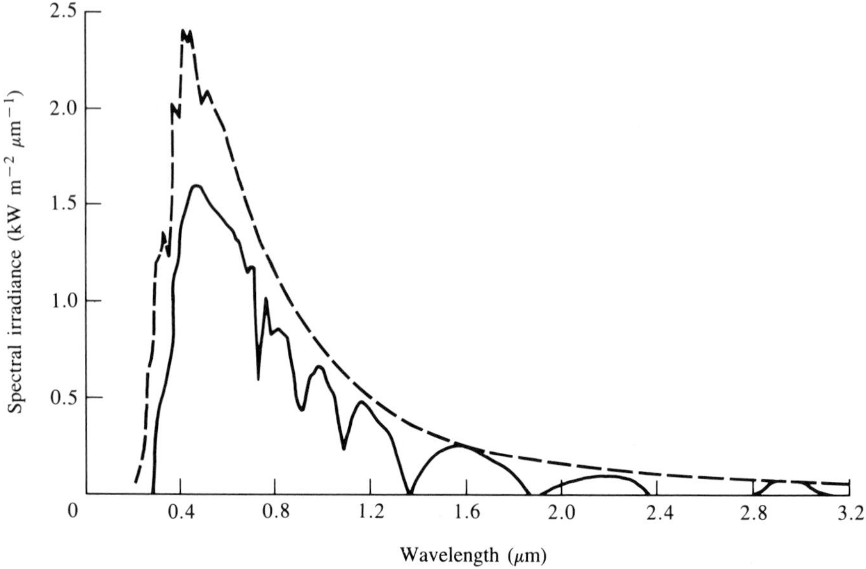

Figure 4.2 Spectral irradiance outside the earth's atmosphere (--) and at the earth's surface (——). In the ultraviolet region, atmospheric absorption is primarily attributable to O_3 in the stratosphere, whereas in the infrared region, water vapor absorbs intensely in several bands. (From Campbell, 1977.)

character that demarcates lower altitudinal limits and a warmer temperature regime that allows winter survival at higher altitude.

While the presence of heavy cloud cover can decrease global solar radiation by more than 80% (Gates, 1980), scattered cumulus clouds can actually increase instantaneous values of global solar radiation from an unobscured sun by reflecting sunlight toward an observer. Light regimes of partly cloudy days are especially dynamic, with wide oscillations in global radiation over intervals of only a few seconds. Such short-term fluctuations can induce large changes in leaf temperature, transpiration, plant water status, and photosynthesis (Stansell *et al.*, 1973; Pereira *et al.*, 1986).

Of the total energy incident upon the earth's surface, only a small fraction is captured by photosynthesis. There are a number of reasons for this lack of efficiency (Slatyer, 1967, p. 36). First, only 44% of the energy in solar irradiance is in the wavelengths used in photosynthesis (400–700 nm) and the leaf will reflect and transmit some of this. Second, only 20% of the energy in 10 moles of quanta required to fix a mole of CO_2 in photosynthesis is stored in the product. Further reductions result from the lower efficiency of energy conversion of shorter and longer wavelengths of light in the solar spectrum. The total effect of these factors reduces the maximum attainable efficiency of conversion to 8 or 9% (Slatyer, 1967). Light saturation of photosynthesis, leaf aging, and a myriad of biotic and abiotic factors reducing the capacity of the photosynthetic machinery diminish the efficiency still further.

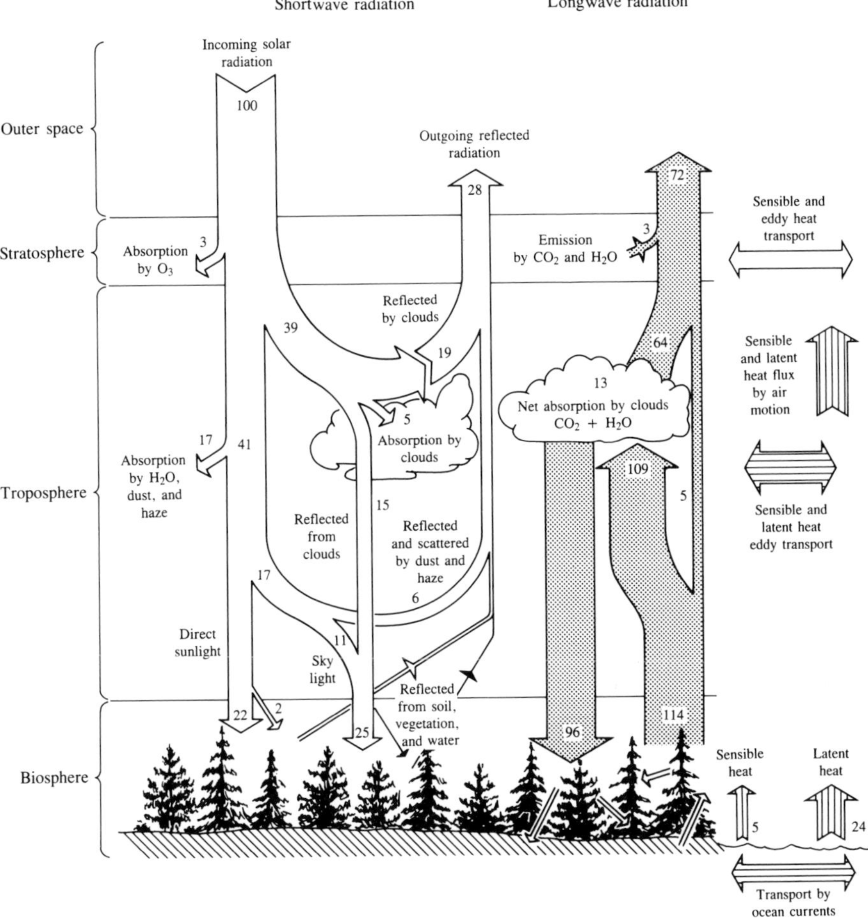

Figure 4.3 The exchange of radiation in the atmosphere and at the earth's surface. Numbers indicate relative magnitude of radiant transfers, with incoming short-wave radiation equaling 100. (From Schneider and Londer, 1984.)

It has been estimated that, for the entire earth, only 0.21% of the incident solar radiation can be accounted for in gross primary production (GPP) (Gates, 1980), and the maximum efficiency of photosynthesis rarely exceeds 2 to 3% of the incident solar radiation (Bonner, 1962, cited in Slatyer, 1967). Conversion efficiencies of solar radiation based on net primary production (NPP) of 0.5% for a northeastern North American pine forest and 0.9% for a tropical forest also emphasize the relative inefficiency of photosynthesis as a light-harvesting process (Gates, 1980).

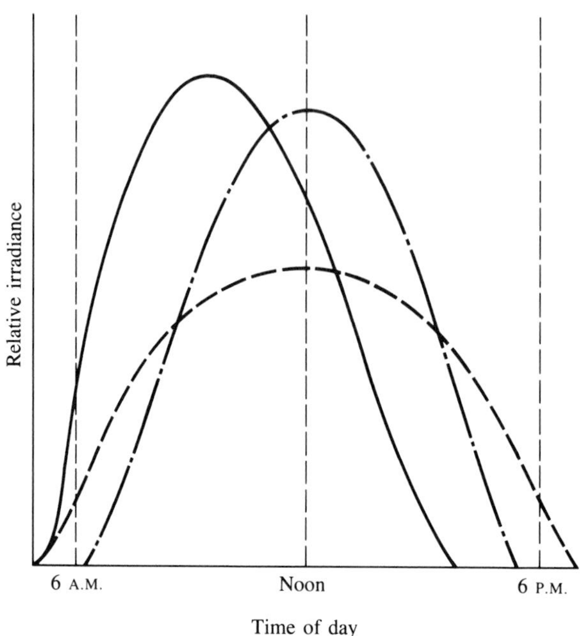

Figure 4.4 Relative irradiance received on June 21 on a 45° slope in the southern Appalachian Mountains in the eastern United States. North, ––; south, –––; east, ——. The curve for the west slope is the mirror image of that shown for the east-facing slope, with a maximum in the afternoon. (From Spurr and Barnes, 1980. Copyright © 1980 by John Wiley & Sons.)

The Solar Spectrum

The sun behaves nearly like a blackbody at 6000°K (Fig. 4.6), emitting radiation continuously from about 300 to 4000 nm with peak irradiance at approximately 500 nm (Campbell, 1977) (cf. Figs. 4.2 and 4.6). About half of the solar energy incident at the earth's surface is in the visible wavelengths (less than 700 nm) and the rest is in the infrared. There is little overlap in wavelength between the radiation emitted from the sun at 6000°K and that emitted by the earth at an air temperature of 288°K (Fig. 4.6) (Campbell, 1977). Because of minimal overlap in spectra, solar radiation and terrestrial thermal radiation can usually be considered separately. Solar radiation is often characterized as "short-wave" even though it includes some infrared radiation; terrestrial radiation is frequently called "long-wave" radiation (Slatyer, 1967).

Comparison of the spectrum of extraterrestrial solar radiation and that at ground level shows several bands of intense absorption by the atmosphere (Fig. 4.2). In the ultraviolet region (<400 nm), and to some extent in the visible region, absorption is attributable to O_2 and O_3 present in the upper region of the atmosphere, while these two gases plus water vapor and CO_2 account for most absorption in the infrared (Nobel, 1983). This same capacity for absorption of infrared light by CO_2 [and by methane, nitrous oxide, chlorofluoromethane compounds (CFCs), and a few other

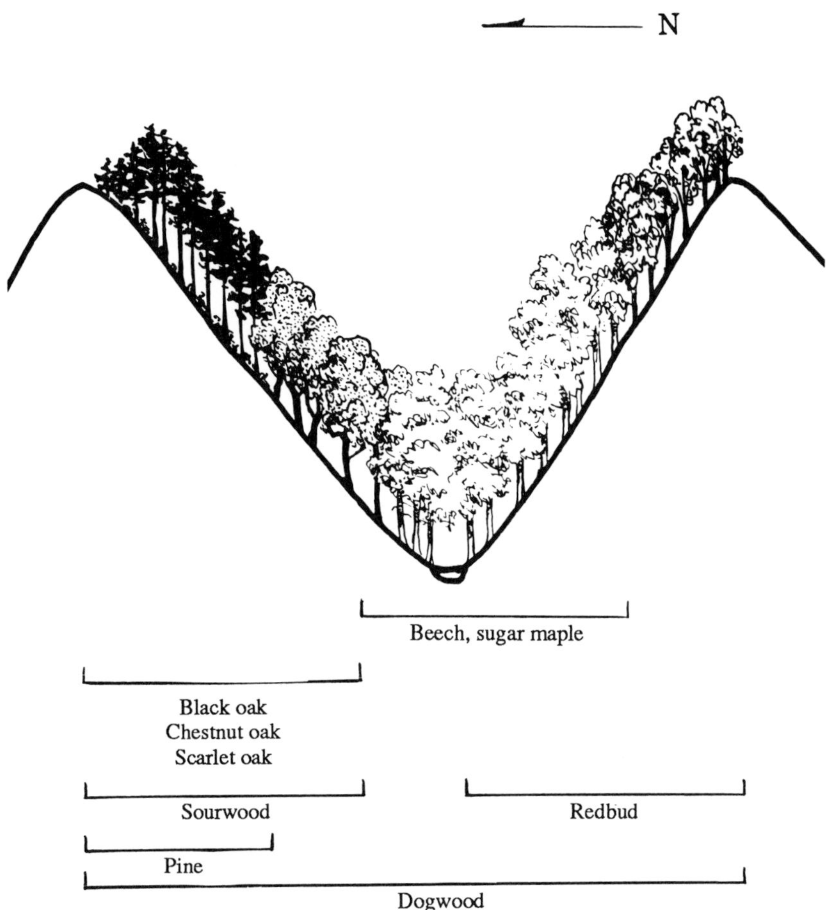

Figure 4.5 Influence of microclimate on the distribution of trees on north- and south-facing slopes in the Cumberland Plateau of West Virginia. (Figure from "Ecology and Field Biology," 3d. ed. by Robert L. Smith. Copyright © 1980 by Robert Leo Smith. Reprinted by permission of Harper & Row, Publishers, Inc.)

trace gases] is responsible for the "greenhouse effect" whereby outgoing long-wave radiation emitted by the earth is absorbed in the atmosphere, where it is reradiated in all directions and about half is returned to the earth (Gates, 1980). Hence increases in the level of atmospheric gases that absorb long-wave radiation, primarily CO_2 and methane, may have an important impact on the energy balance and temperature relations of the earth (see Chapter 10).

Optical Properties of Leaves

Because of similarity in composition of pigments, other organic constituents, and water, leaves from a wide variety of plants possess similar optical properties. The leaf absorptance, reflectance, and transmittance data for a typical broad-leaved plant

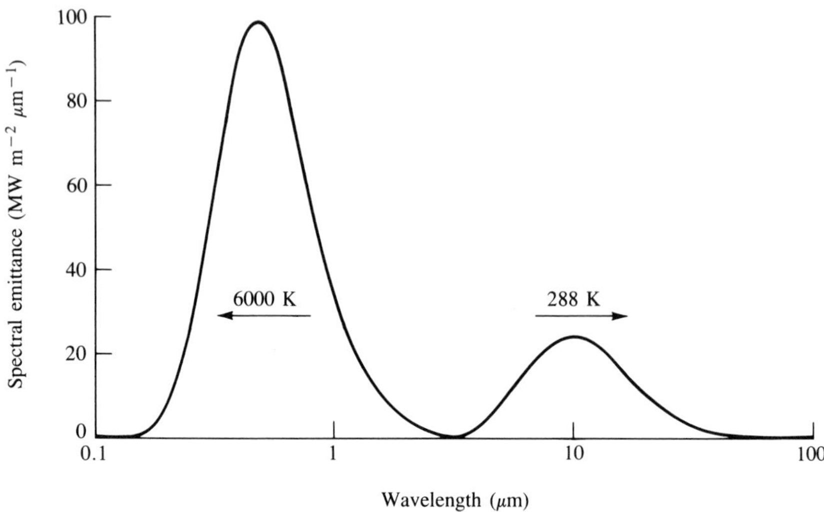

Figure 4.6 Blackbody spectral emittance for a 6000 K source (similar to the sun) and a 288 K source (similar to the earth). (From Campbell, 1977.)

such as eastern cottonwood (Fig. 4.7) show that light is strongly absorbed in most of the visible range between 400 and 700 nm except for a "green drop" region for which leaf constituents (mainly photosynthetic pigments and water) lack the capacity to absorb radiation. In this region some incident radiation is transmitted and some is reflected, giving leaves their green color. Integrated leaf absorptance often increases somewhat during expansion and maturation, and sun leaves tend to show higher absorptances than do those developed in shade. Gates (1980) noted that leaves for a wide range of broad-leaved species had mean absorptances of 0.48 to 0.56 under high sun illumination. Leaf absorptance generally decreases under low-angle solar illumination because of increased reflectance and the depletion of blue and enrichment of red wavelength components of the incoming radiation. Conifers show very high absorptance, particularly in the visible region, where values as high as 0.97 have been measured (Gates *et al.*, 1965). This capacity is attributable to the dark pigmentation of conifer foliage, which offsets to some extent the reduced absorption per unit surface area of the nonplanar leaves of these plants.

Particularly interesting and relevant to leaf energy exchange are the optical properties of leaves beyond the visible range in the infrared, where about half of the energy of solar radiation resides. There is an abrupt decline in leaf absorptance above 700 nm and extending to about 1400 nm, which results in a "decoupling" of the leaf from much of the nonphotosynthetically active energy of solar radiation. This trough in absorptance is attributable to a lack of overlap between the wavelengths of light capable of causing electronic energy transitions upon absorption by molecules and those wavelengths that can induce molecular vibrational and rotational transitions. As a result of this decoupling, a substantial amount of incoming solar energy is not absorbed by leaves, but is both transmitted and reflected (Fig. 4.7). It has been estimated that absorption of this excess energy could result in an elevation of leaf temperature of several degrees Celsius, suggesting that this gap in absorption may be beneficial under extreme environmental conditions (Gates, 1980).

At wavelengths beyond 1400 nm, absorptance increases rapidly, mostly because of absorption by liquid water, and absorptance is nearly total beyond 2000 nm. Because Kirchoff's Law states that a good absorber of radiation at any wavelength is also an equally effective emitter, leaves also emit radiation very efficiently in the far infrared. This is important in considering the energy exchange of leaves, as Figs. 4.6 and 4.7 show that radiation emission from blackbodies at terrestrial temperatures spans much of the region across which leaves *behave* like blackbodies. This pattern of emissivity allows leaves to reradiate much absorbed energy in the far infrared.

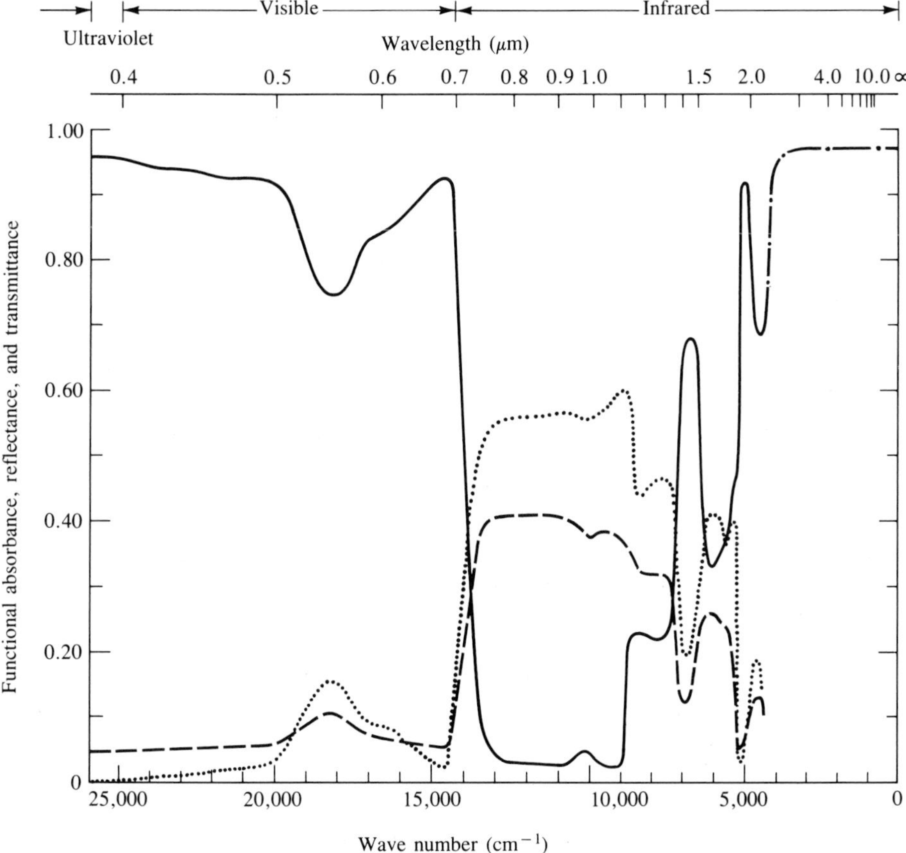

Figure 4.7 Spectral properties of a leaf of eastern cottonwood as a function of wave number and wavelength; absorbance (——), reflectance of upper surface (– – –), and transmittance (····). (From Gates, 1980.)

Although most leaves have similar optical properties, they can be modified by environmental and plant factors. For example, Waring *et al.* (1986) noted that stresses that cause degradation in chlorophyll or increased chlorophyll fluorescence induce shifts in the apparent reflectance spectrum of a canopy. Healthy leaves show strong reflectance in the near infrared (NIR, 0.75–1.3 μm) and a rapid rise in reflectance between 0.68 μm and the NIR band. The latter, termed the "red edge," shows characteristic shifts in position and slope that can be related to chlorophyll concentrations (Horler *et al.*, 1980, 1983, cited in Rock *et al.*, 1986). In other work, reduced reflectance in the NIR band has been attributed to poor foliage vigor, and increased reflectance in the 1.65- to 2.2-μm region has been linked with reduced leaf water content (Rock *et al.*, 1986, Waring *et al.*, 1986).

Utilization of these spectral properties of plants holds promise in assessing the response of vegetation to environmental stresses by remote sensing with satellites and aircraft. Such techniques have already been employed in geobotanical mineral prospecting because the presence of high concentrations of trace metals causes a "blue shift" in the red edge portion of the reflectance spectrum of plants (Collins *et al.*, 1983, cited in Rock *et al.*, 1986). Waring *et al.* (1986) and Rock *et al.* (1986) summarized the status and potential of remote sensing as a tool in monitoring terrestrial ecosystems.

The Energy Balance of Leaves

Absorption of radiation by leaves results in an increase in energy, a small part of which is captured in the products of photosynthesis and the remainder (1) heats up the leaf, (2) is transferred to the surrounding environment by reradiation, or as sensible heat by conduction and convection, or (3) is consumed in evaporation of water and is lost as latent heat from leaves through transpiration (Monteith, 1973). The disposition of energy from all sources can be described by developing an energy budget for a leaf (Table 4.1). Such budgets emphasize the complexity and consequences of energy exchange by leaves. Of primary significance is the dependence of leaf temperature on energy balance. If the energy absorbed from incoming solar radiation is not converted into chemical energy or transferred to the surroundings, it will result in increased leaf temperature.

As noted previously, the energy consumed by photosynthesis usually comprises less than 5% of the energy in solar radiation, and the production of heat energy associated with respiration of leaves is less than 1%. Thus most of the energy transfer in leaves is purely physical and does not directly influence plant metabolism. Under daytime conditions, at the same time that leaves absorb radiation from the surroundings they emit a large proportion of it back (Nobel, 1983). However, if a leaf at 25°C in the dark under a clear sky were to be suddenly illuminated by global solar radiation of 840 W m^{-2}, the leaf temperature would have to increase to nearly 55°C to totally reradiate the additional energy input. Such temperature increases rarely occur because a substantial part of the solar energy absorbed by leaves normally is transferred back to the surroundings through conduction and convection or is consumed in latent heat transfer associated with transpiration.

Table 4.1
Energy Balance Equation for a Leaf[a]

$$R_n + C + \lambda E = S_{leaf} \cdot \rho_{leaf} \cdot A \cdot dz \cdot dT_{leaf}/dt + P_s + R_s$$

Factor	Description	Units	Value
R_n	Net radiation (balance between incoming and outgoing long-wave and short-wave radiation)	cal min^{-1}	+0.460
C	Transfer of sensible heat by conduction	cal min^{-1}	−0.100
λE	Transfer of latent heat through transpiration	cal min^{-1}	−0.378
S_{leaf}	Specific heat of leaf	cal gm^{-1} °C^{-1}	0.73
ρ_{leaf}	Leaf density	g cm^{-3}	0.75
A	Leaf area	cm^2	1
dz	Leaf thickness	cm	0.015
dT_{leaf}/dt	Rate of change of leaf temperature	°C min^{-1}	0
P_s	Energy consumption by photosynthesis	cal min^{-1}	−0.023
R_s	Energy production by respiration	cal min^{-1}	+0.005

[a] Approximate values for factors are given for a typical leaf of a well-watered plant under full sunlight and moderate wind and stable leaf temperature.

Sensible heat is transferred by conduction across the zone of still air that forms a boundary layer around the leaf, then moves by convection into turbulent air surrounding the leaf, and eventually crosses a canopy boundary layer to the bulk atmosphere. Convective transfer from leaves is of two types: (1) forced convection, which involves transfer by a horizontally moving airstream, and (2) free convection, which depends on energy transfer by ascending or descending air currents. Forced convection is most common, with free convection important only where wind speeds are less than 0.1 m sec^{-1} (Nobel, 1983). Some authors suggest that free convection may be important within dense crop and forest canopies (Monteith, 1973), but the close coupling of aerodynamically "rough" forest canopies to the bulk atmosphere (see Jarvis, 1985) suggests that forced convection is more important in most forests and orchards. Convective energy transfer is proportional to temperature differences between the leaves and air and inversely proportional to thermal resistance and boundary layer thickness. Boundary layers are thinner around small and narrow leaves and with higher wind speeds.

Latent heat transfer associated with transpiration is a function of the difference in vapor pressure between the leaf mesophyll spaces and that in the atmosphere and is inversely related to the leaf (stomatal and cuticular) and boundary layer resistances to water vapor loss. The latent heat of vaporization for water is quite high (2.442×10^6 J kg^{-2}) because of strong hydrogen bonds that must be broken as liquid water evaporates, and energy flux equal to several hundred W m^{-2} can be dissipated by transpiration. If relative humidities are very low and stomatal conductance is high, energy in excess of that gained in net radiation can be dissipated by rapid evaporation of water, resulting in leaf temperatures of well-watered plants that are lower than air temperatures, despite a high leaf radiation load. In fact, Idso *et al.* (1980) used remote sensing of decreasing difference between leaf and air temperature as an indicator of increasing water stress in the arid southwestern United States. In areas where bare, dry soil alternates with areas of transpiring plant canopies, lateral transfer of sensible heat energy by wind movement from bare areas, a process called "advection," can cause latent heat transfer from vegetation that is far in excess of net radiation (Slatyer, 1967, pp. 38–41). This is sometimes called the oasis effect. In the absence of significant advection, and given the normal range of stomatal conductances in woody plants, in humid regions the leaf to air vapor pressure differences are seldom great enough to cause undercooling, and even well-watered plants, when fully illuminated, often have leaf temperatures that exceed air temperature.

Partitioning of net radiation into latent and sensible heat transfer varies with season (Fig. 4.8) (Rauner, 1976). In temperate mixed-deciduous forest dominated by European white birch and European aspen, early- and late-season disposition of net radiation was dominated by sensible heat transfer because of the leafless condition of deciduous trees in the early spring and late fall. From May to September,

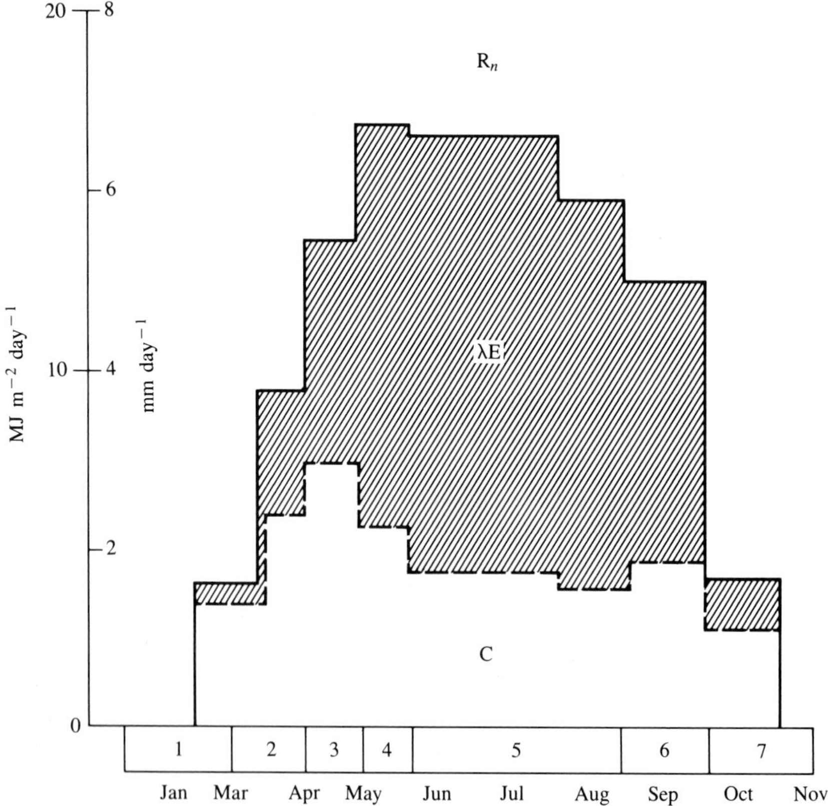

Figure 4.8 Annual courses of fluxes of net radiation (R_n), sensible heat (C), and latent heat (λE) from European birch–European aspen forest near Moscow. Numerical references indicate (1) presence of snow cover, (2) snowmelt period, (3) post-snowmelt leafless period, (4) leaf-out, (5) summer foliage period, (6) leaf senescence and abscission, and (7) postabscission leafless period. (From Rauner, 1976.)

latent heat transfer associated with transpiration exceeded that of sensible heat by a wide margin. In a variety of well-watered deciduous forests, latent heat transfer accounted for at least two-thirds and often for more than 90% of net radiation (Rauner, 1976).

The influence of climate on energy partitioning has been evident in studies of energy balance in conifers (Jarvis *et al.*, 1976). In maritime regions, where vapor pressure differences between needles and the air are relatively small (e.g., about 5 mbar), transpiration rates are relatively low and more of the incoming radiation is transferred from the needles by conduction and convection of sensible heat than by transpiration. For conifers growing in continental climates where vapor pressure deficits may be much higher (≥20 mbar), transpiration and attendant latent heat transfer are promoted. These influences explain the relatively high ratio of sensible to latent heat transfer (Bowen ratio) observed in a Scotch pine forest at Thetford, U.K., compared with that for Norway spruce on the European mainland (Jarvis *et al.*, 1976).

Partitioning of energy between latent heat and conductive transfer also is very sensitive to plant water status (Fig. 4.9). When plants are dehydrated, little transfer of latent heat is possible because stomatal closure greatly increases leaf resistance to water vapor loss. This condition can cause large temperature increases in the leaf as sensible heat transfer predominates. Thus, leaf temperatures of dehydrated plants tend to be higher than those of turgid plants because conduction and convection (which transfer sensible heat energy in proportion to leaf–air temperature differences) must dissipate more of the incoming radiant energy when transpiration is reduced.

Photosynthesis

Basic Responses of Photosynthesis to Light Intensity

In photosynthesis, light is harvested by chlorophyll–protein complexes associated with photosystem reaction centers located on chloroplast thylakoid membranes. Energy from absorbed photons is transferred twice to electrons split from water molecules as they pass through a chain of electron carriers. Ultimately this electron transport produces both $NADPH_2$ and ATP, which are then consumed in the production of carbohydrates. More extensive discussions of photosynthesis in woody plants can be found in Chapter 2, and in Larcher (1969), Schaedle (1975), Kramer and Kozlowski (1979), and Nelson (1984).

Light-driven photosynthesis within the leaf mesophyll creates the diffusion gradient for movement of substrate CO_2 inward, but the stomata also influence the actual flow of CO_2. Hence the response of stomata to light (e.g., Fig. 4.10) and other environmental and endogenous factors also influence photosynthetic responses of the mesophyll (Davies and Kozlowski, 1974). Stomata may respond to light directly, they may be influenced indirectly through photosynthetic depletion of

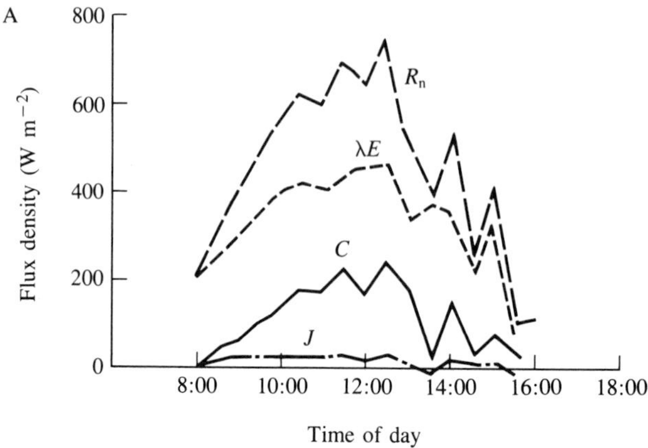

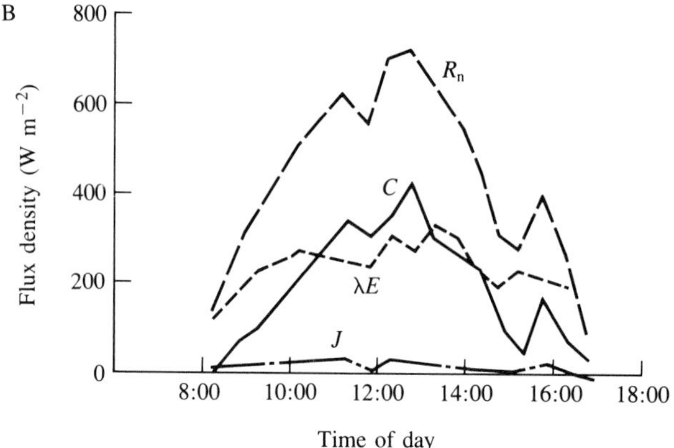

Figure 4.9 Diurnal patterns of energy balance of a mature tea plantation in Kenya during the (A) wet and (B) dry season. R_n, net radiation (-–); *C*, sensible heat (——); λ*E*, latent heat (---); *J*, below-canopy fluxes (-–-). (From Squire and Callander, 1981.)

CO_2 in the mesophyll, or both responses may occur simultaneously (Sharkey and Ogawa, 1987). Action spectra of stomatal responses indicate that a number of different responses of stomata to light are possible. Stomata often respond to light in a fashion similar to the action spectrum of photosynthesis whether the mesophyll is present or not, suggesting that guard cell photosystems may drive the ion movement necessary for stomatal opening. Also there is an accentuated opening response to weak blue light, suggesting that another pigment system, possibly flavin based, may be involved in light-mediated stomatal movements (Zeiger *et al.*, 1987b). Stomatal function was recently reviewed by Zeiger *et al.* (1987a).

For most plants, photosynthetic response to light intensity is roughly hyperbolic, with 90% of maximum net photosynthesis (P_n) occurring between one-third and

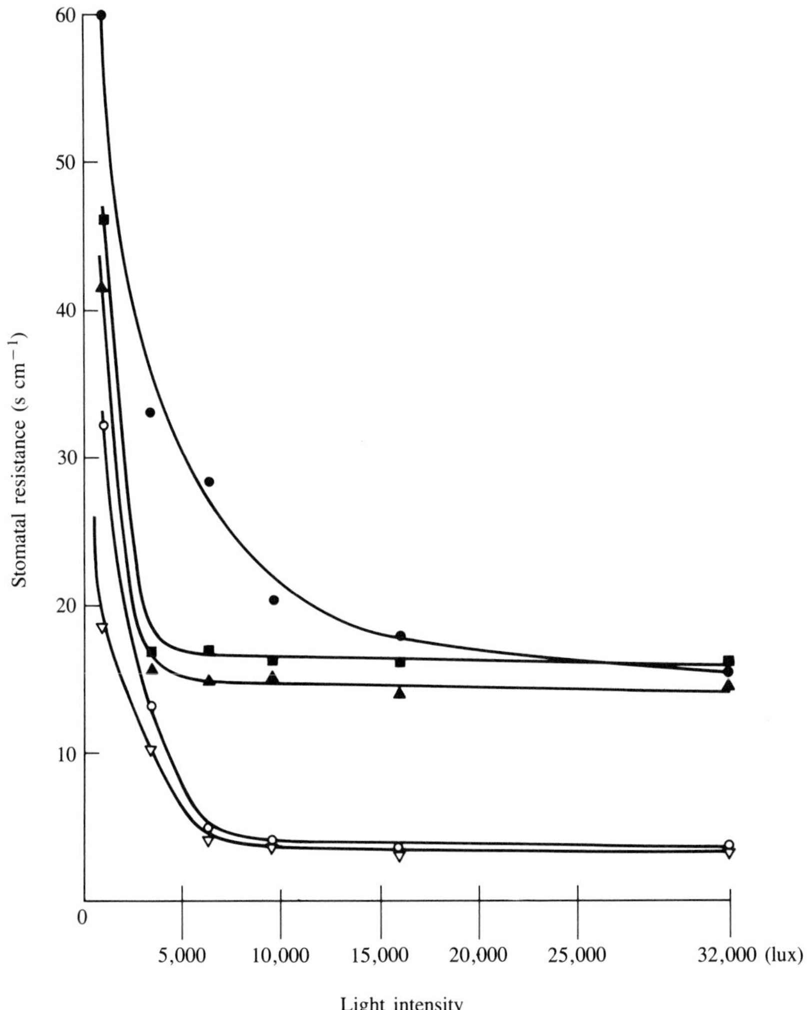

Figure 4.10 Relationship between stomatal resistance and incident light intensity for five species of woody angiosperms: sugar maple (●), eastern redbud (▲), orange (■), white ash (○), and bur oak (▽). (From Davies and Kozlowski, 1974.)

two-thirds of full sunlight (Fig. 4.11). Certain species show P_n saturation at lower light intensities, whereas P_n of others does not appear to be light saturated at all. For example, many shade-tolerant tree species, such as red maple and American beech, show maximal photosynthesis at light intensities as low as 5 to 10% of full sunlight (Loach, 1967). In contrast, shade-intolerant species show light saturation at much higher light levels (yellow poplar), or not at all (trembling aspen) (Loach, 1967).

Light response curves, obtained under conditions where levels of other

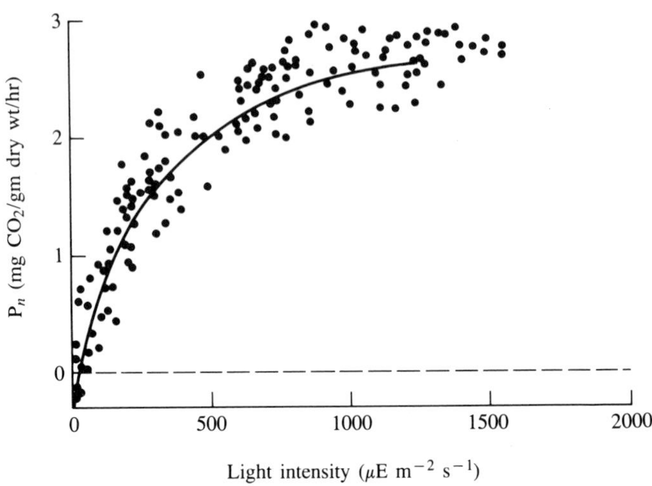

Figure 4.11 Rate of net photosynthesis as a function of light intensity for sunlit lateral branches of black spruce trees at air temperatures between 13 and 20°C. (After Vowinckel *et al.*, 1975.)

environmental factors are not limiting, provide much useful information. Each curve illustrates the light compensation point (i.e., when light intensity is reduced to the point at which net photosynthesis equals zero), which depends on dark respiration in the plant. The quantum yield (the initial slope of the light versus P_n relationship) indicates the efficiency of the photosynthetic apparatus under light-limiting conditions. Finally, the maximum photosynthetic capacity under saturating light provides a relative measure of investment in photosynthetic electron transport capacity and Calvin cycle enzymes and intermediates.

Light responses of single leaves may be quite different from those of shoots and tree canopies because of mutual shading within canopies, with individual leaves usually showing light saturation at lower light intensities (Kramer and Decker, 1944; Kramer and Clark, 1947). Within the crowns of large forest trees, such as Douglas-fir, there are substantial reductions in photosynthetic rate with increasing depth of light penetration (Helms, 1964). Woodman (1971) noted that photosynthetic rates of leaves of a canopy Douglas-fir tree varied widely with time of day, season, vertical and horizontal location within the crown, and needle age. Within the crown, the highest rates of photosynthesis were always associated with needles located at an identifiable boundary between fully sunlit and shaded conditions (Fig. 2.7).

Light has additional regulatory effects on photosynthesis beyond those associated with providing the energy transduced in the electron transport chain. For instance, light induction (i.e., a period of time in the light) is required for full development of photosynthetic capacity (Perchorowicz *et al.*, 1981), most likely through light regulation of key enzymes in the photosynthetic carbon metabolism, particularly Rubisco (Portis *et al.*, 1986; Salvucci *et al.*, 1986).

Sun–Shade Phenomena and the Concept of "Shade Tolerance"

The concept of "shade tolerance" has persisted in forestry and horticulture despite its uncertain exact physiological nature and its apparent lack of independence from other aspects of competition. Despite this ambiguity, silviculture systems and planting recommendations for ornamental species based on empirically determined shade tolerances have proven successful. Tables and texts denoting relative shade tolerances of forest trees have been developed by several authors (e.g., Baker, 1950; Franklin and Dyrness, 1973; Daniel *et al.*, 1979; Harlow *et al.*, 1979) (Table 4.2). Perhaps the earliest classification can be traced to field observations by Hayer in 1852 (cited in Burns, 1923). Most authors acknowledge that placement of species in such groups is subject to modification by influences of plant age, genotype, site, climate, and microclimate (e.g., Daniel *et al.*, 1979). Hence relative shade tolerance varies across the geographic ranges of co-occurring species. It is nearly impossible to make close comparisons of shade tolerance between species that are native to different regions.

The influence of factors other than light on capacity of young trees to persist in an understory was clearly shown in the trenching experiments in stands of Scotch pine by Fricke, reported by Zon (1907), in which colonization of seedlings in areas delimited by soil trenching was greatly stimulated. This study has been supported by similar experiments with eastern white pine in the northeastern United States (Toumey and Kienholz, 1931) and with loblolly pine in North Carolina (Korstian and Coile, 1938) among others. Decreased rates of photosynthesis in shade limit root growth and may result in reduced capacity to survive drought (Kramer and Decker, 1944; Bourdeau and Laverick, 1958). Although the immediate cause of death is desiccation, the ultimate cause is reduced photosynthesis in shade—sometimes related to self-shading of leaves (Kramer and Clark, 1947) or to declining photosynthetic efficiency of leaves during seedling ontogeny (Bormann, 1956).

Table 4.2
Shade Tolerances of Some Common Conifer Species of the Pacific Northwest in North America[a]

Intolerant	Intermediate	Tolerant	Very tolerant
Noble fir	California red fir	White fir	Silver fir
Western larch	Incense cedar	Grand fir	Western hemlock
Lodgepole pine	Sugar pine	Alpine fir	
Ponderosa pine	Western white pine	Port Orford cedar	
Douglas-fir		Engelmann spruce	
Red alder		Sitka spruce	
Black cottonwood		Coast redwood	
Oregon white oak		Western red cedar	
		Mountain hemlock	
		Bigleaf maple	

[a] Adapted from Franklin and Dyrness (1973).

Further, Givnish (1988) pointed out that although respiration and construction costs of nonphotosynthetic tissues will have a substantial influence on whole-plant carbon balance and survival, the impact of these demands on a plant's carbon budget is virtually never considered in studies of shade tolerance. When a more comprehensive approach to calculating net carbon gain was employed, which incorporated costs of night respiration, leaf development, roots, and supporting tissues, the effective light compensation points for yellow poplar were elevated by 140 μmol m^{-2} sec^{-1} for trees 1 m in height and by 1350 μmol m^{-2} sec^{-1} for 30-m-tall trees.

As discussed in Chapter 3, the death of trees under low-light conditions sometimes can be related to increased susceptibility to biotic factors such as insects and diseases. For example, Augsberger (1984) found that fungal disease (primarily caused by damping-off organisms) was the most frequent cause of death of shaded tree seedlings in Panama. Hence it is appropriate to consider the possibility that persistence in the understory is indicative of the presence of several adaptations, an important one being the capacity to maintain adequate photosynthetic rates and hence whole-plant carbon gain in the shade, but others such as drought tolerance and disease resistance also may be involved (Fenner, 1987).

Responses of the Photosynthetic Apparatus to Shade

In addition to effects of light on gross morphology and anatomy, which will be discussed later, the photosynthetic apparatus shows several responses to the light regime during leaf development (Boardman, 1977). Further, as mentioned previously, there are inherent differences among species in capacity to tolerate shading. Adaptation of the photosynthetic apparatus to low light intensity has been studied extensively in herbaceous plants, but less in trees (Berry, 1975; Boardman, 1977). Shade-tolerant plants generally have lower dark respiration rates and, consequently, lower light compensation points (Boysen-Jensen, 1929; Bourdeau and Laverick, 1958; Grime, 1965; Boardman, 1977; Field, 1988). For example, light compensation points and dark respiration rates of shade-tolerant species such as American beech and red maple were lower than those of intolerant species such as yellow poplar and trembling aspen regardless of irradiation level during growth (Loach, 1967). Shade-tolerant plants also generally have lower light saturation points for photosynthesis (Kramer and Decker, 1944).

The physiological mechanisms responsible for these differences in responses to light are not well characterized in woody plants. Leaves of shade-adapted species generally contain more chlorophyll on a weight basis, but less on an area basis because they usually are thinner. The richer content of chlorophyll in chloroplasts of shade-adapted leaves combined with the reduction in leaf thickness may allow for more efficient light utilization (Boardman, 1977). The chlorophyll $a:b$ ratio decreases in shade-grown leaves of many woody species, including Sitka spruce and European beech (Lewandowska *et al.*, 1976; Lichtenthaler *et al.*, 1981).

Chloroplast ultrastructure of many woody species is very responsive to light regime. Chloroplasts from European beech leaves grown under low-light conditions had substantially greater numbers of thylakoids per granum and greater thylakoid stacking compared with sun-type chloroplasts (Lichtenthaler *et al.*, 1981). Shade-type chloroplast grana also were wider and did not possess large starch grains characteristic of the chloroplasts of sun leaves. Similar differences in granal width and thickness and starch grain content were found between chloroplasts of sun and shade leaves of *Euphorbia forbesii,* a woody C_4 species found in shaded, mesic forests of Hawaii (Pearcy and Franceschi, 1986). In contrast, a C_3 species (*Claoxylon sandwicense*) showed little variation in chloroplast ultrastructure with light regime during growth, but accumulated less starch in shade-type chloroplasts.

Leaves of shade-adapted species generally contain lower levels of Rubisco, ATP synthase, and electron transport chain constituents such as cytochromes than do sun-adapted species (Lewandowska *et al.*, 1976; Nasrulhaq-Boyce and Mohamed, 1987; Anderson *et al.*, 1988; Chow *et al.*, 1988, Evans, 1988). The low levels of these constituents are consistent with a reduced capacity for electron transport (as reported, e.g., in Sitka spruce; Lewandowska *et al.*, 1976) and carbon fixation (e.g., in grapefruit and orange; Syvertsen, 1984). Hence, synthesis and maintenance of a large pool of Rubisco and electron carriers in sun-adapted species allow electron transport and carbon fixation to proceed rapidly at high light intensity; however, in shade-adapted plants this would be wasteful as light capture and photosynthetic electron transport can only proceed at much reduced levels.

Maximum rates of photosynthesis generally are higher for plants grown at high light intensity, with shade-intolerant species showing the greatest plasticity in response. For example, when seedlings of the light-demanding West African species idigbo (*Terminalia ivorensis*), afara (*T. superba*), and obeche (*Triplochiton scleroxylon*) were grown at high light intensity, they showed at least twofold increases in light-saturated rates of photosynthesis over shaded seedlings (Kwesiga *et al.*, 1986). However, seedlings of the shade-tolerant kuka (*Khaya senegalensis*) showed little difference in light-saturated rates of photosynthesis whether grown at high or low light intensity. In other shade-tolerant species (e.g., iripilbark tree, *Pentaclethra macroloba*) the increase in light-saturated photosynthesis at high light during development may be moderate (25%) (Oberbauer and Strain, 1986; Field, 1988) or occasionally substantial (e.g., coffee; Friend, 1984).

Light compensation points of sun leaves may be higher than those of leaves developed in the shade because of higher respiration rates that are associated with increased costs of synthesis and maintenance of an expanded complement of photosynthetic machinery (Boardman, 1977). Decreases in light compensation point with increased shading during growth have been shown for a large number of woody species (e.g., Loach, 1967; Langenheim *et al.*, 1984; Friend, 1984; Oberbauer and Strain, 1986). Loach (1967) showed that the light compensation point decreased by 50 to 75% as the light regime during growth decreased from 100% of full sunlight to 3% in American beech, red maple, northern red oak, yellow poplar, and trembling aspen. Dark respiration rates (on a leaf area basis) decreased in close correspondence with light compensation point in these species.

At low light intensities, quantum efficiency of shade-grown leaves frequently may be higher than that in sun leaves (Table 4.3). This type of response to light level during growth has been observed in a variety of species, for example, coffee (Friend, 1984), iripilbark tree (Oberbauer and Strain, 1986), black walnut (Dean *et al.*, 1982), sugar maple (Logan and Krotkov, 1968), American beech, red maple, northern red oak, yellow poplar, and trembling aspen (Loach, 1967), and Venezuelan copaltree (*Copaifera venezuelana*) (Langenheim *et al.*, 1984).

Thus the lower dark respiration rates, reduced light compensation points for photosynthesis, and greater quantum efficiency of many shade-tolerant species apparently help them maintain adequate whole-plant carbon balance despite low light intensity.

Utilization of Sunflecks

The light regime beneath a woody plant canopy is very heterogeneous with respect to both space and time. A substantial amount of the total radiant energy flux at the forest floor is received as short-lived episodes of high intensity, called sunflecks. For instance, between 55 and 80% of total daily photosynthetically active radiation on the floor of rain forests is associated with sunflecks (Anderson and Osmond, 1987).

The capacity for photosynthetic use of sunflecks by shade-tolerant woody plants has been debated. While Woods and Turner (1971) noted that stomata of several shade-tolerant woody species would open quickly enough to allow increased gas exchange, Pereira and Kozlowski (1977b) could identify no close relationship between shade tolerance and stomatal responsiveness to light.

Recent work with tropical plants (Chazdon and Pearcy, 1986a,b) has shown that light induction of the photosynthetic apparatus is required to obtain significant amounts of carbon fixation during sunflecks. However, once photosynthesis was induced, leaves of the shade-tolerant rain forest understory species *Alocasia macrorrhiza* (elephant's-ear plant) showed high rates of carbon fixation that persisted for several seconds after passage of a 20-sec sunfleck. Once induction has occurred, the photosynthetic rate increases substantially for a given light intensity not only in response to the initial light event but also to subsequent events. This is ecologically significant because the best predictor of occurrence for a sunfleck appears to be a prior sunfleck. Pearcy (1988) noted that in an Australian tropical forest 70% of sunflecks occurred within 1 min of the preceding sunfleck. In contrast, only 5% of sunflecks were preceded by the absence of sunflecks for an hour or more. Hence photosynthetic induction results in elevated capacity for photosynthesis when probabilities of sunflecks in the understory are high.

For sunflecks of fairly long duration (at least 3 min), Harbinson and Woodward (1984) showed that photosynthesis was essentially light saturated for shade leaves of a number of British woodland species, including English holly and European beech. Photosynthetic rates during sunflecks were only 20 to 40% higher than those in deep shade for these species, suggesting that the influence of sunflecks on carbon balance was modest. In contrast, substantial increases in photosynthetic rates during sunflecks were found in European ash, English ivy, European filbert, and Wych elm,

Table 4.3
Photosynthetic and Anatomical Characteristics of Leaves of Black Walnut Seedlings Grown under Several Shading Regimes[a]

Shading treatment[b]	Light transmission (% of PAR)	Stomatal density (no. mm^{-2})	Palisade layer ratio[c]	Leaf thickness (μm)	Quantum efficiency (mol CO_2 fixed per mol PAR absorbed)
Control	100.0	290.0a	1.00a	141.3a	0.023ab
GL1	50.0	293.1a	1.44b	108.3b	0.023a
	20.2				
ND1	15.7	264.0b	1.25ab	104.9b	0.026ab
GL2	20.9	209.2c	1.85c	89.9c	0.030bc
	8.3				
ND2	3.3	187.0d	1.93c	88.2c	0.033c

[a] From Dean *et al.* (1982).

[b] The GL treatments consisted of green celluloid film with holes to simulate a canopy that had sunflecks. For GL1 and GL2, upper value indicates the sum of 100% transmission through holes and transmission through remaining shaded area; lower value indicates transmission through shade material only (GL1, one layer; GL2, two layers). ND1 and ND2 represent treatments with two levels of shading by neutral density shade cloth. Mean values within a column followed by the same letter are not significantly different ($P \leq 0.05$).

[c] Palisade layer ratio equals the cross-sectional area of palisade layer of control leaves divided by that of leaves of other treatments.

indicating a potential for substantial stimulation of total photosynthesis by sunflecks. Efficient utilization of sunflecks may depend on the establishment of a balance among components of the processes occurring in chloroplasts, particularly the pools of electron transport carriers and intermediate compounds of carbon assimilation, that allows swift response of electron transport to increasing light and prompt consumption of products after passages of sunflecks (Anderson and Osmond, 1987).

Photoinhibition

Commonly, if plants grown in shade are suddenly exposed to high light intensity, the photosynthetic apparatus shows a substantial reduction in response. Shade-tolerant species show greater tendency than intolerant species for photoinhibition and greater injury when it occurs. Quantum efficiency is reduced, as is the maximum capacity for photosynthesis (Osmond, 1987). For example, in *Claoxylon sandwicense,* a shade-tolerant subcanopy species of tropical forests, leaves grown under high light showed reduced quantum efficiency and maximum photosynthesis compared with leaves grown under low light (Pearcy and Franceschi, 1986).

Although the exact nature of photoinhibition remains uncertain, some evidence suggests that it results from damage to the photosystem reaction centers (primarily PS II) (Kyle, 1987; Cleland, 1988; Critchley, 1988). At the same time that photoinhibitory damage occurs, repair processes at the molecular level can also take place. Hence the development of photoinhibition depends on both the rate of injury and repair.

Apart from repair mechanisms, avoidance of photoinhibition appears to be related to the capacity to minimize light absorption by paraheliotropic leaf movements (orientation of the leaf blade parallel to incoming solar radiation) and wilting (Ludlow and Björkman, 1984), and through thermal dissipation of excess absorbed light energy. For example, in sun leaves of balsam poplar and English ivy, a carotenoid pigment, zeaxanthin, is produced on exposure to intense light, and the eventual onset of photoinhibitory damage after prolonged illumination is closely correlated with cessation of further zeaxanthin accumulation (Demmig *et al.*, 1987). At chilling temperatures, accumulation of zeaxanthin in American mangrove leaves is suppressed and recovery of photosynthesis from inhibition caused by chilling and high light is delayed (Demmig-Adams *et al.*, 1989). Hence, the production of zeaxanthin may facilitate diversion of light energy from excited chlorophyll molecules and eventual thermal dissipation.

Photoinhibition may be ecologically important in forest stands and plantings of other woody species subjected to drastic changes in canopy cover caused by events such as clear-cutting and heavy thinning. Anderson and Osmond (1987) suggested that effective acclimation of plants to bright light was associated with the capacity to maintain efficient photosynthesis to very high light intensities (i.e., to maintain a linear relationship between P_n and absorbed light with unaltered slope). Incomplete acclimation occurs when the light-saturated rate of photosynthesis cannot be elevated sufficiently, or if reductions in quantum yield occur, and these responses are nearly always indicators of photoinhibited plants. Hence established understory individuals of shade-tolerant species, which frequently lack acclimation capacity, would be at a serious disadvantage following canopy removal. In this situation, existing understory and subdominant plants would undergo long-term injury from changes in light regime that place existing plants at a disadvantage compared to colonizing vegetation and resistant competitors already present on the site. Additionally, the foliage of many shade-tolerant species that develops after canopy opening may show persistent inclination toward photoinhibition, indicating an inherently limited capacity for acclimation to light (Osmond, 1987). Mortality of advance regeneration in stands harvested by clear-cutting and heavy shelterwood cuts and of shaded ornamental plants suddenly exposed to full sunlight may be partly attributable to such influences.

Interactive Effects of Light and Other Environmental Factors on Photosynthesis

Light interacts with numerous other environmental factors in the control of photosynthesis. Some of these effects are indirect. For example, illuminated leaves in arctic plants are warmed to temperatures nearer the optimum for photosynthesis, despite low air temperatures (Chapin and Shaver, 1985). Conversely, water deficits can be induced by high transpiration rates in sunlit leaves if air temperature is high. This often causes a transient midday drop in P_n and stomatal conductance that disappears with a decline in afternoon light intensity (Kramer and Kozlowski, 1979). Light also plays a substantial role in determining vapor pressure differences between leaves and the air because illuminated leaves tend to be warmer and thus have a higher saturation vapor pressure in the mesophyll air spaces than shaded, cooler leaves. This results in a greater vapor pressure difference between the leaf and air in illuminated leaves, which increases transpiration and may influence stomatal aperture.

Light may exaggerate the inhibitory effects of many other environmental factors on photosynthesis. For example, high light intensity may enhance thermal injury effects on photosynthesis (Al-Khatib and Paulsen, 1989). There is some evidence that water-stressed plants may show greater photoinhibition than those that are well watered (Powles, 1984). In oleander this accentuated effect was attributed to a reduction in intercellular CO_2 in leaves with tightly closed stomata (Björkman and Powles, 1984). Lacking the normal substrate for photosynthesis, the capacity for utilization of light energy was greatly reduced, causing a more severe photoinhibitory response. Other researchers have not associated photoinhibition with depletion of internal CO_2 (Sharp and Boyer, 1986; Boyer *et al.*, 1987), suggesting instead that water stress may more directly predispose the photosynthetic mechanism to photoinhibition.

In chilling-sensitive plants, low above-freezing temperatures can cause accentuated photoinhibition under subsequent exposure to high light intensity.

Photooxidation of lipids in these plants is also increased, and substantial destruction of chloroplast structure can be observed when low temperatures are combined with high light intensity (Wise and Naylor, 1987). Repair of photoinhibitory damage requires protein synthesis and hence photoinhibition is aggravated by constraints on metabolism at low temperature. Leaves of some tropical and subtropical woody plants (e.g., coffee and *Citrus*) are chilling sensitive (Kramer and Kozlowski, 1979) (see Chapter 5).

Freezing temperatures in combination with high light intensity can also induce photosynthetic injury in evergreen conifers, even if they are winter-hardy. In Scotch pine, low-temperature inhibition of photosynthesis was light dependent (Leverenz and Öquist, 1987), with evidence of injury to PS II. Prolonged light exposure at chilling or freezing temperatures frequently results in pigment bleaching caused by photooxidation, causing destruction of chlorophyll and accessory pigments (Öquist *et al.,* 1987). Photooxidation or metabolic destruction of chlorophyll *a* in reaction centers and of light-harvesting chlorophyll may be especially sensitive to high light intensity and low temperatures in Scotch pine.

Biomass Productivity and Radiation Interception

Despite the obvious link between total photosynthesis and growth of plants, there often is a lack of close coupling between measured photosynthetic rates and productivity (see also Chapter 2). This is partly because it is usually feasible only to estimate P_n by brief measurements on a small fraction of the total leaf surface area (Ledig, 1976; Helms, 1976; Kramer and Kozlowski, 1979). In one of the few instances where P_n of an entire tree beyond the seedling stage was monitored over a long period of time, the correlation between season growth of a young apple tree and measured photosynthesis was moderately good (Heinicke and Childers, 1937). Bate and Canvin (1971) also found reasonably good agreement between increment in carbon mass and integrated CO_2 uptake measured with infrared gas analyzers for seedlings of trembling aspen over a period of 56 days. However, these measurements are extremely difficult and expensive to conduct, particularly when they must be done in the field. Because of the problems inherent in long-term measurements of canopy photosynthesis, it has been proposed that measurements of leaf area increase and duration offer an attractive alternative method of assessing the photosynthetic potential of a tree canopy (Landsberg, 1986). Recent study and modeling of the interaction between dynamic radiation regimes and developing canopies have focused on the total interception of radiant energy as a valuable predictor of dry matter production.

On a given site where environmental conditions are favorable for photosynthesis, the amount of photosynthetic surface produced and its efficiency in light capture account for most of the variation in productivity (Linder *et al.*, 1985; Cannell *et al.*, 1987). The situation becomes more complex when variations among sites with respect to environmental limitations such as temperature, mineral nutrients, and water supply are involved because they frequently have impacts on both leaf area development and photosynthetic rates. These environmental constraints modify the slope of the expected linear relationship between intercepted radiation and dry matter production.

Canopy Density and Architecture and Interception of Solar Radiation

The vertical distribution of canopy leaf area can often be described by a normal curve (Fig. 4.12), which indicates maximum leaf area at intermediate positions within the crown and gradually decreasing leaf area upward and downward. However, the point of maximum leaf area is not always symmetrically located between the crown base and top. Maximum rates of photosynthesis are usually found in the upper part of the crown and may be lower in leaves at the extreme outer edge of the crown than those somewhat inside (Woodman, 1971; see Fig. 2.7).

Leaf area index (LAI) in natural forests varies widely with species, stand density, and environment (Anderson, 1981), with values for temperate deciduous forests ranging from 3 to 9. However, lower LAIs (1.5–1.9) have been reported for more arid eucalyptus forests in southeastern Australia (Anderson, 1981). Temperate coniferous forest LAI values may reach 11 to 12 (see also Chapter 3).

Capacity for leaf retention under environmental stress also is an important factor in growth of woody plants. It is well known that extensive leaf shedding occurs when trees such as poplars, yellow poplar, black walnut, and willows are subjected to drought (e.g., Addicott and Lyon, 1973; Parker and Pallardy, 1985; Kozlowski, 1991). A eucalyptus (*E. maculata*) forest in Australia lost more than half its leaf

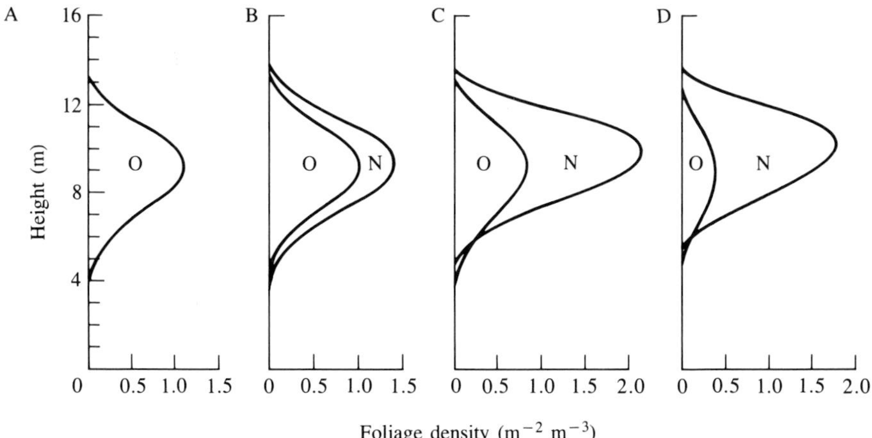

Figure 4.12 Vertical distribution of old (O) and new (N) foliage of loblolly pine sampled at four times during 30 weeks of the growing season. (A) Week 0, (B) week 10, (C) week 20, and (D) week 30. (From Kinerson *et al.*, 1974.)

area in a severe drought in 1980 (Pook, 1986). Leaf loss will result in reduced leaf area duration (LAI × time) and growth potential, depending on the severity of drought and how early in the growing season leaf abscission begins.

Leaves in a tree canopy interact in a very complex manner with solar radiation throughout the diurnal traverse of the sun. The amount of solar radiation absorbed by a canopy varies with the amount of leaf area, leaf angle distribution, clumping characteristics of the foliage, the position of the sun, and the fractions of sunlight that are direct and diffuse. To predict average irradiance at any level of a continuous canopy, a form of Beer's Law is often employed (Monsi and Saeki, 1953; Landsberg, 1986):

$$\phi(z) = \phi(0) \exp{(-k_{\phi})}\cdot \mathrm{LAI}(z)$$

where $\phi(z)$ = shortwave irradiance at level z, $\phi(0)$ = irradiance at the top of the canopy, k_{ϕ} = an empirically determined extinction coefficient, and LAI(z) = cumulative LAI to point z in the canopy. This relationship can be extended to noncontinuous canopies such as those of fruit orchards if modified to account for the fraction of unintercepted radiation (Jackson, 1985) (see Chapter 3).

Light and Biomass Production

Canopy photosynthesis can be estimated from time-averaged radiation interception functions and light response curves for photosynthesis. When combined with measurements of leaf area duration and estimates of growth- and maintenance respiration, estimates of growth and total production of tree crops can be obtained. Alternatively, measurements of canopy light interception can be related empirically to dry matter production over time to produce predictive models of productivity as a function of intercepted radiation (Landsberg, 1986) (Fig. 4.13):

$$W(t) = \epsilon_{\phi} \sum \phi_{abs} \, \Delta t$$

where $W(t)$ = dry matter mass at time t, ϵ_{ϕ} = conversion efficiency of light energy to dry matter, ϕ_{abs} = absorbed radiation, and Δt = time interval.

The relationship between dry matter production and intercepted radiation under relatively constant environmental conditions at a single site usually is linear or at least monotonic (e.g., Monteith, 1977; Linder *et al.*, 1985; Eckersten, 1986; Cannell *et al.*, 1987). The inclusion of a conversion efficiency term attempts to account for variations among sites and seasons in temperature, water, and nutrient regimes, which alter the slope of the relationship between absorbed radiation and production. Short-term influences such as transient water stress should reduce ϵ_{ϕ}, while chronic stresses such as prolonged drought or infertile soil would likely affect both conversion efficiency and absorbed radiation (through reduction in leaf area production).

Development of this "top-down" approach to viewing light harvesting (so called presumably because it is based on analysis of light and plant growth at a high level of biological organization and not at the molecular or cellular level) holds particular promise for modeling purposes, where the extreme complexity of physiological

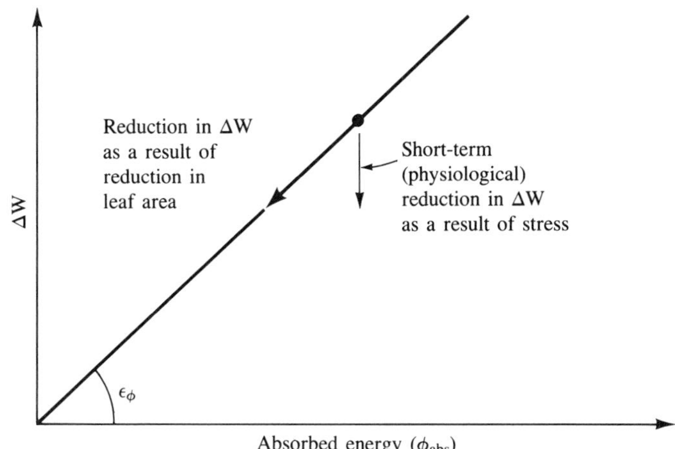

Figure 4.13 The basic relationship underlying "top-down" models that allows calculation of dry matter production (Δt) during which an amount of radiant energy (ϕ_{abs}) is absorbed by the canopy. The slope of the relationship (ϵ_ϵ) represents an "efficiency" factor that estimates the efficiency of conversion of light into chemical energy. (From Landsberg, 1986.)

processes has hindered effective development of models (Landsberg, 1986). Because of their simplicity and the relative ease in collection of required data, absorbed radiation models, if considered carefully in the context of other limiting factors, should be of great benefit in analyzing productivity in forest and orchard management.

Light and Plant Development

Light Quality in Forest Canopies

In passing through vegetation the spectral quality of solar radiation is substantially altered. Evolutionary responses of plants include the development of photoreceptor pigments that greatly influence plant growth and morphology by employing perception of light quality and quantity. Many responses to light quality are ecologically significant, as they alter plant form and function in such a way as to obtain more light and use it more efficiently. Specific examples will be given later.

The light that penetrates a forest canopy is a combination of daylight that has passed through gaps (sunflecks) and that which has been attenuated by passage through the canopy (Smith and Morgan, 1981). In passage through leaves, daylight is depleted of red, blue, and, to some extent, green wavelengths because of absorption by chlorophyll (primarily) and light scattering (Fig. 4.14). In contrast, almost no far-red (730 nm) radiation is absorbed by leaves, and usually only a small fraction is reflected. The result is an enrichment in the proportion of far-red radiation on a forest or orchard floor. Except near sunrise and sunset (when the angle of

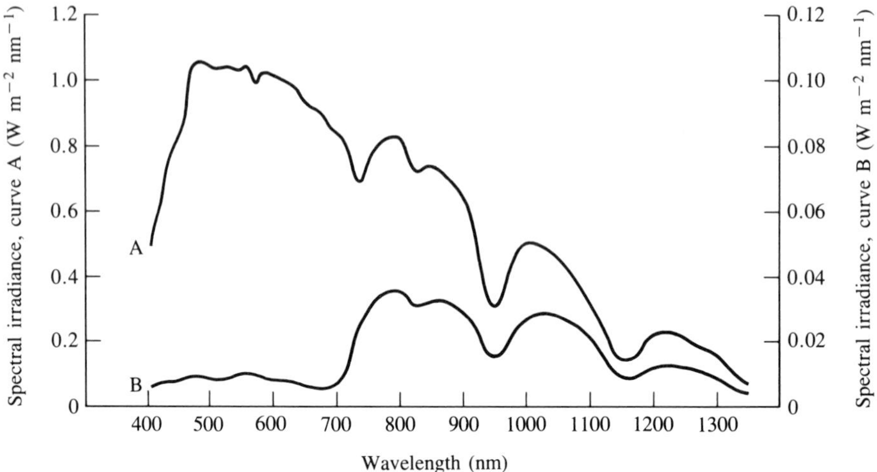

Figure 4.14 Spectral distribution of noontime July light in the open (A) compared to that within a woodland (B). (From Kendrick and Frankland, 1983.)

the sun is <10° and accentuated light scattering in the atmosphere lowers the relative amount of red in sunlight), the red : far-red ratio (R : FR) of sunlight is approximately 1.1. Measurements in various forests throughout the world have shown that a forest canopy significantly decreases R : FR, with temperate deciduous forests showing a generally greater reduction than evergreen coniferous forests (Table 4.4). For example, while R : FR ratios under a closed sugar maple forest in the northeastern United States ranged from 0.08 to 0.11 under clear skies, they were between 0.25 and 0.26 for a forest of eastern white pine under the same conditions (Federer and Tanner, 1966). It must be kept in mind that light quality beneath a forest canopy is quite heterogeneous in space and time, especially where canopy gaps, and resulting sunflecks, are abundant. However, there is general agreement that the R : FR ratio is a reliable index of the degree of shading.

Pigment Systems Involved in Light Responses

The primary pigment involved in plant responses to light quality is phytochrome, a protein weighing about 124,000 g mole^{-1} and possessing a light-responsive portion (Vierstra and Quail, 1983). Phytochrome is present in highest concentrations in dark-grown etiolated tissue (Hart, 1988). Exposure to light inhibits synthesis of phytochrome to some extent and hastens its destruction (Salisbury and Russ, 1985). The pigment is present throughout the cytoplasm and is also found within or associated with the plasmalemma and chloroplast membranes. This remarkable protein undergoes reversible conformational changes in structure when illuminated with red (660 nm) and far-red (730 nm) light, with red light converting a red-absorbing form (P_r) to a form that absorbs far-red light (P_{fr}) and far-red causing the reverse reaction

Table 4.4
Estimation of R : FR Ratio for Shadelight beneath Various Vegetation Canopies[a]

Canopy	Sky conditions	R : FR[b]
Broadleaf Deciduous Woodland		
Beech	Overcast	0.36–0.97
	All conditions	0.16–0.64
Oak	Clear	0.12–0.17
	Hazy	0.32
	All conditions	0.37–0.77
Chestnut	Clear	0.12
Sugar maple	Clear	0.14–0.28
	Clear	0.08–0.11
	Overcast	0.21
Birch	All conditions	0.56–0.78
Mixed hawthorn shrub	Unknown	0.12
	Unknown	0.30
Alder shrub	Unknown	0.35
Coniferous Evergreen Woodland		
Spruce	Clear	0.33
	Clear	0.15
	Overcast	0.46
Red pine	Clear	0.47–0.76
	Clear	0.33
	Partially overcast	0.55
	Hazy	0.61
White pine	Clear	0.25–0.26
	Hazy	0.49
Jack pine	Clear	0.32
	Overcast	0.76
Tropical Rain Forest		
Montane lowland	Bright	0.22–0.30
	Unknown	0.26

[a] From Smith and Morgan (1981).
[b] In daylight, R : FR is 1.1.

(Fig. 4.15). In pines these peaks are shifted to somewhat shorter wavelengths (656 and 714 nm, respectively) (Kendrick and Frankland, 1983). In the light, phytochrome is never entirely present in one form or the other, as the absorption bands of the two forms overlap. Rather a photoequilibrium concentration ratio is established [$\phi = ([P_{fr}]/[P_r]+[P_{fr}])$] in which the extremes in ϕ range from 0.8 in red light (660 nm) to 0.01 in far-red light (720 nm) (Hart, 1988). The process of conversion between P_r and P_{fr} is not understood, although several intermediate forms of the pigment are thought to exist.

In darkness, P_{fr} may gradually revert to P_r or it may be destroyed, possibly by denaturation (Salisbury and Ross, 1985) or by enzyme-catalyzed conversion to an inactive form (Hart, 1988). Despite the profound difference in biological activity of

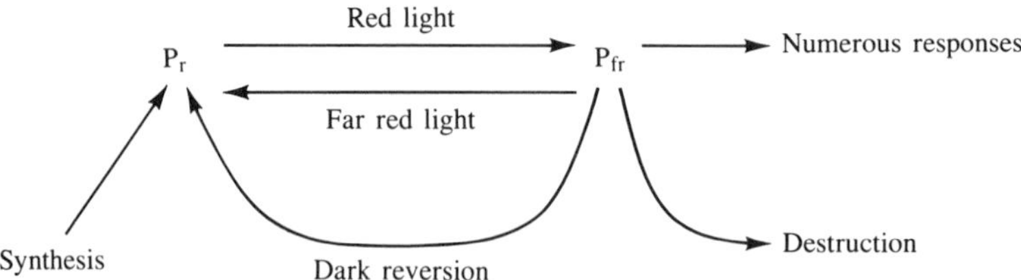

Figure 4.15 Scheme showing photoconversion of phytochrome by red and far-red light into far-red-absorbing (P_{fr}) and red-absorbing (P_r) forms, respectively. P_{fr} may gradually revert to P_r in the dark, or it may be destroyed. (From *Plant Physiology,* 2E, by Frank B. Salisbury and Cleon W. Ross. © 1978 by Wadsworth Publishing Company Inc. Reprinted by permission of the publisher.)

the two forms of phytochrome there are few detectable physical and chemical differences between them (Hart, 1988). Whatever the differences, the P_{fr} molecule induces changes in metabolism that result in exceedingly diverse physiological changes and extreme alteration in growth and development, involving photoperiodism, photodormancy, and photomorphogenesis. Some phytochrome effects are very rapid, such as changes in membrane potentials, whereas others are slower and require gene expression and protein synthesis (Hart, 1988).

Classic phytochrome-mediated responses, such as photoreversible seed germination, respond to very low levels of light energy. Another class of responses, called high irradiance responses (HIR), can also be observed (Mohr and Shropshire, 1983; Hart, 1988). These responses, which include seed germination in some species, stem growth, leaf expansion, and pigment synthesis, differ from low-energy phytochrome responses in that (1) they require high irradiance and long exposure times (on the order of hours) or repeated exposures for elicitation, (2) they do not exhibit R–FR reversibility nor reciprocal relationships between time and intensity (i.e., do not respond similarly to short, high-intensity episodes as compared with long, low-intensity irradiation), and (3) they exhibit one or more response peaks in the blue, red, and far-red regions. High irradiance responses in the red to far-red region can be related to creation of specific ϕ levels, while the relative importance of phytochrome and the "blue-light receptor" associated with processes responding only in the blue region is not known (see the following discussion).

In addition to phytochrome, there appears to be an as yet unidentified photoreceptor pigment, often called the "blue-light receptor" or "cryptochrome," that shows an action spectrum with peak absorption in the UV/blue region, suggesting that it may be a flavinlike compound (Mohr and Shropshire, 1983). In some cases phytochrome and the blue-light receptor pigment appear to have similar effects, whereas in other situations their actions apparently are independent (Salisbury and Ross, 1985). Hart (1988) noted that specific blue-light responses are characterized by (1) responsiveness to a wide range of light intensities (<1 to >400 μmol m^{-2} sec^{-1}), (2) relatively rapid response time (on the order of seconds), and (3) effectiveness mainly on the irradiated tissue. Nastic movements, as in mimosa leaves, and the blue-light responses of stomata (Zeiger *et al.*, 1987b) are examples of rapid responses mediated by the blue-light receptor.

Light Effects on Seed Dormancy

Germination of seeds of woody species exhibits three patterns of light sensitivity: (1) stimulation under continuous or interrupted light, (2) improved germination under transient illumination (phytochrome linked), and (3) light insensitivity (USDA Forest Service, 1974). A number of woody species show germination patterns that indicate the presence of a phytochrome-linked germination system (Kramer and Kozlowski, 1979). Woody genera possessing some species that are known to exhibit wavelength sensitivity include *Alnus, Abies, Betula, Fraxinus, Picea, Pinus,* and *Ulmus* (Asakawa and Inokuma, 1961; Nagao and Asakawa, 1963; Hatano and Asakawa, 1964). Table 4.5 illustrates the photoreversibility of red-light-stimulated germination of Virginia pine seeds. Far-red inhibition reduced germination, respiratory activity, and cell division in Scotch pine seeds (Nyman, 1961). Ecologically, phytochrome-mediated seed dormancy would appear to be of most value to shade-intolerant species. For example, seeds of red alder, a shade-intolerant pioneer species of the northwestern United States, did not germinate under low irradiance (simulating seed burial) or when broad-leaved plants were used to filter incoming incident light (Bormann, 1983). Seed dormancy in this species would thus be maintained until opening of the canopy by logging or fire occurred, signaled by increased R : FR of unfiltered direct light.

Some species germinate better under continuous light (e.g., Douglas-fir and jack pine) even if it is weak, but alternate light and dark periods can also substitute for continuous light (e.g., in Douglas-fir) (Jones, 1961; Ackermann and Farrar, 1965). The ecological significance of stimulation of germination by continuous light is doubtful, as seeds are not subjected to such conditions in nature. High temperatures

Table 4.5
Influence of Alternating Red and Far-Red Irradiation on Germination of Virginia Pine Seeds[a]

Character of irradiation[b]	Percentage germination
Dark control	4
R	92
R + FR	4
R + FR + R	94
R + FR + R + FR	3
R + FR + R + FR + R	93

[a] From Toole *et al.* (1961).
[b] R = red; FR = far red.

will lengthen the light period that promotes germination in eastern hemlock (Stearns and Olson, 1958), and temperatures of 20°C negated the influence of alternate light and dark periods in stimulating germination of hairy birch seeds, while stimulation was observed at 15°C (Black and Wareing, 1955)

Photomorphogenesis

In dicotyledonous angiosperms, seedlings growing in the dark or in low light show etiolation, increased internode elongation, and a suppression of leaf growth. Ecologically, this pattern of growth allocates reserve material in the most efficient manner to attain a sunlit position. Upon illumination, the first influence of P_{fr} in a germinating seedling occurs when the epicotyl (epigeal species) or hypocotyl (hypogeal species) hook emerges from the soil. This hook functions to protect the stem apex from mechanical damage by the soil, but upon emergence, exposure to sunlight rich in red wavelengths produces P_{fr}, inducing a series of events leading to differential growth and straightening of the hook to produce an erect stem.

Another early event in angiosperm development, attributable to light-induced conversion of phytochrome to the P_{fr} form, is the triggering of an early step in the biochemical pathway through which chlorophyll is synthesized. In contrast, formation of chlorophyll *a* and *b* in cotyledons of conifers appears to be independent of light, and chloroplasts of such species developed in darkness may possess prolamellar bodies and granal and stromal thylakoids (Lewandowska and Öquist, 1980). After photosynthesis begins in plants, stem elongation is inhibited by light, particularly when the R : FR ratio is high.

Leaves of shade-tolerant plants are less affected by the light regime during development than those of intolerant species (Hart, 1988). For plants that show responses, leaves of shade-grown plants are thinner, have reduced stomatal density, greater specific leaf area, and more chlorophyll per unit fresh weight, and show much reduced tendency to develop multiple palisade layers. For example, shade-intolerant black walnut seedlings, whether subjected to shading with neutral density screening or discontinuous green celluloid film designed to stimulate a natural canopy, had fewer stomata per unit area, and both palisade layer development and leaf thickness were reduced as the plants were increasingly shaded (Table 4.3) (Dean *et al.*, 1982).

While leaves of red maple, American beech, and flowering dogwood usually had only one layer of palisade tissue regardless of the light environment during growth, shade-intolerant species such as yellow poplar, black cherry, and sweet gum had two or three layers when grown in full sunlight, but only one layer when grown in shade (Jackson, 1967). In contrast, leaves of shade-tolerant European beech had one palisade layer when developed in shade and two layers when developed in full sun (Eschrich *et al.*, 1989). Interestingly, differentiation into sun- and shade-leaf primordia was predetermined to some degree by early August of the year of primordium formation. When shaded branches were exposed to high light during the next growing season, shade buds produced either shade leaves or leaves with characteristics that were intermediate between sun and shade leaves.

Increases in specific leaf area (the amount of leaf area per gram leaf dry weight) in response to shade have been shown in a large number of genera, including *Betula, Acer, Liriodendron, Quercus, Pseudotsuga, Fagus, Coffea, Pentaclethra,* and *Castanospermum* (Loach, 1967; Drew and Ferrell, 1977; Nygren and Kellomaki, 1983; Friend, 1984; Jurik, 1986; Oberbauer and Strain, 1986; Myers *et al.*, 1987). Increases in specific leaf area frequently are accompanied by increased amounts of chlorophyll per unit of *dry weight,* but since shade leaves are considerably thinner than sun leaves, the amount of chlorophyll per unit *leaf area* decreases.

Shade needles of conifers generally show responses similar to those of angiosperms, being thinner, richer in chlorophyll, and having reduced stomatal density compared with sun-grown needles (Lassoie *et al.*, 1985). Orientation of foliage in the upper branches of many conifers is distinctly vertical, giving branches a brush- or comblike (pectinate) appearance. This reduces the effective light intensity for the needles and decreases light interception, but permits deeper penetration of light into the canopy.

Some morphological responses of woody plants to shade beneath the canopy are mediated by phytochrome. The stimulation of shoot elongation and suppression of lateral branching observed in many woody plants, particularly shade-intolerant species, when grown under shade are promoted by the low R : FR ratio of woodland understory light (Warrington *et al.*, 1988). For example, when cuttings of Monterey pine taken from mature and juvenile trees were exposed to radiation regimes varying widely in R : FR ratio, both types of cuttings showed greater internode elongation and stem length, a decreased number of needles per unit of stem weight, and increased stem dry weight when R : FR ratios were low (Morgan *et al.*, 1983). Stimulation of stem elongation by low R : FR ratios also has been observed in tropical angiosperms (e.g., idigbo, *Terminalia ivorensis;* Kwesiga and Grace, 1986). Ecologically, this response presumably allows a better-lit position in the understory and greater prospect of survival. Increases in shoot growth under shade may occur at the expense of belowground growth, resulting in decreased root–shoot ratios under shade, and total dry weight of shade-grown plants is nearly always less than that of sun-grown plants (Logan and Krotkov, 1968). Anatomical and physiological differences between sun and shade leaves of plants appear to be correlated better with the total amount of light received than with peak light intensities or light quality (Hart, 1988).

Chloroplasts show very different alignments in leaves exposed to bright or weak light, with those of intensely illuminated leaves arranged in tiers along radial walls of mesophyll cells, a response that may promote more efficient absorption of light under low irradiance and provide protection from photodestructive effects at high light intensity (Salisbury and Ross, 1985; Chow *et al.*, 1988). The chloroplasts of shaded plants are arranged in monolayers along the upper and lower cell walls.

These alignment patterns may be attributable to responses mediated by the blue-light receptor or by phytochrome (Mohr and Shropshire, 1983).

Photoperiodic Effects

Seasonal fluctuation in day length is zero at the equator (Fig. 4.16). The difference between the shortest winter day and the longest day in summer increases with latitude, and at the poles deep winter and summer are characterized by continuous night and day, respectively. In middle to high latitudes, the increasing photoperiod in late winter and decrease in late summer provide reliable environmental cues that signal the impending onset and end of the growing season. Phenological events of many temperate woody plants, particularly vegetative and reproductive growth, are synchronized to growing season by responses to day length and other interacting environmental factors. Even within a species, photoperiod-sensitive ecotypes may develop if there is broad distribution across latitudes (Salisbury and Ross, 1985).

The mechanism of perception of day length-sensitive plants involves phytochrome, although phytochrome action is modified by factors such as low and high temperature, water stress, and the physiological condition of the plant. In addition to the familiar responses of flowering to photoperiod, growth and phenological phenomena as diverse as the timing of initiation and cessation of growth, internode extension growth, dormancy induction and release, coldhardiness, autumn leaf abscission, and cambial activity respond to day length in at least some woody species (Vince-Prue, 1975).

Shoot Growth

The timing and duration of shoot growth in woody plants, as well as autumn leaf abscission, are influenced by several environmental factors, including day length and temperature. Day length influences shoot growth patterns and the timing of bud dormancy in many species, particularly those of temperate regions. This response has obvious ecological value, as changing photoperiod length is a certain indicator of the coming of autumn and winter. Bud formation may be attributable simply to telescoping of the shoot as internode extension is greatly suppressed, resulting in buds in which ensheathing leaf stipules serve as bud scales (e.g., in *Betula*). Alternatively, specialized bud scales may form as internode extension is inhibited and shoot telescoping develops (*Ribes* sp.) (Vince-Prue, 1975; Kramer and Kozlowski, 1979). In plants that have sympodial growth, such as *Robinia, Rhus,* and *Syringa,* the shoot apex simply aborts and subsequent shoot elongation occurs from the first lateral bud (Kramer and Kozlowski, 1979).

In many temperate trees and shrubs (e.g., black locust, yellow poplar, weigela, and red maple), shoot growth ceases and resting buds develop in late summer in response to short days. However, if long days and moderate temperatures are maintained, shoot growth will continue. For example, Kramer (1936) showed that

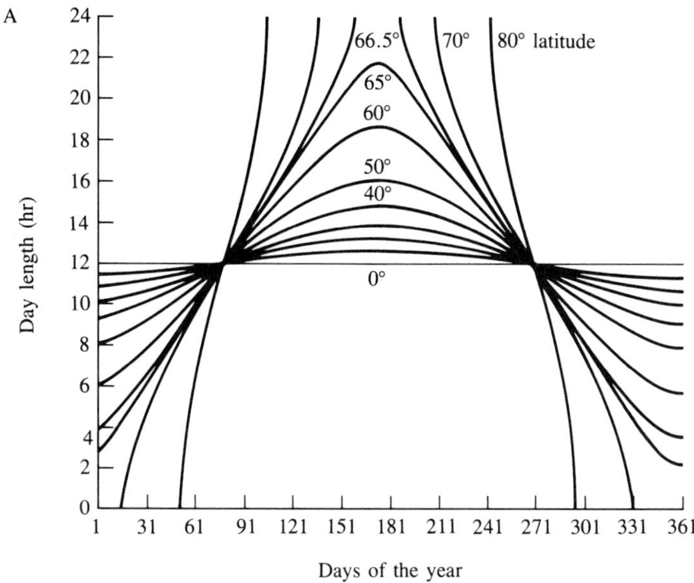

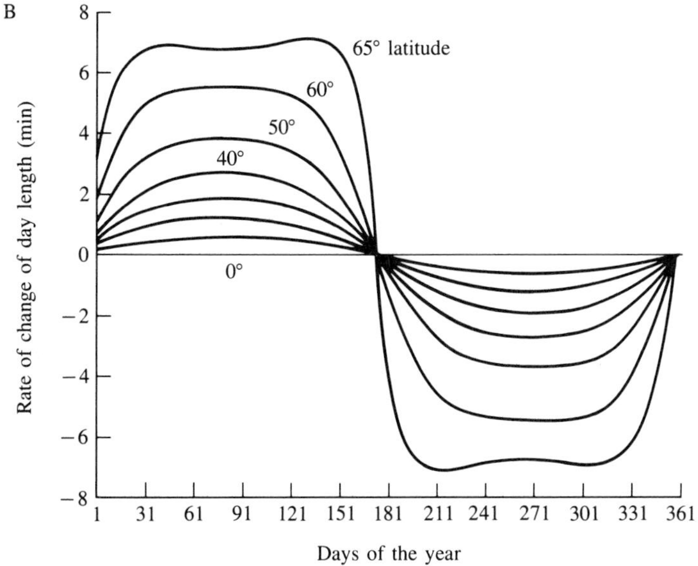

Figure 4.16 Annual cycle of day length and rates of change in day length at various latitudes. From *Plant Physiology,* 2E, by Frank B. Salisbury and Cleon W. Ross. © 1978 by Wadsworth Publishing Company Inc. Reprinted by permission of the publisher.)

shoot growth persists through the winter in yellow poplar and sweet gum plants kept in a greenhouse under 16-hr days. An example of the stimulation of shoot growth by long days in slash pine is shown in Fig. 4.17.

In other species (e.g., royal paulownia, horse chestnut, and sycamore maple), short days also cause cessation of shoot growth, and the period of shoot growth can be prolonged by long days, but not indefinitely (Vince-Prue, 1975). Scotch pine and sycamore maple will eventually cease shoot growth even in continuous light (Wareing, 1956). Long days or continuous illumination can prolong the flushing behavior of species, such as many oak species, that are predisposed to show this pattern of growth, and can induce repeated flushing in certain trees (e.g., pines) that normally have only a single flush of shoot growth (Wareing, 1956). For some species, such as European mountain ash, European ash, and apple, shoot growth lacks sensitivity to day length (Wareing, 1956; Vince-Prue, 1975).

The influence of day length on breaking of dormancy varies with species. In species that have developed resting buds during the summer, such as Scotch pine and oaks, exposure to long days or continuous light will cause renewed shoot growth. If buds of these trees are allowed to develop full winter dormancy, only chilling will break this dormancy. In contrast, European beech can be induced to grow by long-day treatment regardless of the degree of dormancy (Wareing, 1956).

Photoperiod effects also are subject to modification by other factors. For example, Nitsch (1962) noted that black locust plants subjected to short days at 10°C stopped growing, but subsequently resumed growing if exposed to long days. In contrast, plants subjected to short days at 14°C became dormant and subsequently did not respond to long days, indicating that higher temperatures influence dormancy, perhaps through an effect on respiration. Also, shoot growth pattern will influence responses to short days. Although it is possible to experimentally induce shoot growth cessation in many woody species by exposure to short days, this does not mean that short days are invariably responsible for growth cessation in the field. For example, those plants that show a fixed shoot growth pattern complete shoot development and cease shoot growth long before the onset of short days late in the growing season (Kramer and Kozlowski, 1979).

Leaf abscission in some species is directly influenced by photoperiod; in other species, temperature appears to control leaf fall more strongly. For example, in smooth sumac and yellow poplar, the leaves are retained under long days and shed under short days (Garner and Allard, 1923). The leaves of hairy birch, black locust, sweet gum, and white oak remain on trees exposed to short days as long as the temperature does not drop too low (Kramer 1936; Wareing, 1954). Black locust seedlings kept out-of-doors under long days retain leaves until they are killed by frost (Jester and Kramer, 1939).

Cambial growth may be indirectly controlled by photoperiod because of its dependence on photosynthate from the leaves and hormones from active apical meristems (e.g., in red pine; Larson, 1962). There is also evidence (e.g., in black locust and Scotch pine) that long days may extend cambial growth independently of shoot extension (Wareing, 1951).

Figure 4.17 Height growth of slash pine after 15 months under 12-, 14-, and 16-hr day lengths (*left* to *right*). (From Downs, 1962.)

Street and parking area lights may delay dormancy induction in trees and shrubs planted nearby. It is a common observation that leaves persist in the autumn around branches near the lamps of street lights (e.g., in silver maple). Kramer (1937) demonstrated that winter injury after several hard frosts to a hedge of glossy abelia was localized to a portion of the hedge near nighttime electric lighting. Hence, depending on the sensitivity of the species to photoperiod, the delay in winter dormancy caused by artificial lighting may increase susceptibility to injury from cold weather in the late autumn and winter. The significance of this problem relative to other environmental stresses that are characteristic of urban areas is not known.

Flowering

Everyday observations tell one that irises bloom in the spring and chrysanthemums in the autumn, that most trees flower in the spring, and that witch hazel flowers in the late autumn and early winter. Certain plants require long days to flower, whereas others respond to short days or are insensitive to day length. In long-day plants (LDP), P_{fr} promotes flowering and, during the spring, increasing day length results in nights becoming too short to permit reversion of this flower-promoting form of the pigment to P_r. In short-day plants (SDP), low P_{fr} at night is required for flowering and the same mechanism works to inhibit flowering under long days (although there is a requirement for P_{fr} during the daylight period *before* the critical dark night in SDP; Vince-Prue, 1975). Hence it is *night length* that is most critical to flowering to both LDP and SDP plants.

Detection of photoperiod-sensitive flowering responses in woody plants is far more difficult than it is in herbaceous species because (1) flowering of trees and shrubs does not occur until after a variable juvenile period of several years regardless of day length patterns, (2) there are considerable physical and logistical problems in controlling light regimes of large plants for experiments that could determine the presence of sensitivity to day length, and (3) there is a substantial time lag between primordium initiation and flowering (USDA Forest Service, 1974). Nevertheless, it appears that floral initiation is generally less sensitive to photoperiod in woody species than in many herbaceous plants. Exceptions include Wych elm, Scotch pine, and *Betula* sp., which requires long days for initiation of flower buds (Wareing, 1956; Wareing and Longman, 1960). On the other hand, young lodgepole pine trees initiated more female strobili when subjected to short days. Day length and flower bud initiation often may *appear* to be related because of the natural timing of flower bud initiation near the end of seasonal vegetative growth in many temperate woody species and the tendency for shoot growth to cease under short days (Kramer and Kozlowski, 1979).

In addition to the effect of photoperiod, it is a common observation that flowering is more profuse under full sunlight. In fact, flowering in forest stands is often essentially limited to dominant trees (Kramer and Kozlowski, 1979). A relationship between flowering and light intensity has been quantitatively confirmed in apple, Douglas-fir, and white spruce, among others (Silen, 1973; Jackson and Palmer,

1977; Marquard and Hanover, 1984). Stimulation of flowering under high light intensity may be linked with elevated bud temperatures, which leads to increased concentrations of growth regulators, particularly gibberellins. Application of gibberellins is known to stimulate flowering (Pharis and Kuo, 1977; Ross *et al.*, 1983).

Genetic Variation in Photoperiodic Responses

As mentioned previously, genetic differentiation in photoperiodic response of growth and dormancy induction is commonly reported in species with large north-to-south ranges in the temperate zones and constitutes an important adaptive response in seasonal synchronization of phenological events. Photoperiodic ecotypes and clines of many woody species have been identified and a few examples will be given here. In black cottonwood, eastern cottonwood, and balsam poplar sampled from an extensive latitudinal gradient in North America and grown in Massachusetts, shoot growth of plants of more northern origin ceased earlier in the summer than it did in plants of southern origin (Pauley and Perry, 1954). Similar results were observed for several other woody species (Vaartaja, 1959). The ecological significance of these differences is obvious if photoperiodic ecotypes are planted outside their natural range. In the Northern Hemisphere, if southern-latitude plants are moved northward, they often show prolonged late-summer growth and delayed dormancy because of the longer photoperiods there, predisposing them to injury by early frosts (e.g., in European aspen; Moshkov, 1935). Conversely, shoot growth of northern sources of day length-sensitive species growing in southern locations will cease early in the growing season because of shorter days (see also Chapter 5 and Fig. 1.8).

Ultraviolet Radiation and Plants

In recent years, interest in the potentially injurious effects of ultraviolet radiation on plants has increased because of predictions of increased exposure resulting from reduction in upper-atmosphere O_3 levels. Most O_3 formation occurs 10 to 40 km high in the atmosphere and is initiated by photodissociation of O_2 and subsequent reaction of an activated atomic oxygen atom with O_2 (Klein, 1978). Ozone absorbs essentially all radiation below 280 nm, with absorption declining to 50% at 310 nm and zero at 330 nm (Klein, 1978). Predictions of reductions in the amount of O_3 at high altitude are based on consequences of the release to the atmosphere of large amounts of CFCs, primarily from spray propellants and refrigerants (Caldwell, 1981). When these compounds eventually reach the stratosphere, they photodissociate under UV-C light. The elemental chlorine thereby released reacts with O_3, converting it to O_2. Depletion of stratospheric O_3 during the Antarctic spring in the 1980s has been documented, although it has not been definitively linked with increased CFCs (Brasseur *et al.*, 1988). The damage caused to plants by UV light is discussed in more detail later.

The Ultraviolet Spectrum

The solar spectrum reaching the earth's surface contains about 7% ultraviolet radiation (<400 nm) on an energy basis (Caldwell, 1981). Ultraviolet light is divided for convenience into UV-A (320–390 nm), UV-B (280–320 nm), and UV-C (200–280 nm) bands. Extraterrestrial solar radiation is much richer in UV light than that reaching the ground, but intense absorption in the upper atmosphere, primarily by O_3, removes essentially all UV-C light and depletes sunlight of UV-B wavelengths (Fig. 4.18). There is also some attenuation of UV light by scattering, but much of this radiation eventually reaches the earth in the diffuse component of solar radiation. Because of incomplete absorption by O_3 and because of its impact on constituents of plants (Fig. 4.8), solar UV-B radiation is the most important portion of the UV spectrum.

The intensity of the UV radiation spectrum varies widely with latitude, elevation, solar angle, and cloud conditions. Attenuation is greater in the UV-B region when the optical path lengthens, moving the solar irradiance "shoulder" to longer wavelengths (Fig. 4.18). Hence UV-B is depleted more at the ends of the day and at higher latitudes. In addition to a greater path length, UV intensity decreases at higher latitudes because atmospheric O_3 increases in abundance toward the poles through stratospheric migration from a band of highest production in the tropics.

Direct beam UV increases with elevation because of reduced optical path length; diffuse UV may show a small increase with elevation or may decrease because of reduced scattering at high altitude. Clouds across the sun attenuate direct beam UV, but usually reduce global UV-B less than global solar radiation because of the high proportion of diffuse radiation in the former (Caldwell, 1981).

Influences of Ultraviolet Light on Plants

Numerous compounds absorb UV light, including purines, pyrimidines, aromatic amino acids, certain lipids, chlorophyll, abscisic acid, and indole-acetic acid (Fig. 4.18). The biological impact of UV light is not strictly dependent on UV flux, as each plant response has a different action spectrum. For example, damage to DNA occurs under light of longer wavelength than that required for photosynthetic inhibition (Caldwell, 1981). Hence the complex interactions between individual biological processes and UV light require careful weighting of responses.

The most significant photoreaction induced by UV light occurs with the pyrimidine base thymine, in which light absorption results in the formation of thymine dimers and several other products. Because thymine is a constituent of DNA, formation of abnormal thymine products inserts an "unreadable" segment into the genetic code (Hader and Tevini, 1987). Leaves of higher plants appear to be much less sensitive to nucleic acid damage than bacteria and algae. For example, the amount of 254-nm radiation required to kill the leaves of very sensitive higher plant species is four orders of magnitude greater than that necessary to kill cells of highly UV-resistant bacterial species (Caldwell, 1981). In contrast, pollen may be quite susceptible to injury and death from UV-B radiation after shedding.

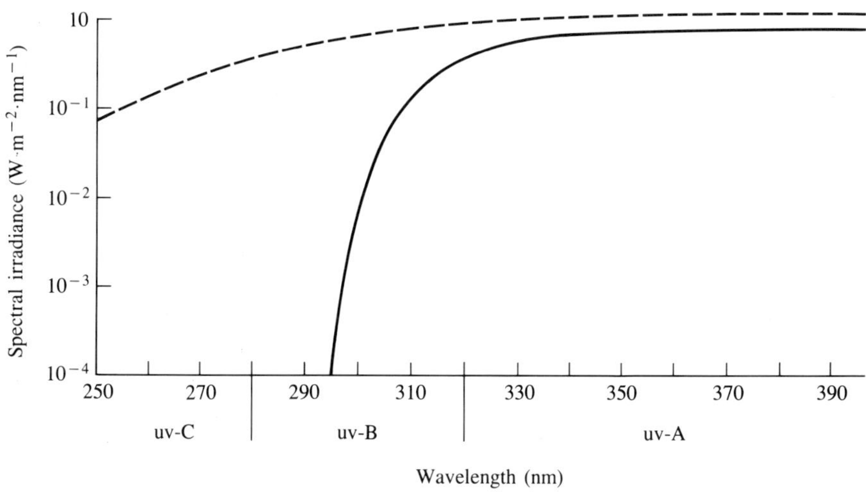

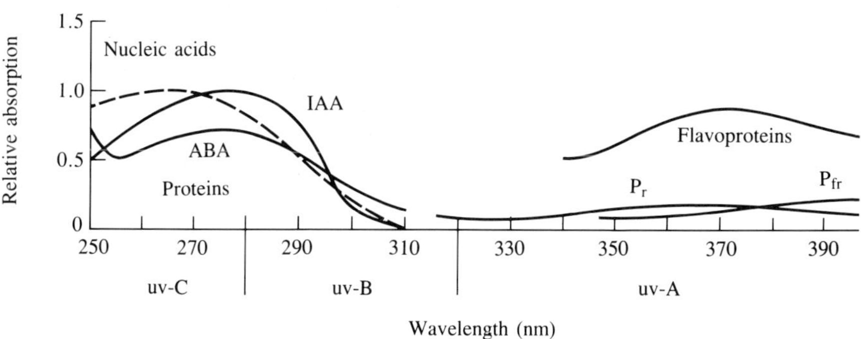

Figure 4.18 *Top,* Solar spectral irradiance before (––) and after (——) attenuation by the earth's atmosphere, as would be received at midday during the summer in temperate latitudes at sea level. *Bottom,* Absorption spectra of several chromophore molecules. (From Caldwell, 1981.)

Proteins containing disulfide bridges and aromatic amino acids also may react with UV light to produce a variety of products. Lipid molecules with isolated or conjugated double bonds, which are abundant in the phospho- and glycolipids that constitute cell membranes, will react with UV light and other compounds to form highly reactive peroxides and free radicals that may disrupt the structural integrity of membranes and destroy transport proteins (Murphy, 1983). However, injury to membranes from solar UV radiation is unlikely to cause mortality, as the intensity required is far higher than that necessary for extensive DNA damage. It is more likely that membrane-mediated effects of UV radiation exacerbate the effects of other biotic and abiotic stresses (Murphy, 1983).

Photosynthetic rates are depressed when plants are exposed to UV-B light. Photosystem II appears to be more sensitive than PS I, and ultrastructural damage to chloroplasts, particularly thylakoid membranes, may result from UV-B irradiation (Brandle *et al.*, 1977; Caldwell, 1981). Another effect of UV-B light on the photosynthetic apparatus is the destruction of plastoquinone (Mantai and Bishop, 1967). Stomatal conductance is usually decreased under UV-B light, but sometimes the reductions are small (Teramura, 1983) and may vary between stomata located on upper and lower leaf surfaces (Bennett, 1981).

Inhibition of photosynthesis by UV-B light is cumulative and not dependent on dose, but if UV-B is supplied in the presence of high levels of visible light flux, less photosynthetic inhibition occurs (Caldwell, 1981). Light in the UV-A region is absorbed by photosynthetic pigments and hence might drive photosynthesis, but it is uncertain how important this light might be because of the highly effective absorption of UV-A light by the leaf epidermis (see the following discussion). McCree and Keener (1974, cited in Caldwell, 1981) noted that stripping of the epidermis resulted in an increase in photosynthesis in wild cabbage (*Brassica oleracea*), which they attributed to increased UV-A penetration into the leaf.

Caldwell (1981) summarized the effects of UV light on growth and morphology of higher plants, noting that UV light can increase the root–shoot ratio, depress flower development, and induce leaf abscission and a loss of apical dominance. The absorption of UV by growth regulators such as IAA and ABA (Fig. 4.18) suggests that UV light might influence the hormone balance of plants, a result that would have profound effects on plant growth and development. However, the relative ineffectiveness of UV-B and UV-A in causing molecular changes in these compounds suggests that this influence would be small.

There is considerable evidence that UV light affects competition among plants. For example, Gold and Caldwell (1983), employing data of Bruzek, reported that removal of UV-B region light from solar radiation increased total shoot biomass and altered the relative distribution of shoot biomass between paired European ash and sycamore maple seedlings. Other experiments using a variety of herbaceous species in replacement series (pots comprising a continuum of reciprocal density changes between two species and constant total density), where solar UV-B was supplemented, indicated little change in total biomass production under UV enhancement, but significant alteration of biomass distribution between the members of species pairs (Gold and Caldwell, 1983). Performance of an individual species was highly dependent on the species with which it was paired. These data indicate that changes in UV-B in the natural solar radiation regime might have a significant impact on species distribution and community composition.

Avoidance and Repair Mechanisms

Plants possess protection and repair mechanisms that can shield or restore the integrity of the genetic code. The leaf epidermis prevents a large fraction of the incident UV flux from penetrating the mesophyll and hence functions as an UV avoidance mechanism. Accumulation of intensely UV-absorbing flavonoid and related phenolic compounds, particularly flavonols and anthocyanins, in the leaf epidermis and outer cell layers is well established. Flavonoid synthesis in the epidermis can be induced by UV-B illumination, indicating an apparent facultative UV avoidance response (Caldwell *et al.*, 1983). Species with leaves that are densely pubescent or very glaucous reflect a substantial fraction of incoming solar radiation, with some species (e.g., Colorado blue spruce) even showing increased reflectance in the UV region. Cuticular waxes also absorb some UV light.

As a result of differences in these leaf properties the mesophyll may be subjected to UV fluxes very much less than those incident to the leaf. For example, Robberecht *et al.* (1980) showed that mean epidermal transmittance of UV-B light ranged from 0.1 to 0.7% for several species of eucalyptus and was 1.1% for willow. A latitudinal and altitudinal gradient in epidermal effectiveness was clearly evident (Fig. 4.19), with plants growing at high altitudes in equatorial and tropical latitudes showing much less epidermal transmittance of UV than arctic plants. The latter also exhibited greater variability in transmittance properties of the epidermis. Exotics from low elevations at high latitude established at high-elevation sites at low latitude

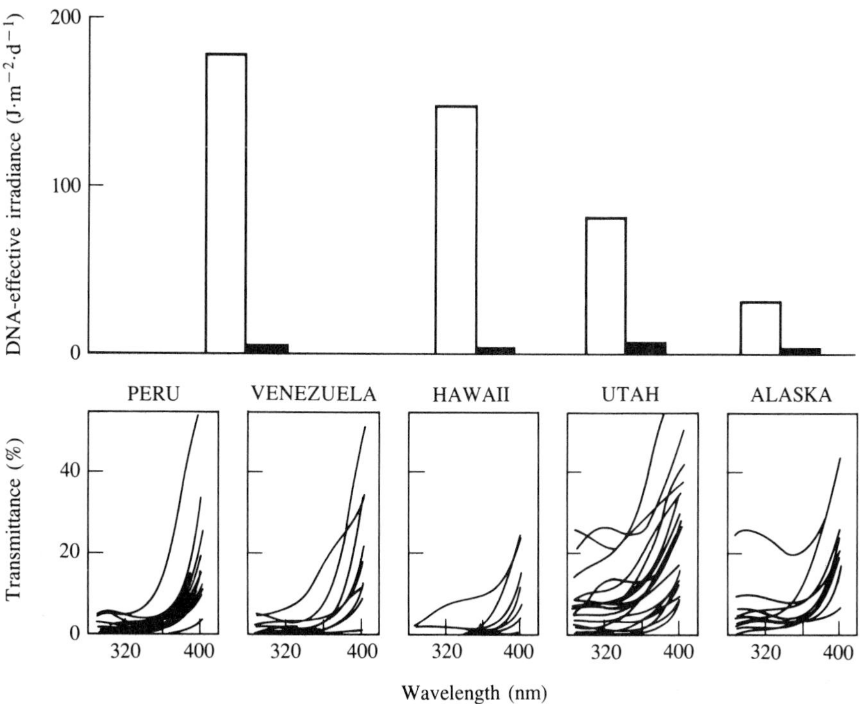

Figure 4.19 *Top,* Daily UV-B irradiance weighted for effectiveness in damaging DNA at the seasonal maximum at several locations along a latitudinal and altitudinal gradient (open bars), and mean effective UV-B flux at the mesophyll (filled bars). *Bottom,* Spectra of epidermal UV transmittance for a variety of species at each location. (From "Leaf ultraviolet optical properties along a leaf latitudinal gradient in the artic-alpine life zone" by R. Robberecht, M. M. Caldwell, and W. D. Billings, *Ecology,* 1980, *61,* 612–619. Copyright © 1980 by the Ecological Society of America. Reprinted by permission.)

showed remarkable acclimation capacity, successfully enduring solar UV-B levels equivalent to three times that associated with a 16% reduction in O_3 layer thickness in temperate latitudes. However, Caldwell *et al.* (1983) cautioned that these results do not allow generalization about plant response to elevated UV-B solar radiation, as they do not represent the performance of *all* high-latitude species, but only the successful genotypes of a few species.

Repair mechanisms have been most widely studied in bacteria, but many are also present in higher plants (Caldwell, 1981). These mechanisms involve several biochemical reactions, some of which are promoted by UV-A and visible light, through which DNA base sequences are reconstituted to their original state. Hader and Tevini (1987) provided a detailed discussion of repair mechanisms in plants.

Summary

The influence of solar radiation on plants is characterized by a diversity of effects and significant impacts on physiological processes and plant morphology. Much of the interaction between woody plants and the radiation environment occurs in leaves, where most of the energy in solar radiation is absorbed and converted through physical processes to long-wave radiation, sensible heat, and latent heat. Only a small fraction of solar energy is captured in photosynthesis, but in the absence of other limiting factors the productivity of woody plant communities is closely related to the amount of radiation absorbed by the canopy.

Plants show adaptation and capacity for acclimation to various solar radiation regimes, particularly adaptation and acclimation of the photosynthetic apparatus to sun or shade. Shaded forest understories have quite distinctive light quality compared with direct sunlight, and woody plants possess sensitive photoreceptor systems that induce adaptive physiological and morphological responses to these differences, including maintenance of seed dormancy and changes in photosynthetic characteristics, leaf anatomy, and root and shoot growth patterns. Woody plants also exhibit sensitivity to day length. Photoperiodic responses of flowering, shoot and cambial growth, leaf senescence and abscission, and dormancy have been shown.

Recent interest in depletion of upper-atmosphere O_3 has stimulated research on the effects of ultraviolet light on plants. Ultraviolet light affect nucleic acid metabolism, photosynthesis, growth patterns, and competition. Similar to microorganisms, plants possess several mechanisms to repair damage to nucleic acids. Leaves of plants avoid much of the UV light in incoming radiation because the epidermis is especially effective in absorbing UV light before it enters the mesophyll, particularly if epidermal pigments such as anthocyanins accumulate.

General References

Berry, J. A. (1975). Adaptation of photosynthetic processes to stress. *Science* **188,** 644–650.

Boardman, K. (1977). Comparative photosynthesis of sun and shade plants. *Ann. Rev. Plant Physiol.* **28,** 355–377.

Campbell, G. S. (1977). "An Introduction to Environmental Biophysics." Springer-Verlag, New York and Berlin.

Gates, D. M. (1980). "Biophysical Ecology." Springer-Verlag, New York and Berlin.

Hart. J. W. (1988). "Light and Plant Growth." Unwin Hyman, London.

Kendrick, R. E., and Frankland, B. (1983). "Phytochrome and Plant Growth," 2nd ed. Edward Arnold, London.

Kramer, P. J., and Kozlowski, T. T. (1979). "Physiology of Woody Plants." Academic Press, New York.

Kyle, D. J., Osmond, C. B., and Arntzen, C. J. (1987). "Photoinhibition." Elsevier, Amsterdam.

Landsberg, J. J. (1986). "Physiological Ecology of Forest Production." Academic Press, London.

Monteith, J. L. (1973). "Principles of Environmental Physics." Edward Arnold, London.

Monteith, J. L. (1976). "Vegetation and the Atmosphere. Vol. 2: Case Studies." Academic Press, London.

Nobel, P. S. (1983). "Biophysical Plant Physiology and Ecology." W. H. Freeman, San Francisco.

Shropshire, W., and Mohr, H., eds. (1983). Photomorphogenesis. *Encycl. Plant Physiol.* **16A.**

Slatyer, R. O. (1967). "Plant–Water Relationships." Academic Press, London.

Smith, H., ed. (1981). "Plants and the Daylight Spectrum." Academic Press, London.

Vince-Prue, D. (1975). "Photoperiodism and Plants." McGraw-Hill, Maidenhead, England.

Zeiger, E., Farquhar, G. D., and Cowan, I. R. (1987). "Stomatal Function." Stanford University Press, Stanford, CA.

Chapter 5

Temperature

Introduction

Temperature has profound effects on distribution and growth of woody plants. For example, climatic zones, which are determined largely by temperature variations related to altitude and latitude, support vegetation types that are determined by natural selection for various temperature regimes (Grace, 1987).

Altitudinal limits for tree growth are largely determined by low temperature. Although "timberline" and "tree line" often are used synonymously, it is useful to retain a distinction between them. Timberline refers to the altitudinal limit of commercial timber, which usually is the upper limit of closed forest. However, for several hundred meters above true timberline (to the tree line), scattered and stunted trees (called krummholz) of negligible commercial value often occur (Kimmins, 1987).

Low winter temperatures are chiefly responsible for the northern and altitudinal limits of broad-leaved evergreen woody plants and for the northern limit of temperate deciduous forests in North America and Eurasia (Larcher and Bauer, 1981). However, the distribution of certain species appears to be influenced more by low summer temperatures. According to Wareing (1985), the rise in mean temperature in the spring in boreal forests lags behind improvement in light conditions by several weeks. Growth processes such as leaf initiation and leaf expansion usually require higher minimum temperatures than does photosynthesis. Under cool conditions the growth of plants often is limited more by the rate at which photosynthetic products are used in growth than by the rate of photosynthesis. Hence, in temperate regions, low temperature often limits plant growth even during late spring. Low summer temperatures may also adversely affect reproductive growth. For example, littleleaf linden trees at the northern limits of the species range do not set seeds except in unusually warm summers (Pigott and Huntley, 1981).

When changes in the boundaries of the Arctic air mass occur over periods of hundreds of years, the limits of the boreal forest in North America also change. Records of such shifts in ranges of species exist in the podzolic soils developed under these forests. North of the present boreal forest, podzolic soils alternate with the gray-brown soils of the treeless tundra, providing a long-term record of changes in the boundary of the boreal forest (Bryson and Murray, 1977). At the alpine timberline, forests give way abruptly to heath-grassland, a change associated with low photosynthetic capacity, incomplete maturation of tissues, and severe desiccation of tree foliage (Tranquillini, 1969). In the forest–tundra ecotone region of the Washington and Oregon Cascades, invasion of alpine fir and hemlock into subalpine meadows during 1928–1937 was associated with a period of warm climate (Franklin *et al.*, 1971), further emphasizing the importance of temperature regimes on distribution of woody plants.

Growth of woody plants is very sensitive to temperature, and a change of only a few degress Celcius often leads to a change in the rate of growth. For example, increasing the day temperature during the growing season from 23 to 30°C increased shoot growth of loblolly pine seedlings about 1.2 cm per degree (Kramer, 1957). Small temperature changes affect growth indirectly, and often slowly, by altering the rates of a variety of physiological processes. In contrast, very large temperature changes may injure plants directly and rapidly, by freezing or severe dehydration, or occasionally through direct injury by overheating. Temperature effects are modified by other environmental factors, including water supply, light intensity, daylength, and mineral supply. Interactions between temperature and other stress factors often are complex.

Cardinal Temperatures

The cardinal temperatures are the minimum below which a physiological process, including growth, is not measurable; the maximum above which it is not measurable; and the optimum at which it proceeds at the highest rate. The rates of physiological processes generally increase as the temperature is raised from near 0°C and become maximal in the range of 20 to 35°C, depending on the process and the species. With further increase in temperature the rates of physiological processes rapidly decrease. This pattern resembles the effect of temperature on enzyme activity (Kramer and Kozlowski, 1979, p. 647).

Individual plant processes have different optimal temperatures and various organs have different cardinal temperatures for a given process. Cardinal temperatures also vary with the age of plant tissue, duration of temperature regimes, and other environmental factors.

Variations in temperature requirements for growth of different organs and tissues are well known. In the temperate zone, for example, seasonal growth of roots begins at a lower temperature than is required for bud expansion (Kramer and Kozlowski, 1979). Young red pine seedlings grew rapidly at 20°C. At 30°C, growth of shoots was reduced by 35% and roots by 56% (Kozlowski, 1968a).

The optimal temperatures were near 33°C for dry weight increment of leaves of cacao seedlings and 22°C for dry weight increase of stems or roots (Sena Gomes and Kozlowski, 1987). The partitioning of dry matter also was affected by the temperature regime, with proportionally more photosynthate retained by shoots and less translocated to roots over a 60-day period at high temperatures. Similar effects of temperature on partitioning of carbohydrates have been shown for macadamia (Trochoulias and Lahav, 1983).

Energy Exchange

The temperature of a plant depends on its rate of energy exchange with the environment (Fig. 5.1). The temperatures of roots, stems, and leaves of the same plant often differ appreciably because of variations in the rate of energy transfer between various parts of the plant and the air. Small branches and twigs are closely coupled to air temperatures, hence their temperatures do not deviate much from that of air. By comparison, the temperatures of large branches and stems may vary considerably from air temperatures. The energy relations of plants are discussed in more detail in Chapter 4 of this volume, by Gates (1976, 1980), and by Nobel (1983).

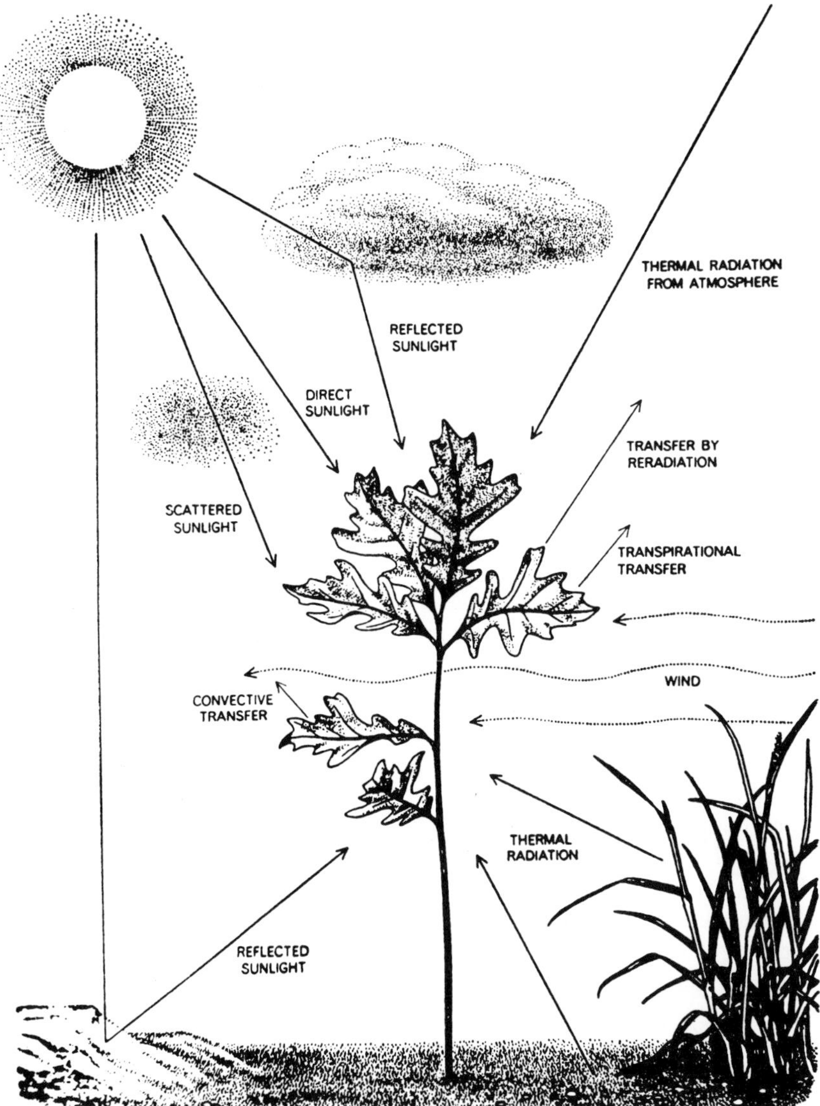

Figure 5.1 Energy flow between a young oak tree and its environment. (From Gates, 1973.)

Growth

The effects of temperature on growth have been studied with a variety of objectives in mind. Some investigators determined minimum or optimum temperatures required for growth of plants, and others emphasized the importance of day versus night temperatures (Went, 1953). For example, growth of loblolly pine seedlings was more closely related to the difference between day and night temperature than to the actual day or night temperature (Kramer, 1957, 1958). Perry (1962) found that the optimum day and night temperatures for growth of red maple seedlings from different geographic sources varied with the source. However, differences in day and night temperatures are not always as important as originally supposed. Redwood seedlings grew vigorously with little or no difference between day and night temperatures (Hellmers, 1966), and growth of Engelmann spruce seedlings was regulated almost entirely by night temperatures (Hellmers *et al.*, 1970). Apparently if the mean temperature is unfavorable, a high day temperature may compensate for a too-low night temperature, or vice versa. For example, in greenhouses, low night temperatures can compensate for too-high day temperatures (Warrington *et al.*, 1977).

Some studies emphasize the importance of soil temperature in regulating growth. For example, roots of redwood seedlings grew best in soil at 18°C under both high and low air temperature conditions (Hellmers, 1963), and loblolly pine seedling roots at 25°C (Barney, 1951).

Several investigators correlated heat sums or degree-days with tree growth. Hellmers (1962) defined heat sum as total daily degree-hours. The temperature was the numerical value above 0°C times the number of hours that plants were at that temperature during a 24-hr day. Growth of eastern hemlock and Jeffrey pine seedlings was closely related to the total daily heat sum (Olson *et al.*, 1959; Hellmers, 1962). Cleary and Waring (1969) developed a heat sum index that related the effects of field temperature to distribution of Douglas-fir. However, they acknowledged that their procedure did not account for interactions of temperature with other environmental factors such as light intensity and water supply. Thomson and Moncrief (1982) developed a model, based on degree-day accumulation, for predicting dates of seasonal initiation of shoot expansion from observations of dates of bud opening and weather records. A limitation of several of the heat sum studies is that they assume a constant relationship between temperature and growth during all stages of plant growth and during the day and night. Effects of temperature on vegetative and reproductive growth will be discussed in more detail in the following sections.

Shoot Growth

Temperature influences shoot growth through its effects on bud development, release of bud dormancy, expansion of internodes and leaves, and induction of bud dormancy toward the end of the growing season (Kramer and Kozlowski, 1979).

Both the temperature of the year of bud formation and temperature of the year of bud expansion into a shoot may influence the amount of shoot growth (Kozlowski, 1983b), especially in trees with fixed growth.

Bud Dormancy

In plants of the temperate zone, bud dormancy can be induced by low temperatures and likewise broken by winter cold. Although temperatures near 5°C are effective and freezing temperatures are not required, bud dormancy may be broken faster by freezing temperatures than by low temperatures above freezing. Bud dormancy also can be broken by alternating warm and cold temperatures, but continuous chilling usually breaks dormancy faster.

The amount of chilling needed for release of bud dormancy varies with species, genotype, location of buds on the tree, and even with weather of the previous season (Kramer and Kozlowski, 1979). About 1200 to 1600 hr of chilling were needed to break bud dormancy of sweet gum (Farmer, 1968). More than 2000 hr of chilling were required for sugar maple in southern Canada, but much less chilling was needed for seedlings from more southerly sources of this species (Kriebel and Wang, 1962; Taylor and Dumbroff, 1975). When seedlings of several red maple provenances were brought to Gainesville, Florida, and exposed to chilling temperatures, both racial and local variations in chilling requirements were found. Plants from very cold parts of the species range required up to a few weeks of exposure to low temperature to release bud dormancy; those from warm parts of the range needed little or no chilling (Perry and Wang, 1960). Chilling of loblolly pine seedlings in storage was as effective as natural chilling, but there were significant differences among families and provenances in the amount of chilling required (Fig. 5.2).

As mentioned, buds on different locations of the same trees often have different chilling requirements. For example, the terminal vegetative buds of peach had a shorter chilling requirement and were less affected by temperature level than the lateral vegetative buds or flower buds (Fig. 5.3).

Temperature often interacts with photoperiod in breaking bud dormancy. Chilled European beech trees do not break bud dormancy after prolonged chilling under short days because their long-day requirements are not altered by chilling (Wareing, 1953). By comparison, dormancy of European white birch buds is readily broken by chilling even under short days (Wareing, 1956).

The effects of temperature on development of bud dormancy as well as on subsequent release from dormancy appear to be mediated by a balance between growth-inhibiting hormones (e.g., abscisic acid) and growth-promoting hormones (e.g., gibberellins) (Kramer and Kozlowski, 1979; Walton, 1980). At least three major and several minor phases of bud development are recognized, each controlled by balances of growth hormones (Fig. 5.4). In sycamore maple the greatest amounts of inhibitors were found in buds and leaves during dormancy, and the lowest amounts when the apex was expanding (Phillips and Wareing, 1958).

Release from bud dormancy is accompanied by an increase in growth promoters,

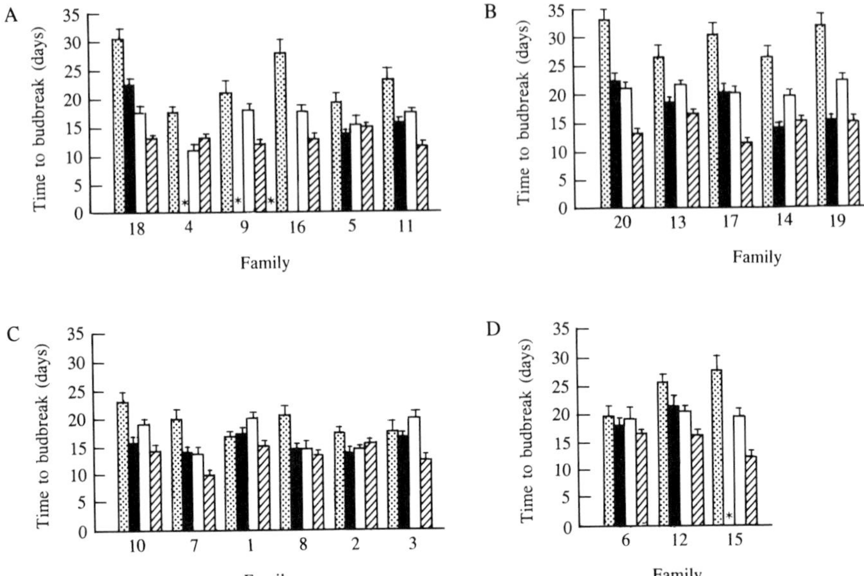

Figure 5.2 Number of days to budbreak of families of loblolly pine seedlings subjected to chilling sums of 207 hr in nature (▩), 207 hr in nature plus 500 hr in storage (■), 734 hr in nature (□), or 734 hr in nature plus 500 hr in storage (▨). Asterisks indicate seedlings destroyed. Families are from (A) the North Carolina coastal plain, (B) the North Carolina piedmont, (C) the Mississippi–Alabama coastal plain, or (D) the Arkansas–Oklahoma region. (From Carlson, 1985a.)

a decrease in inhibitors, or both. In sycamore maple the release of bud dormancy was correlated more with an increase in gibberellin content of buds than with a decrease in growth inhibitors (Eagles and Wareing, 1964). Growth promoters other than gibberellins also appear to be involved in the mechanism of dormancy release. For example, in dormant buds of paper birch and balsam poplar, kinetinlike activity increased progressively until shortly after buds opened, and decreased thereafter (Domanski and Kozlowski, 1968).

Shoot Expansion

After bud dormancy is broken by chilling temperatures, exposure to warm temperatures for a critical period is necessary for normal shoot expansion. After bud dormancy of sugar maple seedlings was broken by chilling, exposure to at least 140 hr of warm temperatures was required for shoot elongation to begin (Taylor and Dumbroff, 1975). In most deciduous trees of the temperate zone, initial leaf emergence usually depends on the cumulative thermal sum (degree-hours or degree-days) to which buds are exposed after a prerequisite cold period (Lechowicz, 1984). The importance of temperature for resumption of shoot extension is shown by variations in growth initiation of the same species at different latitudes. Whereas shoots of

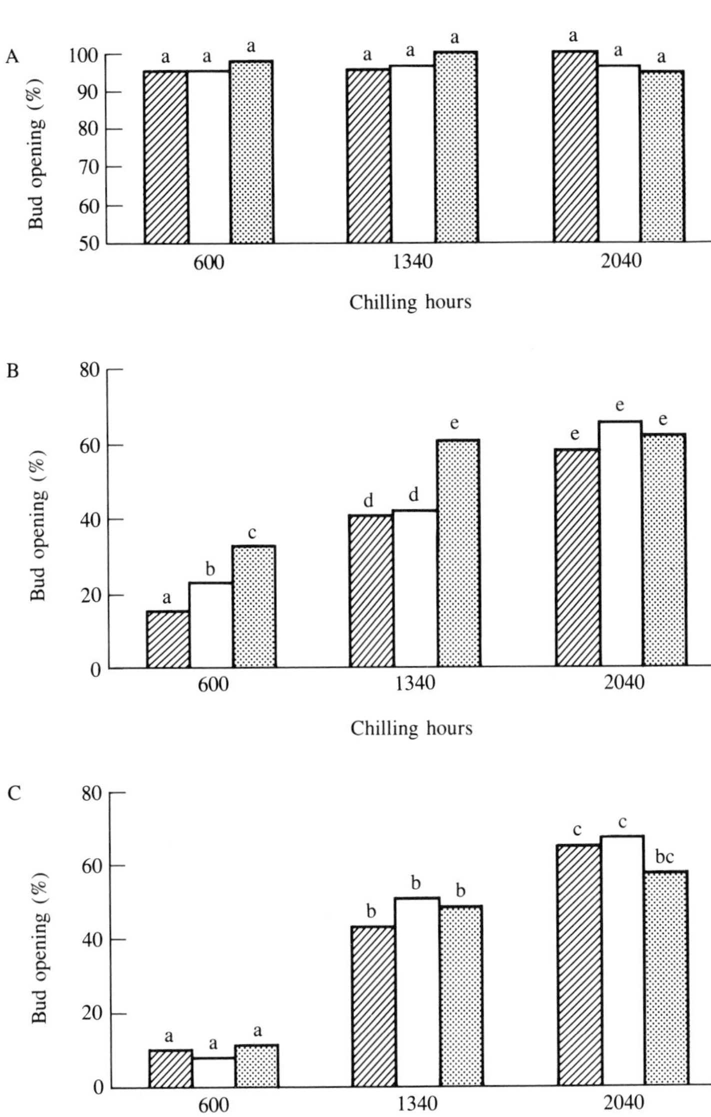

Figure 5.3 The effect of temperature level and duration of chilling on budbreak of peach. Mean separation by Duncan's multiple range test, 5% level. (A) Terminal buds; (B) lateral vegetative buds; (C) flower buds. (From Scalabrelli and Couvillon, 1986.)

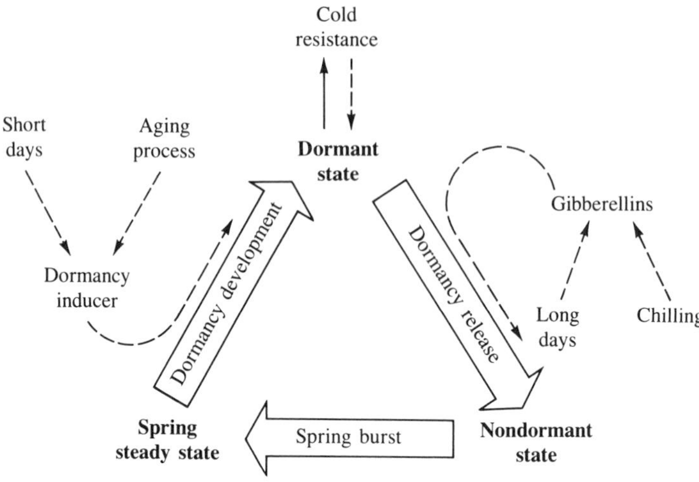

Figure 5.4 The relationship of the dormancy phases of bud development to the annual cycle. The three steady states are shown in heavy lettering and the transitional phases in enclosed arrows. The possible mediation of environmental and endogenous factors by postulated substances is represented by dashed arrows. (After Smith and Kefford, 1964; from Kramer and Kozlowski, 1979, by permission of Academic Press.)

eastern white pine and red pine begin expanding in North Carolina in late March, they do not do so in New Hampshire until about the first week of May (Fig. 5.5).

Temperature also influences shoot growth by regulating both the rate and duration of internode expansion. The amount of shoot growth of European larch was progressively reduced at high altitudes, largely because of delay in initiation of growth in the spring. Growth of Norway spruce also started progressively later with increase in altitude (Fig. 5.6). Height growth of bristlecone pine decreased rapidly near timberline. Tree height averaged 5 m at an elevation of 3400 m but only 2 m at 3487 m (La Marche and Mooney, 1972). Temperature also influences formation and expansion of conifer needles. The optimum temperature for needle initiation in black spruce and white spruce seedlings was 25°C, with needle initiation inhibited by both higher and lower temperatures (Pollard and Logan, 1977). The needles of Swiss stone pine were 7.6 cm long on trees at an elevation of 1300 m, and only 5.2 cm long at 2000 m near timberline (Tranquillini, 1965). When late summer temperatures were low, the period for bud differentiation in Norway spruce was shortened and the shoots developing from such buds were short (Heide, 1974).

In species exhibiting fixed growth of shoots, the total length of shoots is correlated with the size of the winter bud, with small buds producing shorter shoots than large buds (Kozlowski *et al.*, 1973). Hence, in many species the amount of shoot growth is better correlated with the temperature regime of the year of bud formation than with temperature during the year of bud expansion into a shoot. For example, environmental conditions during bud development greatly influenced the subsequent year's growth of shoots of red pine (Olofinboba and Kozlowski, 1973) and Scotch

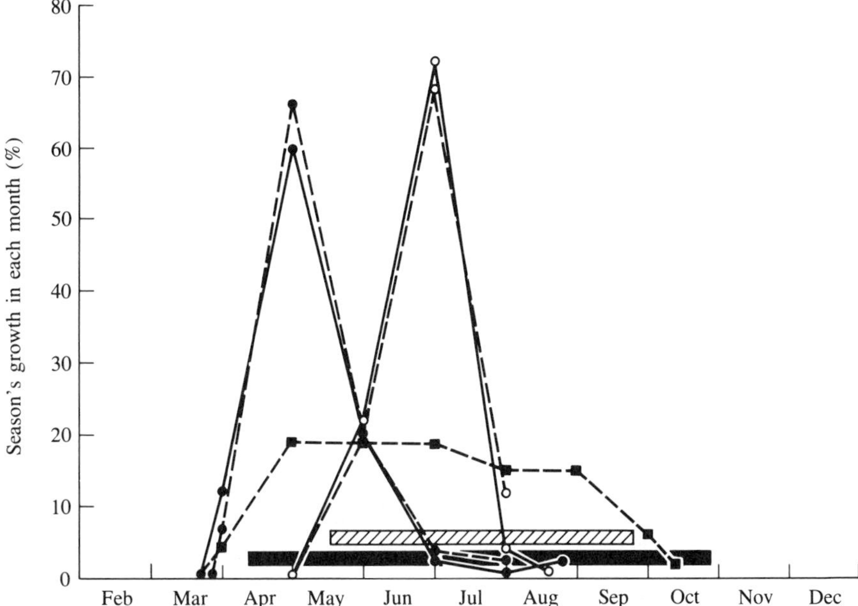

Figure 5.5 Variations in seasonal height growth patterns of red pine (——) and white pine (---) in North Carolina (●) and New Hampshire (○), and of three southern pines (loblolly, shortleaf, and slash pines; ■) in North Carolina. Hatched box, frost-free season in New Hampshire; filled box, frost-free season in North Carolina. (After Kramer, 1943; from Kramer and Kozlowski, 1979, by permission of Academic Press.)

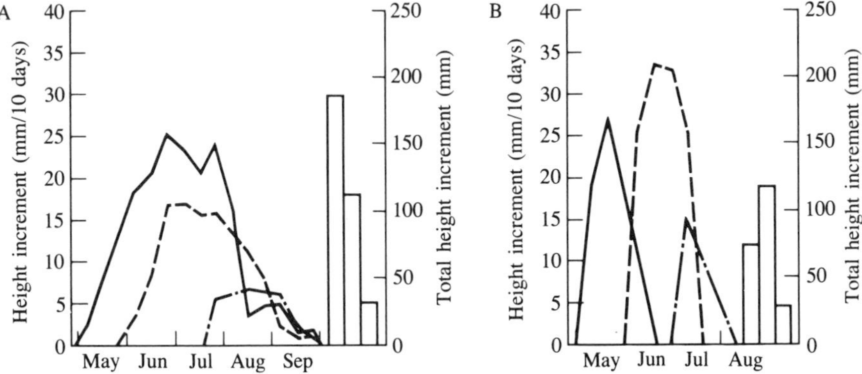

Figure 5.6 Effect of altitude on the amount and seasonal duration of height growth of young European larch (A) and Norway spruce (B) trees in Austria. 700 m (——); 1300 m (--); 1950 m (---). Adapted from Tranquillini and Unterholzner, 1968, and Oberarzbacher, 1977; from Kozlowski, 1983b, by permission of Westview Press.)

pine (Junttila, 1986). Both leader length and number of needle fascicles per leader of Scotch pine trees in northern Norway and Finland were positively correlated with June to August temperatures of the previous season (Junttila and Heide, 1981).

Cambial Growth

Temperature regulates cambial growth by influencing the time of seasonal initiation of division of fusiform cambial cells to produce xylem and phloem, as well as the subsequent rate and duration of production of xylem and phloem tissues (Kramer and Kozlowski, 1979, p. 650).

The effects of temperature on cambial growth are particularly evident in subarctic regions, where temperature is more limiting than water supply. Mikola (1950) considered temperature to be the decisive environmental factor controlling diameter growth in northern Finland. In that region, low rates of cambial growth from 1902 to 1911, and increases thereafter, were closely correlated with mean July temperature (Mikola, 1962). Siren (1963) attributed at least three-fourths of the effect of climate on xylem increment in a subarctic area to air temperature of both the current and previous summer. Cambial growth of Norway spruce on the east coast of Hudson Bay lasted for less than 8 weeks (Marr, 1948), but about 300 miles farther south at Chalk River, Ontario, it continued for about 12 weeks (Fraser, 1962).

The effects of both latitude and altitude on seasonal duration of cambial growth, which are mediated through temperature, have been well documented. Whereas cambial growth of some temperate zone species in cold regions may continue for only a few weeks, in many tropical and subtropical areas cambial growth may be more or less continuous throughout the year, resulting in indistinct growth rings or absence of rings (Kozlowski, 1971b, Fig. 1.5).

Some insight into the predominating effect of temperature on cambial growth can be gained from studies of forest trees planted at different altitudes. For example, yearly diameter growth of mature Norway spruce trees in the Austrian Alps was 6 mm at low altitudes and 3 mm near timberline (Fig. 5.7). The reduced growth at high altitudes was attributed to a shorter growing season as a result of late initiation of cambial growth and low physiological activity during the growth period (Holzer, 1973). Near the Austrian timberline (1950 m), seasonal cambial growth of European larch began 2 weeks later than at 1300 m, and 10 weeks later than at 700 m (Tranquillini and Unterholzner, 1968).

Some studies show a fairly rapid response of cambial growth to a change in temperature. In Wisconsin, daily radial growth of northern pin oak early in the growing season was correlated with daily maximum and minimum temperatures (Kozlowski *et al.*, 1962). There also was a 1-day lag effect of temperature on cambial growth. This was consistent with data of Fritts (1959), who found that maximum temperature of the preceding day influenced cambial growth of white oak and sugar maple in Indiana. The rates of basipetal transport of carbohydrates often vary from 0.5 to 1 m per hour (Zimmermann and Brown, 1971). Thus, in tall trees a time lag between synthesis of carbohydrates in the shoots, their translocation to the

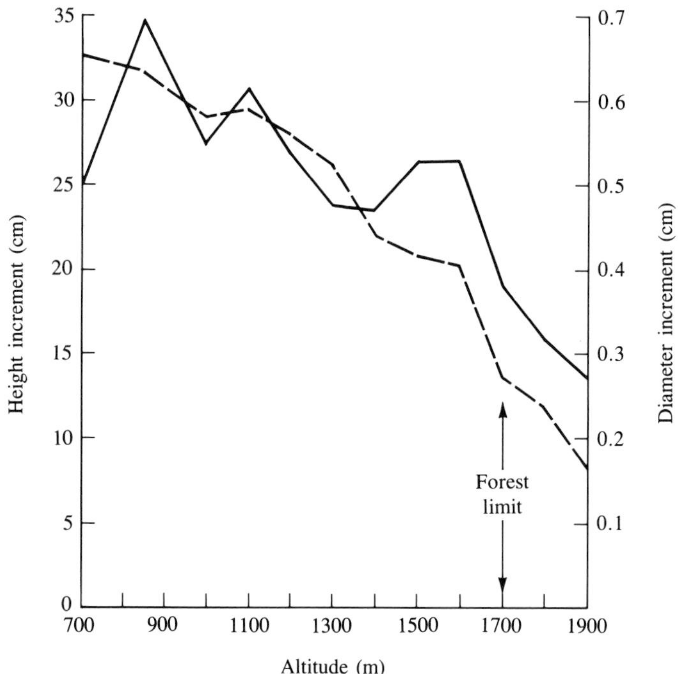

Figure 5.7 Effect of altitude on height (– –) and diameter (——) growth of mature (70- to 140-year-old) Norway spruce trees in the Austrian alps. (After Holzer, 1973; from Kozlowski, 1983b, by permission of Westview Press.)

lower stems, and eventual utilization in cambial growth of the lower stem may be expected. There also is a time lag in temperature changes in the plant body.

Thermal Shrinkage and Expansion

Variations in stem diameter result from changes in production of wood and bark tissues as well as from superimposed, reversible hydration and thermal changes. Shrinkage and swelling of tree stems associated with variations in hydration are discussed in Chapter 7.

As early as 1906, Wiegand reported that stems of woody plants shrank considerably when exposed to cold. Daubenmire and Deters (1947) showed that decreases in air temperature during the growing season not only decreased cambial growth of forest trees but also induced stem shrinkage.

Most thermal contraction and expansion of tree stems and branches occur during the winter (Kozlowski, 1965). In Wisconsin, the stems of yellow birch, sugar maple, paper birch, trembling aspen, and American basswood showed a large diameter decrease in mid-December as the air temperature dropped rapidly (Winget and Kozlowski, 1965). The diameters of many trees decreased two to three times as much in winter as they increased while growing during the previous summer, and some shrank over five times as much (Fig. 5.8). Maximum stem contraction occurred in the coldest part of the winter. The amount of winter shrinkage of stems was greater in American basswood than in yellow birch, sugar maple, or eastern hemlock.

Much of the contraction and expansion of stems and twigs with temperature changes is localized in the bark. When Wiegand (1906) measured expansion on thawing of willow twigs with and without bark, he found more than half of the expansion in the bark. McCracken and Kozlowski (1965) reported that after transfer from a warm (20°C) to a cold (–20°C) environment, twigs with bark of English oak, Norway maple, and European beech shrank much more than those without bark.

Thermal shrinkage and expansion of tree stems and branches occur very rapidly. Shortly after a cold spell in North Carolina, stems of shortleaf pine shrank appreciably but expanded immediately after the freezing weather ended (Byram and Doolittle, 1950). Diameters of white ash trees varied diurnally with temperature changes during the winter (Small and Monk, 1959). The highest temperatures occurred near noon when the largest increases in stem diameter were recorded. Most stem contraction occurred near 6:30 A.M. when the daily temperature was lowest.

Root Growth

Soil temperature influences both initiation of roots and growth of existing roots. The optimum temperature for root growth varies with the species and genotype, stage of development, and supply of soil moisture and oxygen.

Roots usually begin to grow shortly after the soil becomes free of frost. Root growth therefore begins earlier in trees growing at low than at high altitudes. Roots probably have no inherent dormancy and in the southeastern United States some growth of tree roots occurs in every month of the year. The minimum temperature for root growth of many woody plants of the temperate zone is between 0 and 5°C; the optimum temperature is between 20 and 25°C (Kramer and Kozlowski, 1979). Root regeneration of Norway spruce and European larch seedlings was low at a soil temperature of 4°C and optimal at 20°C (Tranquillini, 1973). In Missouri, growth of black walnut roots in well-watered soil began at a soil temperature of 4°C, increased slowly to 13°C, and then increased more rapidly to a maximum at 17 to 19°C. The number of growing roots peaked at 21°C, the highest temperature encountered (Kuhns *et al.*, 1985). Barney (1951) reported an average daily growth of 5.2 mm per day of roots of loblolly pine seedlings at a soil temperature of 25°C, but only 0.17 mm per day at 5°C. There are significant differences among families of loblolly pine with respect to root growth in cold soil (Carlson, 1985b; Kramer and Rose, 1986) and differences might be found in other species also.

Growth of seedlings in the cotyledon stage of development is particularly sensitive to temperature extremes, with root growth being reduced more than shoot growth (Kozlowski and Borger, 1971). Over a 7-week period, dry weight increment of red pine seedlings in the cotyledon stage of development was reduced by

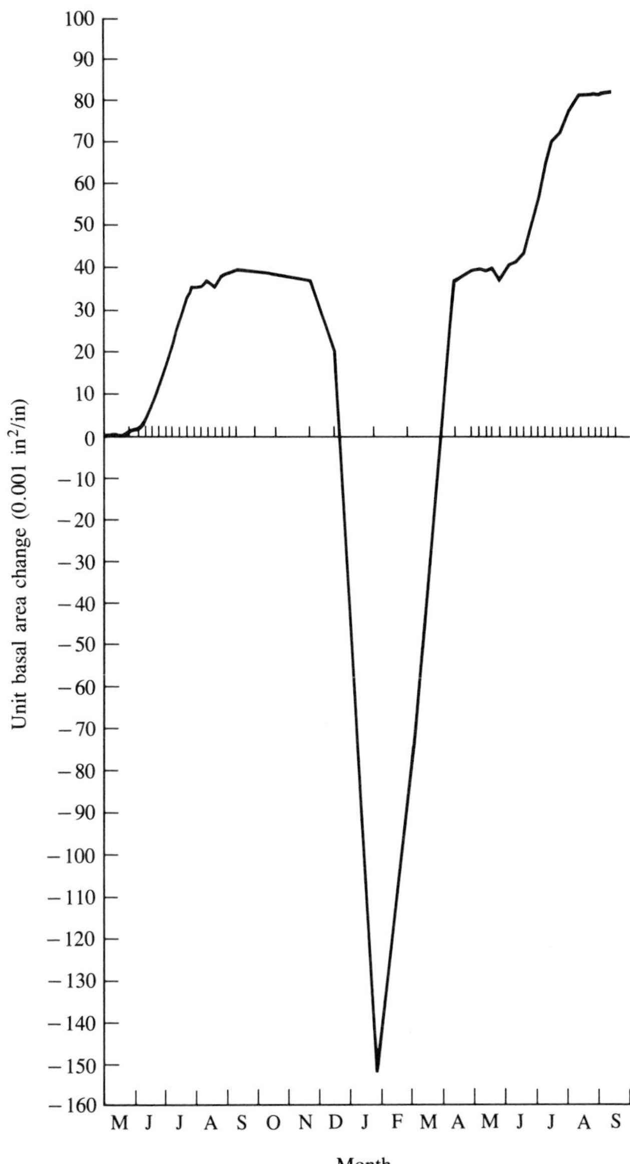

Figure 5.8 Seasonal changes in stem size of dominant American basswood trees in Wisconsin. Data are average change in basal area per inch of stem circumference. (From Winget and Kozlowski, 1965.)

temperatures above or below 20°C. At 10°C, root growth was reduced by 61% and shoot growth by 26%. At 30°C root growth was reduced by 56% and shoot growth by 35% (Kozlowski, 1967a).

Studies of effects of temperature on root growth are complicated by interactions of temperature with soil moisture supply. In North Carolina, the periods of slowest root growth of loblolly and shortleaf pines in the winter were correlated with low soil temperature. However, the periods of slowest root growth in the summer coincided with periods of lowest soil moisture content (Reed, 1939). In Missouri, root growth of white oak was controlled by a balance between temperature conditions and availability of soil moisture. At soil temperatures below 17°C, temperature was the important factor in influencing root elongation, but at temperatures above 17°C, soil water potential became the dominant factor (Teskey and Hinckley, 1981).

Reproductive Growth

Temperature affects flowering and fruit production in several ways, including the chilling required to break bud dormancy, control of time of flowering and fruit set, control of fruit growth, and killing of flowers and fruits by untimely freezes (Kramer and Kozlowski, 1979, pp. 690–692).

In most instances, dormancy of flower buds is broken long before temperatures are warm enough to permit flowering. However, buds sometimes open prematurely during a short period of warm weather, only to be killed later by freezing weather. This is a common problem with fruits, especially with peaches and to a lesser degree apples, and citrus fruits maturing during the winter in Florida often are injured by untimely freezes. In contrast, in mild climates, lack of sufficient cold weather to break bud dormacy results in failure of flowers to open properly. For example, the peach crop was a failure in most of the southeastern United States following the mild winter of 1931–1932. Failure of fruit trees to flower is a common problem when trees of the temperate zone are transplanted to subtropical climates.

The amount of chilling required for breaking of dormancy of flower buds varies among species and cultivars. Although 1000 hr of chilling will break dormancy of most varieties of peach, the requirement varies from near 50 to more than 1100 hr (Table 5.1).

Dogwood does not flower until early May in Columbus, Ohio (Wyman, 1950), although dormancy probably was broken earlier, as it flowers much earlier in the southeastern states. The time of flowering of a given species also varies greatly from year to year because of temperature differences. For example, over a 43-year period the date of full bloom of Cox's Orange Pippin apple trees at East Malling, England, varied from April 15 to May 23rd (Tydeman, 1964). The effects of elevation and latitude on flowering dates of pines are mediated through temperature influences. In the United States, most western pines show a delay in flowering of about 5 days for each degree of latitude northward, and flowering of western white pine was delayed by 5 days for each 300 m of elevation (Bingham and Squillace, 1957).

Table 5.1
February 15th Chilling Requirements of Flower Buds of Peach Cultivars[a]

Cultivar	Hours of chilling required at 7.2°C (45° F)
Mayflower	1150
Raritan Rose, Dixired, Fairhaven	950
Sullivan Early Elberta, Trigem, Elberta, Dixired, Georgia Belle, Halehaven, Redhaven, Candor, Rio Oso Gem, Golden Jubilee, Shippers Late Red	850
Afterflow, July Elberta (Burbank), Hiland, Hiley, Redcap, Redskin	750
Maygold, Bonanza (dwarf), Suwannee, June Gold, Springtime	650
Flordaqueen	550
Bonita	500
Rochon	450
Flordahome (double flowers)	400
Early Amber	350
Jewel, White Knight No. 1, Sunred Nectarine, Flordasun	300
White Knight No. 2	250
Flordawon	200
Flordabelle	150
Okinawa	100
Ceylon	50–100

[a] From Teskey and Shoemaker (1978).

Some trees and shrubs ordinarily bloom during the winter. Witch hazel blooms in the winter and winged elm and red maple flower and fruit in late winter or early spring, while frosts are still common. Some varieties of Japanese quince bloom all winter and the flowers are not injured by severe freezes. Camellias also flower all winter, even in the colder part of their range, but their flowers often are injured by freezing weather.

Physiological Processes

Both soil and air temperatures affect growth of plants by altering rates and balances among physiological processes such as photosynthesis, respiration, water relations, hormone synthesis, enzymatic activity, mineral absorption, and cell division and enlargement.

Water Relations

As mentioned in Chapters 2 and 7, plants tend to dehydrate whenever the rate of transpiration exceeds the rate of water absorption. Water stress (dehydration) of plants may result from rapid transpiration, slow absorption of water, or both. The water balance of plants is influenced by effects of soil temperature on water absorption and of air temperature on transpiration.

Soil Temperature

Plants from warm climates often are influenced by decreased absorption of water at low temperatures. For example, absorption of water of loblolly pine, a southern species, was reduced more than that of eastern white pine, a northern species (Kozlowski, 1943). The decrease in absorption of water at low temperatures is caused largely by increased resistance to water flow through roots (Kramer, 1983a). Furthermore, if stems are chilled to a few degrees below 0°C, the water in the xylem elements freezes, preventing water transport to the leaves even if some of the roots are in unfrozen soil (Zimmermann, 1964). When this occurs, winter desiccation injury may result. This will be discussed later.

Air Temperature

The difference in water vapor pressure between the leaves and air causes the movement of water vapor out of leaves, largely through stomatal pores. An increase in air temperature without an increase in water content or absolute humidity of the air tends to increase the rate of transpiration because it increases the vapor pressure gradient from the evaporating surfaces to the air. However, this may be counteracted by decreasing stomatal conductance with increasing vapor pressure deficit. (See Chapter 7 and Sena Gomes *et al.*, 1987). For example, increase in temperature from 10 to 20°C almost doubles the vapor pressure gradient (Table 5.2 and Fig. 5.9).

Food Relations

Photosynthesis of woody plants occurs over a wide temperature range from near freezing to over 40°C, the specific range depending on plant age and origin and season (Kramer and Kozlowski, 1979, pp. 196–199). Net photosynthesis usually

Table 5.2
Effect of Temperature on the Saturation Vapor Pressure of Water and the Vapor Pressure Gradient (Δe) from Water Surface to Air

Temperature (°C)	Vapor pressure at saturation (kPa)	Vapor pressure of air at 70% relative humidity (kPa)	Δe (kPa)
0	0.610	0.427	0.183
10	1.227	0.858	0.369
20	2.337	1.635	0.701
30	4.243	2.970	1.273
40	7.377	5.163	2.214

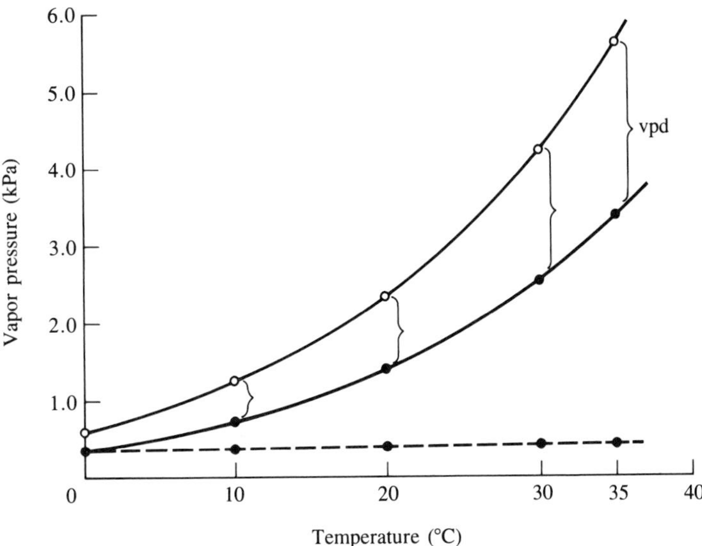

Figure 5.9 Effect of increasing temperature on the vapor pressure difference (vpd) between the leaf (○) and air (●) if the air in the leaf is saturated, leaf and air temperatures are similar, and the external air is at 60% relative humidity at each temperature. The dashed line shows water vapor pressure in the atmosphere if the absolute humidity is the same at all temperatures. (From Kramer, 1983a, by permission of Academic Press.)

increases with rising temperature, up to some critical temperature, above which it begins to decline rapidly. In most temperate zone woody plants, the rate of photosynthesis increases from near freezing until it attains a maximum at a temperature between 15 and 25°C. In tropical species the minimum temperature for photosynthesis generally is several degrees above freezing and the optimum well above 25°C (Fig. 5.10). Soil temperatures as well as air temperatures affect photosynthesis, with the rate of CO_2 uptake decreasing at low temperatures (Table 5.3). The effect of temperature often is modified by light intensity, CO_2 availability, water supply, and the preconditioning effects of environmental factors.

Woody plants often show photosynthetic acclimation to the temperature regime in which they are grown. For example, the optimum temperature for net photosynthesis of Sitka spruce was about 10°C in the spring and 20°C in September (Neilson *et al.*, 1972). The optimum temperature also varies with the altitudinal temperature gradient. For example, balsam fir seedlings showed a clinal pattern of adaptation, with the optimum temperature for photosynthesis decreasing 2.7°C for each 300 m of elevation (Fryer and Ledig, 1972). The time required for plants to acclimate to the temperature optimum of photosynthesis in response to temperature change varies with species, ontogeny, and nutritional status. Photosynthesis of California encelia acclimated to a new temperature regime within 24 hr (Mooney and Shropshire, 1967), but 7 to 30 days were required for acclimation of different altitudinal populations of snowgum in Australia (Slatyer and Ferrar, 1977).

Table 5.3
Effect of Soil Temperature and Soil Water Potential on Net Photosynthesis of Monterey Pine Seedlings[a]

Soil temperature (°C)	Soil water potential (bars)	Net CO_2 uptake (10^{-5} cm^3 cm^{-2} sec^{-1})
10.0	−0.35	2.90
15.6	−0.35	3.46
21.1	−0.35	4.06
26.7	−0.35	4.28
10.0	−2.50	1.66
15.6	−2.50	1.99
21.1	−2.50	2.23
26.7	−2.50	2.36

[a] After Babalola *et al.* (1968); from Kramer and Kozlowski (1979), by permission of Academic Press.

Preconditioning Effects of Temperature on Photosynthesis

Subjecting plants to either a low or high temperature affects the subsequent rate of photosynthesis at another temperature (Pearcy, 1977). Although subfreezing temperatures injure the photosynthetic mechanism, the damage generally is reversible

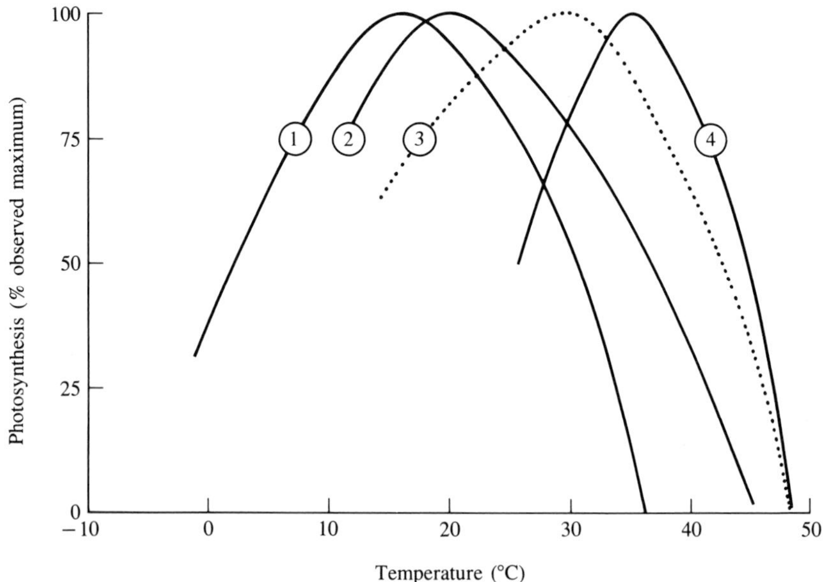

Figure 5.10 Effect of temperature on photosynthesis of temperate [*Pinus cembra* (1) and *Fagus sylvatica* (2)] and tropical [*Ficus retusa* (3) and *Acacia craspedocarpa* (4)] trees. (After Larcher, 1969; from Kramer and Kozlowski, 1979, by permission of Academic Press.)

with time at temperatures above freezing. When Douglas-fir and ponderosa pine seedlings were subjected to subfreezing temperatures for varying numbers of nights, photosynthesis on the following day was depressed by an amount that varied with the number of nights at a given subfreezing temperature and with the coldness of the night. Recovery of photosynthesis generally occurred for most plants at 3°C, but the time of recovery depended on species and the subfreezing pretreatment (Fig. 5.11). After a single night at −4°C, photosynthesis of ponderosa pine recovered completely within 6 days, whereas in Douglas-fir it recovered only partially in 60 days. After exposure to −6 or −8°C, photosynthesis of ponderosa pine recovered only partly and, after −10°C treatment , neither species showed photosynthetic recovery when returned to 3°C. When recovery from subfreezing treatment did occur, plants that were warmed slowly subsequently had higher rates of photosynthesis than plants that were warmed rapidly (Pharis *et al.*, 1970).

The inhibitory effect of high-temperature preconditioning may last for many days, with the aftereffect much greater on photosynthesis than on respiration. For example, when European silver fir and sycamore maple leaves were exposed to nearly lethal heat for 60 min, CO_2 uptake at moderate temperatures was inhibited for many days, whereas respiration rates were not appreciably altered (Larcher, 1969). This situation probably results in depletion of food reserves.

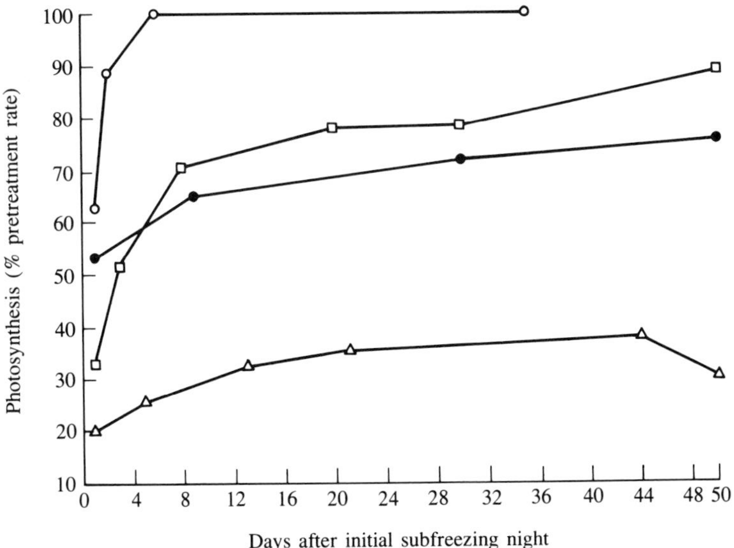

Figure 5.11 Effect of a single 16-hr night of −4(○), −6(●), −8(□), or −10°C(△) on photosynthesis of ponderosa pine seedlings. The plants were maintained and photosynthesis was measured at 3°C. (After Pharis *et al.*, 1970; from Kramer and Kozlowski, 1979, by permission of Academic Press.)

Mechanisms of Photosynthetic Inhibition

The ways in which temperature influences photosynthesis are complex and may involve direct effects on synthesis and activity of carboxylating enzymes as well as indirect effects on stomatal aperture, dark respiration, and increased photoinhibition. The mechanisms of photosynthetic inhibition by unusually low or high temperatures may differ greatly.

Low-temperature inhibition Low air temperatures in the chilling range (see the next section) reduce photosynthesis by both stomatal (Öquist, 1983) and nonstomatal inhibition. The latter includes effects on the photosynthetic machinery, such as decrease in carboxylation and electron transport and destruction of chlorophyll. The latter is especially important when plants are exposed simultaneously to low temperature and bright light. Nevertheless, much of the reduction in photosynthesis at very low temperatures involves adverse effects on the photosynthetic mechanism. After photosynthesis of Scotch pine was reduced by frost, recovery on return to high temperature occurred some time after the rate of transpiration increased, emphasizing a direct effect of low temperature on the photosynthetic process (Zelawski and Kucharska, 1969). Much of the reduction in photosynthesis of conifers during the winter is associated with changes in the structure and function of the chloroplast thylakoids (Berry and Björkman, 1980; Larcher and Bauer, 1981). Following exposure of olive leaves to 5°C, the photosynthetic capacity recovered fully if the leaves had been chilled in a high light intensity for 8 hr or less, but did not recover fully from longer chilling, which induced loss of chlorophyll (Bongi and Long, 1987).

When roots are exposed to temperatures near freezing or below and leaves are at temperatures favorable for photosynthesis, absorption of CO_2 is reduced by both stomatal and nonstomatal inhibition. Lawrence and Oechel (1983) attributed the reduction in CO_2 absorption by seedlings of plants of subarctic coniferous forests in cold soil to decreased stomatal conductance, associated with leaf dehydration caused by reduced absorption of water (as discussed earlier), and decreased mesophyll conductance for CO_2. DeLucia (1986) showed that during the first several hours of root chilling of Engelmann spruce seedlings, the decrease in photosynthesis was caused by a decrease in stomatal conductance, but in the longer term it was caused largely by reduced carboxylation efficiency.

High-temperature inhibition A decrease in net photosynthesis following exposure to high temperature may involve effects on respiration, stomatal closure, or injury to the photosynthetic mechanism. Respiration usually continues to increase above a critical high temperature at which photosynthesis begins to decrease, thus reducing net photosynthesis (Figs. 2.10 and 5.12). Although stomatal conductance may be reduced by high temperature, it occurs chiefly because increasing leaf temperature increases the vapor pressure deficit from leaf to air. In several studies the photosynthetic machinery appeared to be injured by increasing temperature while stomatal conductance remained high.

Reduction in rates of photosynthesis at high temperatures usually is associated with changes in properties of thylakoid membranes, inactivation of enzymes of

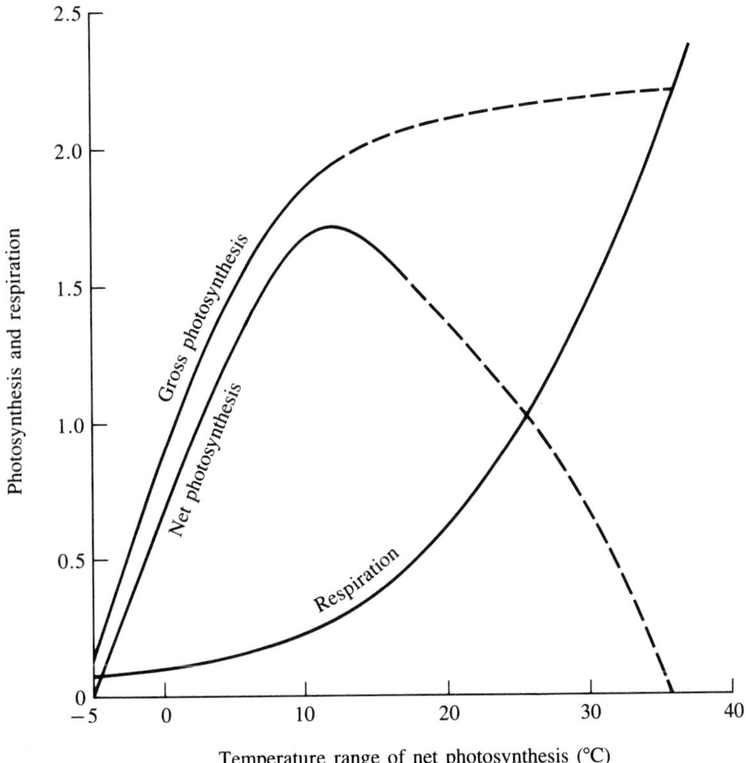

Figure 5.12 Effects of temperature on photosynthesis, respiration, and net or apparent photosynthesis of Swiss stone pine seedlings. Solid lines are from actual measurements; broken lines are estimated. (After Tranquillini, 1955; from Kramer and Kozlowski, 1979, by permission of Academic Press.)

photosynthetic carbon metabolism, and a decrease in soluble leaf proteins as a result of protein denaturation and precipitation (Berry and Björkman, 1980; Björkman *et al.*, 1980). Injury to the photosynthetic mechanism by heat can be repaired by a lowered temperature if cell membranes have not been severely injured. Often the extent of recovery depends on the severity of the heat stress and time for recovery. When English ivy leaves were exposed for 30 min to 44°C they lost about half their capacity for photosynthesis, which they subsequently recovered on exposure to 20°C for 7 days. Exposure to 48°C resulted in loss of about three-fourths of the photosynthetic capacity and recovery at 20°C required 8 weeks (Berry and Björkman, 1980).

Effects of Low Temperatures

Low air temperatures injure plants by both chilling and freezing. Many tropical and subtropical plants suffer chilling injury at temperatures several degrees above freezing. Such injury is distinct from freezing injury, which occurs at temperatures low enough to form ice. Low soil temperatures often injure plants by reducing water absorption and causing shoot dehydration.

Chilling Injury

Vegetative and reproductive tissues of many tropical and subtropical plants, and a few temperate zone plants, exhibit injury after exposure to temperatures below approximately 15°C but above freezing (McWilliam, 1983). For example, necrotic lesions developed on coffee leaves after a 12-hr night exposure to 5°C (Bauer *et al.*, 1985). Other responses to chilling include reduced photosynthesis, increased susceptibility to decay organisms, ultrastructural changes, reduced growth, and ultimate death of plants (Wise *et al.*, 1981; Wang, 1982). Chilling stress and light stress often interact to injure plants. Chilling injury generally occurs under clear night skies, especially if followed by bright days. Injury to the photosynthetic apparatus by chilling is increased by high light intensity (Powles *et al.*, 1983)(see also Chapter 4).

Although little information is available for tree seeds, germinating seeds and young seedlings generally are especially susceptible to chilling injury. The first indication of sensitivity to chilling is evident as soon as imbibition of water by seeds begins. Some species also show increased sensitivity 1 to 4 days later. When cotton seeds were chilled at the beginning of imbibition of water, the tips of radicles aborted. However, if seeds were exposed to chilling 24 hr after the start of imbibition, the root cortex was injured (Christiansen, 1964). Injury to soybean seeds caused by imbibition of water at low temperatures was studied by Leopold (1980) and Chabot and Leopold (1982). Symptoms of chilling injury in seedlings may include wilting, desiccation and necrosis of leaves, decreased growth, and death of plants (Wolk and Herner, 1982).

Chilling Injury of Fruits

Injury of fruits by chilling is of particular interest because many fruits are shipped, stored, and displayed at low temperatures to extend their life.

There is considerable variation among fruits with respect to sensitivity to chilling and to symptom expression. Symptoms include surface lesions (Fig. 5.13), surface and internal discoloration, breakdown of tissues, failure to ripen, accelerated senescence, increased susceptibility to decay organisms, and shortened storage and shelf life (Kramer and Kozlowski, 1979, p. 691). Bananas are very sensitive. With mild chilling, green bananas may not show any symptoms, but when they ripen the color of the peel often varies from dull yellow to grayish-yellow or smoky gray (Fig. 5.14). More severe symptoms include a dull yellow shine, browning of the skin, failure to ripen, loss of flavor, and increased susceptibility to mechanical injury. Chilling in the field often causes brown epidermal streaks and softening of the pulp during ripening as well as predisposition to storage rots and blemishes (Tai, 1977).

Among citrus fruits, grapefruits and limes are most sensitive. Grapefruits develop

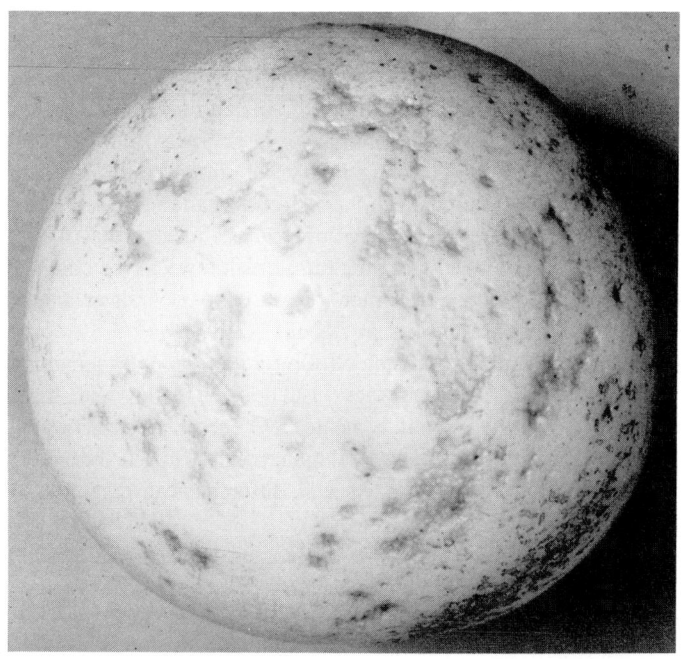

Figure 5.13 Surface pitting of grapefruit following chilling. (U.S. Department of Agriculture photo.)

Figure 5.14 Discoloration of peel and flesh of banana following chilling. Chilled fruit (top); unchilled fruit (bottom). (U.S. Department of Agriculture photo.)

pits or depressions of the rind when stored at 4 to 5°C (Fig. 5.13). However, at 0 to 1°C the rind shows superficial brown staining. Chilled oranges develop small sunken spots in the rind and uniform browning over large areas. Lemons generally do not show external symptoms of chilling injury but, like oranges and grapefruits, exhibit softening of the rind and flesh, a condition called "watery breakdown." Apples develop "soft-scald," a brown discoloration as a result of death of patches of epidermis and cortex, and also low-temperature breakdown in the cortex. Peaches, plums, and nectarines develop a dry, mealy texture, loss of flavor and color, and browning of the flesh. Often they fail to ripen.

The critical low temperature at which chilling injury becomes apparent in tropical fruits is near 10 to 12°C, but species susceptibility varies with the region of origin. The lower temperature limit is near 12°C for many varieties of banana; between 8 and 12°C for citrus, avocado, and mango; and 0 to 4°C for temperate zone fruits such as apples. There also are varietal differences in response to chilling (Simmonds, 1982).

For any fruit at any temperature down to freezing there appears to be a critical exposure time for induction of irreversible injury. This varies from a few hours to several weeks depending on the fruit and the temperature (Couey, 1982). Fruits that are chilled for short periods may appear to be uninjured when removed from low-temperature regimes, but injury symptoms often appear in a few days at higher temperatures.

The effects of chilling are cumulative and low temperatures in the field before harvest and during transport can accentuate the effects of chilling in storage. Grapefruits are injured most by chilling temperatures in the autumn and spring but are resistant to chilling injury in the winter (Kawada *et al.*, 1979). The field temperatures during the few weeks before harvesting are particularly important. For example, pears that mature in a cool preharvest temperature ripen better when stored at 0°C than pears that mature at higher temperatures. However, fruits on the tree may be injured if the temperatures drop to the chilling range shortly before harvest (Mellenthin and Wang, 1976).

Susceptibility to chilling injury also is influenced by maturity of fruits. Bananas are most sensitive at the stage at which they normally are harvested. Immature, green grapefruits are very susceptible to chilling and become less susceptible as the green color develops. Sensitivity of avocados is relatively high in the respiratory preclimacteric stage, increases to a maximum at the climacteric respiratory peak before the fruit is ripe, and decreases rapidly as the fruit ripens (Kosiyachinda and Young, 1976). Development of low-temperature breakdown of apples occurred early after harvesting, whereas browning of the core appeared only during advanced senescence (Fidler *et al.*, 1973).

Mechanism of Chilling Injury

Some investigators claim that chilling temperatures induce a primary physical phase transition of the lipids in cell and organelle membranes from a liquid crystalline to a solid gel state (see reviews by Levitt, 1980a; Wang, 1982). This change presumably induces contraction that causes cracks or channels in membranes, thereby increasing their permeability. The alteration in membrane structure may or may not be followed by metabolic changes (Fig. 5.15), accumulation of toxic products, a range of visible symptoms, and ultimately death of plants. Considerable recent evidence suggests, however, that phase transition of membranes is not as important in chilling injury as originally proposed. For example, O'Neill and Leopold (1982) did not find phase transitions in membranes of chilling-sensitive soybeans in the temperature range in which chilling injury occurs. Markhart *et al.* (1980) reported that changes in membrane fatty acid unsaturation occurred during acclimation of soybean plants to chilling, but the degree of unsaturation was not a good indicator of differences in chilling tolerance among species. Kenrick and Bishop (1986) found wide variations in chilling sensitivity of 27 species of herbaceous and woody plants. However, the

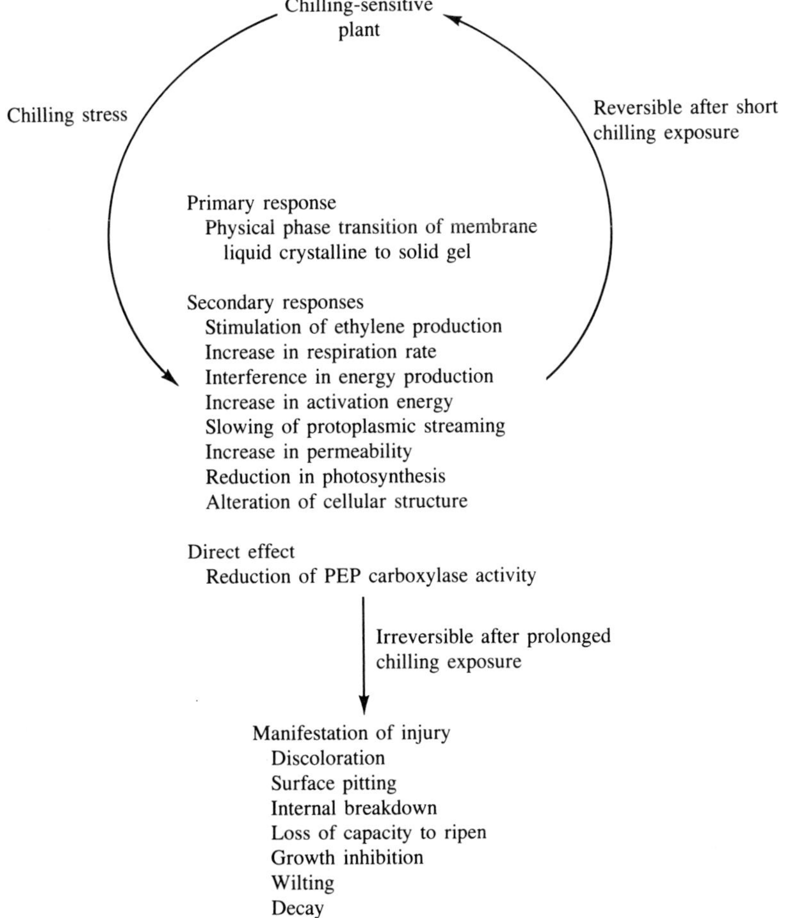

Figure 5.15 Responses of sensitive plants to chilling. (From Wang, 1982.)

content of high-melting-point fatty acids was not related to chilling sensitivity. Wang (1982) and Steponkus (1984) reviewed in more detail the sequential events leading to chilling injury, and there is a useful discussion by Graham and Patterson (1982) of chilling injury at the subcellular level.

Freezing Injury

When subfreezing temperatures occur, most injury is caused by ice formation in plant tissues. Ice usually forms first in the large-diameter vessels and rapidly spreads both inter- and intracellularly throughout the shoot system. The pattern of freezing varies with the anatomy of the plant, its degree of hardening, water content, and rate of freezing. In some species, freezing is postponed by extreme supercooling (undercooling). A decreasing water content and increasing concentration of osmotically active compounds in the cell sap not only lower the freezing point but also decrease the threshold subcooling temperature and prolong supercooling. Hence, some tissues do not freeze for a long time even when exposed to very low temperatures.

Deep supercooling has been reported in more than 240 species of woody plants. Generally, species that possess a capacity for deep supercooling are distributed at lower latitudes, where the air temperature is unlikely to fall below the homogeneous nucleation point of supercooled water within cells. Species that survive by tolerating extracellular freezing inhabit colder regions than do species that experience deep supercooling (Quamme, 1985).

The capacity for supercooling varies appreciably in different organs and tissues of the same plant. For example, hardy xylem ray parenchyma and axial parenchyma cells, floral tissues, phloem, and buds of some woody plants and some seeds supercool and do not exhibit typical extracellular freezing. Although the bark of several species freezes at –10°C, xylem parenchyma cells supercool to much lower temperatures. Supercooling of xylem ray parenchyma cells of deciduous trees down to temperatures as low as –38 to –47°C has been reported (Quamme, 1985).

Freezing injury can be caused directly by intracellular freezing of the cell contents or indirectly by dehydration of tissues resulting from extracellular freezing. Plants are generally killed if ice crystals form within cells, but formation of ice crystals in the intercellular spaces between cells usually is not fatal. When frost-hardened plants are cooled slowly, ice initially forms in the intercellular spaces and, as the temperature decreases, water moves out of the cells to these ice nuclei. Hence, the concentration of the cell sap is increased and its freezing point is lowered. If cold-hardened plants are cooled rapidly they may be injured, possibly because the water does not move out of the cells fast enough and intracellular freezing suddenly occurs. Killing by intracellular freezing often has been attributed to mechanical disruption of cell membranes. Freezing of needles, buds, and shoots of Norway spruce injured cellular membranes, causing increased efflux of cellular electrolytes (Pukacki and Pukacki, 1987). However, most freezing injury probably is caused by dehydration, and freezing tolerance depends chiefly on the capacity of protoplasm to survive such dehydration (Levitt, 1980a; Sakai and Larcher, 1987).

Symptoms of Freezing Injury

Freezing damage is expressed as discoloration, bleaching, shrinkage, and dieback of tissues, rupturing associated with the mechanical effects of frost, malformation resulting from injury to meristematic and incompletely differentiated tissues, and delay in development (e.g., late sprouting) (Sakai and Larcher, 1987).

Injury to shoots Shoots often are injured by temperatures at or below freezing (Fig. 5.16). The amount of tissue killed varies with the severity and duration of the frost as well as the degree of coldhardiness of the shoots at the time of the frost.

Frosthardiness of shoots disappears rapidly in the spring before the buds open. Spring frosts that occur shortly before the buds open usually injure only the buds because they are in a more advanced stage of dehardening than are fully expanded shoots. Spring frosts that occur after seasonal bud opening generally injure both the leaves and succulent shoots (Kozlowski, 1971a). Frost damage to needles of white spruce consisted of collapse of mesophyll cells and coagulation of cytoplasm (Fig. 5.17). When buds or young leaves are killed by spring freezes, a new crop of leaves usually emerges from previously dormant buds.

In Scotland, Sitka spruce plantations commonly are injured in the spring, when temperatures of the expanding shoots drop to between –3 and –5°C (Cannell, 1984). Some shoots are injured more than others because of variations among buds in the rate of loss of coldhardiness. For example, lateral buds that expand before terminal buds do are injured most severely (Cannell and Sheppard, 1982).

Sudden temperature drops in the autumn may injure shoots that have stopped growing but have not had sufficient time to develop coldhardiness before frosts occur. Early autumn frosts injure only succulent shoots, hence injury most often occurs to late-season growth flushes (e.g., lammas shoots or late-season growth flushes of recurrently flushing species) (Kozlowski, 1971a). In Scotland, autumn frost damage to Sitka spruce after annual shoot elongation has ceased is characterized by browning and death of needles (Redfern and Cannell, 1982). Ornamental plants subjected to long photoperiods from artificial lights also are subject to autumn frost injury (Kramer, 1937; Sakai and Larcher, 1987).

Sunscald On cold winter days, the sun shining on one side of a tree stem may increase its temperature by 20 to 30°C over that on the shaded side, resulting in thawing. After sundown, or after the sun disappears behind a cloud, the temperature of the thawed tissue may decrease so rapidly that freezing and death of cambial tissues follow. Sometimes drastic fluctuations in stem temperatures can be avoided by painting stems with white paint or by wrapping them in straw, matting, white paper, or tape (Martsolf *et al.*, 1975; Harris, 1983).

Winter sunscald or sunscorch lesions sometimes form on thin-barked trees when freezing and thawing alternate during the winter and early spring. Such lesions are characterized by dead patches of bark, which may peel, or by sunken cankers,

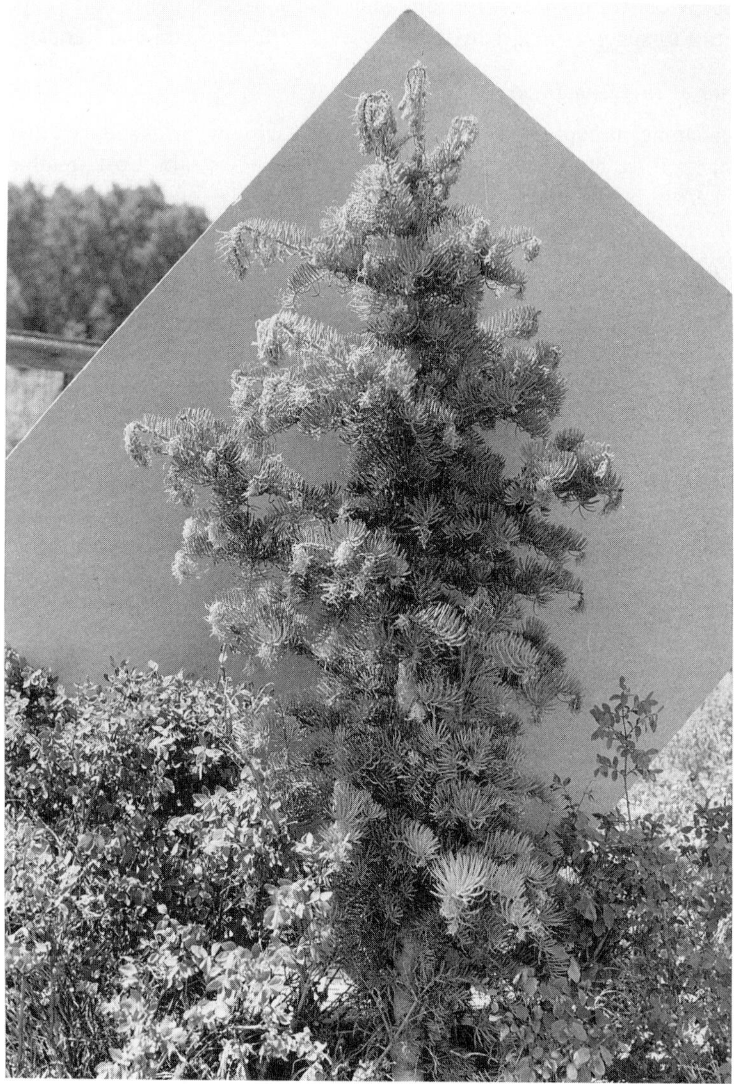

Figure 5.16 Frost damage to white fir foliage. (U.S. Forest Service photo.)

which become sites for invasion by insects and fungi. These lesions also injure the phloem, thereby impeding translocation of organic solutes.

Frost cracks During periods of freezing temperatures both stems and branches of some trees may develop vertical frost cracks (Fig. 5.18). Such cracks tend to close and reopen as the air temperature increases or decreases. In the spring, vertically oriented protrusions of callus tissue, called "frost ribs," often develop

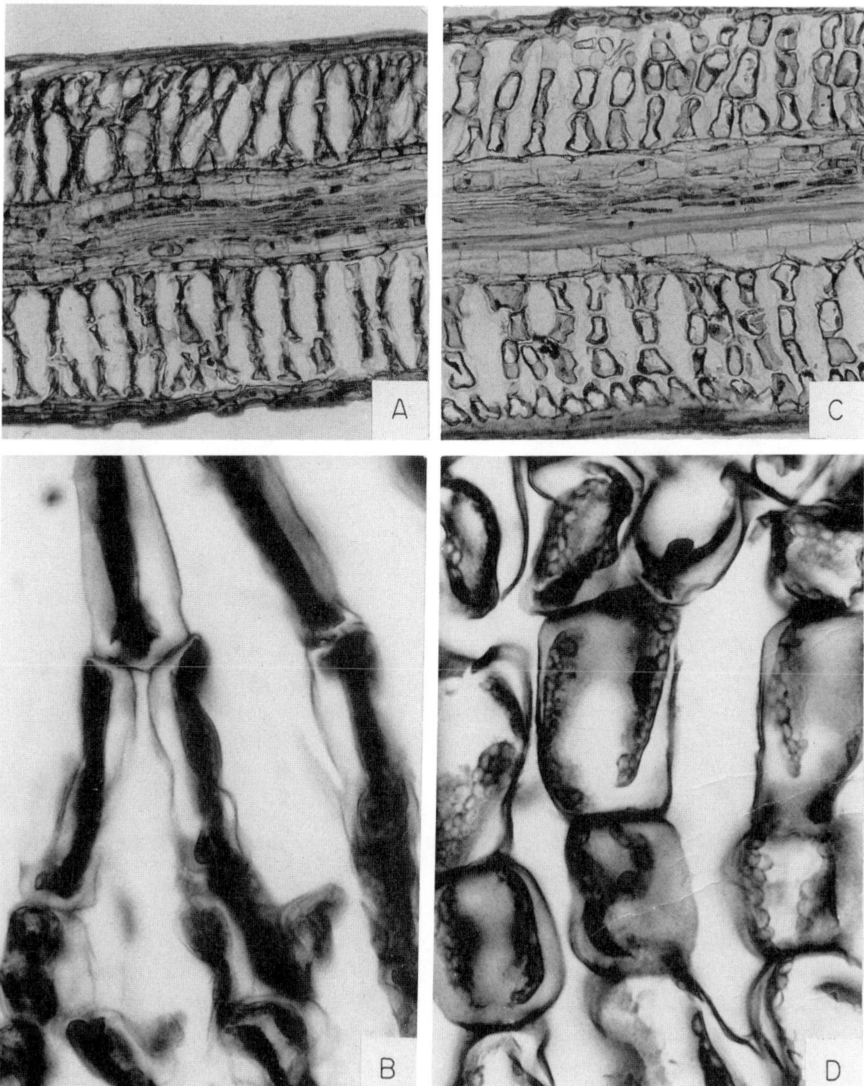

Figure 5.17 Frost damage to white spruce needles. (A) Longitudinal section through a frost-damaged needle; mesophyll cells are collapsed and cytoplasm in coagulated; ×80, (B) Longitudinal section through frost-damaged needles showing a few damaged mesophyll cells; ×380. (C) Longitudinal section through a normal needle; ×100. (D) Longitudinal section through a normal needle showing arrangement of a few mesophyll cells; note cell turgidity; ×440. (After Glerum and Farrar, 1965; from Kramer and Kozlowski, 1979, by permission of Academic Press.)

Figure 5.18 Frost cracks (arrows) on stems of 13-year-old Golden Delicious apple trees on East Malling VII rootstock. *Left,* Extreme splitting from the main scaffold branches to the stock–scion union. *Right,* Tree that has recovered from previous injury. (After Simons, 1970; from Kramer and Kozlowski, 1979, by permission of Academic Press.)

along the edges of frost cracks. During a mild winter the callus tissue may grow together and imbed the frost crack.

Frost cracks form because wood tends to shrink more tangentially than radially during freezing. They are caused primarily by freezing out of water from the wood cell walls into the lumens of wood cells. The higher vapor pressure of some of the bound water in the cell walls causes diffusion of cell wall water to ice in the cell lumens. Other factors contributing to formation of frost cracks include the faster cooling of the outer wood than the inner wood, expansion of freezing water in cell lumens, and formation of lenses of ice in wood (Kubler, 1983, 1987, 1988).

Cracking of stems often occurs instantaneously and is associated with a loud noise resembling a gunshot; the sound sometimes is heard in forests on very cold nights. Frost cracks may form in only a few trees of a forest stand, even though all the trees are exposed to similar freezing regimes. This probably reflects individual differences in moisture contents of trees. Stems of trees with very wet wood tend to crack more than dry stems at freezing temperatures. Stem wounds also trigger frost cracks (Butin and Shigo, 1981).

Frost rings Frost during the growing season often injures the cambium, causing abnormal "frost rings" to form. These consist of an inner part made up of cells killed by the frost (usually xylem mother cells and differentiating cambial derivatives) and an outer part of abnormal xylem cells produced after the frost (Kramer and Kozlowski, 1979, p. 652). Frost rings of conifers are characterized by underlignified, abnormal tracheids, collapsed cells, and traumatic parenchyma cells. The xylem rays usually are laterally displaced and expanded (Fig. 5.19).

Frost rings occur more frequently in trees with thin bark than in those with thick bark, probably because of the insulating effect of the thick bark. On the same tree twigs are injured more than large branches, and the latter are injured more than the main stem. Formation of frost rings was discussed by Glerum and Farrar (1966).

Root injury and mortality Freezing tolerance of roots is particularly important for survival of container-grown plants. Although roots are much less cold-hardy than stems, they usually suffer less injury in winter because the soil and snow cover protect them from exposure to very low temperatures. Nevertheless, when the soil freezes many of the small physiologically active roots often are killed and freezing injury to roots of plants in containers can be severe. During the winter, roots freeze at temperatures in the general range of –5 to –20°C, with the critical temperature varying appreciably among species (Table 5.4). In Sweden, roots of seedlings of Scotch pine, Norway spruce, and lodgepole pine that were grown in containers showed a progressive increase in coldhardiness toward midwinter and loss of hardiness as spring approached. Roots of Norway spruce were less sensitive to low temperatures than roots of Scotch pine or lodgepole pine. Northern provenances of Scotch pine and Norway spruce had the hardiest roots during the autumn. Although the root systems survived temperatures down to –15°C to –25°C, during late autumn and midwinter the capacity to produce new roots declined at considerably higher temperatures (Lindstrom and Nystrom, 1987). The coldhardiness of roots

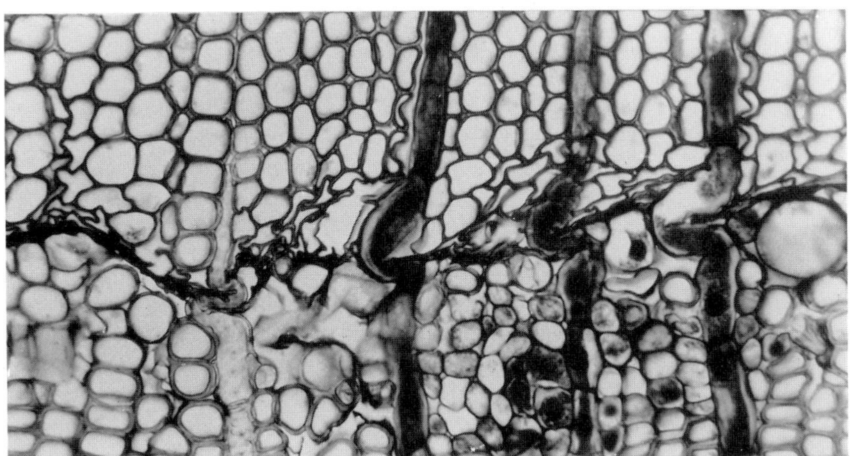

Figure 5.19 Frost ring in a conifer. The effects of frost are indicated by collapsed tracheids, displacement of rays, and excessive production of parenchyma tissue. (U.S. Forest Service photo.)

Table 5.4
Frost Tolerance of Roots of Woody Plants[a]

Species	LT_{50} (°C)
Magnolia × soulangeana	− 5.0
Magnolia stellata	− 5.0
Cornus florida	− 6.7
Daphne cneorum	− 6.7
Ilex crenata	− 6.7
Ilex opaca	− 6.7
Pyracantha coccinea	− 7.8
Cryptomeria japonica	− 8.9
Cotoneaster horizontalis	− 9.4
Viburnum carlesii	− 9.4
Cytisus × praecox	− 9.4
Buxus sempervirens	− 9.4
Ilex glabra	− 9.4
Euonymus fortunei	− 9.4
Hedera helix	− 9.4
Pachysandra terminalis	− 9.4
Vinca minor	− 9.4
Pieris japonica	− 9.4
Acer palmatum cv. Atropurpureum	−10.0
Cotoneaster adpressa praecox	−12.2
Taxus × media cv. Nigra	−12.2
Rhododendron cv. Gibraltar	−12.2
Rhododendron cv. Hinodegiri	−12.2
Pieris japonica	−12.2
Leucothoe fontanesiana	−15.0
Pieris floribunda	−15.0
Euonymus fortunei cv. Colorata	−15.0
Juniperus horizontalis	−17.8
Rhododendron carolinianum	−17.8
Rhododendron catawbiense	−17.8
Rhododendron P.J.M. hybrids	−23.3
Potentilla fruticosa	−23.3
Picea glauca	−23.3
Picea omorika	−23.3

[a] From Havis (1976).

also varies with rootstocks, and appears to be greater in sandy than in clay soils. Injury also appears to be greater in dry than in wet soils.

It has been questioned whether the relatively low tolerance of roots to freezing is an inherent characteristic or results from failure to develop freezing tolerance in the warm soil. By storing plants at −7°C the cold tolerance of roots of seedlings of honeysuckle, cotoneaster, and euonymus was greatly increased. However, the tolerance of privet roots to cold was not increased (Pellett, 1971), and Havis (1976) could not induce cold-hardening in roots of magnolia, holly, dogwood, or cotoneaster. Sakai and Larcher (1987) concluded that the low resistance of roots to freezing is an inherent characteristic, but the differences in timing of growth of roots and shoots, and the lesser impact of frost action in the soil, also are important.

Injury to roots by "frost heaving" often causes loss of trees in nurseries and young plantations (Kramer and Kozlowski, 1979, p. 656). Such injury may occur when roots are frozen in a mass of ice and subsequently additional layers of ice form below. As lifting occurs, some of the deeper, fine roots may be broken. The larger, deeper roots are less prone to injury because the frozen surface soil layer slides upward along the stems and down again on thawing. Smaller seedlings tend to remain lifted on thawing because their lower roots are more likely to be on a frozen layer. During subsequent freezing the partially lifted plants tend to be pushed out even more. If the roots are broken or the shoots and exposed roots are dehydrated, injury or death usually follows (Kramer and Kozlowski, 1979, p. 656).

Coldhardiness

Trees and shrubs native to warm regions usually cannot be moved to very cold regions because they do not develop enough hardiness to cold, do not acclimate fast enough to survive early cold weather, deharden too quickly, or are killed by subfreezing temperatures (Kramer and Kozlowski, 1979). For example, freezing tolerance of twigs of species collected in midwinter from various parts of North America varied greatly (Table 5.5). Northern species such as trembling aspen, balsam poplar, paper birch, and tamarack resisted freezing to −80°C, and some survived temperatures down to −196°C after prefreezing to −15°C. Rocky and western mountain conifers such as ponderosa pine, western white pine, lodgepole pine, Jeffrey pine,

Table 5.5
Variations in Freezing Tolerance of North American Tree Species and Minimum Temperatures at Northern Limits of Natural Ranges or Artificial Plantings[a]

Relative hardiness classification	Representative species	Average minimum temperatures at northern limits of growth (°C)		Observed freezing resistance (°C)
		Natural range	Artificial plantings	
Tender evergreen species	*Quercus virginiana*	−3.9 to −6.7	−9 to −12	−7 to −8
Hardy evergreen species	*Magnolia grandiflora*	−9 to −12	−18 to −20	−15 to −20
Hardy deciduous species	*Liquidambar styraciflua*	−18 to −20	−26 to −29	−25 to −30
Very hardy deciduous species	*Ulmus americana*	−37 to −46	−40 to −43	−40 to −50
Extremely hardy deciduous species	*Betula papyrifera*	Below −46	Below −46	Below −80
	Populus deltoides	−32 to −34	−37 to −45	Below −80
	Salix nigra	−32 to −34	−37 to −45	Below −80

[a] After Sakai and Weiser (1973); from Kramer and Kozlowski (1979), by permission of Academic Press.

blue spruce, Engelmann spruce, alpine fir, white fir, and western larch survived freezing at temperatures between –60 and–80°C. However, southern species such as slash pine, longleaf pine, live oak, and southern magnolia survived temperatures only down to –15°C (Sakai and Weiser, 1973).

Much interest has been shown in within-species differences in freezing tolerance. Climatic races have become adapted to particular environments and those of mild climates may not be sufficiently cold-hardy when moved northward (Kramer and Kozlowski, 1979). For example, sugar maple, sweet gum, and ponderosa pine from northern provenances are much more tolerant of low temperature than those from southern sources (Kriebel, 1957; Squillace and Silen, 1962; Williams and McMillan, 1971). Flint (1972) grew red oak seedlings at Weston, Massachusetts, from acorns collected over a wide geographic range and found large differences in coldhardiness. Acclimation of northern provenances to cold often is related to their phenology, with seed sources that set buds early exhibiting less injury from early frosts. For example, southern sources of Douglas-fir set buds later and were injured more by early frosts than were northern sources when both were grown at Corvallis, Oregon (Campbell and Sorenson, 1973). For planting in cold regions it often is useful to select planting stock for late bud opening. Hence, late-flushing provenances of Norway spruce often are planted at very cold sites in England and Scandinavia, and the late-flushing black spruce may be planted in preference to white spruce at unusually cold sites in Canada (Dormling, 1982; O'Reilly and Parker, 1982). However, selection of Sitka spruce clones for late bud opening was not very useful in upland England. Although bud opening could be delayed 7 days by selecting late-flushing clones, this only altered the probability of damage from once every 3 to once every 4 years, because of the high incidence of late spring frosts. To avoid frost damage altogether, bud opening would have to be delayed by at least 2 weeks (Cannell *et al.*, 1985).

Provenance variations in coldhardiness have not been demonstrated for certain species. For example, cottonwood and black willow from various parts of North America survived freezing down to at least –50°C, regardless of their native habitats (Sakai and Weiser, 1973). Northern clones of red-osier dogwood hardened earlier than southern clones. Nevertheless, there were no real differences in their capacity to develop hardiness, and all of them withstood temperatures of –90°C by early December (Smithberg and Weiser, 1968).

Hardiness of Different Organs and Tissues

As mentioned in preceding sections, coldhardiness of different organs and tissues on the same plant often varies greatly, with reproductive tissues being more sensitive than vegetative tissues to cold. Roots are especially sensitive to cold and do not harden to temperatures more than a few degrees below 0°C. Shoot tips and cambium cells also are sensitive. Expanded leaves of deciduous trees never develop substantial hardiness to cold. In rhododendron, the leaf midrib and petiole as well as the stem cambium, phloem, and cortex were much less cold-hardy than the stem xylem and interveinal leaf tissues (Holt and Pellett, 1981). Terminal buds usually develop less coldhardiness than lateral buds. However, vegetative buds lose coldhardiness as they begin to open in the spring, and unopened buds may also be killed by large and sudden temperature drops. Complete defoliation of trees by killing of young leaves by late spring frosts is well known.

Seasonal development of cold-hardening and dehardening often varies in different tissues and organs. For example, in the spring, loss of coldhardiness begins earlier in the roots than in the shoots, and earlier in flower buds than in vegetative buds (Sakai and Larcher, 1987).

Severe winters often eliminate potential fruit and seed crops because of the relatively high susceptibility of flower buds and young conelets to frost injury. Flower primordia in winter buds may be 10 to 20°C more sensitive than vegetative buds. In general, the more advanced the bloom the greater is the injury from a given low temperature. Apricots bloom before peaches, which in turn bloom before apples, and spring frost injury is therefore likely to be greatest in apricot and least in apple. Considerable variability among cultivars in susceptibility to spring frosts also has been demonstrated. For example, the peach cultivars Elberta, Sunhigh, Redskin, and Golden Jubilee are injured more than Veteran, Vedette, Early Red Free, or Sunrise (Teskey and Shoemaker, 1978). Frost injury to flowers or young fruits often results in mature fruits that are abnormal and of low quality (Simons and Lott, 1963).

In some years freezing in the spring is a major cause of mortality and abortion of conelets of conifers, although often only part of the conelet crop is killed by freezing. Sensitivity to frost varies with the stage of conelet development, and injury usually is greater in ovulate than in staminate conelets. Ovulate conelets at the stage of maximum pollen receptivity are very sensitive to frost damage, but generally are more resistant after pollination. In addition to killing first-year conelets, freezing temperatures may reduce the number of sound seeds in the cones that survive. Some ovules are killed or pollen is rendered sterile, prior to ovule abortion, even if the cones are not destroyed.

Coldhardiness in Relation to Dormancy

There has been considerable debate about the importance of a state of deep dormancy in woody plants for development of coldhardiness. Most woody plants of the temperate zone become dormant before they develop coldhardiness. For example, eastern white pine trees in eastern Canada were dormant by September and the dormant condition could only be broken by prolonged chilling. During chilling, coldhardiness increased to a maximum by early December, by which time the chilling requirement also was satisfied (Glerum, 1976). In black spruce seedlings, development of coldhardiness was correlated with a low mitotic index of the embryonic shoot (Colombo *et al.*, 1989). Reduction in coldhardiness of some species (e.g., box elder) brought indoors during the winter occurs only after they are no longer dormant (Irving and Lanphier, 1967). However, some hardy woody plants develop coldhardiness independently of bud dormancy. For example, cold tolerance of red-osier dogwood can be maintained long after dormancy has been broken. Other species become dormant but never acquire coldhardiness.

Development of Coldhardiness

Coldhardiness of northern trees appears to develop in two stages. The first stage occurs at 10 to 20°C in the autumn when reserve carbohydrates and lipids accumulate. These compounds appear to be substrates and energy sources for metabolic changes occurring in the next stage. The second stage, which is promoted by freezing temperatures, is characterized by synthesis of proteins and membrane lipids as well as by structural changes. Only healthy plants can become fully hardened to very low temperatures. Diseased plants, mineral-deficient plants, and those with low amounts of reserve carbohydrates do not become fully cold-hardened (Sakai and Larcher, 1987). Six species of undefoliated broad-leaved evergreens survived the winter, whereas defoliation in the autumn, which prevented accumulation of carbohydrates, resulted in winter killing of most plants (Kramer and Wetmore, 1943).

There has been controversy about whether two or three distinct steps are necessary for full development of coldhardiness in northern trees. Tumanov (1967) suggested that three sequential conditions were required, including (1) a temperature slightly above 0°C, (2) a temperature slightly below 0°C (–3°C to –5°C), and (3) slow cooling for several weeks at –10°C to –60°C. However, subzero temperatures are not always prerequisites for inducing coldhardiness to –70°C or below, especially in very hardy species. Sakai and Larcher (1987) reviewed the evidence for the view that deep coldhardiness of most northern species of woody plants develops in two stages.

Sakai and Larcher (1987) described patterns of seasonal development of coldhardiness for species of warm climates and those of cold climates as follows: Most woody plants growing in mild climates acquire coldhardiness as a direct consequence of progressively decreasing temperatures. In these species coldhardiness, which stops at the first stage, can be achieved within a day or two and may be lost just as quickly, without appreciable reduction in metabolism or development. Species in this group, such as the recurrently flushing tropical pines, olive, and species of *Eucalyptus*, are relatively insensitive to photoperiod with respect to hardening (Paton, 1982; Sakai and Larcher, 1987).

Woody plants of the cool temperate zone exhibit a decrease in growth and metabolism and gradually develop coldhardiness, long before the onset of low temperatures. Plants of such species only slowly resume the capacity for metabolic activity in the late winter and early spring. Species in which coldhardiness is associated with dormancy acquire some degree of coldhardiness even when grown in a greenhouse. A state of dormancy also is important for retention of coldhardiness during warm spells in midwinter. In many species of this group, coldhardiness appears to be induced by short days. For example, considerable injury to nonhardened pine seedlings occurred at –8°C, whereas in trees previously exposed to short days for 3 weeks at 2°C, comparable injury occurred only when the temperature was decreased to −19°C (Bervaes, *et al.*, 1978). In some species the development of coldhardiness can be prevented or delayed by long days. This is emphasized by extensive injury to plants growing near street lamps (Kramer, 1937). In Sapporo, Japan, poplar trees normally defoliate in late October, but those near street lights may retain their leaves to mid-November and exhibit more winter injury than trees a short distance away (Sakai and Larcher, 1987).

In some species of cold climates a combination of short days and low temperatures induces coldhardiness, whereas short days alone do not, an example being cloudberry (Kaurin *et al.*, 1982). Cessation of shoot growth and coldhardiness of Scotch pine were induced by exposure to short days and low temperatures. Short days and warm temperatures resulted in intermediate hardiness of shoots and dormancy, but long days plus low temperatures induced more coldhardiness than short days plus warm temperatures (Smit-Spinks *et al.*, 1985).

Biochemical Changes Associated with Coldhardiness

During development of coldhardiness, changes occur in enzyme activity and in concentrations of sugars, proteins, amino acids, nucleic acids, and lipids. Tissue hydration is reduced and permeability of membranes increases. Some of these changes, alone or in combination, have been causally related to acclimation to frost (Sakai and Larcher, 1987).

Sugars increase in the autumn as plants become cold-hardy (Fege and Brown, 1984). The increase largely reflects starch hydrolysis, but sugars may also accumulate directly from photosynthesis. Decreases in growth and respiration also reduce the use of sugars, resulting in their accumulation. In the spring the sugar content usually decreases as plants deharden and respiration increases (Siminovitch, 1981). The specific sugars that accumulate during hardening vary with the carbohydrate metabolism of various species, but an increase in sucrose is most common.

Sugars might increase coldhardiness in any of three ways (Sakai and Larcher, 1987).

1. Accumulation of sugars in vacuoles decreases the amount of extracellular ice that forms and may thereby decrease freeze-induced dehydration.
2. When sugars are metabolized in the protoplasm they may cause changes that can increase tolerance of freeze-induced dehydration.
3. Sugars may dilute compounds such as electrolytes that are potentially toxic to membranes in high concentrations.

Some investigators reported a strong correlation between coldhardiness and the soluble protein content of plant tissues (Li and Weiser, 1967; Brown and Bixby, 1975). Changes in soluble proteins may involve an increase in water-binding proteins that could decrease free cellular water. This would tend to decrease formation of intracellular ice during the latter stages of acclimation to cold.

An autumn increase and a spring decrease in lipids, particularly phospholipids, also have been shown. For example,development of coldhardiness of black locust and poplar was consistently accompanied by an increase in bark phospholipids (Siminovitch *et al.*, 1968; Yoshida and Sakai, 1973; Yoshida, 1974). During cold-hardening, lipids also increased in apple shoots (Ketchie and Burts, 1973) and chloroplasts of eastern white pine (de Yoe and Brown, 1979). The membrane lipid content of winter-hardened Norway spruce needles was almost twice as high as that in unhardened needles (Senser, 1982).

It is inevitable that some biochemical changes occur in plants subjected to decreasing temperature because (1) decreasing temperature affects various enzyme-mediated processes differently, (2) the decrease in or cessation of growth results in decreased use of soluble carbohydrates and nitrogen compounds, and (3) the lowering of respiration rates also decreases the use of soluble carbohydrates. Thus, it is not surprising that sugars and soluble proteins accumulate in cold-hardened plants, but such accumulation may or may not be causally related to increase in coldhardiness even when the two processes are well correlated.

Winter Desiccation Injury

Winter injury to evergreens often is caused by desiccation of leaves rather than by direct thermal effects. Such injury is well known in forest trees (Kozlowski, 1982a) as well as ornamental trees and shrubs (Havis, 1971). Usually the leaves are killed and eventually shed and the buds survive, but some trees may be killed. Winter desiccation injury occurs because the rate of transpirational water loss greatly exceeds the rate of absorption of water. Transpiration increases as the leaf temperature rises above freezing during sunny winter or spring days and increases the vapor pressure gradient between the leaves and surrounding air. Because the surface soil is cold or even frozen, water cannot be absorbed through the roots fast enough to replace transpirational losses; hence the leaves become dehydrated. The extent of winter desiccation injury varies with plant species, soil water content, time and depth of freezing of soil, depth of snowfall, air humidity, and wind velocity (Kozlowski, 1976a).

Winter desiccation is one of the most important factors limiting the altitudinal and latitudinal range of conifers in forests in North America (Kozlowski, 1982a), Europe (Larcher, 1983), and Japan (Sakai and Larcher, 1987). Sometimes leaves of trees on whole mountainsides appear as though scorched by fire as a result of winter desiccation. Winter drying injury, sometimes called "red belt" when it occurs in horizontal bands on mountainsides, results when the soil is frozen at certain elevations or when warm winds follow belts. When lower branches are buried in snow they generally are not injured. In red belt injury the older needles are shed first (MacHattie, 1963). Desiccation injury often is most severe on south and southwest sides of individual trees. In central and eastern Japan on northern mountain slopes at elevations of 400 to 600 m, the soil usually freezes for 1 to 2 months to a depth of 10 to 30 cm. Here, evergreens show extensive desiccation injury. However, on south- and east-facing slopes, where the soil remains unfrozen, desiccation injury is negligible. In northern Honshu and Hokkaido, where the soil freezes to a great depth, winter desiccation injury occurs in windswept areas regardless of the direction of slope (Sakai and Larcher, 1987).

In general, temperate conifers are more susceptible than boreal conifers to winter desiccation. Such winter-hardy species as Scotch pine and jack pine show little injury even under extreme desiccation (Sakai, 1970). In the Adirondack Mountains of New York State, the order of winter desiccation injury was red spruce > eastern hemlock > eastern white pine > balsam fir. Injury was negligible on black spruce, white spruce, and red pine (Curry and Church, 1952). In northern Japan, tea plants show severe desiccation injury, characterized by death of leaves, twigs, and stems (Fuchinoue, 1982). Injury is especially severe during extended winter droughts, when winds are strong, and when temperatures are unusually low. Damage to fine roots by soil temperatures of –2 to –4°C accentuates injury to shoots. The extent of winter desiccation injury differs among cultivars and with age of plants, being especially severe in plants 1 to 3 years old.

Symptoms of winter desiccation often resemble those of freezing injury. Distinguishing between these two types of injury may require close monitoring of the water balance of plants throughout the winter. Dehydration of leaves precedes development of winter desiccation injury, and accelerated loss of water follows the development of symptoms after freezing injury (Sakai and Larcher, 1987). Sometimes, however, leaves are injured by both desiccation and freezing (Wardle, 1981).

According to Tranquillini (1979), death of conifer needles at the alpine timberline in Europe often results from dehydration caused by lack of water absorption from frozen soil. However, in the Rocky Mountains of eastern Wyoming, winter death of timberline conifers appears to be caused chiefly by severe abrasion of needle cuticles by wind-driven snow, which leads to dehydration (Hadley and Smith, 1986).

Effects of High Temperatures

Dangerously high temperatures of plant tissues sometimes are reached under natural conditions. Exposed tree stems, for example, may become very hot. Temperatures higher than 50°C are not uncommon in leaves of some desert plants (Kappen, 1981).

The thermal death point of most active plant cells varies from 50 to 60°C depending on species, age of tissue, and duration of exposure to high temperature. Physiologically inactive, dehydrated tissues tolerate high temperatures better than hydrated, growing tissues do. For example, dry seeds may survive temperatures up to 120°C whereas hydrated tissues often are killed by temperatures of 50 to 60°C (Levitt, 1980a).

Woody plants often are injured by high temperatures below the thermal death point. Symptoms of heat injury include scorching of leaves and fruits, sunscald, sunburn, abscission of leaves, and growth inhibition. In exposed locations the temperature of the surface layer of soil may exceed 65°C, and temperatures of desert soils may reach 70 to 80°C. Excessive heating of stems of tree seedlings near the soil surface sometimes causes death of cells and stem lesions. Such injury often is most prevalent on the south sides of the stems and resembles injury caused by "damping-off" fungi, except that the injury is localized in aboveground tissues. Another type of lesion caused by dehydration of the cambium and inner bark is called sunscald or "bark scorch," as discussed earlier.

High air temperatures often injure fruits, including apples, avocados, cherries, grapes, peaches, and oranges. Common symptoms include burns and lesions, which vary somewhat among species. Symptoms of heat injury to orange fruits include sunburned peels, dried flesh, and increased granulation (Ketchie and Ballard, 1968). Prunes develop translucent areas in the mesocarp next to the pit cavity. These subsequently turn brown, and the injured cells generally disintegrate. When injury is severe the flesh may develop a brown, mushy appearance (Maxie, 1957). These symptoms are different from the high-temperature injury of Kelsey plums called "Kelsey spot." In mild cases of Kelsey spot, the distal end of the fruit develops a slight, reddish depression, which is underlaid by necrotic tissue. In more severe cases the injured areas develop a purplish red color and the injury may extend through the flesh to the pit (Proebsting, 1937).

Because of the increased interest in growing trees in containers, much attention has been given to effects of high temperatures on roots. The high ratio of surface area to volume of plant containers and the absorption of solar radiation by container walls often result in increases in soil temperatures of 4 to 10°C per hour, and maxima approaching 50°C. Such high soil temperatures often injure the roots and decrease their growth. For example, high temperatures sequentially inhibited growth, induced death of root tips and whole roots, and eventually led to death of shoots of five species of woody plants (Wong *et al.*, 1971). The critical temperature for injury to root cell membranes of two species of *Ilex* decreased linearly as duration of exposure increased exponentially. The critical temperature for injury was higher for *Ilex vomitoria* 'Helleri' than for *Ilex vomitoria* 'Schellings' (Ingram, 1986), indicating differences among varieties.

Mechanisms of Heat Injury

Often it is difficult to distinguish between heat injury and desiccation injury caused by excessive water loss at high temperatures, or when soils are cold.

The mechanism of heat injury is very complex. According to Levitt (1980a), the several types of primary indirect injury (e.g., growth inhibition, starvation, toxicity, biochemical lesions, and protein breakdown) in response to moderately high temperatures for hours to days are metabolic in nature and may be traceable to a single central mechanism. This may involve deficiency of an essential metabolite or an increase in a toxic compound or compounds. However, direct "heat shock" injury, following exposure to high temperatures for seconds or minutes, is largely the result of damage to cell membranes, as shown by leakage of ions. Both liquefaction of protein lipids and denaturation of proteins may be associated with membrane damage.

According to Levitt (1980a), variations in heat tolerance of plants are largely due to differences in protein thermostability. The rate of inactivation of enzymes increases rapidly at high temperatures, and most plant enzymes are inactivated above 70°C. Enzyme inactivation by heat, which almost always results from denaturation of proteins, is influenced by pH, ionic strength, protein concentration, and the protective action of substrates and inhibitors (Dixon *et al.*, 1979).

Plants often are injured by dehydration associated with high transpiration rates caused by high temperature. As temperature rises, a rapid increase in transpiration occurs, largely because of an increase in the vapor pressure gradient between the leaf and surrounding air. Winds also reduce boundary layer thickness, which influences transpiration. Hence, injury and shedding of leaves often are associated with hot, dry winds (Kozlowski, 1976a, 1991). Reed and Bartholomew (1930) distinguished between two types of injury by hot winds in California, namely, "windburn" and "scorch." Leaves killed by windburn first wilt, then dehydrate rapidly, and become brittle within a day. If the wind stops within a few hours the wilted leaves sometimes regain turgidity. However, if the leaves are severely dehydrated they are shed within a few days. By comparison, "scorched" leaves turn brown and become brittle without an intermediate wilting stage, but are not shed for several weeks despite their exposure to strong winds.

Adaptation to Heat Stress

Woody plants can adapt to heat as shown by increasing tolerance from the lowest in the spring to highest in the summer. Heat tolerance also is higher in unusually hot summers than in normal ones (Kappen, 1981).

Acclimation to heat can be achieved by brief exposure to high temperature. For example, hardening of European beech occurred following a heat shock of 55°C. However, exposure to 50°C actually lowered heat tolerance (Wagenbreth, 1965). Specific proteins, called "heat shock" proteins, appear in tissues subjected to high temperatures. For example, heat shock proteins formed rapidly when the temperature of soybean seedling tissue was increased from 28 to 40°C. When the tissue was returned to 28°C there was a progressive decrease in synthesis of heat shock proteins, followed by a normal pattern of protein synthesis (Key *et al.*, 1981). Burke and Orzech (1988) developed a model for the potential mechanism for triggering certain biochemical responses to high-temperature-induced leakage of extracellular and vacuolar ions into the cytoplasm. They suggested that many heat shock proteins may be involved in protecting enzymes from inactivation and nucleic acids from cleavage induced by elevated levels of specific metals. Heat shock proteins may result simply from perturbations in protein synthesis and may not necessarily be protective.

Summary

Distribution and growth of woody plants are influenced by temperature regimes. Small temperature changes affect plant growth indirectly by altering physiological processes. Large temperature changes injure plants directly and rapidly by freezing, overheating, and dehydration.

Temperature affects shoot growth by influencing bud development and dormancy as well as expansion of internodes and leaves. Shoot growth is variously influenced by temperature of the year of bud formation and temperature of the year of bud expansion into a shoot. In subarctic regions, temperature predominates in control of cambial growth. Soil temperature influences growth of existing roots and initiation of new ones. Yields of seeds and fruits are influenced by effects of temperature on floral initiation, regulation of bud dormancy, anthesis, fruit set, and growth of fruits.

Vegetative and reproductive growth of tropical plants are adversely affected by temperatures below approximately 15°C but above freezing. Symptoms of such chilling injury to fruits include surface lesions, discoloration, tissue breakdown, failure to ripen, early senescence, susceptibility to disease, and short storage and shelf life.

Freezing injury, which occurs commonly, includes discoloration, bleaching, shrinkage and dieback of tissues, rupturing by mechanical effects of frost, malformations, and delay in development. Freezing injury is caused directly by intracellular freezing of cell contents or indirectly by tissue dehydration resulting from extracellular freezing.

Coldhardiness, which develops in two stages, varies greatly among species, genotypes, and different organs and tissues. During development of coldhardiness changes occur in enzyme activity and concentrations of sugars, proteins, amino acids, nucleic acids, and lipids.

Winter desiccation injury to shoots limits the range of conifers. Such injury, which occurs because transpiration exceeds absorption of water when the soil water is cold or frozen, varies with species, soil water content, time and depth of freezing of snow, depth of snow, air humidity, and wind velocity.

Woody plants are killed when temperatures climb to the thermal death point and often are injured by high temperatures below the thermal death point. Heat injury includes scorching of leaves and fruits, sunscald, sunburn, abscission of leaves, and growth inhibition. Indirect heat injury may involve deficiency of essential metabolites or increases in toxic compounds. Direct "heat shock" injury results from damage to cell membranes associated with liquefaction of protein lipids and denaturation of proteins. Heat injury may also result from dehydration because of high transpiration rates.

General References

Gates, D. M. (1980). "Biophysical Ecology." Springer-Verlag, New York.

Grant, B. W. W., and Morris, G. J., eds. (1987). "The Effects of Low Temperatures on Biological Systems." Edward Arnold, London.

Kappen, L. (1981). Ecological significance of resistance to high temperature. *Encycl. Plant Physiol.* **12A,** 439–474.

Kramer, P. J., and Kozlowski, T. T. (1979). "Physiology of Woody Plants." Academic Press, New York.

Larcher, W. (1983). "Physiological Plant Ecology," 2d ed. Springer-Verlag, New York.

Levitt, J. (1980). "Responses of Plants to Environmental Stresses. I. Chilling, Freezing, and High Temperature Stresses." Academic Press, New York.

Li, P. H., and Sakai, A., eds. (1978). "Plant Cold Hardiness and Freezing Stress," Vol. I. Academic Press, New York.

Li, P. H., and Sakai, A., eds. (1982). "Plant Cold Hardiness and Freezing Stress," Vol. II. Academic Press, New York.

Lieth, H., ed. (1974). "Phenology and Seasonality Modeling." Springer-Verlag, New York

Long, S. P., and Woodward, F. I., eds. (1988). "Plants and Temperature." S.E.B. Symposium No. 42. Company of Biologists Ltd., Cambridge Univ., England.

Lyons, J. M., Graham, D., and Raison, J. K., eds. (1979). "Low Temperature Stress in Crop Plants. The Role of the Membrane." Academic Press, London.

Morris, G. J., and Clarke, A., eds. (1981). "Effects of Low Temperature on Biological Membranes." Academic Press, London.

Nobel, P. (1983). "Biophysical Plant Physiology and Ecology." Freeman, San Francisco.

Olien, C. R., and Smith, M. N., eds. (1981). "Analysis and Improvment of Plant Cold Hardiness." CRC Press, Boca Raton, FL.

Pantastico, E. B. (1975). "Postharvest Physiology, Handling, and Utilization of Tropical and Subtropical Fruits and Vegetables." AVI, Westport, CT.

Ryall, A. L., and Lipton, W. J. (1979). "Handling, Transportation, and Storage of Fruits and Vegetables," 2d ed. AVI, Westport, CT.

Sakai, A., and Larcher, W. (1987). "Frost Survival of Plants." Springer-Verlag, New York.

Steponkus, P. L. (1984). Role of the plasma membrane in freezing injury and cold acclimation. *Ann Rev. Plant Physiol.* **35,** 543–584.

Tranquillini, W. (1979). "Physiological Ecology of the Alpine Timberline." Springer-Verlag, New York.

Underwood, L. S., Tieszen, L. L., and Folk, G. E., eds. (1979). "Comparative Methods of Cold Adaptation." Academic Press, New York.

Woodward, F. I., ed. (1987). "Climate and Plant Distribution." Cambridge University Press, New York.

Chapter 6

Soil Properties and Mineral Nutrition

Introduction

Growth of trees, shrubs, and other plants varies widely within their geographic range because of differences in the productive capacity or site quality of land. This varies with soil conditions, such as its capacity to supply water, minerals, and oxygen, its physical properties that affect root growth, and the topography and slope exposure. Although farmers and gardeners probably knew of the differences in productivity related to soil conditions before the beginning of recorded history, the reasons for the relationship only began to be understood during the nineteenth century. Evelyn's "Sylva," published in 1670, mentioned differences in site requirements for several species of trees, and in 1767 Enderlin discussed forest soils in relation to tree growth. However, it was not until after the middle of the nineteenth century when Liebig had disproved the "humus theory" of plant nutrition that the mineral nutrition of trees began to receive serious attention among French, German, and Scandinavian foresters. In recent decades the increasing cost of establishing forests and the fact that they often are relegated to less productive soils have increased the interest of foresters in analyzing the components of site quality. An understanding of site quality is even more important to horticulturists and arborists because of the high value of fruit and shade trees and ornamental shrubs. Therefore, we will discuss the factors affecting site quality in some detail.

Site Quality and Site Index

Site quality for tree growth has been measured by various methods (Carmean, 1975), but "site index" is used most frequently. It usually is defined as the height of dominant trees at 50 years of age, although a shorter time sometimes is used for rapid-growing trees and a longer time for very-slow-growing trees. Height is used because at the spacing generally employed in managed forests, height is little affected by stand density, except for stagnating conifer stands on especially poor sites (Smith, 1986). Direct estimates of site index are based on measurements of age and height of existing dominant and codominant trees used with site index curves to estimate height of trees at some future age. Such curves are most applicable to even-aged, undamaged stands of fairly uniform growth (Hagglund, 1981).

It is often necessary to estimate future productivity of sites on which no trees of the desired species are growing. This can be done by relating site index to easily measured soil properties that influence the supply of water and the suitability of the soil as a medium for root growth. For example, site quality for loblolly and shortleaf pine on the piedmont of the southeastern United States improves with increasing depth of the A horizon and decreasing plasticity and improved drainage of the subsoil (Coile and Schumacher, 1953). Zahner (1968) also found site quality in southern Arkansas and northern Louisiana to be well correlated with those soil properties that affect water-holding capacity and aeration.

In Massachusetts, increase in silt and clay content of the A horizon was associated with increase in site index for eastern white pine, presumably because it improved the supply of water and mineral nutrients. However, increase in silt and clay in the B horizon decreased site quality, probably because it decreased aeration and hindered deep root growth (Mader, 1976). In Wisconsin, growth of trembling aspen was better correlated with those physical characteristics that increased the water-holding capacity of the soil than with the amount of available mineral nutrients (Stoeckeler, 1960; Fralish and Loucks, 1975). In some soils the concentration of mineral nutrients is too high to be a limiting factor for site quality (Tarrant, 1956), but mineral deficiencies often are limiting on sandy soils and eroded sites, and water supply also is more likely to be limiting on sandy than on clay soils. The relationship of mineral nutrition to tree growth is discussed in detail later in this chapter.

In hilly or mountainous regions, steepness of slope and slope exposure often affect site quality. The site index usually is higher on well-drained bottomland and lower parts of slopes and on northern and eastern exposures than on upper slopes and those with southern and western exposures. Einspahr and McComb (1951) reported that the site index for oak trees in Iowa is 2.4 to 3.6 m greater on north and east than on south and west slopes. Site quality for oak in the Appalachian Mountains seems to be related to the effects of topography on the availability of water (Meiners *et al.*, 1984). However, in the North Carolina Piedmont, topography seems to have less effect on site quality than the physical characteristics of the soil profile (Coile and Schumacher, 1953). The effects of topography and slope exposure are complex, involving differences in irradiance, temperature, and evapotranspiration, which affect tree water stress, and in soil fertility.

Soil as a Medium for Root Growth and a Reservoir for Water and Minerals

Soil is important because it provides the water and minerals essential for plant growth and the anchorage necessary for plants to grow upright. Thus its quality for plant growth depends on its suitability for root growth and its capacity to supply water and essential mineral elements. The characteristics that control these functions will be discussed in detail. Soil consists of varying proportions of rock particles and organic matter forming the solid matrix, and air and water that occupy varying proportions of the pore space in the matrix. In addition, it contains living organisms such as bacteria, fungi, algae, protozoa, earthworms, insects, and roots that directly or indirectly affect its structure and gas composition. A good agricultural or forest soil usually contains 40 to 60% by volume of pore space, of which about half is filled with air and the other half with water containing the dissolved minerals and gases that constitute the soil solution.

The suitability of soil as a substrate for root growth is important because a healthy, deep, wide-spreading root system is essential to provide the anchorage, water, and minerals required to support a large tree canopy. There also is increasing evidence that hormones produced in roots have important effects on shoot growth. Root development depends on soil physical properties such as bulk density (mass per unit of volume), mechanical resistance to root penetration, and aeration; and on chemical properties such as pH, deficiencies in essential minerals, toxic concentrations of elements such as aluminum or manganese, or excessive amounts of salt. The capacity to hold water and mineral nutrients (cation-exchange capacity) is related to soil texture, being greater in fine-textured soils that possess more surface than in coarse-textured soils. Thus the suitability of a soil for tree growth depends largely on soil texture, structure, and depth to a layer or horizon that cannot be penetrated by roots. We will therefore discuss those soil characteristics in more detail before turning to chemical characteristics.

Soil Texture

Soil is classified broadly as sand, loam, or clay depending on the relative proportions of large (2–0.2 mm), intermediate (0.02–0.002 mm), and small (<0.002 mm) particles of which it is composed. A sand contains less than 15% of clay and silt; clay contains over 40% of clay particles and less than 45% of sand and silt; whereas loam soils contain intermediate proportions of sand, silt, and clay. Sandy soils are loose, noncohesive, well drained, and well aerated, but have a low storage capacity for water and minerals. Clay soils are more compact and cohesive, often poorly drained and deficient in aeration, but have a large storage capacity for minerals and water. The differences in water storage capacity of various textural grades of soil are shown in Fig. 6.1.

Soil Structure

Soil structure depends on the degree of aggregation of primary soil particles into aggregates (crumbs, granules, small clods). Structure is important because it influences the amount of pore space and pore size, which in turn affects drainage, aeration, and root penetration. It is particularly important in fine-textured soils, especially those composed chiefly of montmorillonite-type clay minerals, which swell when wet and reduce infiltration, drainage, and aeration. Most soils contain approximately 50% of pore space, but the distribution between capillary and noncapillary space varies widely, as shown in Fig. 6.2. The noncapillary pore space is that fraction from which water drains by gravity, providing air space. Capillary pore space consists of pores less than 60 to 30 μm in diameter that are not drained by gravity and are too small to be penetrated by roots. Soils suitable for root growth

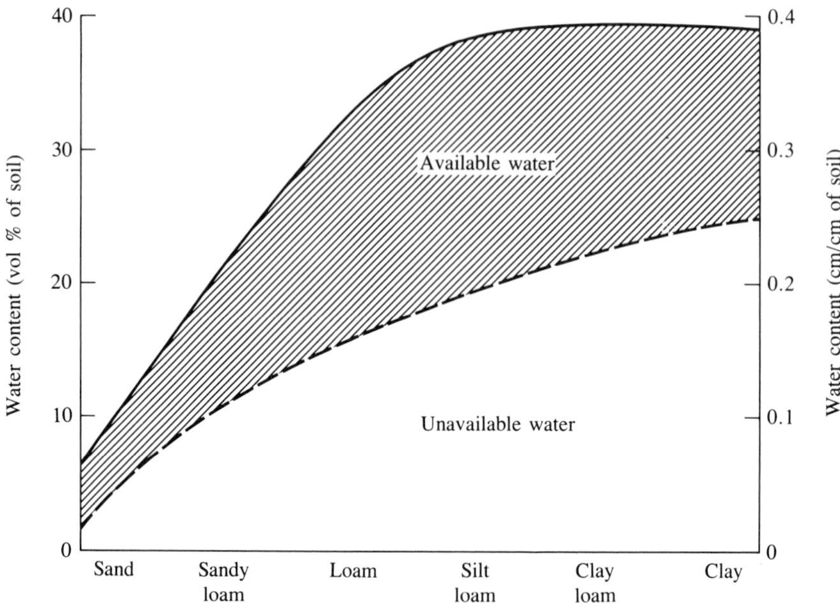

Figure 6.1 Diagram showing the relative amounts of available water in soils ranging in texture from sands to clays. Permanent wilting percentage, – –; *in situ* field capacity, ——. Amounts are expressed as percentages of soil volume and as centimeters of water per centimeter of soil depth. (After Cassell, 1983; from Kramer, 1983a, by permission of Academic Press.)

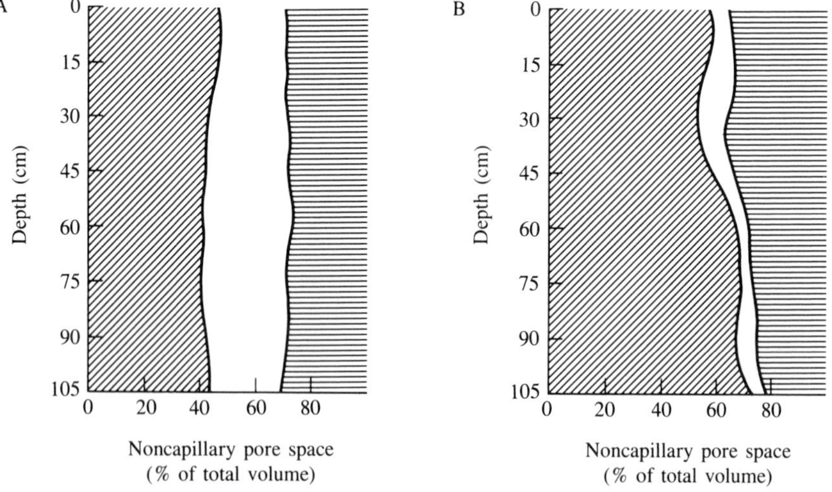

Figure 6.2 Examples of the differences in amounts of capillary (*right*) and noncapillary (*center*) pore space in two dissimilar soils. The Marshall silt loam (A) is more satisfactory for root growth than the Shelby loam (B) because the larger proportion of noncapillary pore space improves drainage and aeration. Soil volume shown on left. (From Kramer, 1983a, by permission of Academic Press.)

contain about equal volumes of capillary and noncapillary pore space. Forest soils usually have a much larger proportion of noncapillary pore space than agricultural soils of the same textural grade, as shown in Fig. 6.3, because they have suffered less mechanical disturbance.

The aggregation of clay particles into larger units appears to be related to the presence of root exudates and organic colloids produced by soil organisms, and the large amounts of organic matter and microbial activity in forest soil probably aid in maintaining its good structure. As pointed out in Chapter 8, the structure of clay soils is easily destroyed by compaction by machinery or by flocculation with an excess of potassium or sodium ions. Vehicular traffic, especially on wet soil, causes

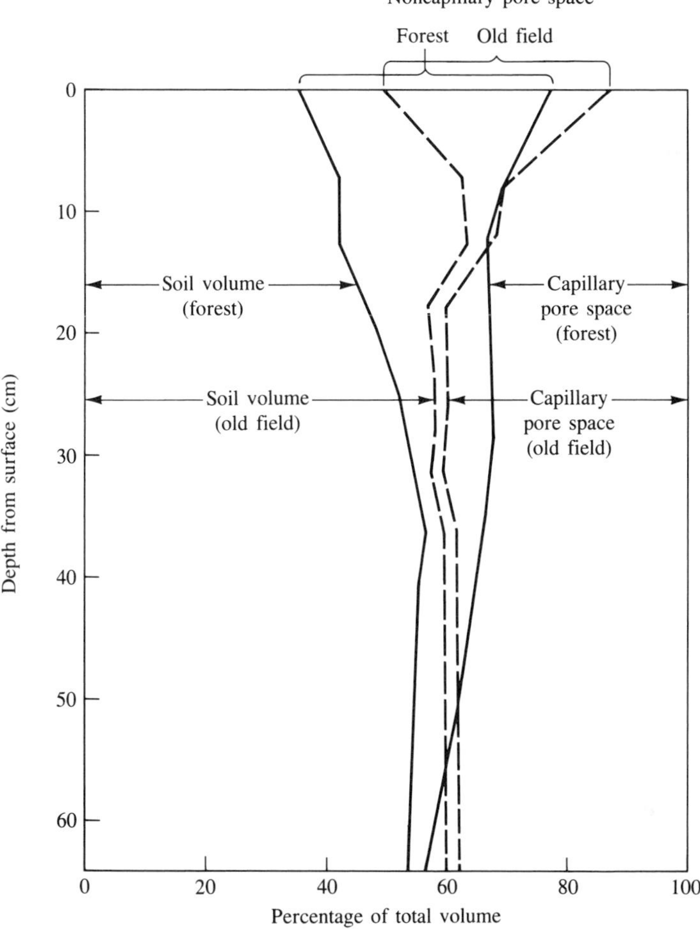

Figure 6.3 Differences in amount of capillary pore space in an old field and in an adjacent forest growing on the same type of soil. The large percentage by volume occupied by noncapillary pore space in the forest soil provides better aeration for roots. It also increases the rate of infiltration, and decreases runoff during heavy rains. (After Hoover, 1944; from Kramer, 1983a, by permission of Academic Press.)

compaction that greatly decreases aeration and mechanically hinders root extension, and the damage often persists for many years (Greacen and Sands, 1980).

It has been known for many years that high soil bulk density or "strength" as measured with a penetrometer affects root growth (Barley, 1963; Taylor and Ratliff, 1969). Apparently, increasing soil density not only affects root extension but also decreases shoot growth, stomatal conductance, and photosynthesis of herbaceous plants (Carmi *et al.*, 1983; Masle and Passioura, 1987), and increases water use efficiency, the ratio of roots to shoots, and carbon isotope discrimination (Masle and Farquhar, 1988). Although this research was done on herbaceous seedlings, it is possible that soil compaction has important physiological effects on shoots of tree seedlings in addition to the mechanical effects on root extension. Soil compaction is discussed further in Chapter 8.

Soil Permeability and Internal Drainage

Heavy clays and badly compacted soils often suffer from slow infiltration of water and poor internal drainage. Rapid infiltration of water from rainfall and irrigation is important with respect to recharging the soil and because slow infiltration results in surface runoff, sheet erosion, and gully formation.

Rapid internal drainage is important in maintaining good aeration. Arborists and homeowners often discover that the topsoil used in holes dug in heavy clay to plant trees and shrubs becomes waterlogged and the root systems are injured by poor aeration. Low permeability and slow water movement make clay soils difficult to drain with tile. Forest soils usually have much higher rates of infiltration and drain more rapidly than cultivated soils because they contain more noncapillary pore space, as shown in Fig. 6.3.

Soil Profiles

Most soils show important changes in texture and structure and related properties with increasing depth that affect their capacity to store water and minerals and support root growth. Figure 6.3 shows the changes in capillary and noncapillary pore space with increasing depth under a forest and in a nearby field on the same soil type, and Fig. 6.4 shows the horizons that might occur in a well-developed soil profile. The large volume of noncapillary pore space in forest soils results in better infiltration of rainwater, less surface runoff and erosion, and better aeration than in cultivated soils.

An important limitation on root development and tree growth is the presence of an impermeable layer near the soil surface that restricts root penetration and limits the size of the water and mineral reservoir available for growth. This is discussed in Chapter 8. In some areas, chemical barriers to root penetration occur in the form of excessive acidity (low pH) and the presence of high concentrations of aluminum and sometimes of manganese or other elements. In regions with low rainfall, carbonates

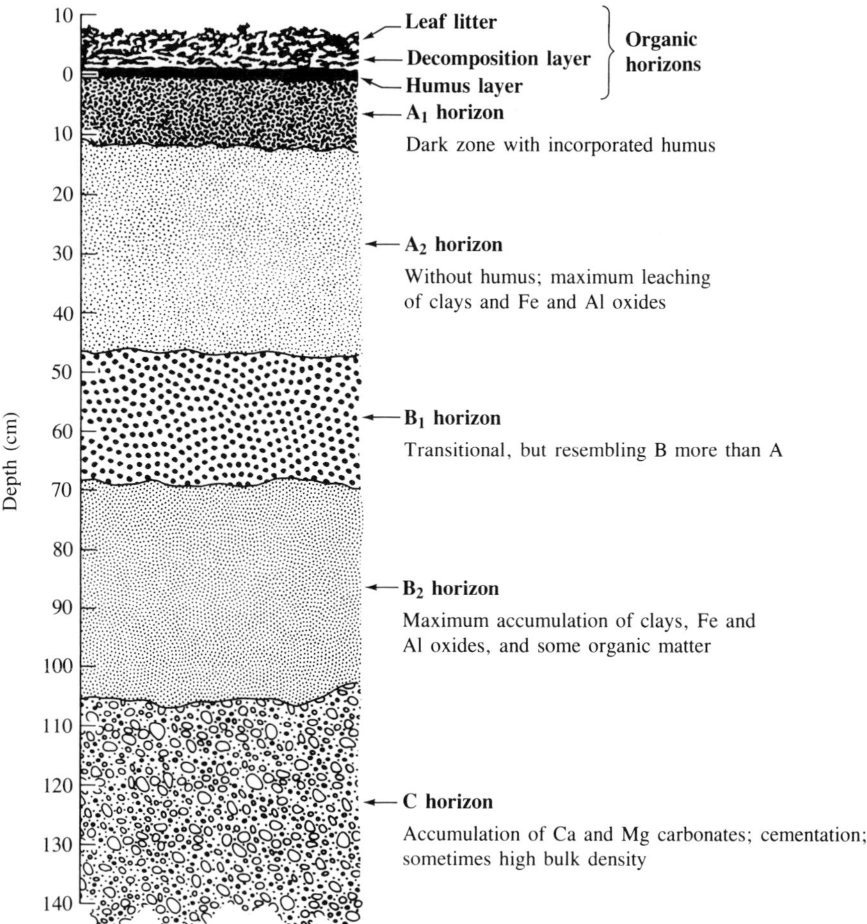

Figure 6.4 A soil profile under an old shortleaf pine stand in the North Carolina Piedmont. (Modified from Billings, 1978, and Kramer, 1983a.)

often accumulate at the deepest point wetted by rain, forming a hardpan layer (caliche) that is difficult for roots to penetrate. Aluminum toxicity is a serious limitation to root growth in the acid soils of the southeastern United States and in many tropical soils. At least part of the injury to forests caused by air pollution probably results from changes in soil chemistry that affect plant nutrition (Schulze, 1989). Cassell (1983) discussed chemical and physical barriers to root growth with respect to crop plants, and the same general principles apply to trees. Perhaps more energetic attempts should be made to find genotypes of important tree species that have roots capable of penetrating compact soils and that are tolerant of low pH and high concentrations of aluminum.

Soil Water Relations and Terminology

Discussions of soil water often use terms such as field capacity, permanent wilting percentage, readily available water, and water potential. Water potential (ψ) is a physical term referring to the free energy status of water in the soil or plant as compared to that of free water. It usually is expressed in pressure terms as bars, or preferably in international units as megapascals (MPa). One MPa equals 10 bars. Water potential of plant tissue is discussed in Chapter 7.

Assuming isothermal conditions, the factors affecting soil water potential can be summarized by the equation

$$\psi_w = \psi_m + \psi_s + \psi_g$$

In this equation, ψ_m is the matric potential produced by capillary and surface forces binding water, ψ_s the osmotic potential resulting from solutes, and ψ_g the gravitational force operating on water in a soil column. The water potential at field capacity is about -0.03 MPa and at permanent wilting percentage about -1.5 MPa, as shown in Fig. 6.5.

The field capacity, or *in situ* field capacity, is the water content after downward drainage by gravity has become very slow and the water content has become relatively stable. The water potential at field capacity is approximately -0.03 MPa or -0.3 bar. It is not a true equilibrium but rather the condition in which water

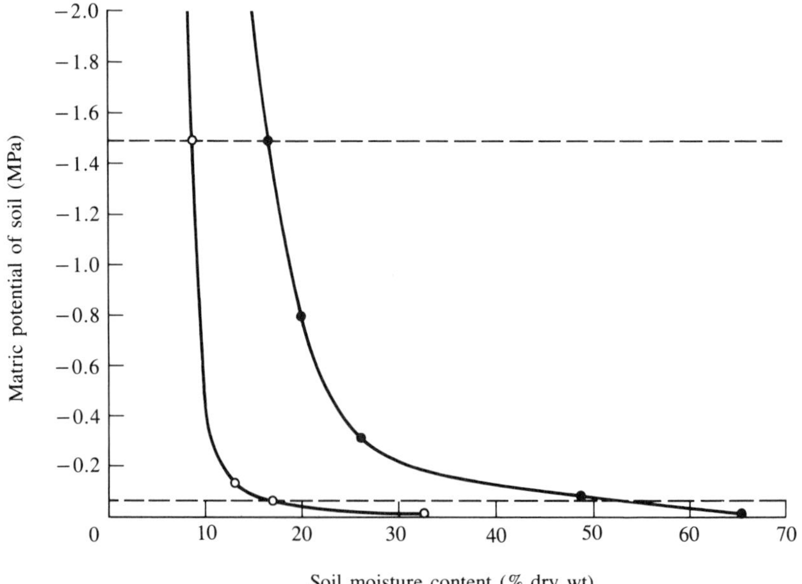

Figure 6.5 Soil matric water potential of clay loam (●) and sandy loam (○) soils plotted over gravimetric water content. Upper dashed line represents permanent wilting percentage. Lower dashed line represents field capacity. (From Kramer, 1983a, by permission of Academic Press.)

content does not change materially from day to day. The field capacity of the surface soil is attained more rapidly in sandy and deep soils than in shallow or clay soils. Permanent wilting percentage is the water content at which plants remain permanently wilted unless water is added to the soil. This occurs at a soil water potential of approximately −1.5 MPa or −15 bars, but varies somewhat depending on the kind of plant and environmental conditions. It is not really a soil constant, but depends on the osmotic potential at which leaves lose their turgor (Slatyer, 1957). Nevertheless, it can be used as an approximate soil constant. The readily available water is the amount retained in the soil between field capacity and the permanent wilting percentage. It varies from a minimum in sandy soils to a maximum in silts and clays, as shown in Fig. 6.1. The water storage capacity in various horizons of an unusually deep, pumice soil in New Zealand is shown in Fig. 6.6. The volumetric water storage capacity can be seriously reduced by the presence of rock fragments. For example, Richter *et al.* (1989) found that forests on certain mountain soils in North Carolina were subjected to water stress because rock fragments occupied about 40% of the total volume of soil.

Soil Mineral Pool

Even when a soil has the physical characteristics favorable for vigorous root growth, it cannot support good tree growth unless it contains adequate amounts of over a dozen essential mineral elements, most of which are listed in Tables 6.1 and 6.2. The original source of most of the minerals found in natural soils is weathering of the parent rock, which provides the solid matrix of all except a few organic soils. The mineral ions released during weathering are adsorbed on the surfaces of the soil particles or exist in the soil solution. The extremely large surface area of colloidal-sized clay particles provides much more ion binding capacity per unit of soil volume than exists in sand, and organic matter also increases the surface. An acre (0.405 ha) of sandy soil 6.6 in. (16.7 cm) deep contains about 5000 acres (2025 ha) of surface, and an acre of clay soil contains 500,000 acres (202,500 ha) of internal surface (Lyon and Buckman, 1943). The cation-exchange capacity is high in organic and clay soils and low in sands. As a result, sandy soils are apt to be low in mineral nutrients because of loss by leaching.

The pool of soil minerals is increased slowly by rock weathering and by deposition from the atmosphere in rainfall and as dust. The increasing pollution of the atmosphere is providing significant amounts of mineral nutrients to the soil, as shown in Table 6.3. In fact, atmospheric deposition of nitrogen is exceeding its use in some northern forest ecosystems and may be contributing to forest decline (Aber *et al.*, 1989; Schulze, 1989). In contrast, Kenk and Fischer (1988) reported increased growth of Norway spruce in Germany, beginning about 1960, which may be caused by increased deposition of nitrogen from the atmosphere, although this is not proven. Air pollution is discussed in detail in Chapter 9.

The largest natural additions come from decomposition of fallen litter, dead roots, dead soil organisms, and other organic matter and from root exudates (Smith,

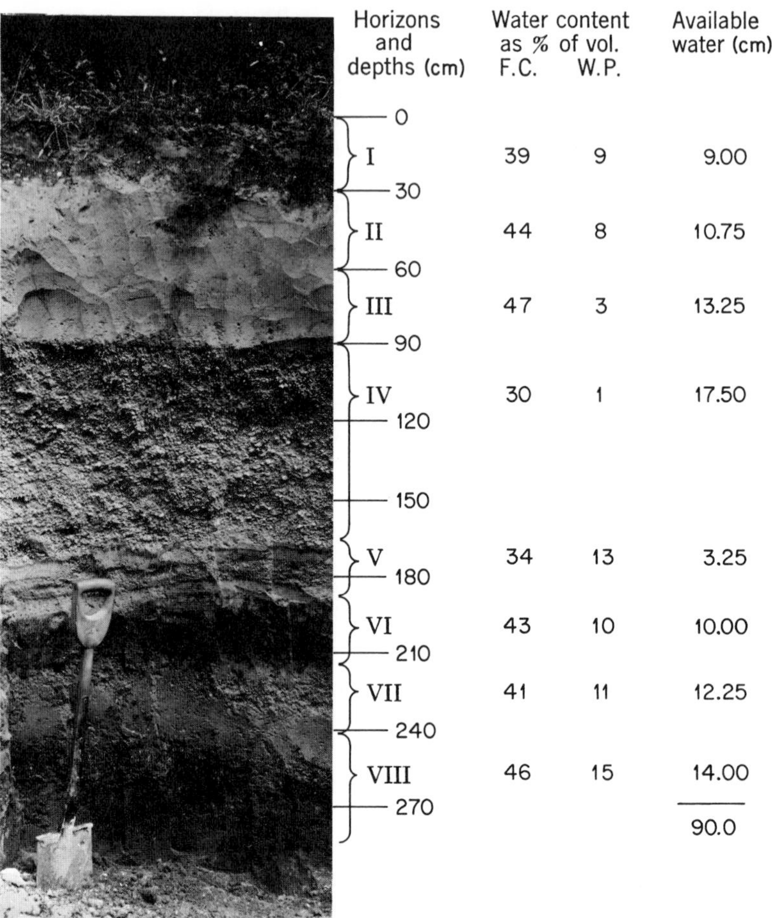

Horizons and depths (cm)
Water content as % of vol.
F.C.
W.P.
Available water (cm)
0
I
39
9
9.00
30
II
44
8
10.75
60
III
47
3
13.25
90
IV
30
1
17.50
120
150
V
34
13
3.25
180
VI
43
10
10.00
210
VII
41
11
12.25
240
VIII
46
15
14.00
270
90.0

Figure 6.6 Profile of a deep pumice soil in New Zealand, formed from volcanic deposits, showing the water-holding capacity of various horizons expressed as percentages by volume. This soil has an extraordinarily high water-holding capacity and forests growing on it are deep rooted and very productive. (From Kramer, 1983a, by permission of Academic Press.)

1976). The turnover is so rapid in warm climates that relatively little organic matter accumulates on or in the soil, but in cool climates decomposition is much slower and so much of the total mineral pool is accumulated in litter that mineral deficiencies often result. In managed forests, additional minerals sometimes are added to the soil pool by fertilization, as discussed in Chapter 13. The pool of soil minerals is depleted by plant absorption and by leaching away in drainage water, and some nitrogen is lost as gas by ammonification and denitrification. According to Mitchell *et al.* (1975), about 80% of the nitrogen in a hardwood forest in southwestern North Carolina is in the soil, and according to Switzer *et al.* (1968), the same is true of a

Table 6.1
The Amounts of the Major or Macronutrient Elements in the Various Components of Loblolly Pine (*Pinus taeda*) Trees in a 16-Year-Old Plantation[a,b]

Component	N	P	K	Ca	Mg
Needles, current	55	6.3	32	8	4.8
Needles, total	82	10.3	48	17	7.9
Branches, living	34	4.5	24	28	6.1
Branches, dead	26	1.5	4	30	3.0
Stem wood	79	10.7	65	74	22.7
Stem bark	36	4.2	24	38	6.5
Aboveground, total	257	30.9	165	187	46.2
Roots	64	16.9	61	52	21.9
Tree total	321	47.8	226	239	68.1

[a]From Wells *et al.* (1975).
[b]Measurements are in kg ha^{-1}.

20-year-old pine stand in Mississippi. In tropical forests a smaller proportion of the nitrogen occurs in the soil and more in the trees.

Mineral Nutrient Recycling

An important factor in the maintenance of a soil mineral pool that is adequate for plant growth is the continual recycling of minerals. This occurs at three levels: (1) inputs and losses in soil that are independent of vegetation, (2) the exchange between soil and vegetation, and (3) retranslocation within plants. The first two levels will be discussed in this section, and retranslocation within plants is discussed in Kramer and Kozlowski (1979, pp. 377–379).

Table 6.2
The Amounts of Minor or Micronutrients in the Various Components of Loblolly Pine (*Pinus taeda*) Trees in a 16-Year-Old Plantation[a,b]

Component	Mn	Zn	Fe	Al	Na	Cu
Needles, current	1.222	0.166	0.334	2.178	0.258	21.5
Needles, total	2.544	0.327	0.650	4.116	0.356	31.6
Branches, living	1.716	0.345	0.915	2.519	1.384	63.7
Branches, dead	—	0.289	1.281	2.902	0.314	69.5
Stem wood	8.445	1.086	1.830	1.790	3.640	275.0
Stem bark	0.951	0.336	1.126	9.705	0.590	59.4
Tree total	13.656	2.383	5.802	21.032	6.284	499.2

[a]From Wells *et al.* (1975).
[b]Measurements are in kg ha^{-1} except for Cu, which is in g ha^{-1}.

Table 6.3
The Average Input of Mineral Nutrients from Precipitation and the Loss in Runoff for a Year from Five Pine-Forested Watersheds in Northern Mississippi[a]

Nutrient	Input ($kg\ ha^{-1}$)	Loss	Gain or loss
NO_3	3.12	0.32	+2.80
NH_4	5.73	3.35	+2.38
PO_4	0.07	0.04	+0.03
K	4.98	3.31	+1.67
Ca	7.72	6.21	+1.51
Mg	3.03	3.05	−0.02

[a]Data from Schreiber *et al.* (1976); from Kramer and Kozlowski (1979), by permission of Academic Press.

Soil Cycling

The weathering of rock provides not only the solid matrix of inorganic soil but also most of the minerals. Some minerals are released mechanically by freeze-fracturing and friction of rocks against one another in streams and under glacial ice. Chemical weathering involves reactions with organic acids, especially H_2CO_3, and is most important in sedimentary rocks, especially limestone. The solution of limestone, resulting in the formation of caves such as Mammoth Cave and Carlsbad Cavern and countless smaller caves, is a dramatic example of chemical weathering. Organic acids released by plant roots are very effective in chemical weathering, and lichens, fungi, and microorganisms also contribute organic acids. The ions released sometimes enter into secondary reactions that reduce their availability, as when phosphorus combines with iron. The ecological significance of rock weathering is discussed by Waring and Schlesinger (1985, pp. 130–137). The return of minerals to the soil by decomposition of organic matter will be discussed in another section.

The elements released by mechanical and chemical weathering and by decomposition of organic matter can be removed from soil by plants, adsorbed on soil particles, or lost by leaching and stream flow. Much of the minerals is lost in solution, but iron and most of the phosphorus are relatively insoluble and so strongly adsorbed that they are lost only as particulate matter. Table 6.4, from a study of a New England watershed, shows the relative amounts of various elements lost in stream flow from a forested watershed. Losses by leaching are more serious in sandy soils than in clay soils, which, as stated earlier, have more cation-binding surface.

Large amounts of nitrogen and some sulfur are volatilized and lost as a result of fires (Chapter 11), but other elements left in the ash are readily available. However, they also are lost more readily in runoff. Nevertheless, pine forests in the southeastern United States have been burned for decades without serious deterioration in fertility (Jorgensen and Wells, 1971; Richter *et al.*, 1982), and Stark (1977) con-

Table 6.4
Mineral Loss in Stream Flow from a Northern Mixed Hardwood Forest Watershed of the Hubbard Brook Experimental Forest in New Hampshire[a]

Element	Particulate + dissolved element (total kg ha^{-1})	Particulate		Dissolved	
		kg ha^{-1}	%	kg ha^{-1}	%
Al	3.37	1.38	40.9	1.99	59.1
Ca	13.93	0.21	1.7	13.7	98.3
Cl	4.58	—	0	4.58	100
Fe	0.64	0.64	100	—	0
Mg	3.34	0.19	5.7	3.15	94.3
N	4.01	0.11	2.7	3.90	97.3
P	0.019	0.012	63.2	0.007	36.8
K	2.40	0.52	21.7	1.88	78.3
Si	23.8	6.19	26.0	17.6	74.0
Na	7.48	0.25	3.3	7.23	96.7
S	17.63	0.03	0.2	17.6	99.8
C	12.3	3.98	32.4	8.35	67.5

[a]From Likens *et al.* (1977); after Waring and Schlesinger (1985), by permission of Academic Press.

cluded that prescribed burning could be practiced indefinitely in western Montana without serious loss of soil fertility.

Most of the nitrogen and sulfur in forest soils comes from the decomposition of organic matter, although some nitrogen is supplied by both symbiotic and nonsymbiotic nitrogen fixation and by deposition from the atmosphere. Industrial pollution of the air has increased the amounts of nitrogen and sulfur supplied from the air, both in rainfall and as dry particulate matter. Volcanoes also supply some sulfur and some nitrogen is fixed by electrical storms. As a result, the input of nitrogen often significantly exceeds the loss in stream flow, as shown in Table 6.3. In fact it is reaching injuriously high concentrations in some soils (Aber *et al.*, 1989). Table 6.3 also shows the amount of mineral nutrients added by rainfall compared to the loss in runoff from pine forests in northern Mississippi. It appears that the addition of minerals in rainfall often exceeds the losses by leaching. The importance of atmospheric sources of minerals is discussed by Lindberg *et al.* (1986), Lovett *et al.* (1982), Miller (1984), Waring and Schlesinger (1985, pp. 123–128), and Aber *et al.* (1989). Floodplains also receive considerable amounts of minerals in the soil deposited from floodwater.

Cycling between Trees and Soil

Large amounts of mineral nutrients are returned to the soil by decay of litter and a small amount by leaching from the foliage. The latter reaches the soil either as direct throughfall or as stemflow down the tree trunks (Patterson, 1975). Leaching from foliage of deciduous trees naturally is most important during the summer and as leaves begin to senesce, and it is likely to be lower for evergreen conifers than for deciduous trees (Waring and Schlesinger, 1985, pp. 171–173). The earlier literature on leaching from foliage was summarized by Tukey (1970).

A large fraction of recycled minerals is in fallen leaves, twigs, and fine roots that die and decay, and in the debris left after logging. Table 6.5 summarizes the amounts of several elements returned to the soil by leaf fall in coniferous and broad-leaved trees in New York and South Carolina. The large amount of minerals returned by death and decay of fine roots is discussed by Waring and Schlesinger (1985, pp. 172–173) and Bowen (1984, pp. 157–160). The rate of decomposition of litter is much slower in coniferous than in deciduous forests and in cool than in warm climates, as shown in Table 6.6. As a result of the slower decomposition, a much larger fraction of the total mineral content of the soil–plant system occurs in the soil and litter in temperate zone forests than in tropical forests. In Ghana over 40% of the total N, P, K, Mg, and Ca in the vegetation and top 30 cm of soil occurs in the vegetation (Greenland and Kowal, 1960), but in a northern hardwood forest less than 10% of the nitrogen is in the vegetation (Bormann *et al.*, 1977). In cool climates the litter often is sharply differentiated from the underlying soil, giving rise to a mor forest floor. In warmer climates, where earthworms, bacteria, and fungi are more active, the decaying organic matter becomes mixed with the mineral soil, producing a mull type of forest floor (Waring and Schlesinger, 1985, pp. 205–208). According to Miller (1986), even in temperate zone forests recycling within trees and through decomposition of the litter is so efficient that after canopy closure the demand for minerals from the soil is much reduced. Mineral cycling is discussed in more detail in Bowen and Nambiar (1984, Chapter 3) and in Waring and Schlesinger (1985, Chapters 6, 7, and 8).

Table 6.5

Amounts of Minerals Returned to the Soil by Litter and Leaf Fall in a Year[a]

Forest	Annual litter fall	Minerals returned to soil by leaf fall ($kg\ ha^{-1}$)					
		N	P	K	Ca	Mg	Total
New York							
Hardwood[b]	3042	18.6	3.7	14.8	73.5	10.3	120.9
Conifer[c]	2759	26.4	2.0	7.1	29.7	5.0	70.2
South Carolina[d]							
Hardwood	4144	25.8			99.0	23.4	
Pine–hardwood	4284	29.7			57.7	12.5	
Pine	3291	14.8			19.4	6.25	

[a]From Kramer and Kozlowski (1979), by permission of Academic Press.
[b]From Chandler (1941).
[c]Average of seven species of conifers, from Chandler (1944).
[d]From Metz (1952).

Table 6.6
Rate of Decomposition of Litter and Release of Mineral Elements on the Forest Floor under Various Climatic Conditions[a,b]

		Mean residence time (years)					
Forest region	No. of sites	Organic matter	N	K	Ca	Mg	P
Boreal coniferous	3	353	230	94	149	455	324
Boreal deciduous	1	26	27.1	10.0	13.8	14.2	15.2
Temperate coniferous	13	17	17.9	2.2	5.9	12.9	15.3
Temperate deciduous	14	4.0	5.5	1.3	3.0	3.4	5.8
Tropical rain forest	4	0.4	2.0	0.7	1.5	1.1	1.6

[a]From Waring and Schlesinger (1985), by permission of Academic Press.
[b]Boreal and temperate values are from Cole and Rapp (1981); tropical values are calculated from Edwards and Grubb (1982) and Edwards (1977, 1982).

The Absorbing System

Any discussion of mineral and water uptake requires consideration of the characteristics of root systems. Although small amounts of water are absorbed by the foliage from dew, fog, and rain (Azevedo and Morgan, 1974; Kramer, 1983a, p. 216) and measurable amounts of minerals are deposited on foliage in both wet and dry states (Lovett *et al.*, 1982; Lindberg *et al.*, 1986), the major portion of the water and minerals absorbed by plants enters through roots.

Bowen (1985) has a good discussion of the importance of roots for vigorous tree growth. He pointed out that there is a relationship between extensive root systems, adaptability to a wide range of soil conditions, and ability to succeed in competition with other species. Fitter (1987) regards the architecture of the root system as more important than root morphology with respect to absorption of water and minerals and suggests that the architecture of root systems has been a basic factor in natural selection. For example, desert plants that are active during the rainy season have shallow, branched roots that intercept water from summer showers, but plants that are active through the dry season have deeper roots with "herringbone" branching that are more efficient in absorbing water deep in the soil, but require more photosynthate for growth and maintenance. There obviously is a trade-off between the cost of producing an extensive root system and the disadvantage of a smaller absorbing surface, which, however, requires less total photosynthate. Factors affecting the development of root systems were discussed in Kramer (1983a, Chapter 6), and Caldwell and Virginia (in Pearcy *et al.*, 1989) reviewed methods of studying root system development and functioning in the field.

Because of the low mobility of nutrients such as phosphate, and the relatively low concentration of most minerals in the soil solution, extensively branched root systems are important for the effective extraction of mineral nutrients from the soil. The little-leaf disease of shortleaf pine develops where soil conditions are unfavorable for root growth and damage to roots by *Phytophthora cinnamomi* reduces the absorbing surface so much that severe nitrogen deficiency occurs (Campbell and Copeland, 1954). Barley (1970) discussed the importance of the configuration of root systems in relation to mineral absorption by crop plants and similar principles apply to woody plants. Clarkson (1985) reviewed the absorption process and factors affecting the absorption of minerals. Huck (1983) suggests that in agroforestry, growing shallow- and deep-rooted plants together minimizes competition for minerals and water, and this also is true in forests.

Root Density

The relative abundance or density of roots often is expressed as centimeters of roots per cubic centimeter of soil. A root density of 1 to 2 cm cm^{-3} of soil is regarded as adequate for water and mineral absorption by crop plants. Bowen's (1984, Table 1) summary of root abundance for a number of tree species indicated densities ranging from less than 0.2 cm to over 8 cm cm^{-3} of soil. Data of Coile (1937) indicate root length densities of 0.5 to 2.0 cm cm^{-3} for roots less than 2.5 mm in diameter in the surface soil under various forest stands on various soil types in the North Carolina Piedmont. The concentration of roots generally decreases rapidly with soil depth and over 90% of the small roots occur in the top 12.5 cm under pine stands on heavy clay soils (Coile, 1937). As the canopy closes, the density of roots in the surface soil stabilizes and they form a dense mat that must be very efficient in intercepting minerals released by decomposition of litter. This may be particularly important in tropical rain forests (Jeffrey, 1967; Stark and Jordan, 1978).

Root Turnover

One of the remarkable characteristics of root systems of trees is the rapid turnover of fine roots that typically are short-lived. The growth and death of roots of trees are discussed in detail in Chapter 5 of Kozlowski (1971b), by Head in Kozlowski (1973), and by Lyford in Torrey and Clarkson (1975). Bowen (1984, pp. 157–160) cites considerable literature indicating that as much photosynthate often is used for root growth as for shoot growth, and Harris *et al.* (1977) reported that root growth was 2.8 times greater than aboveground wood production in both pine and hardwood forests in the southeastern United States (also see Kramer, 1983, p. 154). Many of the small roots of fruit trees also are short-lived. However, McClaugherty *et al.* (1982) suggest that the methods used to calculate root production of temperate forests may significantly overestimate the amount of roots produced. In any case, the amount of photosynthate used in growth and respiration of roots represents an important diversion from use in growth of aboveground tissue. It is claimed that reoccupation of soil by new roots improves the exploitation of less mobile elements in the soil and is more efficient than the cost of maintenance of longer-lived roots during the dormant season (Reynolds, in Torrey and Clarkson, 1975). However, it seems doubtful that the benefits compensate for the costs of rapid turnover (Caldwell, 1976). The biological costs of root systems are discussed in Givnish (1986, Chapters 6, 8, and 9).

Root–Shoot Interrelationships

The successful growth of trees depends on maintenance of a balance in growth and functions between roots and shoots such that neither suffers serious deficiencies in supplies of essential substances contributed by the other. Shoots are dependent on roots for water, minerals, and hormones such as cytokinins and gibberellins, whereas roots depend on shoots for carbohydrates, other organic nutrients, and hormones. In apple and many other trees, nitrate reduction occurs chiefly in roots (Bollard, 1958; Barnes, 1963), but Smirnoff and Stewart (1985) cite cases where NO_3 is translocated to the leaves of woody plants. If damage to the root system seriously reduces the absorbing surface, shoot growth is reduced by lack of water, minerals, and hormones. If a tree is defoliated by disease or insects, root growth is reduced because of the decreased supply of carbohydrates and other organic compounds. Thus the ratio of roots to shoots can be very important and it is not surprising to find that there appears to be an inherent tendency toward homeostasis by repair of imbalances. For example, partial defoliation or pruning usually results in increased shoot growth and reduced root growth (Wareing *et al.*, 1968; Fordham, 1972). Santantonio *et al.* (1977) reported a consistent relationship between weight of root systems and tree diameter for a wide range of trunk diameters. Carlson and Harrington (1987) reported that the sum of root cross-sectional area is closely related to stem area near the ground level in southern pines.

Some data on root–shoot ratios indicate ratios of about 1 : 5 in temperate forests and somewhat less in tropical forests. Such data are not very reliable because different investigators use methods that recover varying proportions of the root surface and static root weights are misleading because they do not account for the large turnover in fine roots. Also root weight is less important than root surface because most absorption occurs through fine roots, which contribute little to dry weight. Perhaps the ratio of root surface to leaf surface or land area would be a more meaningful measurement. The latter was attempted for some herbaceous species by Newman (1969). Competition among organs for carbohydrates was mentioned in Chapter 2 and discussed by Cannell (1985).

Mineral Uptake

The uptake of mineral nutrients depends on movement through the soil to root surfaces, the amount of root surface in contact with the soil, the absorptive capacity of the roots, and the rate at which minerals are being used in the plant. The soil and root systems have been discussed and we will now turn to root absorbing surfaces and plant demand for minerals.

Root Absorbing Surfaces

The classic view is that mineral and water absorption occurs chiefly through the unsuberized regions near the tips of roots. However, it also has been noted that few or no unsuberized root tips can be found in cold or dry soil, and Kramer and Bullock (1966) found that in midsummer less than 1% of the root surface in the upper 10 cm

Figure 6.7 Various kinds of mycorrhizal root development on Douglas-fir (*Pseudotsuga menziesii*). (A) Typical cluster, (B) gray pinnately branched type, (C) orange mycorrhiza, (D) yellow mycorrhiza with rhizomorphs extending into the soil. (From Kramer and Kozlowski, 1979, by permission of Academic Press.)

of soil under stands of loblolly pine and yellow poplar was unsuberized. Head (1967) also reported a large reduction in root growth of apple and plum in midsummer and concluded that a considerable proportion of the water and mineral uptake must occur through suberized roots. Kramer and Bullock (1966), Chung and Kramer (1975), and Queen (1967) all observed significant uptake of water and phosphorus through suberized roots and Sands *et al.* (1982) found that the resistance of suberized roots of loblolly pine to water movement was only about twice that of unsuberized roots. Bowen (1984, p. 150) gives some additional data, but reminds readers that most of these measurements were made on roots in solution and do not necessarily represent the field situation in which root–soil contact is an important factor. However, Kramer (1946) reported water absorption from moist soil through suberized roots and MacFall *et al.* (1990) observed depletion of soil water in the vicinity of woody roots. The writers think that a substantial part of the mineral and water absorption by woody plants must occur through suberized roots because the unsuberized surfaces often are too limited in extent to supply the amounts required.

Mycorrhizae

The absorbing surface of most root systems is greatly increased by the presence of mycorrhizal roots from which fungal hyphae extend out into soil not penetrated by roots or root hairs. The hyphae also can enter spaces between soil particles that are too small to be penetrated by roots. This is very important for absorption of relatively immobile elements such as phosphorus, and Bowen (1984, p. 170) suggests that the presence of mycorrhizal roots greatly increases the competitive capacity of trees. He also suggests that they may compensate for environmental conditions such as high acidity, salt, aluminum or other toxic minerals, and poor aeration that restrict root growth. For example, inoculation with mycorrhizal fungi improved seedling growth on mine dumps (Marx, 1980; Walker *et al.*, 1989). Bowen suggests that it should be possible to find mycorrhizal fungi that are tolerant of various unfavorable conditions and thus improve tree growth on a variety of poor sites. Janos (1980) reported that endotrophic mycorrhizae (vesicular-arbuscular or VAM mycorrhizae) are essential to the continued growth of seedlings of many species of trees in tropical rain forests. The importance of mycorrhizae in forestry is discussed briefly by Bowen (1984, pp. 172–174) and the role of mycorrhizal roots is discussed in more detail in Marks and Kozlowski (1973) and in Harley and Smith (1983). An example of mycorrhizal roots is shown in Fig. 6.7. The role of mycorrhizal roots is discussed later in this chapter in connection with management practices.

Functions of Minerals

Mineral nutrients have many roles in plants, including functions as constituents of plant tissues, regulators of osmotic potential, constituents of buffer systems, activators of enzymes, and regulators of membrane permeability. For example, Ca is found in the middle lamella of cell walls, K is involved in guard cell activity and is an activator of enzymes involved in starch and protein synthesis, Mg occurs in chlorophyll, P in phospholipids and nucleotides, and S in some proteins, N is an important constituent of protein, chlorophyll, and many other compounds, and P is a constituent of important organic compounds involved in metabolism and energy transfer.

Essential Elements

Because of their relative abundance in plant tissues (see Table 6.1), N, P, K, Ca, Mg, and S are known as macronutrients. Another group of elements, including Fe, Cu, Zn, Mn, B, Mo, and Cl, are required in such small quantities (Table 6.2) that they are known as micronutrients. For instance, less than 1 ppm of Mo is required for plant growth. Iron, Cu, Zn, and Mn act as coenzymes or prosthetic groups with enzymes. Molybdenum is involved in nitrogen metabolism and Cl and Mn in photosynthesis, although their roles are not entirely clear. Although Al and Na are present in significant quantities they are not known to be essential. More detailed discussions of the functions of the various elements can be found in plant physiology texts, in Epstein (1972), in Marschner (1986), and in a review by Clarkson and Hanson (1980).

What Makes an Element Essential?

An essential element (1) is necessary for completion of the life cycle as part of a molecule in some essential constituent and cannot be replaced by another similar element or (2) is directly involved in plant metabolism and does not play some secondary role. An example of an element playing a secondary role is cobalt, which is essential for the nitrogen-fixing bacteria found in nodules on legumes, but it is not directly essential for the host plant. Although sodium resembles potassium in many respects and is found in some plants in large quantities, it cannot replace potassium. Determination of the essentiality of an element requires great care to exclude contamination from water, air, containers, and reagents used in the experiments. Some of these problems are discussed by Epstein (1972) and Tinker and Läuchli (1986).

Many elements in addition to those regarded as essential are found in plants, and it seems probable that any element in the root environment may be absorbed. This has even led to the use of analyses of plant tissue as a guide in prospecting for various minerals. Examples of this are described by Carlisle and Cleveland (1958), Cannon (1960), Epstein (1972), and Warren (1972). Large quantities of Al, Si, and Na occur in plants and traces of more exotic elements such as Pt, Ag, and Au have been found in trees, but none of these elements is known to have any essential function in plants. However, Epstein *et al.* (1988) reported that maize grows better when supplied with Si than without it. *Astragalus* accumulates so much Se when growing on some soils that it is toxic to animals grazing on it, but Se probably is not essential even for *Astragalus*.

Toxic Concentrations of Minerals

Excessive soil concentrations of even the most essential elements can be as injurious as deficiencies. Injury to meristematic regions as a result of B deficiency is common, but a small excess of B is toxic, especially to conifers. A trace of Cu is essential, but an excess is very toxic, and most other micronutrients are toxic when present in high concentrations. It is claimed that excessive use of pesticide sprays containing lead, copper, and arsenic on orchards has occasionally resulted in toxic accumulations of those compounds in the soil, and mine dumps often support little vegetation because of high concentrations of heavy metals (Antonovics *et al.*, 1971). However, plants of some species growing in such situations acquire tolerance to toxic concentrations of various elements. Steiner *et al.* (1980) observed birch growing on acid spoil banks and they found that some provenances of paper birch tolerate high concentrations of aluminum. Such situations encourage the search for genotypes that are tolerant of various stresses.

Until recently it was assumed that forest soils rarely contain an excess of mineral nutrients and tree growth usually should benefit from added nitrogen. However, it appears that some forests in the northern temperate zone are receiving an excess of nitrogen from polluted air, which is modifying soil chemistry and mineral uptake and possibly contributing to the forest decline often attributed to acid rain (Aber *et al.*, 1989; Schulze, 1989). If air pollution continues to increase, problems with several other elements can be expected.

Salinity

Accumulation of excessive amounts of salt in the soil excludes trees and most kinds of shrubs from large areas, partly because of osmotic inhibition of water absorption and partly because of toxic effects of high concentrations of ions. Only a few woody halophytes such as greasewood and saltbush can tolerate highly saline soils. Mangroves grow on salty tidal flats, but salt spray often damages coastal trees and shrubs (Boyce, 1954), and salt spray from highways treated with salt to control ice formation often injures adjacent vegetation. Some species are more tolerant of soil salinity than others because their root systems tend to prevent it from reaching the shoots. For example, Cleopatra mandarin orange rootstocks exclude Cl and trifoliate orange rootstocks exclude Na from the shoots of Valencia oranges grafted on them, apparently because they sequester ions from the xylem sap in the cells of wood and bark in the root–shoot transition region (Walker, 1986; Lloyd *et al.*, 1987). Such a protection has only limited capacity and is effective only for a short time or in soil containing a relatively low concentration of salt.

Mineral Deficiencies

Mineral deficiencies are common and often limit the growth of trees and shrubs. They usually are chronic rather than catastrophic and therefore often go undetected.

Most of the research on mineral nutrition has been done on annual crop plants that must absorb their requirements of mineral nutrients within a growing season that is measured in weeks, whereas absorption by trees and shrubs continues over many years. Vigorous tree growth therefore often occurs on soil that is too low in essential mineral elements to produce profitable yields of annual crops without adding fertilizers. However, as mentioned earlier, the increasing cost of establishing and maintaining forest plantations is resulting in increasing interest in the detection and elimination of mineral deficiencies.

Nitrogen deficiency occurs worldwide and phosphorus deficiency is common in some areas such as Australia, New Zealand, and the southeastern United States and in many tropical soils. However, deficiencies of most elements occur only in limited areas, such as soils with excessively high or low soil pH or soils derived from rocks unusually high or low in certain elements. For example, trees grow poorly or fail on serpentine soils formed from rocks that yield excessive amounts of Mg and Fe relative to Ca (Proctor and Woodell, 1975). Flooding and other unfavorable soil conditions can also cause deficiencies or produce symptoms resembling those of mineral deficiencies (Kozlowski and Pallardy, 1984). This is discussed in Chapter 8.

Detection of Mineral Deficiencies

A deficiency of any of the essential elements can cause disturbance of the normal course of biochemical and physiological processes, resulting in reduced growth. Deficiencies are most likely to develop early in the growth cycle (Miller, 1986). Unfortunately reduction in growth caused by mild mineral deficiencies often goes undetected for many years unless more obvious symptoms develop. In addition to visual symptoms such as chlorosis, malformed leaves, and dieback of shoots, the nutritional status of trees can be diagnosed by plant tissue analysis, soil analysis, or testing the effects of fertilizing with specific elements. No one method is entirely satisfactory and a combination of methods often is desirable (Mead, 1984, p. 260).

Visual Symptoms

Leaves and stem and root apices are particularly sensitive to mineral deficiencies. Leaves of mineral-deficient plants tend to be small and pale in color (chlorotic) and sometimes have dead areas at the tips and margins or between the veins. Sometimes they develop in tufts or rosettes, needles of conifers become fused, and various other abnormalities in shape and color develop that enable experienced observers to diagnose the cause. Other visible symptoms include dieback of stem tips and twigs, lesions in the bark, and excessive gum formation. Some examples of deficiency symptoms are shown in Fig. 6.8, and many examples are shown in Hacskaylo *et al.* (1969). One of the most common symptoms is loss of green color caused by breakdown of or interference with synthesis of chlorophyll, which is commonly caused by deficiency of N but is also produced by deficiencies of Fe, Mn, Mg, and other elements. Chlorosis can be caused by deficient soil aeration, water stress, air

Figure 6.8 Examples of visual symptoms of mineral deficiencies. (A) Apple leaves developing magnesium deficiency. (B) Manganese-deficient (*left*) and normal tung leaves (*right*). [(A) Courtesy of Crops Research Division, U.S. Department of Agriculture; (B) courtesy of R. D. Dickey, Florida Agricultural Experiment Station.]

pollution, and an excess of minerals. Genetic factors also produce chlorosis, ranging from mottling to albino seedlings. A troublesome type of chlorosis develops in trees and shrubs growing on calcareous soil, usually resulting from Fe or Mn deficiency. Stone (1968) discussed iron chlorosis in trees. Bowen (1984, p. 262) suggested that remote sensing might be developed to detect symptoms of mineral deficiencies over large areas.

Tissue Analysis

Foliar analysis has been used extensively in horticulture and forestry to determine the level of mineral nutrients on the assumption that a low concentration of an element in leaves indicates a deficiency. However, it is necessary to establish the leaf concentrations that are necessary for vigorous growth (e.g., Wells, 1968) and to decide on a satisfactory sampling procedure. Nutrient levels vary with the season, leaf age, and position in the crown, and variations among trees often make sampling of a large number of trees necessary. These problems are discussed in detail by Bowen (1984, pp. 263–279). Ingestad (1979) regarded any deviation from the usual proportions of elements to be an indicator of deficiency. He reported a ratio of 100 : 13 : 65 : 6 : 8.5 for N:P:K:Ca:Mg for several conifers in Norway, and Adams and Allen (1985) reported a ratio of 100:9.3:36.5:17.2:9.2 for loblolly pine in North Carolina. McNeil *et al.* (1988) found a good correlation between foliar P and response to fertilization with P in loblolly pine, especially when fertilizer was applied in late summer and winter.

Some attempts have been made to use analyses of buds, phloem, roots, or fresh litter as indicators of mineral nutrition status. It has also been suggested that analysis of xylem sap would be a good indicator of mineral nutrition status (Stark and Spitzner, 1985). However, there are large seasonal and species variations in the mineral composition of xylem sap that complicate the use of this method (Barnes, 1963).

Biological Tests

Biological tests include fertilizer trials in the field and pot tests in the greenhouse. Field trials are desirable and even necessary for calibrating other methods. The design of such experiments is discussed in Chapter 16 of Bowen and Nambiar (1984). Sometimes tree seedlings are grown in pots of a problem soil with various fertilizer treatments in order to learn which element or elements produce the most improvement in growth. Ingestad (1982) and others have used solution cultures for this purpose. Ingestad and his coworkers emphasize the necessity of adjusting the supply of mineral nutrients to the rate of growth to prevent development of deficiencies as seedlings grow larger.

Soil Analysis

Analyses of soil have been used widely by farmers and gardeners to determine the kinds and amounts of fertilizer to be added. They also are used to indicate the existence of mineral nutrition problems before plantations are started. However, there are two major problems relating to their use in forestry. The first is the uncertainty involved in sampling because of the wide variations occurring in mineral nutrients over short distances laterally and with depth in forest soils. Metz *et al.* (1966) reported that the variation among soil samples was greater than the variation among tissue samples from different trees. The second problem is uncertainty concerning methods of extracting N and P that can be related to their availability to trees. Overall, soil analysis probably is less satisfactory than tissue analysis as an indicator of mineral deficiencies.

Effects on Growth and Reproduction

This section deals with the economically important effects of mineral deficiencies expressed in terms of reduced production of wood, flowers, fruit, and seed.

Vegetative Growth

It is well known that nitrogen deficiency reduces biomass production, but there is some uncertainty as to how the reduction is brought about. This uncertainty can best be answered by considering how nitrogen fertilization increases growth. There is much evidence that the most important effect of nitrogen is to produce an increase in leaf area or photosynthetic surface, per unit of land (leaf area index), resulting in more photosynthate for use in growth and reproduction. For example, nitrogen fertilization of young, N-deficient loblolly pine caused a 50% increase in leaf area index and a proportional increase in volume of wood produced (Vose and Allen, 1988). Such fertilization increased leaf area most in the middle and lower part of the crown, as shown in Fig. 6.9 (Vose, 1988). An increase in the amount of photosynthate can result from increase in number, size, or longevity of leaves, and the length of time over which they remain active in photosynthesis (Linder and Rook, 1984, pp. 214–217). Nitrogen fertilization sometimes also increases the amount of palisade tissue, the specific leaf weight (weight of leaf tissue per unit of area), and stomatal frequency, all of which are favorable for high rates of photosynthesis (Kozlowski and Keller, 1966). The effects on leaf area are most important in young stands of trees and become less important after the canopy is closed and the benefit of additional leaf surface is counterbalanced by self-shading. Chlorophyll deficiency is a fair indicator of mineral deficiency and it often is at least moderately correlated with decreased photosynthesis and biomass production. Although nitrogen deficiency sometimes decreases the rate of photosynthesis, it seems likely that the most important effect of mineral deficiency, at least in conifers, is exerted through reduction in photosynthetic surface (leaf area) rather than through decrease in rate of photosynthesis (Brix, 1983). The seasonal duration of photosynthesis may be reduced in deciduous species by the shorter life of mineral-deficient leaves.

Reproduction

In general, mineral and especially nitrogen deficiency reduces flowering and seed production and fertilization usually increases seed production in seed orchards (Sweet, 1975; Gregory *et al.*, 1982). However, exactly why this occurs is uncertain, although gibberellic acid probably plays a role, at least in conifers (Pharis and King, 1985). A high ratio of carbohydrate to nitrogen and retarded shoot growth are regarded as favorable to flowering, and girdling of stems and branches has been used to increase flower and fruit production of apples and pine trees (Ross *et al.*, 1985). Fruit and seed production requires quantities of both carbohydrate and

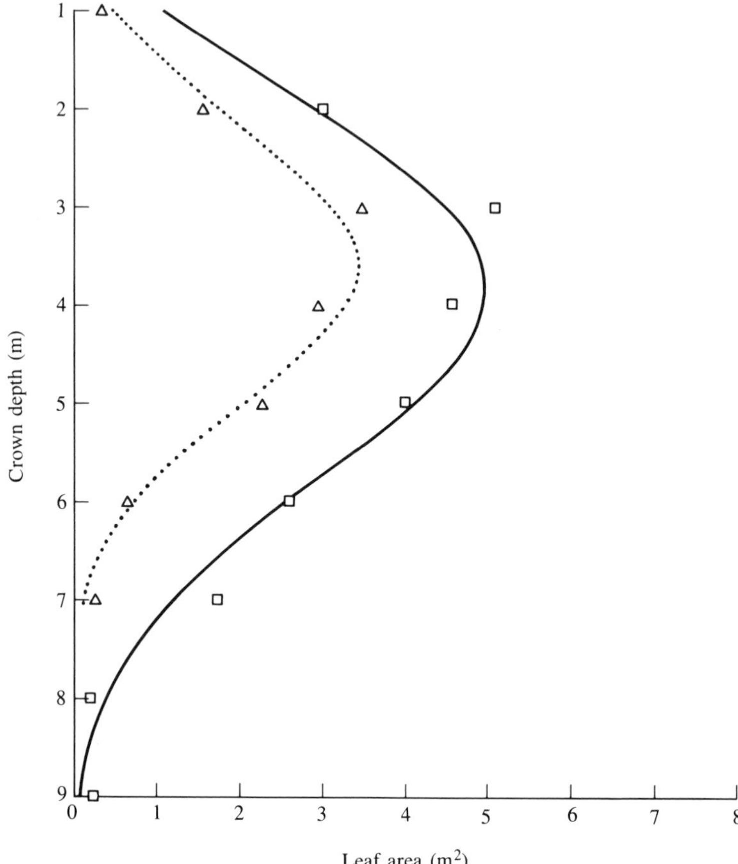

Figure 6.9 Effect of nitrogen fertilization on foliage area of crown of young fertilized (□) and unfertilized (△) loblolly pine trees. Note that most of the increase in leaf area occurred in the middle and lower part of the crown. (From Vose, 1988.)

nitrogen and both are increased by fertilization with nitrogen. According to Cromack and Monk (1975), over 10% of the minerals in the total litter fall of an oak stand was in the acorns, and a crop of apples is said to remove 80 kg ha^{-1} and a crop of oranges 200 kg ha^{-1} of minerals (Labanauskas and Handy, 1972).

Partitioning of Photosynthate

Mineral-deficient plants growing on infertile soils tend to have a higher proportion of roots to shoots than plants growing on fertile soils. For example, Linder and Axelsson (1982) reported that in young Scotch pine trees on infertile soil, more than 60% of the photosynthate was allocated to root growth, but less than 40% was allocated to roots on fertile soil. In a study of Douglas-fir, root production accounted for 23% of the total annual biomass production (17.8 tons ha^{-1}) on a good site and 53% of the total (15.4 tons ha^{-1}) on a poor site (Keyes and Grier, 1981). Thus mineral deficiencies not only reduce the total production of photosynthate but also affect its partitioning. In fact, Axelsson and Axelsson (1986) suggested that the decrease in allocation of dry matter to production of fine roots is an important reason for the increase in production of stem wood resulting from improvement in mineral nutrition. However, in contrast, Nadelhoffer *et al.* (1985) concluded from experiments on deciduous trees in Wisconsin that increase in soil nitrogen had little effect on the ratio of above- to belowground components. Such relationships probably vary with age, species, soil fertility, and degree of water stress.

Species Differences in Tolerance of Deficiencies

Tree species differ significantly in their capacity to absorb various mineral elements and in their tolerance of limited supplies of essential elements. For example, leaves of flowering dogwood and white oak contain over twice as much Ca as leaves of post oak and loblolly pine growing on the same soil (Coile, 1937). Incidentally, a high concentration of Ca is supposed to hasten decomposition of litter. On the infertile sand plains of New York, Norway and white spruce trees showed severe symptoms of mineral deficiency or even died, whereas Scotch and jack pine showed only slight symptoms and mugo pine showed none. There were differences among individual trees of eastern white pine, with some trees dying while others grew well (Heiberg and White, 1951). It was reported by Kahdr *et al.* (1965) that orange trees growing on trifoliate orange rootstock are more susceptible to zinc and iron deficiency on calcareous soils than when they are grown on rough lemon rootstocks. This situation emphasizes the value of identifying individuals or families of trees with unusual tolerance of low fertility for planting on infertile soils. Some of these problems were discussed by Goddard *et al.* (1976) with reference to loblolly and slash pine. They concluded that there was more likelihood of finding families of slash than of loblolly pine that responded well to phosphorus, and considered it unlikely that strains of either species unusually responsive to nitrogen would be found.

Management Practices in Relation to Soil Fertility

Although cultural and management practices will be discussed in detail in Chapter 13, it is useful to discuss briefly some problems related to soil fertility and the uptake of minerals. These include the use of mycorrhizae and nitrogen-fixing understory plants, the effects of whole-tree harvesting, and the special problems of tropical soils. Fertilization is an important part of soil fertility management, but it will be treated in Chapter 13 on cultural practices.

Use of Mycorrhizae-Forming Fungi

The roots of most trees and shrubs and many herbaceous plants are invaded by fungi that form symbiotic associations called mycorrhizae. The most common type, called ectotrophic mycorrhizae, form mycelial mats on root surfaces, penetrate between the root cells, and cause dichotomous branching and development of the characteristic coralloid clusters of roots shown in Fig. 6.7. Some genera of trees, including *Acer, Liquidambar, Liriodendron,* and *Citrus,* and many herbaceous plants develop endotrophic mycorrhizae, also known as vesicular-arbuscular mycorrhizae (VAM). In this type the hyphae penetrate the host cells but do not form mycelial mats on root surfaces nor modify root morphology. Sometimes both types occur on the same plant. Mycorrhizal development is a complex process affected by a variety of internal and external factors (Slankis and Hacskaylo, in Marks and Kozlowski, 1973; Harley and Smith, 1983). Growth of mycorrhizae seems to require a surplus of carbohydrates in roots and is most abundant where moderate mineral deficiencies inhibit growth, resulting in carbohydrate accumulation. Very fertile soil and heavy shading both suppress mycorrhizal formation.

The beneficial effects of mycorrhizae result largely from the increased mineral-absorbing surface provided by the hyphae and rhizomorphs (bundles of hyphae) extending out into the soil. They extend farther out in the soil than root hairs, live longer, and may function for months. This is particularly important in phosphorus-deficient soils and it has proven impossible to grow trees in some areas unless they are inoculated with mycorrhizae-forming fungi. Mycorrhizae also increase uptake of many other mineral nutrients (Bowen, 1984, p. 167) and water, and probably improve drought tolerance (Dixon *et al.*, 1980; Duddridge *et al.*, 1980). It has been demonstrated that inoculation in nursery seedbeds usually results in improved growth of pine seedlings (Marx *et al.*, 1984). Inoculation is important because nursery soil often loses its naturally occurring flora during fumigation and spraying for disease control (Marx and Cordell, 1985). Dixon *et al.* (1983) found that inoculation of bare-rooted and container-grown black oak seedlings with mycorrhizae-forming fungi decreased their water stress after outplanting. In addition to their direct effects through increasing the absorbing surface, mycorrhizal roots can absorb a wide range of nitrogen compounds, convert P to more readily mobile forms, and increase formation of soil aggregates (Bowen, 1984, p. 168). It also has been suggested that the mycorrhizal mat developed in the litter under tropical trees intercepts minerals released by decaying litter and returns them promptly to the trees (Went and Stark, 1968; Stark and Jordan, 1978).

The benefits and energy costs of mycorrhizal development are discussed in more detail in Bowen and Nambiar (1984, Chapter 6), Harley and Smith (1983), and Marks and Kozlowski (1973). Their role in water absorption is discussed in Chapter 7.

Understory Vegetation

The amount of understory vegetation varies widely, depending chiefly on the amount of light penetrating the canopy, and generally is greater under deciduous forests than under coniferous forests. For example, there may be practically no understory vegetation under a well-stocked stand of loblolly pine, but Gosz (1980) found that the litter fall from understory vegetation in a trembling aspen stand in New Mexico amounted to 30% of the total litter fall. MacLean and Wein (1977) reported that the understory vegetation in an open 16-year-old stand of jack pine contained 25% of the N and Ca, 30% of the P, and over 65% of the K in the aboveground biomass. Thus litter from deciduous understory vegetation sometimes plays an important part in mineral recycling. It usually decays more rapidly than coniferous vegetation, thus increasing the rate of release of minerals into the soil pool.

It has been known for centuries that legumes are beneficial to other plants, and Davey and Wollum (1984, p. 375) have quoted English poetry dated 1613 stating that alder nourishes other plants growing near it. Broom (*Cytisus* spp.) was used with conifers in England and lupine (*Lupinus* sp.) in Germany, but with mixed results. Alder is used extensively as an understory plant in forests in northern Europe and Japan and its value is now appreciated in the Pacific Northwest. On the Pacific Coast of the United States, *Ceanothus* also contributes substantial amounts of nitrogen (Binkley *et al.*, 1982). Two forms of nitrogen-fixing organisms are important: bacteria of the genus *Rhizobium*, which grow and form nodules on the roots of legumes, and actinomycetes of the genus *Frankia,* which grow on the roots of a number of other genera, including *Alnus, Myrica,* and *Casuarina.* The latter sometimes are called actinorhizal plants. Much of the research on nitrogen fixation by actinorhizal plants was summarized by Bond (1976) and Youngberg and Wollum (1970).

Davey and Wollum (1984) suggested that nitrogen-fixing plants may be used in plantations by interplanting, underplanting, or in rotation. For example, red alder was interplanted with Douglas-fir, oleaster with black walnut, and black locust with various other species. In those experiments the presence of nitrogen-fixing vegetation resulted in measurable increases in growth of the more valuable species. It has been suggested that various tropical leguminous trees might be interplanted with other more valuable timber species, and this deserves investigation. The chief problem with interplanting of nitrogen-fixing species with other species is that the former sometimes outgrow and suppress the latter. This can be avoided by the choice of shrubs such as alder or by using slow-growing trees.

Underplanting with leguminous herbs such as clovers provides protection from erosion in addition to supplying nitrogen. When *Trifolium subterraneum* and *T. incarnatum* were planted under sycamore, height growth was more than doubled in four years (Haines *et al.*, 1978), and the use of *Trifolium* in loblolly pine plantations in the southeastern United States also is being investigated (Davey and Wollum, 1984, p. 371). In both interplanting and underplanting, root competition for water and minerals can be a problem. This is less serious if one kind of plant is deeper rooted than the other so that it obtains much of its water from a deeper soil horizon. It is even possible that deep-rooted plants may transfer water from deep to shallow soil horizons, thus benefiting the shallow-rooted plants after the surface soil dries (Corak *et al.*, 1987; Field and Goulden, 1988; Caldwell and Richards, 1989).

Rotation of trees with nitrogen-fixing plants ranging from annual legumes to shrubs or trees has been proposed, however, Binkley (1986, p. 225) regards mixed plantings as more profitable. This is already widely used in agroforestry systems in the tropics, where woody species such as *Leucana* supply both nitrogen and much needed firewood. The role of trees in agroforestry is discussed in Chapter 13 and by Huxley and others in a book dealing with many aspects of attempts to grow herbaceous crops with trees (Huxley, 1983).

Whole-Tree Harvesting

The method of harvest used and the post-harvest treatment affect the degree of disturbance of mineral cycling in a harvested forest ecosystem. In recent years there has been increasing interest in harvesting practically the entire tree, sometimes including as much of the root system as can be conveniently removed. It seems obvious that the more biomass that is removed in a harvest, the more minerals will be removed. However, the amount of minerals removed is not proportional to the amount of biomass because different parts of a tree contain different concentrations of various elements, a point emphasized by Binkley (1986, Chapter 7) and McColl and Powers (1984, pp. 397–399). The differences in amounts of the major elements in various components of a 16-year-old pine stand are shown in Table 6.1, for the microelements in Table 6.2. In that plantation, 32% of the nitrogen in the aboveground biomass was in the needles, 23% in the branches, 30% in stem wood, and 14% in the stem bark. Thus removal of only the stems for pulpwood would leave over half the nitrogen and substantial amounts of the other elements in the slash, all of which would be removed by harvesting the entire aboveground portion. About 20% of the N, 27% of the K, and 35% of the P in the entire trees was in the roots, hence harvesting the roots would remove considerably more minerals. Some data for a hardwood forest, summarized by Binkley (1986, p. 170) and given in Table 6.7, show the amounts of biomass and principal mineral elements removed by stem and whole-tree harvesting of an oak–hickory forest. In general, whole-tree harvesting probably removes two or three times as much minerals as harvesting sawlogs and leaving the slash.

Perhaps the important question is not how much mineral is removed, but how the amount removed compares with the amount available in the soil for regrowth after harvesting. This depends on the original soil fertility, the relative amounts of minerals in the ecosystem that occur in the soil and in the biomass, and the rate at which the soil mineral pool is replenished by decomposition of slash, wet and dry deposition from the atmosphere, and weathering of rocks. As mentioned earlier, a considerably larger fraction of the total nitrogen and other mineral elements in the ecosystem occurs in the biomass of tropical forests than in temperate zone forests, and the least occurs in boreal forests. Also more minerals are found in the biomass of temperate zone deciduous trees than in the biomass of conifers (McColl and Powers, 1984, pp. 371 and 397). Thus whole-tree harvesting probably will remove relatively more minerals from tropical than from temperate forest ecosystems and

Table 6.7
Amounts of Biomass and Mineral Nutrients in an Oak–Hickory Forest in Tennessee, and the Amounts Removed in Sawlogs and by Whole-Tree Harvesting[a,b]

Component	Biomass	N	P	K	Ca
Tree					
Foliage	3,900	60	4	50	40
Branch	35,300	85	7	35	200
Stem	133,800	40	16	90	910
Stump	14,700	30	2	10	100
Total tree	187,700	215	29	185	1,250
Forest floor	13,700	150	12	20	160
Soil (0 to 45 cm)					
Extractable	—	15	40	280	500
Total	78,700	3,000	1,370	24,200	6,070
Total ecosystem	280,100	3,380	1,451	24,685	7,980
Harvest removal					
Stem only	57,300	99	6	32	370
Whole tree					
Branches	31,700	75	6	30	180
Stems	133,800	240	16	90	910
Total	165,500	315	22	120	1,090

[a]From Binkley (1986), by permission of John Wiley & Sons.
[b]Calculated from Johnson *et al.* (1982).

more from deciduous than from coniferous forests of similar biomass (Phillips and Van Loon, 1984). Also, such removal is more injurious on infertile soil, where deficiencies already exist. Some data on losses of nutrients by harvesting and time required for replacement are summarized in Waring and Schlesinger (1985, p. 152). Binkley (1986, Chapter 7) discussed the consequences and economics of whole-tree harvesting in detail.

Slash management also affects the amount of mineral nutrients lost by harvesting trees. Much nitrogen is lost during slash burning, which typically creates hot fires (see Chapters 11 and 13), and the losses are particularly serious in tropical forests. Pushing slash into heaps or rows before burning often is accompanied by transfer of considerable topsoil and redistribution of minerals, which result in an overall decrease in site quality. The problems of site preparation in relation to mineral nutrition are discussed in Chapter 13 and in more detail by Binkley (1986, Chapter 7).

Short Rotations

The increasing interest in shorter rotations also should be considered with respect to its effect on mineral nutrition of trees. Mitchell (in McColl and Powers, 1984, p. 398) reported that two 20-year rotations of Monterey pine in Australia would increase the loss of N and P by 18% over a single 40-year rotation, and Crane and Raison (1980) concluded that fertilization will be necessary to sustain yields of

Monterey pine on short rotations. Converting from one 30-year rotation for trembling aspen to three 10-year rotations was estimated to increase depletion of N, P, and K by 345, 239, and 234%, respectively (Boyle, 1975). This is largely because young trees contain higher concentrations of minerals than that in the larger proportion of mature tissue found in older trees. During a long rotation there is more likelihood of the soil mineral pool being replenished by natural processes, but this will not occur during short rotations and fertilization will become increasingly important. The loss of nutrients that occurs during site preparation for replanting also will be increased by two harvests instead of one.

Tree harvesting and subsequent site preparation often cause considerable damage to the environment and this is increased by shorter rotations. For example, logging machinery often causes soil compaction, thus decreasing infiltration of water and soil aeration and reducing root growth, and the effects sometimes persist for decades. The problem of soil compaction is discussed in Chapter 8. Although clear-cutting usually does not seriously increase loss of minerals in drainage water (McColl and Powers, 1984, pp. 399–401), there are notable exceptions (Likens *et al.*, 1970). The amount of nutrient loss depends on the soil type, rainfall pattern, and rapidity of regrowth of new vegetation. In one experiment, heavy loss of minerals occurred because regrowth of vegetation was prevented by the application of herbicides (Likens *et al.*, 1970), but this is abnormal. Where rapid regrowth of vegetation is occurring, loss of minerals in the drainage water often decreases after 2 or 3 years because they are increasingly bound in the biomass.

Site Deterioration

There are conflicting reports concerning decline in site quality and rate of tree growth during a second or third rotation, but this has been reported from South Africa, Australia, and New Zealand, especially for Monterey pine on infertile soils (McColl and Powers, 1984, pp. 405–407). Binkley (1986, p. 184) attributes this decline chiefly to mineral deficiencies, but deterioration of soil structure, decrease in mycorrhizal fungi, and accumulation of toxic substances also have been suggested as causes. The reported decline in site quality resembles the "replant problem" encountered in horticulture, in which difficulty is encountered in successfully replanting apple, peach, and citrus orchards and vineyards (Yadava and Doud, 1980). The failure to obtain good survival and vigorous growth has been attributed to accumulation of allelochems, nematodes, pathogenic fungi, and deteriorating soil structure and aeration. It is likely that as forestry adopts the use of shorter rotations, it too will encounter more of the problems already existing in horticulture.

Problems of Tropical Soils

Climatic conditions in the tropics vary from those of deserts to cold mountain plateaus, but most of the trees occur in lowland rain forests. Although there is considerable variation in the characteristics of tropical soils (Vitousek, 1984), most are fairly deep and well drained, but they are commonly acidic and infertile, and often high in aluminum, deficient in phosphorus, and low in exchangeable calcium. Because of the heavy rainfall and low cation-exchange capacity, most of the silica and cations are leached out, but contrary to popular belief, tropical soils are not necessarily very low in organic matter (Sanchez, 1973).

Marbut and Manifold (1926) stated that most of the soils in the Amazon Basin resemble the principal soils of the southeastern United States. According to Sanchez *et al.* (1982), the surface soils of the Amazon Basin generally have favorable physical properties, but the acidic and infertile subsoils present a chemical barrier to root penetration. Thus tropical forests tend to be shallow rooted, with most of the roots in the top 30 cm, as Greenland and Kowal (1960) reported for the forests of Ghana. Of course Coile (1937) found that over 90% of the roots less than 2.5 mm in diameter were in the top 12.5 cm of soil under pine and oak forests on the clay soils of the North Carolina Piedmont. According to Sanchez *et al.* (1982), only a small percentage of tropical soils are subject to laterization and harden when exposed. However, because of the heavy rainfall typical of those regions, after harvesting they are subject to sheet erosion and need to be protected by cover crops. The percentage of phosphorus in the litter generally is higher in tropical than in temperate forests and phosphorus and other minerals usually are cycled more efficiently by tropical forests (Vitousek, 1984). Tropical forests and tropical soils obviously present different problems from temperate zone forests and deserve more study.

Summary

Growth and yield of all kinds of plants vary widely because of differences in the site quality or productive capacity of the soil in which they are growing. The site quality for trees often is expressed as actual or calculated height at age 50 years. This depends on the capacity of the soil to supply water and minerals and on those physical properties that determine its suitability for root growth. A good soil consists of about 50% solids and 50% pore space, the latter being occupied by approximately equal volumes of water and air in well-drained soils. Sandy soils are loose, noncohesive, and well aerated, but with a low capacity to store water and minerals. Clay soils are compact, cohesive, and often poorly drained and aerated, but with a large capacity to store water and minerals. Loam soils are intermediate in properties. Most soils show important changes in properties with increasing depth that affect their capacity to store water and minerals and allow root growth.

Soils most favorable for plant growth contain relatively large amounts of readily available water between field capacity and the permanent wilting percentage, and also contain over a dozen minerals essential for plant growth. The pool of minerals in the soil is derived from rock weathering, deposition from the atmosphere, and recycling by decomposition of dead organic matter. Minerals are depleted chiefly by plant absorption and by leaching in drainage water. Fire also vaporizes much nitrogen and some is lost by denitrification. Mineral and water uptake requires extensive, much branched root systems, and the efficiency with which minerals are absorbed often is increased by the presence of mycorrhizal roots.

The most common mineral deficiencies are N and P, but deficiencies of other essential elements are common locally. Occasionally toxic concentrations of essential elements occur, and high concentrations of salt reduce growth or exclude plants from some soils. Mineral deficiencies reduce growth by hindering important physiological and biochemical processes. Deficiencies usually can be recognized by visual symptoms such as chlorosis or malformation of leaves or twigs or by tissue or soil analysis.

Current forest practices such as short rotations and whole-tree harvest deplete the soil mineral pool rapidly and are likely to increase the need for fertilization. Tropical soils present special problems because a larger percentage of the total mineral pool occurs in the vegetation in the tropics than in temperate forests, hence deforestation is more apt to result in depletion of minerals in the tropics. Nitrogen-fixing understory crops may be one answer to this problem in some circumstances. In general there is need for much more research on the detection and cure of mineral deficiencies in woody plants.

General References

Binkley, D. (1986). "Forest Nutrition Management." Wiley, New York.

Bowen, G. D., and Nambiar, E. K. S., eds. (1984). "Nutrition of Forest Plantations." Academic Press, London.

Epstein, E. (1972). "Mineral Nutrition of Plants: Principles and Perspectives." Wiley, New York.

Gregory, P. J., Lake, J. V., and Rose, D. A., eds. (1987). "Root Development and Function." Soc. Exp. Biol. Seminar Series 30. Cambridge University Press, Cambridge.

Harley, J. L., and Smith, S. E. (1983). "Mycorrhizal Symbioses." Academic Press, London.

Hillel, D. (1982). "Introduction to Soil Physics." Academic Press, New York.

Marks, G. C., and Kozlowski, T. T. (1973). "Ectomycorrhizae." Academic Press, New York.

Marschner, H. (1986). "The Mineral Nutrition of Higher Plants." Academic Press, London.

Mengel, K. and Kirby, E. A. (1982). "Principles of Plant Nutrition," 3d ed. International Potash Institute, Berne, Switzerland.

Pritchett, W. G. (1979). "Properties and Management of Forest Soils." Wiley, New York.

Tinker, B., and Läuchli, A. (1986). "Advances in Plant Nutrition." Vol. 2. Praeger, New York.

Waring, R. H., and Schlesinger, W. H. (1985). "Forest Ecosystems." Academic Press, Orlando, FL.

Wild, A. (1989). "Russell's Soil Conditions and Plant Growth." Wiley, New York.

Chapter 7

Water Stress

Introduction

Everyone who has any experience with plants is aware of the importance of an adequate supply of water and has observed the injurious effects of drought. Water, temperature, and nitrogen usually are the most common limiting environmental factors for plant growth, and wherever temperature and mineral supply permit plant growth, the quantity and quality depend chiefly on the water supply. Regions with abundant, well-distributed rainfall develop luxuriant forests containing a variety of species, as in tropical rain forests, the mixed mesophytic forests of the southern Appalachian mountains, and the coniferous forests of the Pacific Northwest. Regions with consistently severe seasonal droughts tend to be covered with chaparral or grasslands and those with the least rainfall produce only desert scrub (Chabot and Mooney, 1985). This correlation between water supply and plant distribution and growth (Currie and Paquin, 1987) exists because water stress affects most of the physiological processes involved in growth. As a result, successful competition for water often becomes a dominant factor in success, both in dry Mediterranean climates (Schlesinger *et al.*, 1982) and in humid climates (Bunce *et al.*, 1977; Bahari *et al.*, 1985).

Frequency of Droughts

Drought is difficult to define objectively (McWilliam, 1986), but it will be treated in this book as a meteorological phenomenon (May and Milthorpe, 1962), namely, a period without rainfall of sufficient duration that it produces prolonged plant water deficits and reduces growth. Even in humid regions, droughts of sufficient severity to affect growth and crop yield occur frequently enough to be regarded as a major problem. Decker (1983) discussed several methods of determining the probability and intensity of droughts, including the amount of rainfall, the impact on crop yields, and the amount of available water stored in the soil. He concluded that it is most reasonable to base a definition of drought on the extent to which soil moisture is depleted. Van Bavel calculated the probability of drought from rainfall and estimated evapotranspiration and the capacity of the soil in the root zone to hold water (Van Bavel and Verlinden, 1956). Using this formula and assuming a soil water-holding capacity of 7.5 cm (3 in.) of available water in the root zone, the probable number of drought days for crop plants 5 years out of 10 is 20 in central North Carolina, 40 to 50 in central Georgia, and 70 to 80 in southern Arkansas and northern Louisiana, but only 30 to 40 in central Minnesota (Table 7.1). Because of their deeper rooting (1 m or more), the frequency of drought days for most kinds of trees should be lower than for most annual crops, but no comparable data are available for trees.

The frequency of drought depends not only on the amount of precipitation, but also on the rate of soil water depletion by evaporation and transpiration. Thus an

Table 7.1
Probable Number of Drought Days in 5 Years out of 10, Assuming 7.6 cm of Available Water in the Root Zone[a]

Region	Drought days
Central Virginia	20
Central North Carolina	20
Central South Carolina	40–50
Central Georgia	40–50
Northern Louisiana	70–80
Central Minnesota	30–40

[a]After Kramer (1982).

amount of rainfall that can support a forest in a cool climate such as that of Minnesota can only support grasslands in the warmer climate of Texas. This fact led Transeau (1905) to develop the concept of the precipitation–evaporation ratio, which was refined by Thornthwaite (1948) and others and is incorporated in climatic diagrams such as those of Walter (1979, pp. 25–30). A simple diagram showing the precipitation–evaporation ratio for the eastern half of the United States is shown in Fig. 7.1.

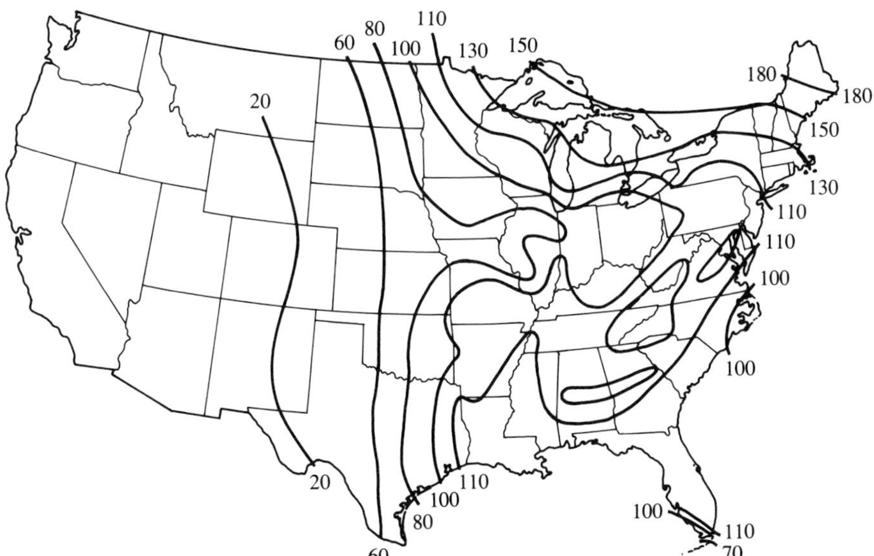

Figure 7.1 Ratios of precipitation to evaporation for the eastern and central United States, as calculated by Transeau (1905). (From "The Study of Plant Communities," 2d ed. by Henry J. Oosting. Copyright © 1956 by Henry J. Oosting. Reprinted with permission by W. H. Freeman and Company.)

As stated in Chapter 2, the close relationship between water supply and plant growth exists because plant water deficits affect almost every aspect of tree physiology and morphology. Thus it is important to examine the causes of plant water deficits and the mechanisms by which water stress reduces growth. We will first discuss the nature of water deficits and their causes, and later examine some effects on specific physiological processes.

What Constitutes Plant Water Stress

The definition of plant water stress depends somewhat on the objectives of the observer. Farmers, foresters, horticulturists, and others who grow plants evaluate stress in terms of reduction in quantity and sometimes quality of economic yield, whereas ecologists and foresters also are interested in the effects of water stress on the growth of forests and other plant communities. Physiologists evaluate water stress in terms of loss of turgor, reduction in growth, closure of stomata, inhibition of processes such as photosynthesis, and disturbance of the normal course of other processes such as nitrogen and carbohydrate metabolism. For example, starch tends to be hydrolyzed to sugars, protein synthesis is reduced, and proline and other intermediate compounds accumulate in water-stressed plants. The causes of some of these reactions will be discussed later.

Causes of Water Stress

Water stress usually is attributed to drought, but it develops whenever water loss exceeds absorption long enough to cause a decrease in plant water content and sufficient loss of turgor to cause a decrease in cell enlargement and perturbation of various essential physiological processes. For instance, cold soil or high salinity can reduce water absorption and cause plant water stress. The plant water balance can be treated as analogous to a bank balance that depends on the relationship between deposits and withdrawals. Thus, using economic terminology, plant water stress can be caused by either rapid transpiration (excessive withdrawals) or slow absorption (inadequate deposits), or in hot, dry weather a combination of the two. In the absence of irrigation, drought eventually results in sustained plant water stress, but transient water stress often develops on hot sunny days in the absence of drought (Figs. 7.2, 7.4, 7.6, and 7.7). For example, Sucoff (1972) concluded that red pine growth on a sandy soil in Minnesota was limited by water stress every day during the summer of 1969. Although water relations are commonly treated as a continuum of soil, plant, and atmospheric water, absorption and transpiration, which control plant water balance, are not tightly coupled, particularly in large woody plants. This is partly because of the capacitance factor provided by the readily available water stored in the leaves and stems of herbaceous plants and the sapwood and inner bark of trees and partly because absorption and transpiration are controlled by different factors. The importance of water storage is discussed later in connection with drought tolerance.

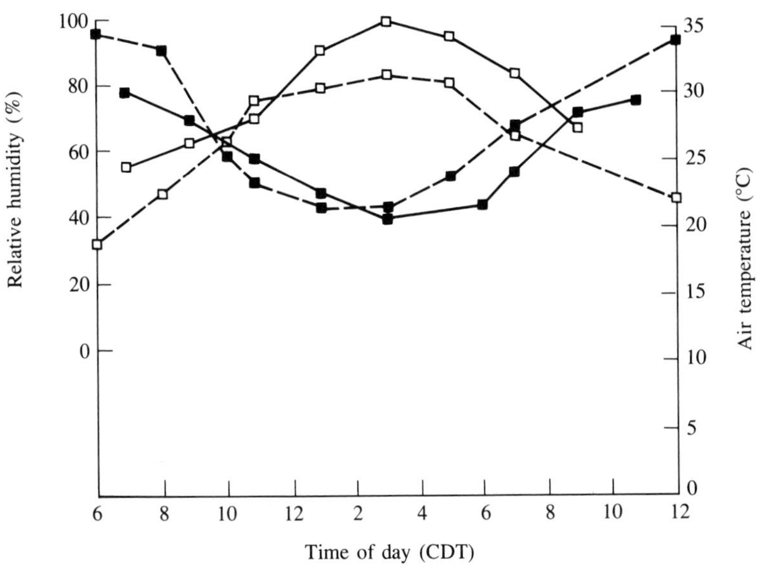

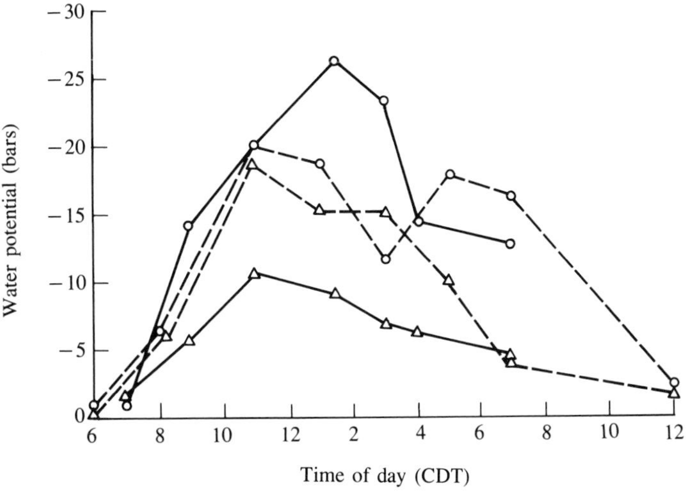

Figure 7.2 *Top,* Change in relative humidity (■) and air temperature (□) over a day in mid-July (– –) and mid-August (——). *Bottom,* Afternoon decrease in leaf water potential of seedlings of two tree species (*Ulmus alata,* ○; *Diospyros virginiana,* △) in mid-July (---) and mid-August (——) in an open field in southern Illinois. (From Bacone *et al.*, 1976.)

In moist soil, water absorption is controlled largely by the rate of transpiration, but in drying soil it is gradually reduced by the decreasing difference in water potential between roots and soil and the increasing resistance to water movement toward roots through drying soil (Kramer, 1983a, Chap. 9). In hot, sunny weather as transpiration increases rapidly in the morning, water usually is removed first from leaves and adjacent sapwood because the resistance to removal of water from the cells of those tissues is lower than the resistance to intake through the roots. With increasing stress, resistance to movement through the xylem may be increased by cavitation (Tyree *et al.*, 1984; Tyree and Sperry, 1989). As a result, water absorption of trees and forests often lags behind transpiration during the morning and early afternoon, as shown in Fig. 7.3. The resulting midday water deficits are often severe enough to cause temporary wilting, stomatal closure, reduction in photosynthesis, and even shrinkage of stems and fruits (Fig. 7.4), but in moist soil, recovery usually occurs in the late afternoon or overnight, accompanied by swelling of plant organs.

The severity and duration of midday water stress therefore vary considerably in different parts of a tree. The exposed leaves are subject to significant water deficits on almost every hot, sunny day, but this deficit may not reach the lower part of the trunk and roots until several hours later and usually disappears sooner each day in

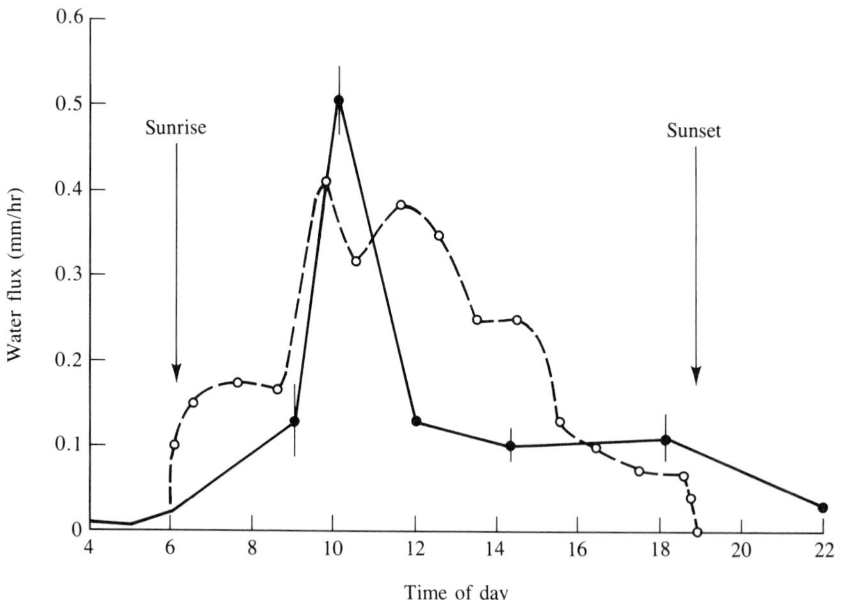

Figure 7.3 The lag of absorption (●) behind transpiration (○) during daylight hours in a 36-year-old Scotch pine forest. The daytime lag was about 30%, but it was compensated for by excess absorption overnight. Uptake was estimated from radioactive tracers and transpiration from leaf area, stomatal conductance, and meteorological conditions. (After Waring *et al.*, 1980; from Waring and Schlesinger, 1985.)

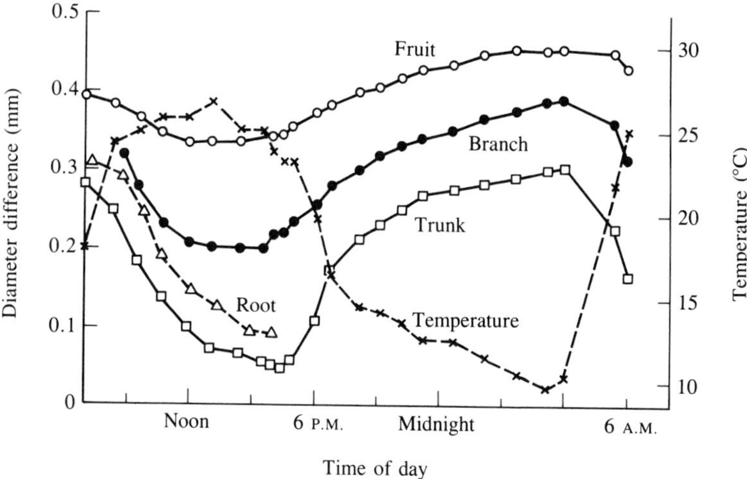

Figure 7.4 Midday shrinkage of various parts of an avocado tree caused by water deficits produced by rapid transpiration. (After Schroeder and Wieland, 1956; from Kramer, 1983.)

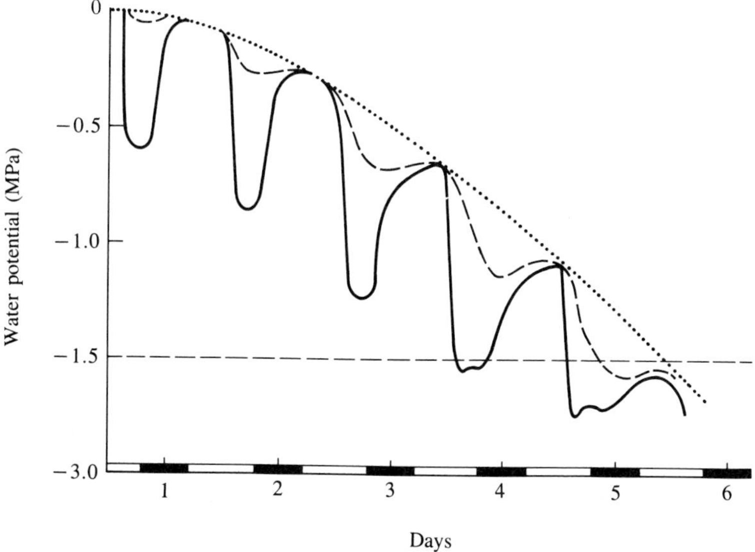

Figure 7.5 Diagram showing probable daily changes in leaf (——) and root (−−) water potential of a transpiring plant rooted in drying soil (····). The dark bars indicate darkness. (After Slatyer, 1967; from Kramer, 1983.)

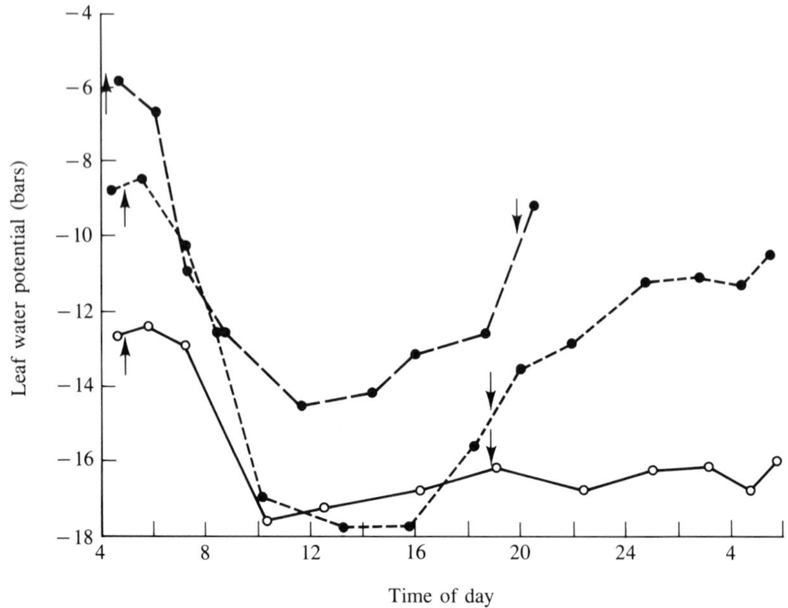

Figure 7.6 Daily cycle in leaf water potential of red pine trees showing slowing recovery with decreasing soil moisture during a drought. Tree W, July 1 (—–), was in moist soil; tree R, August 20 (---), was in intermediate soil; tree W, August 20 (——), was in dry soil. Arrows denote sunrise (*left*) and sunset (*right*). (See Sucoff, 1972, for more details.)

those regions than in the upper part of trees (Jarvis, 1981b). Hinckley *et al.* (1974) found that development of stem water deficits lagged 0.5 to 3 hr behind leaf water deficits in white oak seedlings less than 2 m in height and the lag was up to 6 hr in taller trees. Schulze *et al.* (1985) reported lags of 2 to 3 hr in larch and spruce trees. Thus leaves are likely to be subjected to more severe water stress for longer periods of time than stem or root tissues. As soil dries, resistance to water movement toward roots increases and the soil water potential decreases until it approaches the plant water potential. As a result, absorption becomes too slow to replace the water loss of the previous day and permanent wilting develops, as shown in Fig. 7.5. An example of the effect of drying soil on the daily cycle of leaf water potential of red pine is shown in Fig. 7.6.

Plant Water Relations

The water relations of trees and shrubs can be considered in terms of both whole plants and stands of plants, as discussed in Chapter 2 and later in this chapter. However, most of the water is found in cells, which are deeply involved in absorption and transpiration, and plant water status is determined by cell and tissue water status. It therefore seems desirable to discuss cell and tissue water relations in more detail before discussing the absorption, translocation, and loss of water by plants. This will be followed by discussion of some specific effects of water stress.

Cell Water Relations

The water relations of cells depend on their water potential, which in turn depends on the osmotic potential of the vacuolar sap and the turgor or pressure potential. Water potential refers to the free energy level or the capacity of the water to do work, and the relationship among these terms can be shown by the following equation:

$$\Psi_w = \Psi_s + \Psi_p$$

where Ψ_w is the total water potential of the cell, Ψ_s the osmotic potential of the cell sap, and Ψ_p the turgor potential or turgor pressure. Ψ_s is negative because the presence of solutes lowers the potential of cell sap below that of pure water, whereas Ψ_p ranges from zero in a flaccid cell to a value equal to Ψ_s, but opposite in sign, in a fully turgid cell. The relationships can be seen in the following tabulation, which for convenience disregards the accompanying change in cell volume. The pascal (Pa) is the international unit for pressure and 10^5 Pa equal 1 bar or 0.987 atmosphere and 10 bars equal 1 MPa or 10^6 Pa.

	Ψ_w	=	$\Psi_s + \Psi_p$	
Fully turgid	0	=	−2.0 + (+2.0)	MPa
Partly turgid	−1.0	=	−2.0 + (+1.0)	MPa
Flaccid	−2.0	=	−2.0 + 0	MPa

The role of cell water potential in plant water relations can be illustrated as follows. During the morning the turgid leaf cells of a well-watered tree or shrub lose water by transpiration, reducing the pressure term and cell volume and decreasing the cell water potential below zero. This causes flow of water from the xylem into the leaf cells, producing tension in the xylem sap, which is transmitted in the sap stream to the roots (see the section Ascent of Sap). The tension lowers the potential of the xylem sap, causing water to move into the xylem from root cells, while the reduced potential in the root cells causes inflow of water from the soil. Thus water movement from soil to leaves through a transpiring plant can be regarded as occurring along a gradient of decreasing water potential produced by the loss of water in transpiration.

Growth requires increase in cell volume, which depends on maintenance of a positive water balance, that is, high cell turgor, which requires that water absorption exceed water loss. Thus the water status of a plant can range from fully turgid through temporary to permanent wilting, to death from desiccation, depending on the extent to which water loss exceeds absorption over periods of several to many days.

Transpiration

Transpiration can be defined as the loss of water from plants in the form of vapor. Small amounts are lost through the bark of tree trunks and twigs (Geurten, 1950;

Huber, 1956; Schönherr and Ziegler, 1980), but most of the water is lost from leaves. Generally, except in dry soil, plant water relations are dominated by transpiration. A deciduous forest in humid southwestern North Carolina, where the average annual precipitation is 158 cm, transpires 40 to 55 cm of water per year, and a single exposed tree may lose 200 to 400 liters (50 to 100 gal) on a hot summer day. Such losses often result in temporary midday wilting of plants even in moist soil, and during droughts they cause reduction in growth and finally death from desiccation. Although high rates of transpiration often cause injury, transpiration is unavoidable because a leaf structure favorable for the entrance of CO_2 also is favorable for the loss of water vapor, and in general a high rate of photosynthesis has been more important for survival during the evolution of plants than a low rate of water loss. A few plants such as cacti have structural and physiological adaptations that minimize water loss, but such plants generally grow slowly and are confined to dry habitats.

The rate of transpiration is controlled by the energy supply, the vapor pressure gradient from leaves to air, the boundary layer resistance around the leaves or plant canopy, leaf resistance, and the water supply to the roots. The boundary layer as used here refers to the layer of water vapor and other gases surrounding leaf surfaces in quiet air. The energy required to evaporate water from transpiring structures comes chiefly from incident solar radiation, to a small extent from reflected radiation, and, especially in desert regions, from advective flow of heat from the surroundings. Thus the rate of transpiration is quite sensitive to light intensity and is much reduced by shading and by cloudy weather (Fig. 2.3). The rate increases as the steepness of the vapor pressure gradient increases from plant to air, which depends on temperature and the humidity of the atmosphere (Fig. 7.2). Another significant environmental factor is wind, which would be expected to increase transpiration by reducing the thickness of the boundary layer around leaves and plant canopies, but at the same time it decreases the rate by cooling leaves, which reduces the vapor pressure gradient (Knoerr, 1967; Dixon and Grace, 1984). Most of the cooling effect occurs at low wind velocities and transpiration often is reduced at higher velocities because of stomatal closure (Tranquillini, 1969; Caldwell, 1970). The effects of wind are discussed in more detail in Chapter 12.

Trees and shrubs go through daily and seasonal transpiration cycles related to changes in solar energy supply and leaf area. The rate of transpiration of trees that are well supplied with water goes through a daily cycle with a maximum near midday and a minimum at night (Figs. 2.1 and 7.7). This is because the stomata are usually wide open, the irradiance and energy supply are high, and the vapor pressure gradient from leaf to air is steepest near midday, but all of these conditions are reversed at night, when transpiration becomes very slow.

Even when environmental conditions are favorable for evaporation, the rate of transpiration at the leaf level is controlled largely by stomatal conductance, which depends on the degree of stomatal opening. The leaves of most woody plants are enclosed in a relatively waterproof cuticle so little water escapes, except through stomata. Readers who wish to learn more about stomata are referred to the section on stomatal behavior and to papers by Jarvis and Mansfield (1981), Mansfield

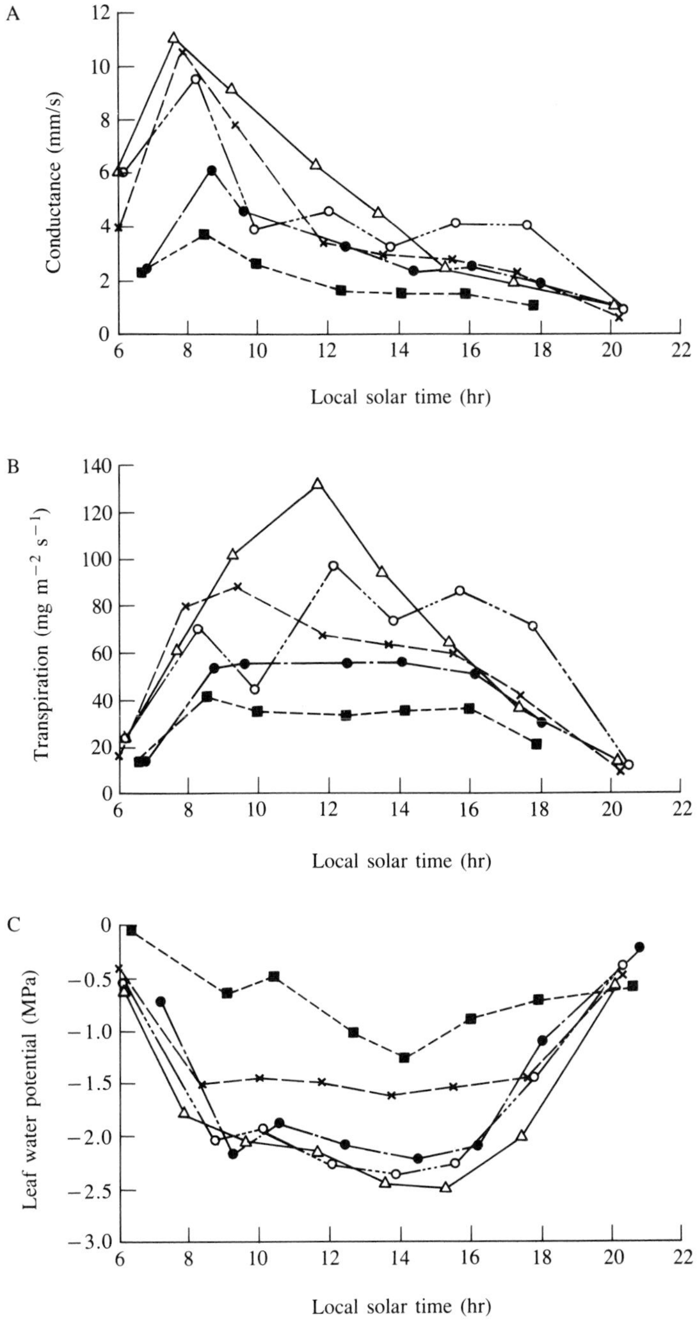

A

Conductance (mm/s)

Local solar time (hr)

B

Transpiration (mg m^{-2} s^{-1})

Local solar time (hr)

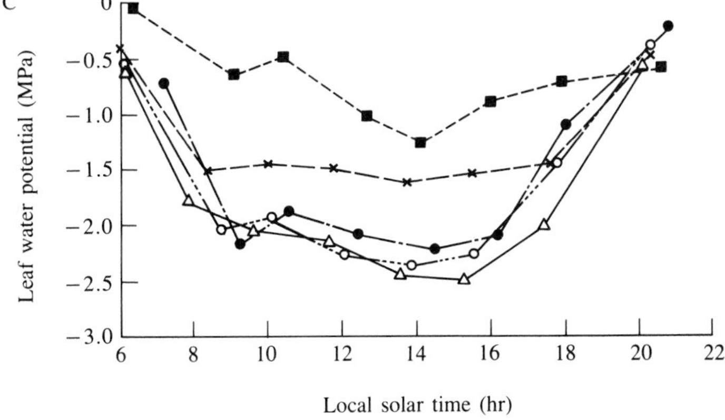

Figure 7.7 Daily cycle in (A) leaf conductance, (B) transpiration, and (C) leaf water potential of bigtooth aspen (△), paper birch (x), red oak (○, ●), and sugar maple (■) on July 20 in northern Michigan. Leaves of aspen and birch were at the top of the canopy, and those of red oak were at the top (○) and 1.5 m above the soil surface (●). Sugar maple leaves were at the top of a 5-m sapling in the understory. (From Jurik *et al.*, 1985.)

(1986), and Zeiger *et al.* (1987b). The complex relationships between the environment and water loss in tree canopies are discussed by McNaughton and Jarvis in Kozlowski (1983a) and by Jarvis (1986).

In moist soil the rate of transpiration is controlled largely by atmospheric conditions, but as the soil dries and plants dehydrate, stomatal opening decreases and transpiration is greatly reduced even at midday. The accompanying water stress reduces cell enlargement and shoot growth, causes closure of stomata, inhibits photosynthesis, and affects other physiological processes. As mentioned earlier, water stress results in narrower annual rings of wood, often with a smaller proportion of earlywood in years when drought occurs (Fig. 7.8).

Water Absorption

Most of the water used by plants is absorbed from the soil, but in special situations such as fog belts, water supplied by clouds and fog is important. For example, the

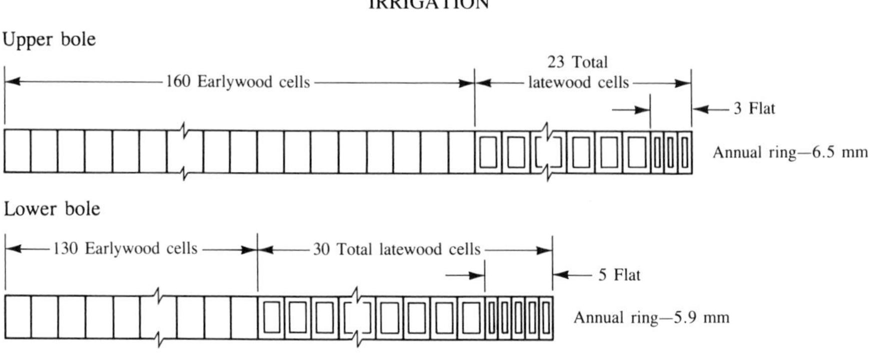

Figure 7.8 Difference in width of tracheids and in proportion of latewood in xylem rings of red pine grown with irrigation and under water stress. Also note differences between upper and lower bole. (From Zahner, 1968.)

redwood forest of the Pacific Coast of North America is confined to the fog belt. There has been considerable discussion concerning the amount of water absorbed directly from moist air or from water deposited on leaves as dew. Rundel (1982) concluded that direct absorption from the air is important only for epiphytes, but dew, fog, and cloud drip may replenish soil water. The importance of these sources of water was reviewed by Chaney (1981). It should also be noted that air pollutants tend to become concentrated in fog and mist (Lovett *et al.*, 1982) and CO_2 concentration in the air is also higher on foggy mornings than later in the day (Wilson, 1948).

Root systems of slowly transpiring plants in moist, well-aerated soil sometimes behave like osmometers because the accumulation of solutes in the root xylem sap lowers its potential below that of the soil water. This results in the development of root pressure and the "bleeding" observed from wounds in many herbaceous plants and from birch, grape, and a few other kinds of woody plants in the spring. Later in the season, rapid sap flow sweeps most of the solutes out, rendering the osmotic mechanism largely inactive, and water absorption is brought about principally by the tension or reduced water potential in the xylem sap. Thus in rapidly transpiring plants, water is pulled in through the roots and up to the transpiring leaves.

As water is removed from the soil by transpiring plants, the soil water potential approaches the lowest water potential that can be developed in the plant. Absorption decreases because of the decreasing water potential gradient between soil and roots, the water lost by transpiration is not replaced, and permanent wilting occurs (Slatyer, 1957). Also, as the soil dries, resistance to water movement toward the roots increases because of loss of continuity in capillary columns and possibly because of decreasing contact between soil and roots (Huck *et al.*, 1970; Tinker, 1976). High salinity reduces water absorption by decreasing the gradient in water potential from the substrate to the roots and cold soil increases the resistance to water flow through the roots. The absorption of water was reviewed by Passioura (1988).

An extensive root system that occupies a large volume of soil is important for the survival of plants in drying soil. As roots grow older they become increasingly suberized and therefore less permeable to minerals and water. However, it seems probable that much of the mineral and water absorption by woody perennials occurs through suberized roots because the unsuberized root surface usually is inadequate (Kramer and Bullock, 1966; Chung and Kramer, 1975). Mycorrhizae also increase the absorbing surface of roots and this is particularly important with respect to uptake of less mobile elements such as phosphorus. Their role in mineral absorption is discussed further in Chapter 6. Their role in water absorption was discussed by Duddridge *et al.* (1980), Sands *et al.* (1982), and Parke *et. al.* (1983), and root conductivity was discussed by Colombo and Asseltine (1989).

Ascent of Sap

The mechanism by which water moves from the roots to the tops of tall trees once was a mystery that caused considerable controversy. However, it is now generally agreed that water is pulled up through the xylem by the decrease in water potential developed in leaf cells by loss of water during transpiration. The conducting system of angiosperms consists of dead xylem elements from which the protoplasts and many end walls have partially or totally disappeared, resulting in conducting elements 20 to 800 μm in diameter and several centimeters to several meters in length, the longest xylem elements occurring in ring-porous trees and in vines. Water conduction usually is confined to a few outer rings because in older wood, vessels often are blocked by tyloses and gas bubbles. According to Ellmore and Ewers (1985), most of the water conduction in American elm occurs in the outermost annual ring. The restricted conducting system of trees with ring-porous xylem makes them more susceptible to injury from vascular diseases such as Dutch elm disease or oak blight than trees with diffuse-porous xylem. In gymnosperms, movement is through tracheids, which are single cells about 30 μm in diameter and 3 to 5 mm in length.

Most trees have more conducting tissue than is necessary for survival and live after considerable fractions of the cross section of the trunk are inactivated by injury or cavitation (Kramer, 1983, pp. 269–273), although leaf water potential may be reduced (Richter, 1974). It is claimed that there is a decrease in tracheid and vessel size from base to top in some trees that probably contributes to increased water stress toward the top and may limit height growth (Richter, 1974; Rundel and Stecker, 1977; Zimmermann, 1978). In tall trees, the effect of gravity on the water potential of the xylem sap combined with greater exposure to sun and moving air also may affect growth and result in more xeromorphic leaves near the top. The essential facts are that continuous water columns exist in the xylem from the evaporating surfaces in leaves to roots and this water has sufficient internal cohesive force to permit water to be pulled to the transpiring tops.

In trees of a given species there appears to be a fairly consistent relationship between leaf area and the cross section of the sapwood that supplies it with water. However, the ratio varies among species, presumably because of differences in the efficiency of the conducting system (Kaufmann and Troendle, 1981) and in transpiration rate per unit of leaf area. Zimmermann (1978) made numerous measurements of the conductivity in various parts of trees of several species and many of these are summarized in his book (Zimmermann, 1983). An example is shown in Fig. 7.9. In general, xylem structure seems to be an evolutionary compromise between efficient conduction and protection against blockage by gas bubbles (embolisms), with long vessels having large diameters being most efficient for water conduction but more vulnerable to blockage by gas bubbles formed by cavitation. Tracheids offer more resistance to flow than vessels, but would seem less subject to blockage by cavitation and they form the water-conducting system of many of the earth's tallest living trees. However, Tyree and Dixon (1986) reported that maple, which has vessels, is less subject to cavitation than cedar and hemlock, which have only tracheids, and Tyree and Sperry (1989) claim that embolism formation is not directly correlated with conduit diameter. Cavitation refers to the rupture of stressed water columns, resulting in formation of gas embolisms in the xylem that block sap flow (Tyree *et al.*, 1984; Tyree and Dixon, 1986; Tyree and Sperry, 1989).

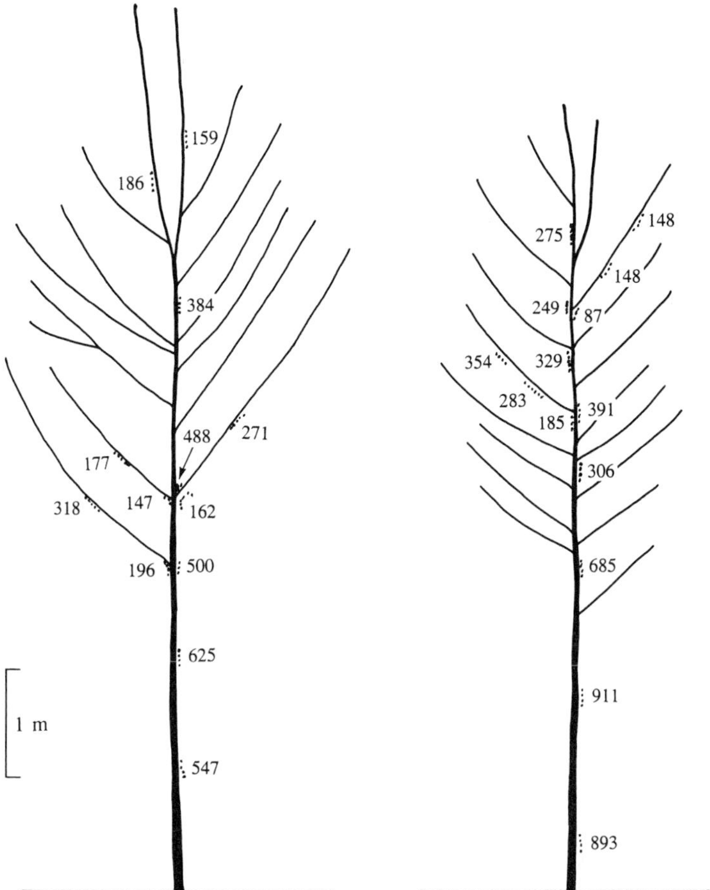

Figure 7.9 The leaf specific conductivity in microliters per hour, per gram fresh weight of leaves supplied, in two paper birch trees. Conductivity is lower in the branches than in the trunk and much lower at points where branches are attached to the trunk. (After Zimmermann, 1978; from Kramer, 1983.)

Tyree and Sperry (1988) suggest that woody plants often approach a situation where blockage of the xylem by cavitation seriously decreases xylem conductivity. For example, sugar maple lost over 30% of its conductance during a wet summer (Sperry *et al.*, 1988). Schultz and Matthews (1988) reported that some cavitation occurs even in well-watered grapevines during periods of rapid transpiration and the amount of blockage increases greatly as the soil dries. Vapor blockage also is very extensive during the winter, but root pressure probably aids in refilling gas-filled xylem vessels in the spring in species such as grape and maple (Sperry *et al.*, 1988). The occurrence and importance of cavitation in the xylem were reviewed by Tyree and Sperry (1989). Jones (1989) discussed stem conductance and cavitation, but suggested that root resistance generally is more important than stem resistance.

The cohesive water columns in the xylem that connect leaves and roots provide a feedback control mechanism that is essential for plant survival because it coordi-

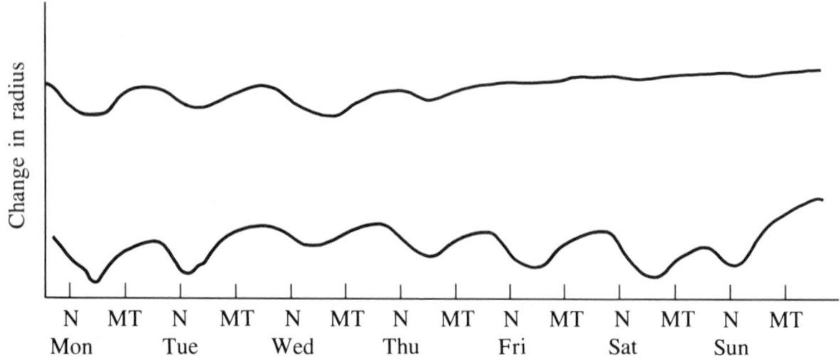

Figure 7.10 Dendrograph traces showing afternoon shrinkage and night swelling of red pine stems during the summer. The upper tracing (July 10–17) shows no daily shrinking and swelling during the latter part of the week, but only continuous diameter growth because cloudy, rainy weather reduced transpiration. The lower trace is for a week of sunny weather, August 21–28. (From Kozlowski, 1968c.)

nates the rate of absorption with the rate of transpiration. This is important for plants that undergo rapid changes in rate of transpiration from hour to hour. Water stored in the xylem of tree trunks also provides a limited capacitance effect that somewhat reduces the impact of rapid increases in transpiration on leaf water potential (Waring and Running, 1978). As mentioned earlier, when transpiration starts in the morning, water is removed first from tissues in the upper part of the stem and then from those in the lower part before appreciable increase in absorption occurs through the roots, because the resistance to water movement out of stem tissue is lower than the resistance to inward movement across roots. As a result, sap flow starts in the upper part of the stem before it begins in the lower part, and stem shrinkage also begins first in the upper part of rapidly transpiring trees (Zaerr, 1971; Lassoie, 1973; Schultze *et al.*, 1985; and others). Measurable midday shrinkage has been observed in the stems of many kinds of plants as a result of this lag of absorption behind water loss (Kozlowski, 1968b). Examples are shown in Figs. 7.10 and 7.22. Seasonal decreases in water content also occur, and trunks of birch and poplar containing so much water that they cannot float when cut in the spring will dry to a specific gravity of 0.60 to 0.75 and float readily if cut late in the season (Gibbs, 1935).

Some Effects of Water Stress

Having discussed the processes involved in plant water relations and the development of plant water stress, we will now turn to a more detailed discussion of the effects of water stress.

General Effects

The primary effects of water deficits are (1) decrease in water content and cell turgor of plant tissue and (2) decrease in the free energy status or potential of the remaining water. Reduction in turgor decreases cell enlargement and growth, while decreases in water potential affect water movement, especially water movement from the soil into roots and water movement into growing regions.

In general, cell enlargement and growth are very sensitive to water deficits, because some minimum degree of turgor is necessary for cell expansion (Boyer, 1985a, p. 492). However, several investigators have failed to find a clear relationship between turgor and cell growth (see Turner, 1986, pp. 13–15). This is not surprising because cell enlargement is a complex process requiring simultaneous intake of water, extension of cell walls, and a sustained supply of solutes necessary to maintain turgor in the expanding cells (Boyer, 1985a). Thus cell enlargement can be inhibited by the physical processes associated with the water supply or by metabolic processes associated with cell wall extensibility and the supply of solutes, or by a combination of the two. The complexity of the situation in the field was discussed by Wenkert *et al.* (1978), who concluded that within the range of water deficits usually encountered in the field, soybean leaf growth is more likely to be inhibited by metabolic factors than by low turgor. Turner (1986) attributes some discrepancies to difficulties in accurately measuring turgor and it seems possible that changes in cell wall extensibility also might cause changes in the relationship between turgor and cell enlargement (Shackel and Matthews, 1986). Cosgrove (1986) claimed that water stress has no important effect on the yielding properties of cell walls, but this is debatable.

Another problem is the relationship between water potential and various enzyme-mediated steps in processes such as photosynthesis, respiration, and nitrogen and carbohydrate metabolism. As Hsiao (1973) and Hsiao and Bradford (1983) point out, there is no obvious means (transducer) by which a decrease in water potential of -1.0 or -1.5 MPa can affect enzyme-mediated processes, yet both increases and decreases in enzyme activity occur. Proline accumulates in water-stressed tissue because its synthesis from glutamate is increased by loss of feedback inhibition and the rates of oxidation and incorporation into new tissue are decreased. Amylase activity is increased in leaves of water-stressed plants (Eaton and Ergle, 1948), resulting in a decrease in starch in a wide variety of plants. Because of the difficulties in relating changes in metabolic processes to changes in water potential, Sinclair and Ludlow (1985) proposed that plant water status be expressed in terms of relative water content (RWC). They cited experiments indicating that photosynthesis, protein synthesis, NO_3 reduction, and leaf senescence are, at least in some experiments, better correlated with changes in cell volume and relative water content than with water potential. It is unlikely, however, that expression of plant water status in terms of water potential will be abandoned because water movement is controlled by differences in water potential. Also the relative water content units are not applicable to soil water, while water potential units have a thermodynamic basis and are equally applicable for expressing the water status of both plants and the soil in which they are growing.

As stated earlier, water deficits affect almost every aspect of plant physiology and morphology, but only a few specific effects can be discussed. They will be treated under two general headings: (1) growth and (2) other physiological processes. The extensive literature on effects of water deficits on woody plants was reviewed in detail by Kozlowski (1982a). Table 7.2 shows the differing sensitivity of various processes to water stress.

Growth

Growth usually implies permanent increase in size and results from cell division followed by enlargement and differentiation. Cell enlargement generally is regarded as more sensitive than cell division to water deficit (Kramer, 1983a, p. 355), although McCree and Davis (1974) reported no difference in sensitivity of the two processes in sorghum. There also is some evidence that cell size affects cell division because cells do not divide until they have attained a certain size (Doley and Leyton, 1968). It is generally accepted that some degree of turgor is essential for permanent cell enlargement and growth and many investigators have reported inhibition of cell and organ enlargement by water deficits (Boyer, 1985a; Bunce, 1977; Cleland, 1971; Hsiao, 1973; Hsiao *et al.*, 1976; Michelena and Boyer, 1982; Turner and Begg, 1981; and others). However, Turner (1986) cited several cases in which leaf growth was inhibited although there was no loss of turgor and concluded that leaf growth can be affected by the root water status as well as by leaf turgor. This also was discussed by Davies *et al.* (1986). This may be true if the roots are in dry or flooded soil, which affects the hormone supply from roots to shoots, but can scarcely apply to plants rooted in moist soil but exposed to hot sun and dry air (Kramer, 1988b).

In trees, extension growth occurs in the apical meristems of roots and stems, diameter growth occurs in the cambia, and much cell enlargement occurs during leaf expansion. Overall, tree growth is greatly reduced on dry sites by prolonged water stress and even on good sites by the random droughts characteristic of many temperate regions. The effects of water deficits on shoot and root growth will be discussed separately.

Shoot and Leaf Growth

Shoot growth of seedlings is very sensitive to water stress and is much slower if soil is periodically allowed to dry below field capacity than if it is kept near field capacity (Zahner, 1968, pp. 197–200). Irrigation, therefore, is desirable and profitable in forest nurseries. However, growth of many species is inhibited by an excess of water sufficient to cause soil saturation, a problem discussed in Chapter 8 and by Kozlowski (1984b).

After the seedling stage the effects of water deficits on shoot growth become more complex and depend in part on the growth habit. If stem elongation occurs largely in a single flush from buds formed the previous season (fixed growth) as in many northern conifers, both the water supply during bud development the previous year and that of the current year during bud expansion may be important. The current rainfall is most important for growth of trees that make several flushes of growth each season (free growth), such as tulip poplar and the southern pines. Drought and resulting tree water stress during any of these flushes can reduce

Table 7.2

Relative Sensitivity to Water Stress of Various Plant Processes[a]

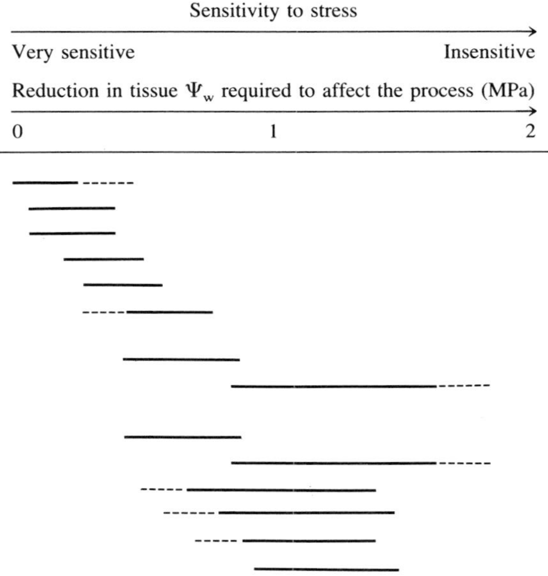

Process affected	Sensitivity to stress → Very sensitive … Insensitive; Reduction in tissue Ψ_w required to affect the process (MPa) → 0 … 1 … 2	References
Cell growth (−)		Acevedo *et al.*, 1971; Boyer, 1968
Wall synthesis[b] (−)		Cleland, 1967
Protein synthesis[b] (−)		Hsiao, 1970
Protochlorophyll formation[c] (−)		Virgin, 1965
Nitrate reductase level (−)		Huffaker *et al.*, 1970
ABA synthesis (+)		Zabadal, 1974; Beardsell and Cohen, 1974
Stomatal opening[c] (−)		
mesophytes		Hsiao, 1973 (review)
some xerophytes		van den Driessche *et al.*, 1971
CO_2 assimilation[c] (−)		
mesophytes		Hsiao, 1973 (review)
some xerophytes		van den Driessche *et al.*, 1971
Respiration (−)		
Xylem conductance[d] (−)		Boyer, 1971; Milburn, 1966
Proline accumulation (+)		
Sugar level (+)		

[a]After Hsiao *et al.*, 1976. Length of solid horizontal bars represents the range of stress within which a process is first affected; the dashed portion is the portion of the water potential range in which the response is not well established. In the left column, (+) indicates an increase and (−) indicates a decrease in a process.

[b]Fast growing tissue.

[c]Etiolated leaves.

[d]Should depend on xylem dimension.

growth, and late summer droughts often reduce the number of flushes. S. G. Pallardy observed that bud set of some poplar clones is very sensitive to water stress. Much relevant literature is cited by Zahner (1968) and by Kozlowski (1982a).

At the community level, tree height often is limited by the availability of water and trees usually grow taller in valleys than in the shallower, drier soil of the adjacent uplands. Redwood grows to a height of 100 m on the deep soils of alluvial flats in northwestern California, but only to 30 m on the upper edge of the fog belt. Waring and Schlesinger (1985, p. 91) suggested that decreasing predawn water potential is well correlated with decreasing tree height at maturity, as shown in Fig. 7.11.

In addition to the long-term effects of drought, midday reduction in shoot growth occurs because morning transpiration often exceeds water absorption. For example, Fielding (1955) reported that shoot elongation of Monterey pine trees often ceased at midday, at the same time that shrinkage was occurring in the lower part of the stem.

One of the damaging effects of water stress is reduction in leaf area, which reduces the loss of water, but unfortunately also reduces the surface that carries on photosynthesis, thus decreasing the amount of photosynthate available for growth.

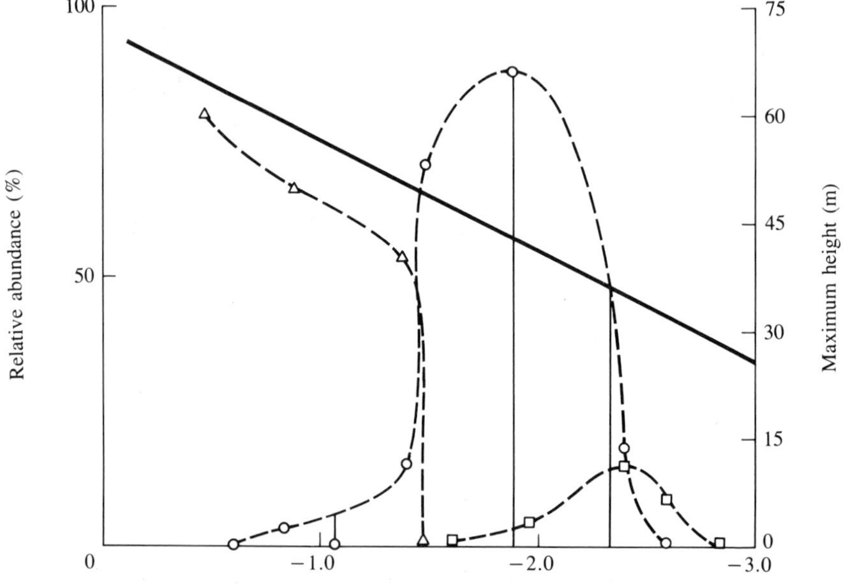

Figure 7.11 Relationship between predawn water stress of seedlings during summer drought and maximum height of Douglas-fir trees in southwestern Oregon (bold line). Also shown is the relative abundance of trees of three other species: *Abies concolor* (△), *Pinus ponderosa* (○), and *Quercus garryana* (□). (After Waring, 1970; from Waring and Schlesinger, 1985, by permission of Academic Press.)

Experiments with farm crops led to the conclusion that cultural treatments such as irrigation and fertilization increase yield chiefly by increasing leaf area (Watson, 1952). There is some evidence that this also is true of forest trees, as several investigators have reported that increase in leaf area following fertilization of forest trees is more important than increase in rate of photosynthesis per unit of leaf area (Brix, 1983). Other research indicates that there is a good correlation between leaf area and stem biomass (Isebrands and Nelson, 1982; Ridge *et al.*, 1986) or at least between leaf area and sapwood area (Kaufmann and Troendle, 1981). The history and application of this relationship were reviewed by Maguire and Hann (1987) and are discussed further in Chapter 4.

Water stress not only reduces leaf size, but often increases the ratio of mesophyll to external leaf surface (Nobel, 1980). The leaves at the top of a tall tree are subjected to considerably more water stress than those in the lower part of the crown because there is a decrease of water potential of 0.1 bar per meter of height due to the weight of the water columns in the xylem, plus the additional potential gradient required to overcome resistance to flow (Fig. 7.12). Leaves in the top of a canopy also are exposed to more xeric conditions than those within the canopy. Long ago Huber (1923) reported that leaves at the top of a tall tree are more xeromorphic than those near the base of the crown, and Connor *et al.* (1977) reported that in eucalyptus there is a decrease in area and increase in thickness of leaves with increasing height. There also is said to be a decrease in tracheid diameter in branches with increase in height in giant sequoia (Rundel and Stecker, 1977), and Waring and Schlesinger (1985, p. 91) suggested that limitation on size and number of water-conducting elements may limit tree height by reducing the water supply to the upper leaves and branches. Ginter-Whitehouse *et al.* (1983) reported that leaf water potential of eastern red cedar was lower than that of white oak or black walnut growing on the same site, probably because of the higher resistance to sap flow in gymnosperm than in angiosperm stems.

Leaf expansion, like other processes dependent on cell enlargement, is a complex process. For example, the leaves of European white birch grow more rapidly when illuminated, but those of sycamore maple grow better in darkness (Taylor and Davies, 1986). It was concluded that growth of maple leaves is limited by leaf turgor, whereas growth of birch leaves is limited by cell wall extensibility, which is increased by light (Taylor and Davies, 1988). It is reported that leaf expansion of black cottonwood is stimulated by light, while leaves of eastern cottonwood expand in darkness, but leaves of hybrids grow well both in light and in darkness. Incidentally, the larger size of the leaves of the hybrids of these two species seems to result from inheritance of large cell size from black cottonwood and large number of cells from eastern cottonwood, thus providing a larger photosynthetic surface (Ridge *et al.*, 1986). It should be reiterated that the cell expansion necessary for growth requires both supplies of water and solutes and cell walls capable of plastic extension (Boyer, 1985a). Growth, therefore, can be affected both directly and indirectly by a number of internal and external factors in addition to water stress.

The difference in anatomy between sun and shade leaves is well known (Chapter

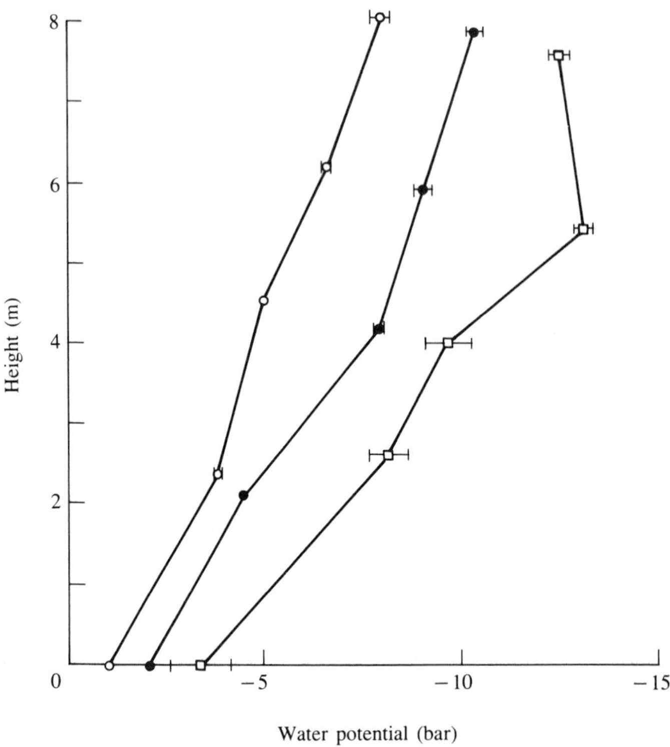

Figure 7.12 Vertical gradients in midday water potential in a Sitka spruce at midday on three different days in August. August 6, warm, quiet, overcast (○); August 7, warm, quiet, intermittent sun (●); August 4, warm, sunny, breezy (□). Root water potential was measured about 1 m from base of trunk. (From Hellkvist *et al.*, 1974.)

4). Where leaf quality is important, as in tea production, droughts often reduce or stop leaf production and injure young leaves. This results in cyclical variations in quality and quantity of yield. Thus irrigation of tea plantations often is profitable (Squire and Callander, in Kozlowski, 1981), and tea and tobacco sometimes are shaded to decrease water stress and improve leaf quality.

Diameter Growth

The relationship between rainfall and diameter growth, determined by the width of the annual rings laid down each growing season, has been known for centuries (Glock, 1955) and the science of dendrochronology is based on the relationship between rainfall and width of tree rings (Fritts, 1976). Both the quantity and quality of wood produced by a tree are affected directly and indirectly by water supply (Zobel and Van Buijtenen, 1989). The change from large- to small-diameter xylem elements can be hastened by water stress or postponed by irrigation. In years with an abundance of rainfall, tree rings are not only wider but contain a larger proportion of earlywood-containing xylem elements having larger diameters and thinner walls than in dry years (see Figure 7.8 and Kozlowski, 1971b, pp. 164–165). These differences were attributed by Larson (1964) to a decreased supply of auxin to the differentiating xylem derivatives, but another study showed little change in auxin during the growing season (Zahner, 1968, p. 224). There also may be a decrease in the supply of metabolites to the cambial region. Lack of turgor probably inhibits enlargement of xylem initials and water stress seems to have direct effects on the development of cell walls of xylem derivatives (Zahner, 1968, 210–215; Kozlowski, 1982a; Sheriff and Whitehead, 1984) and the incorporation of labeled glucose into tracheid cell walls (Whitmore and Zahner, 1967). Differences in proportions of earlywood and latewood have important effects on the specific gravity and other properties of wood.

In addition to the long-term effects of drought on diameter growth, there are short-term daily effects that can only be observed by sensitive dendrographs (Bormann and Kozlowski, 1962). For example, tree stems often show midday shrinkage (Figs. 7.10 and 7.22), even when growing in moist soil, with the daily amplitude increasing toward midsummer then transpiration is most rapid and decreasing late in the season as transpiration decreases. This daily shrinking and swelling is imposed over the permanent increase in diameter that occurs during the growing season and can cause errors in measuring diameter growth (Kozlowski, 1972b, Chapter 1; Braekke and Kozlowski, 1975; Kozlowski, 1982a). Long-term increase in diameter results from increase in number and size of xylem cells, but short-term variations are caused by reversible shrinking and swelling of existing cells, chiefly those of the inner bark and cambial region (see Fig. 7.22).

Root Growth

The shoot water deficits that develop on hot, sunny days eventually are transmitted to the roots through the sap stream. However, because of the time lag in transmission of water stress from leaves to roots, the roots are the last tissues to be stressed, and because they are closest to the source of water they are the first to recover in the evening. According to data shown in Fig. 7.12, midday root water potentials of sitka spruce were much higher than those in the upper part of the trees. The physiological effects of these transient stresses on roots have not been evaluated fully and deserve further study. According to Sharp and Davies (1979) and Molyneaux and Davies (1983), root growth of several herbaceous plants is maintained longer than shoot growth in drying soil. Kaufmann (1968) reported that root growth of loblolly and Scotch pine seedlings in slowly drying soil was reduced to about 25% of the rate at field capacity by a soil water potential of −0.6 or −0.7 MPa. He also reported that less root growth occurred after the second or third drying cycle than after the first cycle, and that shoot growth was reduced more than root growth in drying soil. Waring and Schlesinger (1985) cited several experiments suggesting that tree roots do not grow much at a soil water potential below −7 bars. Additional data are presented by Kuhns *et al.* (1985) for walnut and by Teskey and Hinckley (1981) for white oak. Roots usually resume growth within a day or two after the soil is rewetted by rain or irrigation.

Slowing or cessation of root growth in drying soil decreases water absorption because it reduces the invasion of previously unoccupied soil and is accompanied by increase in the proportion of root surface that is suberized. However, considerable absorption occurs through suberized roots (Chung and Kramer, 1975). In recent years it has been suggested that water stress modifies the amount and kind of hormones exported from the roots to the shoots and that much of the perturbation in shoot metabolism of water-stressed plants is caused by chemical signals from the roots (Davies *et al.*, 1986; Zhang and Davies, 1989a, b). These include decrease in the amount of cytokinin and increase in the amount of ABA, although major emphasis is placed on ABA by Zhang *et al.* (1987). This may be true for trees growing in slowly drying soil, but it probably does not occur until root growth is slowed and probably is superimposed on the hydraulic effects of shoot water stress.

Development of root systems is affected by soil properties and by rainfall patterns, as well as by heredity. In heavy, poorly aerated soils, most root development occurs near the surface. For example, Coile (1937) found that over 90% of the roots less than 2.5 mm in diameter occurred in the top 12.5 cm under pine and oak forests on the heavy clay soils of the North Carolina Piedmont. However, other studies showed enough deep roots to extract water from more than a meter below the soil surface under a loblolly pine stand (Hoover *et al.*, 1953). Even in sandy soils the trees often develop numerous roots near the surface, possibly because the surface soil contains more nutrients (Woods, 1957). It also seems likely that shallow root development is stimulated by summer showers that wet only the surface soil. It appears that at night, deep roots can transfer water from deep, moist soil horizons to the dry surface soil, where it can be absorbed later by shallow-rooted plants (Corak *et al.*, 1987; Richards and Caldwell, 1987; Caldwell and Richards, 1989) or even by shallow roots on the same plant (Field and Goulden, 1988).

Urban trees and shrubs and those grown in containers often suffer from restricted root growth caused by deficiencies in aeration, water supply, and growing space (Craul, 1985). The restricted root development may also reduce the supply of hormones such as cytokinins and gibberellins that are produced in the roots and translocated to the shoots. The effects of restricted root systems were reviewed by Krizek and Dubik (1987). Experiments with herbaceous plants indicate that if roots have difficulty penetrating soil, shoot physiology may be affected and the root–shoot ratio reduced (Masle and Passioura, 1987).

Reproductive Growth

The reproductive stage of growth usually is more sensitive to water stress than vegetative growth (Kramer, 1983a, p. 354), but the effects vary among species and with the stage of development. Reproduction includes flower bud initiation and development, flowering, pollination and fruit setting, and fruit and seed maturation, and all of these stages are susceptible to injury from water deficits. Most of the information comes from research on fruit trees in dry areas where the water supply can be controlled by irrigation. Water stress inhibits flower bud initiation in some trees, such as apricots (Uriu, 1964) and peaches (Proebsting and Middleton, 1980), but the information concerning forest trees is somewhat contradictory (Kozlowski, 1982a, p. 76). This probably results from the difficulty in correlating rainfall or irrigation of seed orchards or forest stands with the limiting stage in the reproductive process, which may extend over 2 or 3 years. Also reproduction in forest trees operates in various time frames. Development of acorns of the white oak subgenus occurs the year of pollination, acorns of the red oak subgenus the next year, and growth and maturation of the cones of some pines extend through three growing seasons (Kozlowski, 1971b, Chapter 8). Thus the weather of the preceding year may affect the seed crop of a given year. For example, Wenger (1957) and Shoulders (1967) reported that seed production of southern pine stands varied inversely with the seed crop of the previous year and directly with the May to July rainfall of the second preceding summer. Other examples of these complex relationships are summarized by Kozlowski (1982a, p. 76), and the effects of water stress on several kinds of fruits are discussed in Kozlowski (1983a).

In a few cases, a period of water stress seems to be necessary to break dormancy before flowering can be brought about by rain or irrigation. Examples are coffee (Alvim, 1973) and cacao (Alvim, 1977), and Southwick and Davenport (1986) reported that severe water stress down to −3.5 MPa induced flowering of Tahiti lime (*Citrus latifolia*). Jones (1987) reported that both dehydration and defoliation often cause repeat flowering of apple if they occur after flower buds are differentiated and suggested that dehydration causes repeat flowering because it results in defoliation. Dewers and Moehring (1970) reported that loblolly pine trees irrigated from April to June, but subjected to late summer drought, bore more conelets than those irrigated in late summer.

In dry climates, irrigation of fruit trees during fruit setting and development is beneficial or even essential for high yield and quality (Uriu and Magness, 1967; Kriedemann and Barrs, in Kozlowski, 1981; Kozlowski, 1981). Even in humid regions such as England, irrigation increases apple yield by increasing the number of flower clusters and decreasing fruit drop (Table 7.3) (Goode and Hyrycz, 1964). However, moderately low humidity during the time of pollination usually is beneficial because pollen distribution is improved. Severe water stress during maturation reduces fruit size and sometimes causes shriveling of the kernels of nuts. However, Richards and Wadleigh (1952) cited research indicating that moderate water stress improves the quality of apples, peaches, pears, and prunes. This view was strongly supported by the older generation of California fruit growers. An excess of water sometimes causes fruit cracking or splitting, while a deficiency sometimes causes "corking" or water-soaked areas near the core of apples. Prolonged wetting of leaf and fruit surfaces increases the likelihood of infection by pathogenic organisms. Much literature on the effects of water deficits on reproductive growth was summarized by Kaufmann in Kozlowski (1972a), and the problem in tropical forests is discussed by Doley (p. 295), for apples by Landsberg and Jones (p. 452), and for

Table 7.3
Effect of Water Deficits on Various Aspects of Reproductive Growth in Cox's Orange Pippin Apple Trees[a]

		Treatment		
	Date	Droughted	Control	Irrigated
Flowers per tree	May 12	3228	3431	2464
Fruitlets per tree	June 9	275	540	591
Fruit set as percentage of flower number	June 9	8.5	15.7	24.0
Fruit clusters per tree	June 9	144	252	219

[a]From Powell (1974).

citrus by Kriedemann and Barrs (p. 377), all in Kozlowski (1981). The effects of water stress on citrus and apples also are discussed in Kozlowski (1983a).

Stomatal Behavior

Stomata are tiny pores with controllable apertures in the epidermis of leaves and other structures through which gas exchange occurs. They are most common on leaves, especially on the lower surface of leaves of woody plants, but they can occur on almost all aerial parts, including green stems, flower parts, and fruits. The number per unit of area varies widely among species, ranging from 11,500 per cm^2 in yew and 13,000 per cm^2 in American basswood to 68,000 in northern red oak and over 100,000 in scarlet oak (Meyer *et al.*, 1973, pp. 74–75). The actual number varies widely within a species, depending on the conditions under which the leaves develop. Stomata are very small, the pores averaging 10 to 20 μm in length and 5 to 10 μm in width when fully open, and the open pore space usually is less than 1% of the leaf area. Yet because of the physical principles governing diffusion through small pores in a membrane (Nobel, 1983, pp. 395–398), diffusion of water vapor through stomata sometimes exceeds 50% of the rate of evaporation from a free water surface having the same area as a leaf. Stomata are equally effective as entrances for CO_2, although the rate of diffusion of H_2O is 1.56 times that for CO_2, because of the higher density of the latter.

The size of the stomatal aperture is controlled by the two guard cells bordering the aperture, as shown in Fig. 7.13. It increases as guard cell turgor increases and decreases as their turgor decreases. In general, stomata of woody plants open in the light or in response to a low concentration of CO_2 and they close in darkness or when dehydration causes a loss of turgor. Stomata of different species vary considerably in their reaction to light. For example, Woods and Turner (1971) found that stomata of shade-tolerant beech opened more quickly in bright light (3 sec) than those of shade-intolerant tulip poplar (20 sec), while red maple and northern red oak were intermediate. Presumably rapid opening is beneficial to shade plants illumi-

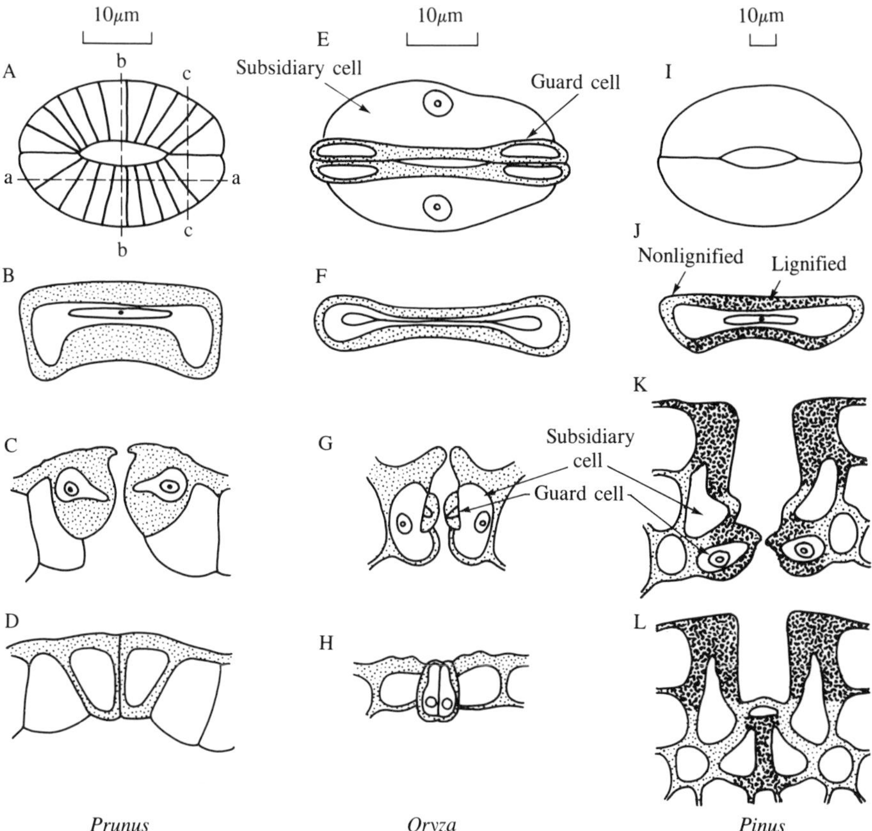

Figure 7.13 Surface and sectional views of three kinds of stomata. Diagrams A, E, and I are surface views. The other diagrams show sections made in the planes indicated on diagram A. B, F, and J in plane aa; C, G, and K in plane bb; D, H, and L in plane cc. (From Esau, 1977.)

nated chiefly by sunflecks. In general they found that stomata opened more quickly than they closed, taking 3 to 20 sec to open and 12 to 36 sec to close. However, as pointed out in Chapter 4, Pereira and Kozlowski (1976) found no relationship between the speed of stomatal response to change in light intensity and shade tolerance. As leaves grow older the stomatal activity often decreases, and Tazaki *et al.* (1980) reported that by late summer the stomata of mulberry remained permanently open.

In recent years it has been shown that exposure to dry air causes stomatal closure in turgid leaves of many but not all kinds of plants (Lange *et al.*, 1971; Schulze and Kuppers, 1979; Marshall and Waring, 1984; Sandford and Jarvis, 1986; Sena Gomes *et al.*, 1987). Sandford and Jarvis (1986) reported that in the four species of conifers they studied (3-year-old seedlings of Sitka spruce, lodgepole and Scotch pine, and a hybrid larch), stomatal closure caused by dry air varied from 7.5% for

Scotch pine to 65% for Sitka spruce. Marshall and Waring (1984) reported that conifer stomata were more sensitive to a high vapor pressure deficit than those of several broad-leaved species. Pallardy and Kozlowski (1979c) reported that differences in sensitivity of the stomata of certain poplar clones to humidity were correlated with differences in drought tolerance of their parents (Fig. 7.21). They also pointed out that partial closure of stomata in dry air compensates somewhat for the higher rate of transpiration that would otherwise occur, moderating the effect of a high vapor pressure deficit on plant water status. Wind or mechanical shaking of trees causes closure of stomata of some species, but not all (Kramer, 1983a, p. 304; and Chapter 12).

Temperature usually has only a moderate effect on stomatal aperture, and Kaufmann (1982) concluded that it could be disregarded for the four species of subalpine trees in the narrow temperature range that he studied. However, stomatal closure of plants of some tropical herbaceous species is greatly slowed at low temperatures (McWilliam *et al.*, 1982). Wuenscher and Kozlowski (1971) found that the leaf conductance of several kinds of tree seedlings decreased as the temperature was increased from 20 to 40°C, and the increase was greatest for species growing on dry sites. However, this may have been at least partly a response to increasing vapor pressure deficit. It is difficult to generalize concerning stomatal behavior because it is greatly affected by conditions that existed prior to the experiments. Also, several investigators report that stomatal closure is better correlated with soil water status than with leaf water status (Morrow and Mooney, 1974; Bates and Hall, 1981; Gollan *et al.*, 1985; Pereira *et al.*, 1986; Zhang and Davies, 1989a), but this probably is only true for plants growing in soil so dry that turgor is not recovered overnight.

The response of stomata to water stress is modified by their past treatment. For example, stomata of plants previously subjected to water stress generally close at a lower leaf water potential than those of plants not previously stressed. Schulte and Hinckley (1987) reported that stomata of black cottonwood not previously water stressed remain open when stressed, but if the leaves have been stressed previously the stomata close. The effect of water stress on stomatal behavior probably is a factor in the better survival of outplanted tree seedlings that have been subjected to water stress by wrenching, undercutting, or controlled irrigation in the seedbed (Rook, 1973; Johnson *et al.*, 1986). The physiology of stomatal opening is complex and the details are beyond the scope of this book. For example, puzzling cycling or oscillation between the open and closed condition with a periodicity ranging from 15 to 120 min is sometimes observed in water-stressed plants (Kramer, 1983a, p. 321). Readers are referred to physiology texts or reviews such as those of Mansfield (1986) and Zeiger *et al.* (1987a) for more details.

Optimization

The response of stomata in water-stressed leaves is critical to the success of trees and other plants because they control both the entrance of CO_2 necessary for photosynthesis and the escape of water vapor that threatens the plant water balance.

This has led to attempts to explain stomatal behavior in terms of optimization of costs to benefits, benefit being decreased loss of water and cost being decreased uptake of CO_2 used in photosynthesis. Optimum efficiency should result when stomatal aperture varies during the day in a manner that results in minimum transpiration (E) for maximum photosynthesis (A). This would require a constant ratio of transpiration to photosynthesis, that is daily dE/dA = a constant.

Wong *et al.* (1979) claim that a feedback mechanism relating stomatal aperture to the photosynthetic capacity of the mesophyll cells maintains a fairly constant ratio of internal to external concentration of CO_2. The optimization principle was developed in detail by Cowan (1982), by Farquhar and Sharkey (1982), and in a book edited by Givnish (1986). However, it is doubtful if optimal stomatal response saves enough water to be an important factor in the success of plants in stands and it is not certain that it is of common occurrence. Sandford and Jarvis (1986) cited some of the contradictory literature and reported that several woody species fail to show a constant dE/dA. Anyway, conservation of water is beneficial only if the plants with optimal behavior have no competitors for the soil water that is conserved by them. Usually competing plants benefit from the water not used by more efficient plans (Bunce *et al.*, 1977).

Photosynthesis

As stated in Chapter 2, although growth is not necessarily closely correlated with the rate of photosynthesis, it is dependent on food supplied by photosynthesis. Thus any serious interference with that process by water deficit is likely to significantly reduce growth. Numerous measurements indicate that increasing water deficits are accompanied by decrease in rate of photosynthesis (Bormann, 1953; Kozlowski, 1949; Brix, 1962, 1979; Kriedemann, 1971; and others). An example is shown in Fig. 7.14. This is generally believed to be brought about by both stomatal and nonstomatal inhibition of the process (Teskey *et al.*, 1986).

The leaf water potential at which photosynthesis is materially reduced varies widely among species and families of a species, and with past treatment. Usually plants native to dry areas carry on photosynthesis to a lower water potential than those native to humid areas. For example, photosynthesis of alder and green ash was reduced at −1.0 MPa but that of creosote bush, a desert shrub, was not reduced until −2.0 MPa (Chabot and Bunce, 1979), and some plants native to Death Valley were little affected at −2.9 MPa. Morrow and Mooney (1974) reported photosynthesis in two California evergreen sclerophylls at −2.5 MPa. Other examples of species and ecotype differences are cited by Kozlowski (1982a, pp. 70–71). Bunce (1977) found large differences in sensitivity to water stress among 12 woody plant species ranging from streamside species to desert shrubs, and nonstomatal inhibition of photosynthesis began at the same water potential as decrease in stomatal conductance.

On a typical hot, sunny, summer day, plants rooted in moist soil show a rapid increase in stomatal conductance, photosynthesis, and transpiration during the early

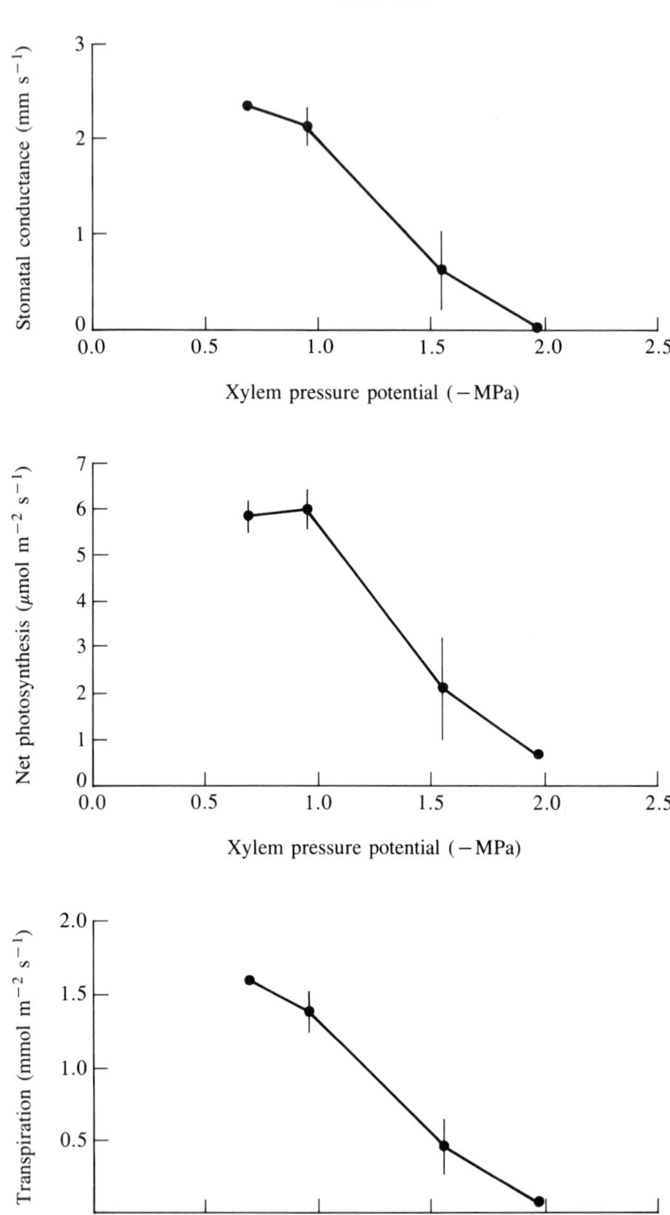

Figure 7.14 Relationship between xylem water potential, net photosynthesis, transpiration, and stomatal conductance in loblolly pine. Vertical lines = ±1 SEM. (From Teskey *et al.*, 1986.)

morning, but often show a decrease toward midday, often followed by at least partial recovery by midafternoon and a decrease in late afternoon with decreasing irradiance. As the soil dries, the midday decrease becomes more severe, the stomata close earlier in the day or fail to open, and the rate of photosynthesis falls to a very low level (Fig. 7.15). There usually is a close correlation between reduction in stomatal aperture and reduction in rate of photosynthesis, both during midday water stress and in drying soil. This was shown for citrus by Kriedemann (1971), for apple by Lakso (1979), for eastern cottonwood by Regehr *et al.* (1975), and for loblolly pine by Kozlowski (1962), and Teskey *et al.* (1986) (see Figs. 7.14 and 7.15).

Stomatal versus Nonstomatal Inhibition of Photosynthesis

The close correlation often observed between stomatal conductance and rate of photosynthesis resulted in the assumption that a decrease in photosynthesis, especially the midday decrease, is caused by stomatal closure. However, during dehydration, photosynthesis frequently is reduced more than can be explained solely by stomatal closure (Beadle and Jarvis, 1977; Luukkanen, 1978; Scarascia-Mugnozza *et al.*, 1986; and others). There is considerable evidence that water deficits severe enough to cause stomatal closure also inhibit operation of the photosynthetic process (Boyer, 1971; Farquhar and Sharkey, 1982). For example, Teskey *et al.* (1986) concluded that although stomatal response of loblolly pine to several environmental variables is closely correlated with change in rate of photosynthesis, internal limitations actually are the major cause of the reduction. De Lucia (1986) reported that the decrease in rate of photosynthesis of Englemann spruce during the first 10 hr after the roots are chilled was caused by decreased stomatal conductance, but after 24 hr it resulted from nonstomatal limitation. However, this is not always true. Comstock and Ehleringer (1984) found that as water stress increased in *Encelia,* the stomatal conductance decreased more rapidly than photosynthetic capacity and accounted for 40% of the reduction in photosynthesis of severely stressed plants. Gollan *et al.* (1985) concluded that photosynthesis of oleander in drying soil is controlled chiefly by stomatal conductance rather than by reduction in capacity of the photosynthesis apparatus. Ni and Pallardy (1987) found that water stress increased the stomatal limitation on photosynthesis of post oak, but produced little change in white oak.

Most of the claims for nonstomatal inhibition of photosynthesis in water-stressed plants are based on observations that the internal CO_2 concentration remains high in stressed leaves after the stomata are closed. The internal concentration usually is calculated from stomatal conductance to water vapor loss determined from transpiration data, the rate of net photosynthesis measured by gas exchange with an IRGA, and the external CO_2 concentration. However, Downton *et al.* (1988) and Terashima *et al.* (1988) reported that gas exchange measurements made with infrared gas analyzers do not provide reliable estimates of the average internal CO_2 concentration in amphistomatous or stressed leaves because the stomata do not close uniformly over the surface of the leaves. Nonuniform closure of stomata has been

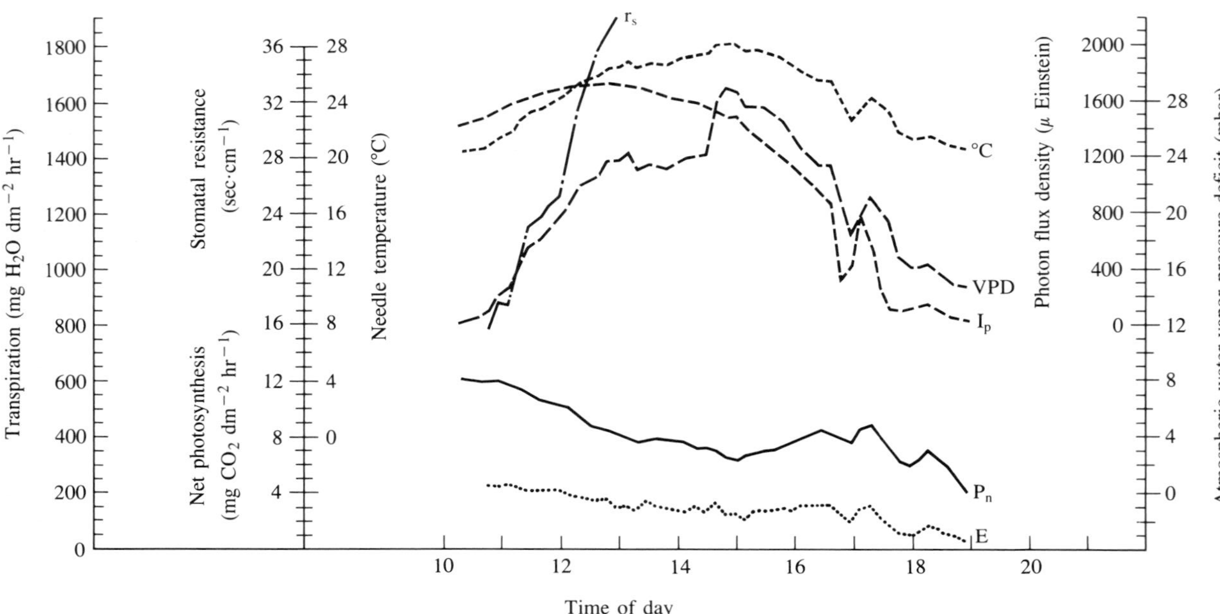

Figure 7.15 Gas exchange of Monterey pine needles at 4.9 m above the ground in a 9-year-old stand after a long dry period with predawn water potential of −1.05 MPa and soil water potential of −0.82 MPa. Ambient temperature, °C; atmospheric vapor pressure deficit, VPD; photon flux density at 400–700 nm, I_p; net photosynthesis, P_n; transpiration, E; stomatal diffusion resistance to water vapor, r_s. (From Benecke, 1980.)

reported by others, including Daley *et al.* (1989) and Omasa *et al.* (1985), which results in nonuniform occurrence of transpiration and photosynthesis over the leaf surface. Measurement of photosynthesis by the oxygen electrode or by chlorophyll fluorescence in leaves treated with exogenous ABA led to the conclusion that the presumed nonstomatal inhibition of photosynthesis caused by ABA does not exist. In fact Downton *et al.* (1988) stated that it seems likely that stomatal closure can fully account for inhibition of photosynthesis caused by water stress and possibly by other stresses. If correct, these observations require reevaluation of several papers cited earlier in this section.

In the long term, nonstomatal limitation probably is important, including decrease in carboxylating enzyme, electron transfer, photophosphorylation, and chlorophyll. There is sometimes an increase in dark respiration in stressed plants that decreases net photosynthesis. The increase in temperature and respiration of wilted leaves also may be a factor in the midday decrease in photosynthesis. A temporary increase in respiration of water-stressed conifers was observed by Parker (1952) and Brix (1962) and in broadleaf evergreens by Mooney (1969). This may be caused by hydrolysis of starch to sugar, but needs further investigation.

Premature Senescence and Leaf Shedding

Another effect of water stress is to cause premature senescence and shedding of leaves. This may involve the normal process of abscission or simply withering. If dehydration occurs rapidly, as by exposure to dry, hot winds, leaves may wither and die but remain attached. Sometimes abscission of leaves injured by dehydration does not occur until the plants are rewatered, suggesting that resumption of metabolic activity is necessary for formation of the abscission layer. The amount of premature abscission varies with the season and among species. It is more common in older leaves, and the youngest leaves often are retained after all the older leaves are shed. There also are significant differences among species, with leaves of buckeye, tulip poplar, and black walnut showing premature senescence more frequently than those of oaks. In pines and some other conifers the branches bearing bundles of needles are shed and in some other species entire twigs are shed. In tropical and subtropical areas having well-defined dry seasons, some species lose some of their leaves and others all, and the total leaf area of the vegetation is greatly decreased during the dry season. Occasional exceptions to this pattern of behavior are reported. For instance, Radwanski and Wickens (1967) observed that *Acacia albida* and *Salix subserrata* growing in the Sudan bear leaves during the dry season and are leafless during the rainy season. Leaf shedding caused by water stress is discussed in more detail by Kozlowski (1991).

Leaf shedding of Christmas trees in storage and transit is a serious problem (Mitcham-Butler *et al.*, 1987). However, leaf shedding of water-stressed plants usually is regarded as a beneficial adaptation that reduces water loss and prolongs survival. Borchert (1980) reported that loss of leaves from the tropical tree *Erythrina peoppigiana* results in decrease in tree water stress, and this is followed by production of new leaves. He stated that a transition often occurs from the evergreen to the deciduous habit along altitudinal gradients of increasing water stress. Actually the stomata on water-stressed plants often are closed and transpiration already is so low that only a small additional reduction in transpiration may result from loss of the older, lower leaves in a tree canopy. However, any reduction in water loss is beneficial during droughts, especially if it decreases the probability of injury to meristematic tissue. Incidentally, meristematic tissue seems to have a high priority during internal competition for water, and stem tips and fruits often continue to enlarge while older tissues are shrinking (Kramer, 1983a, pp. 351–352).

Premature leaf senescence involves complex interactions of hormones and nutritional balance. Water stress probably reduces the supply of carbohydrates and nitrogen both to very young leaves and to older, lower, heavily shaded leaves in the forest canopy. This may be aggravated by decrease in the supply of cytokinins from the roots and increase in the production of ethylene. The relative importance of hormones, metabolites, and water stress on leaf senescence and shedding needs more study.

Seed Germination

There are wide differences in the tolerance of tree seeds to desiccation. Seeds of pines, sweet gum, and eucalyptus are tolerant, whereas seeds of oak, willow, bald cypress, and elm are killed by desiccation, but still other seeds can be wetted and dried repeatedly without injury. For example, seeds of *Eucalyptus sieberi* can survive several periods of imbibition separated by periods of dehydration (Gibson and Bachelard, 1986). Regardless of the type of seed, the first step in germination is imbibition of enough water to activate the physiological processes essential for germination. This may amount to two or three times the dry weight of the seed and in the early stages most of the water is bound tightly by imbibitional forces. Generally there is rapid increase in the rate of respiration after a certain amount of water is imbibed (Vertucci and Leopold, 1987a) (Fig. 7.16), stored food is digested by enzyme activity and translocated to the growing regions, existing embyronic tissues expand, and new tissue is produced.

An assured supply of water therefore is necessary for successful germination and establishment of seedlings. In nature, most seedlings fail to become established because of death from desiccation and only after wet springs do we see large crops of pine and hardwood seedlings surviving in and adjacent to forests. Even on good sites only a few seedlings survive to grow into trees and on dry sites only seedlings of the more drought-tolerant species become established. The need for water necessitates the installation of irrigation equipment in forest nurseries, but care also must be taken to avoid overwatering. Because of their high level of physiological activity, germinating seeds have a high oxygen requirement and those of most species are injured by saturating the soil. Even the seeds of bald cypress germinate better on exposed soil than under water.

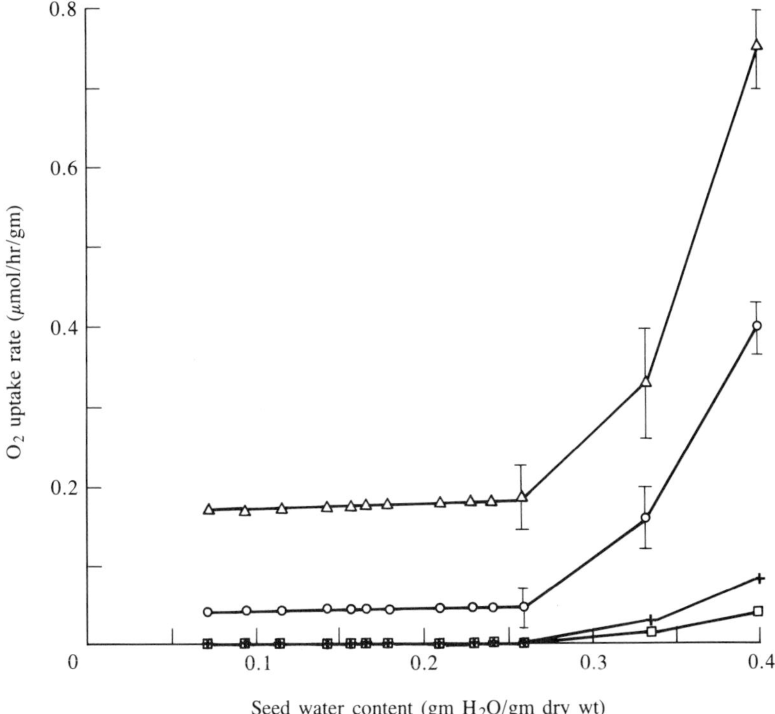

Figure 7.16 Rate of oxygen uptake by pea seeds with increasing water content at 10 (□), 15 (+), 25 (○), and 35°C (△). Water content must exceed about 25% of dry weight before oxygen consumption begins to increase. Vertical bars indicate standard errors. (From Vertucci and Leopold, 1987a.)

Nearly all seeds can absorb sufficient water for germination from soil at or near field capacity, but germination of seeds of some species decreases rapidly as the soil dries. Satoo (1966b) reported that germination of seed of hinoki cypress was reduced about 6% for each 0.1 MPa decrease in soil water potential and that of Japanese red pine about 2.5%. Kaufmann (1969) found that 88% of citrus seed germinated in soil at 0 MPa water potential, 23% at −0.23 MPa, but only 3% at −0.47 MPa, and germination of seed of eastern cottonwood also decreased with decreasing water potential of the substrate.

There also are interactions between temperature and water supply that vary among species. For eastern cottonwood, low temperature seems to aggravate the effect of low soil water potential (Farmer and Bonner, 1967). Considerable injury has been observed when seeds of some kinds of plants, especially those of legumes, are allowed to imbibe water rapidly at low temperatures. This apparently involves injury to cell membranes (Leopold, 1980; Chabot and Leopold, 1982). It has not been reported for seeds of woody plants, but probably occurs. High salt content of the substrate generally is injurious to seed germination independent of direct osmotic effects, but few data are available for seeds of woody plants. The role of water in seed germination was discussed in detail by Koller and Hadas (1982).

Disease and Insect Injury

Injury to plants from diseases and insects depends on interactions among the host plants, the pests, and the environment. Environmental conditions such as the amount of rainfall can strongly affect both the inherent resistance of the host plants and the vigor with which the pests can attack. Using a broad definition of disease as any abnormality in form or function not caused by mechanical injury, water stress itself is a disease. Water stress usually increases the susceptibility of plants to attack by insects and pathogenic fungi but sometimes decreases susceptibility to injury from air pollutants. In addition, some direct effects of water stress can be regarded as physiological or abiotic diseases (Durbin, 1978). Examples are the premature shedding, cracking, or pitting of young fruits and internal necrosis such as the bitter pit of apple. Blossom end rot of tomato and blackheart of head lettuce seem to involve the interaction of water stress and calcium deficiency (Durbin, 1978; Tibbits, 1979).

There is some evidence that water stress favors invasion by certain fungi (Fig. 7.17). Bier (1959) reported that the fungus causing bark cankers in willow and poplar invades the bark only when the relative water content is below 80%, and

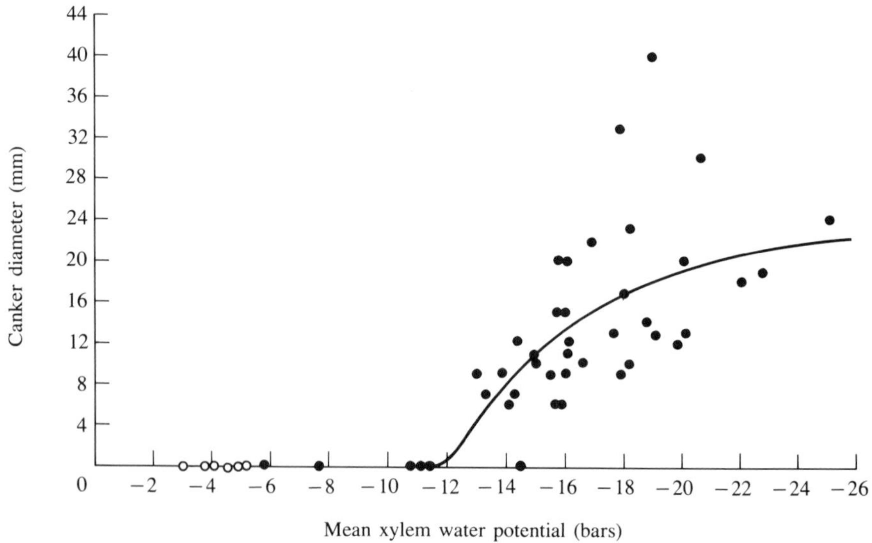

Figure 7.17 Size of bark cankers on stems of white birch inoculated with the causal organism, *Botryosphaeria dothidea*, in relation to twig water stress. (From Crist and Schoeneweiss, 1975.)

Parker (1961) concluded that development of bark cankers usually is related to low water content. Appel and Stipes (1986) reported that water stress makes pin oak more susceptible to attack by blight fungus and it also makes beech more susceptible to attack by fungi (Chapela and Boddy, 1988). Schoeneweiss (1978a, b) reported several cases where water stress increased canker formation in woody plants. Perhaps changes in biochemistry caused by water stress, such as the increase in proline observed in quaking aspen, are favorable for the growth of fungi (Griffin *et al.*, 1986). Damage by *Fomes annosus* also is said to be increased by dry weather, but it is probable that the root systems already have been injured and dry weather merely makes the effects of reduction in absorbing surface more evident. Likewise a decrease in water transport to shoots caused by vascular disease becomes more evident in dry weather when plants are subjected to water stress. Vascular diseases were discussed in detail by Talboys (1968). The interrelationship between water status and various "decline" diseases is complex and may also involve other factors such as injury to roots by poor soil aeration, mineral deficiencies, and air pollution, in addition to attack by pathogens.

Powdery mildews are more troublesome in dry weather but most diseases are more common in wet weather because leaf and fruit surfaces must remain wet for many hours to permit spore germination. The relationship of water stress to plant diseases was discussed in detail by Ayres and by Schoeneweiss in Kozlowski's "Water Deficits and Plant Growth," Vol. 5 (1978a), and some of the effects of diseases on plant water relations were discussed in Chapter 5 of Misaghi (1982).

Water stress sometimes promotes outbreaks of leaf-eating insects because it produces chemical changes that improve the nutritional quality of leaves and environmental conditions favorable for insect reproduction (Mattson and Haack, 1987b). However, extreme stress is likely to be as unfavorable for insects as it is for the host plants. Scriber (1977), for example, reported that growth of insect larvae was slowed by feeding on leaves low in water content and such leaves are likely to be tough, low in nitrogen, and unpalatable. In contrast to the view that weak trees are most susceptible to attack, Amman *et al.* (1988) stated that there is considerable evidence that mountain pine beetle prefers the largest trees in a lodgepole pine stand. They attributed the low infestation of poorly growing trees in recently thinned stands to changes in microclimate such as increased light and decreased humidity. This suggests that at least in some cases, infestation depends as much on the environment as on the condition of the host.

There is some difference of opinion concerning the effects of water deficits on attacks by boring insects. Vité (1961) stated that infestation of ponderosa pine by bark beetles is favored by dry weather because it lowers oleoresin exudation pressure. In contrast, Lorio and Sommers (1986) concluded that moderate water stress is unfavorable for attacks of southern pines by the southern pine beetle because slowing growth makes more food available for conversion into oleoresin, and abundant resin production is unfavorable to survival of southern pine beetles. Ferrell (1978) reported that engraver beetles cannot establish themselves in white fir unless the water potential is lower than −1.5 MPa because resin flow is too strong in unstressed trees. Lorio *et al.* (1990) reviewed the literature on conifer physiology in relation to borer resistance and suggested some ideas for future research.

Beneficial Effects of Water Stress

Not all the effects of water stress are detrimental. For instance, water stress tends to decrease injury from air pollution because it causes stomatal closure and stomata are the principal pathway by which air pollutants such as SO_2 enter leaves (e.g., Norby and Kozlowski, 1982). The extensive literature on air pollution in relation to woody plants will be discussed in Chapter 9. Even though water stress reduces vegetative growth, it sometimes has other effects that are beneficial. For example, moderate water stress increases the rubber content of the desert shrub guayule enough to increase the yield of rubber, although the total yield of plant material is decreased (Wadleigh *et al.*, 1946) (Fig. 7.18). Richards and Wadleigh (1952) stated that the quality of apples, pears, peaches, and plums is improved by mild water stress, although the size of fruits often is decreased. The oil content of olives also is said to be increased (Evenari, 1960). The benefit from subjecting seedlings to water stress before transplanting is discussed later in this chapter under Drought Hardening. It also can be argued that the increased proportion of thick-walled xylem elements produced in water-stressed trees (Zahner, 1968; Zobel and van Buijtenen, 1989) is beneficial because it increases wood density. More research on the effects of water stress on wood quality would clearly be useful.

Drought Tolerance

Over one-third of the earth's surface is classified as arid or semiarid because of deficiencies in rainfall and even the humid areas often are subjected to random droughts of varying severity. Thus deficiency in water supply is an important limitation on plant distribution over most of the world. Zahner (1968, p. 230) stated that 80% of the variation in width of annual rings of trees in humid climates and 90% of the variation of those in semiarid climates can be explained by variations in rainfall. For example, in humid Michigan over a 10-year period, 14% of the variation in radial growth of young red pine trees was associated with variation in water stress during the preceding July to September and 68% with water stress during May to September of the current year. Bassett (1964a) found a very high correlation between basal area production per acre of shortleaf and loblolly pine and available soil moisture in Arkansas. Both the establishment of seedlings and the growth of trees and shrubs often are limited by water deficits and their success depends to a large extent on their capacity to survive drought and resume growth when rewatered. Differences in tolerance of drought often determine species distribution, with those species with low tolerance being confined to moist sites while those with greater tolerance occupy drier slopes and ridges. Some of the earlier literature on causes of

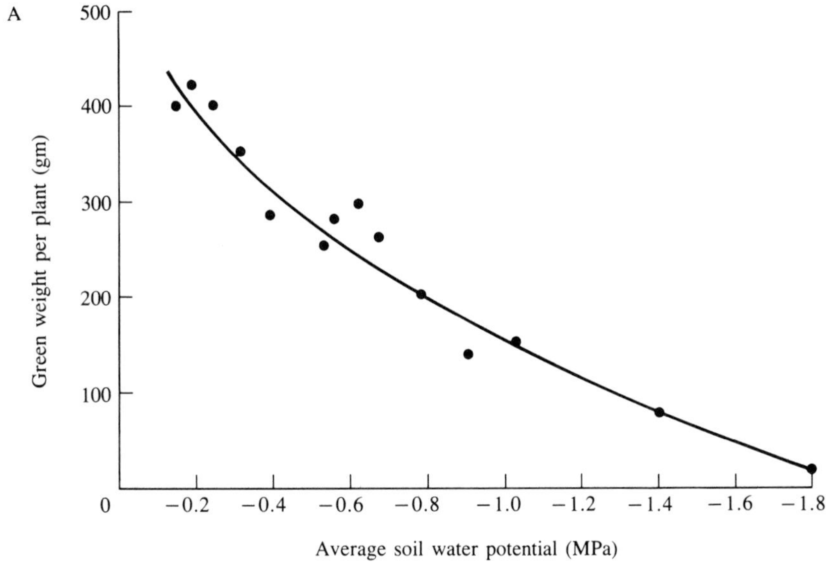

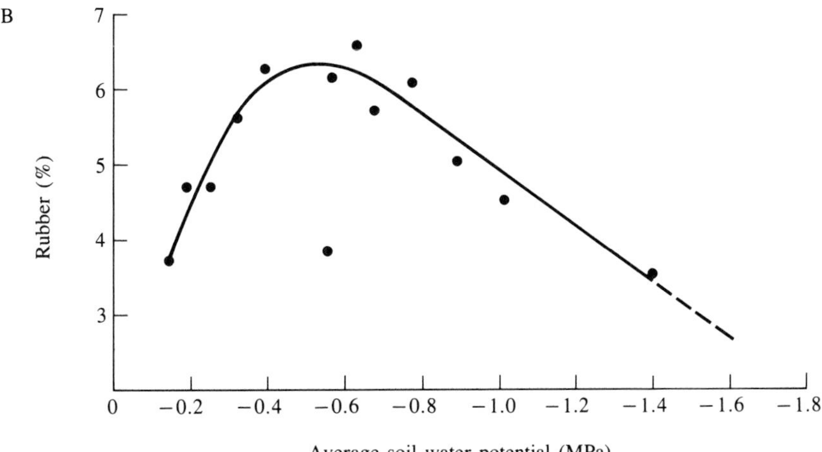

Figure 7.18 Relationship between average soil water potential developed between irrigations and (A) fresh weight of guayule plants and (B) rubber content as a percentage of dry weight. The 275 g of plant containing over 6% of rubber is just as productive as 400 g of plant containing slightly over 4% of rubber and requires much less water. (After Wadleigh *et al.*, 1946; from Kramer, 1983.)

drought tolerance was reviewed by Parker in Kozlowski (1968d) and there is a more recent review by Kozlowski (1982a).

Terminology

Because of differences among writers in terminology, some terms will be defined. As stated earlier, drought will be treated as a meteorological event, as a period without precipitation of sufficient duration that results in depletion of soil water and injury to plants (May and Milthorpe, 1962; McWilliam, 1986). Definition of drought by one author as the "occurrence of substantial water deficit either in the soil, atmosphere, or plant" creates confusion that can be avoided by treating drought exclusively as a meteorological factor. Droughts are permanent in arid regions, seasonal where well-defined wet and dry seasons occur, and unpredictable or random in many humid regions, where they can occur during any season. Table 7.4 shows the variations in rainfall during the growing season in three consecutive years in a humid climate. We prefer to use the term "drought tolerance" rather than "drought resistance," because tolerance more accurately describes the reaction of plants to environmental stresses such as drought, high salinity, and unusually low or high temperatures.

In general terms, droughts are avoided or tolerated to various degrees by plants with the help of structural or physiological adaptations that postpone dehydration or that enable plants to tolerate dehydration without serious injury.

Drought Avoidance

The best known examples of drought avoidance are found in the ephemeral desert annuals that complete their life cycle within a few weeks after winter rains have wetted the soil and produce mature seed by the time the soil dries. Drought avoidance is of little importance in forestry or horticulture because the life cycles of trees are too long for avoidance to occur. Most of what is sometimes called "drought avoidance" (Levitt, 1980b, p. 96) really is postponement of dehydration. Assuming

Table 7.4
Monthly Rainfall during the Summer at Letcombe Laboratory, Oxfordshire, England[a,b]

Year	May	June	July	August	September
1975	54	9	36	29	104
1976	38	24	16	18	99
1977	51	89	19	150	20
7-year avg.	54	60	49	64	66

[a]Rainfall measured in millimeters.
[b]From *Annu. Rep. Agri. Res. Counc. Letcombe Lab. 1977*, p. 92, 1978.

that droughts are periods without rainfall, perennial plants in natural habitats cannot "avoid" droughts.

Postponement of Dehydration

The most important type of drought tolerance involves postponement of dehydration, by either effective root systems, very efficient water transport systems, control of water loss, or all three.

Efficient Absorbing Systems

One of the most effective safeguards against drought injury is a deep, extensively branched root system that can absorb water from a large volume of soil. In the absence of rainfall or irrigation, soil dries from the surface downward because of evaporation and the high concentration of roots common near the surface (Coile, 1937). Thus shallow-rooted plants suffer severe water deficits long before deeper-rooted plants. There is some evidence that significant amounts of water can be transferred from deep, moist soil horizons to the drier surface soil (Corak *et al.*, 1987; Baker and van Bavel, 1986). Caldwell and Richards (1989) demonstrated that labeled water supplied to the deep roots of sagebrush appeared in neighboring shallow-rooted bunchgrass within 24 hr. They considered the transfer of water through sagebrush roots to be beneficial to associated shallow-rooted plants. However, it may also increase water loss from the surface soil by evaporation. Ecologists and foresters have given considerable attention to differences in the initial root development of tree seedlings and other plants as a cause of differences in plant distribution. For instance, the deeper roots of northern red oak enable it to survive where the shallower-rooted American basswood cannot (Holch, 1931) (Fig. 7.19), and shallow-rooted birch seedlings can become established only in permanently moist soil, whereas seedlings of deeper-rooted species can survive summer droughts on drier upland soils. Red maple is very adaptable because it develops deep roots in dry soils and shallow roots in wet soils and so occurs in a wide range of habitats (Toumey, 1929).

Albertson and Weaver (1945) concluded that depth of rooting was an important factor in the survival of trees in the prairie region during the great drought of the 1930s. Oak and hickory seedlings survive better than pine seedlings under forest stands because the deep roots of the former quickly penetrate below the surface horizon to a depth where competition for water is less intense (Coile, 1940). Differences in depth of rooting were important with respect to survival on dry sites in Australia (Grieve and Hellmuth, 1970; Parsons, 1969), in Israel (Oppenheimer, 1951), and Missouri (Bahari *et al.*, 1985; Ginter-Whitehouse *et al.*, 1983), and in Texas (van Buijtenen *et al.*, 1976). However, Kummerow (1980) claimed that leaf adaptations that reduce transpiration are more important than depth of rooting for drought tolerance of shrubs in the California chaparral community. This probably is because the shallow soil of the chaparral region prevents development of deep root systems even when the hereditary potential for deep rooting exists.

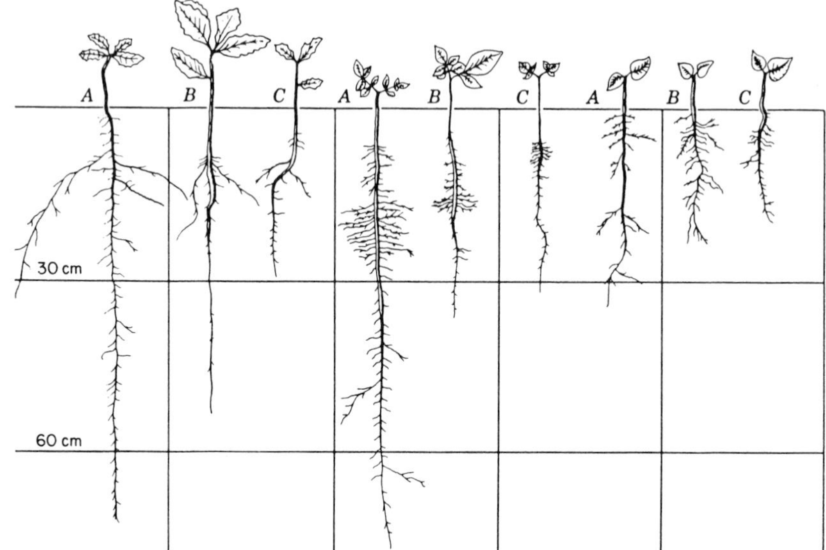

Figure 7.19 Differences in seedling rooting habit of three species in three environments. Northern red oak, *left*; shagbark hickory, *center*; American basswood, *right*. Seedlings A were growing in full sun in open prairie, seedlings B in an oak forest, and seedlings C in the deep shade of a moist basswood forest. It was necessary to water the basswood seedlings in full sun to prevent death by desiccation. Oak developed the deepest and largest root systems in all three habitats, and basswood the smallest and shallowest. (After Holch, 1931; from Kramer, 1983.)

The role of mycorrhizae in mineral absorption is well known, but, as mentioned in the section on water absorption, mycorrhizae may improve water absorption and increase drought tolerance. Dixon *et al.* (1980) reported that inoculation of roots of white oak seedlings with a mycorrhizal-forming fungus increased drought tolerance and speeded up recovery after a drying cycle. Fitter (1987) provides a useful review of literature dealing with the efficiency of various types of root systems and Caldwell and Virginia (in Pearcy *et al.*, 1989) discuss methods of studying root growth and function.

Establishment of outplanted seedlings depends on possession of vigorous root systems that promptly resume growth. Satoo (1956) observed that the size of root systems was important to seedling survival in Japan and Lopushinsky and Beebe (1976) found large root systems to be important to survival of outplanted seedlings of ponderosa pine and Douglas-fir in the Pacific Northwest. Pallardy and Kozlowski (1979b) found differences in root growth among poplar clones and Othieno (1978) found differences among strains of tea. Other observations emphasizing the importance of root systems in seedling survival are cited by Kozlowski (1982a), Pallardy (1981), Mazzoleni and Dickmann (1988), and South (1986). It has even been suggested that infection of root systems by *Agrobacterium rhizogenes,* the cause of hairy root disease, might increase drought tolerance by increasing root surface (Vartanian and Berkaloff, 1989).

The physiological quality of planting stock is as important as its morphological characteristics (Wakeley, 1948), because the capacity to generate new roots (root growth potential) is essential to survival. The physiological characteristics of pine seedlings with respect to survival after outplanting were discussed in detail by Kramer and Rose (1986), and production of drought-tolerant seedlings was discussed by several other authors in South (1986), in Duryea and Brown (1984), and in Chapter 13.

Although a high ratio of roots to shoots is desirable in seedlings and during droughts, it seems possible that many and perhaps most mature trees possess more roots than are necessary for survival. Most readers probably have observed cases where a considerable fraction of the root system of a tree was removed during construction work, yet the tree survived. Teskey *et al.* (1985) noted that removal of one-third of the roots from 15- to 18-year-old trees of Pacific silver fir did not affect leaf conductance or xylem pressure potential, but removal of one-half of the roots decreased both. However, data of Carlson *et al.* (1988) indicate that the entire root system of young loblolly pine trees is used in water absorption. Perhaps the difference in results from the two experiments is related to differences in atmospheric conditions and soil moisture between a subalpine forest in the Pacific Northwest and a pine forest in southeastern Oklahoma. Although large root systems often are beneficial during droughts, readers are reminded that roots use photosynthate that might have gone into trunk or fruit growth. Thus where severe droughts are rare or irrigation is practiced, the large root systems so desirable for survival in nature may be undesirable. Perhaps more attention should be given to selection for the most efficient root systems for particular environments.

Rooting Habit, Vegetation Cover, and Watershed Yield

Although deep root systems have obvious advantages they also have disadvantages that should not be overlooked. One is the large amount of photosynthate used in maintenance respiration at the expense of shoot growth. More important in some situations is the fact that deep-rooted vegetation removes more water from the soil than shallow-rooted types, thus decreasing stream flow from watersheds (Hellmers *et al.*, 1955; Davis and Pase, 1977). Thus in regions with limited rainfall and where maximum yield of water from watersheds is desired, deep-rooted vegetation may be undesirable. Another problem is presented by phreatophytes such as salt cedar, mesquite, and other trees and shrubs whose deep roots draw on shallow underground water tables, and vegetation such as cottonwood and willow that grow along streams and irrigation canals (Robinson, 1952). Daily peaks in transpiration sometimes remove enough water from the soil to cause corresponding decreases in stream flow and in the water level in shallow wells. Removal of phreatophytic vegetation often results in significant increases in water yield. However, maintenance of enough plant cover to prevent surface runoff and erosion is essential.

In general, grasslands yield more water than forested watersheds, and removal of forests often results in an increase in water yield (Dunford and Niederhof, 1944; Hoover, 1944; Trimble *et al.*, 1963). Deciduous forests also usually yield more water than evergreen forests, which waste more water by interception and by transpiration during the dormant season (Dunford and Niederhof, 1944; Swank and Douglass, 1974). Removal of evergreen shrubs such as laurel and rhododendron from beneath a deciduous forest also is likely to increase stream flow (Johnson and Kovner, 1956). The relationship between vegetation, evapotranspiration, and watershed yield was discussed by NcNaughton and Jarvis in Kozlowski (1983a). Removal of forests occasionally results in an undesirable rise in the water table, converting formerly dry areas into swamp (Wilde, 1958; Trousdell and Hoover, 1955). This is most likely to occur where an impermeable layer hinders downward percolation and causes a "perched" water table.

Control of Transpiration

Most plants native to arid and semiarid regions have thick, heavily cutinized, sclerophyllous leaves with low cuticular transpiration and stomata that close promptly in dry air or when leaves are water stressed. The amount of cuticular transpiration probably depends less on the thickness of the cuticle than on the amount of wax deposited in and on it (Schönherr, 1976) and on whether the wax occurs in platelets or is amorphous. The amount of wax depends on both the genetic potential and the environment of the plant, usually increasing in sunshine and dry air in both herbaceous (Van Volkenburgh and Davies, 1977) and woody plants (Pallardy and Kozlowski, 1980). (Differences in wax accumulation are shown in Fig. 7.20.) Accumulation of wax in stomatal antechambers is particularly effective in reducing transpiration (Kozlowski, 1982a, pp. 78–79). Water stress and bright light also cause increases in leaf thickness and in the ratio of mesophyll surface to epidermal surface seen in sun leaves as compared with shade leaves. This tends to increase photosynthesis more than transpiration and improves water use efficiency (Nobel, 1980).

The number and size of stomata vary widely among species and even in strains of the same species (Pallardy and Kozlowski, 1979a; Sena Gomes and Kozlowski, 1988a). However, their responsiveness to leaf water stress probably is more important than their number and size. For example, the differences in transpiration rate and leaf water potential among six poplar clones were more closely related to speed of stomatal closure than to differences in leaf anatomy. There also are differences among the stomata of poplar hybrids in sensitivity to light and the vapor pressure deficit of the atmosphere (Pallardy and Kozlowski, 1979c). Kaufmann (1982) observed the effects of temperature, radiation, humidity, and water stress on leaf conductance of trees of a subalpine forest and concluded that irradiance and humidity or vapor pressure deficit were most important. Teskey *et al.* (1984) considered the low leaf conductance of Pacific silver fir to be an important factor in preventing dehydration damage to trees growing in cold soil.

As mentioned earlier, there often is at least partial closure of stomata at midday, resulting in reduction in exchange of CO_2 and water vapor (Figs. 7.7 and 7.15). As the soil dries and plant water stress becomes more severe, the stomata tend to close earlier, more remain closed, and eventually they fail to open. This results in a large reduction in transpiration because the cuticular resistance is much greater than the

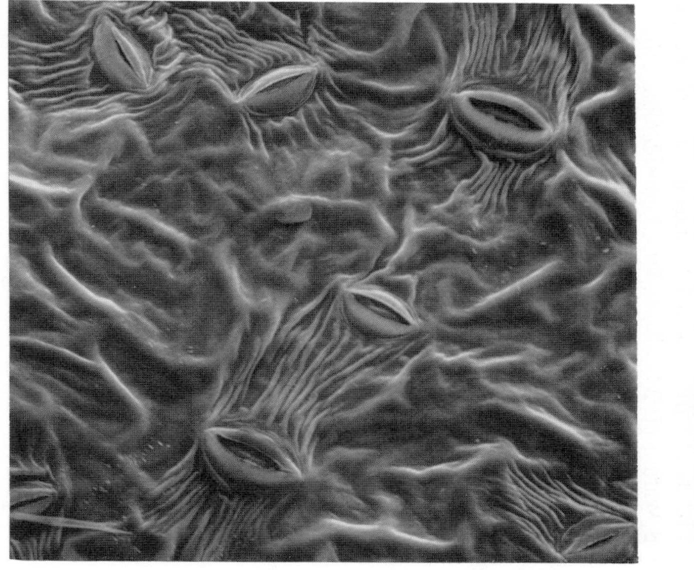

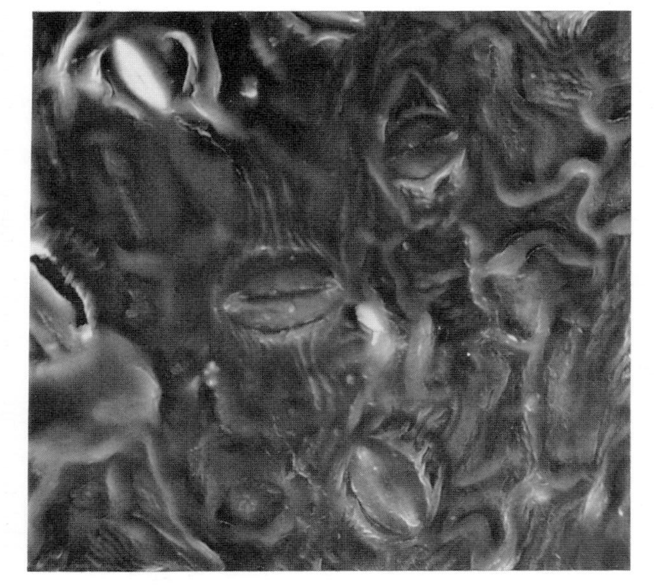

Figure 7.20 Effect of environment on accumulation of wax on leaves of a poplar hybrid (*Populus candicans* Ait. × *P. berolinensis* Dipp.). (A) Grown in a growth chamber; (B) grown in the field. Magnification, ×500. (From Pallardy and Kozlowski, 1980.)

resistance of open stomata. At one time it was supposed that stomata close at some critical or threshold leaf water potential, but it now seems that the relationship varies widely among species, depending on the rate of drying and on whether the plants have been dehydrated previously (Turner, 1986, pp. 13 and 31). Another factor affecting stomatal closure is the increase in concentration of ABA in water-stressed tissue, which may act in addition to and even in advance of the hydraulic effect of decrease in water content and turgor (Zhang and Davies, 1989a). However, the importance of ABA is debatable (Downton *et al.*, 1988; Mansfield, 1986, pp. 202–206; Trewavas, 1981).

Recently it has been claimed that hormonal signals from water-stressed roots may play an important role in causing reduced shoot growth and stomatal closure independently of the hydraulic effects of water stress (Davies *et al.*, 1986; Turner, 1986; Zhang *et al.*, 1987; Zhang and Davies, 1989b). However, in sunny climates, shoot water stress often develops before roots are stressed (Kramer, 1988b).

As stated earlier, it is now well established that the stomata of some, but not all, plants also close in response to decrease in vapor pressure of the atmosphere. This can even result in a decrease in transpiration while decreasing atmospheric humidity favors an increase. Humidity effects were shown for apricot by Schulze *et al.* (1972), for Engelmann spruce by Kaufmann (1976), and for cacao for Sena Gomes *et al.* (1987). Pallardy and Kozlowski (1981) found considerable differences in responsiveness of stomata of various poplar hybrids to atmospheric humidity (Fig. 7.21). Marshall and Waring (1984) also reported considerable differences among species in response to a high vapor pressure deficit. In spite of uncertainties concerning the mechanisms by which stomata respond to water stress there is no doubt that they are the most important control over water loss from most woody and herbaceous plants. The use of antitranspirants that cause stomatal closure is discussed later in this chapter.

Reduction in leaf expansion and consequently in leaf area is caused by water stress and has the beneficial effect of reducing transpiration, but the undesirable effect of also reducing the photosynthetic surface. Exposure to radiation is reduced in mature leaves by wilting and by the rolling characteristic of grass leaves. Stålfelt (1956, p. 336) cited work indicating that leaf wilting and rolling reduce transpiration about 35% for plants of moist habitats, 55% for Mediterranean species, and as much as 75% for some desert xerophytes. Begg (1980) attempted to use severity of leaf rolling as a quantitative measure of water stress. Pine needles that occur in bundles also shade one another and have a lower rate of photosynthesis, and presumably a lower rate of transpiration, than single, fully exposed needles (Kramer and Clark, 1947).

The leaf surface exposed to radiation sometimes is reduced by changes in orientation and angle of exposure. An example of orientation is found in the desert shrub jojoba, which has vertical leaves that intercept only about half as much radiation at noon as in the morning or evening. Seedling leaves of turkey oak also are oriented approximately parallel to incident radiation. It is claimed that this aids survival in the North Carolina sand hills, where there is strong reflection from the white sand,

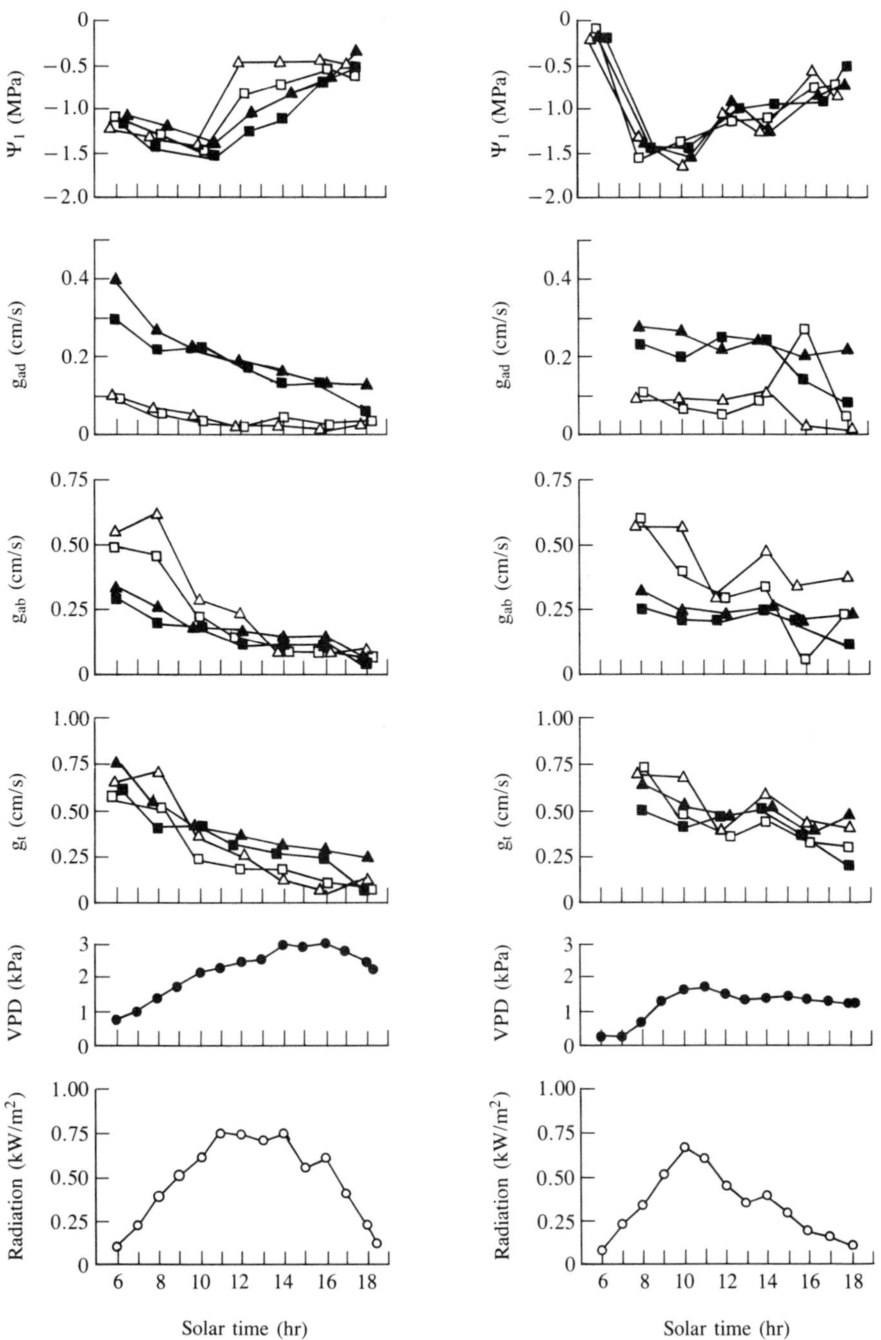

Figure 7.21 Diurnal trends in leaf water potential (Ψ_1), stomatal conductance of upper (g_{ad}) and lower (g_{ab}) leaf surfaces, and total stomatal conductance (g_t) of two fast-growing (□, △) and two slow-growing (■, ▲) poplar clones. *Left*, July 19, a sunny day. *Right*, July 27, a partly cloudy day. (From Pallardy and Kozlowski, 1981.)

but this has never been proven. Leaf orientation was discussed by Begg (1980), Ehleringer (1980), and Ehleringer and Werk in Givinish (1986b). The latter consider solar tracking and leaf angle to be important with respect to the amount of solar energy received.

Leaf area and transpiring surface also are decreased by the premature senescence and shedding of leaves that are common in water-stressed plants as mentioned earlier in this chapter. This usually begins with the oldest leaves and progresses toward stem tips. As the first leaves to die are often shaded and have low rates of transpiration, the reduction in water loss is not necessarily closely correlated with the reduction in leaf area. However, insect defoliation decreased transpiration of a hardwood forest enough to decrease water stress (Stephens *et al.*, 1972).

Plants from regions with well-defined wet and dry seasons often shed part or all of their leaves during the dry season (Reich and Borchert, 1982). Mooney and Dunn (1970) regard leaf shedding as important in increasing drought tolerance of shrubs and it seems to be important for some tress. Usually new leaves appear when soil moisture is replenished by rain, or sometimes even sooner (Richards, 1966). Premature abscission of leaves of water-stressed plants was discussed earlier in this chapter and in detail by Kozlowski (1991). Although premature leaf shedding usually is treated as an important adaptation that increases tolerance of drought, it also is a symptom of severe drought injury. It is rather expensive in terms of the photosynthate and nitrogen required to replace leaves and there have been lively discussions on the pros and cons of the deciduous versus the evergreen habit (e.g., Field *et al.*, 1983; Gower *et al.*, 1989).

The relative importance of control of transpiration by responsive stomata and leaf structure versus a large absorbing system probably varies with soil and atmospheric conditions. Van Buijtenen *et al.* (1976) rated stomatal control first, deep root systems second, and leaf morphology third with respect to drought tolerance of loblolly pine in West Texas, but few such comparisons have been made.

Water Storage

Storage of water in plants is another way of postponing dehydration during drought. Use of stored water to replace loss by transpiration is responsible for the midday decrease in water content and shrinkage often observed in plant organs (see Figs. 7.4 and 7.10). The daily transpiration of most herbaceous plants exceeds their total water content, but according to Waring and Running (1978), the water stored in the sapwood of large Douglas-fir trees is equivalent to several days of transpiration and it is reported that water stored in the sapwood supplies about one-third of the daily transpiration of a Scotch pine forest (Waring *et al.*, 1979). Schulze *et al.* (1985) estimated water storage in the crown to be 24% of daily transpiration for a 20-m spruce tree and 14% for a 25-m tree 28 cm in diameter. They also reported that very little water was used from the trunks below the crown in those trees. Landsberg *et al.* (1976) reported that the water stored in a 9-year-old apple tree is equal to 2 hr of loss by transpiration.

It is difficult to understand how so much water can be removed from the xylem, except as a result of cavitation, but water released by cavitation cannot be replaced overnight. Brough *et al.* (1986) discussed the problem of water storage in stems in detail. They estimated a 10% decrease in xylem water content from densitometer measurements of trunks of 7-year-old apple trees growing in soil near field capacity in Kent, England. There was only a decrease of 0.1% in xylem radius but a large decrease in bark thickness, suggesting that the bark is an important water storage tissue, as proposed by others (Kozlowski, 1973; Powell and Thorpe, 1977). Data on diurnal changes in radius of the xylem and xylem plus bark as related to changes in relation to xylem water content and leaf water potential of apple are shown in Fig. 7.22. Water storage also is discussed by Waring and Schlesinger (1985, pp. 78–80) and by Waring and Running in Lange *et al.* (1976). Presumably, this stored water reduces the severity of midday water deficits, and Waring and Franklin (1979) regarded it as very important to the success of the forests of the Pacific Northwest. If, as they suggest, it permits photosynthesis to continue at the normal rate for an additional hour each day, it would be highly beneficial.

Water storage is important in desert succulents such as euphorbias and cacti, and a large saguaro can store tons of water in its stem (Meyer, 1956). The loss of water from most of these plants is slow because of their thick cuticle and daytime closure of stomata and they can survive for a year or more on stored water. The ecological importance of water storage in the stems of seven giant, rosette species of *Espeletia* was studied by Goldstein *et al.* (1984). Considerable water is stored in the trunk of the baobab tree (*Adansonia digitata*), which sometimes attains a diameter of several meters. Numerous adaptations to dry habitats are discussed in chapters by Troll and Killian and Lemée in Volume 3 of Ruhland's "Encyclopedia of Plant Physiology" (1956). The numerous structural modifications found in plants of dry habitats support the statement attributed to the German morphologist Goebel that the variety of structures in plants is much greater than the variety of environments in which they live. We can add to this that a variety of physiological and morphological modifications contribute to survival. Examples can be found in papers by Abrams and Knapp (1986), Ginter-Whitehouse *et al.* (1983), LoGullo and Salleo (1988), Lucier and Hinckley (1982), Martin *et al.* (1987), and Wuenscher and Kozlowski (1971).

Antitranspirants

Antitranspirants are used to reduce transpiration, either by causing closure of stomata or by coating the foliage with a film more or less impermeable to water. Unfortunately, treatments that reduce efflux of water vapor usually decrease influx of CO_2, although partial closure of stomata should reduce photosynthesis less than transpiration, as was demonstrated for cotton by Slatyer and Bierhuizen (1964a, b). Some reasons for this are given in Chapter 10 and Table 10.2. Waggoner and Bravdo (1967) reduced transpiration and increased runoff from a pine stand by spraying it with phenylmercuric acetate, which causes stomatal closure. Unfortunately that substance is somewhat toxic to trees (Keller, 1966) and anyway it is unwise to release mercury compounds into the environment. Davies and Kozlowski (1975a, b)

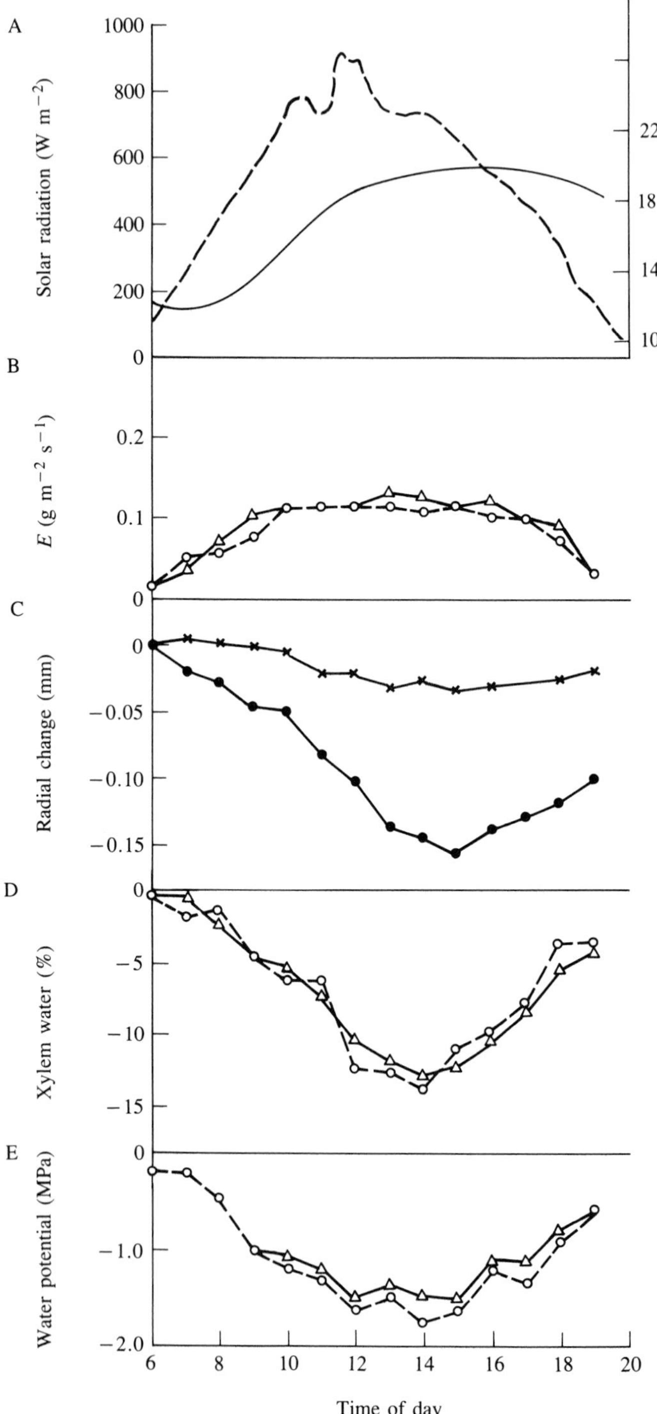

A
B
C
D
E
Solar radiation (W m⁻²)
1000
800
600
400
200
0
22
18
14
10
E (g m⁻² s⁻¹)
0.2
0.1
0
Radial change (mm)
0
−0.05
−0.10
−0.15
Xylem water (%)
0
−5
−10
−15
Water potential (MPa)
0
−1.0
−2.0
6
8
10
12
14
16
18
20
Time of day

used ABA to cause stomatal closure on tree seedlings, but Davies *et al.* (1979) are pessimistic about its practical usefulness. Films such as latex emulsion, polyvinyl chloride, and hexadecanol also have been applied with variable results. Film antitranspirants have been used on fruit trees to increase fruit size (Uriu *et al.*, 1975) and on seedlings when transplanted. The problems encountered in using films on tree seedlings were discussed by Davies and Kozlowski (1975b). The concept is attractive, but Solarova *et al.* (1981) concluded that antitranspirants have a very limited usefulness, chiefly in getting plants over the shock of transplanting. A major problem is that no film-forming substance is available that is more permeable to CO_2 than to water vapor.

Dehydration Tolerance

During severe droughts the capacity of the mechanisms that postpone dehydration is exhausted and the protoplasm ultimately is subjected to severe dehydration that often results in irreversible injury and death to tissue.

Differences in Tolerance of Dehydration

There are wide variations among plants in the tolerance of dehydration, ranging from mesophytes such as sunflower leaves that are irreversibly injured at -2.0 MPa (Fellows and Boyer, 1978) to certain lichens, mosses, ferns, and other "resurrection plants" that can tolerate the air-dry condition for months and resume physiological activity in a few hours or days after rehydration (Gaff, 1980). There are considerable differences in tolerance of dehydration between plants of dry habitats [e.g., *Acacia aneura* of the Australian desert (Slatyer, 1960), ceanothus of the California chaparral (Schlesinger *et al.*, 1982), and creosote bush of the North American desert (Odening *et al.*, 1974)] and dogwood and redbud, which often are killed by midsummer drought. Various trees such as willow, bald cypress, and water tupelo grow well only on permanently wet sites.

Leaf and root tissues of post oak, a dry site species, survive dehydration better than those of white, black, and northern red oak, which grow in moister habitats (Seidel, 1972). Bourdeau (1954) also concluded that post oak and blackjack oak have more protoplasmic tolerance of dehydration than white or northern red oak. The roots of the oaks seem to have more tolerance of dehydration than the shoots, resulting in the ability to sprout from the root crown after the shoots are killed by desiccation. However, Brix (1960) and Leshem (1965) concluded that roots of pine are less tolerant of dehydration than the shoots. Pharis and Ferrell (1966) reported that Douglas-fir seedlings from the interior have more tolerance of dehydration than

Figure 7.22 Light, transpiration, water status, and radial growth of apple trees at East Malling, Kent, England, on a sunny July day. (A) Solar radiation (dashed line) and air temperature (solid line); (B) transpiration of irrigated (△) and unirrigated (○) trees; (C) shrinkage of xylem (X) and of xylem plus bark (●); (D) xylem water content and (E) leaf water potential of irrigated (△) and unirrigated (○) trees. (Modified from Brough *et al.*, 1986.)

coastal seedlings, and Parker (1968) regarded differences in tolerance of dehydration to be important in trees. According to Oppenheimer (1932), leaves of almond could be dried to a saturation deficit of 70% before injury occurred, and those of olive to 60%, but leaves of fig to only 25%. There also are seasonal differences, with tolerance being greatest in the winter in some plants (Pisek and Larcher, 1954). According to Runyon (1936), the small leaves produced by creosote bush during dry weather are more tolerant of dehydration than the larger leaves produced during the rainy season.

Martin *et al.* (1987) used the amount of leakage of electrolytes from leaves (Leopold *et al.*, 1981) as an indicator of dehydration tolerance in several kinds of hardwoods. Leakage increased as leaf water potential decreased until late in the growing season, but leaves of three species of oaks showed less leakage than those of sugar maple or black walnut over the growing season, and leaves of dogwood showed most leakage until late in the growing season. The oaks showed some reduction and dogwood a strong reduction in leakage late in the growing season, suggesting drought hardening, but there was little or no decrease in maple and walnut. Failure to develop dehydration tolerance may be a factor in the early leaf abscission of walnut and trees of some other species. The investigators concluded that dehydration tolerance may play a significant role in plant distribution.

Dehydration injury usually is attributed to physical disorganization of the fine structure of membranes and organelles (e.g., Gaff, 1980; Leopold, 1983; Hincha *et al.*, 1987), but some tolerant plants show considerable disorganization under the electron microscope. There also is considerable interest in the effects of reduction in protoplast and chloroplast volume on processes such as photosynthesis (Kaiser, 1982; Sinclair and Ludlow, 1985; Sen Gupta and Berkowitz, 1988). The causes of injury and reasons for differences in tolerance of dehydration among various kinds of plants were discussed by Bewley (1979), Gaff (1980), and Lee-Stadelmann and Stadelmann (1976). However, as yet there is no adequate explanation of differences in dehydration tolerance at the cellular level.

Although the capacity to survive dehydration and resume growth is important to the survival of seedlings and natural vegetation, it is less important in forestry and horticulture than postponement of dehydration. Neither forest nor fruit trees are profitable if severely dehydrated and neither forestry nor agriculture is likely to be profitable where prolonged droughts occur frequently, unless irrigation is feasible.

Osmotic Adjustment

In some plants, tolerance of dehydration is increased by osmotic adjustment. This refers to a decrease in osmotic potential greater than can be explained by concentration of solutes during dehydration, indicating accumulation of additional solutes (Turner and Jones, 1980, p. 89). For example, loss of 10% of the osmotic or solvent water from plant tissue with an initial osmotic potential of −1.0 MPa would decrease the osmotic potential to −1.11 MPa. However, accumulation of additional solutes might decrease the osmotic potential to −1.2 or −1.5 MPa, permitting maintenance of turgor-dependent processes to a lower water potential than would otherwise be possible. Some decrease in osmotic potential inevitably accompanies dehydration, but true osmotic adjustment may or may not occur. For example, Bowman and Roberts (1985) observed osmotic adjustment by increase in solutes in two species of ceanothus but in a third species the changes in osmotic potential were caused largely by changes in solvent water volume.

Reviews by Morgan (1984), Radin (1983), and Turner (1986, pp. 32–34) cited much of the literature on osmotic adjustment. It was reported in apple leaves by Goode and Higgs (1973) and Osonubi and Davies (1978) attributed the greater drought tolerance of English oak compared to European white birch seedlings to the greater osmotic adjustment of the former. Osmotic adjustment was reported in water-stressed pine seedlings by Johnson *et al.* (1986) and Augé *et al.* (1986) reported that roses subjected to repeated water stress showed osmotic adjustment of 0.4 to 0.6 MPa, and as much as 0.9 MPa when inoculated with mycorrhizal fungi. Clayton-Greene (1983) reported that osmotic adjustment was three times greater in *Callitris columellaris* than in two species of eucalyptus, and Parker and Pallardy (1985) reported significant differences in osmotic adjustment in both leaves and roots of seedlings of black walnut from different sources. Myers and Neales (1986) reported significant osmotic adjustment in water-stressed eucalyptus seedlings. Other examples of osmotic adjustment in woody plants are cited by these writers.

Osmotic adjustment is supposed to maintain turgor and permit stomatal opening, leaf expansion, photosynthesis, root growth, and water absorption to lower plant water potentials than would otherwise be possible. It may also aid in postponing leaf senescence and in protecting meristems in water-stressed plants. However, the range of adjustment is small, usually less than 1.0 MPa, and it usually lasts a week or less (Turner and Jones, 1980). Although Hinckley *et al.* (1980) doubted if osmotic adjustment is very important for woody plants, it currently is receiving considerable attention. It is reported by Sen Gupta and Berkowitz (1988) that osmotic adjustment in the chloroplasts of spinach permits photosynthesis to continue to low leaf water potential.

Bound Water

Earlier in this century there was wide interest in "bound" water as a factor involved in the tolerance to dehydration and low temperature. The early work was reviewed by Kramer in Ruhland's "Encyclopedia of Plant Physiology," Volume 1 (1955). Interest in bound water seems to be reviving at least partly because of improvements in measuring it (Vertucci and Leopold, 1984). Bound water refers to water so firmly fixed or bound to a matrix that its thermodynamic properties such as its freezing point differ from those of bulk water. Development of new methods of estimating bound water such as the water sorption isotherm (Vertucci and Leopold, 1987b), NMR spectroscopy (Roberts, 1984), and nuclear magnetic resonance imaging (Johnson, *et al.*, 1987) probably will result in further study of the importance of bound water as a factor in tolerance of dehydration and freezing (Vertucci and Leopold, 1987b). Vertucci and Leopold (1987c) suggest that tightly bound water makes an important contribution to desiccation tolerance.

Cell Wall Properties

Occasionally it is claimed that thick cell walls containing a large amount of water contribute to dehydration tolerance (Gaff and Carr, 1961; Teoh *et al.*, 1967). However, Slatyer (1967, pp. 177) pointed out that the half-time for equilibration between wall and vacuolar water is only about 10 sec, so cell wall water should have little buffer effect. Furthermore, fully hydrated cell walls contain less than half of the water occurring in mature leaves (Kramer, 1983b, p. 265), hence only a limited volume is available to the vacuoles. Elasticity of cell walls, usually expressed as the bulk modulus of elasticity (ϵ), may be important (Nobel, 1983, pp. 40–44). If cell walls are inelastic (have a high modulus of elasticity), a small decrease in water content produces a greater decrease in cell water potential and turgor pressure than if they are more elastic. An increase in cell wall elasticity will lower the water potential at which turgor is lost as effectively as osmotic adjustment of solutes. Joly and Zaerr (1987) reported that water stress causes reversible changes in the bulk modulus of elasticity of cell walls in needles of Douglas-fir, which regulate turgor. Some investigators claim that elastic cell walls contribute to drought tolerance (Sanchez-Diaz and Kramer, 1971; Kramer, 1983b) and some of the ecological implications of cell wall elasticity are discussed by Osmond *et al.* (1980, pp. 271–272) and Tyree and Jarvis (1982, pp. 60–62).

Although the success of plants in nature usually is closely related to their drought tolerance, it often is difficult to distinguish the causes of differences in tolerance. This is because drought tolerance depends on a number of morphological and physiological characteristics, the relative importance of which vary in different environments. As Pallardy (1981) pointed out, the potential for deep rooting is less important on shallow soils than responsive stomata, and stomata that close promptly do not confer the maximum protection unless the leaf structure minimizes cuticular transpiration. Thus the search for drought tolerance must include consideration of the environment in which a plant will be grown.

Drought Hardening

It is well established that plants previously subjected to water stress usually suffer less injury from transplanting and drought than plants not previously stressed. Growers of both herbaceous and woody seedlings intended for transplanting to the field find it advantageous to "harden" their seedlings. Often this consists simply of exposing seedlings to full sun and decreased water supply. Augé *et al.* (1986) reported that roses subjected to four drying cycles sufficiently severe to cause stomatal closure showed more osmotic adjustment during a subsequent period of water stress than those not previously stressed. Clemens and Jones (1978) found that seedlings of acacia and eucalyptus that had been repeatedly stressed had better control of water loss and were more drought tolerant than seedlings not previously stressed. Operators of forest nurseries often undercut or "wrench" the roots to produce more compact root systems (Van Dorsser and Rook, 1972). These treatments also produce temporary water stress in the shoots. Several experiments on "hardening" tree seedlings by undercutting and wrenching are reported in South (1986) and root pruning is discussed in Chapter 13. The chief effects are on root–shoot ratio, cutinization, stomatal behavior, and osmotic adjustment, but there also is some evidence of increased protoplasmic tolerance of water stress. For example, Mooney *et al.* (1977) indicated that the photosynthetic apparatus of creosote bush was more tolerant of water stress when produced in a dry habitat then when produced in a moist habitat. Martin *et al.* (1987) also found that leaves of several kinds of forest trees developed increased tolerance of dehydration toward the end of the growing season.

The importance of subcellular changes such as the development of special proteins has not been explored in woody plants. Water stress, like heat stress, changes the pattern of protein synthesis in plant tissue (Bewley *et al.*, 1983; Heikilla *et al.*, 1984; Bozarth *et al.*, 1987; Singh *et al.*, 1989). However, it is uncertain whether these changes significantly alter the tolerance of dehydration.

Summary

Plant distribution over the earth's surface, wherever temperature permits plants to grow, is controlled chiefly by the availability of water. Even in many humid regions, more or less random droughts during the growing season cause sufficient water stress to reduce growth and yield. Although water stress usually is caused by drought, it develops whenever transpiration exceeds water absorption long enough to cause loss of plant turgor. Thus plant water stress can be caused by slow water absorption resulting from drying or cold soil or high salinity, by excessive water loss caused by warm, sunny weather and low humidity, or by a combination of the two sets of conditions.

The primary effects of water stress are a decrease in water content and turgor, which reduces cell enlargement and growth, and decrease in the free energy status or potential of the water, which affects water movement, especially movement of water from soil to roots. Plant water stress affects almost every aspect of plant morphology and physiology. In addition to reducing plant size, it often increases root–shoot ratio, wood density, the ratio of leaf weight to surface, and cutinization and wax formation on leaf surfaces. Water stress reduces photosynthesis, often causes changes in amounts of secondary compounds such as terpenes, rubber, and essential oils, causes premature shedding of leaves and fruits, and finally results in death of plants from desiccation. It sometimes increases susceptibility to attacks by pathogens and insects, but mild water stress can increase the content of rubber in guayule and of oil in olives, improve the quality of fruits, and increase the ability of seedlings to survive transplanting.

Differences in drought tolerance often determine the distribution of species within an ecosystem, those with lower tolerance being confined to moister sites. Among the factors affecting drought tolerance are the size of root systems, the control of water loss by responsive stomata and low cuticular transpiration, water storage in stems, and tolerance of dehydration. Thus the specific reasons for drought and dehydration tolerance differ among species and often are difficult to identify.

General References

Boyer, J. S. (1985). Water transport. *Ann. Rev. Plant Physiol.* **36,** 473–516.

Givnish, T. J., ed. (1986). "On the Economy of Plant Form and Function." Cambridge University Press, Cambridge.

Hinckley, T. M., Lassoie, J. P., and Running, S. W. (1978). Temporal and spatial variations in the water status of forest trees. *For. Sci. Monogr.* **20,** 1–72.

Jarvis, P. G., and Mansfield, T. A., eds. (1981). "Stomatal Physiology." Cambridge University Press, Cambridge.

Kozlowski, T. T., ed. (1968–1983). "Water Deficits and Plant Growth," Vols. 1–7. Academic Press, New York.

Kozlowski, T. T. (1982). Water supply and tree growth. I. Water deficits. *For. Abstr.* **43,** 57–95.

Kramer, P. J. (1983). "Water Relations of Plants." Academic Press, New York.

Landsberg, J. J. (1986). "Physiological Ecology of Forest Production." Academic Press, London.

Lange, O. L., Kappen, L., and Schulze, E.–D., eds. (1976). "Water and Plant Life." Springer-Verlag, Berlin.

Lange, O. G., Nobel, P. S., Osmond, C. B., and Ziegler, H., eds. (1982). "Water Relations and Carbon Assimilation," *Encycl. Plant Physiol.* **12B.** Springer-Verlag, Berlin.

Larcher, W. (1980). "Physiological Plant Ecology," 2d ed. Springer-Verlag, Berlin.

Levitt, J. (1980). "Responses of Plants to Environmental Stress," 2d ed., Vol. 2. Academic Press, New York.

Turner, N. C. (1986). Crop water deficits: A decade of progress. *Adv. Agron.* **39,** 1–51.

Waring, R. H., and Schlesinger, W. H. (1985). "Forest Ecosystems." Academic Press, Orlando, FL.

Westgate, M. E., and Boyer, J. S. (1985). Osmotic adjustment and the inhibition of leaf, root, stem, and silk growth at low water potentials in maize. *Planta* **164,** 540–549.

Chapter 8

Soil Aeration, Compaction, and Flooding

Introduction

Respiration of roots and soil organisms uses large amounts of soil oxygen (O_2). A constant supply of O_2 is required to maintain the healthy roots necessary to provide plants with adequate amounts of water, mineral nutrients, and certain hormonal growth regulators. However, the soil does not contain enough O_2 at any one time to maintain normal root respiration for more than a few days (Drew, 1983). Furthermore, under a crop, as much as half of the soil O_2 may be consumed by microorganisms (Clark and Kemper, 1967). Howard (1925) claimed that in India several species of trees were killed by dense growth of grass over their roots, which produced anaerobic soil conditions, and McComb and Loomis (1944) suggested that deficient aeration under various prairie grasses hindered invasion by trees. As much as 2.5 to 17 liters of O_2 per square meter of land area should diffuse into well-drained, porous soils of the temperate zone each day to replace O_2 depletion by respiration of roots and microorganisms (Russell, 1973). Unfortunately gas exchange between the soil and the air is impeded on many sites by compaction of soil, a high or perched water table, impermeable layers such as pavements, or flooding of soil. Inadequate soil aeration also is common in heavy-textured soils in which the drainable pore space occupied by air is low.

Soil Compaction

Soils often are compacted by pedestrian traffic, grazing animals, and heavy machinery. Compacted soils are common on golf courses, campsites, parks, construction sites, timber-harvesting areas, and fruit orchards. Tractors and skidders are very heavy and together with skidding of logs may greatly compact forest soils. For example, a rubber-tired skidder reduced macropores in a forest soil by 38 to 68% (Dickerson, 1976). Once compacted by logging machinery, forest soils may not recover for many years. For example, soils of the Boreal Forest of Canada remained in a compacted condition for several decades, even though they were subjected to annual cycles of freezing and thawing (Corns, 1988). The prolonged compaction of these soils was associated with the presence of hydrous mica clays characterized by a nonexpanding lattice structure and frequently by low moisture contents in late summer and winter. Compaction of forest soils by heavy machinery was reviewed by Greacen and Sands (1980).

Forest soils are particularly susceptible to compaction because of their loose, friable structure and high porosity. Furthermore, as tree roots increase in size they compact the soil, and also transmit the weight of the trees onto the soil. Clay soils and those with a high proportion of exchangeable Ca are very susceptible to compaction, especially when wet. Dry and coarse-textured soils are less prone to compaction.

Effects of Compaction on Soils

Compaction changes the physical properties of soils by increasing their bulk density and mechanical resistance. Soil compaction decreases total pore space, especially the number and size of macropores (those greater than 50 μm), and increases the proportion of micropore space. The very large pores (50–150 μm) are affected first (Hartage, 1968). Decrease in macropore space limits draining and diffusion of O_2 into and diffusion of CO_2 out of the soil. The decrease in O_2 diffusion rate in compacted soils is largely attributed to the decrease in cross-sectional area of soil pore space across which the process can occur.

There are many examples of increase in the bulk density of compacted soil. For example, the bulk density of soil at a campsite in Michigan increased by 21% (Legg and Schneider, 1977). In the Washington, D.C., area, compaction of several soils by pedestrian traffic increased the bulk density from a range of 1.20 to 1.60 g cm^{-3} to a range of 1.70 to 2.20 g cm^{-3} (Patterson, 1976). In a lodgepole pine forest the bulk density of soil was 28% higher on skid trails than in undisturbed soil (Froehlich *et al.*, 1986). Trampling of soil associated with continuous grazing and intensive rotation grazing by livestock appreciably increases the bulk density of soil (Willatt and Pullar, 1983; Warren *et al.*, 1986). Cultivated soils usually have a greater bulk density and less macropore or noncapillary pore space than forest soils (see Fig. 6.3).

When soil is compacted and bulk density increased, the amount of water available for uptake by plants is reduced and a greater proportion of the water moves through the capillary or micropore space. Associated with reduction in total pore space is a decrease in saturated hydraulic conductivity. Furthermore, soil compaction often results in development of a soil crust that reduces infiltration of water, thus increasing surface runoff of water and erosion.

Effects of Compaction on Tree Growth

Growth of woody plants in compacted soils is inhibited. For example, approximately 60% of seeded loblolly pines died in clay loam soils because the radicles could not penetrate the compacted surface (Pomeroy, 1949). Penetration of soil by roots of Douglas-fir seedlings declined linearly with increasing bulk density of soil (Heilman, 1981). A bulk density of 1.8 g cm^{-3} reduced establishment of pitch pine, Austrian pine, and Norway spruce seedlings. Root penetration was restricted at a bulk density of 1.4 g cm^{-3} on a silt loam soil and at 1.6 g cm^{-3} on a sandy loam (Fig. 8.1). Root growth was reduced when soil compaction decreased soil air space below 15% and roots stopped growing when the air space dropped to 2% (Richards and Cockcroft, 1974). Growth of Douglas-fir seedlings was reduced in areas subjected to heavy logging traffic (Youngberg, 1959), and compaction of soil reduced the diameter growth of young red maple and black oak trees (Donnelly and Shane, 1986) and the volume growth of young ponderosa pine trees (Froehlich *et al.*, 1986). Compaction of soil also reduced dry weight increment of ailanthus seedlings within a month, and shifted growth from taproots to lateral roots (Pan and Bassuk, 1985).

Sands and Bowen (1978) grew Monterey pine seedlings in pots in sandy soil with different degrees of compaction over the range of bulk densities normally found in the field. At harvest (151 days after sowing seeds) the dry weights of roots and shoots, seedling heights, and root volumes were highest in the least compacted soil (bulk density of 1.35 g cm^{-3}) and lowest in the most compacted soil (bulk density of 1.60 g cm^{-3}). The dry weights of seedlings in the most compacted soil were only about half of those in the least compacted soil (Table 8.1).

Mechanisms of Growth Inhibition

Compaction of soil reduces growth of woody plants in complex ways. Contributing to growth reductions are changes in soil aeration, infiltration and retention of water, and soil strength (Greacen and Sands, 1980).

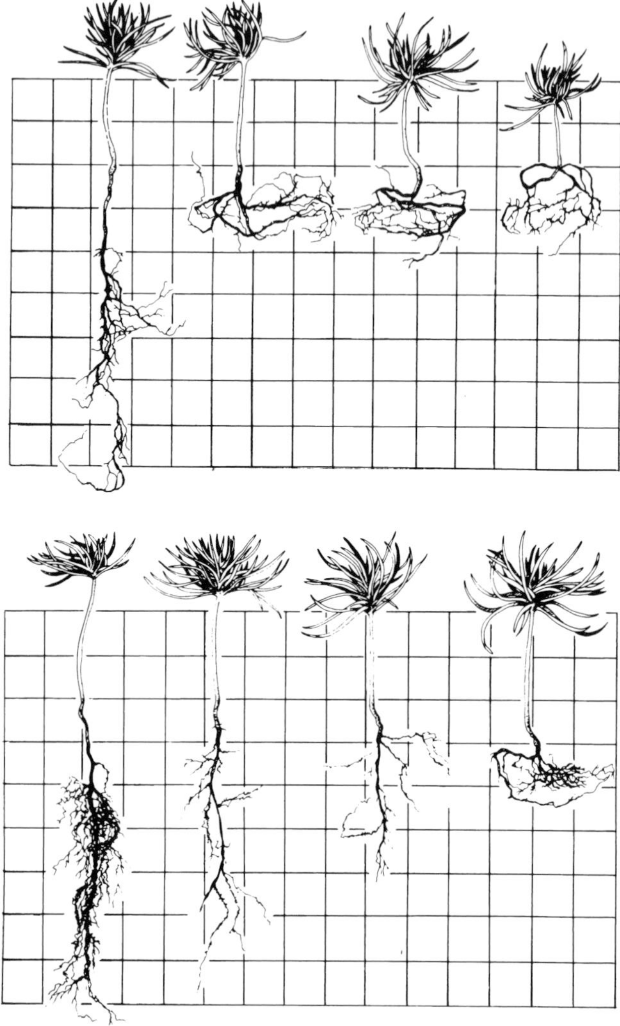

Figure 8.1 Effect of increasing soil compaction on growth of Austrian pine seedlings on a silt loam soil (*top*) and a sandy loam soil (*bottom*). Bulk densities from left to right of each soil are 1.2, 1.4, 1.6, and 1.8 g cm^{-3}. (From Zisa *et al.*, 1980.)

Depletion of soil O_2 often decreases root growth. It is difficult, however, to establish a critical minimal O_2 concentration for root growth because of the complex interactions involved. Nevertheless, some data are available for growth of various species on specific sites. For example, root growth of lodgepole pine was drastically reduced when the soil O_2 content dropped below 10% (Lees, 1972). Root growth of apple trees in Australia was inhibited when compaction reduced soil air space to less

Table 8.1

Dry Weight of Roots and Shoots, Root Volume, and Height of Monterey Pine Seedlings Grown in Soil with Different Bulk Densities[a,b]

Soil bulk density ($g\ cm^{-3}$)		Dry weight (g)		Root volume	Height
Before planting	At harvest	Shoots	Roots	(cm^3)	(cm)
1.60	1.58	3.6	3.0	24.7	20.5
	***	***	*	***	***
1.48	1.48	5.9	4.9	39.3	29.2
	***	***	NS	*	***
1.35	1.37	7.0	5.6	47.3	32.8
n =	20	20	9	15	20

[a]From Sands and Bowen (1978).

[b]NS, not significant; *, $p < 0.01$; ***, $p < 0.001$; levels of significance refer to adjacent treatments.

than 15% and was negligible when the air space dropped to 2% (Richards and Cockcroft, 1974). In the United States, soil O_2 concentrations greater than 10% were required for vigorous growth of apple roots. However, roots grew slowly with 3 to 5% O_2 and survived even with 0.1 to 0.3% O_2. When the O_2 concentration dropped below 10% and CO_2 increased above 5%, formation of new roots was negligible (Boynton *et al.*, 1938). A critical limit of 9 to 10% air space at field capacity was found for the 25- to 75-cm soil layers of citrus orchards in Israel (Patt *et al.*, 1966). Figure 8.2 shows the correlation between air capacity of the soil and growth of fine roots of citrus.

The capacity of roots to penetrate soils has an important effect on their growth. Root elongation and penetration into soil of high bulk density become confined to cracks rather than occurring throughout the soil matrix (Patterson, 1976). Establishment of loblolly pine seedlings was reduced as much as 24% because the emerging radicles could not penetrate the compacted soil (Foil and Ralston, 1967). Reduction of root extension into compacted soil makes less water available as that around the root is depleted. Capillary movement of water from moist to dry regions in soil at or below field capacity is very slow, hence continuous root extension is necessary for absorption of enough water to meet transpiration requirements and maintenance of a favorable internal water balance. Reduced water uptake in compacted soils also has been attributed to effects of high CO_2 content in reducing permeability of roots (Kramer and Kozlowski, 1979, p. 549).

Impermeable Layers

During construction, O_2 deficiency of soil often results from compaction and placement of fill around shade trees (Fig. 8.3). Yelenosky (1964) studied the composition

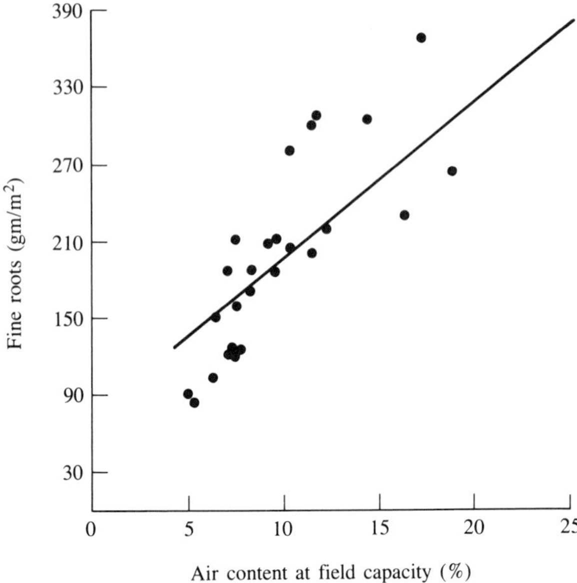

Figure 8.2 Influence of soil pore air content on development of fine roots of citrus in the 25- to 75-cm soil horizon. [From J. Patt, D. Caimeli, and T. Zafrir, Influence of soil physical condition on root development and on production of citrus trees, *Soil Science,* **102,** 82–84, © by Williams & Wilkins (1986).]

of soil air under a newly paved road, an unpaved parking lot, and in an adjacent undisturbed forest. The poorest soil aeration was found in an area where clay fill had been placed around tree stems. Oxygen concentrations were as low as 1% and CO_2 concentration was above 20% in the soil air under this fill. By comparison, the air above ground contained approximately 21% O_2 and 0.03% CO_2. The next most poorly aerated soil was present where layers of clay and gravel were laid down, rolled, and packed in preparation for construction of a driveway. Within 2 weeks the soil O_2 content dropped from 20 to 4% and CO_2 content increased from near 0 to 16%. The most favorable soil atmosphere for tree growth was found in the undisturbed forest. Here, the soil O_2 content was never lower than 18% and CO_2 never higher than 2%.

Flooding

Temporary or continuous flooding of soils is very common as a result of overflowing rivers, storms, overirrigation, seepage from irrigation channels, movement of water by impounded aquifers, and impoundment by flood control dams. Although most woody plants are poorly adapted for growth and survival in flooded soils, some can adapt morphologically and physiologically to grow in waterlogged soils.

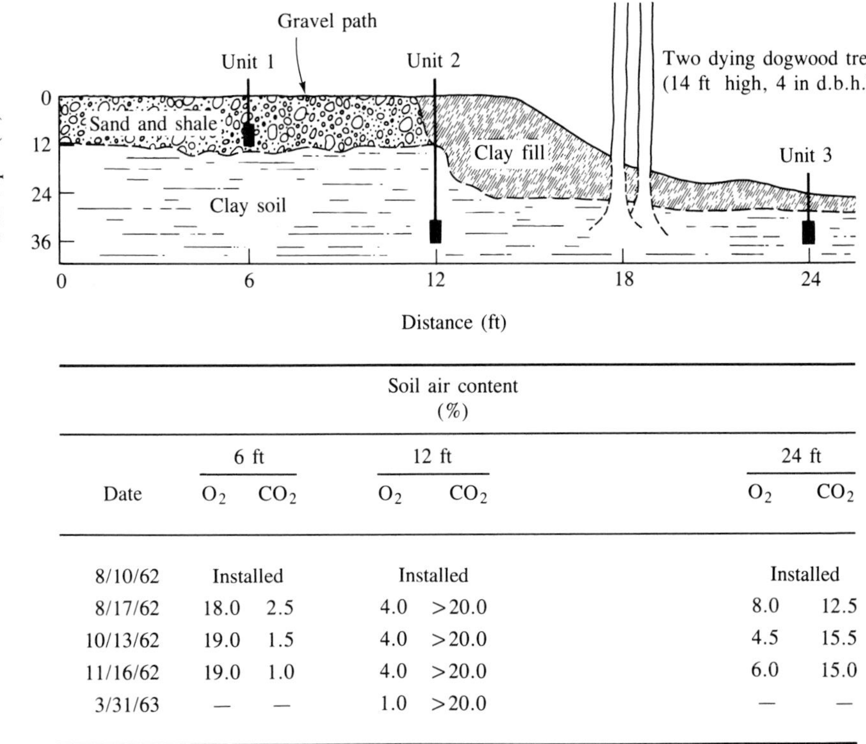

	Soil air content (%)					
	6 ft		12 ft		24 ft	
Date	O_2	CO_2	O_2	CO_2	O_2	CO_2
8/10/62	Installed		Installed		Installed	
8/17/62	18.0	2.5	4.0	>20.0	8.0	12.5
10/13/62	19.0	1.5	4.0	>20.0	4.5	15.5
11/16/62	19.0	1.0	4.0	>20.0	6.0	15.0
3/31/63	—	—	1.0	>20.0	—	—

Figure 8.3 Change in oxygen and carbon dioxide concentrations beneath soil fills of sand and clay. (After Yelenosky, 1964; from Kramer and Kozlowski, 1979, by permission of Academic Press.)

Wetlands

Although wetlands are variously defined, they generally are characterized as ecosystems whose formation was dominated by water. They usually have been wet enough for a sufficient time to support only those plants that are adapted for life in saturated soil conditions. Wetlands include swamps, marshes, bogs, and similar areas. Although wetlands may contain salt or fresh-water, in the United States fresh water wetlands comprise nearly 90% of the total wetland areas.

The 49 continental states contain about 80 million acres of commercially forested wetlands. At least 40 species of trees can grow to commercial size in wetlands of the southern United States alone (Johnson, 1978). The most important species occur in two types of forests.

1. Oak–gum–cypress forests in which water tupelo, black gum, sweet gum, various oaks (overcup, laurel, Nuttal), or bald cypress, alone or in combination, predominate. Major associated species include eastern cottonwood, black willow, green ash, elms, hackberry, sugarberry, and red maple.

2. Elm–ash–cottonwood forests in which American elm, green ash, eastern cottonwood, or silver maple, alone or in combination, dominate. Major associated species include black willow, American sycamore, and American beech.

In the southeastern United States, wetland forests can be found under several wetland forest conditions (Boyce and Cost, 1974). These include (1) stream margins; (2) swamps, bogs, and wet pocosins (poorly drained depressions, often flooded for long periods, usually with peat soils and boggy conditions); (3) flatwoods and dry pocosins (relatively level areas usually with sandy soils, wet in winter and dry from late spring to fall); and (4) cypress forests and channels (depressions filled with water most of the year and with a fluctuating water table).

In the western United States, black cottonwood, western red cedar, willows, and ashes grow on wetland sites as well as on uplands. In Alaska, spruce–poplar forests occur on approximately 18 million acres of bottomland, primarily level floodplains and low river terraces (Johnson, 1978).

Effects of Flooding on Soils

Flooding results in immediate poor soil aeration whereas compaction of soil depletes soil O_2 much more gradually. When a soil is flooded, the water occupies the previously gas-filled pores and gas exchange between the soil and air is then limited to molecular diffusion in the soil water. Such diffusion is very slow. For example, O_2 diffusion through air-filled pores is about 10,000 times as fast as through water, hence the supply of O_2 to flooded soil is severely limited. In addition, almost all the remaining O_2 in standing water and flooded soil is consumed by microorganisms within a few hours. The O_2 concentration remains high only in the few millimeters of the surface soil that are in contact with oxygenated water. Drastic reduction of gas exchange between flooded soil and the atmosphere leads to accumulation of gases such as nitrogen, CO_2, methane, and hydrogen (Ponnamperuma, 1984).

The structure of soil is greatly altered by soil inundation. Important changes include breakdown of aggregates as a result of reduced cohesion, deflocculation of clay because of dilution of the soil solution, and destruction of cementing agents.

When a soil is flooded the aerobic organisms are replaced by anaerobic organisms, primarily bacteria, that can survive without free oxygen. These bacteria are active in metabolism that causes denitrification and reduction of Mn, Fe, and S.

Flooding of soil also reduces the soil redox potential (Fig. 8.4), increases the pH of acid soils, mainly because of change of Fe^{3+} to Fe^{2+}, and decreases the pH of alkaline soils, largely because of accumulation of CO_2, which eventually forms carbonic acid (H_2CO_3). Soil inundation also decreases the rate of decomposition of organic matter. In unflooded soil the decomposing leaf litter binds cations and traps mineral nutrients, preventing their leaching into the deeper layers of soil. In these soils the organic matter is decomposed through the activities of many organisms, including bacteria, fungi, and soil fauna. Rapid decomposition of organic matter is

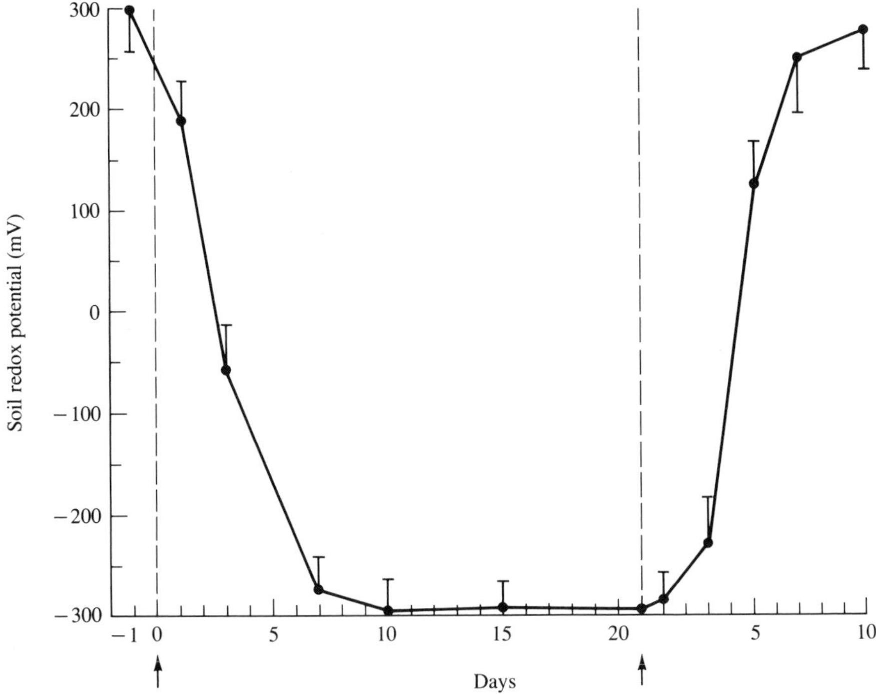

Figure 8.4 Effect of flooding of soil on soil redox potential. The arrows indicate the beginning and end of flooding. (From Pezeshki and Chambers, 1986.)

associated with respiration of aerobic organisms. In flooded soils, however, decomposition of organic matter is limited to activity of anaerobic bacteria, which are less diverse than aerobic microorganisms and consume organic matter at a much slower rate. The rate of decomposition of organic matter in flooded soil may be only half that in an unflooded soil. Thus thick layers of partially decayed organic matter known as peat accumulate.

The end products of decomposition of organic matter in unflooded soils are CO_2 and humic materials. The CO_2 escapes into the air and the humic materials are bound to clay and hydrous oxides of Al and Fe, thus improving aggregation and soil structure. The nitrogen that is released as ammonia is converted to nitrate, and sulfur compounds are oxidized to sulfate. The major end products of decomposition of organic matter in flooded soils are CO_2, methane, and humic materials. The anaerobic bacteria in flooded soils produce many compounds that are not found in well-drained soils. These include gases (including nitrogen, CO_2, methane, and hydrogen), hydrocarbons, alcohols, carbonyls, volatile fatty acids, nonvolatile acids, phenolic acids, and volatile sulfur compounds. The gases build up pressure and escape as bubbles (Ponnamperuma, 1984).

Effects of Flooding on Plants

Inundation of soil during the growing season affects plants at all stages of development. Responses of unadapted plants include inhibition of seed germination, reduction in vegetative and reproductive growth, changes in anatomy and morphology, injury, and often death (Fig. 8.5). Flowing water is less harmful than stagnant water to plant growth. Because of its internal turbulence and turbulent interaction with the atmosphere, flowing water effectively increases the transfer rate of O_2 to the rhizosphere by mass flow, which is a much faster process than diffusion.

Seed Germination and Seedling Establishment

Activation of biochemical processes involved in seed germination depends on availability of water and oxygen. Seed respiration usually is stimulated by an increase in hydration above some critical level (Kozlowski and Gentile, 1958) (see Fig. 7.16). However, the supply of O_2 to respiratory enzymes of seeds is reduced when the soil pores become filled with water.

In general, soaking of air-dry seeds for short periods stimulates germination whereas prolonged soaking leads to loss of viability (Kozlowski, 1984b). However, there are wide variations in tolerance of seeds of different species to flooding. For example, seeds of tupelo gum, swamp tupelo, and bald cypress remain viable even when submerged for long periods. The seeds of these species usually do not germinate under water but do so when the flood waters recede. The seeds of eastern cottonwood, black willow, and American sycamore often germinate under water, hence these species can become established on sandbars and sediments as the flood waters drain away. By comparison, the seeds of green ash and box elder lose viability when submerged for more than short periods (Hosner, 1957, 1962).

Flood tolerance varies appreciably among seedlings of different species and influences regeneration on wet sites. Seedlings of bald cypress can withstand long periods of flooding, and those of silver maple and buttonbush are more flood tolerant than those of hackberry or cherrybark oak (Hosner, 1958, 1960). Flood tolerance of seedlings of closely related species also varies widely. For example, seedlings of paper birch are much less flood tolerant than seedlings of river birch (Norby and Kozlowski, 1983).

Shoot Growth

Soil anaerobiosis adversely affects shoot growth of many upland plants by inhibiting internode elongation and formation and expansion of leaves, and by inducing chlorosis, premature senescence, and abscission (Fig. 8.6).

Reduced height growth of many flooded conifers and broad-leaved trees has been shown in experiments with potted plants. For example, flooding reduced height growth of white spruce, black spruce, eastern white pine, balsam fir (Ahlgren and Hansen, 1957), shortleaf pine, loblolly pine, pond pine (Hunt, 1951), slash pine (McMinn and McNab, 1971), jack pine, and red pine (Tang and Kozlowski, 1983). Soil inundation also decreased height growth of broad-leaved species such as speck-

Figure 8.5 Trees killed by flooding as a result of dam construction. (U.S. Forest Service photograph.)

Figure 8.6 Effect of flooding of soil on shoot growth of seedlings of red gum (*Eucalyptus camaldulensis)* (*top*) and blue gum (*E. globulus)* (*bottom*). *Left*, Unflooded seedlings. *Right*, Flooded seedlings (Photo by A. R. Sena Gomes.)

led alder, American sycamore, river birch, American elm, winged elm, red maple (McDermott, 1954), bur oak (Tang and Kozlowski, 1982a), and *Eucalyptus camaldulensis* and *E. globulus* (Sena Gomes and Kozlowski, 1980c). Height growth of some flood-tolerant species may be increased by flooding if the water is flowing (Dickson *et al.*, 1965).

Leaf initiation and expansion Waterlogging of soil often reduces leaf area by inhibiting leaf formation and expansion (Table 8.2). For example, flooding of soil reduced both formation and expansion of leaves of seedlings of paper birch (Tang and Kozlowski, 1982c), American elm (Newsome *et al.*, 1982), and Japanese larch (Tsukahara and Kozlowski, 1984). American sycamore seedlings were flooded when the leaves were one-third expanded or left unflooded. The leaves of unflooded seedlings expanded fully within 110 days; those of flooded seedlings expanded by only 10% (Tsukahara and Kozlowski, 1985). Pear, apple, and peach trees flooded for 20 months produced fewer and smaller leaves than unflooded trees (Andersen *et al.*, 1984).

Leaf abscission The leaves of many flooded plants turn yellow, senesce early, and are shed, with abscission occurring first in the older, basal leaves and subsequently in the younger, upper leaves. For example, premature leaf shedding was induced by flooding of southern red oak, overcup oak, swamp chestnut oak (Parker, 1950), peach (Marth and Gardner, 1939), pecan (Alben, 1958), and pear (Andersen *et al.*, 1984). Tulip poplar seedlings shed all their leaves within 2 weeks of flooding, those of white oak and silver maple within 3 weeks, honey locust within 4 weeks, and American elm within 8 weeks or more (Yelenosky, 1964). Fewer leaves were present on flooded paper birch seedlings after 60 days of flooding than before flooding was initiated (Tang and Kozlowski, 1982c).

Cambial Growth

Much interest has been shown in the effects of flooding on production and differentiation of cambial derivatives, because of obvious implications in wood production. Quantifying such effects often has been complicated by swelling of tree stems, differences in increment of wood and bark tissues, and variations in cambial growth at different stem heights.

Flooding may increase or decrease diameter growth of tree stems. Kozlowski (1984b) cited several examples of increase in diameter growth of flood-tolerant trees after temporary flooding. Usually, however, early acceleration of growth of flooded trees is followed by decreased diameter growth and subsequent death of trees. For example, in a Mississippi River Valley stand that had been flooded continuously for 4 years with less than 30 cm of water, diameter growth of several species of trees was increased during the first year (Broadfoot and Williston, 1973). However, during the second growing season, American elm, sugarberry, honey locust, and persimmon trees died. After 3 years, most willow oak, black oak, overcup oak, green ash, and sweet gum trees were dead.

Diameter growth of most flood-intolerant species usually is reduced by prolonged flooding of soil during the growing season (Kozlowski, 1984b,c). Such

Table 8.2
Effect of Flooding of Soil on Growth of Cacao Seedlings[a,b]

		Days of flooding					
		15			60		
	Before flooding	Unflooded	Flooded	Change by flooding (%)	Unflooded	Flooded	Change by flooding (%)
Leaves							
Number	4.3	6.4	4.4[c]	−31.2	10.6	6.1[c]	−42.5
Dry wt. (g)	0.15	0.37	0.33	−10.8	2.14	0.64[c]	−70.1
$\overline{RGR}$	—	0.061	0.052	−14.7	0.045	0.024[c]	−46.7
Area per leaf (dm^2)	0.17	0.23	0.24	+ 4.3	0.56	0.26[c]	−53.6
Area per plant (dm^2)	0.73	1.47	1.03[c]	−29.9	5.94	1.60[c]	−73.1
Stems							
Diameter (mm)	3.18	3.40	3.54	+ 4.1	5.07	3.73[c]	−26.4
Dry wt. (g)	0.11	0.21	0.26[b]	+23.8	0.94	0.42[c]	−55.3
$\overline{RGR}$	—	0.044	0.058[b]	+31.8	0.036	0.023[c]	−36.1
Shoots							
Height (cm)	10.0	11.7	10.3	−12.0	17.2	10.8[c]	−37.2
Dry wt. (g)	0.26	0.58	0.57	− 1.7	3.08	1.06[c]	−65.6
$\overline{RGR}$	—	0.055	0.054	− 1.8	0.042	0.024[c]	−42.9
Roots							
Dry wt. (g)	0.07	0.11	0.10	− 9.1	0.57	0.11[c]	−80.7
$\overline{RGR}$	—	0.029	0.020	−31.0	0.035	0.008[c]	−77.1
Root–shoot ratio	0.28	0.19	0.17	− 8.9	0.18	0.11[c]	−38.9
Whole plants							
Dry wt. (g)	0.33	0.69	0.67	− 2.9	3.65	1.17[c]	−67.9
$\overline{RGR}$	—	0.050	0.048	− 4.0	0.040	0.021[c]	−47.5

[a]From Sena Gomes and Kozlowski (1986).
[b]Significantly different from unflooded plants, at $P \leq 0.05$.
[c]Significantly different at $P \leq 0.01$.

growth inhibition is mediated through physiological changes and actual growth reduction may not be apparent for a long time after initiation of flooding. Sometimes the decrease in growth occurs so long after flooding that other causes are sought. Diameter growth of seedlings of flood-intolerant species commonly is rapidly reduced, particularly by stagnant water. Examples are paper birch (Tang and Kozlowski, 1982c), box elder, Norway maple (Yamamoto and Kozlowski, 1987e,f), rubber (Sena Gomes and Kozlowski, 1988b), jack pine, red pine (Tang and Kozlowski, 1983), and Japanese larch (Tsukahara and Kozlowski, 1984).

Flooding of soil not only affects the size of the xylem and phloem increments but may also greatly alter the anatomy of xylem and phloem tissues. For example, the stems of flooded Aleppo pine seedlings contained proportionally more parenchyma tissue in the form of abundant xylem rays, enlarged ray cells, and numerous resin ducts, as well as more phloem parenchyma cells, than did stems of unflooded seedlings (Yamamoto *et al.*, 1987). Because of the abundance of such low-density cells and extensive intercellular spaces in the bark, the basic density of stem segments was lower in flooded than in unflooded seedlings.

Sometimes it is difficult to make valid estimates of wood production in flooded trees by measuring changes in stem diameter, especially in young trees. For example, short-term flooding increased stem diameter of Aleppo pine, Japanese red pine, Japanese hiba (*Platycladus orientalis*), and cryptomeria seedlings, largely because of an increase in bark thickness rather than increased wood production (Fig. 8.7). This was the result of proliferation of phloem parenchyma cells and increase in intercellular space (Yamamoto *et al.*, 1987; Yamamoto and Kozlowski, 1987c,d). However, the effects on xylem increment of these four species were much less consistent. Flooding accelerated tracheid production in the upper stems of Aleppo pine and Japanese hiba seedlings but not in those of cryptomeria or Japanese red pine.

Whereas the tracheids produced following flooding of cryptomeria and Japanese hiba seedlings had normal rectangular shapes, those of Aleppo pine and Japanese red pine were short, round, thick-walled, and surrounded by intercellular spaces (Fig. 8.8). In cross section these tracheids resembled those of compression wood. Tracheids of both unflooded and flooded seedlings of these two species had developed three cell wall layers, including an outer S_1 layer, a middle S_2 layer, and an S_3 layer adjacent to the lumen (Yamamoto and Kozlowski, 1987b,c). However, an S_3 wall layer usually is absent in well-developed compression wood (Coté and Day, 1965).

Root Growth

Growth of existing roots and formation of new ones in flood-intolerant trees are greatly reduced or stopped in poorly aerated soils (Kozlowski, 1984a,b). Rooting depth often is correlated with depth to the water table, with trees on sites with high water tables having shallow, spreading root systems (Lieffers and Rothwell, 1986a). On wet peatland sites the roots of black spruce and tamarack were confined to

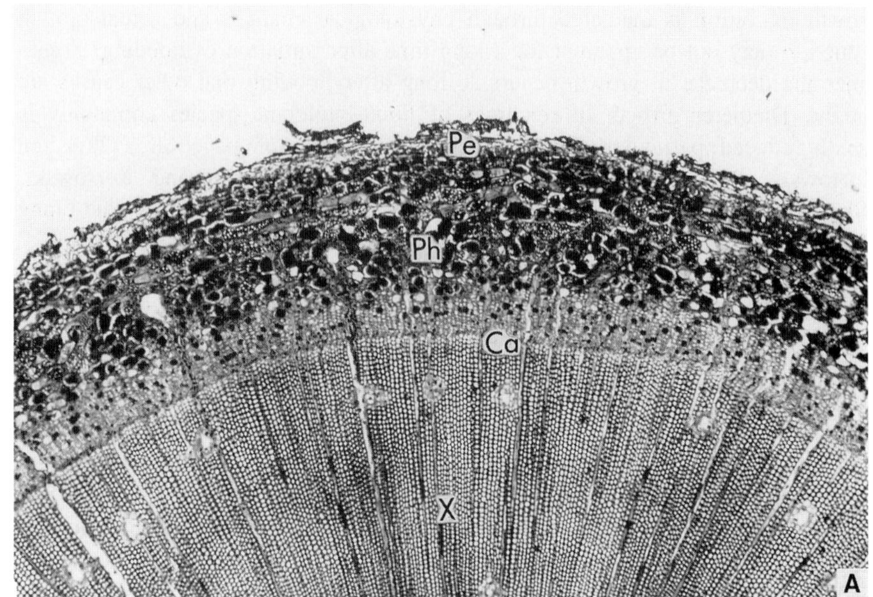

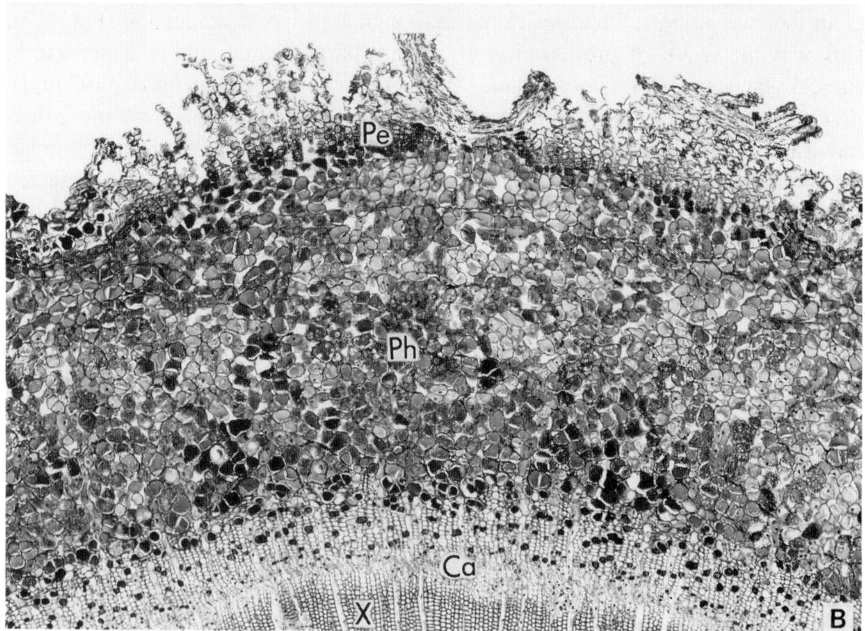

Figure 8.7 Effect of flooding on growth of bark tissues of Aleppo pine. (A) Unflooded seedling; (B) flooded seedling. Ca, cambium; Pe, periderm; Ph, phloem; X, xylem. (From Yamamoto *et al.*, 1987.)

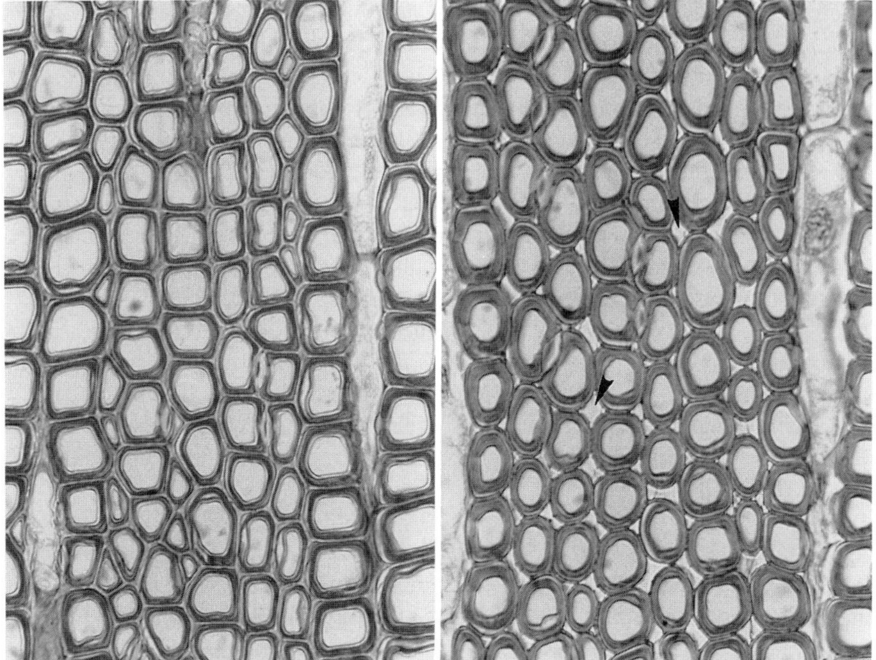

Figure 8.8 Effect of flooding on xylem anatomy of Aleppo pine. *Left*, Unflooded seedling. *Right*, Flooded seedling. Note rounded tracheids with thick walls and extensive intercellular spaces (arrow) in the xylem of the flooded seedling. (From Yamamoto *et al.*, 1987.)

hummocks, whereas on dry sites the roots grew to a depth of 60 cm (Lieffers and Rothwell, 1986b).

There are many examples of drastic reduction in the size and health of the root system of flooded plants and a few will be given. Large and rapid reductions in root growth following flooding of seedlings of forest trees of the temperate zone have been reported for bur oak (Tang and Kozlowski, 1982a), American sycamore (Tang and Kozlowski, 1982b), paper birch (Tang and Kozlowski, 1982c), American elm (Newsome *et al.*, 1982), and Aleppo pine (Sena Gomes and Kozlowski, 1980d; Yamamoto *et al.*, 1987). Similar results were reported for fruit trees, including apple (Childers and White, 1942), pear (Valoras *et al.*, 1964), and citrus species (Stolzy *et al.*, 1965). Soil inundation greatly inhibited root growth of tropical species such as cacao (Sena Gomes and Kozlowski, 1986) and rubber (Sena Gomes and Kozlowski, 1988b).

Flooding of soil also leads to death and decay of large portions of the existing root system, primarily as a result of increased activity of *Phytophthora* fungi, which can tolerate low soil O_2 content. The activity of fungi is stimulated by the low vigor of host trees as well as by attraction of fungal zoospores to root exudates such as sugars, amino acids, and ethanol (Stolzy and Sojka, 1984). Whereas all the woody roots of lodgepole pine seedlings survived flooding, all the nonwoody roots died (Coutts, 1982). Very few new roots were produced by overirrigated citrus trees, and the lack of absorbing roots was attributed to root death caused by *Phytophthora parasitica*. Of much practical importance was the observation that new absorbing roots formed when the soil was allowed to dry between irrigations (Feld, 1982).

The shallow root systems of trees growing in poorly aerated soils often make them prone to windthrow (Fraser and Gardiner, 1967). Because root growth usually is reduced more than shoot growth, the drought tolerance of flooded trees may be decreased after the flood waters recede because the rate of absorption of water by the small root systems will be too low to meet transpiration requirements.

Flooding of soil with stagnant water reduced the rate of dry weight increase of root systems of Aleppo pine seedlings within 8 days. Most of the small roots of flooded plants had decayed within 43 days, the duration of the experiment (Fig. 8.9). Some studies showed that the weight of the root system was appreciably lower after a period of flooding than before flooding began, emphasizing extensive decay of roots (Fig. 8.10).

Mortality

Complete soil anaerobiosis eventually kills most woody plants. Trees may grow for some time after they are first flooded and then they die suddenly. In southern Illinois

Figure 8.9 Flooded (*left*) and unflooded (*right*) Aleppo pine seedlings. Note the sparsely branched roots of flooded plants. (From Sena Gomes and Kozlowski, 1980d.)

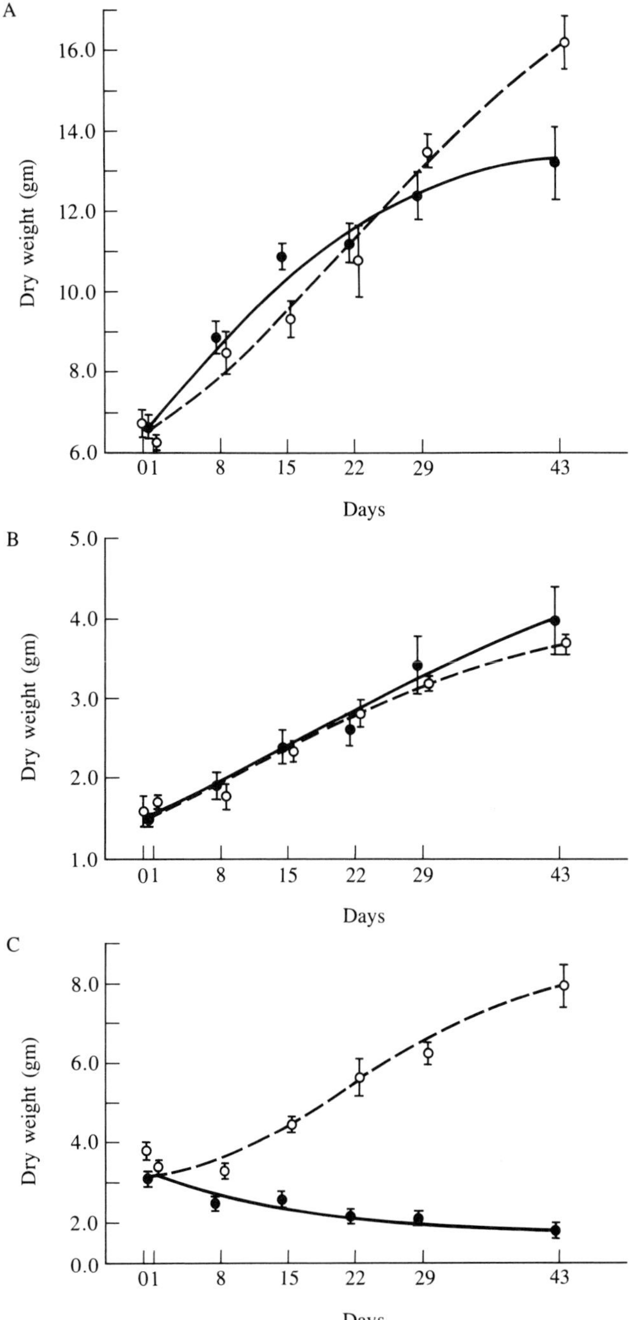

Figure 8.10 Effects of flooding on dry weight changes of (A) needles, (B) stems, and (C) roots of Aleppo pine seedlings. Unflooded seedlings (○); flooded seedlings (●). (From Yamamoto *et al.*, 1987.)

the mortality rate was high in canopy-level bottomland trees that were flooded by impounded water continuously for 83 days after the buds opened (129 days after the beginning of flooding in mid-March). Species affected included silver maple, eastern cottonwood, swamp white oak, pin oak, and American elm (Dellinger *et al.*, 1976). In an impounded flood-control lake in Oklahoma, mortality was much higher in upland species such as southern red oak, chestnut oak, post oak, and black locust than in bottomland species such as eastern cottonwood, green ash, and box elder (Harris, 1975).

Brink (1954) noted wide variations in mortality of flooded forest trees as well as cultivated species. Sitka spruce survived flooding; western red cedar became chlorotic when flooded but survived; and Douglas-fir and western hemlock succumbed readily to flooding. Mortality of broad-leaved species also differed, with red-osier dogwood readily surviving flooding and red alder rarely surviving even short periods of soil inundation.

Mechanisms of Growth Inhibition and Injury

The mechanisms by which flooding alters growth and injures plants are complex. The basic injury occurs at the cellular level, probably to the oxidative system, which in turn affects cell membranes, leading to disruption of many physiological processes. There is evidence that changes in carbohydrate, mineral, hormone, and water relations are involved. In addition, waterlogged soils contain many phytotoxic compounds.

Photosynthesis Flooding of soil usually leads to rapid and drastic reduction in the rate of photosynthesis of upland species. For example, a few days of soil inundation greatly reduced the rate of photosynthesis of eastern cottonwood (Regehr *et al.*, 1975), silver maple (Bazzaz and Peterson, 1984), citrus (Phung and Knipling, 1976), and blueberry (Davies and Flore, 1986). The rate of photosynthesis of Douglas-fir seedlings dropped significantly within 5 hr after flooding (Zaerr, 1983). Soil inundation is followed by rapid stomatal closure (Fig. 8.11) in many species of forest trees (Pereira and Kozlowski, 1977a; Kozlowski and Pallardy, 1979; Sena Gomes and Kozlowski, 1980a, 1986; Kozlowski, 1982a, 1984a,b, 1985a; Pezeshki and Chambers, 1986). Photosynthesis of flooded plants appears to be reduced at first largely because of stomatal closure, resulting in reduced CO_2 absorption by leaves, and later by a reduction in photosynthetic capacity as well (Bradford, 1983a,b). Slow recovery of photosynthesis after floodwaters drain away has been linked to slow reopening of stomata (Davies and Flore, 1986),but it may also be related to slow recovery of the photosynthetic mechanism.

Long-term reduction in photosynthesis is related to inhibition of photosynthetic capacity caused by decrease in activity of carboxylation enzymes, loss of chlorophyll, and reduced leaf area due to inhibition of leaf formation and expansion as well as premature leaf abscission (Rowe and Beardsell, 1973; Kozlowski and Pallardy, 1984).

Mineral nutrition Injury to the root systems of flood-intolerant plants decreases uptake of macronutrients. Also the amount of energy released by anaerobic

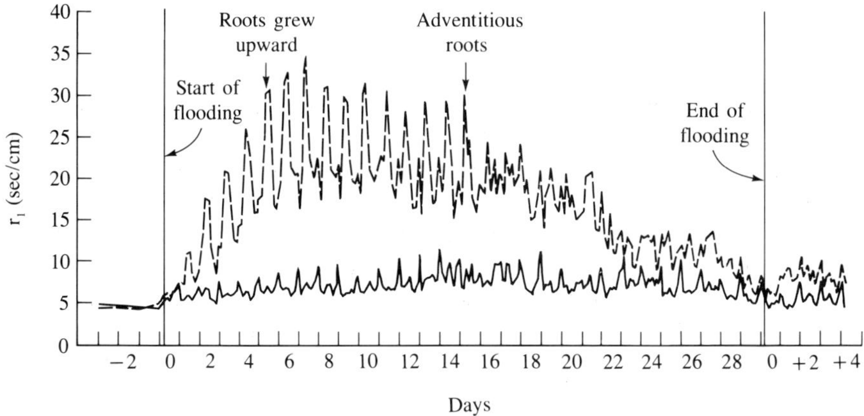

Figure 8.11 Effect of flooding on stomatal aperture of cacao seedlings. An increase in leaf resistance (r_l) indicates stomatal closure. Unflooded (——); flooded (---). (From Sena Gomes and Kozlowski, 1986.)

respiration in soils may be inadequate to sustain uptake of macronutrients by flood-intolerant plants in the amounts needed for growth (Kozlowski and Pallardy, 1984). Furthermore, under anaerobic conditions the permeability of root cell membranes may be affected, resulting in loss of ions by leaching (Rosen and Carlson, 1984).

Flooding reduces both N concentration and total N content of tissues of flood-intolerant plants. A contributory factor is the rapid depletion of nitrate, which is very unstable in the anaerobic zone and is lost from the soil after conversion to nitrous oxide or N_2 by denitrification. The low N concentrations in flooded plants also reflect reduced absorption of nitrate because of the effect of low O_2 supply on root metabolism. Uptake of K and P also is reduced in flooded soils. Decreased uptake of mineral nutrients is partly the result of suppression of mycorrhizae, which accelerate absorption of minerals. Mycorrhizal fungi are strongly aerobic, hence their numbers are reduced in flooded soils (Wilde, 1954; Filer, 1975).

By comparison to reduced uptake of macronutrients, the absorption of Fe and Mn by plants in flooded soils is increased as ferric and manganic forms are converted to the more reduced soluble ferrous and manganous forms (Ponnamperuma, 1972). Although concentrations of Fe and Mn are increased in plant tissues, total uptake is decreased as a result of reduced growth of plants (Kozlowski and Pallardy, 1984).

In contrast to reducing absorption of macronutrients by flood-intolerant plants, flooding often has little effect on mineral relations of flood-tolerant species such as green ash, tupelo gum, black willow, and bald cypress. Sometimes more mineral nutrients are absorbed by flood-tolerant plants when compared with well-watered (but not waterlogged) controls (Kozlowski and Pallardy, 1984).

Hormone relations Flooding of soil rapidly alters the amounts of hormonal growth regulators in plants as a result of differences in synthesis, destruction, and translocation of these compounds. For example, flooding increases levels of auxins, ethylene, and abscisic acid (ABA) but decreases levels of gibberellins and cytokinins in stems (Reid and Bradford, 1984).

Several flood-induced responses, including leaf epinasty, senescence and abscission, stem hypertrophy, and production of adventitious roots, have been attributed to increased ethylene production. Although there is strong evidence that ethylene causes leaf epinasty and abscission (Osborne, 1973), it is unlikely that other morphological changes are caused by ethylene alone. For example, the weight of evidence indicates that production of adventitious roots involves subtle relations among several compounds (Haissig, 1974, 1982). The essentiality of auxins and cofactors for initiation and development of such roots is well established (Kramer and Kozlowski, 1979, pp. 580–581). The effects of ethylene on hypertrophy of tissues are modified by associated enzymes and growth hormones (Tang and Kozlowski, 1984).

Phytotoxic compounds Waterlogged soils contain many phytotoxic compounds, including sulfides, CO_2, soluble Fe and Mn, ethanol, acetaldehyde, and cyanogenic compounds. For a long time the adverse effects of flooding were attributed to the presence of excess ethanol in plant tissues, however, ethanol is readily eliminated from plants. Furthermore, when ethanol was supplied to nutrient solutions at concentrations approximately 100 times those in flooded soils, plants were not injured. Hence, the adverse effects of ethanol on flooded plants seem to have been exaggerated (Jackson *et al.*, 1982). Nevertheless, Crawford and Finegan (1989) suggested that removal of ethanol is beneficial to flooded plants. Although toxic products such as reduced Fe, fatty acids, and ethylene inhibited root growth of conifer seedlings, the adverse effects of O_2 deficiency were much more substantial (Sanderson and Armstrong, 1980a,b). The weight of evidence indicates that oxygen deficiency is the most important environmental factor that triggers growth inhibition and injury in flooded plants.

Factors Affecting Flooding Tolerance

The responses of woody plants to flooding vary appreciably with species and genotype, age of plants, condition of the floodwater, and time and duration of flooding.

Species and genotype The flood tolerance of different species and genotypes varies greatly (Kozlowski, 1982b, 1984b,c). Most broad-leaved trees tolerate flooding better than most conifers do, but wide variations occur within each of these groups. Whereas black willow, tupelo gum, black gum, and mangroves are rated as very flood tolerant, flowering dogwood, yellow poplar, sweet gum, and paper birch are considered intolerant. Hall *et al.* (1946) classified forest trees of the Tennessee Valley in three broad groups of flood tolerance (Table 8.3): (1) tolerant species that survived continuous flooding for up to two growing seasons; (2) moderately tolerant species that usually survived one season of flooding but succumbed during the second growing season; and (3) intolerant species that did not survive continuous flooding for one growing season. Many of the latter group succumbed to 4 weeks or less of flooding.

Table 8.3
Variations in Flood Tolerance of Trees of the Tennessee Valley[a]

Tolerant	Moderately tolerant	Intolerant
Silver maple	Red mulberry	American ash
Sweet gum	Swamp chestnut oak	Chinkapin oak
Black maple	Hackberry	Mockernut hickory
Persimmon	Winged elm	Shagbark hickory
Green ash	Hawthorn	Black locust
Honey locust	Osage orange	Sassafras
Overcup oak	Box elder	Flowering dogwood
Eastern cottonwood	Loblolly pine	Sourwood
Water hickory	River birch	Southern red oak
Black willow	American elm	American basswood
Tupelo gum	Sycamore	Blackjack oak
Bald cypress	American holly	Black cherry
		Shortleaf pine
		Virginia pine
		Eastern red cedar
		Eastern redbud
		Black walnut
		Swamp hickory
		American beech
		Tulip poplar
		Yellow buckeye
		Sugar maple
		Post oak

[a]From Hall *et al.* (1946).

As mentioned, a few species of conifers are very flood tolerant. Bald cypress and tamarack tolerate flooding well (Fowells, 1965), and redwood grows vigorously on seasonally flooded alluvial flats (Stone and Vasey, 1968). Lodgepole pine tolerates soil inundation better than Sitka spruce (Coutts and Philipson, 1978a,b). In another study, flood tolerance varied in the following order: balsam fir > black spruce > white spruce > eastern white pine > red pine (Ahlgren and Hansen, 1957).

Rowe and Beardsell (1973) ranked fruit trees in the following order of flood tolerance: quince, extremely tolerant; pear, very tolerant; apple, tolerant; citrus and plum, intermediately tolerant; cherry, intermediately sensitive; apricot, peach, and almond, sensitive; and olive, very sensitive. Such ratings are complicated by differences in flood tolerance among rootstocks on which fruit trees are grown (Table 8.4). For example, Red Delicious apple scions on MM.111 and seedling rootstocks were more sensitive to flooding than those on M.27 or MM.106 rootstocks. Scions on M.26 rootstocks were least affected (Rom and Brown, 1970). Rowe and Beardsell (1973) compiled the following rating of sensitivity of apple rootstocks to flooding of soil: fairly resistant—M.2, M.3, M.6, M.7, M.13, M.14, M.15, M.16,

Table 8.4
Sensitivity of *Prunus* spp. Rootstocks to Waterlogging[a]

Sensitivity to waterlogging	Rootstock	Species
Resistant	Damas de Toulouse	*P. domestica*
	Damas GF1869	*P. domestica*
	GF8-1	*P. cerasifera* cv. Marianna
	S2544-2	
Moderately resistant	GF31 hybrid	*P. cerasifera* × *P. salicina*
	Myrob B	*P. cerasifera*
	P936	*P. cerasifera*
	P938	*P. cerasifera*
	P855	*P. cerasifera*
	P34	*P. cerasifera*
	St. Julian A	*P. domestica*
	St. Julian GF355-2	*P. domestica*
	Brompton	*P. domestica*
	Ciruelo 43	*P. domestica*
Moderately sensitive	S37	*P. salicina*
	S2540	
	S2541	
	S300	
Sensitive	S2514	
	S2508	
	S763	
	S2538	
Extremely sensitive	Apricot	*P. armeniaca*
	St. Lucie 39	*P. mahaleb*
	Cherry	*P. avium*
	Peach	*P. persica*
	GF305	*P. davidiana*

[a]After Rowe and Beardsell (1973); from Kozlowski (1984b), by permission of Academic Press.

Crab C, Jonathon; moderately sensitive—M.4, M.9, M.26; very sensitive—M.2, MM.104, MM.109; and extremely sensitive—M.779, M.789, M.793, and Northern Spy.

Sensitivity to flooding may vary greatly among closely related woody plants. For example, *Eucalyptus* species varied in the following order from most to least flood tolerant: *Eucalyptus grandis*, *E. robusta*, and *E. saligna* (Clemens *et al.*, 1978). *Prunus japonica* was very flood tolerant; *P. salicina* and *P. cerasifera* were tolerant; and *P. persica*, *P. mume*, *P. tomentosa*, *P. pauciflora (pseudocerasus)*, and *P. subhirtella* were less tolerant. *Prunus armeniaca* was less flood tolerant than any of the aforementioned *Prunus* species (Mizutani *et al.*, 1979). Both jack pine and red pine seedlings adapted poorly to flooding but jack pine was more adversely affected, as shown by earlier and greater growth reduction (Fig. 8.12). Pond pine and loblolly pine adapted much better than sand pine to soil anaerobiosis, with extensive shoot dieback and chlorosis developing only in the latter species (Topa and McLeod, 1986a).

Age of plants Adult trees tolerate flooding better than overmature trees or seedlings of the same species. Hence, some species rated as flood tolerant may be quite sensitive in the seedling stage. Kennedy and Krinard (1974) reported that a major flood during the spring and summer killed seedlings of several bottomland species. By comparison, trees 1 year old and older survived when flooding occurred in the first 2 months of the growing season. Flooding of black poplar trees 1 to 4, 5 to 6, and 8 to 15 years old for 150 days injured the youngest trees most (Popescu and Necsulescu, 1967). Because the crowns of adult trees remain above water when

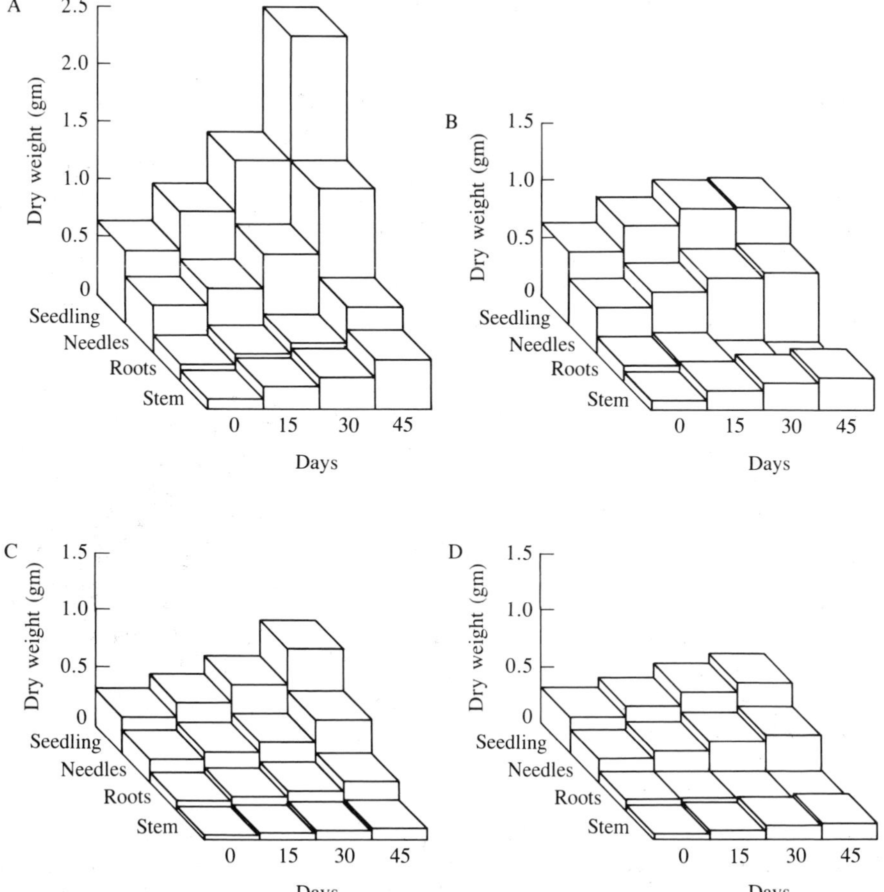

Figure 8.12 Effects of flooding on dry weight changes of seedlings, needles, roots, and stems of jack pine and eastern white pine seedlings. (A) Jack pine, unflooded. (B) Jack pine, flooded. (C) Red pine, unflooded. (D) Red pine, flooded. (From Tang and Kozlowski, 1983.)

flooded, they obviously are subjected to less severe conditions than young seedlings are. Seedlings often die because they are pushed over, buried in mud, or uprooted by floods.

Condition of the floodwater Flooded plants are injured more by stagnant water (which contains less O_2) than by moving water in the absence of debris impact effects, with even the most flood-tolerant species being injured by standing water. For example, flooding with stagnant water of seedlings of the very flood-tolerant species bald cypress decreased the rate of height growth, needle initiation and expansion, and dry weight increment of needles and roots. After 8 weeks the leaf area of flooded seedlings was only 45% of that of unflooded seedlings (Table 8.5). Height growth of swamp tupelo and tupelo gum seedlings growing in moving water was about twice that of seedlings in standing water, and dry weight increment was up to five times greater (Hook *et al.*, 1970b).

Time and duration of flooding Flooding of woody plants during the growing season usually is much more harmful than flooding during the dormant season. For example, flooding of pecan trees in September and October when temperatures were high caused leaf scorching and abcission, but flooding in April and May did not (Alben, 1958). The greater growth inhibition and more extensive injury by flooding during the warm season are associated with the high O_2 requirements of growing roots with high respiration rates. Also a higher transpiration rate is more likely to cause dehydration. The lack of foliage during the winter also contributes to reduction in injury attributable to dormant-season flooding. In the absence of foliage, flooded trees do not place great transpirational demands on the weakened or injured root system.

The duration of flooding on many sites varies greatly between and within years. For example, on Mississippi floodplain sites flooded annually, the duration of flooding varied from 6 to 40% of the year (Bedinger, 1981). Such differences may greatly influence seedling establishment as well as growth and survival of established trees.

Some species of trees can survive continuous flooding for two growing seasons and others are killed by less than a month of flooding. Near a reservoir margin all of the 39 species of woody plants studied were killed when the root crowns were flooded for half the growing season or more (Hall and Smith, 1955). Of six bottomland species only black willow survived 32 days of flooding in laboratory experiments. When completely submerged for 16 days, many green ash, some sweet gum, and a few box elder seedlings survived. Cottonwood and silver maple seedlings did not survive flooding for 16 days or more (Hosner, 1958).

Adaptations to Flooding

Many wetland species have become morphologically and metabolically adapted to soil anaerobiosis. Various morphological and physiological adaptations make it possible for plants rooted in anaerobic soils to maintain more normal physiological processes.

Oxygen transport Many wetland plants absorb O_2 through aerial tissues from which it moves through the stem to the roots (Kramer *et al.*, 1952). Subsequent

Table 8.5
Effects of Flooding of Soil with Standing Water for 8 or 14 Weeks on Growth of Bald Cypress Seedlings[a]

Weeks	Treatment	No. of leaves	Height (cm)	Leaf area (cm^2)	Dry weight (g)				Root–shoot ratio
					Leaves	Stems	Roots	Whole plant	
0	—	19.0	31.0	115	0.58	0.24	0.37	1.19	0.45
8	Unflooded	42.9	76.0	903	5.12	3.46	4.71	7.84	0.55
	Flooded	40.6	62.6	410	2.43	2.94	2.15	7.52	0.40
14	Unflooded	49.1	84.3	988	6.00	7.05	10.06	23.32	0.78
	Flooded	41.0	59.5	437	2.50	4.17	4.69	11.35	0.70

[a]From Shanklin and Kozlowski (1985a).

diffusion of O_2 out of the roots to the rhizosphere benefits plants by oxidizing reduced soil compounds such as ferrous and manganous ions, which are toxic to roots (Opik, 1980). Rhizospheric oxidation may occur directly by reaction with molecular O_2 and enzymatically by the influence of microbes associated with roots (Armstrong, 1975).

Entry of O_2 through leaves has been demonstrated for *Populus* × *petrowskiana*, *Salix alba* (Chirkova, 1968), *Salix fragilis*, *Salix atrocinerea* (Armstrong, 1968), and lodgepole pine (Philipson and Coutts, 1978), as well as through lenticels of twigs, stems, and roots (Hook, 1984). Internal O_2 transport in the flood-tolerant lodgepole pine occurred in both the wood and the bark, whereas in Sitka spruce it was restricted to the bark. The deeper penetration of waterlogged soil by roots and greater flood tolerance of lodgepole pine were attributed to more efficient O_2 transport (Philipson and Coutts, 1980).

Formation of lenticels Many flood tolerant plants produce hypertrophied lenticels on the submerged stem and roots (Fig. 8.13). Such lenticels may facilitate aeration of the stem and permit release of potentially toxic compounds (e.g., ethanol, acetaldehyde, ethylene) produced by flooded plants, for lenticels are connected by continuous intercellular spaces in the cortex and phloem. Armstrong (1968) showed that O_2 entered stems of woody plants through the lenticels located

Figure 8.13 Hypertrophied lenticels and adventitious roots on the submerged stem of a green ash seedling. (From Pereira and Kozlowski, 1977a.)

just above the waterline. The hypertrophied lenticels of flooded plants are more pervious to gas exchange than lenticels of nonflooded plants because they contain larger intercellular spaces and lack the many closing layers associated with the former (Hook, 1984).

Formation of hypertrophied lenticels following flooding has been demonstrated for many species of broad-leaved trees, including black willow, eastern cottonwood, green ash, bur oak, and American sycamore (Kozlowski, 1984b), and conifers such as ponderosa pine, red pine, jack pine, Virginia pine, and tamarack (Hahn *et al.*, 1970). The formation of hypertrophied lenticels in flooded American elm seedlings involved stimulation of phellogen activity and elongation of cork cells (Angeles *et al.*, 1986).

Root structure Mangroves have morphologically adapted root systems that assist in gas exchange of submerged roots. *Avicennia* produces numerous vertical air roots that protrude from the mud around the base of the tree. A single tree may produce as many as 10,000 of these "pneumatophores." Enough air diffuses in through lenticels on the vertical roots when the tide rises to maintain aerobic conditions in the root sytem without appreciably reducing respiration during tidal inundation (Curran *et al.*, 1986). The branched, looping aerial "stilt" roots of *Rhizophora* bear lenticels that are connected by aerenchyma tissue to roots buried in the mud. Plugging of the lenticels with grease was followed by reduction in O_2 content of roots buried in the mud, emphasizing that the stilt roots assisted in aeration of the submerged roots (Scholander *et al.*, 1955). Some investigators claimed that the conical vertical "knees" of bald cypress serve as aerating organs that supply O_2 to submerged roots. The knees develop because of localized cambial activity on the upper surfaces of roots, which are better aerated than the lower surfaces. Most of the oxygen is utilized locally and the knees do not appear to be important for supplying O_2 to the roots (Kramer *et al.*, 1952).

Aerenchyma The stems and roots of many wetland plants are permeated by aerenchyma tissues (soft tissues characterized by large intercellular spaces). Aerenchyma tissues form either by dissolution of entire cells or by separation of cell walls. Kawase (1979) demonstrated that O_2 deficits in flooded plants increased production of ethylene, which stimulated cellulase activity, leading to formation of aerenchyma tissue.

Aerenchyma tissues, which often are better developed in flood-tolerant than flood-intolerant woody plants, occur in mangroves (Gill and Tomlinson, 1975), lodgepole pine (Coutts and Philipson, 1978b), and loblolly pine (Hook, 1984). Aerenchyma tissues in the roots of anaerobically grown pond pine and loblolly pine seedlings were much more extensive than in the roots of the less flood-tolerant sand pine. Furthermore, the capacity for O_2 transport and rhizospheric oxidation was lower in sand pine than in the other two species (Topa and McLeod, 1986b).

Adventitious roots Many flood-tolerant species grow adventitious roots on the submerged portion of the stem, on the original root system, or both (Figs. 8.14 and 8.15). When roots of flooded plants die back to the major secondary roots or the primary roots, new roots, which seem more tolerant of anaerobic conditions, often

Figure 8.14 Extensive formation of adventitious roots on the stem of a previously flooded American sycamore seedling. The horizontal line shows the height to which the seedling was flooded. (From Kozlowski, 1986d.)

Figure 8.15 Flood-induced adventitious root emerging from the stem of a seedling of American elm. The root originated in the ray parenchyma of the secondary phloem. (From Angeles *et al.*, 1986.)

emerge from these points. The development of adventitious roots on flooded plants varies considerably with the condition of the floodwater. For example, many adventitious roots formed on tupelo gum flooded with moving water, but very few or none developed on trees flooded with stagnant water (Hook *et al.*, 1970b).

Flood-induced adventitious roots usually are thicker, have larger cells and more intercellular space, and have fewer root hairs (often none) than do roots growing in well-aerated soils. Often the new roots of flooded plants are more succulent and permeable than the original roots (Hook *et al.*, 1971).

The flood-induced adventitious roots may or may not emerge through lenticels (Figs 8.13 and 8.15). Both the origin and development of adventitious roots of flooded American elm seedlings were similar whether they emerged from lenticels or through the bark (Angeles *et al.*, 1986). The primordia of the adventitious roots originated in the ray parenchyma of the secondary phloem (Fig. 8.15). However, in cryptomeria, flood-induced adventitious roots arose in the xylem parenchyma (Yamamoto and Kozlowski, 1987d).

There has been some controversy about whether flood-induced adventitious roots are physiologically important or whether they merely represent nonfunctional expressions of flooding injury. Gill (1975) noted that excision of flood-induced adventitious roots had relatively little effect on leaf growth of European alder and attributed only minor physiological importance to such roots. Tripepi and Mitchell (1984) concluded that the adventitious roots of flooded red maple and river birch seedlings were not important for their survival in anaerobic soils. Nevertheless, the following lines of evidence indicate that flood-induced adventitious roots are physiologically important and confer some degree of flood tolerance to plants that possess the capacity to form such roots.

1. Production of adventitious roots and flood tolerance often are related. Many flood-intolerant woody plants (e.g., Aleppo pine, jack pine, red pine, and paper birch) have limited capacity to produce adventitious roots when flooded (Sena Gomes and Kozlowski, 1980d; Tang and Kozlowski, 1982b, 1983). By comparison, abundant production of adventitious roots by melaleuca was correlated with its high degree of flood tolerance (Sena Gomes and Kozlowski, 1980b). Flood-tolerant species such as green ash and river birch also produce many adventitious roots when flooded (Kozlowski, 1984b; Krasny *et al.*, 1988). The relative flood tolerance of three species of *Eucalyptus* was *E. grandis* $\geq$ *E. robusta* $\geq$ *E. saligna* and was correlated with their capacities for producing adventitious roots (Clemens *et al.*, 1978). The more flood-tolerant *Eucalyptus camaldulensis* produced abundant adventitious roots on submerged stems, whereas the less flood-tolerant *E. globulus* did not (Sena Gomes and Kozlowski, 1980c).

2. Adventitious roots of flood-tolerant plants increase their capacity for absorption of water and mineral nutrients, thereby compensating for the loss of absorbing capacity as a result of decay of a portion of the original root system. Transpiration, hence absorption of water, by green ash seedlings was up to 90% higher in seedlings with adventitious roots on submerged portions of stems than in seedlings from which such roots had been excised (Fig. 8.16). Furthermore, increased production

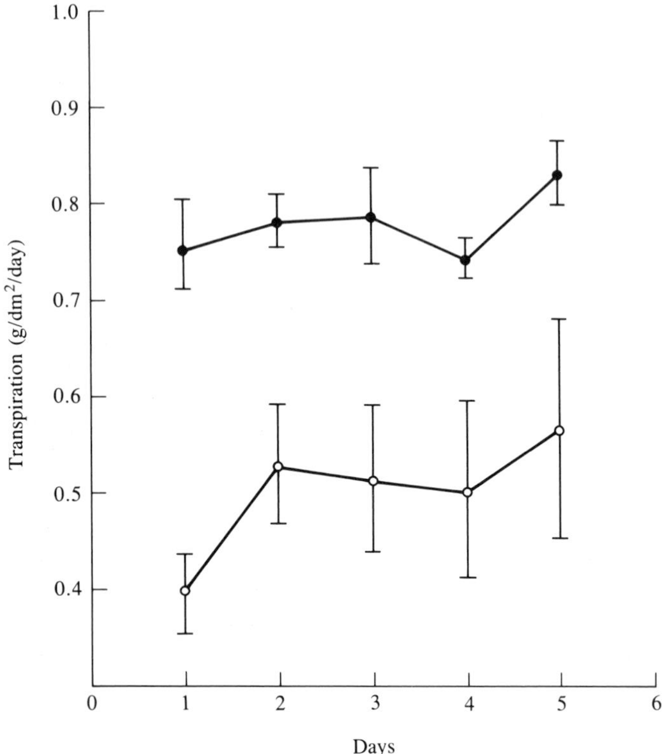

Figure 8.16 Transpiration rates of flooded green ash seedlings with adventitious roots (●) and seedlings without adventitious roots (○). (From Sena Gomes and Kozlowski, 1980a.)

of adventitious roots of flooded seedlings was correlated with eventual reopening of stomata, which had closed shortly after the soil was flooded (Sena Gomes and Kozlowski, 1980a; Kozlowski, 1985a). When flood-induced adventitious roots were excised from the submerged portions of American sycamore stems, subsequent height growth, diameter growth, and relative growth rates of leaves and roots were reduced (Tsukahara and Kozlowski, 1985). Shoot growth also was greater in herbaceous plants with adventitious roots than in plants from which the flood-induced adventitious roots were removed (Jackson, 1942, 1943).

3. Flood-induced adventitious roots play a role in oxidizing the rhizosphere and detoxifying soil toxins. The flood-induced adventitious roots of swamp tupelo oxidized the rhizosphere, but roots of unflooded plants did not (Hook *et al.*, 1970a; Hook, 1984). In addition, the capacity for oxidizing the rhizosphere was higher in flood-induced roots of tupelo gum, a flood-tolerant species, than in the new roots of the less flood-tolerant American sycamore (Hook and Brown, 1973).

4. Flood-induced adventitious roots increase the availability of root-synthesized

growth hormones, especially gibberellins and cytokinins, to shoots (Reid and Bradford, 1984).

Often several adaptations are involved in flood tolerance. The greater flood tolerance of *Leptospermum scoparium* over that of *L. ericoides* was attributed to better development of aerenchyma tissue, greater root growth, and higher redox potentials in the rhizosphere of waterlogged plants (Cook *et al.*, 1980).

Relative flood tolerance among seedlings of five species varied in the following order: tupelo gum > green ash > American sycamore > sweet gum > tulip poplar. When these species were examined for specific physiological adaptations (capacity to tolerate high CO_2 concentrations, oxidize the rhizosphere, accelerate anaerobic respiration) and morphological adaptations (capacity to develop adventitious roots and new secondary roots, capacity of secondary roots to survive), all species except tulip poplar exhibited at least two of these adaptations, and the two most flood-tolerant species exhibited four (Hook and Brown, 1973). Coutts and Armstrong (1976) emphasized that whenever root growth continued in flooded soils, such growth was invariably associated with adaptive mechanisms for internal transport of oxygen to the roots.

Metabolic adaptations In addition to the morphological adaptations already discussed, some plants exhibit biochemical adaptations that enable them to survive even with prolonged soil anaerobiosis. The major mechanisms involve maintaining a glucose supply and adjusting carbon metabolism to avoid accumulation of toxic products. Davies, D. D., *et al.* (1974) advanced a theory that flood tolerant plants can be distinguished from intolerant plants by permeability of the tonoplast to certain metabolites. Under anaerobic conditions severe acidification may occur in root tips. Roberts *et al.* (1985) concluded that if acid leaks from cell vacuoles to the cytoplasm the acidification may cause injury or death. Metabolic adaptations to flooding are discussed by Davies (1980).

Adaptations to flooding with salt water Successful growth of halophytes requires adaptations to salinity stress as well as to anaerobiosis. Adaptation to salinity is achieved either by tolerance of salinity or avoidance of it by excluding salt passively, excluding salt actively, or diluting the entering salt (Levitt, 1980b).

The salt concentration in the xylem of mangroves is much lower than in the surrounding water. For example, the concentration of NaCl in the bleeding sap of *Avicennia* was only 15% as high as in seawater; in *Rhizophora* and *Sonneratia* it was only 1.5% of that of seawater. These low salt concentrations are traceable to the capacity of mangroves to exclude or excrete salt.

The root membranes of such mangroves as *Rhizophora mangle*, *Ceriops tagal*, and *Bruguiera gymnorhiza* exclude salt by separating fresh water from salt water by a nonmetabolic ultrafiltration process in the xylem (Scholander, 1968). Other mangroves eliminate salt through glands located on leaves, as in *Acanthus*, *Aegiceras*, *Aegialitis*, and *Avicennia*. Salt is excreted by an active mechanism located within the gland itself as shown by excretion of brine by glands of detached leaves and even by glands of isolated leaf discs (Waisel, 1972). Some mangroves (e.g.,

Languncularia and *Conocarpus*) have epidermal structures that resemble salt glands. However, the capacity of these structures to excrete salts has not been conclusively demonstrated (Tomlinson, 1986).

Summary

Inadequate soil aeration is common in compacted soils, soils with a high or perched water table, soils with impermeable layers, flooded soils, and heavy soils with small amounts of pore space.

Many soils are compacted by pedestrian traffic, grazing animals, and heavy machines. Compacted soils are common in golf courses, campsites, parks, construction sites, timber harvesting areas, and fruit orchards. Compaction increases bulk density of the soil and decreases the number and size of macropores, thereby impeding diffusion of O_2 into the soil and diffusion of CO_2 out of it. Reduced growth of plants in compacted soils is associated with decreased soil aeration and mechanical resistance as well as reduced infiltration and retention of water.

Temporary or continuous flooding of soil decreases soil O_2 and alters soil structure. The anaerobic bacteria in flooded soils produce potentially toxic compounds, including gases, hydrocarbons, alcohols, carbonyls, volatile fatty acids, nonvolatile acids, phenolic acids, and volatile sulfur compounds.

Flooding adversely affects shoot growth of upland trees by inhibiting internode elongation and leaf formation and expansion, and by inducing leaf senscence, injury, and abscission. Although flooding often increases diameter growth of tree seedlings, largely by stimulating growth of bark tissue, it decreases long-term diameter growth of mature upland trees. Flooding also alters xylem and phloem anatomy. Growth of roots and formation of new ones are drastically reduced in flooded plants. In addition flooding leads to death and decay of many roots as a result of accelerated activity of *Phytophthora* fungi.

Effects of flooding on plant growth are mediated by changes in carbohydrate, hormone, mineral, and water relations. Flooding reduces the rate of photosynthesis by inducing stomatal closure, altering carboxylating enzymes, and reducing chlorophyll content. Absorption of mineral nutrients is decreased because of the small root systems and because energy released by anaerobic respiration is inadequate for root functions. In addition, ions are lost from roots by leaching.

Synthesis, destruction, and translocation of hormonal growth regulators are altered in flooded plants. Levels of auxins, ethylene, and abscisic acid are increased in stems while gibberellins and cytokinins are decreased.

Flood tolerance varies greatly among species and genotypes, time and duration of flooding, and condition of the floodwater. Morphological adaptations to flooding include: (1) O_2 transport from shoots to roots, and subsequent release of O_2 and oxidation of reduced soil compounds; (2) production of lenticels on roots and submerged stems; (3) production of aerenchyma tissues; and (4) production of adventitious roots, which replace loss of the original roots. An important metabolic adaptation involves adjustment of carbon metabolism to avoid accumulation of toxic compounds and maintenance of a low metabolic rate.

General References

Cannell, R. Q. (1977). Soil aeration and compaction in relation to root growth and soil management. *Appl. Biol.* **2,** 1–86.

Clark, J. R., and Benforado, J., eds. (1981). "Wetlands of Bottomland Hardwood Forests." Elsevier, New York.

Crawford, R. M. M. (1989). "Studies in Plant Survival." Blackwell, Oxford.

Drew, M. C., and Lynch, J. M. (1980). Soil anaerobiosis, microorganisms and root function. *Ann. Rev. Phytopathol.* **18,** 37–66.

Etherington, J. R. (1983). "Wetland Ecology." Arnold, London.

Glinski, J., and Stepniewski, W. (1985). "Soil Aeration and Its Role for Plants." CRC Press, Boca Raton, FL.

Good, R. E., Whigham, D. F., and Simpson, R. L., eds. (1978). "Freshwater Wetlands: Ecological Processes and Management Potential." Academic Press, New York.

Greacen, E. L., and Sands, R. (1980). Compaction of forest soils: A review. *Austr. J. Soil Res.* **18,** 163–189.

Greeson, P. E., Clark, J. R., and Clark, J. E., eds. (1979). "Wetland Functions and Values: The State of Our Understanding." American Water Resources Association, Minneapolis.

Hook, D. D., and Crawford, R. M. M., eds. (1978). "Plant Life in Anaerobic Environments." Ann Arbor Science, Woburn, MA.

Jackson, M. B. (1985). Ethylene and responses of plants to soil waterlogging and submergence. *Ann. Rev. Plant Physiol.* **36,** 145–174.

Kozlowski, T. T., ed. (1984). "Flooding and Plant Growth." Academic Press, New York.

Maltby, E. (1986). "Waterlogged Wealth." International Institute for Environment and Development, London and Washington, D.C.

Mitch, W. J., and Gosselink, J. G. (1986). "Wetlands." Van Nostrand, New York.

Patterson, J. C. (1976). Soil compaction and its effects upon urban vegetation. *In* "Better Trees for Metropolitan Landscapes" (F.S. Santamour, H.D. Gerhold, and S. Little, eds.), pp. 91–101. U.S.D.A. Forest Service, Gen. Tech. Rept. NE–22. Upper Darby, PA.

Richardson, C. J. (1981). "Pocosin Wetlands." Hutchinson Ross, Stroudsburg, PA.

Ruark, G. A., Mader, D. L., and Tattar, T. A. (1980). The influence of soil compaction and aeration on the root growth and vigor of trees—A literature review. Part 1. *Arboric. J.* **6,** 251–265.

Tomlinson, P. B. (1986). "The Biology of Mangroves." Cambridge University Press, Cambridge.

Chapter 9

Air Pollution

Introduction

Plants are exposed to a wide variety of chemicals that contaminate both the air and the soil. Some chemicals affect growth directly by causing foliar injury and reduction in photosynthesis and the supply of carbohydrates and growth regulators, thereby inhibiting both shoot and root growth. Chemicals entering the soil can injure roots directly and reduce their efficiency in absorbing water and minerals, or affect the soil pH and the availability of various mineral elements such as Mg, which in turn inhibit growth. Air pollutants[1] can also act indirectly, such as by destruction of the ozone layer, resulting in increasing exposure to ultraviolet radiation, or by increasing temperature as a result of increasing CO_2 concentration. Simultaneous pollution stress on both roots and tree crowns is likely to be more injurious than stress exerted on only one or the other. As suggested in Chapter 1, plants are normally subjected to a variety of co-occurring natural stresses on which air pollution is superimposed, often with disastrous results.

Types and Sources of Air Pollutants

The important air pollutants that are known to influence growth of woody plants include sulfur dioxide (SO_2), ozone (O_3), fluorides, oxides of nitrogen (NO_x), peroxyacetylnitrate (PAN), and several particulates. Pollutants are classified as primary or secondary. Whereas primary pollutants (e.g., SO_2, HF) originate in a toxic form at the source, secondary pollutants are produced by interactions between primary pollutants. For example, the secondary pollutants O_3 and PAN are formed by the action of sunlight on products of fuel combustion.

Sulfur Dioxide

A variety of sources emit SO_2. These include smelters, refineries, paper mills, power plants, furnaces and volcanic eruptions. Ores of copper, zinc, lead, nickel, and iron often form as sulfides in minerals, which may contain as much as 10% sulfur. During smelting this waste product combines with O_2 in the air at high temperatures to produce SO_2.

In the United States, emissions of SO_2 increased by 18% between 1940 and 1982, but there was a slight decrease from 1970 to 1982. Despite efforts at reducing the amount of SO_2 released, it has been estimated that emissions of SO_2 would be appreciably higher in 1990 than they were in 1975 (Environmental Protection Agency, 1978a,b). Rubin (1981) estimated that use of coal would increase several times by the year 2000 in industrial countries, thereby increasing the amount of SO_2 released into the atmosphere.

Acid Precipitation

There is much concern with the fact that SO_2 and nitrogen oxides, together with HCl and some other compounds, mix in the atmosphere with oxygen and water to form solutions of mineral acids that are deposited on plants and soils as acid rain, snow, or fog. Rain formed in a nonpolluting area should have a pH of about 4.6 (Galloway *et al.*, 1984). However, rains with low pH values fall in the eastern United States

[1]In this chapter the concentrations of air pollutants are given in the units reported in individual studies. Amounts of different pollutants in $\mu g\ m^{-3}$ equivalent to 1 vpm (1000 ppbv) at different temperatures and standard atmospheric pressure are given in Table 9.1, and in ppbv equivalent to 1000 $\mu g\ m^{-3}$ in Table 9.2.

Table 9.1
Amounts of Different Pollutants in $\mu g\ m^{-3}$ Equivalent to 1 ppm (1000 ppb) at Different Temperatures but Standard Atmospheric Pressure[a]

Temperature (°C)	SO_2	NO_2	NO	NH_3	O_3	CO_2	CO	H_2S	HF
−5	2914	2092	1365	774	2183	2001	1274	1550	910
0	2860	2054	1340	760	2143	1965	1250	1521	893
5	2809	2017	1316	747	2104	1929	1228	1494	877
10	2759	1981	1292	733	2067	1895	1206	1468	862
15	2711	1947	1270	721	2031	1862	1185	1442	847
20	2665	1914	1248	708	1997	1831	1165	1417	832
25	2620	1882	1227	696	1963	1800	1146	1394	818
30	2577	1850	1207	685	1931	1770	1127	1371	805
35	2535	1821	1187	674	1899	1742	1108	1348	792

[a]From Wellburn (1988). Copyright © 1988 by Longman Group UK Ltd.

and Canada, often hundreds of kilometers from major sources of air pollution (Likens and Bormann, 1974).

The average pH of rainwater at the Hubbard Brook Experimental Forest in New Hampshire varied from 4.03 to 4.09 during 1956 to 1971. In 1956 zones of very acid rains were found in states with high sulfur emissions (parts of Ohio, Pennsylvania, West Virginia, New York, and New England). By 1973, the area with an average pH of rain below 4.5 had extended to include parts of Mississippi, Alabama, Georgia, South Carolina, North Carolina, Kentucky, Virginia, and north into New England and Canada. On an annual basis, rain and snow over large regions of the world are 5 to 30 times more acid than would be expected in an unpolluted

Table 9.2
Amounts of Different Pollutants in ppb Equivalent to 1000 $\mu g\ m^{-3}$ at Different Temperatures but Standard Atmospheric Pressure[a]

Temperature (°C)	SO_2	NO_2	NO	NH_3	O_3	CO_2	CO	H_2S	HF
−5	343	478	733	1291	458	500	785	645	1099
0	349	487	746	1315	467	509	800	657	1120
5	356	496	760	1339	475	518	814	669	1140
10	362	505	774	1364	484	528	829	681	1161
15	368	514	787	1388	492	537	844	693	1181
20	375	523	801	1412	501	546	858	705	1202
25	382	531	815	1436	509	556	873	718	1222
30	388	540	828	1460	518	565	888	730	1243
35	394	549	842	1484	527	574	902	742	1263

[a]From Wellburn (1988). Copyright © 1988 by Longman Group UK Ltd.

atmosphere. Furthermore, the rain of individual storms may be several hundred times more acid than expected (Likens *et al.*, 1979)

Photochemical Oxidants

The most important pollutants of this group are O_3, NO_2, and PAN. Most secondary pollutants form from the action of sunlight on products of fuel combustion, such as nitrogen oxides, hydrocarbons, aldehydes, and carbon monoxide. Oxidants are important components of photochemical smog, which contains O_3, PAN, NO_2, hydrogen peroxide, formaldehyde, higher aldehydes, acrolein, and formic acid (National Academy of Sciences, 1977b).

Throughout most of the eastern and southwestern United States, during sunny weather O_3 levels regularly exceed natural background concentrations (0.02 to 0.03 ppm) over the growing season. Concentrations of O_3 in central and northwestern Europe are comparable to those in the United States and southern Canada (Reich, 1987).

Several oxides of N occur in polluted air. NO and NO_2 are interconvertible and are collectively designated NO_x. In addition to NO and NO_2, polluted air contains very small amounts of nitrous oxide (N_2O), nitrogen trioxide (NO_3), and dinitrogen pentoxide (N_2O_5).

Nitrogen oxides arise from both natural and human-made sources. In the United States, emissions of NO_x approximately doubled between 1940 and 1982. From 1970 to 1982 the rate of increase was reduced, primarily through controlling emission from motor vehicles. Emissions of NO_x by industrial processes were rather constant between 1970 and 1982 (Environmental Protection Agency, 1984a).

Most NO_x injury is indirect through involvement in photochemical reactions that produce atmospheric oxidants such as O_3 and PAN, but primarily O_3.

Fluorides

The major industrial sources of F are aluminum smelting; manufacture of steel; conversion of fluoroapatite to phosphate and phosphorus; and production of glass, bricks, and ceramics. Natural sources of F include soils, fumaroles, and volcanoes.

Although SO_2, O_3, and NO_x cause more injury to plants than fluorides do, the latter are the most phytotoxic air pollutants and may injure plants at much lower concentrations than other pollutants (Weinstein and Alscher-Herman, 1982). The leaves of highly susceptible species such as apricot and young needles of some conifers may develop lesions from accumulation of less than 0.1 to 0.2 parts per hundred million (pphm) F. Gaseous fluorides are much more phytotoxic than those containing particulate compounds.

Particulates

Particulate air pollutants consist of finely divided particles of solid or liquid matter (dust, smoke, aerosols). The important particulates include

1. *Cement Kiln Dusts* Thick crusts of dusts in waste gases from cement kilns often cover leaves in adjacent areas.

2. *Fluorides* Dust-formed fluoride compounds are emitted from various industrial activities.
3. *Soot* Forest fires often produce carbonaceous particulate matter in addition to large amounts of O_3 and NO_2.
4. *Heavy Metals* The major sources of heavy metals in the environment include the iron and steel industry, primary base metal smelters, secondary base metal smelters that reprocess discarded products into their constituent metals, base metal refineries that convert crude concentrates from smelters into almost pure metals, mine spoils and tailings, pesticides that contain metals as active ingredients, contaminated sewage sludge, solid wastes and fertilizers, electric power stations, municipal incinerators, motor vehicles, and sulfuric acid aerosols.

In the United States, emissions of particulates decreased progressively between 1940 and 1982 as a result of installation of control equipment on stationary sources, obsolescence of coal-burning locomotives, reduction in wildfires, and decreased burning of solid wastes (Environmental Protection Agency, 1984a).

Responses of Woody Plants to Air Pollution

Many air pollutants, alone and in combination, induce injuries and physiological changes in plants that are later expressed in growth changes. Pollution-induced biochemical changes in cells are sequentially followed by physiological changes in individual cells, plants, species, and plant communities (Fig. 9.1).

Growth

A voluminous literature shows that various air and soil pollutants, alone and in combination, reduce shoot growth, cambial growth, root growth, and reproductive growth of trees near point sources of pollution. Most studies have been conducted with seedlings.

Shoot Growth

Both gaseous pollutants and particulates reduce shoot growth of seedlings, the amount of reduction varying with the pollutant as well as its concentration and duration of exposure (Table 9.3). Some studies show that height growth of older trees may be reduced as well. For example, a general decline in height growth of dominant and codominant white ash and red oak trees was attributed to increases in several industrial pollutants, including O_3, SO_2, NO_x, and particulates (McClenahen, 1983).

Leaf growth The leaf area of individual trees and stands of trees exposed to air pollutants may be reduced because of inhibition of leaf formation, arrested leaf

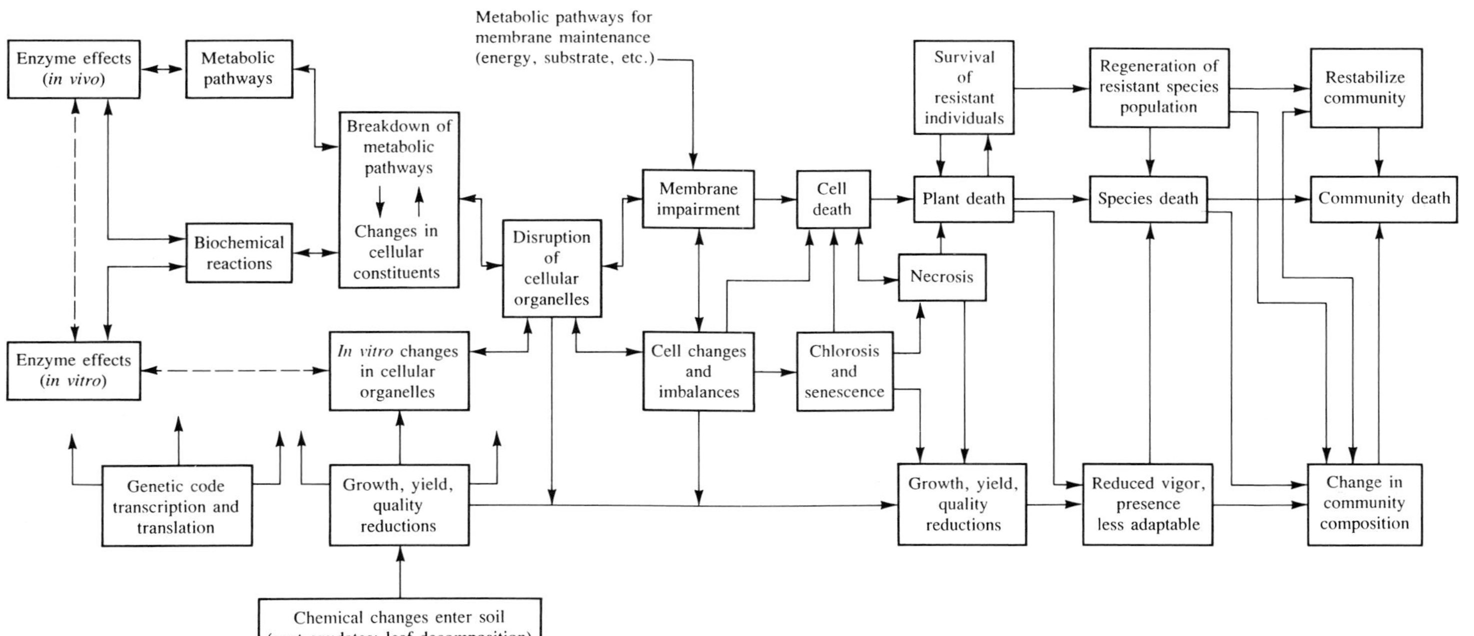

Figure 9.1 Sequential responses of plants to pollutants at the biochemical, cellular, whole-plant, species, and community levels. (From Heck, 1973.)

Table 9.3
Representative Studies That Show Reduction in Height Growth of Seedlings Exposed to Various Pollutants

Species	Pollutant	Source
Green ash	SO_2	Jensen (1977)
Sycamore	SO_2	Jensen (1977)
Eastern cottonwood	SO_2	Jensen (1977)
Silver maple	SO_2	Jensen (1977)
Tulip poplar	SO_2	Jensen (1977)
Black locust	SO_2	Suwannapinunt and Kozlowski (1980)
Ailanthus	SO_2	Marshall and Furnier (1981)
Paper birch	SO_2	Norby and Kozlowski (1981c)
Red gum (*Eucalyptus camaldulensis*)	SO_2	Norby and Kozlowski (1981c)
Blue gum (*Eucalyptus globulus*)	SO_2	Norby and Kozlowski (1981c)
Red pine	SO_2	Norby and Kozlowski (1981c)
Scotch pine	SO_2	Farrar *et al.* (1977)
Japanese larch	SO_2	Tsukahara *et al.* (1985)
Japanese red pine	SO_2	Tsukahara *et al.* (1985)
Japanese black pine	SO_2	Tsukahara *et al.* (1985)
Sugar maple	O_3	Jensen (1973)
Silver maple	O_3	Jensen (1973)
Black alder	O_3	Jensen (1973)
Green ash	O_3	Jensen (1973)
Black walnut	O_3	Jensen (1973)
Sycamore	O_3	Jensen (1973)
Green Ash	O_3	Duchelle *et al.* (1982)
Tulip poplar	O_3	Duchelle *et al.* (1982)
Virginia pine	O_3	Duchelle *et al.* (1982)
Loblolly pine	O_3	Kress *et al.* (1982)
Silver maple	$CdCl_2$ in soil	Lamoreaux and Chaney (1977)
Tulip poplar	$CdCl_2$ in soil	Kelly *et al.* (1979)
Yellow birch	$CdCl_2$ in soil	Kelly *et al.* (1979)
Silver maple	$CdCl_2$ in soil	Kelly *et al.* (1979)
Chokecherry	$CdCl_2$ in soil	Kelly *et al.* (1979)
Eastern white pine	$CdCl_2$ in soil	Kelly *et al.* (1979)
Loblolly pine	$CdCl_2$ in soil	Kelly *et al.* (1979)
Black alder	Industrial dusts	Greszta *et al.* (1982)
Scotch pine	Industrial dusts	Greszta *et al.* (1982)
Austrian pine	Industrial dusts	Greszta *et al.* (1982)
Norway spruce	Industrial dusts	Greszta *et al.* (1982)

expansion, and accelerated leaf abcission. For example, SO_2 reduced leaf weight and leaf area of Norway maple (Garsed *et al.*, 1979). Rates of leaf expansion of American elm seedlings exposed to SO_2 were only 40% of those of leaves of control plants (Constantinidou and Kozlowski, 1979a). SO_2 also inhibited needle expansion of conifers, reflecting an inhibitory effect on both cell size and cell number (Bucher and Keller, 1978; Halbwachs, 1984). Inhibition of leaf expansion by O_3 was shown for American elm (Constantinidou and Kozlowski, 1979a), green ash (Jensen, 1981), silver maple, and eastern cottonwood (Jensen, 1982). Arrested leaf expansion also has been attributed to effects of fluorides (Ferlin *et al.*, 1982) and particulates (Darley, 1966).

Abscission Premature defoliation is a common response to air pollution (Usher and Williams, 1982; Carlson, 1980; Reich and Lassoie, 1985). For example, the leaf area index of ponderosa pine stands was greatly lowered by oxidants. Severely injured trees carried essentially only 1-year-old needles (Axelrod *et al.*, 1980).

Cambial Growth

Pollutants often reduce diameter growth of seedlings (Table 9.4), saplings, and mature trees, commonly without visible leaf symptoms.

Stem radial growth of apple, sweet cherry, sour cherry, and plum trees was reduced by SO_2 by as much as 40% (Guderian and Stratmann, 1968). Treatment of soil with Pb or Cd decreased radial growth of sycamore saplings by up to 65% (Carlson and Bazzaz, 1977).

There are many examples of reduced diameter growth of mature trees located near sources of pollution; for example, eastern white pine trees near a smelter in Sudbury, Ontario (Navratil and McLaughlin, 1979); Norway spruce and Scotch pine near a sulfite plant in Sweden (Westman, 1974); balsam fir near a copper smelter (Robitaille, 1981); singleleaf pinyon pine near a copper smelter (Thompson, 1981); and Austrian pine near a brickworks complex in England (Gilbert, 1983). Fluorine dust from a fertilizer plant in Finland reduced wood production in Scotch pine trees (Havas and Huttunen, 1972).

Using analysis of tree rings, several investigators showed that air pollution reduced xylem increment of mature trees (e.g., Ashby and Fritts, 1972; Thompson, 1981). In the United States, dendroecological techniques were used to study effects

Table 9.4
Representative Studies That Show Reduction in Diameter Growth of Seedlings following Exposure to Various Pollutants

Species	Pollutant	Source
Eastern white pine	SO_2	Linzon (1978)
Japanese larch	SO_2	Tsukahara *et al.* (1985)
Japanese red pine	SO_2	Tsukahara *et al.* (1985)
Japanese black pine	SO_2	Tsukahara *et al.* (1985)
Sophora	SO_2	Shanklin and Kozlowski (1985b)
Hinoki cypress	SO_2	Tsukahara *et al.* (1986)
Ponderosa pine	Oxidants	Ohmart and Williams (1979)
White fir	Oxidants	Ohmart and Williams (1979)
Scotch pine	Fluorine dust	Havas and Huttunen (1972)
Douglas-fir	F	Treshow and Anderson (1982)

of climatic factors, age of trees, soil type, and air pollution on tree growth from Maine to North Carolina and as far west as Missouri (McLaughlin *et al.*, 1983). Nash *et al.* (1975) described a method for separating effects of air pollution from those of climatic factors on width of xylem rings.

Air pollutants not only decrease xylem increment but sometimes also alter wood structure. For example, SO_2 reduced the amount of latewood formed in annual rings of Norway spruce (Keller, 1980a). Scotch pine, Norway spruce, and European larch trees that had been exposed to air pollution had short tracheids, and English oak, European aspen, and European white birch had short vessels, tracheids, and fibers. The number of vessels in English oak also was reduced by air pollution (Grill *et al.*, 1979).

Root Growth and Rhizosphere Symbionts

Because the aboveground parts of woody plants are directly exposed to air pollutants and the roots are not, it sometimes has been assumed that shoot growth is reduced more than root growth by pollution. However, as emphasized in Chapter 3, when photosynthesis is reduced by environmental stresses, proportionally less of the carbohydrate pool is allocated to the roots than to growing shoots. For example, the amounts of starch, sucrose, and reducing sugars were lower in roots of green ash seedlings fumigated with O_3 than in roots of nonfumigated controls (Jensen, 1981). Similar results have been reported for conifers (McLaughlin *et al.*, 1982). Hence, growth of both shoots and roots of polluted trees is reduced, but root growth generally is reduced more.

Several studies showed that SO_2 significantly reduced root growth of seedlings. Examples are American elm (Constantinidou and Kozlowski, 1979a), paper birch, red gum (*Eucalyptus camaldulensis*), blue gum (*E. globulus*), red pine (Norby and Kozlowski, 1981a,b), and hinoki cypress (Tsukahara *et al.*, 1986). Freer-Smith (1984) reported large decreases in root–shoot ratios of European white birch seedlings that were fumigated with SO_2, or with a combination of SO_2 and NO_2. Lead contamination of the soil close to a motor road in Sweden was associated with decreased growth of pine roots and of root mortality of Norway spruce trees (Majdi and Perrson, 1989).

Air pollutants may also reduce mycorrhizal populations, often because of the reduced flow of assimilates from leaves to roots (Kasana and Mansfield, 1986). For example, mycorrhizal populations on citrus roots were reduced after chronic exposure to O_3 (McCool *et al.*, 1979). SO_2 reduced the numbers of infected roots and the percentage of infection of red oak seedlings with mycorrhizal fungi (Reich *et al.*, 1985). Low concentrations of O_3 stimulated mycorrhizal infections of red oak seedlings but high concentrations reduced infection (Reich *et al.*, 1986). Acid rain decreased mycorrhizal infection of eastern white pine seedlings. The decrease in the number of mycorrhizal roots was the result of fewer short roots available for infection and a lower percentage of roots infected (Stroo *et al.*, 1988).

The effects of pollution on root growth and mycorrhizal infection are of enormous practical importance and deserve further study. Lowering of the root–shoot ratio by pollution may decrease drought tolerance because the small root systems may be unable to absorb water fast enough to keep up with transpirational losses. Inhibition of mycorrhizal development by air pollutants may also be expected to decrease absorption of mineral nutrients (Chapter 6).

Reproductive Growth

Air pollution often decreases the yield and quality of fruits. For example, near Los Angeles, California, ambient levels of photochemical oxides reduced yield of citrus fruits by as much as half (Thompson and Taylor, 1969). Following application of 15 sprays of 0.0025 or 0.00125 *N* NaF solutions, yield of citrus fruits over a 6-year period was reduced 15% by annual accumulation of 75 ppm F and 22% by accumulation of 150 ppm F in the leaves (Brewer *et al.*, 1967).

Air pollutants adversely influence several stages of reproductive growth (Fig. 9.2). Inhibition of reproductive growth may result from less partitioning of carbohydrates to reproductive organs or from a smaller carbohydrate pool. By injuring leaves, air pollutants decrease the amounts of available carbohydrates and hormonal growth regulators. Hence, yield may be reduced following a decrease in tree vigor and in physiological efficiency of the foliage. Reproductive growth may also be inhibited by the effects of pollutants on mechanisms of flowering and fruiting as well as by direct injury to reproductive structures. In addition, certain pollutants are toxic to pollinating insects. Decreased reproductive growth in O_3-affected plants is

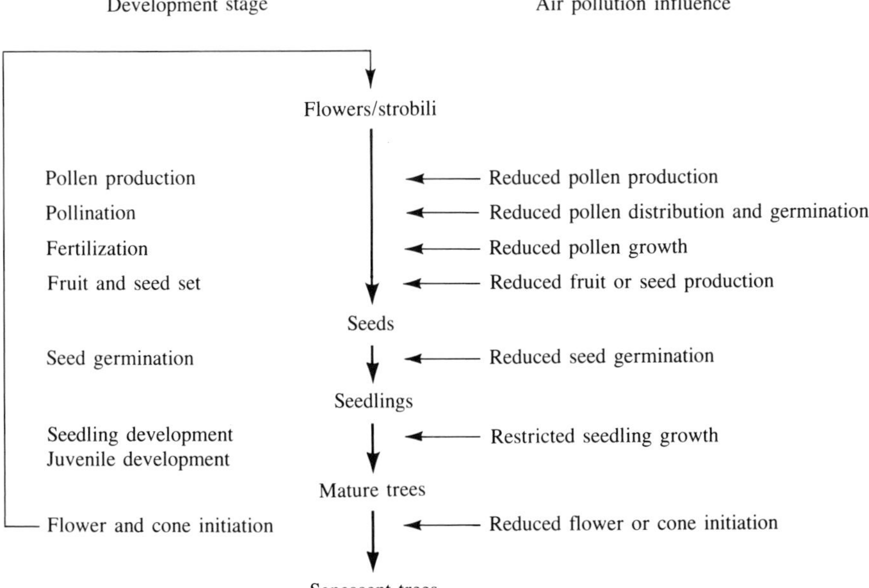

Figure 9.2 Effects of air pollutants on various developmental stages of woody plants. (From Smith, 1981.)

mediated primarily through reduced physiological efficiency of injured foliage (Bonte, 1982).

In Ohio the average weight of red pine seeds, percentage of filled seeds, and seed germination capacity were reduced by air pollution (Houston and Dochinger, 1977). In France air pollution reduced cone size and the number of seeds per cone, and increased the rate of cone abortion (Roques *et al.*, 1980). Germination of red pine seeds decreased by as much as half following exposure of seeds to 5 pphm SO_2 for 8 hr, 7 days a week (Riding and Boyer, 1983).

Reduction in seed production often has been attributed to inhibitory effects of individual and combined air pollutants on germination of pollen and growth of pollen tubes. Germination of pollen of mugo pine, Austrian pine, Scotch pine, and European silver fir was reduced by 0.225 ppm SO_2. The pollen of Austrian pine and Scotch pine was more sensitive than that of mugo pine, which responded in the same way as pollen of European silver fir (Keller and Beda, 1984). Both germination of pollen and growth of pollen tubes of red pine and eastern white pine trees on SO_2-polluted sites were lower than in trees on unpolluted sites (Houston and Dochinger, 1977). Van Ryn *et al.* (1988) studied effects of acid mist (pH range of 5.6 to 2.6) on pollen tubes of red maple in the field. Both pollen germination and growth of pollen tubes were reduced by low pH, but the effects on germination were not as severe as on pollen germinated *in vitro* (Van Ryn *et al.*, 1986). Heavy metals also inhibited pollen germination, with Cd^{2+} being the most toxic ion studied, followed by Cu^{2+}, Hg^{2+}, Pb^{2+}, Zn^{2+}, and Ba^{2+} (Chaney and Strickland, 1974, 1984).

Growth of pollen tubes of apricot and sweet cherry was inhibited by HF during anthesis. The results were similar whether fumigation occurred before or after pollination, indicating an effect on pollen tube growth rather than on stigma receptivity or inhibition of pollen germination (Facteau *et al.*, 1973; Facteau and Rowe, 1977, 1981).

Injury

Several forms of foliar injury in forest, ornamental, and fruit trees have been attributed to air pollutants. Gaseous air pollutants injure leaves after being absorbed through stomatal pores (Fig. 9.3). Acute injury is severe and follows rapid absorption of enough pollutant to kill tissues. During or soon after exposure, leaf cells collapse and necrotic patterns subsequently appear. In broad-leaved trees both SO_2 and HF cause collapse of spongy mesophyll cells and those of the lower stomata-bearing epidermis, followed by distortion of chloroplasts and injury to palisade cells. Vascular tissues are injured least (Ormrod, 1978). Chronic injury caused by rapid absorption of sublethal amounts of pollutants over a long time is characterized by chlorosis, which develops slowly. Sometimes chronic injury is accompanied by necrotic markings. Acute injury by pollution is relatively rare, whereas nonvisible effects of pollution occur commonly.

When only a single pollutant induces injury, the symptoms may be distinct. For

Figure 9.3 Injury to (A) poplar and (B) ash leaves by SO_2. (U.S. Department of Agriculture photos.)

example, SO_2 injury to broad-leaved trees is characterized by areas of injured leaf tissue located between the healthy tissue around the veins. O_3 injury to leaves is expressed variously in different species (e.g., flecks or stipples of dead tissue in sugar maple; bronzing in clematis, blackening in poplar). The symptoms may be dose dependent or related to factors affecting stomatal aperture.

Symptoms of injury by individual pollutants are less definite for conifers than for broad-leaved trees. Nevertheless, some distinctions can be made between acute and chronic injury in conifers. High doses of SO_2 cause brown tipburn of needles, with the color change progressing downward from the tip as more pollutant is absorbed. Chronic injury is expressed by chlorosis, especially in the older needles, which often are shed prematurely. Acute O_3 injury is characterized by discoloration of needle tips or whole needles. Eventually, all except the current-year needles may be shed, giving the branches a tufted appearance (Fig. 9.4). Mild O_3 injury sometimes appears as chlorotic mottling of needles. Several pollutants, including SO_2, O_3, and fluoride, cause tipburn, depending on the dosage and plant species. Tipburn of conifers is also caused by some herbicides, deicing salts, excess fertilizers, and natural winter injury, often making it difficult to ascertain the specific cause (Kozlowski, 1980a,b).

The waxes on leaf surfaces are physiologically important because they restrict loss of water vapor, control gaseous exchange, reduce leaching of nutrients and metabolites, and act as a barrier to reactive pollutants such as SO_2, NO_2, and O_3. Leaf waxes are subject to natural weathering, which may be accelerated by wind damage, mechanical abrasion, and chemical interaction with pollutants. Thus older or damaged leaves are likely to be more susceptible to injury.

Both the morphology and distribution of leaf waxes often are modified by air pollutants. Responses include inhibition of wax formation (Grill, 1973), degrada-

Figure 9.4 Effects of oxidant on ponderosa pine needles. The branches are from three trees ranging from oxidant tolerant to oxidant sensitive (*left to right*). (U.S. Forest Service photo.)

tion of surface waxes (particularly around stomatal pores), and occlusion of stomatal chambers (Percy and Riding, 1978). Exposure of Scotch pine needles to SO_2 accelerated weathering of needle waxes and caused thickening of the wax tubes (Crossley and Fowler, 1986).

Ecosystem Structure

Accumulation of sufficient toxic substances in the environment can induce changes in the structure and function of communities of woody plants. During normal plant succession in an unpolluted atmosphere, the number of plant species, productivity, biomass, community height, and structural complexity increase over time and tend to maximize in the climax stage (Chapter 3). When a severe pollution stress is imposed on a forest its capacity to select for resistance often is overwhelmed. Hence, extreme pollution sets in motion a retrogression that is characterized by reduction in structural complexity, biomass, productivity, and species diversity (Whittaker, 1975).

Reduced photosynthesis and visible leaf symptoms often precede growth inhibition and mortality of the more pollution-sensitive plants, leading to alteration of ecosystem structure. Community structure is also influenced by changes in reproductive growth because plant vigor and reproductive capacity are highly correlated and exposure of plants to pollution decreases their vigor (Kramer and Kozlowski, 1979, pp. 680–681). Pollutants also influence plant community structure by their interactions with plant diseases and insects. For example, exposure to air pollution may decrease or intensify disease responses. Pollutants, alone or in combination, may act directly on fungi or bacteria, thereby inhibiting parasites, or they may change the physiology of the host so as to render it more or less sensitive to a specific pathogen. If the host is weakened by air pollution, it usually becomes more sensitive to weak pathogens but less sensitive to obligate parasites. However, if pollutants physically injure the host, infection generally is facilitated (Treshow, 1980).

Responses of forest ecosystems vary from total destruction under high pollution dosages to negligible changes under low dosages (Table 9.5). High dosages of pollutants can induce severe and dramatic degradation of forest stands, as found near major point sources of pollution such as smelters. The effects usually are confined to a zone a few kilometers around the pollution source and several kilometers downwind. The most sensitive species of the tree layer are affected first and canopy trees may be killed. Shrubs and herbs are then destroyed in order. Simplification of the ecosystem is accompanied by reduced energy fixation, lowered biomass, and increased loss of mineral nutrients.

A classic example of the destructive effect of a high SO_2 dosage on a forest stand was shown around an iron-sintering plant near Wawa, Ontario, Canada (Gordon and Gorham, 1963). Severe SO_2 injury to plants was confined to a narrow strip northeast from the point source because southwest winds predominated. SO_2 caused successive deterioration of trees, shrubs, and herbaceous plants, and floristic variety

Table 9.5
Influence of Air Pollution on Forest Ecosystems[a]

Air pollution dosage	Response of vegetation	Impact on ecosystem
High	1. Acute morbidity	1. Simplification; increased erodibility, nutrient attrition, altered microclimate and hydrology
	2. Mortality	2. Reduced stability
Intermediate	1. Reduced growth (a) decreased nutrient availability (i) depressed litter decomposition (ii) acid rain leaching (b) suppressed photosynthesis, enhanced respiration	1. Reduced productivity, lessened biomass
	2. Reduced reproduction (a) pollinator interference (b) abnormal pollen, flower, seed, or seedling development	2. Altered species composition
	3. Increased morbidity (a) predisposition to entomological or microbial stress (b) direct disease induction	3. Increased insect outbreaks, microbial epidemics Reduced vigor
Low	1. Act as a sink for contaminants	1. Pollutants shifted from atmospheric to organic or available nutrient compartment
	2. No or minimal physiological alteration	2. Undetectable influence, fertilizing effect

[a]From Smith (1974).

declined sharply close to the pollution source. The number of macrophyte species varied from 43 at distances beyond 16 km from the source to 2 to 4 within 5 km of the source. The number of species declined from 28 beyond 16 km to 0.2 within 5 km (Fig. 9.5). Other specific examples of destruction of forest ecosystems by high dosages of air pollutants from point sources are given by Miller and McBride (1975), Linzon (1978), Winner and Bewley, (1978), Rosenberg *et al.* (1979), and Guderian and Kueppers (1980).

Under intermediate pollution dosages, tree growth is inhibited and forest biomass decreased as a result of a reduction in photosynthesis, a decrease in mineral supply, and lowered reproductive capacity, as well as an increase in stress from insects and diseases. Changes in physiological processes vary from subtle to severe for different components of the ecosystem, resulting in changes in competitive ability and hence changes in succession and species composition.

Under low SO_2 dosages (1 to 4 $\mu g\ m^{-3}$) forests act as sinks for pollutants, are not adversely affected, and occasionally may even benefit from pollution. For exam-

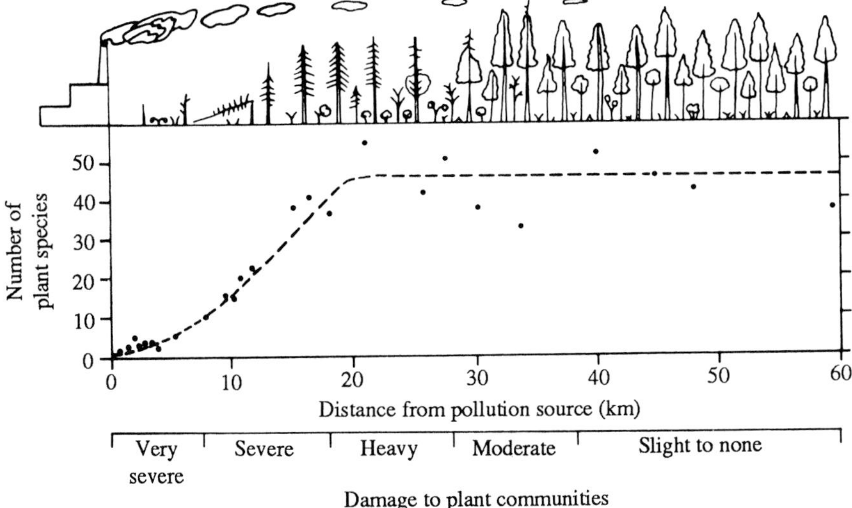

Figure 9.5 Effects of SO_2 produced by an iron-sintering plant on numbers of plant species. (Reproduced from Whittaker, 1975; after Gordon and Gorham, 1963.)

ple, in limited areas where the soil is deficient in S, growth of plants is increased by very slight SO_2 pollution (Faller *et al.*, 1970; Bennett *et al.*, 1974; Cowling and Lockyer, 1976).

The response of forests in polluted air is an integrated response to the pollutant as well as to other environmental stresses present, such as those associated with plant competition. In a mixed forest, some species may actually benefit from pollution if they are given a competitive advantage by the relatively greater adverse effect of the pollutant on other species in the ecosystem. Growth of still other species in the same ecosystem may be reduced because of their lowered competitive potential (Kozlowski, 1985c). For example, in the mixed conifer forest of the Sierra Nevada, SO_2 pollution reduced basal area of ponderosa pine but increased basal area of white fir (Kercher *et al.*, 1980).

Effects of Acid Rain

Evidence that acid rain causes economic damage to woody plants has been vigorously debated (Morrison, 1984). A number of short-term studies, mostly with seedlings, reported such potentially deleterious effects of acid rain as inhibition of pollen germination (Van Ryn *et al.*, 1988) and seed germination (Teigen, 1975), injury to leaf cells (Adams *et al.*, 1984), leaching of mineral nutrients from leaves (Puckett, 1982; Reich *et al.*, 1988) and soil (Johnson *et al.*, 1983), decrease in cation-exchange capacity (Klein and Perkins, 1988), and inhibition of height and diameter growth of trees (Matziris and Nakos, 1978; Jonsson and Sundberg, 1980). However, other studies showed no deleterious effects of acid rain on tree seedlings. Short-term studies have not shown important interaction between O_3 and acid deposition on photosynthesis or growth of conifers (Reich *et al.*, 1988) or broad-leaved trees (Reich *et al.*, 1986). In fact, Wood and Bormann (1977) and Reich *et al.* (1987) showed that photosynthesis and growth of eastern white pine seedlings were increased by simulated acid rain as a result of increased nitrogen fertilization. Hanson *et al.* (1988a) found that after 13 weeks of exposure, pH 3.3 and 4.5 rain treatments increased absorption of CO_2 by loblolly pine seedlings by 52% over seedlings exposed to pH 5.2 treatments. Amthor (1984) concluded that acid rain alone will not be a major problem in most forests.

Despite the lack of negative effects of acid rain in the short term, harmful effects might occur over long periods because of altered mycorrhizal and nutrient relations. Simulated acid rain decreased mycorrhizal infection in eastern white pine and red oak seedlings (Reich *et al.*, 1986; Stroo *et al.*, 1988). In eastern white pine the percentage of short roots infected with mycorrhizal fungi decreased by more than 10%, and in red oak by 20%, as the pH of the rain treatment decreased from 5.6 to 3.0. As emphasized by Reich *et al.* (1988), acid rain might have adverse long-term effects on forests because of possible accumulation of toxic elements in soils and because of deficiencies of important major elements in plant tissues, perhaps partly because of decrease in mycorrhizal roots.

European studies show that the effects of acid rain on growth of forest trees change over time. Initially the effect may be beneficial from increased availability of S and N if cations in the soil are ample but N is limiting growth. Subsequently a nutritional imbalance may develop as acidic deposition accelerates leaching of cations from the soil while growth of trees is accelerated by a high N supply. Eventually, as nutritional deficiencies develop and physiological processes are adversely affected, tree growth may be inhibited (Oren *et al.*, 1988b). Reduction in growth of a declining Norway spruce stand that had been exposed to acidic deposition was limited by low sink strength for photosynthate (Oren *et al.*, 1988a). The rate of photosynthesis was reduced by severe Mg deficiency, which was related to Mg availability in the soil and root development (Meyer *et al.*, 1988), uptake into the xylem (Osonubi *et al.*, 1988), and leaching from the needles, all of which are affected by acid deposition. Oren *et al.* (1988b) concluded that imbalanced nutrition was more important than deficiency of a particular element in reducing the vigor of declining trees that had been exposed to acidic deposition for some time. Much more research is needed on the long-term effects of acid rain on forest ecosystems.

Physiological Processes

Reduction in growth of woody plants by air pollutants is mediated by changes in physiological processes. In addition to causing nutrient imbalances, several air pollutants inhibit chlorophyll synthesis and photosynthesis and alter stomatal aperture, permeability of cell membranes, amounts and types of stored carbohydrates and proteins, and activity of enzymes (Kozlowski and Constantinidou, 1986a; Darrall, 1989).

Photosynthesis

Much attention has been given to effects of air pollutants on photosynthesis because the rate is correlated with biomass production of woody plants. Reduction in the rate of photosynthesis would be expected when leaves are injured or shed, but the rate often is reduced by several pollutants long before visible injury or growth reduction occurs. The amount of reduction varies with specific pollutants and dosage (Table 9.6) and with species, clones, cultivars, and environmental conditions (Kozlowski and Constantinidou, 1986b).

Sulfur dioxide Most studies show that high dosages of SO_2 rapidly and substantially reduce the rate of photosynthesis. For example, exposure to 20 pphm (524 g m^{-3}) SO_2 for up to 70 days greatly reduced the rate of photosynthesis of 3-year-old grafts of European silver fir, Norway spruce, and Scotch pine (Keller, 1977b,c, 1978). Keller (1983) also studied the effect of a 3-month winter fumigation with 10 or 20 pphm SO_2 on photosynthesis of two Norway spruce clones. In spite of low photosynthetic activity during the dormant season, the rate was substantially reduced.

Wide variations in photosynthetic inhibition by SO_2 have been shown for different plant species and genetic materials. For example, the amount of reduction of photosynthesis by SO_2 varied in the following order: sycamore maple > English oak, horse chestnut, European ash > European white birch (Piskornik, 1969). SO_2 at 5 pphm for 2 hr decreased photosynthesis of sensitive eastern white pine clones by 27% and tolerant clones by 10% (Eckert and Houston, 1980). Variations also were shown among Scotch pine clones and provenances in photosynthetic responses to SO_2 (Lorenc-Plucinska, 1982; Oleksyn and Bialobok, 1986).

Ozone Many examples are available of reduction of photosynthesis by relatively low levels of O_3 (Kozlowski and Constantinidou, 1986a) and only a few will be cited. Exposure of ponderosa pine seedlings to 15 pphm O_3 for 30 days decreased photosynthesis by 10% (Miller *et al.*, 1969). Reich (1983) emphasized that chronic exposure of hybrid poplar leaves to low levels of O_3 decreased photosynthesis and chlorophyll contents while increasing dark respiration. During the first 7 days, photosynthesis of leaves chronically exposed to 0.125 μl liter^{-1} O_3 differed only slightly from that of control leaves. However, once leaves were fully expanded (within 14 days) the rate of photosynthesis was greatly reduced (Fig. 9.6). Accelerated leaf aging by O_3 was partly responsible for decreased photosynthetic capacity. The control leaves lived 10 to 15% longer than those exposed to O_3. Concentrations of O_3 representative of those found in clean and mildly to moderately polluted air appreciably reduced photosynthesis of eastern white pine, red oak, sugar maple, and poplar seedlings. The reductions in rates were linear with respect to O_3 concentrations, and no visible O_3 injury was detected.

Fluorides Inhibition of photosynthesis by F is common near smelting, aluminum, and fertilizer plants. The amount of reduction of photosynthesis near a

Table 9.6
Threshold Pollutant Doses for Suppression of Photosynthesis of Forest Tree Seedlings and Saplings[a]

Pollutant	Concentration	Time	Experiment duration	Species	Reference
			Seedlings		
SO_2	600 pphm (15.7×10^3 μg m^{-3})	4–6 hr	Single treatment	Red maple	Roberts *et al.* (1971)
	100 pphm (2620 μg m^{-3})	2–4 hr	Single treatment	Quaking aspen White ash	Jensen and Kozlowski (1974)
	10 pphm (262 μg m^{-3})	Continuous	2 weeks	White fir	Keller (1977c)
	20 pphm (524 μg m^{-3})	Continuous	2 weeks	Norway spruce Scotch pine	Keller (1977c)
O_3	30 pphm (588 μg m^{-3})	9 hr day^{-1}	10 days	Ponderosa pine	Miller *et al.* (1969)
	15 pphm (294 μg m^{-3})	Continuous	19 days	Eastern white pine	Barnes (1972)
	15 pphm (294 μg m^{-3})	Continuous	84 days	Slash pine Pond pine Loblolly pine	Barnes (1972)
F	30 μg g^{-1} dry wt. basis foliar tissue			Pines (various)	Keller (1977a)

Pb	<10 μg g^{-1} dry wt. basis foliar tissue			American sycamore	Carlson and Bazzaz (1977)
Cd	<10 μg g^{-1} dry wt. basis foliar tissue			American sycamore	Carlson and Bazzaz (1977)
			Saplings		
SO_2	100 pphm (2620 μg m^{-3})	30 min	Single treatment	Red maple (excised leaves)	Lamoreaux and Chaney (1978)
	50 pphm (1310 μg m^{-3})	7–11 hr	1–2 days	Black oak	
	50 pphm (1310 μg m^{-3})	7–11 hr	1–2 days	Sugar maple	Carlson (1979)
	50 pphm (1310 μg m^{-3})	7–11 hr	1–2 days	White ash	
O_3	50 pphm (980 μg m^{-3})	4 hr	Single treatment	Eastern white pine	Botkin *et al.* (1972)
	50 pphm (980 μg m^{-3})	7–11 hr	1–2 days	Black oak	Carlson (1979)
	50 pphm (980 μg m^{-3})	7–11 hr	1–2 days	Sugar maple	Carlson (1979)
Cd	100 μg g^{-1} dry wt.	45 hr	Single treatment	Silver maple (excised leaves)	Lamoreaux and Chaney (1978)

[a]From Smith (1981).

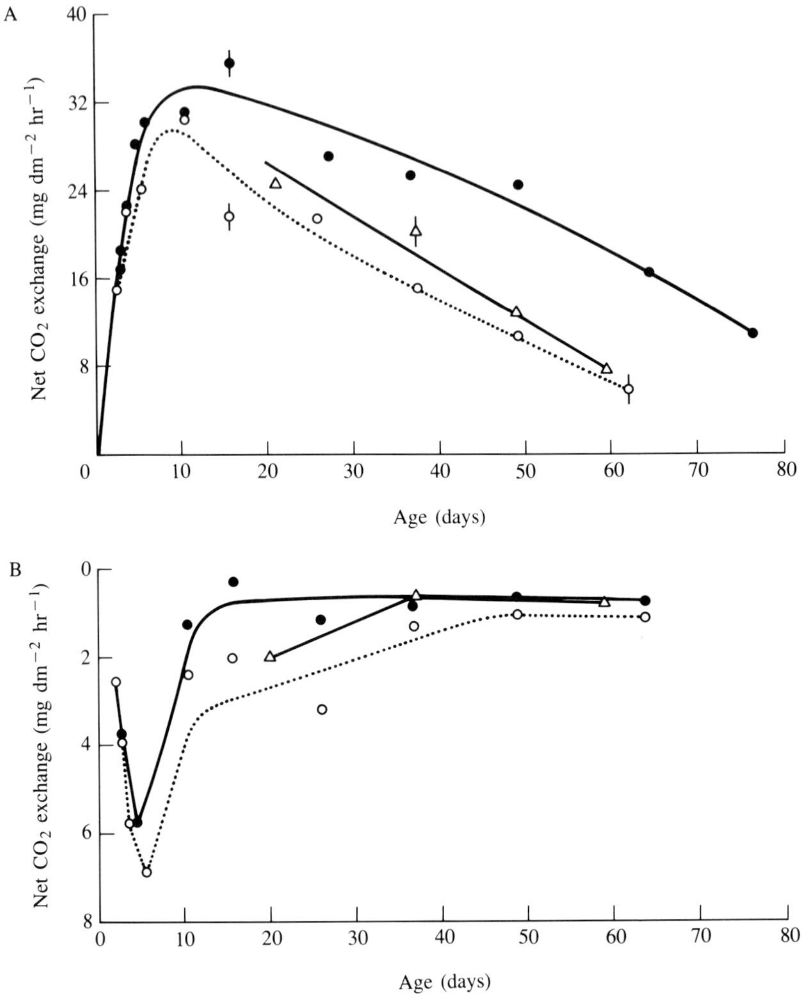

Figure 9.6 (A) Net photosynthesis and (B) dark respiration of variously aged hybrid poplar leaves chronically exposed to 0.085 (△) or 0.125 μl liter^{-1} (○) ozone. Control (●). (From Reich, 1983.)

fluoride-emitting aluminum smelter varied widely among species, with Scotch pine being much more sensitive than Norway spruce or Douglas-fir (Keller, 1973b). Keller (1977a) positioned seven species of conifers (Scotch pine, Austrian pine, eastern white pine, Japanese larch, Douglas-fir, Norway spruce, and European larch) and four broad-leaved species (red oak, gray alder, white beam mountain-ash, and sycamore maple) at various distances from a source of airborne F. Exposure to F caused foliar injury and abscission and reduced total photosynthesis primarily because of loss of leaves. However, the rate of photosynthesis of leaves remaining on the plants was as high as it was on unpolluted plants.

Particulates Substances such as cement kiln dusts, some fluorides, soot, magnesium oxide, iron oxide, foundry dusts, and sulfuric acid aerosols inhibit photosynthesis, often by plugging stomata and intercepting light. For example, cement and coal dust reduced photosynthesis of Norway spruce seedlings (Auclair, 1976) and coal dust reduced photosynthesis of Scotch pine and poplar (*Populus euramericana*) seedlings (Auclair, 1977). Dustlike fluoride compounds (NaF, CaF_2, and cryolite) severely reduced the rate of photosynthesis of Scotch pine and European white birch, often before visible injury was evident (Keller, 1973a).

Combined pollutants Synergistic photosynthetic responses of silver maple and white ash leaves were reported by Carlson (1979). Also a synergistic reduction of CO_2 absorption by Norway spruce due to 7.5 pphm SO_2 and increased tissue F levels resulting from an increase in F content in the rooting medium have been shown (Keller, 1980b). Photosynthetic rates of silver maple leaves were reduced by Cd, but the decrease was greater when 100 to 200 pphm SO_2 was present (Lamoreaux and Chaney, 1978). When 1-year-old silver maple seedlings were exposed for up to 60 days to 10 pphm SO_2, net photosynthesis decreased by 75% (Jensen, 1983).

Mechanisms of photosynthetic inhibition Air pollutants may alter the rate of photosynthesis by several mechanisms, including (1) clogging of stomatal pores; (2) influencing stomatal aperture; (3) altering optical properties of leaves by changing reflectance and decreasing light intensity; (4) altering the heat balance of leaves; (5) inhibiting the photosynthetic process through breakdown of chlorophyll as well as changing activity of carbon-fixing and chloroplast enzymes, phosphorylation rate, and pH buffering capacity; (6) disrupting integrity of membranes and ultrastructure of organelles; and (7) inducing changes in leaf anatomy. In the long term, total photosynthesis per plant is lowered by decreased leaf formation and expansion, necrosis, and abscission (Kozlowski and Constantinidou, 1986a).

Metabolism

Exposure of plants to air pollutants often results in disturbances in metabolic reactions involving carbohydrates, proteins, amino acids, lipids, and enzymes. Such changes may adversely affect growth of woody plants.

Air pollutants appear to be relatively nonspecific agents with several sites of action. SO_2 released into the air normally is oxidized by a series of chemical reactions. If SO_2 enters leaves through stomatal pores, the oxidation reactions occur in mesophyll tissues and often injure plants. If SO_2 is adsorbed on leaf surfaces, the oxidation of SO_2 may be influenced by the other chemicals present and by environmental conditions. When oxidation reactions proceed with SO_2 in the gas phase, it becomes SO_3 gas, which is short-lived and goes into solution with water to form H_2SO_4. SO_2 in air may also go directly into solution with liquid-phase water and form sulfate ions, which are then oxidized to sulfate (Winner *et al.*, 1985).

The impact of SO_2 on plants may be mediated through effects on cellular pH, oxidation–reduction reactions, osmotic alterations, and specific inhibition of key metabolic steps. Ozone affects many aspects of cell metabolism, including synthesis and concentration of metabolites across cell membranes, membrane integrity, and energy metabolism of cells (Kozlowski and Constantinidou, 1986a). Fluorides are toxic to plant cells in millimolar concentrations. The many metabolic events disrupted by fluorides eventually lead to tissue destruction (Chang, 1975).

Carbohydrates Because the major air pollutants inhibit photosynthesis they would be expected to decrease total carbohydrate reserves. However, in plants exposed to air pollutants some individual sugars increase in amount while others decrease (Table 9.7).

Exposure of seedlings to SO_2 reduced soluble sugars in pines (Wilkinson and Barnes, 1973); decreased the amount of starch, reducing sugars, and sucrose in green ash roots (Jensen, 1981); and reduced nonstructural carbohydrates in American elm seedlings (Kozlowski and Constantinidou, 1986b). Decreases in starch and increases in soluble sugars following exposure to O_3 have been attributed to inhibi-

Table 9.7
Effects of Air Pollutants on Carbohydrate Pools of Woody Plants[a]

Species	Dosage and pollutant	Effect
	Sulfur dioxide	
Jack pine	890–1340 $\mu g\ m^{-3}$, 96 hr	Increase in reducing sugars; decrease in nonreducing sugars
Norway spruce	10 $\mu g\ m^{-3}$ (average), 6 successive growing seasons	Decrease in sucrose and starch
American elm	200 pphm, 6 hr, assayed 24 hr after exposure	Decrease in nonstructural carbohydrates in all leaves, stems, and roots
	Ozone	
Eastern white pine Loblolly pine	10, 20, 30, 60, 100 pphm for 10 min, 7 or 21 days	Decrease in label of photoassimilated ^{14}C in soluble sugars
Sugar maple	20–30 pphm for 2 hr or 10 pphm, 2 hr day^{-1}, 14 days	No change in starch concentration
Lemon	25 pphm, 8 hr day^{-1}, 5 days $week^{-1}$, 9 weeks	Increase in reducing sugars; decrease in starch
Ponderosa pine	200 $\mu g\ m^{-3}$, 6 hr day^{-1}, up to 20 weeks	Increase in soluble sugars and starch in shoots, decrease in roots
Eastern white pine	5 pphm; 4, 9, 17, 22 weeks	Increase in soluble and reducing sugars in primary needles
	Fluoride	
European silver fir	450 μg F g^{-1} needle tissue; trees growing near an aluminum smelter	Decrease in fructose, glucose, sucrose, and raffinose; no effect on galactose

[a]From Kozlowski and Constantinidou (1986a).

Table 9.8
Effects of Air Pollutants on Amino Acid and Protein Pools of Woody Plants[a]

Species	Dosage and pollutant	Effect
	Sulfur dioxide	
Eastern white pine	132 μg m^{-3}, 11 weeks	Decrease in total proteins and protein-containing disulfide bonds
Norway spruce	10 μg m^{-3} (average), 6 successive growing seasons	Increase in amino acid content
European beech	0.075 μl $liter^{-1}$	Increase in alanine and cysteine; decrease in serine
	Ozone	
Ponderosa pine	200 μg m^{-3}, 6 hr day^{-1}, up to 20 weeks	No effect on protein and amino acid content of tops; increased levels in roots

[a]From Kozlowski and Constantinidou (1986a).

tion of starch synthesis as well as stimulation of starch hydrolysis (Malhotra and Sarkar, 1979). However, some investigators associated accumulation of soluble sugars in O_3-treated plants with reduced incorporation of glucose into cell walls, arrested translocation of sugars, and retention of sugars in shoots due to formation of glucosides that rendered sugars unavailable for growth or translocation to the roots (Tingey *et al.*, 1976).

Exposure of woody plants to increasingly higher dosages of F reduced total carbohydrates in European silver fir. The effect appeared to be mediated principally through reduction in photosynthesis by inactivation of enzymes of the pentose phosphate pathway, which regulate glucose synthesis. Fluorides may also inhibit enzymes involved in starch hydrolysis (Garrec *et al.*, 1981).

Proteins and amino acids Exposure of plants to SO_2 or O_3 usually results in a decrease in proteins and an increase in amino acids (Table 9.8). The increase in amino acids may result from stimulation of protein hydrolysis or acceleration of amino acid synthesis.

Exposure to SO_2 or O_3 decreased total protein content in American elm seedlings (Constantinidou and Kozlowski, 1979b) and eastern white pine seedlings (Percy and Riding, 1981). SO_2 caused shifts in a number of amino acids in jack pine seedlings. Alanine, valine, glycine, isoleucine, leucine, threonine, methionine, tyrosine, and lysine increased, whereas proline, serine, and phenylalanine decreased (Malhotra and Sarkar, 1979). These changes apparently involved increased conversion of amino acids into related ones.

Lipids Exposure to SO_2 inhibits synthesis of certain classes of lipids that are important constituents of cell membranes, leading to their disorganization (Malhotra and Khan, 1978). SO_2 decreased unsaturated fatty acids and increased saturated fatty acids in lodgepole pine needles (Khan and Malhotra, 1977). Ozone reacts not only with the proteinaceous sulfhydryl and aromatic groups of cell membranes but also with their unsaturated fatty acid moieties (Heath, 1980). However, the membrane proteins are more sensitive to O_3 attack than are the lipid components (Mudd *et al.*, 1984). In most cases the loss of lipids following exposure to O_3 is small. Fumigation of American elm seedlings with O_3 caused only insignificant changes in lipids (Constantinidou and Kozlowski, 1979b).

Enzymes Air pollutants stimulate activity of some enzymes and inhibit that of others. One day after exposure of red spruce seedlings to NO_2 began, nitrate reductase activity was greatly increased over that in control plants (Norby *et al.*, 1989). Peroxidase activity was increased following exposure of woody plants to SO_2 (Keller, 1977b), O_3 (Evans and Miller, 1972), and fluoride (Keller and Schwager, 1971). Peroxidases have been considered to be good indicators of "hidden" pollution injury. For example, fumigation of European silver fir seedlings with SO_2 increased peroxidase activity by as much as 70% even before the appearance of visible symptoms (Keller, 1977b). In contrast to its effect on peroxidase, SO_2 decreased activity of malic acid dehydrogenase (which catalyzes an important step in the oxidation of carbohydrates) and of acid phosphatase (which decomposes phosphate esters and releases inorganic phosphate in cells) (Sarkar and Malhotra, 1979; Malhotra and Khan, 1980).

Factors Affecting Responses to Pollution

Quantifying the effects of pollution on forest ecosystems is very difficult because of the complex nature of pollution stress, variations in stability and growth of forest stands, and influences of a variety of superimposed interacting plant and environmental factors. The responses of plants to pollution vary with plant species and genotype, pollutant dosage, types and combinations of pollutants, plant responses measured, developmental stages of plants, environmental regimes, and interactions of pollutants with plant diseases and insects (Kozlowski and Constantinidou, 1986b).

Injury has been assessed by different investigators as leaf necrosis, shoot dieback, tree mortality, or destruction of forest stands. On the basis of injury, plants have simply been classified by some investigators as sensitive, intermediate, or tolerant to pollution, and divided by other investigators into many classes of pollution tolerance. Injury to leaves has been reported as extent of leaf necrosis in broad-leaved trees, amount of tip necrosis in conifers, and leaf abscission. Growth reponses have been reported as height growth, cambial growth, root growth, and yield of harvested products.

Some investigators studied physiological and biochemical responses to pollution, with a view toward providing a basis for explaining development of pollution symptoms, explaining "hidden" injury, studying the mechanism of action of pollutants, and providing a rapid and sensitive assessment of plant response. Among the processes measured were rates of photosynthesis, respiration, membrane permeability, and enzymatic activity.

Investigators should be aware of limitations in use of certain plant responses as indicators of pollution effects. Use of height growth often is complicated by variations in growth patterns among species, clones, and cultivars. Seasonal duration of height growth of different species of woody plants varies from a few weeks to several months. Hence a pollution episode late in the summer may decrease height growth of species that continue to increase in height for much of the summer, but may not affect growth of species that complete height growth early.

Cambial growth often is reduced by pollutants without associated visible injury. However, in a short-term study, leaf injury occurred following exposure of seedlings of Japanese white birch seedlings to SO_2 although cambial growth was not significantly reduced (Tsukahara *et al.*, 1987). Nevertheless, at high dosages, pollutants are more likely to inhibit cambial growth in seedlings than in adult trees. Whereas leaf injury, which adversely affects photosynthesis, should eventually result in decreased xylem increment, such reduction often is not evident for a long time. Decreased xylem increment in the lower stem is preceded by inhibition of photosynthesis and synthesis of hormonal growth regulators, as well as by decreased transport of both to the lower stem.

Changes in dry weight or biomass of woody plants are often used to assess effects of pollution dosage or environmental stresses on growth (Constantinidou and Kozlowski, 1979a). However, relative growth rate may be more useful for comparing pollution tolerance of plants that vary in size (see Chapter 3).

Ratings of pollution tolerance of woody plants may also vary with experimental methods (Kozlowski, 1985c,d). Observations of effects of pollutants have been based on field studies of native plants or plants transplanted to a polluted area, plants exposed to pollutants in chambers in the laboratory or in the field, or plants artificially exposed to pollution in the field without use of chambers.

Foresters are interested in wood production per unit area of land. In contrast to the ease of establishing pollution as the cause of reduced forest productivity near point sources of pollution, it is much more difficult to obtain rigorous proof of regional reduction in productivity by specific air pollutants. Symptoms commonly induced by air pollutants (chlorosis, premature leaf shedding) resemble those caused by drought, freezing, disease, and some nutrient deficiencies. Furthermore, there is considerable genetic diversity within forests in susceptibility to air pollution, and productivity of forests is rather stable over a fairly wide range of tree densities. As mortality of sensitive trees is increased by air pollution, growth of the more resistant residual trees may be accelerated because of their improved competitive potential. Hence pollution-induced changes in growth or mortality of some trees in a forest may have little effect on total productivity, unless the injury is so severe that the forest is understocked (Woodman and Cowling, 1987).

To establish unequivocal proof of the effects of pollution on forest trees, Woodman and Cowling (1987) advocated adherence to adaptations of Koch's (1882) postulates, which involved the following rules:

1. The injury or dysfunction symptoms observed in individual trees in the forest should be associated consistently with the presence of the suspected causal factors.

2. The same injury or dysfunction symptoms must be seen when healthy trees are exposed to the suspected causal factors under controlled conditions.
3. Natural variations in resistance and susceptibility observed in forest trees must be seen when clones of the same species are exposed to the suspected causal factors under controlled conditions.

Woodman and Cowling (1987) emphasized the need for fulfilling the first two rules for drawing a firm inference about cause and effect. Fulfillment of the third rule increases the rigor of the second rule. After applying these rules to many published sets of data, Woodman (1987a) concluded that O_3 was the only regionally dispersed air pollutant that had been clearly shown to injure and kill forest trees over a wide geographic range. He presented evidence for O_3 injury to eastern white pine in various parts of its range and to ponderosa pine, Jeffrey pine, white fir, sugar pine, and incense cedar in southern California. Oxidant-type injury symptoms also have been shown in the central Appalachian Mountains on tulip poplar, green ash, hickories, black locust, table mountain pine, Virginia pine, pitch pine, and eastern hemlock (Duchelle *et al.*, 1982). Sulfur dioxide has injured forests at distances of over 100 km from point sources, but on a regional scale it is more important as a precursor pollutant that leads to formation of acid rain (Linzon, 1986).

Considerable circumstantial evidence links airborne chemicals to regional injury to forests. Examples are decline of red spruce; reduced growth of pitch pine, shortleaf pine, loblolly pine, and eastern white pine in New Jersey; sugar maple decline; and growth inhibition in pine forests of the southeastern Piedmont Plateau. Nevertheless, Woodman (1987a) considered that even in these cases the evidence was insufficient to prove that the observed changes in forest health were caused by air pollution alone.

The selection of meaningful plant responses should be carefully determined by investigators on the basis of research objectives, types and doses of pollutants, and age of plants, as well as available facilities, time, and funds. Data of various investigators on pollution tolerance of different plants have been more consistent when based on injury than on some aspect of growth or physiological response. Growth responses have differed appreciably because of variations in species growth characteristics, age of plants, experimental conditions, and lack of uniform methods for measuring growth. Use of physiological processes as response parameters is complicated by their extreme sensitivity to stress other than pollution and the difficulty in relating rates of physiological processes to growth. For example, both high and low positive correlations as well as negative correlations between rates of photosynthesis and growth of woody plants have been demonstrated (Kramer and Kozlowski, 1979, pp. 179–180). Selection of the appropriate plant process to measure for evaluating pollution tolerance is another problem.

Species and Genotype

The tolerance of woody plants to pollution varies greatly among species and genetic materials. Several lists of pollution tolerance (based on visible injury) to various air pollutants are available (e.g., Davis and Gerhold, 1976; Smith, 1981; Last, 1982). However, the pollution tolerance of a given species may be ranked somewhat differently by various investigators because the criteria for derivation of the rankings, as well as conditions under which plant responses were observed, have varied greatly. Rankings of species at the extremes (very sensitive or very tolerant to a pollutant) generally are more useful than those for species in an intermediate class. Davis and Gerhold (1976) emphasized that species shown to be very tolerant or very sensitive to an air pollutant in a growth chamber usually also are very tolerant or sensitive, respectively, to the same air pollutant in the field. By comparison, sensitivity of species listed as intermediate in pollution tolerance may differ greatly.

Several investigators found considerable variation in within-species tolerance to air pollution. Differences in pollution tolerance were detected among clones of trembling aspen after exposure to SO_2 or O_3 (Karnosky, 1976, 1977; Kimmerer and Kozlowski, 1981) and among clones of eastern white pine exposed to SO_2 (Genys and Heggestad, 1978). Studies on genotypic and environmental variations in responses of Norway spruce families to HF traced most of the interfamily differences to genetic effects (Scholz *et al.*, 1980). In some species, interpopulation differences in pollution tolerance are comparable to differences among species (Mejnartowicz, 1984).

Mechanisms of Pollution Tolerance

Differences in pollution tolerance are traceable largely to variations in avoidance of uptake of pollutants by leaves or in biochemical tolerance of pollutants. In addition variations in capacity to metabolize pollutants into less toxic substances and in dilution of pollutants by rapid redistribution in plants also may be involved in tolerance mechanisms.

The rate of absorption of air pollutants depends on the concentration gradient from the leaf exterior to the leaf interior as well as the resistance of gaseous flow along the diffusion pathway. Hence, the effects of air pollutants often vary appreciably with differences in stomatal conductance, which is a function of stomatal size, stomatal frequency, and control of stomatal aperture. Stomatal conductance varies widely among species of plants (Reich, 1987) as well as clones and cultivars (Siwecki and Kozlowski, 1973; Pallardy and Kozlowski, 1979d). For a given plant, stomatal conductance also varies diurnally and seasonally.

Variations in pollution tolerance of different species and genotypes often are closely related to stomatal conductance. For example, more SO_2 was absorbed by white ash leaves, with large stomata (high stomatal conductance), than by leaves of sugar maple, with small stomata (low stomatal conductance) (Jensen and Kozlowski, 1975). Leaf conductance values of 10 species of trees and shrubs in California provided a good index of SO_2 uptake and tolerance (Winner *et al.*, 1982). River birch seedlings were more sensitive than paper birch seedlings to SO_2, partly because of the greater stomatal conductance of the former (Norby and Kozlowski, 1983). Susceptibility of some poplar clones to SO_2 was correlated with stomatal conductance (Kimmerer and Kozlowski, 1981), further emphasizing the importance of stomatal characteristics to pollution tolerance.

Reich (1987) observed a similar linear decline in photosynthesis and growth of both broad-leaved trees and conifers in response to O_3 dosage at ambient or near ambient levels. The broad-leaved trees were more sensitive than conifers to O_3 at equivalent total doses. This was because stomatal conductance of the conifers, and hence O_3 uptake, was lower. When the amount of O_3 absorbed was used as a basis for comparison, the sensitivity of both groups differed only slightly.

Certain pollutants themselves cause stomatal opening or closing and thus affect the amounts of pollutants absorbed. Several studies show that stomatal opening occurs at low concentrations of air pollutants, below the threshold for effects on photosynthesis, and closure occurs at injurious concentrations, the latter response often following inhibition of photosynthesis (Darrall, 1989). Low dosages of SO_2 induced stomatal opening in American elm seedlings; higher dosages caused stomatal closure (Noland and Kozlowski, 1979). The initial opening effect of SO_2 probably is related to reduced turgor of subsidiary cells. Stomatal closure at high SO_2 dosages may be associated with accumulation of CO_2 in substomatal cavities following SO_2 inhibition of photosynthesis, or changes in permeability of guard cell membranes. When European white birch trees were exposed to 7 pphm SO_2, photosynthesis was reduced by 19%. It appeared that stomatal closure induced by SO_2 provided a mechanism for avoiding injury without major interference with the CO_2 supply (Mansfield and Freer-Smith, 1984). Reich (1987) emphasized, however, that stomatal conductance varies much more among species of unpolluted plants than it does in a given species in response to low concentrations of O_3.

Some investigators emphasized the importance of biochemical tolerance of pollutants. For example, some Norway spruce clones fixed more S in organic fractions and maintained higher buffering capacity following exposure to SO_2 than did SO_2-susceptible clones (Braun, 1977a,b). Important seasonal differences were identified among clones in their capacity to produce thiols (cysteine, glutathione) after exposure to SO_2 (Schindlbeck, 1977). More SO_2 was absorbed by grafted clones of SO_2-tolerant eastern white pine trees than by grafted clones of susceptible trees (Roberts, 1976). Activation of superoxide dismutase, catalase, and peroxidase enzymes, which could protect cells from O_3-induced oxidation of unsaturated fatty acids of membranes, provided a mitigating mechanism in some O_3-tolerant species (Bennett *et al.*, 1984). Differences were found in enzymatic and isoenzymatic patterns in trees of varying pollution tolerance and were attributed to liberation from disintegrated organs and/or to inactivation of some alleles by pollutants (Mejnartowicz, 1984).

Pollutant Dosage

Accurate information on the amounts of individual and combined pollutants that injure plants and decrease their growth is needed for coping with the pollution problem. With a view toward establishing air quality standards, extensive data on critical dosages for various plant species have been obtained, primarily from two types of studies: (1) exposure of plants to known concentrations of pollutants in controlled environments for specified periods and (2) evaluation of plant responses at varying distances from pollution sources such as industrial plants.

Many investigators have reported threshold dosages (concentration × duration of exposure) of air pollutants that induce foliar injury (Table 9.9). A number of species are sensitive to SO_2 and O_3 at concentrations below air quality standards in the United States. The critical dosage of SO_2 for acute injury may vary from 70 pphm for 1 hr to 18 pphm for 8 hr of exposure, or to many weeks at lower concentrations (10 to 20 pphm) (Environmental Protection Agency, 1984a). However, leaves may become chlorotic when the average SO_2 concentration is in the range of 8 to 17 pphm for a whole growing season during which the SO_2 concentration fluctuates (Linzon, 1978). Threshold concentrations of O_3 for broad-leaved trees range from 20 to 30 pphm for 2 to 4 hr (National Academy of Sciences, 1977b). The critical dosage for NO_x has been reported as 160 to 260 pphm for up to 48 hr. For 1-hr exposures 2000 ppm was critical; for 20-hr exposures, 100 pphm was critical (National Academy of Sciences, 1977a). Threshold concentrations of fluorides are 10 to 1000 times lower than those for SO_2, O_3, or NO_x (Weinstein, 1977).

Several studies showed that exposure of woody plants to very low concentrations of O_3 for prolonged periods adversely affected them. For example, fumigation with low levels of O_3 (<0.15 μl liter^{-1}) for several weeks reduced photosynthesis and growth in hybrid poplar, sugar maple, and red oak seedlings (Reich, 1983;

Table 9.9

Injury Responses to Various Air Pollutants and Threshold Doses[a]

Pollutant	Symptoms	Threshold dose
Sulfur dioxide	Angiosperms: interveinal necrotic blotches Gymnosperms: red-brown dieback or banding	70 pphm (1820 μg m^{-3}) for 1 hr; 18 pphm (468 μg m^{-3}) for 8 hr; 0.8–1.7 pphm (21–44 μg m^{-3}) for growing season
Nitrogen dioxide	Angiosperms: interveinal necrotic blotches similar to SO_2 injury Gymnosperms: red-brown distal necrosis	2000 pphm (38 × 10^3 μg m^{-3}) for 1 hr; 160–260 pphm (3000–5000 μg m^{-3}) for 48 hr; 100 pphm (1900 μg m^{-3}) for 100 hr
Ozone	Angiosperms: upper surface flecks Gymnosperms: distal necrosis, stunted needles	20–30 pphm (392–588 μg m^{-3}) for 2–4 hr; some conifers 8 pphm (157 μg m^{-3}) for 12–13 hr
Peroxyacetylnitrate	Angiosperms: lower surface bronzing Gymnosperms: chlorosis, early senescence	20–80 pphm (989–3958 μg m^{-3}) for 8 hr
Fluoride	Angiosperms: tip and margin necrosis Gymnosperms: distal necrosis	<100 μg g^{-1} fluoride, dry wt. basis

[a] From Smith (1981).

Reich *et al.*, 1984, 1985, 1986; Reich and Amundson, 1985). Low O_3 concentrations in ambient air had a more negative impact than similar concentrations of SO_2. Because elevated levels of O_3 occur more often and over much larger areas, Reich *et al.* (1984) concluded that current levels of O_3, but not SO_2, are toxic to woody plants.

There may be considerable lag response by plants to low concentrations of O_3. Young hybrid poplar cuttings did not show responses to low O_3 levels (<0.15 μl liter^{-1}) during the first 6 weeks of exposure. During the next 4 weeks, however, O_3 reduced height and diameter growth and increased leaf senescence (Reich and Lassoie, 1985).

Identifying precise dosages of pollutants that adversely affect plants is difficult because plant responses to pollutants differ with methods of evaluation, environmental conditions, the plant response observed or measured, frequency of pollutant exposures, length of time between exposures, extent of fluctuations in concentrations, time of day of exposures, their sequence and pattern, and total flux of pollutants as affected by canopy characteristics and leaf boundary layers.

Adding constant amounts of pollutants into chambers for set periods results in very artificial exposure patterns (Jacobson, 1982). Nevertheless, experiments in controlled-environment chambers may be very useful in studies designed to show mechanisms of air pollution effects. In the field the washing of pollutants off plant tissues during emission-free periods results in reduction of dosage and alleviation of injury, except for O_3 (Halbwachs, 1984). Another problem in the field is that the response of various species of trees in a forest stand may differ greatly from predictions based on responses determined without plant competition. Still another problem is that the dose to which a plant is exposed (ambient dose) is not a direct measure of the effective dose (that which induces a physiological response). The effective dose is a function of the rate at which pollutant molecules arrive at affected cells in the leaf interior. This rate is regulated primarily by the gas-to-liquid pathway, which varies appreciably with the environmental regime and genotype (Taylor *et al.*, 1982). In recognizing such difficulties, Last (1982) concluded that if SO_2 were the only atmospheric pollutant present, the critical SO_2 concentration might vary for different plant responses as follows: possibly beneficial, 50 g m^{-3}; beneficial in some instances, detrimental in others, 50 to 100 g m^{-3}; usually inducing a decrease in growth, 100 to 200 g m^{-3}; significantly decreasing growth within 1 to 3 months, 200 to 400 g m^{-3}.

Combinations of Pollutants

Only rarely do air pollutants exist singly in the atmosphere. The effects of combined pollutants may be synergistic, additive, or antagonistic (Kozlowski and Constantinidou, 1986b). Mechanisms of synergism and antagonism are complex and may be related to direct reactivity between pollutants, effects of either or both pollutants on photosynthesis and/or stomatal aperture, competition for reaction sites, changes in sensitivity of reaction sites to pollutants, or combinations of these (Heagle and

Johnston, 1979). The extent of injury by mixtures of gaseous pollutants varies with the concentration of each gas in the mixture with respect to the injury thresholds of the individual pollutants, the relative proportion of the gases, whether there is simultaneous or intermittent application of the combined pollutant stress, and the age and physiological condition of the exposed plant tissues (Reinert *et al.*, 1975).

Most studies of effects of pollutant mixtures have been conducted with SO_2 and O_3. Constantinidou and Kozlowski (1979a) demonstrated variations in susceptibility of seedlings of American elm to SO_2 and O_3 and a greater effect on growth of these combined pollutants compared with either pollutant alone (Fig. 9.7). For example, injury to leaves from SO_2–O_3 mixtures was evident within 24 hr; from SO_2 alone within 20 to 48 hr; and from O_3 alone from 36 to 48 hr after fumigation ceased. Expansion of young leaves and dry weight increase of seedlings were reduced more by SO_2–O_3 mixtures than by either pollutant alone. As indicated by expansion of leaves that emerged after fumigation, actively growing seedlings began recovering from the acute pollutant poisoning. In contrast, quiescent seedlings responded to the SO_2–O_3 mixture by accelerated leaf abscission. It appeared that the recovery mechanisms of the rapidly metabolizing seedlings were not operative in quiescent seedlings, and this led to leaf abscission. Other examples of responses of woody plants to mixtures of pollutants are given in Table 9.10.

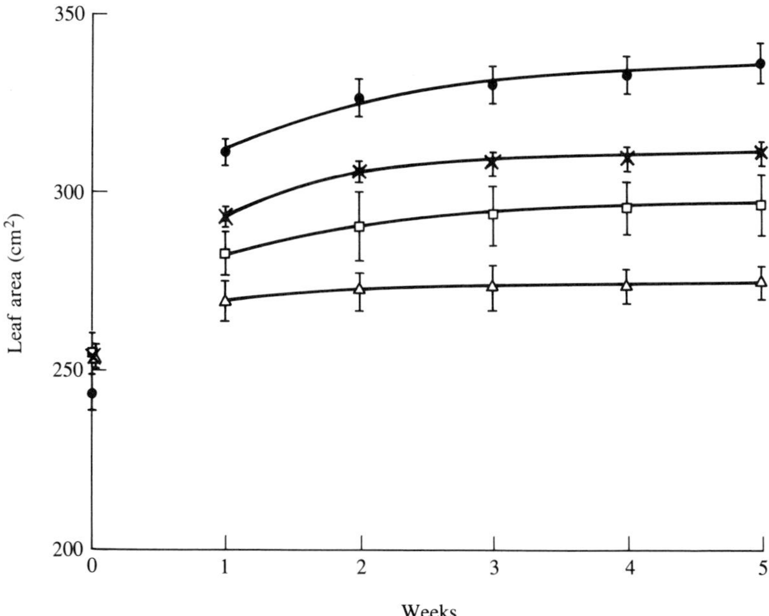

Figure 9.7 Effects of SO_2 (□; 2 ppm for 6 hr), O_3 (x; 0.9 ppm for 5 hr), and SO_2–O_3 mixtures (△; 2 ppm SO_2 and 0.9 ppm O_3 for 5 hr followed by SO_2 for 1 hr) on leaf growth of American elm seedlings. Control (●). (From Constantinidou and Kozlowski, 1979a.)

Table 9.10
Responses of Woody Plants to Pollutant Mixtures

Species	Response measured	Type of response	Source
	$SO_2 + O_3$		
Eastern white pine	Foliar injury	Synergistic	Dochinger *et al.* (1970)
Scotch pine	Foliar injury	Less than additive	Nielsen *et al.* (1977)
American elm	Leaf formation	Additive	Constantinidou and Kozlowski (1979)
	Root growth	Additive	
Apple	Shoot elongation	Synergistic	Kender and Spierings (1975)
	Foliar injury	Synergistic	
	$SO_2 + NO_2$		
European white birch	Dry weight increase	Synergistic	Freer-Smith (1984)
Hairy birch	Dry weight increase	Synergistic	Freer-Smith (1984)
Black poplar	Dry weight increase	Synergistic	Freer-Smith (1984)
Alder	Dry weight increase	Synergistic	Freer-Smith (1984)
	$SO_2 + HF$		
Sweet orange	Foliar injury	Additive	Matsushima and Brewer (1972)
	Shoot elongation	Additive	
	$O_3 + PAN$		
Hybrid poplar	Foliar injury	Synergistic	Kohut *et al.* (1976)
Ponderosa pine	Foliar injury	Less than additive	Davis (1977)

Antagonistic responses of woody plants to SO_2–O_3 mixtures were shown for eastern white pine ramets. Lesion development and total injury to new leaves were less after a 2-hr simultaneous fumigation with 5 pphm of each pollutant than with either SO_2 or O_3 alone (Costonis, 1973). An antagonistic relation in which SO_2 reduced the toxic effects of O_3 on leaf growth also was reported for *Populus deltoides* × *P. trichocarpa* (Noble and Jensen, 1980).

As the complexity of pollutants in the environment increases, entire forest ecosystems can be adversely affected. McClenahen (1979) studied community changes in a deciduous forest spread over 50 km of the upper Ohio River Valley along gradients of chronic exposure to airborne chloride, SO_2, fluorides, and possibly other pollutants. Species diversity decreased in all except the shrub layer. Sugar maple was sensitive in all strata, but yellow buckeye appeared to be tolerant to the mixture of pollutants. Hence, stand composition changed toward predominance of species that were better suited to the polluted environment, but of lower timber value than the species they replaced. Hutchinson and Whitby (1977) reported that SO_2, Ni, Cu, Fe, and Co emitted from a smelter in Ontario almost totally destroyed a mixed deciduous forest over an area of 100 square miles.

Age and Stage of Plant Development

Very young seedlings usually are more susceptible than old seedlings or adult trees to air pollution. Seedlings in the cotyledon stage of development often grow at threshold levels of available carbohydrates, hormones, and mineral nutrients (Kozlowski, 1976b) and are especially sensitive to air pollution. When red pine seedlings in the cotyledon stage of development were fumigated with SO_2, the adverse effects were evident early. SO_2 caused chlorophyll breakdown, inhibited expansion of primary needles and dry weight increase of seedlings, and caused death of needle tips. Seedlings fumigated with SO_2 with 1.3 or 4 ppm SO_2 for only 15 min eventually became chlorotic (Constantinidou *et al.*, 1976). Conifers in the cotyledon and primary needle stages appear to be more sensitive to air pollution than those with secondary needles. Despite the high sensitivity of young seedlings to air pollutants, the older trees in forests around point sources of pollution often are killed faster than young trees, probably because the crown canopy serves as a filter and the young trees are exposed to less of the pollutant. Furthermore, the low stomatal conductance of the shaded, understory plants results in low rates of absorption of gaseous pollutants.

In conifer seedlings the immature secondary needles are very susceptible to pollution (Treshow and Pack, 1970), the elongating needles being sensitive to F but the mature needles very resistant. However, latent responses sometimes occur. For example, necrosis may occur on fully elongated pine needles after they accumulate enough F to be injured when other environmental stresses such as drought or high temperature occur late in the growing season (Weinstein, 1977).

Sensitivity of broad-leaved species to air pollutants varies with leaf age. In cotton, neither very young nor very old leaves are as sensitive to O_3 as leaves that are at least three-fourths expanded (when the amount of intercellular space in the palisade layer is high). O_3 injury of cotton leaves at the susceptible stage was associated with depletion of carbohydrate and amino acid reserves (Ting and Mukerji, 1971). Differences in sensitivity of young and old leaves may also be attributed to variations in stomatal activity.

Environmental Factors

The effects of air pollutants on plants are influenced by prevailing environmental conditions as well as those occurring before and after a pollution episode. Environmental influences on responses to pollutants may be mediated by regulation of stomatal aperture, plant metabolism, and rate of plant development (Kozlowski and Constantinidou, 1986b).

Prevailing Environmental Conditions

Several environmental factors such as soil moisture supply, air humidity, and temperature influence responses of plants to air pollution, largely by regulating stomatal aperture and hence the amount of pollutant absorbed. More gaseous air pollutants are absorbed when the soil is wet than when it is dry because the stomata are more open (Chapter 7). Witholding irrigation from *Diplacus aurianticus* plants induced stomatal closure and decreased SO_2 avoidance. An irrigation treatment may be likened to an unseasonal rain that increases SO_2 absorption (Atkinson *et al.*, 1988).

The stomatal pores of many species of woody plants open or close in response to high or low air humidity (Chapter 7), even though the water content of leaves does not change appreciably (Sena Gomes *et al.*, 1987). For example, the stomata of paper birch seedlings were more open at high humidity than at low humidity and responded rapidly to changes in vapor pressure deficit. The seedlings absorbed more SO_2 and were more sensitive to SO_2 when fumigated at high humidity, as shown by more leaf necrosis and abscission as well as by greater inhibition of growth, than when fumigated at low humidity (Norby and Kozlowski, 1982).

Prevailing air temperature often influences plant responses to air pollution by altering mechanisms of pollutant avoidance, pollutant tolerance, or both. Seedlings of paper birch, white ash, and red pine exposed to SO_2 at 30°C had wider open stomata and absorbed more S than seedlings fumigated at 12°C. The effect was modified somewhat in birch and ash by partial stomatal closure at 12°C in response to the presence of SO_2 (Norby and Kozlowski, 1981b).

Prefumigation and Postfumigation Environment

Environmental conditions before or after a pollution episode often influence plant responses to pollution. Hence, in assessing the effects of environmental factors on pollutant toxicity, investigators should consider the influence of not only prevailing environmental conditions but also of pre- and postfumigation environmental regimes.

Exposure of plants to prefumigation stresses such as drought, flooding, or low temperature affects responses to gaseous pollutants, largely by affecting stomatal aperture and the amount of pollutant absorbed. For example, green ash seedlings grown for 4 weeks at 15°C had fewer leaves and more closed stomata than those grown at 25°C. When both groups were exposed to SO_2 at 22°C, the seedlings preconditioned at 15°C absorbed less S and were injured less (Shanklin and Kozlowski, 1984).

Flooding of soil prior to a pollution episode induces stomatal closure (Chapter 8) and hence decreases absorption of gaseous pollutants. For example, SO_2 reduced growth more in unflooded bald cypress seedlings than in those that had been flooded for 8 weeks (Shanklin and Kozlowski, 1985a). Similarly, flooding of soil induced stomatal closure and reduced SO_2 uptake in river birch seedlings (Norby and Kozlowski, 1983).

The influences of postfumigation environmental regimes on responses to air pollutants often are mediated through effects on metabolism. At high postfumigation temperatures (22 or 32°C), red pine seedlings that previously had been exposed to SO_2 recovered partially and formed new needles that provided an additional source of photosynthate, thereby reducing growth inhibition and decreasing the proportion of injured needles. At a low postfumigation temperature (12°C), the shoots grew slowly and did not replace the injured needles (Norby and Kozlowski, 1981a).

Interactions of Air Pollutants with Disease and Insects

The effects of air pollutants on host–disease and host–insect interactions are rather complex. Air pollutants may decrease or increase the severity of diseases depending on whether the effects of the pollutant are exerted primarily on the pathogen or on the host (Treshow, 1975). Disease severity commonly is influenced by changes in metabolism of the host plant. Pollutants may also alter the habitat for the pathogen through their effects on the structure and microenvironment of plant surfaces (Saunders, 1971, 1973). In addition, pollutants may affect disease severity by impeding mycorrhizal development and thereby influencing absorption of mineral nutrients by the host (Parmeter and Cobb, 1972).

A prevailing view is that by reducing tree vigor, air pollutants, even at chronic levels, predispose trees to certain diseases, particularly those caused by nonobligate pathogens. For example, increased injury by *Armillaria mellea* was associated with trees weakened by air pollutants (Sinclair, 1969). Trees weakened by SO_2 also were susceptible to needle blight caused by *Rhizosphaera kalkoffii* (Chiba and Tanaka, 1968) and to wood rots induced by *Trametes serialis* and *T. heteromorpha* (Jancarik, 1961). Ozone-induced injury favored infection by *Lophodermium pinastri* (Costonis and Sinclair, 1972) and *Fomes annosus* (James *et al.*, 1980).

Single or combined air pollutants may also influence the impact of disease by acting directly on the pathogen. In most cases, disease severity is reduced because of the greater toxicity of pollutants to the pathogen than to the host, especially at high concentrations of pollutants. For example, near a smelter emitting SO_2, rust fungi were absent on trees in the region of greatest SO_2 injury, and abundant where injury was negligible (Scheffer and Hedgcock, 1955). Heart rot and blister rust symptoms were reduced near a SO_2-producing smelter (Linzon, 1958, 1966).

The effects of pollutants on insect injury are exerted largely on insects attacking weak or injured trees. For example, bark beetles attacked weakened ponderosa pine trees more often than they attacked vigorous trees (Larsson *et al.*, 1983). Increased infestations of primary insect pests (those attacking vigorous trees) also have been reported in polluted areas. Such attacks have been attributed to suppression of insect predators of pests (Wentzel and Ohnesorge, 1961). Increased infestations of primary insects in polluted trees may also involve changes in water balance of the host (Templin, 1962) and other changes in the host that attract insects (Huttunen, 1984). For example, emission of terpenes, which serve as chemical messengers for many pests of conifers, increased after balsam fir trees were exposed to SO_2 (Renwick and Potter, 1981). By comparison, the ethane produced by SO_2-injured tissues

(Kimmerer and Kozlowski, 1982) may repel some wood-boring beetles (Sumimoto *et al.*, 1975).

Summary

The important air pollutants that affect woody plants include sulfur dioxide (SO_2), ozone (O_3), fluoride, oxides of nitrogen (NO_x), peroxyacetylnitrate (PAN), and particulates (e.g., cement kiln dusts, fluorides, soot, and heavy metals). Primary pollutants (e.g., SO_2, HF) originate in a toxic form at the source of emission; secondary pollutants (e.g., O_3) are produced by interactions between primary pollutants.

Various pollutants, alone or in combination, reduce plant growth, especially near sources of emissions. Pollutants decrease leaf area by inhibiting leaf formation and expansion, injuring leaves, and hastening abscission. Pollutants also decrease height growth, cambial growth, and reproductive growth of woody plants. Yield of fruits and seeds may be decreased by reduction in physiological efficiency of foliage and direct injury to reproductive organs. Reduction in seed production often is associated with inhibitory effects of pollutants on germination of pollen and growth of pollen tubes.

Responses of forests vary from total destruction under very high pollution dosages to negligible changes under low dosages. Pollutants may alter ecosystem structure by arresting plant succession and causing reduction in structural complexity, biomass, productivity, and species diversity.

Effects of pollutants on growth are mediated by changes in physiological processes and metabolic pools. Pollutants inhibit chlorophyll synthesis and photosynthesis and alter stomatal aperture, permeability of cell membrane, amounts of stored carbohydrates and proteins, and activity of enzymes.

Quantifying the effects of pollutants is complicated by variations in stability and growth of forest stands and influences of interacting plant and environmental factors. Responses to pollution vary with plant species and genotype, pollutant dosage, types and combinations of pollutants, plant responses measured, developmental stages of plants, environmental regimes, and interactions of pollutants with plant diseases and insects.

Differences in pollution tolerance among plant species and genotypes are associated with variations in avoidance of uptake of pollutants or in biochemical tolerance of pollutants, with the former more important. Variations among plants in pollutant tolerance are often closely related to stomatal conductance.

Identifying precise dosages of pollutants that adversely affect various species and genotypes is complicated because plant responses differ with methods of evaluation, environmental conditions, plant responses measured, frequency of pollutant exposures, time between exposures, fluctuations in pollutant concentrations, time of day of exposures as well as their sequence, and total flux of pollutants as influenced by canopy characteristics and leaf boundary layers.

General References

Adams, D. D., and Page, W. P., eds. (1985). "Acid Deposition." Plenum Press, New York.

Guderian, R. (1977). "Air Pollution: Phytotoxicity of Acidic Gases and Its Significance in Air Pollution Control." Springer-Verlag, Berlin.

Heath, R. L. (1980). Initial events in injury to plants by air pollutants. *Ann. Rev. Plant Physiol.* **31,** 395–431.

Hutchinson, T. C., and Meema, K. M. (1985). "Effects of Atmospheric Pollutants on Forests, Wetlands, and Agricultural Ecosystems." Springer-Verlag, Heidelberg and New York.

Koziol, M. J., and Whatley, F. R., eds. (1984). "Gaseous Air Pollutants and Plant Metabolism." Butterworth, London.

Kozlowski, T. T. (1980). Impacts of air pollution on forest ecosystems. *BioScience* **30,** 88–93.

MacKenzie, J. J., and El-Ashry, M. T. (1989). "Air Pollutant's Toll on Forests and Crops." Yale University Press, New Haven, CT.

Mansfield, T. A., ed. (1976). "Effects of Air Pollution on Plants." Cambridge University Press, New York.

Mansfield, T. A., and Freer-Smith, P. H. (1981). Effects of urban air pollution on plant growth. *Biol. Rev.* **56,** 343–368.

Miller, P. R., ed. (1980). Effects of Air Pollutants on Mediterranean and Temperate Forest Ecosystems. U.S. Forest Service Gen. Tech. Rept: PSW–43. Berkeley, CA.

Mellanby, K. (1980). "The Biology of Pollution." Arnold, London.

Moriarty, F. (1983). "Ecotoxicology." Academic Press, London.

Mudd, J. B., and Kozlowski, T. T., eds. (1975). "Responses of Plants to Air Pollution." Academic Press, New York.

National Academy of Sciences (1977). "Ozone and Other Photochemical Oxidants." National Academy of Sciences, Washington, D.C.

National Academy of Sciences (1977). "Nitrogen Oxides." National Academy of Sciences, Washington, D.C.

Nurnberg, H. W., ed. (1985). "Pollutants and Their Ecological Significance." Wiley, Chichester.

Reygens, J. L., and Rycroft, R. W. (1985). "The Acid Rain Controversy." University of Pittsburgh Press, Pittsburgh, PA.

Schulte-Hostede, S., Darrall, N. M., Blank, L. W., and Wellburn, A. R., eds. (1988). "Air Pollution and Plant Metabolism." Elsevier, London and New York.

Sheehan, P. J., ed. (1984). "Effects of Pollutants at the Ecosystem Level." Wiley, Chichester.

Smith, W. H. (1990). "Air Pollution and Forests," 2d ed. Springer-Verlag, New York.

Ulrich, B., and Pankrath, J., eds. (1982). "Effects of Accumulation of Air Pollutants in Forest Ecosystems." Reidel, Dordrecht, Holland.

Unsworth, M. H., and Ormrod, D. P., eds. (1982). "Effects of Gaseous Air Pollution on Agriculture and Horticulture." Butterworth, London.

Winner, W. E., Mooney, H. A., and Goldstein, R. A. (1985). "Sulfur Dioxide and Vegetation. Physiology, Ecology, and Policy Issues." Stanford University Press, Palo Alto, CA.

Chapter 10

Carbon Dioxide

Introduction

Forty to fifty percent of the dry weight of plants is carbon, which comes from atmospheric CO_2 converted into carbohydrates by the process of photosynthesis. However, the atmosphere contains only about 0.035% of CO_2 by volume (350 ppm or 350 μl liter^{-1}) and this low concentration often limits the rate of photosynthesis. Thus it would seem that the prospect of a substantial increase in concentration of atmospheric CO_2 should be welcomed by farmers, foresters, and others interested in plants and plant yields. However, this increase also is a source of concern because, in addition to its direct "fertilization" effects on plants, it may cause an increase in temperature, the "greenhouse effect," and a change in rainfall patterns that indirectly will affect plant growth and human activities unfavorably (Lemon, 1983; Hoffman, 1984; Crane, 1985; Strain and Cure, 1985).

Foresters, farmers, horticulturists, and gardeners are accustomed to coping with drought, flooded soil, mineral deficiencies, and unfavorable temperatures, but the problems resulting from an increase in CO_2 concentration of the atmosphere were recognized only recently. Older textbooks gave the CO_2 concentration of the atmosphere as about 300 ppm by volume and did not mention any change in concentration. However, analyses of air trapped in ice indicate that the concentration has been increasing for many centuries. The rate of increase began to accelerate with increased burning of fossil fuels in the nineteenth century and rose from about 260 or 270 ppm in the middle of the nineteenth century to 300 ppm early in this century. The most accurate measurements of worldwide concentration of CO_2, made at the Mauna Loa observatory in Hawaii, gave a concentration of 315 ppm in 1958 and 350 ppm in 1988 (Fig. 10.1), and a concentration of 600 ppm is predicted by 2050 (Strain, 1987). Thus most of the trees growing today developed during a period of increasing CO_2 concentration and those growing during the next half century will be subjected to even higher concentrations. La Marche *et al.* (1985) reported that tree ring data indicate an increase in diameter growth of subalpine conifers in the western United States since the middle nineteenth century, which they attributed to the increasing concentration of CO_2. An increase in CO_2 can affect growth and yield directly through effects on physiological processes and indirectly through their effects on such climatic factors as temperature and rainfall, which in turn affect physiological processes.

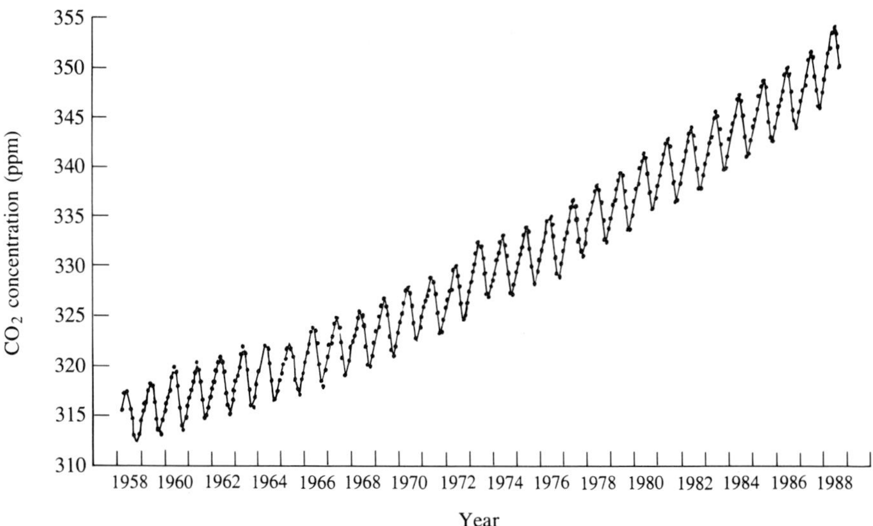

Figure 10.1 Increase in atmospheric concentration of CO_2 at the Mauna Loa observatory in Hawaii and the seasonal cycle in concentration each year since 1958. (From Keeling, 1986; updated courtesy of C. D. Keeling, 1988.)

To develop a better understanding of how increasing CO_2 concentration is affecting woody and herbaceous vegetation, we will discuss the effects of above ambient CO_2 concentration on plant growth and the physiological processes involved, as well as the interaction between CO_2 concentration and environmental factors such as light, water, and mineral nutrient supply. Finally, we will discuss the effects of CO_2 concentration on competition and the development of plant communities.

Effects of CO_2 Concentration on Growth

Much research has been done on the effects of CO_2 concentration on the growth of herbaceous plants, and greenhouse crops such as lettuce, tomatoes, roses, and other flowers have been "fertilized" with CO_2 for many years (Miller, 1938, pp. 585–587). Kimball (1983, 1986) surveyed the literature and collected observations on 38 species of crop plants, which led him to conclude that the marketable yield of herbaceous plants could be increased an average of 33% by doubling the ambient CO_2 concentration if other conditions were favorable. Idso (1984) also expected increases in yield, but Waggoner (1984) expected the beneficial effects of high CO_2 to be nullified by less favorable climatic conditions. The data for woody plants are more limited and naturally are based largely on experiments with seedlings that lasted only a few months or at most a few years.

Some relatively long-term experiments were performed on seedlings of loblolly pine and sweet gum in the Duke University phytotron (Sionit *et al.*, 1985). Seedlings were grown from seed in controlled environment greenhouses in 350, 500, and 650 ppm of CO_2 with natural photoperiods and light, and with temperature regimes following those out-of-doors. At the end of the growing season the total dry weights

Table 10.1
Effects of Atmospheric CO_2 Enrichment on Growth of Sweet Gum and Loblolly Pine at 32 Weeks after Planting (End of First Growing Season)[a,b]

CO_2 concentration (ppm)	Stem weight (gm)	Leaf weight (gm)	Root weight (gm)	Root–shoot ratio	Total weight (gm)	Percentage increase in total weight
			Sweet Gum			
350	2.1a	1.5a	6.3a	1.7ab	9.9a	
500	3.4b	1.6ab	9.2b	1.8b	14.2b	43.4
650	3.7b	2.0b	8.6ab	1.5a	14.3b	44.4
			Loblolly Pine			
350	0.4a	1.4a	2.1a	1.2a	3.9a	
500	1.0b	2.3b	2.9b	0.9a	6.2b	58.9
650	0.8b	2.1b	3.2b	1.1a	6.1b	56.4

[a]From data of Sionit *et al.* (1985).

[b]Means followed by the same letter within a column for each species are not significantly different at the 5% level of probability. Values are means of measurements on 15 seedlings.

of sweet gum seedlings were over 40% greater and those of loblolly pine over 55% greater at 500 than at 350 ppm of CO_2, but there was no increase in dry weight at 650 over that at 500 ppm for either species (Table 10.1). Heights of seedlings of both species also were increased at 650 over 500 ppm (Fig. 10.2). Sweet gum produced more branches, and both species produced more leaves and roots in 500

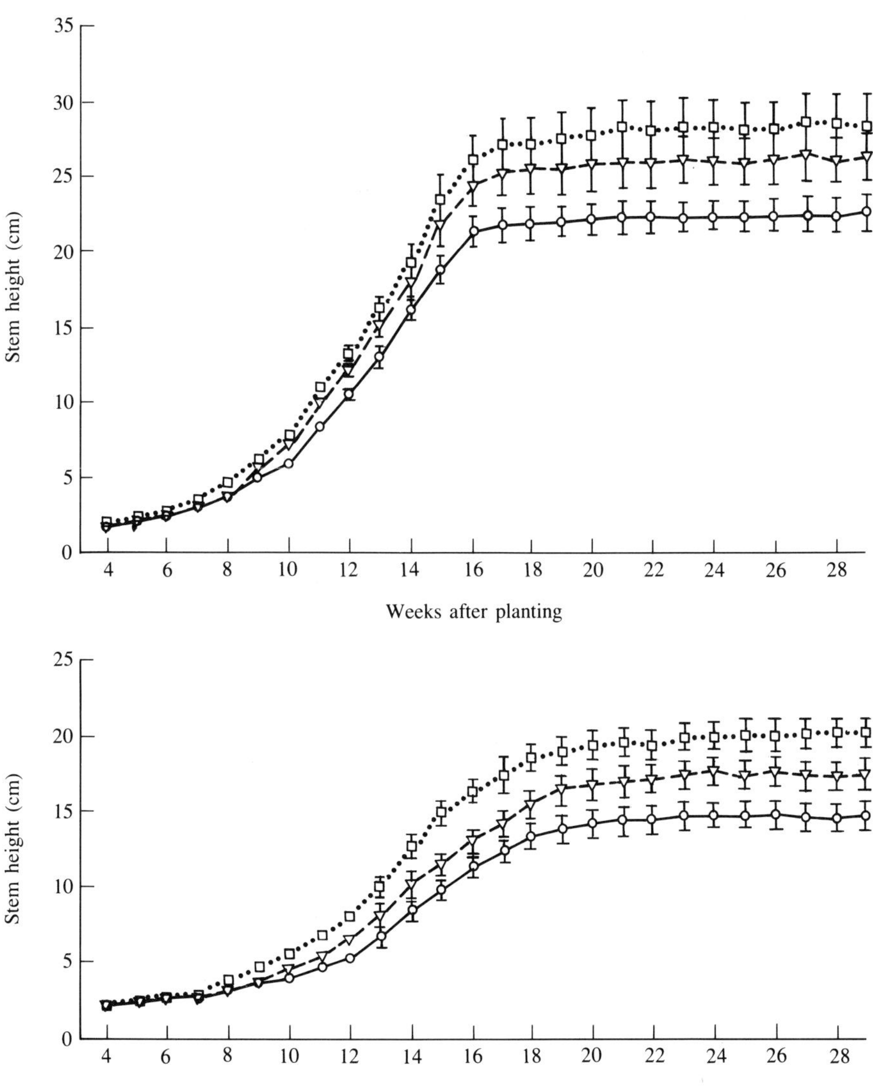

Figure 10.2 Height growth of sweet gum (*top*) and loblolly pine (*bottom*) seedlings grown for 6 months with 350 (——), 500 (–––), and 650 μl liter^{-1} (···) of CO_2 in an air-conditioned greenhouse with natural photoperiod and temperature simulating average outdoor temperatures. The temperature regime is shown in Fig. 10.3. (From Sionit *et al.*, 1985.)

ppm CO_2, as shown in Table 10.1. However, sweet gum produced more leaves at 650 ppm (Fig. 10.3). Leaf senescence on sweet gum began about the same time in all concentrations of CO_2, as shown in Fig. 10.3. All loblolly pine seedlings showed a substantial increase in dry weight over winter, but surprisingly there was no greater gain at 600 than at 350 ppm of CO_2. Perhaps light was the most limiting factor during the winter, although high CO_2 often compensates for low light in vegetable crops. Some of the results from the experiments on loblolly pine were reported in more detail by Sionit *et al.* (1985).

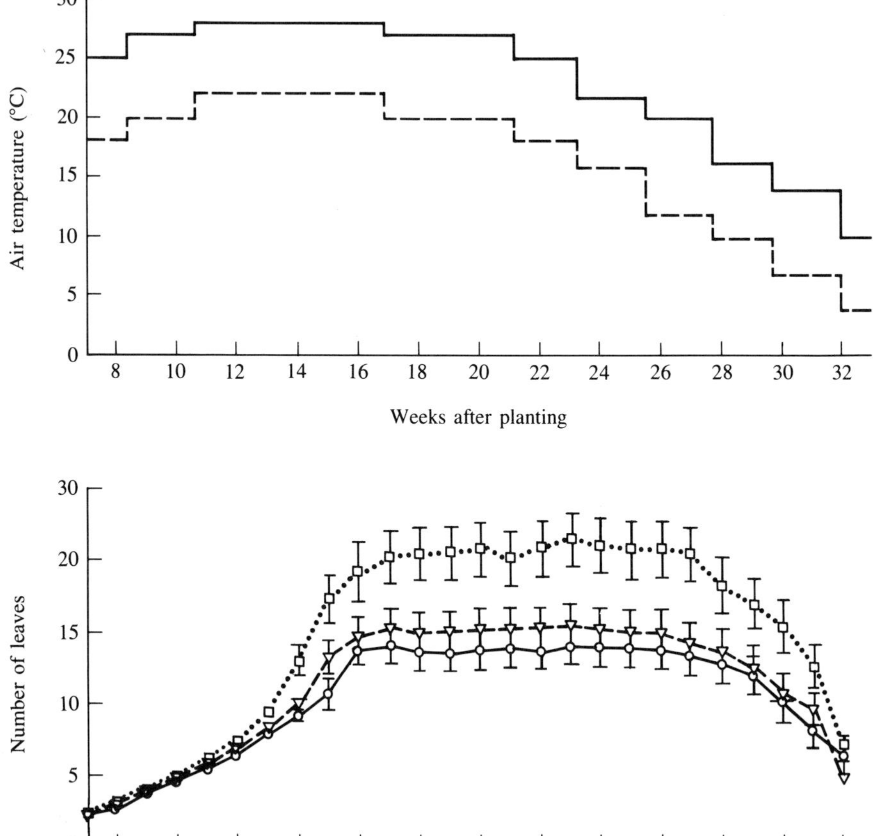

Figure 10.3 *Top,* The temperature regime used in experiments shown in Fig. 10.2 and Tables 10.1 and 10.4. *Bottom,* The effect of 350 (——), 500 (– – –), and 650 μl liter $^{-1}$ (···) of CO_2 on number and duration of leaves of sweet gum. The larger number of leaves at 650 ppm was related to increased number of branches. This experiment was terminated late in November after most leaves had fallen. (From Sionit *et al.*, 1985.)

In another series of experiments with the same species in CO_2 concentrations of 675 and 1000 ppm, height, basal stem diameter, leaf area, and total dry weight were significantly increased at high light intensity, but less increase occurred at a low light intensity in sweet gum and none in loblolly pine seedlings (Tolley and Strain, 1984b). The early increase in dry matter production in high CO_2 resulted from an increased rate of photosynthesis, but the increase in leaf area became an increasingly important factor as the seedlings grew older.

Observations on loblolly pine and sweet gum seedlings after 14 months at 350 or 500 ml liter^{-1} of CO_2 indicated an increase in stomatal limitation on photosynthesis at the higher concentration, but an increase in the rate of photosynthesis of loblolly pine. There was a small decrease in rate in sweet gum, probably caused by a decrease in Rubisco activity (Fetcher *et al.*, 1988). In another experiment, differences in growth rate between seedlings of two tropical tree species were partly attributed to differences in the amounts of photosynthate allocated to leaf growth (Oberbauer *et al.*, 1985). Surano *et al.* (1986) reported that 1-, 2-, and 7-year-old seedlings of ponderosa pine grown for 2.5 years at concentrations of 350, 425, 500, and 650 ppm of CO_2 in open-top chambers showed increases in stem diameter and volume and in height with increasing CO_2 concentration, although the older trees appeared to be saturated at 500 ppm. Norby *et al.* (1986) reported an 85% increase in dry weight of seedlings of white oak grown at 690 ppm of CO_2 for 40 weeks, in spite of nitrogen deficiency. Rogers *et al.* (1983) reported a 40% increase in dry weight of loblolly pine and sweet gum seedlings grown out-of-doors in open-top chambers at a CO_2 concentration of 910 ppm for 3 and 2 months, respectively.

Sionit and Kramer (1986) summarized the results of 15 experiments on various species of woody plants as follows:

1. Dry weight, stem diameter, and height of seedlings are increased by CO_2 concentrations in the range of 400 to 700 ppm.
2. CO_2 concentrations above 500 to 600 ppm usually produce little additional growth and occasionally even cause a decrease.
3. A high concentration of CO_2 seems to have morphological effects such as increased branching, leaf area, and leaf thickness on seedlings of some but not all species.
4. Seedlings of different species react differently and each species should be studied individually.
5. The competitive capacity of seedlings of various species is likely to be affected differently by increasing CO_2 concentration.

Effects on Physiological Processes

As stated in Chapter 1, environmental factors affect growth through modification of physiological processes (see Fig. 1.1) and CO_2 concentration is no exception. Information concerning its direct effects on stomatal conductance and processes such as photosynthesis, respiration, carbohydrate partitioning, and plant water economy, especially water use efficiency, is important to understanding its effects on growth and competitive capacity. The effects on photosynthesis and water use efficiency are related to the decrease in stomatal aperture and increases in leaf area often found in plants grown with high concentrations of CO_2.

Stomatal Conductance

Both inward diffusion of CO_2 and outward diffusion of water occur chiefly through the stomata, hence the effects of increasing CO_2 concentration on stomatal conductance, or its reciprocal, stomatal resistance, are very important. Numerous experiments on herbaceous plants and a few on woody plants indicate that above ambient concentrations of CO_2 usually cause partial closure of stomata and reduce conductance. In one experiment in open-top chambers out-of-doors, doubling the CO_2 concentrations decreased conductance 40% for sweet gum (Rogers *et al.*, 1983), and in another experiment in a phytotron it was reduced 50% for sweet gum, but not at all for loblolly pine (Tolley, 1982). Beadle *et al.* (1979) found that stomatal conductance of Sitka spruce was unaffected by CO_2 concentration over the range from 20 to 600 ppm, but Surano *et al.* (1986) reported that stomatal conductance in midafternoon of needles of ponderosa pine at 500 to 650 ppm was less than half that of needles at lower concentrations. The reasons for differences among species are unknown and the mechanism by which high ambient concentrations of CO_2 cause stomatal closure needs further investigation (Mansfield, 1986; Raschke, 1986). Morison (1985) suggested that stomatal reaction to high concentrations of CO_2 might increase water use efficiency by 30–50%, although probably less at the crop level than at the single-leaf level.

Figure 10.4 shows the complex relationship between stomatal conductance, transpiration, and CO_2 uptake of eastern cottonwood seedlings at several concentrations of CO_2. Stomatal conductance and transpiration were about 25% lower at 600 than at 300 ppm, but photosynthesis increased about 40%. This apparent paradox depends partly on the differences in resistance to diffusion of water vapor and CO_2, and partly on characteristics of the carbon fixation process to be discussed in the next paragraph.

The escape of water vapor from leaves depends largely on stomatal conductance, hence partial closure of stomata causes a more or less proportional decrease in transpiration, the relationship depending on the importance of boundary layer conductance relative to stomatal conductance. However, the entrance and fixation of carbon depend not only on stomatal conductance but on conductance in the liquid pathway through the cell walls, membranes, and cytosol to reaction sites in the chloroplasts, as well as on carboxylation capacity and photochemistry in the chloroplasts. These conductances are known collectively as the mesophyll conductance. In woody plants this often is more limiting than the stomatal conductance (Teskey *et al.*, 1986). Thus a 50% decrease in stomatal conductance has much less effect on the fixation of CO_2 than on the exit of water vapor because stomatal

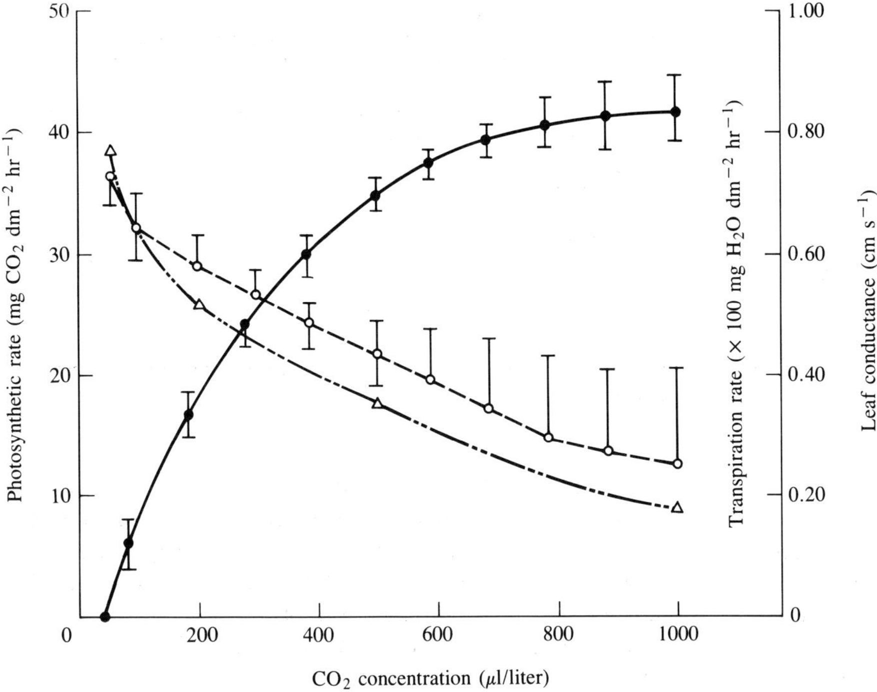

Figure 10.4 Effects of increasing CO_2 concentration on net photosynthesis (●), leaf conductance (△), and transpiration (○) of eastern cottonwood. (Adapted from Regehr *et al.*, 1975; from Sionit and Kramer, 1986; reprinted with permission from "Carbon Dioxide Enrichment of Greenhouse Crops," Volume II. Copyright © 1986 by CRC Press, Inc., Boca Raton, FL.)

conductance is only part of the total limitation on inward diffusion of CO_2, but is the only important limitation on outward diffusion of water vapor (Table 10.2). An increase in ambient CO_2 concentration causes an increase in internal concentration that more than compensates for partial closure of the stomata. This difference in the effect of partial closure of stomata on loss of water and uptake of CO_2 has important effects on water use efficiency and makes possible the use of antitranspirants that operate by decreasing stomatal conductance.

Photosynthesis

The relationships among CO_2 concentration, photosynthesis, and biomass accumulation are somewhat complex (see Chapter 2; Kramer, 1981), which increase the difficulty of predicting the effects on biomass of the increasing concentration in atmospheric CO_2. A much greater increase in photosynthesis can be expected for plants with the C_3 or Calvin cycle than for those with the C_4 or Hatch–Slack cycle. In nearly all woody plants, carbon is fixed by the C_3 pathway in which the

Table 10.2
Approximate Effects of Partial Closure of Stomata on Conductances for Water Vapor and CO_2[a]

	Air and stomatal conductance (cm sec^{-1})	Internal or mesophyll conductance (cm sec^{-1})	Total conductance (cm sec^{-1})	Reduction in total conductance
Water vapor				
Stomata open	0.43	0	0.43	
Stomata closed	0.22	0	0.22	50%
CO_2				
Stomata open	0.27	0.16	0.102	
Stomata closed	0.14	0.16	0.074	27%

[a]From data of Cooke and Rand (1980).

carboxylating enzyme is ribulose bisphosphate carboxylase (RUBP carboxylase, or Rubisco), and carbon is fixed in three-carbon phosphoglyceric acid. Rubisco functions as both a carboxylase and an oxygenase, and its oxygenase activity is responsible for the photorespiration that uses 30 to 50% of the carbon fixed by photosynthesis. Oxygen and CO_2 are competitive substrates for Rubisco and an increase in CO_2 concentration increases the rate of carboxylation and inhibits oxygenase activity, resulting in an increase in net photosynthesis.

In C_4 plants the CO_2 initially is fixed into four-carbon acids by the enzyme phosphoenolpyruvate carboxylase (PEP carboxylase). This enzyme already is saturated at the existing concentration of CO_2 and does not respond significantly to increase in CO_2 concentration. The organic acids produced by carboxylation in the mesophyll cells of C_4 plants diffuse into the bundle sheath cells, where they are decarboxylated, producing a high local concentration of CO_2 that is fixed by enzymes of the Calvin cycle. The overall result is that the apparent mesophyll conductance is high and photosynthesis of C_4 plants is less responsive than that of C_3 plants to an increase in concentration of atmospheric CO_2. Thus C_3 plants are likely to respond much more than C_4 plants to an increase in atmospheric CO_2 concentration (Kramer, 1981; Rogers *et al.*, 1983). For example, when Rogers *et al.* (1983) grew sweet gum, soybeans, and maize in open-top chambers with increased CO_2 concentration, the rate of photosynthesis of the C_3 sweet gum and soybean was increased but not that of the C_4 corn (Fig. 10.5).

Long- versus Short-Term Experiments

Short-term measurements of the effects of CO_2 concentration on photosynthesis are not always good indicators of the effects of prolonged exposure. In a number of long-term experiments with herbaceous species the rate of photosynthesis decreased over time at high CO_2 concentrations (e.g., Kramer, 1981) and this occurred in some experiments with seedlings of loblolly pine and sweet gum (Table 10.3). The rate per unit of leaf surface also was lower in a high CO_2 concentration for the

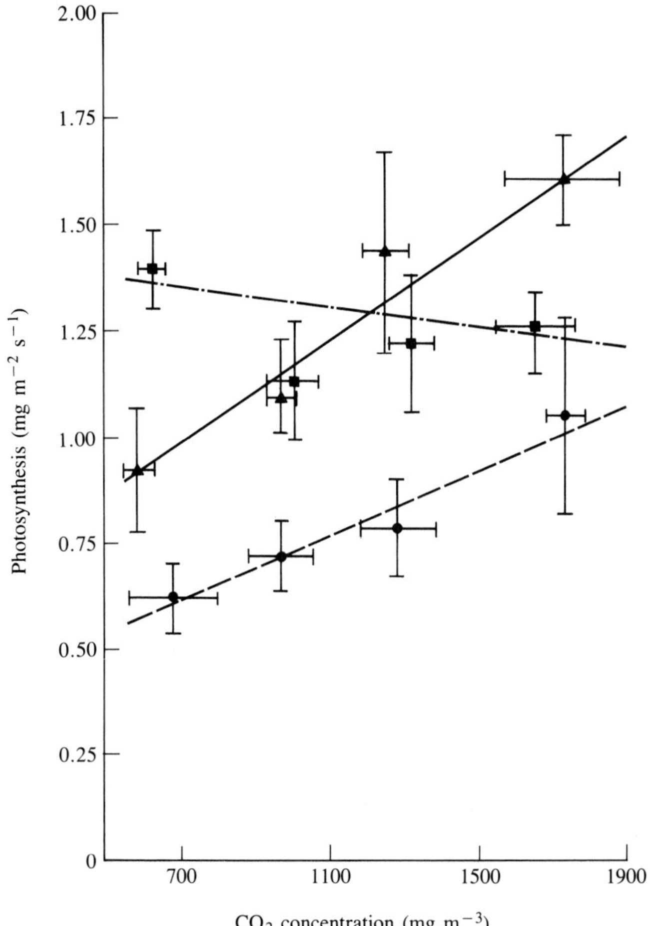

Figure 10.5 Net photosynthesis of soybean (——) and sweet gum (C_3) (– –) and corn (C_4) (– - –) at various concentrations of CO_2. Plants were grown in open-top chambers in the field. Vertical bars indicate one standard deviation in rate of photosynthesis and horizontal bars indicate the same for CO_2 concentration. The C_4 plant, corn, failed to respond to an increase in CO_2 concentration. (Adapted from Rogers *et al.*, 1983.)

seedlings of tropical tree species studied by Oberbauer *et al.* (1985). However, they found that the lower rate was compensated by an increase in leaf area. The reduction in rate might be caused by feedback inhibition resulting from accumulation of carbohydrates (Clough *et al.*, 1981), distortion of chloroplasts caused by accumulation of starch (Wulff and Strain, 1982), reduction in activity of carboxylating enzyme (Wong, 1979), reduction in stomatal conductance, or some combination of these. Sasek *et al.* (1985) discussed reversible inhibition of photosynthesis in cotton and cited some relevant literature.

Table 10.3
Rates of Net Photosynthesis (μmol CO_2 m^{-2} sec^{-1}) of Sweet Gum and Loblolly Pine Seedlings Measured 4 and 6 Weeks after Planting at the Same Concentrations of CO_2 in Which They Had Been Grown[a]

CO_2 concentration (μ $liter^{-1}$)	Sweet gum		Loblolly pine	
	4 weeks	6 weeks	4 weeks	6 weeks
350	11.0 ± 0.8	9.5 ± 0.8	3.5 ± 0.8	3.7 ± 0.8
675	9.8 ± 1.0	9.0 ± 1.1	4.0 ± 0.1	3.5 ± 0.9
1000	9.8 ± 0.8	7.0 ± 0.4	4.0 ± 0.6	2.6 ± 0.4

[a]From data of Tolley (1982).

Perhaps the long-term effects of increased CO_2 concentration are best indicated by the gain in dry matter reported from experiments lasting one or more growing seasons. In experiments by Rogers *et al.* (1983) in open-top chambers in the field, the dry weight of loblolly pine and sweet gum were increased about 40% over a growing season in 910 ppm of CO_2. In growth chamber experiments, Sionit *et al.* (1985) reported dry weight increases at the end of a growing season of 43% for sweet gum and 56% for pine seedlings at 500 ppm, but no further increase occurred at 650 ppm (Table 10.1). Tolley and Strain (1984b) reported similar results. Surano *et al.* (1986) grew 1- and 2-year-old seedlings and 7-year-old saplings of ponderosa pine in an open-top chamber for 2.5 years at 350, 425, 500, and 650 ppm of CO_2 and observed increases in stem diameter, volume, and height. However, the saplings appeared to be saturated at 500 ppm and by the beginning of the third season they showed chlorosis and premature loss of needles at high concentrations of CO_2. This was attributed to excessive needle temperatures caused by reduced transpiration, resulting from decreased stomatal conductance rather than from direct injury by the high concentration of CO_2.

Photosynthesis and Biomass

Although photosynthesis requires CO_2, the relationship between CO_2 concentration and rate is not as predictable as might be expected. For example, doubling the CO_2 concentration decreased the rate of photosynthesis per unit of leaf area of balsa seedlings, but increased leaf area 40% and dry weight 80% over a period of 60 days. The rate of photosynthesis of *Pentaclethra macroloba* was insignificantly reduced in high CO_2, leaf area and specific leaf weights were slightly increased, but total dry weight was increased about 30% (Oberbauer *et al.*, 1985). Perhaps the rate of photosynthesis had been higher earlier in the life of these seedlings as the measurements were made shortly before harvest. Biomass production over time depends on the rate of photosynthesis per unit of leaf area over time, the total leaf area, and leaf duration, as well as the partitioning of photosynthate among such processes as respiration, production of new tissue, and storage in wood, seeds, and other tissues. Leaf duration and the length of time during which trees carry on photosynthesis also may be important with respect to growth and competition between deciduous and evergreen species. Winter photosynthesis is common in evergreen vegetation where winters are mild, as in the Pacific Northwest (Fig. 2.4), the southern United States, and much of western Europe, and an increase in the concentration of atmospheric CO_2 should increase it. However, Sionit *et al.* (1985) reported that although the dry weight of loblolly pine seedlings increased during the winter, the increase was not enhanced by increasing the CO_2 concentration. This is puzzling, unless light was limiting. It seems likely that a shade-intolerant species such as loblolly pine will benefit less from a high concentration of CO_2 at low than at high light intensities (Tolley and Strain, 1984a; Teskey and Shrestha, 1985).

As mentioned in Chapter 2, a serious difficulty in relating photosynthesis to biomass results from the fact that measurements of photosynthesis usually are made more or less instantaneously on single leaves or twigs, whereas biomass is estimated for whole plants or plant stands. As Jarvis (1986) and others have pointed out, one cannot expect short-term measurements on isolated plants or plant organs to indicate how they will behave when growing in a forest. Furthermore, measurements of photosynthesis on a leaf area basis are compromised by the increase in leaf thickness often found in leaves grown with high concentrations of CO_2. This increases the amount of photosynthetic tissue per unit of leaf area.

Partitioning of Photosynthate

The uses made of the products of photosynthesis have a very important effect on the relationship between photosynthesis and biomass (Gifford *et al.*, 1984), including partitioning among processes such as respiration, growth, and reproduction and morphological partitioning among the various structures of plants such as roots, stems, and leaves. The effects of CO_2 on partitioning of photosynthesis in mature trees are largely unknown, but some information is available for seedlings.

Respiration

It was mentioned earlier that increasing the concentration of CO_2 inhibits the oxygenase activity of Rubisco and decreases photorespiration, thereby increasing net production of carbohydrate. This was reported for *Atriplex patula* (Björkman, 1971), grape, and *Leea brunoniana* (Kriedemann *et al.*, 1976). Björkman suggested that the decrease in photorespiration may explain why yields of greenhouse crops are increased by augmenting the CO_2 supply even when irradiance is low.

Dark respiration uses up to 50% of the product of net photosynthesis, but a CO_2 concentration of 5 to 10% usually is needed to significantly reduce dark respiration. It therefore seems unlikely that expected increases in carbon dioxide concentration will significantly affect rates of dark respiration directly. However, if the average temperature increases 3 or 4 degrees as a result of the "greenhouse effect," this might cause a significant increase in use of carbohydrate in maintenance respiration (see Fig. 2.10), which would be undesirable.

Vegetative Growth

The survival of plants is related to maintenance of a functional balance in the partitioning of photosynthate and other metabolites among roots, stems, leaves, and reproductive structures. Their economic success depends on the proportion allocated to merchantable products such as wood, fruit, or seeds. The factors controlling partitioning of food among organs of a plant are poorly understood, but include the role of hormones in controlling sink strength and environmental factors such as nitrogen, water, light intensity, and CO_2 concentration.

Experiments mentioned earlier showed differences among species in effects on branching and leaf area during the first growing season. At 650 ppm of CO_2, the number of branches was doubled on sweet gum seedlings and slightly increased on loblolly pine, but leaf weight was increased 30% on the former and 50% on the latter (Table 10.1). Exposure to 675 ppm of CO_2 increased growth of two tropical trees species, but the increase was twice as great for balsa as for *Pentaclethra macroloba,* probably because more photosynthate was allocated to leaf production in the former (Oberbauer *et al.*, 1985). Increase in leaf area has been reported for most kind of herbaceous plants grown in high concentration of CO_2 (Morison and Gifford, 1984). Differences in response among competing species might eventually lead to differences in composition of plant communities. This will be discussed later.

If an increase in CO_2 concentration results in the production of more photosynthate it might be expected to increase the root–shoot ratio, but this did not occur in sweet gum or loblolly pine. In fact it was decreased in sweet gum in one experiment (Table 10.1). However, O'Neill *et al.* (1987) reported that at 700 ppm of CO_2, mycorrhizal infection and root weight of white oak and shortleaf pine seedlings were greatly increased. High concentrations of CO_2 also usually increase the production of storage organs such as roots, bulbs, corms, and tubers (Bhattacharya *et al.*, 1985; Oechel and Strain, in Strain and Cure, 1985, p. 130).

Reproductive Growth

For obvious reasons there are no experimental data available on the effects of high concentrations of CO_2 on flowering and fruiting of forest trees. The data summarized by Kimball (1983) indicate that reproduction is increased in many annual herbaceous crop plants and this appears to be true of flower production of several greenhouse crops, including roses (Enoch and Kimball, 1986, Chapter 10). A possible reason for the increase in flowering is the increase in branching caused by high CO_2 in some species. Zangerl and Bazazz (1984) found enough differences among plants of different annual species in seed production at above ambient CO_2 concentration to suggest that increasing CO_2 concentration might change their relative competitive capacities. However, because of their long life cycle, this would be a slow process in forest communities.

An increase in CO_2 concentration might be expected to increase the yield of fruits and seeds of woody plants, but few data are available. However, developing cones on red pine (Dickmann and Kozlowski, 1970), grape berries (Hale and Weaver, 1962), pecans (Davis and Sparks, 1974), and other fruits and seeds are strong sinks for photosynthate. There also are reports in the literature indicating that fruit and seed development reduces root growth (Kramer, 1983a, p. 162), suggesting that the supply of photosynthate sometimes is inadequate to maintain high rates of vegetative and reproductive growth simultaneously. The existence of strong sinks should increase the rate of photosynthesis, particularly in a high concentration of CO_2 (Clough *et al.*, 1981). Downton *et al.* (1987) reported that when Valencia oranges were supplied with 800 ppm of Co_2, beginning just before anthesis, fruit retention was increased, resulting in an increase in fruit dry weight yield of over 50%.

Other Uses of Carbohydrates

In addition to the photosynthate used in respiration and in vegetative and reproductive growth, significant amounts are lost by shedding of leaves and branches, by death of roots, in root exudate, and in use by mycorrhizal fungi and symbiotic nitrogen-fixing organisms. Although high CO_2 sometimes hastens senescence of herbaceous plants, there is little information concerning its effects on woody plants. However, there was no evidence that it caused premature leaf abscission in loblolly pine or sweet gum (Tolley and Strain, 1984a; Sionit *et al.*, 1985) (Fig. 10.3).

It is unlikely that doubling the atmospheric CO_2 concentration will directly affect roots or organisms in the soil, where the concentration already is often 5% or more. However, the indirect effects of increased production of carbohydrates in the shoots may have important effects on activities belowground. Luxmoore (1981) predicted that the carbohydrate increase will cause an increase in root exudation and microbial activity in the rhizosphere, and in production of mycorrhizae. Development of mycorrhizal roots seems to be stimulated by an abundance of soluble carbohydrates (Wenger, 1955; Björkman, 1970; Hacskaylo, in Marks and Kozlowski, 1973; Marx *et al.*, 1977), hence their development should be increased by a high concentration of atmospheric CO_2 and the resulting increase in photosynthesis. O'Neill *et al.* (1987) found this to be true for seedlings of white oak and shortleaf pine. The benefits from vigorous development of mycorrhizae often are important (see Chapter 6). A high concentration of CO_2 also increases nitrogen fixation in herbaceous legumes (Hardy and Havelka, 1975). This might be important in forests where alders or other understory plants provide nitrogen (see Chapter 6; Norby, 1987).

Interactions with Other Limiting Factors

In forests, growth often is limited by drought, mineral deficiencies, shade, and unfavorable temperatures, operating singly or in combination. The concept of limiting factors has sometimes been used to argue that if growth already is limited by one factor it will not respond to increase in another suboptimal factor such as CO_2 concentration. This is not true unless the other stresses are extremely severe, because according to Mitscherlich's concept of limiting factors, if growth is limited simultaneously by several factors, the response to an improvement in any one of them is approximately proportional to the degree of deficiency of that factor.

Several experiments have shown that an increase in the concentration of CO_2 causes an increase in the growth of both woody and herbaceous species under stress. For example, as shown in Table 10.4, even with limiting irradiance (250 μmol m^{-2} sec^{-1}) and water stress as severe as -22 to -24 bars, doubling the CO_2 concentration produced significant increases in dry weight of sweet gum and loblolly pine seedlings, although unstressed seedlings showed a much greater increase in dry weight.

Water Stress

Water stress is the most common limitation on growth, beyond the omnipresent suboptimal concentration of CO_2. As mentioned in Chapter 7, it limits growth directly by reducing cell expansion, resulting in decreased leaf area and root and shoot growth. It also causes stomatal and nonstomatal inhibition of photosynthesis, but cell enlargement and growth usually are inhibited before photosynthesis is

Table 10.4
Compensatory Effects of Increased CO_2 Concentration on Water-Stressed Sweet Gum and Loblolly Pine Seedlings[a]

CO_2 (ppm)	Light (μmol m^{-2} sec^{-1})	Water potential (bars)	Total dry weight (gm)	Percentage increase
		Sweet Gum		
350	250	−22	0.81	
675	250	−10	0.95	17
675	1000	−22	1.06	30
675	1000	− 3	1.15	42
		Loblolly Pine		
350	250	−18	0.36	
675	250	−18	0.39	8
675	1000	−24	0.51	41
675	1000	− 7	0.62	72

[a]From Tolley and Strain (1984a,b). n = 10.

significantly reduced (Boyer, 1970). This should result in accumulation of carbohydrates that might contribute to drought tolerance by osmotic adjustment, as reported by Sionit *et al.* (1980) for wheat. Water use efficiency refers to the ratio of units of CO_2 used per unit of water used. In most experiments, increase in CO_2 concentration results in an increase in water use efficiency because, as mentioned earlier, partial closure of stomata reduces transpiration more than it reduces photosynthesis. Rogers *et al.* (1983) observed a large improvement in water use efficiency of sweet gum seedlings as the CO_2 concentration was increased (Fig. 10.6). The improvement in water use efficiency probably will be one of the most beneficial effects of increasing atmospheric CO_2 concentration on plant growth (Morison, 1985).

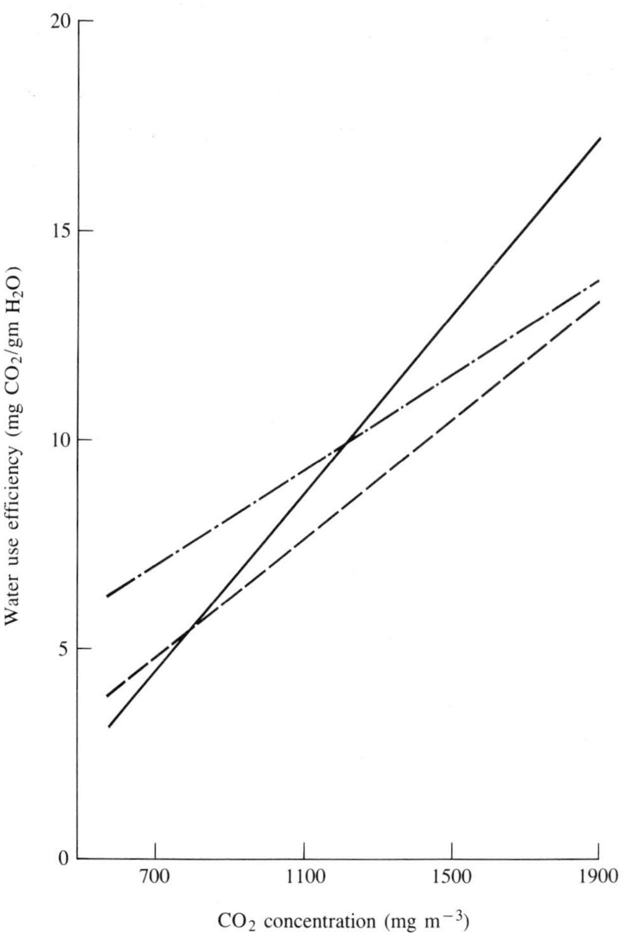

Figure 10.6 Water use efficiencies of sweet gum (–·–), corn (– – –), and soybean (——) with increasing CO_2 concentration. (After Rogers *et al.*, 1983.)

Generally a high concentration of CO_2 seems to compensate to some degree for water stress for both herbaceous and woody seedlings. High CO_2 compensated for water stress to a limited extent for loblolly pine and to a greater extent for sweet gum, leading Tolley and Strain (1984a) to suggest that increasing CO_2 concentration might result in sweet gum replacing loblolly pine on sites too dry for it at present. High CO_2 also compensated for water stress to some degree in wheat (Sionit *et al.*, 1980) and in soybeans (Rogers *et al.*, 1984). Some interactions are shown in Fig. 10.7.

Temperature

The data available indicate that a strong interaction between temperature and CO_2 concentration exists, but the results vary among species. Experiments by Idso *et al.* (1987) on several herbaceous species, including cotton, indicated that high temperature increased the growth enhancement of high CO_2, but lower temperatures at high CO_2 decreased growth. In contrast, Sionit *et al.* (1981) found that okra (*Abelmoschus esculentus*), which is unproductive at day/night temperature regimes below 26/20°C, yields well at 20/14°C in 1000 ppm of CO_2, and Potvin (1985)

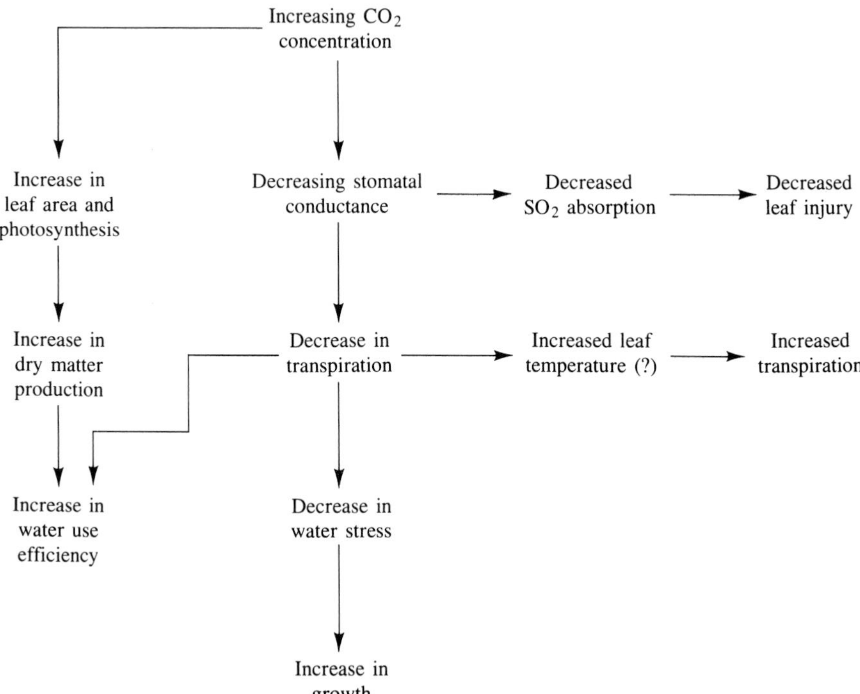

Figure 10.7 Diagram showing interactions between increasing concentration of atmospheric CO_2 and various physiological processes. (After Kramer, 1986.)

reported that a high concentration of CO_2 reduced chilling injury to two species of grasses. Experiments with boreal tree species indicated that high CO_2 concentration moderately increased the growth of quaking aspen but produced only variable increases in white spruce, chiefly in leaf mass (Brown and Higginbotham, 1986). Billings *et al.* (1984) concluded that CO_2 concentration is not a limiting factor for arctic tundra vegetation.

Kriedemann *et al.* (1976) found that a concentration of 1200 to 1300 ppm of CO_2 increased growth and heat tolerance of grape and several other kinds of woody plants subjected to temperatures of 37 to 40°C. Only guava (*Psidium guajava*), among the half dozen species tested, showed no improvement during an exposure of 3 to 6 months. Pearcy and Björkman (1983) suggested that the enhancement of photosynthesis by increasing CO_2 concentration occurs because the increasing concentration compensates for the decreasing affinity of Rubisco for CO_2 at high temperatures.

Light

Generally, a high concentration of CO_2 tends to compensate for low light intensity. Although the largest increase in absolute rate of photosynthesis occurs in high (saturating) intensities, the greatest relative increase occurs at low intensities. This is at least partly because a high concentration of CO_2 increases the quantum yield of C_3 plants by decreasing photorespiration and it also decreases the light compensation point (Heath and Meidner, 1967; Pearcy and Björkman, 1983). Thus enhanced CO_2 concentration of greenhouse air is profitable even in the winter (Hopen and Ries, 1962). Global CO_2 increase should result in increased rates of photosynthesis within tree canopies and even in vegetation of the forest floor. Few data are available for woody plants, but Tolley and Strain (1984a) found that high CO_2 compensated for low light intensity for growth of sweet gum (Fig. 10.8).

It seems that differences among species in the degree to which a higher concentration of CO_2 compensates for shading and low light intensity might have differential effects on the success of plants growing on the forest floor.

Mineral Deficiencies

It would be expected that if dry matter production is increased by high CO_2 there will be increased use of minerals and nitrogen and more fertilization will be required for maximum benefit from the increasing concentration of CO_2. However, Norby *et al.* (1986) found that white oak seedlings made an 85% increase in dry weight at 690 ppm over that at 362 ppm even though they became severely nitrogen deficient. Thus increasing CO_2 concentration apparently can result in increased growth even in mineral-deficient soil, but it is likely to be accompanied by an increase in the carbon–nitrogen ratio. If there is an increase in root development and especially in mycorrhizal roots, as in the experiments reported by O'Neill *et al.* (1987), there should be increased uptake of phosphorus and other minerals. Conroy *et al.* (1986)

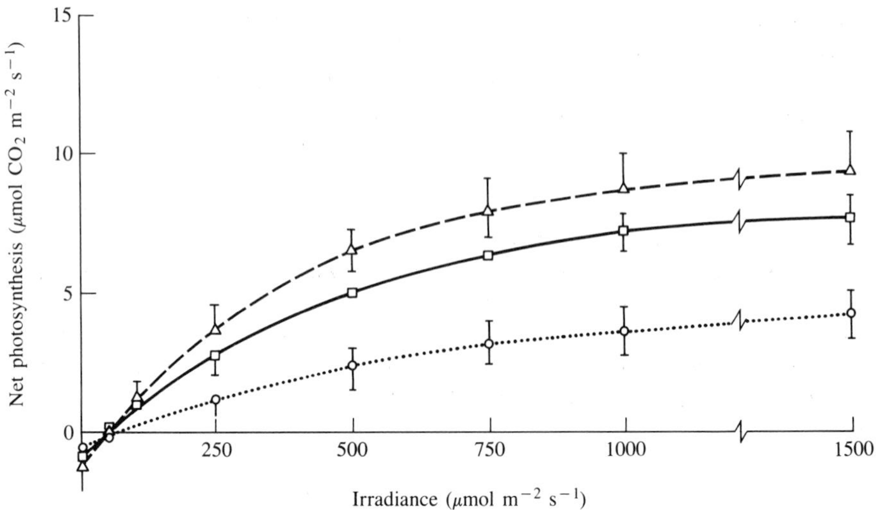

Figure 10.8 Net photosynthesis plotted over irradiance for sweet gum seedlings grown at 350 (···), 675 (−−−), and 1000 ml liter^{-1} (——) of CO_2. (From Sionit and Kramer, 1986; based on data of Tolley and Strain, 1984a. Reprinted with permission from "Carbon Dioxide Enrichment of Greenhouse Crops," Volume II. Copyright © 1986 by CRC Press, Inc., Boca Raton, FL.)

concluded that the level of soil phosphorus adequate for growth of Monterey pine seedlings in 330 ppm of CO_2 would be inadequate for maximum growth at 660 ppm. Growth with adequate phosphorus was 32% greater in 660 than in 330 ppm of CO_2, but in soil deficient in phosphorus it was only 14% greater. More research is needed on the interaction between mineral requirements and increased concentration of atmospheric CO_2.

Effects on Plant Communities and Forests

The preceding discussion of the physiological effects of an increasing concentration of CO_2 serves as background for consideration of how this will affect the future composition of plant communities and the productivity of forests. This may depend as much on the indirect effects of changes in climate as on the direct effects on physiological processes. There are differences in opinion concerning the broad future effects on crops. Idso (1984) and Kimball (1983) claimed that increasing the CO_2 concentration will increase crop yields as much as 30%, but Waggoner (1984) predicted that there will be little change in yield because the favorable effects on physiological processes will be largely nullified by unfavorable changes in temperature and rainfall.

Experiments described by Bazzaz and Carlson (1984) and Zangerl and Bazzaz (1984) suggest that differences in seed production with increasing concentration of

CO_2 may produce changes in composition of some herbaceous plant communities. Wray and Strain (1987) found that a high CO_2 concentration increased the competitive capacity of aster (C_3) versus broomsedge (C_4), suggesting that the usual replacement of aster by broomsedge in old field succession might be delayed. In general, the competitive capacity of C_3 versus C_4 plants should be increased (Lemon, 1983, p. 196; Patterson, 1986). Few data are available for forest trees, but experiments with seedlings suggest that high CO_2 may increase the competitive capacity of sweet gum versus loblolly pine on dry sites (Tolley and Strain, 1984a) and in the shade (Tolley and Strain, 1984b). Sasek and Strain (1990) expect increasing CO_2 concentration and the accompanying global change in climate to increase both the geographic range of kudzu and its effect on other species. Williams *et al.* (1986) found little effect of increased CO_2 for 90 days on seedlings of trees characteristic of upland and lowland forests and concluded that more extensive experiments are needed to predict the effects of increasing CO_2 on the composition of forest communities.

The direct and indirect effects of the increasing concentration of CO_2 and other "greenhouse gases" (Hoffman and Wells, 1987) are of particular importance to foresters because their crops require several decades to mature, hence significant changes in atmospheric CO_2 and climate can occur during a single rotation. Various aspects of this problem are discussed in detail in the book edited by Shands and Hoffman (1987).

It has been predicted that increasing CO_2 concentration will have the greatest effect on biomass accumulation early in secondary succession. Presumably this also would be true of forest plantations. It should speed up canopy closure, intensify competition, and hasten maturity. This probably will be desirable in forestry as it will shorten rotations, but it also will increase the need for fertilization. However, it will have unpredictable results on the composition of naturally occurring plant communities (Strain and Bazzaz, 1983, pp. 209–211; Bazzaz and Garbutt, 1988). As Wong (1979) stated, plant reactions to elevation in atmospheric CO_2 depend on a complex of partially compensating processes, the results of which are not readily predicted. Answers to many of these questions can be obtained only by maintaining experimental plots for several decades.

The extent of the changes in range and productivity of forest trees will depend on the kind and extent of climate changes and the effects of those changes on physiological processes. In general, the "greenhouse effect" of the increasing concentration of CO_2 and methane is expected to produce a warming trend that will be greatest toward the poles, accompanied by a decrease in rainfall at about 30° to 40°N latitude (National Research Council, 1979b; Roberts, 1987). These changes are likely to produce significant changes in the range and productivity of important tree species and require changes in forest management (Hoffman, 1984; Sandenburgh *et al.*, 1987).

Miller *et al.* (1987) predicted that increasing temperature and decreasing rainfall will cause a north and eastward shift in the range of loblolly pine. Although the range may increase somewhat in area, the yield probably will decrease because the soils of the newly occupied areas will be less fertile than those in areas lost by the increasing frequency and duration of drought in the southwestern part of its present range. Solomon and West (1987) predicted from computer models that although changes in species would occur in the forests of east-central Tennessee, the changes will not seriously affect forest management. In contrast, in northwest Michigan, extinction of boreal species such as spruce and birch might create serious management problems in the long term. Leverenz and Lev (1987) predicted changes in western conifers based on probable reactions to increasing temperature and decreasing water supply. One of their models predicts an increase of ponderosa pine in California and western Oregon and a decrease along the east slope of the Rockies, a decrease in the importance of Douglas-fir in the eastern and southern part of its range, and a decline in the area of Engelmann spruce over its entire range as it is restricted to higher altitudes by increasing temperature. Lodgepole pine probably will expand in some areas and be restricted in others. Woodman (1987b) expected no important changes in management of Douglas-fir and western hemlock in the Pacific Northwest. Sasek and Strain (1990) predict a significant northward expansion of kudzu as a pest if winter temperatures increase 3°C.

Predictions concerning changes in plant and animal distribution will remain speculative until it is possible to predict climatic changes more accurately and until we know more about how such changes affect the physiology of various species. There are a few scientists who disagree with the general view that there will be significant warming and changes in rainfall patterns (Roberts, 1989; Kerr, 1989), and it may be a decade or more before reliable predictions concerning climatic change become available. Actually, the climatic changes likely to be produced by increase in CO_2 concentration are no greater than the variations in weather experienced during unfavorable seasons with existing CO_2 concentrations. Thus a better understanding of the response of trees and other plants to environmental stresses will be valuable in ecology, agriculture, and forestry, regardless of the size of the climatic changes produced by increasing CO_2 concentration (Lee and Kramer, 1987).

Miscellaneous Effects

It is impossible to discuss adequately the numerous established and potential effects of increasing CO_2 concentration on plants. The bibliography prepared by Strain and Cure (1986) lists over 1000 titles and there are many additional books and articles that do not fall within the scope of the work. We will discuss a few effects that did not fit into the outline of this chapter but deserve mention.

Tolerance of Air Pollution

It seems likely that the partial closure of stomata caused by high concentrations of CO_2 will reduce entrance of pollutants and reduce injury. Decreased injury was reported for several C_3 species of herbaceous plants fumigated with SO_2 (Carlson and Bazzaz, 1982). Green and Wright (1977) reported that a high concentration of CO_2 increased the rate of photosynthesis of ponderosa pine injured by air pollution. The action of air pollutants is discussed in more detail in Chapter 9 and by Kozlowski and Constantinidou (1986a,b).

Diseases and Insect Pests

An increasing concentration of CO_2 is likely to be accompanied by an increase in the carbon–nitrogen ratio of foliage, which probably will increase herbivory. In one study, larvae of the soybean looper (*Pseudoplusia includens*) ate more tissue per larva on leaves of soybeans grown with 500 to 650 ppm of CO_2 than on those of plants grown at 350 ppm (Lincoln *et al.*, 1984). In another study, insect larvae fed on plantain leaves grown in 700 ppm of CO_2 grew more slowly and experienced greater mortality than those fed on leaves grown with a normal concentration of atmospheric CO_2 (Fajer *et al.*, 1989). The effect of the changing carbon–nitrogen ratio in leaves on attacks by fungi has not been evaluated as yet. However, if significant changes occur in rainfall and temperature patterns, changes in attacks by pests also are likely to occur.

Seedling Production

In most experiments it was reported that an increase in concentration of atmospheric CO_2 increased seedling growth. For example, in one series, growth in 500 ppm of CO_2 resulted in increases in dry weight of over 55% and 40% for loblolly pine and sweet gum at the end of the first growing season (Table 10.1). Tinus (1972) reported that seedlings of blue spruce and ponderosa pine grown for a year in 1200 ppm of CO_2 were taller, heavier, and had more side branches than those grown with 300 ppm. This suggests that it might be profitable to increase the CO_2 concentration in enclosures used for the production of container-grown seedlings. A combination of prolonged photoperiod and enhanced CO_2 concentration should be particularly effective on slow-growing seedlings of northern conifers.

Wood Quality

Although the effects of increased CO_2 on wood quality are important to forestry, the limited data available come from seedlings and show no clear trend. However, it seems unlikely that there will be any important direct effects on wood quality although there might be indirect effects if water stress is increased.

Litter Quality

It seems likely that if the ratio of carbon to nitrogen in leaves increases, litter quality and rate of decomposition will decrease (Melillo, in Strain and Cure, 1986).

However, decrease in quality of litter may be compensated by increase in quantity. There also may be changes in moisture supply and temperature that will affect decomposition rates.

Salinity

According to Schwarz and Gale (1984) and Bowman and Strain (1987), higher concentrations of CO_2 appear to increase tolerance of high salinity, but more data are needed to draw any firm conclusions.

Vegetation and Atmospheric Carbon Dioxide

Emphasis thus far has been on the effects of increase in CO_2 concentration of the atmosphere on plant growth. However, plants, and especially forests, also affect the CO_2 concentration of the atmosphere by removing CO_2 used in photosynthesis and releasing CO_2 produced by respiration during periods when there is little or no photosynthesis. This is shown by the seasonal cycle in CO_2 concentration visible in the ascending concentration curve shown in Fig. 10.1. There is great concern regarding the effects of deforestation, especially removal of tropical rain forest, on the global CO_2 balance. This deforestation not only decreases the amount of CO_2 removed by photosynthesis but also temporarily increases the amount released into the atmosphere by burning and microbial decomposition of slash. The overall effect is regarded as catastrophic by many writers, for example, Woodwell *et al.* (1983). However, it is difficult to accurately estimate the biomass of tropical forests (Brown and Lugo, 1984) and Armentano and Ralston (1980) concluded that temperate zone forests serve as important sinks for CO_2. This is supported by the calculations of Tans *et al.* (1990) which indicate that in the northern hemisphere terrestrial ecosystems must absorb a large amount of CO_2.

In the long run a mature forest releases about the same amount of CO_2 as it fixes, but if a mature forest is replaced by a young forest or other rapidly growing vegetation there will be a temporary increase in the amount of CO_2 fixed. However, taking into account the losses of CO_2 after harvesting from decay and burning of slash and processing byproducts, Harmon *et al.* (1990) concluded that conversion of old growth forests to young forests adds CO_2 to the atmosphere. Jarvis (1989) doubted if expansion of forested area can significantly reduce the increase in atmospheric CO_2 concentration. The global carbon balance and carbon cycle continue to be subjects of debate as indicated in a paper by Hoffman and Wells (1987) and a group of letters in *Science* (**241,** 1736–1739, 1988).

Summary

The CO_2 concentration of the atmosphere has increased from about 300 ppm early in this century to about 350 ppm at present and may increase to 600 ppm by the middle of the next century. This increase is affecting plant growth directly by increasing photosynthesis, the "fertilization effect," and indirectly by causing climatic changes that are likely to affect the growth and distribution of plants, the "greenhouse effect." The direct effects include decrease in stomatal conductance and transpiration and increase in photosynthesis, resulting in increased water use efficiency. With high CO_2, partitioning of photosynthate to leaves and roots often is increased, resulting in a larger leaf area and root system. In addition, photorespiration usually is decreased and there generally is an increase in biomass and sometimes an increase in fruit and seed production.

There are strong interactions between CO_2 supply and environmental stresses such as light, water, and mineral nutrition. There is some indication that a high concentration of CO_2 compensates at least in part for shading, moderate water stress, and mineral deficiencies. However, on most soils, fertilization probably will be required to obtain maximum benefits from the increasing concentration of CO_2. The decrease in stomatal conductance observed at high concentrations of CO_2 tends to decrease injury to leaves from air pollutants, but the increase in the carbon–nitrogen ratio may increase insect feeding on leaves. A high concentration of CO_2 should increase the growth of container-grown seedlings produced in greenhouses, but little information is available concerning seedling quality. There is need for more long-term experiments on the effects of high CO_2 concentrations on trees.

It is likely that differences in competitive capacity at high CO_2 concentrations exist that will cause changes in the composition of plant communities. Such differences may also affect the choice of species and families for future forest plantations. At present the information concerning the extent of the "greenhouse effect" on climate is too uncertain to predict what changes in temperature and rainfall distribution are likely to occur or what the effects will be on the ranges of major forest tree species. To better anticipate the effects of possible climatic changes we need more information concerning the physiological effects of changes in temperature and water supply. There is a special need for more information concerning differences between and within species with respect to their responses to increasing CO_2 and to related changes in other environmental factors such as increase in temperature and decrease in rainfall.

General References

Allen, L. H. (1990). Plant responses to rising carbon dioxide and potential interactions with air pollutants. *J. Environ. Quality* **19,** 15–34.

Enoch, H. Z., and Kimball, B. A., eds. (1986). "Carbon Dioxide Enrichment of Greenhouse Crops, Vol. II. Physiology, Yield, and Economics." CRC Press, Boca Raton, FL.

Hoffman, J. S. (1984). Carbon dioxide and future forests. *J. Forestry* **82,** 164–167.

Jarvis, P. G. (1989). Atmospheric carbon dioxide and forests. *Phil. Trans. R. Soc. B (London)* **324,** 369–392.

Kimball, B. A. (1983). Carbon dioxide and agricultural yield: An assemblage and analysis of 430 prior observations. *Agron. J.* **75,** 779–788.

Kramer, P. J. (1981). Carbon dioxide concentration, photosynthesis, and dry matter production. *Bio-Science* **31,** 29–33.

Lemon, E. R., ed. (1983). "CO_2 and Plants." Westview Press, Boulder, CO.

Shands, W. E., and Hoffman, J. S., eds. (1987). "The Greenhouse Effect, Climate Change, and U.S. Forests." The Conservation Foundation, Washington, D.C.

Sionit, N., Strain, B. R., Hellmers, H., Riechers, G. H., and Jaeger, C. H. (1985). Long-term atmospheric CO_2 enrichment affects the growth and development of *Liquidambar styraciflua* and *Pinus taeda* seedlings. *Can. J. For. Res.* **15,** 468–471.

Strain, B. R., and Cure, J. D., eds. (1985). "Direct Effects of Increasing Carbon Dioxide on Vegetation." U.S. Department of Energy (DOE/ER-0238). Available from National Technical Information Service, Springfield, VA 22161.

Strain, B. R., and Cure, J. D., eds. (1986). "Direct Effects of Atmospheric CO_2-Enrichment on Plants and Ecosystems: A Bibliography with Abstracts." Oak Ridge National Laboratory/CDIC-13. Available from National Technical Information Service, Springfield, VA 22161.

Tolley, L. C., and Strain, B. R. (1984a). Effects of CO_2 enrichment and water stress on growth of *Liquidambar styraciflua* and *Pinus taeda* seedlings levels. *Can. J. Bot.* **62,** 2135–21310.

Tolley, L. C., and Strain, B. R. (1984b). Effects of CO_2 enrichment on growth of *Liquidambar styraciflua* and *Pinus taeda* seedlings under different irradiance level. *Can. J. For. Res.* **14,** 343–350.

Chapter 11

Fire

Introduction

Fire, one of the most dramatic forces that alters forest ecosystems, has been recognized as a prominent factor in shaping the landscape for centuries. Early humans used fire for hunting, clearing of forests for agriculture, improving grazing for livestock, producing ash to fertilize fields, favoring some plants over others, facilitating harvesting of crops, and eliminating undesirable plant materials. Unfortunately, however, reckless burning by early settlers caused so much damage to timber and property that fire became widely regarded as an insidious enemy (Kozlowski and Ahlgren, 1974). In the United States, the size, intensity, frequency, and destructive action of wildfires were greatly increased by settlement. The Miramichi and Maine fires (1825), North Carolina fire (1898), Idaho and Montana fire (1898), and Alaska fire (1957) covered more than 3 million acres each. Several other fires, including those in Wisconsin (1894), Michigan (1871), and Washington and Oregon (1910), exceeded 2 million acres. About 1500 lives were lost in the Peshtigo, Wisconsin, fire in 1871 and 400 people perished in the Hinckley, Minnesota, fire in 1894. All of these catastrophic fires were caused by humans. Holbrook (1944) presented vivid and detailed accounts of these and other catastrophic fires in the United States.

For a long time, suppression of uncontrolled forest fires was considered necessary because of rapid population increases, accelerating demand for forest products, and the apparently harmful effects of wildfires. However, our attitudes have changed drastically in recent decades from insistence on complete exclusion of fire from forests to greater tolerance of natural fires and use of controlled burning because of their importance in reducing fuels for wildfires, disease control, and perpetuation of certain natural plant associations. As emphasized by C.E. Ahlgren (1974a), the growing recognition of the role of fire in maintaining certain desirable natural ecosystems led some investigators to the view that fire could be used to solve most or all problems of forest regeneration. But not all ecosystems are fire-adapted. Hence C.E. Ahlgren (1974a) cautioned against permitting natural fires in all forests. He emphasized that natural fires are beneficial in some plant associations but not others. Furthermore, the response of modern forests to fire differs appreciably from that of the primeval forest. Because of increasing demands for forest products and diminishing forest resources, permitting large uncontrolled fires in many forests may be inadvisable (C.E. Ahlgren, 1974a; Ahlgren and Ahlgren, 1984).

Forest fires usually are classified on the basis of the height to which they burn. The generally flameless *ground fires*, which burn below the surface in thick organic matter (duff, muck, or peat) above the mineral soil, generate high temperatures and may kill most plants with roots growing in organic matter. Ground fires are persistent and difficult to control. *Surface fires*, the most common type of forest fires, burn much of the litter layer and often kill understory vegetation, but usually do not kill trees of the overstory. By comparison, *crown fires* often move from one crown to another and kill most trees. In nature many fires are combinations of surface and crown fires. Sometimes surface fires also start ground fires. Conifers, with highly flammable foliage, are particularly susceptible to crown fires. A major cause of injury by crown fires is reduction in photosynthetic surface by defoliation. This is especially serious in young forest stands, which are more likely to be completely defoliated than older, taller trees. Ground fires often kill meristems, thus preventing regrowth.

Effects of Fire on Vegetation and Site Quality

The effects of fires are very complex and range from catastrophic to beneficial. Fires destroy forests, modify site conditions, and alter physiological processes and growth of plants. The responses of plants to fire vary greatly with the type and timing of the fire as well as the species affected, and with other factors, such as prefire status of vegetation and postfire history (e.g., weather).

Fire alters site quality by influencing vegetation, soil properties, hydrology, and geomorphic processes (Fig. 11.1). Site quality is affected largely by the burning of organic matter and by heating of surface soil layers; the former effect is much more widespread. The amount of organic matter consumed by fire varies greatly among plant ecosystems and generally is greater in evergreen than in deciduous forests (Fig. 11.2).

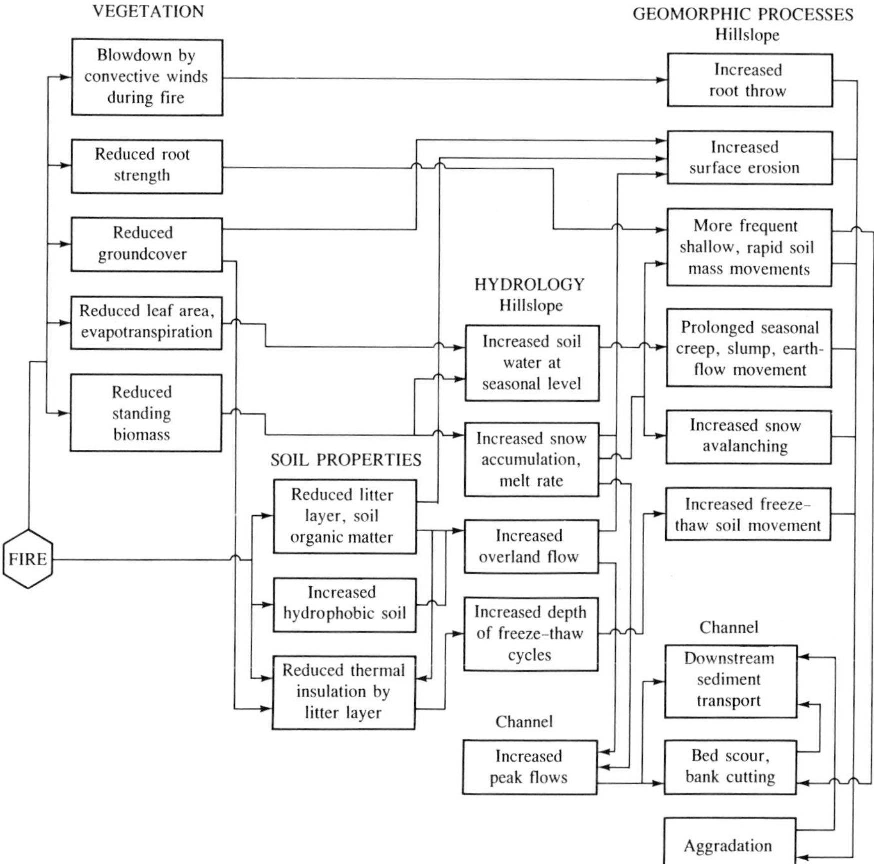

Figure 11.1 Some effects of fire on vegetation, soil properties, hydrology, and geomorphic processes. (From Swanson, 1981.)

Both the physical and biological properties of soil are greatly altered by fire. For example, severe fires sometimes reduce the capacity of soils to store water and cause development of water-repellent layers. They also induce soil compaction, fuse soil surface layers, accelerate soil erosion, and increase fluctuation in soil temperature. Fires are associated with increases in some fungal diseases and insects as well as air pollutants (Rundel, 1981), but may reduce other diseases.

During fires, the surface soil is only briefly exposed to very high temperatures. Mineral soil is a poor heat conductor, hence a steep and transient temperature gradient may be expected with increasing soil depth. Heating of dry soils causes a greater increase in surface temperature but less heat penetration than does heating of wet soils (Raison, 1979). The rise in soil temperature varies with the type of fire. In the southeastern United States, controlled burning seldom results in soil

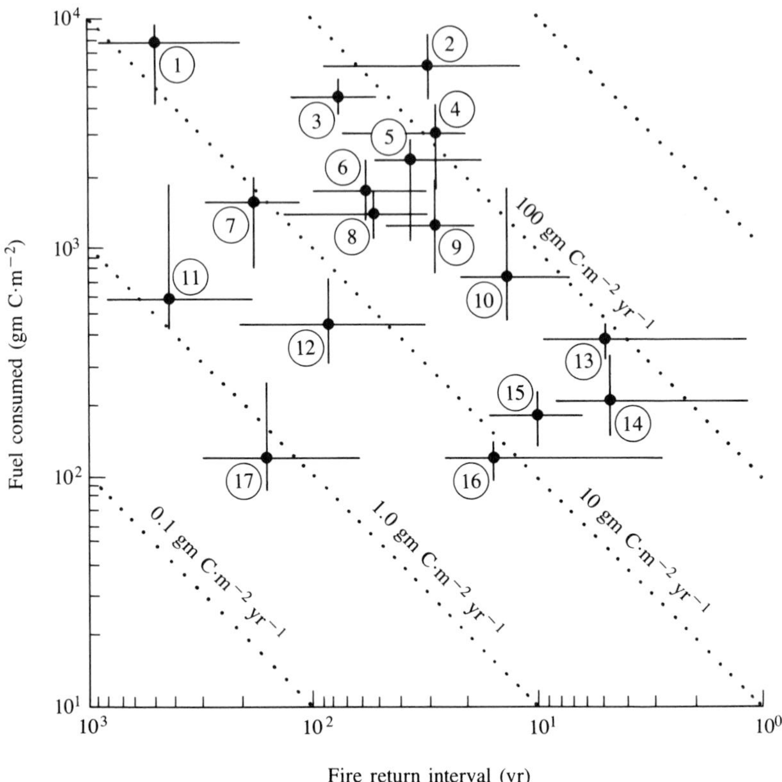

Figure 11.2 Comparison of the amount of fuel consumed and fire return intervals for several plant ecosystems. The diagonal-dotted lines are isopleths of the amount of carbon consumed per square meter per year. (1) Western conifer crown; (2) Western conifer understory; (3) White cedar; (4) Sand pine scrub; (5) Pine barrens; (6) Jack pine; (7) Pinyon juniper; (8) Chaparral; (9) Pocosin; (10) Southern pine flatwoods; (11) Deciduous forest; (12) Scrub steppe; (13) Tall prairie; (14) Southern pine savanna; (15) Annual grassland; (16) Short prairie; (17) Arid desert. (From Christensen, 1987.)

temperatures higher than 52°C for more than 15 min at a 3- to 6-mm depth, and surface temperatures are below 120°C most of the time (Heyward, 1938). However, under slash fires and wildfires the soil may be exposed to very high temperatures for long periods. For example, slash fires in Douglas-fir stands raised the surface temperature to near 1000°C and the soil temperature at a depth of 2.5 cm to 320°C (Isaac and Hopkins, 1937). Surface soil temperatures in chaparral, shrub bogs of the southeastern United States, and British heaths normally vary between 200 and 300°C during fire, but they may exceed 500°C. However, the temperature decreases rapidly with soil depth and rarely exceeds 100°C below a 3-cm depth (Christensen, 1985).

Following fires, daytime temperatures in forests often increase because of greater absorption of light by blackened surfaces and elimination of the insulating effects of vegetation. The rise in temperature is highest in the surface soil, but some increase can occur down to a depth of 18 cm. Beaton (1959) reported that the soil temperature after fire was increased as much as 10°C at a 7.6-cm depth. As might be expected, the temperature increases following fire decline progressively as plant cover becomes reestablished. The higher soil temperatures after fire often stimulate growth of plants directly, as well as indirectly by accelerating microbial activity and mineralization of soil organic matter (Raison, 1979). The release of nutrients immobilized in litter is particularly important in boreal regions, where litter decomposition is very slow.

Organic Matter

Litter and soil organic matter have many benefits as well as some harmful effects in forest ecosystems. They promote development of a soil structure that increases the rate of infiltration and cation-exchange capacity, reduce soil erosion, stabilize soil temperatures, delay soil freezing, and minimize formation of frost. However, they also decrease soil moisture by intercepting rainfall and often impede seed germination and seedling establishment. Furthermore, prolonged exclusion of fire often results in accumulation of combustible organic matter to a dangerous level and leads to catastrophic fires as well as disease and insect problems (Wright and Bailey, 1982). The tendency for mineral nutrients to be tied up in litter in regions with cold climates was cited earlier.

The amount of litter consumed varies with fire intensity. In Douglas-fir–larch forests in Montana, for example, high-intensity fires (300°C at the soil surface) consumed all the litter, medium-intensity fires (180 to 300°C) consumed about half the litter, and light fires only scorched the litter (Stark, 1977).

Mineral Nutrients

Forest fires result in large losses of mineral nutrients from the biomass and debris (Tables 11.1, 11.2, and 11.3). Nutrients are lost by volatilization and ash convection, and subsequently by leaching. Nitrogen and sulfur volatilize at relatively low temperatures and large amounts are lost to the atmosphere during hot fires. Ecosystems with large amounts of mineral nutrients aboveground may experience catastrophic losses of nutrients. For example, 50 to 70% of the N and other nutrients (and occasionally as much as 80%) may be lost during high-intensity fires (Wright and Bailey, 1982; Waring and Schlesinger, 1985).

Mineral nutrients dissolved in ash are available for absorption by plants but may be lost in surface runoff, leached into the soil and retained, or leached through the soil profile. Losses by surface runoff are appreciable when large amounts of ash are exposed and little ground cover is present. However, nutrient losses decrease with time as the vegetation becomes reestablished (Fig. 11.3). Losses by water runoff vary with the amount of ash, slope, infiltration capacity of the soil, and the amount and duration of rainfall after fire. Losses to groundwater are greatest for K^+,

Table 11.1
Percentage Loss of Mineral Nutrients from Fuel (Vegetation and Litter) to the Atmosphere by Volatilization and Ash Convection during Fire[a]

Ecosystem	Nitrogen	Phosphorus	Potassium	Calcium	Magnesium
Longleaf pine savanna, North Carolina	70	46	46	15.5	13.5
Pine plantation, South Carolina	33.3	—	—	—	—
Limestone grasslands, England (2 sites)	73	25, 39	39, 45	9, +1[b]	—
Eucalyptus forest, Tasmania	—	10	9	17	17
Mixed conifer forest, Washington	39	—	35	11	15
Chaparral, California	39	1	16	2	—
Heather moorland, England	100	3	12.5	3	4.5

[a]From Boerner (1982).
[b]Gain of calcium may be due to limestone fragments in samples.

Table 11.2
Effect of Fire on Loss of Sulfur from Forest Foliage and Litter[a,b]

Type of material	Time (min)	Temperature range (°C)			
		375–575	575–775	775–975	975–1175
Pinus ponderosa	5	58	79	81	81
	30	58	77	77	88
	60	54	77	77	92
Pseudotsuga menziesii	5	24	69	80	85
	30	46	67	76	85
	60	50	67	77	81
Alnus sinuata	5	79	56	69	83
	30	79	79	73	84
	60	82	82	81	85
Ceanothus velutinus	5	70	35	73	78
	30	68	68	62	76
	60	63	60	68	63
Forest litter	5	32	49	61	64
	30	38	46	60	68
	60	37	50	59	61

[a]From Tiedemann (1987).
[b]The data are percentages of the weight of sulfur in unburned plant material.

Table 11.3
Transfer of Mineral Nutrients to the Atmosphere from the Litter Layer and *Daviesia mimosoides* Understory of Three *Eucalyptus* Subalpine Forests in Australia during a Prescribed Burn[a,b]

Site	N	P	K	Ca	Mg	Mn	B
				Total transfer			
E. pauciflora	109	3.04	21.0	20.4	9.7	4.3	118
E. dives–E. dalrympleana	83	1.96	12.1	18.7	4.5	1.9	79
E. delegatensis	74	2.71	12.5	29.7	4.6	1.6	81
				% transfer[c]			
E. pauciflora	75	50	66	34	49	43	54
E. dives–E. dalrympleana	57	38	44	31	28	27	39
E. delegatensis	54	37	43	32	25	25	35

[a]From Raison *et al.* (1985).
[b]All weights are kilograms per hectare except for boron (grams per hectare).
[c]Transfer as a percentage of the initial mass of element present in the litter and understory.

followed by NH_4^+, Mg^{2+}, and Ca^{2+}. When the ash contains some NH_4^+ ions and pH at the soil surface is increased, NH_3 may be lost to the atmosphere (Wright and Bailey, 1982).

Pathways of potential N transfer in burned and unburned ecosystems are shown in Fig. 11.4. The rate of N transfer in a burned ecosystem is strongly influenced by physical and chemical properties of soil, the amount of litter present, slope, and

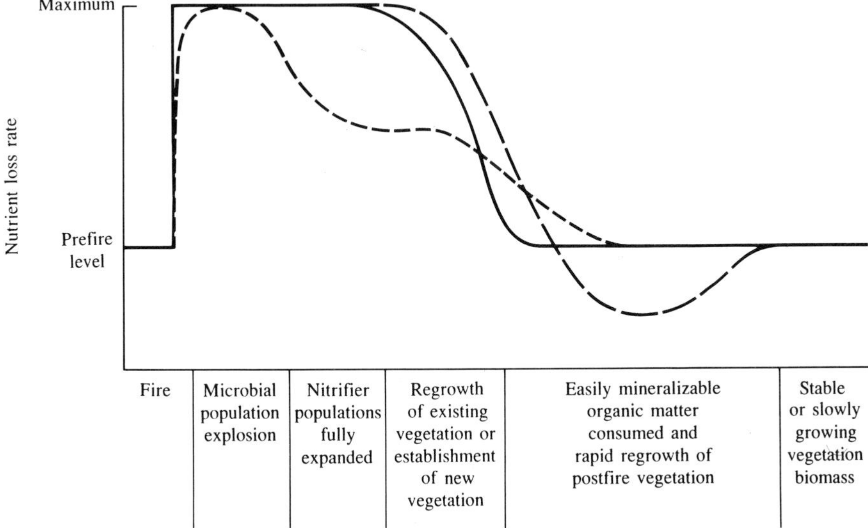

Figure 11.3 Hypothetical potential of losses of mineral nutrients from an ecosystem after fire due to wind erosion (——), leaching (– –), and water erosion (---). (From Woodmansee and Wallach, 1981.)

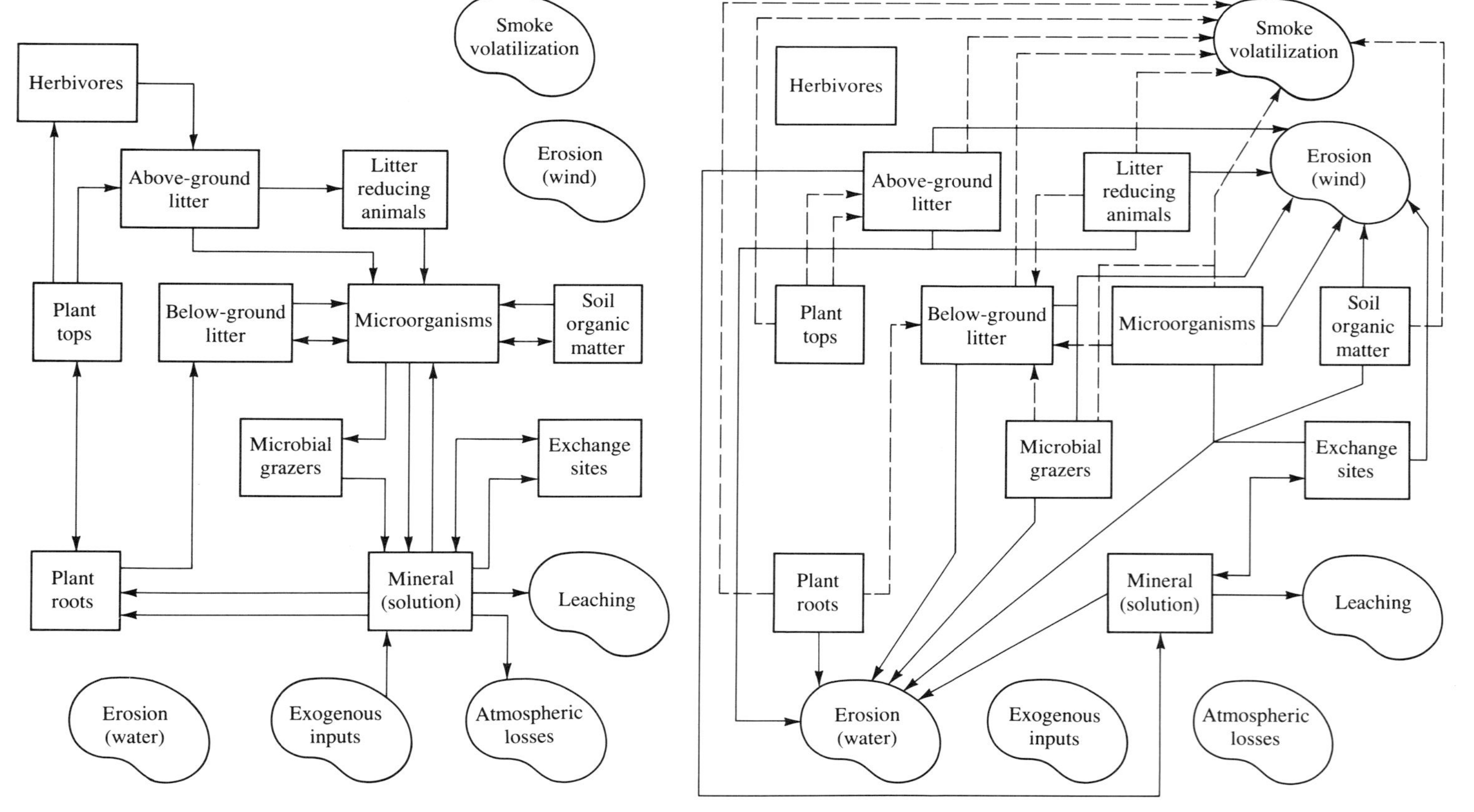

Figure 11.4 Pathways of nitrogen movement in an undisturbed ecosystem (*left*) and in an ecosystem affected by fire (*right*). The solid lines indicate flows of elements that may occur following fire; the dashed lines indicate flows of elements that may occur as a direct result of burning. (From Woodmansee and Wallach, 1981).

weather. The relative contribution of structural components shown in Fig. 11.4 varies among ecosystems, and processes within various ecosystems commonly express unequal rates of activity.

Organic N is volatilized primarily as N_2 or oxides of N. Many estimates have been made of the potential amounts of N lost to ecosystems by burning and only a few will be cited here. Soils heated to 300°C lost 25% of their N, and those heated to 700°C lost 64% (Knight, 1966). Nitrogen losses from burning of young, dense stands of ponderosa pine trees averaged 87% (Klemmedson, 1976). Losses of N from chaparral fuels varied from 40% at 600°C to 80% at 825°C (De Bano *et al.*, 1979). As much as 97% of the N in the forest floor against 66% of that in the A_1 horizon of mineral soil was lost by fire in a conifer stand (Grier, 1975). Whereas total and available N were lost from the burned forest floor, they increased in the unburned layers, probably because of downward transport (Mroz *et al.*, 1985b). As emphasized by Mroz *et al.* (1980), because of appreciable downward movement of N compounds from the forest floor, studies that examine only the burned material tend to overestimate N losses.

When organic matter is burned, CO_2 is released and mineral nutrients are concentrated in the ash. Hence, fire often increases the amount of mineral nutrients available to plants. Even though much N is lost by combustion, it usually is rather rapidly replaced by inputs in rain and by increased microbiological activity and N fixation by free-living organisms (Fig. 11.5). The effect is partly related to a change in soil pH from acidic to neutral conditions. As a result of release of basic cations by

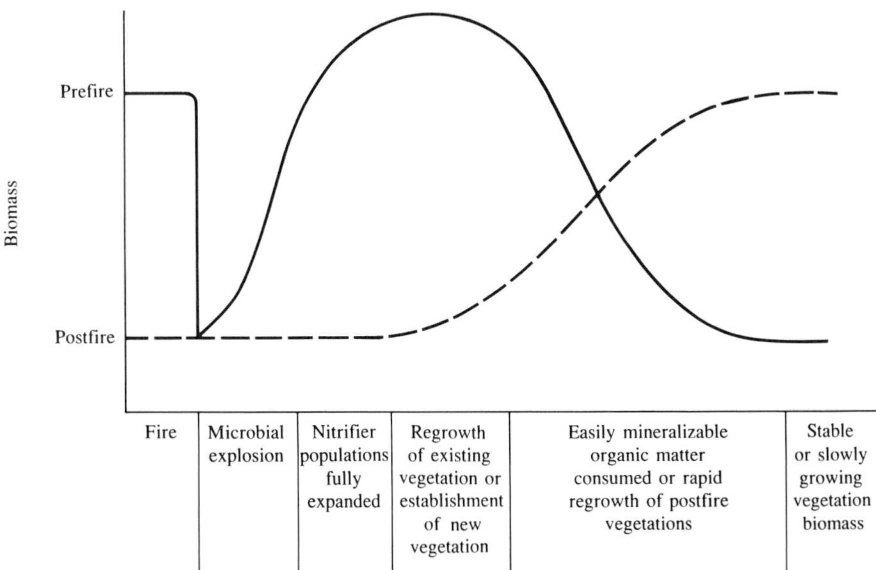

Figure 11.5 Relative time course of nitrogen transformations in soils after fire. Microorganisms (——); vegetation (– – –). (From Woodmansee and Wallach, 1981.)

combustion of organic matter, fire increases soil pH. The amount of pH increase varies with the original pH and organic matter content, amount and composition of ash, and amount and duration of rainfall. Hot slash fires in the western United States may raise the soil pH from 5 or 6 to 7 or more. The increase is smaller where fire intensity is lower, as in pine stands in the southeastern or central United States (Rundel, 1981). However, much of the time the change is in the range of 0.2 to 0.5 pH units.

The extent of increase in N after fire varies among ecosystems and depends to a considerable extent on N fixation by legumes. In many soils, however, N fixation by free-living aerobic *Azotobacter* and anaerobic *Clostridium* bacteria is more important than N fixation in legume nodules. These two species of bacteria increase following fire (I.F. Ahlgren, 1974). By comparison, replacement of P usually is slow because inputs in rain and from weathering of minerals occur slowly (Raison *et al.*, 1985).

Despite potential nutrient losses due to volatilization and increased runoff, the availability of nutrients is generally higher after fire (Christensen, 1985). For example, one year after burning, the concentrations of both ammonium and nitrate N were higher in repeatedly burned ponderosa pine stands than in unburned stands (Covington and Sackett, 1986). Several mechanisms may be involved in such increases, including direct production by burning, leaching into the mineral soil, stimulation of microbial mineralization because of greater availability of soil moisture, and decreased uptake by plants because of root mortality. Both ammonium and organic N increase shortly after fire in chaparral soils. The high nitrate levels, together with increases in other mineral nutrients, create favorable conditions for growth of plants. By comparison, the low nutrient levels in chaparral soils that have not been burned for a long time are correlated with decreased vegetative growth (Christensen and Muller, 1975a,b). In a Mediterranean oak forest the concentration of N in burned surface soil 5 months after a fire was substantially increased (Rashid, 1987).

Concentrations of mineral nutrients after fire tend to be more variable than those before fire, because of local differences in fire intensity and uneven distribution of ash (Christensen and Muller, 1975a; Westman *et al.*, 1981). Whereas peat P concentrations in certain locations of a shrub bog were not affected by fire, they were increased by 10 times in other locations (Wilbur and Christensen, 1983).

Fire-caused increase in nutrient availability does not necessarily improve site quality. If the highly soluble minerals in the ash are leached into the soil and absorbed by roots, site quality is temporarily improved. But if these minerals leach below the root zone or are washed off the surface and removed in the stream flow, site quality may be reduced. Improvement of site by fire is characteristic of sandy to loamy soils on level ground. Site deterioration by loss of minerals is more likely to occur on very coarse sands or heavy soils (Spurr and Barnes, 1980).

Fire may degrade sites where the soil is almost wholly composed of organic matter. For example, when peat bogs are burned after extended droughts, the soil sometimes is destroyed. However, this effect varies with fire intensity. Actually most peatland fires in the southeastern United States and Canada burn only on the surface. Another example of site degradation is the exposure of erodable soil on steep slopes by the burning of organic matter. In contrast, in cold climates, where low temperatures prevent decomposition of organic matter and heavy mats of acid raw humus exist, fires generally improve the site quality for forest species.

Soil Erosion

By removing vegetation and decreasing soil infiltration capacity, forest fires increase both sheet and rill erosion (removal of soil on sideslopes by small channels). The amount of erosion varies with fire intensity, topography, amount of plant cover remaining, and seasonal and long-term climatic conditions (Wright and Bailey, 1982). Whereas wildfires often are intense enough to change the structure of soil and induce appreciable erosion, light fires, such as prescribed burns, generally do not cause significant soil erosion.

Infiltration rates of forest soils often are appreciably reduced after fire (Table 11.4). Much water-induced erosion has been linked to formation of water-repellent soil layers after fire. Water repellency results when soil particles are coated with hydrophobic organic compounds liberated by the burning of litter (De Bano, 1981). Such compounds diffuse downward into the soil and condense on the cooler soil

Table 11.4
Average Infiltration Rates of Seven Forest Soils Unburned for Several Years and Comparable Soils Burned Annually[a]

Soil type	Unburned soil (in.)	Burned soil (in.)	Apparent reduction by annual burning (in.)	(%)
Tilsilt silt loam	1.926	1.079	0.847[b]	44
Tilsilt very fine sandy loam	1.761	0.920	0.841[b]	48
Sabula silt loam	1.672	1.378	0.294[b]	18
Ironton silt loam	2.440	0.929	1.511[b]	62
Ironton stony silt loam	1.716	0.953	0.763[b]	44
Lebanon silt loam	2.040	1.374	0.666[b]	33
Clarksville cherty silt loam	3.288	2.634	0.654[b]	20
All soil types	2.121	1.324	0.797[b]	38

[a]From Arend (1941). Reprinted from *Journal of Forestry*, published by the Society of American Foresters, 5400 Grosvenor Lane, Bethesda, MD 20814-2198.

[b]Significant at the 1% level.

particles (Fig. 11.6). Water-repellent layers are more likely to form in dry, sandy soils than in wet, fine-textured soils.

Erosion of soils following fires has been well documented in temperate forests (Little, 1974), chaparral (Biswell, 1974, 1989), and forests and wooded savannas of Africa (Phillips, 1974). Chaparral soils on steep mountain slopes are especially prone to erosion, with soil losses after wildfire up to 10 times as high as those of the already high rates before fire. In some chaparral ecosystems, erosion occurs primarily in the early winter of the first year after a fire, before the soil has become covered with new vegetation (Biswell, 1974).

Usually the amount of soil eroded away is closely correlated with slope. Losses of soil by erosion following prescribed burning on juniper lands in Texas were negligible on level sites, whereas erosion was severe and its effects evident for 9 to 15 months on moderate slopes and for more than 30 months on steep slopes (Wright *et al.*, 1976). Soil losses by erosion stabilized within 18 months, by which time the plant cover amounted to 63 to 68%.

On many sites the amount of soil erosion after burning is related to the amount of plant cover. Soil erosion after fire on steep prairie slopes invaded by Ashe juniper was extensive (Wright *et al.*, 1982). However, seeding with grasses reduced erosion by as much as 90%. Furthermore, the major impact of fire on erosion was reduced within 3 months on burned and seeded watersheds but not for 15 to 18 months on unseeded watersheds. Soil losses were stabilized when cover (live vegetation plus litter) reached 64 to 72% during normal to wet years and 53 to 60% during dry years. Hence, erosion was strongly affected by the amount of rainfall and cover. There has been some concern that seeding may slow erosion in the short term but, because planted exotics may slow the succession of native species, seeding may eventually increase erosion. Prescribed burning every 4 to 5 years in ponderosa pine

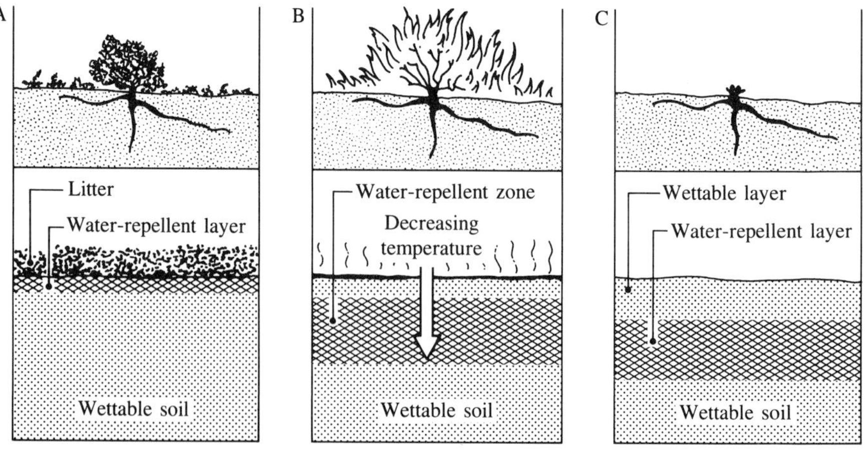

Figure 11.6 Location of water-repellent layer under chaparral; unburned (*left*), intensely burned (*right*), moderately intense fire (*center*). (U.S. Forest Service figure.)

stands in northern California had little effect on surface runoff and soil erosion even after heavy rains. The partially decomposed duff and debris that remained after prescribed burns maintained high infiltration and penetration capacity (Biswell and Schultz, 1957).

Diseases and Insects

Forest fires aid in the control of certain plant diseases and in the spread of others. Some diseases can be controlled because fire eliminates the fungus inoculum. For example, seedlings often thrive on ash beds in part because of the absence of fungi that cause seed decay, damping-off, and root rot. Burning decreases the incidence of brown needle spot caused by *Systremma acicola* on longleaf pine seedlings. Temperatures high enough to kill the needles also kill the fungus. Hence, inoculum usually is negligible on needles that are killed but not consumed by fire. *Nectria* cankers and *Septoria* leaf spot of *Vaccinium* also are controlled by fires (I.F. Ahlgren, 1974).

In other cases, fire is associated with the spread of fungus pathogens by stimulating formation of dense stands of host plants and by creating infection courts for fungi. Spread of powdery mildew (*Microsphaera alni*) and rust (*Pucciniastrum myrtelli*) fungi on *Vaccinium* is favored by thick stands of host plants (Demaree and Wilcox, 1947). Rapid postfire spread of *Rhizina undulata* root rot in Scotch pine plantations led to reduced use of prescribed burning in Finland (I.F. Ahlgren, 1974). Surface roots damaged by fire and fire scars on tree stems are primary infection courts for decay fungi. Almost all decay of timber in Kentucky upland oak stands was associated with injury by fire (Berry, 1969).

Trees weakened by fire are more susceptible than vigorous trees to attack by certain insects, especially wood-boring beetles. Some species of *Melanophila* beetles are attracted to the heat and smoke of forest fires from distances of up to 5 km. Dispersing beetles preferentially land on fire-scarred trees and especially on trees with fungal decay (Gara *et al.*, 1984). Virtually every tree in a coniferous forest may become infected with a variety of wood borers within a few weeks after a fire, even where the burned area extends for hundreds of acres (Evans, 1972). Outbreaks of spruce budworm (*Choristoneura fumiferana*) often start in local environments (epicenters) in mixed conifer–hardwood forests exposed to fire, and then spread into spruce–fir stands. Some insects, including ants, termites, and wood-boring beetles, bore into the wood of dead trees, thereby increasing invasion of wood-decaying fungi and the rate of wood decay. The result is a decreased fuel load, which lowers the likelihood of subsequent severe forest fires (Berryman, 1986).

Air Pollution

Fires often release significant amounts of particulate and gaseous pollutants to the atmosphere. In 1974, forest fires and burning of slash accounted for approximately

15% of the total particulate emissions in Canada (National Research Council, 1979a). Particulates at ground level downwind increased by 10 times as a result of broadcast burning of Douglas-fir slash (Fritschen *et al.*, 1970). Burning also released several pollutants, including low concentrations of ethylene, ethane, propene, propane, methanol, and ethanol. Wildfires produce about 10 times as much particulate matter per acre of fuels burned as do prescribed burns. This is because wildfires consume about three times more fuel per acre than prescribed fires, and because the particulate count per ton of fuel consumed is greater. Furthermore, more green, living tissues are burned by wildfires (Biswell, 1989). According to the Environmental Protection Agency (1977), wildfires and prescribed burning in the United States accounted for only 3% of the particulates and 0.6% of the oxides of N released from 1970 to 1976.

Species Composition

Some species of woody plants are much better adapted than others to complete their life cycle in a fire-prone environment. Except for rain forests, most forested areas have been burned periodically. As a result, succession to a climatic climax stage is arrested in many regions and vegetation is almost indefinitely maintained in a subfinal "fire climax" stage. A few examples will be given.

In the central and northeastern parts of the United States, trembling aspen, jack pine, and black spruce are perpetuated after fire on many sites. By comparison, eastern white and red pines do not reproduce naturally after fire because of lack of seed, increased competition from brush and aspen, and presence of white pine blister rust (Ahlgren and Ahlgren, 1984). The incidence of white pine blister rust increases because germination of its alternate hosts, wild currants and gooseberries, increases after fire removes the duff. Very shade-tolerant species such as sugar maple, red maple, American beech, eastern hemlock, balsam fir, and eastern arborvitae tend to occupy protected and wet sites that are less subject to intense fires. When fire is excluded, they spread to adjacent areas and replace the fire-dependent species (Spurr and Barnes, 1980).

In the southeastern United States, subclimax pine forests are maintained by recurring fires. Without fire, longleaf pine of the coastal plain is succeeded by hardwoods and other pines. This is also true for coastal plain slash pine forests and loblolly pine forests of the piedmont plateau (Komarek, 1974).

In the western and northwestern parts of the United States, several species of trees are dominant over vast acreages because of the effects of fire. For example, Douglas-fir is the principal fire-climax species over much of the Pacific Northwest. When gaps are created in pine forests by death of trees, lightning, disease, or windthrow, the felled and dead trees are consumed by subsequent fires. Seeds of ponderosa pine trees readily germinate in the ash beds, and because seedlings of this species are more resistant to fire than those of associated species, even-aged groups of ponderosa pine are established. Following fire, lodgepole pine rapidly forms dense stands and may occupy areas previously dominated by Douglas-fir. Western larch, another fire-climax species, is commonly associated with Douglas-fir and ponderosa pine, but also occurs in pure stands. Coast redwood also survives fire when other tree species are killed (Weaver, 1974). Chaparral vegetation is especially well adapted to recurring fires (Biswell, 1974).

Adaptations to Fire

Adaptive traits of fire-tolerant trees and shrubs include those that increase survival during fire, such as thick bark, and those that stimulate regeneration of forests, including fire-stimulated seed dispersal and germination, fire-accelerated flowering, and seed storage on the plant or in the soil (Gill, 1981).

Bark Thickness

The insulating properties of bark often determine the extent of fire injury to the phloem, cambium, and buds beneath the bark of some plants. Species with thick or corky bark such as redwood, western larch, ponderosa pine, Douglas-fir, longleaf pine, and bur oak often escape fire injury. In contrast, thin-barked species such as Engelmann spruce, alpine fir, lodgepole pine, and eastern white pine often are killed by ground fires. Young trees of species that have thick bark when mature may be readily killed by fire.

Sprouting

Many species survive fire by sprouting even after most of their foliage has been destroyed. Loss of leaves by fire generally releases the dormancy of buds that survive under the bark or stimulates production of adventitious buds. Dormant buds, originally located in leaf axils of twigs, are connected to the pith by a bud trace. They remain undeveloped under the bark and grow only enough each year for the tip to keep pace with cambial growth. Branching from dormant buds occurs commonly. Adventitious buds form rather irregularly on older portions of a plant rather than at stem tips or in leaf axils. They form on parts of the root or stem that have no connection to apical meristems and they may originate from deep-seated tissues or peripheral ones. The distinction between adventitious and dormant buds is the lack of a bud trace all the way to the pith in the former (Kozlowski, 1971a).

Capacity for sprouting is a much more important adaptation to fire in broad-leaved trees and shrubs than in conifers. Nevertheless, sprouting is crucial for survival of a few species of conifers after fire. For example, pitch pine, one of the most fire-tolerant species of the northeastern United States, often sprouts at the stem base if the stem is killed by fire. Even if all the foliage is scorched or killed, the trees usually refoliate by sprouting. Other conifers with high capacity for sprouting include coast redwood, pond pine, shortleaf pine, and bald cypress.

Sprouting of a few species of conifers after fire has sometimes been attributed to adventitious buds (Mattoon, 1908). At least some such reports were erroneous because Stone and Stone (1954) found that several sprouting species of pines had small buds in the axils of primary needles just above the cotyledons. At that location the needles were closely spaced, and the buds often were clustered after the stem thickened. These buds often produced basal or root collar sprouts. The bud steles were traced back to an origin in the first-year stem, emphasizing that sprouts arose from dormant rather than adventitious buds.

Various broad-leaved trees such as oaks regenerate by sprouts from roots that were not killed by fire. Most of the important species of climax chaparral produce vigorous stump sprouts (Table 11.5), although the amount of sprouting varies among species. For example, nearly all eastwood manzanita plants produce sprouts after fire, but only about 70 to 75% of chamise plants do so. The sprouting species

Table 11.5

Common Sprouting and Nonsprouting Species of Climax Chaparral[a]

Sprouting species	
Chamise (*Adenostema fasciculatum*)	Yerba santa (*Eriodictyon californicus*)
Scrub oak (*Quercus dumosa*)	Chaparral whitethorn (*Ceanothus leucodermis*)
Interior live oak (*Quercus wislizenii* var. *frutescens*)	Bear bush (*Garrya fremontii*)
Leather oak (*Quercus durata*)	Coffee berry (*Rhamnus californica*)
Canyon oak (*Quercus chrysolepis*)	Red berry (*Rhamnus crocea*)
Eastwood manzanita (*Arctostaphylos glandulosa*)	Evergreen cherry (*Prunus ilicifolia*)
Woolyleaf manzanita (*Arctostaphylos tomentosa*)	Black sage (*Salvia mellifera*)
Chaparral pea (*Pickeringia montana*)	Christmas berry or toyon (*Heteromeles arbutifolia*)
California laurel (*Umbellularia californica*)	
Nonsprouting species	
Wedgeleaf ceanothus (*Ceanothus cuneatus*)	Common manzanita (*Arctostaphylos manzanita*)
Hoaryleaf ceanothus (*Ceanothus crassifolius*)	Stanford manzanita (*Arctostaphylos stanfordiana*)
Wavyleaf ceanothus (*Ceanothus foliosus*)	Littleberry manzanita (*Arctostaphylos sensitiva*)
Gregg ceanothus (*Ceanothus greggii*)	Bigberry manzanita (*Arctostaphylos glauca*)
Whiteleaf manzanita (*Arctostaphylos viscida*)	

[a]After Biswell (1974); from Kozlowski and Ahlgren (1974), by permission of Academic Press.

have a distinct competitive advantage over those that reproduce only by seeds because the sprouts grow faster than seedlings. Postfire successions in chaparral are toward strongly sprouting species (Biswell, 1974). On sandy soils in Florida, the scrub oak, *Quercus inopina*, which resprouts from rhizomes after fire, regains and often exceeds its preburn cover within 3 years, whereas seeds of the associated nonsprouting Florida rosemary (*Ceratiola ericoides*) do not even germinate for 2 years after fire. Furthermore, seedlings of *Ceratiola* do not reach reproductive capacity for 10 to 15 years (Johnson *et al.*, 1986).

Shrub species in mediterranean climates vary from obligate sprouters (regenerating primarily from sprouts) to facultative sprouters (reproducing from seeds and sprouts) to obligate nonsprouters (reproducing entirely from seeds) (Christensen, 1985) (Table 11.6).

Following disturbance of forest stands by fire, some species regenerate largely by root suckers, which arise from adventitious buds. An example is trembling aspen.

Serotinous Cones

Seeds may be stored on plants and released after a fire, as in species that produce serotinous (late-to-open) cones. In lodgepole pine, jack pine, pitch pine, sand pine, knobcone pine, black spruce, and Sargent's cypress, cones containing viable seeds often remain unopened on the tree for years. Seeds of jack pine can tolerate a temperature of 150°C for 30 to 45 sec and 370°C for 10 to 15 sec (Beaufait, 1960). When the resinous material on serotinous cones is destroyed by fire, the cone scales open and the seeds are released. The cones of black spruce are borne near the center of the upper crown, where they are unlikely to be severely damaged by fire. Usually the seeds are gradually dispersed during the first year after a fire. Where fires occur frequently, the release of copious amounts of seeds from serotinous cones is likely to determine the dominant species in forest stands.

Seed Germination and Seedling Establishment

Site conditions after burning, such as removal of litter, exposure of mineral soil, increase in soil temperature, and at least temporary increase in availability of mineral nutrients, are conducive to seed germination and seedling establishment. The abundance of vigorous seedlings of coast redwood in burned areas reflects the beneficial effects of reduced duff and debris to expose mineral soil and the reduction in pathogens and potentially toxic compounds in the seedbeds (Stone *et al.*, 1972; Weaver, 1974). By removing and exposing mineral soil, fire also creates suitable sites for germination of seeds of the neotropical savanna shrubs *Miconia albicans* and *Clidemia sericea* in Belize (Miyanishi and Kellman, 1986). In some species the effect of fire on germination is indirect as a result of improving the light regime. Although most species germinate as well in the dark as in the light, some species of birches, pines, and spruces require very low irradiance for germination (Jones, 1961; see also Chapter 4).

Table 11.6
Regeneration Capacity after Fire of some Mediterranean Woody Plants[a]

Name of plant	Resprouting	Spreading by seeds[b]
Trees		
Pinus halepensis	−	+
Quercus calliprinos	+	−
Quercus ithaburensis	+	−
Quercus boisseri	+	−
Ceratonia siliqua	+	−
Styrax officinalis	+	−
Laurus nobilis	+	−
Arbutus andrachne	+	−
Rhamnus alaternus	+	−
Pistacia palaestina	+	−
Phillyrea media	+	−
Cercis siliquastrum	+	−
Shrubs		
Pistacia lentiscus	+	−
Rhamnus palaestina	+	−
Calycotome villosa	+	+
Dwarf shrubs		
Sacropoterium spinosa	+	+
Cistus salvifolius	+	+
Cistus villosus	+	+
Salvia triloba	+	+
Teucrium creticum	+	+
Majorana syriaca	+	+
Satureja thymbra	+	+
Thymus capitatus	+	+
Climbers		
Rubia tenuifolia	+	−
Smilax aspera	+	−
Tamus communis	+	−
Asparagus aphyllus	+	−
Clematis cirrhosa	+	−
Lonicera etrusca	+	−
Prasium majus	+	−

[a]From Naveh (1974), by permission of Academic Press.
[b]Only plants with pronounced postfire germination.

Viable seeds that are buried in the ash layer often germinate readily. Establishment of some species before fire is particularly difficult because, although the seeds may germinate, the rootlets cannot penetrate the leaf litter and reach mineral soil before they dehydrate. By destroying litter, fires also may remove sources of germination inhibitors (Went *et al.*, 1952).

Seeds of certain species require heat or fire to stimulate germination (Keeley and Zedler, 1978). Exposure to high temperatures for short periods stimulated germination of seeds of California red fir and ponderosa pine but not those of lodgepole pine or sugar pine (Wright, 1931). Seeds of several chaparral species are stimulated to germinate by heat. These include Mexican manzanita, pinkbract manzanita, deerbrush, Gregg ceanothus, and *Eriodictyon angustifolium* (Biswell, 1974, 1979). Germination of seeds of snowbrush and greenleaf manzanita was stimulated by temperatures of 60 to 105°C (Gratkowski, 1961a,b). Germination of sugar sumac (*Rhus ovata*) seeds in California chaparral usually is obvious only after fire. Exposure of dry seeds to 80°C for 5 to 60 min resulted in 25% germination, and exposure to 100°C for 5 to 10 min resulted in 33% germination compared with negligible germination without fire. The higher temperatures stimulated germination by cracking the seed coat near the micropyle and increasing permeability to water (Stone and Juhren, 1951). The accelerated germination of seeds of many herbs was attributed to soil heating by increased insolation following fire (Christensen and Muller, 1975a,b).

Regeneration of some species after fire is related to unusual systems of seed storage and stimulation of seed germination. After severe fires, regeneration of mountain ash in Australia depends on the seed stored in canopy trees. Flowering periodicity of this species is strongly biennial and seed formed in a good flowering year is released over a 2-year period. In contrast, if the crown is severely scorched by fire, virtually all the seeds are released within a few days. Hence, the seed supply after fire is much greater than in years between fires (Gill, 1981).

Multiple Adaptations

Some species exhibit more than one adaptation to fire. For example, redwood is favored over other species by its fire-resistant bark and its capacity to sprout prolifically around the root crown and along the stem. In contrast to western hemlock, seedlings of redwood do not readily become established on the undisturbed forest floor (Stone *et al.*, 1972; Biswell, 1974).

Adaptation of chaparral to fire is associated with (1) seed production at an early age; (2) production of seeds that remain viable for decades, are resistant to fire, and, in some cases, fail to germinate in the absence of fire; and (3) sprouting from lignotubers (Biswell, 1974, 1989). About half the species of California chaparral can sprout after the top is killed or consumed by fire (Hanes, 1988). Survival of heathlands following fire has been attributed to sprouting from stems or roots, or to germination of seeds stored on plants or in the soil (Gill and Groves, 1981).

Prescribed Burning

In recent years there has been wide recognition of the benefits of controlled or prescribed burning. When properly used, prescribed burning can result in a variety of benefits to forest ecosystems without many of the harmful effects of wildfires.

The potential benefits include reduction in fuels and fire hazards, restoration of fire to its appropriate role in ecosystem functioning, improvement of wildlife habitat and ranges for livestock, reduction of understory brush, preparation of seedbeds for planting, stabilization of watersheds, decrease in disease and insect infestation, stimulation of seed production or opening of cones, reduction in use of herbicides and insecticides, and facilitating forest and range management (Kayll, 1974; Biswell, 1989).

Although prescribed burning often is carried out with a primary objective in mind, it usually has multiple beneficial effects. For example, burning to reduce brush also decreases the risk of intense wildfires by reducing the amount of fuel and improves habitats for wildlife. Controlled burning alone is not always a panacea for problems, hence other cultural practices may also be advisable. For example, where forests are overbrowsed or soils infertile, reseeding and fertilizing, in addition to prescribed burning, may be beneficial. Biswell (1989, p. 187) suggested that prescribed burning can increase the yield of wood products by as much as 50 to 75% in California forest types that depend on fire, and at a much lower cost than management practices that do not include use of fire and that require large expenditures for wildfire control and suppression. Biswell (1989) emphasized that low-intensity prescribed burns can be beneficial when conducted under fuel and weather conditions that allow burning without excessively heating the soil, and when repeated often enough to prevent accumulation of so much fuel that almost any fire will expose the surface soil to excessive heating. Furthermore, properly conducted prescribed burns do not measurably increase soil erosion and they facilitate nutrient cycling. Whereas high-intensity fires can volatilize nearly all the N, low-intensity prescribed burns volatilize very little.

Even though prescribed burning reduces organic matter on the soil surface, it may increase the amount in the soil when followed by growth of more grasses and legumes that directly add organic matter to the soil. Prescribed burning also can increase water yield by (1) reducing interception losses through fuel reduction; (2) removing deep-rooted plants that consume much water in favor of shallow-rooted plants that use less water; (3) reducing water loss by evaporation and transpiration by growing younger stands, as in chaparral; and (4) reducing interception, evaporation, and transpiration in forests by burning of understory vegetation (Biswell, 1989).

Prescribed burning has been especially well developed in Scandinavian countries, where one of the most serious obstacles to regeneration of forests is a thick layer of humus (Viro, 1974). In Manitoba, Canada, burning of peat on lowland sites improved the seedbed and regeneration of black spruce after harvesting. After five growing seasons, stocking of black spruce was 94% on a moderate burn, 70% on a light burn, and only 35% on an unburned site (Chrosciewicz, 1976).

There also are many examples of beneficial effects of prescribed burning in the United States and only a few will be given here. Major forest management problems ascribed to many years of fire exclusion in ponderosa pine forests in the western United States include overstocked sapling patches, reduced growth, stagnated nutrient cycles, disease and insect infestation, and excess fuel accumulation (Fig. 11.7). These problems can be alleviated by prescribed burning (Fig. 11.8). For example, a controlled burn one year before wildfire reduced the impact of subsequent wildfire on overstory ponderosa pine trees, surface vegetation, and organic soil layers (Wagle and Eakle, 1979). Prescribed burning in a ponderosa pine forest destroyed 76% of the litter and 23% of the duff. However, enough litter and duff remained after the fire to protect the soil from erosion (Sweeney and Biswell, 1961).

Near Flagstaff, Arizona, prescribed burning in a ponderosa pine stand decreased organic matter by 63%, with a 99% reduction of the large, decayed wood. Nutrient storage in the woody debris decreased by 80% for N, 62% for P, 70% for Ca, 71% for Mg, and 74% for K. Forest floor storage was less drastically altered, with organic matter content declining by 37% immediately after burning. However, the nutrient content of the forest floor was not appreciably influenced by burning, partly because mineral nutrients from the woody debris were transferred to the forest floor (Covington and Sackett, 1984). Similarly, N mineralization and nitrification in the forest floor were increased immediately after a prescribed burn in a ponderosa pine stand, and they remained elevated for at least 10 months after the fire. Although these rates in the mineral soil were not changed immediately by burning, they were appreciably increased 6 months after the fire (White, 1986).

Figure 11.7 Heavy accumulation of fuel in a mixed conifer forest in California. A wildfire in this forest would be very destructive and difficult to control. (From Biswell, 1989.)

Figure 11.8 Effects of prescribed burning on fuel reduction in a stand of giant sequoias. *Top*, Before the burn; *Bottom*, After the prescribed burn. Most of the remaining heavy fuels were piled and burned to further reduce the fire hazard. (From Biswell, 1989.)

Prescribed burning is practiced routinely in the southeastern United States. A low-intensity prescribed burn in April in a loblolly pine stand in North Carolina did not reduce the N content of the litter. However, the rate of decomposition of the forest floor more than doubled during the first growing season after burning. This resulted in release of 60 kg N ha^{-1} more than in an unburned portion of the same stand (Schoch and Binkley, 1986).

As emphasized by Biswell (1989), successful prescribed burning requires a sound knowledge of basic fire ecology, very careful planning, patience, experience, and competent supervision. A description of methods of prescribed burning is beyond the scope of this volume. For such information the reader is referred to Brown and Davis (1973), Kayll (1974), Chandler *et al.* (1983), Pyne (1984), and Biswell (1989).

Summary

In recent decades, the policies of forest managers have changed from complete exclusion of forest fires to tolerance of some natural fires because of their importance in perpetuating natural forests.

Effects of forest fires range from catastrophic to beneficial. Fires destroy timber, reduce the capacity of soils to store water, cause development of water-repellent layers, and induce soil compaction, fusion of soil surface layers, and soil erosion. Fires result in losses of mineral nutrients by volatilization, ash convection, and subsequent leaching. Trees weakened by fire are more susceptible than vigorous trees to attacks by certain insects, particularly wood-boring beetles. Burning of forests and forest debris releases some particulate and gaseous pollutants to the atmosphere. Periodic fires also arrest succession of forests to a climax stage. Hence, fires often maintain forests in a subfinal "fire-climax" stage.

Fires may benefit forests by preparing seedbeds and removing organic matter, hence preventing accumulation of combustible fuel. Because mineral nutrients are concentrated in the ash, fires often increase the amounts of mineral nutrients available to plants. However, if highly soluble mineral nutrients leach below the root zone or are washed off the soil surface, site quality may be reduced.

Some diseases are controlled when fire eliminates the fungus inoculum. Fires may also spread pathogens by favoring formation of dense stands of host trees and creating infection courts for pathogens.

Fire tolerance varies appreciably among species. Adaptations of fire-tolerant trees include those that increase survival during fire (e.g., thick bark) and those that stimulate regeneration of forests (e.g., fire-stimulated seed dispersal and germination, fire-accelerated flowering, seed storage on the plant or in the soil, and sprouting capacity).

Some important benefits of prescribed burning include (1) reduction in accumulated fuels and fire hazards, (2) control of insects and diseases, (3) improvement of wildlife habitats and range for livestock, (4) preparation of seedbeds for planting,

(5) stimulation of seed production or opening of cones, and (6) control of undesirable understory vegetation.

General References

Biswell, H. H. (1989). "Prescribed Burning in California Wildlands Vegetation Management." University of California Press, Berkeley.

Booysen, P. de V., and Tainton, M. H. (1984). "Ecological Effects of Fire in South African Ecosystems." Springer-Verlag, New York.

Brown, A.A., and Davis, K.P. (1973). "Forest Fire: Control and Use." McGraw-Hill, New York.

Chandler, C., Cheyney, P., Thomas, P., Trabaud, L., and Williams, D. (1983). "Fire in Forestry. Vol. I. Forest Fire Behavior and Effects." Wiley, New York.

Chandler, C., Cheyney, P., Thomas, P., Trabaud, L., and Williams, D. (1983). "Fire in Forestry. Vol. II. Forest Fire Behavior and Effects." Wiley, New York.

Christensen, N. L. (1985). Shrubland fire regimes and their evolutionary consequences. *In* "The Ecology of Natural Disturbance and Patch Dynamics" (S.T.A. Pickett and P.S. White, eds.), pp. 85–455. Academic Press, Orlando, FL.

Davis, K.P. (1959). "Forest Fire: Control and Use." McGraw-Hill, New York.

Kozlowski, T.T., and Ahlgren, C.E., eds. (1974). "Fire and Ecosystems." Academic Press, New York.

Pyne, S. J. (1982). "Fire in America: A Cultural History of Wildland and Rural Fire." Princeton University Press, Princeton, NJ.

Pyne, S.J. (1984). "Introduction to Wildland Fire. Fire Management in the United States." Wiley, New York.

Spurr, S.H., and Barnes, B. V. (1980). "Forest Ecology." Wiley, New York.

Trabaud, L., ed. (1987). "Fire Ecology." SPB, The Hague, Netherlands.

van Nao, T., ed. (1982). "Forest Fire Prevention and Control." Martinus Nijhoff/W. Junk, The Hague, Boston, and London.

Wright, H.A., and Bailey, A.W. (1982). "Fire Ecology." Wiley, New York.

Chapter 12

Wind

Introduction

Wind is an omnipresent part of the environment and has both harmful and beneficial effects on woody plants. Wind causes injury by toppling trees, breaking stems and branches, uprooting trees, causing stem malformations, and injuring leaves. In addition, wind erodes soils, inhibits tree growth by adversely affecting physiological processes, and increases the risks of air pollution, insect attack, disease, and fire. However, wind also benefits plants by assuring dispersion of pollen and propagules. The complex effects of wind on vegetation are exerted by its physical force as well as by its role in turbulent transfer of heat, water vapor, CO_2, air pollutants, spores, pollen, and seeds.

Injury

Conifers are more susceptible than broad-leaved trees to damage by wind. Pines, spruces, and firs are particularly prone to injury. Sensitivity to wind also varies among fruit trees, with pear trees being more readily damaged than apple trees (Waister, 1972a).

Windthrow and Uprooting of Trees

The stems of trees break when they are subjected to forces that exceed the strength of the stem but are not strong enough to break roots and upturn the root system. Trees are uprooted when subjected to lateral wind forces that exceed the root–soil holding strength but that do not break the stem. Less soil is disturbed by snapped trees than by uprooted trees. Of 310 trees injured by a hurricane on Barro Colorado Island, Panama, 70% snapped, 25% were uprooted, and 5% were broken off at the ground line. The uprooted trees tended to be larger and shorter for a given stem diameter, and had denser and stronger wood than the snapped trees (Putz *et al.*, 1983). During a hurricane on the island of Dominica, large-diameter trees tended to be uprooted more often than small-diameter trees, which tended to break (Lugo *et al.*, 1983).

The windward side of a tree stem is under tension and the leeward side is under compression. Compression failures in stems can occur without breakage of the stem if the wind is strong enough to induce compression failure on the leeward side but without causing tension failure on the windward side. Mergen (1954) described progressive compression failure in stems of pine trees during windstorms. The compression force was greatest in the outer layers of the sapwood and decreased toward the center of the stem, with stresses greatest in the lower stem. When the anchoring strength of the root system was strong enough to prevent windthrow during strong winds, the swinging movement of the crown often bent trees until the compression strength of the wood was exceeded. When fibers were compressed endwise beyond their limit, they buckled and bent. Exposure to continued winds resulted in a progressive increase in the number and extent of failures, with initial failures in the outer rings extending farther into the stem, and the extent of failure increasing in the outer rings. Compression stresses are augmented by diagonal shearing stress during bending of wood.

Uprooting of trees by wind is a serious problem in forests and orchards on windy seacoasts and in other areas subjected to strong winds or hurricanes. The stability of trees is influenced by soil characteristics, the root system, and stem and crown characteristics. Restricted root development of trees growing on shallow or water-logged soils increases the probability of wind damage. In Malaysia, uprooting of rubber trees is associated with shallow root systems caused by impervious rock or clay horizons 1 to 1.2 m below the surface (Savill, 1976). Injury to root systems by flooding, soil compaction, or fungal infection also increases susceptibility to overthrow by wind (Mathew and George, 1967). Trees growing in swamps usually have shallow, flat root systems and are particularly likely to be overthrown. Tall trees are more likely to be damaged than shorter ones.

In hardwood forests of the northeastern United States, species susceptibility to catastrophic winds varied largely with canopy position (Foster, 1988). Fast-growing, pioneer species that were overstory dominants (eastern white pine, red pine, poplars, and paper birch) were injured by wind much more than slower-growing or tolerant species occurring largely in codominant, intermediate, and suppressed canopy positions (hickories, red maple, white oak, black oak, and hemlock). It is said that in Scotland and Ireland, Sitka spruce trees become susceptible to windthrow when they grow to a height of 10 to 12 m. The tendency for damage by windthrow continues to increase with increasing height up to 17 or 18 m. Trees grow taller than 17 to 18 m only on protected sites.

If heavy rains precede high winds and soften the soil, uprooting is more likely than if the soil is dry. In dry soil, stems are more often snapped. The hurricane of 1938 uprooted many trees in New England because it was preceded by rain, but the soil was dry when Hurricane Hazel struck eastern North Carolina in 1954 and many tree stems broke.

Once a stem is deflected by wind, the combined weight of the stem and crown contributes to uprooting. Roots are much stronger than soil and stretch more than soil does before they break. Hence they function as reinforcing rods in the soil mass.

Coutts (1986) studied components of stability of 35-year-old, shallow-rooted Sitka spruce trees on peaty gley soils in England. When the trees were pulled laterally from the crown region with a winch, the soil failed when the crown was deflected about 4 m and when the applied force was about 70% of that necessary for uprooting. By the time the maximum turning moment at the stem base was reached, many roots were broken, the crown was deflected by about 8 m, and the deflected weight of the crown and stem contributed appreciably to the uprooting forces.

The total resistive turning moment afforded by the anchorage consisted of soil resistance (the resistance to uprooting by the soil), resistance of roots under tension on the windward perimeter, the weight of the root soil plate, and resistance to bending at the hinge on the lee side (Fig. 12.1). The importance of these components varied among trees and with time after subjecting trees to lateral forces. The importance of soil resistance was greatest in the early stages and least in the later stages. When the turning moment due to the applied force was at a maximum, the components of anchorage varied in the following order:

windward roots > plate weight > lee hinge > soil resistance.

Critical wind speeds for uprooting of trees are much lower than those predicted from data obtained in tree-pulling studies. This is because swaying of trees by strong winds loosens their roots in the soil, as shown for Scotch pine on sandy soils in England (Oliver and Mayhead, 1974) and Norway spruce trees on gley soils in Germany (Hütte, 1968). Root movement disrupts root–soil contact and decreases water absorption for some time, probably increasing the severity of water stress in trees.

Trees with diseased roots or stems are far more susceptible than healthy trees to

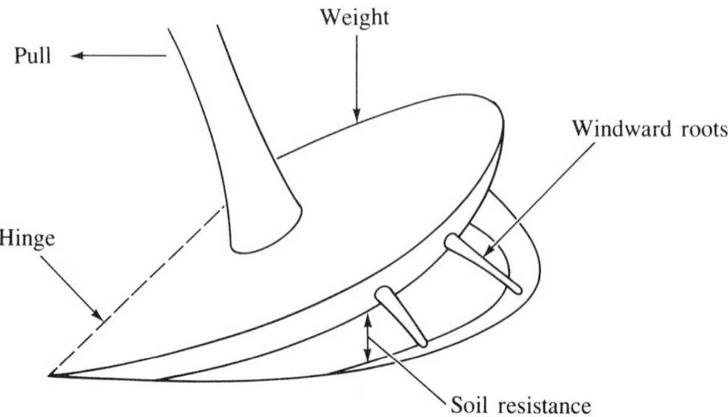

Figure 12.1 Diagram of a shallowly rooted tree, showing four components of the anchorage that resist the horizontal force acting on the stem. These components include (1) weight of the root–soil plate, (2) resistance of the soil to (chiefly) tensile failure, (3) resistance of the roots placed under tension on the windward side of the tree, and (4) resistance to bending at the hinge. (From Coutts, 1986.)

windthrow and windbreak. For example, in New Zealand a high proportion of the Monterey pine trees that were infested with *Armillaria* root rot were blown down by the wind (Shaw and Taes, 1977). After a severe windstorm in Colorado, 86% of the trembling aspen trees that were blown down were found to be infected with the root and butt rotting fungus, *Fomes applanatus*, but only 5% of the trees that remained standing were infected (Landis and Evans, 1974). Fire scars may also predispose trees to wind damage, often because of fungal infection (Gordon, 1973).

Trees growing in dense forest stands become very prone to windthrow when the surrounding trees are cut (Fig. 12.2). Open-grown, strongly tapered trees are most stable (Petty and Worrell, 1981). Cremer *et al.* (1982) considered the height–diameter ratio for dominant Monterey pine trees in Australia to be the most valuable index of risk of wind damage. Trees with height to diameter values of less than 60 were not readily uprooted by wind; those with values greater than 100 were unstable in wind. King (1986) attributed the greater amount of wind breakage in large over small sugar maple trees to loss of flexibility caused by increased stem diameter and to increase in wind speed with height in the canopy.

Trees with buttressed stem bases are less susceptible to windthrow than trees without buttresses. Henwood (1973) showed that buttresses are effective structural members supporting large trees on soils that provide poor anchorage. The buttresses of some species (e.g., *Quararibea asterolepis*) are concentrated on the windward sides of trees and act as tension members that reduce stresses on tree roots, thus rendering the roots less susceptible to breakage (Richter, 1984). In *Pterocarpus officinalis* in Puerto Rico, however, buttresses were distributed independently of the prevailing wind direction (Lewis, 1988). *Pterocarpus officinalis* grows in deep mud, and buttresses provide a broad platform that minimizes toppling. Smith (1972a) suggested that buttressing provides the maximum support for the minimum amount of wood just as an I beam provides more support than a solid beam containing the

Figure 12.2 Heavy blowdown of conifers in Oregon associated with clear-cutting of an adjacent stand.(U.S. Forest Service photo.)

same amount of metal. The physiological control of buttress formation deserves more study.

Both stem breakage and uprooting of trees vary appreciably with the size of tree crowns. For example, snapping of stems by wind is more likely to occur in trees with large crowns than those with small ones. By increasing crown size, application of N fertilizers increased the risk of wind damage to rubber trees (Rosenquist, 1961). Control of leaf diseases also increases susceptibility to wind damage by increasing crown size (de Silva, 1961; Mathew and George, 1967).

Leaf Injury

Tearing of leaves of palm, banana, apple, pear, and citrus trees by wind is common (Waister, 1972a). Wind (3.0 or 6.0 m sec^{-1}) caused tearing and shredding of leaves of cacao seedlings, as well as interveinal browning and curling, necrosis, and abscission of leaves (Sena Gomes and Kozlowski, 1989). A reduction in the leaf area of cacao trees exposed to wind (1 to 5 m sec^{-1}) was attributed to dehydration as a result of a high rate of transpiration (Leite and Alvim, 1978). The usefulness of windbreaks to protect cacao trees is mentioned in a later section.

Injury to leaves of forest trees sometimes occurs along windy seacoasts. In Scotland, exposure to wind in the field or in a wind tunnel induced permanent abrasive damage to sycamore maple leaves in the form of brown lesions, distortions, and tearing of leaf blades. Young expanding leaves were more susceptible than mature leaves to injury, and the amount of injury increased linearly with wind speed up to 6.3 m sec^{-1}. Most of the damage occurred during the first few weeks after budbreak in April, despite the persistence of wind throughout the season (Wilson, 1980). The injury was associated with collapse of adaxial and abaxial epidermal cells, mesophyll collapse, and disruption of the abaxial epicuticular waxes (Wilson, 1984). In northern Ireland, wind damaged up to 46% of the leaf area of individual leaves of sycamore maple trees, but few leaves had damage greater than 10% (Rushton and Toner, 1989).

Winds also injure trees near the coast by transporting salt spray for long distances. For example, trees injured by salt spray during a hurricane were found as far as 45 miles inland (Moss, 1940). Accumulation of sea salt within 2 miles of the west coast of Wales was closely related to the speed and direction of wind (Rutter and Edwards, 1968). Salt spray carried by wind affects the composition and zonation of vegetation on the North Carolina coast (Oosting and Billings, 1942).

As mentioned in Chapter 5, leaf shedding often is associated with hot, dry winds, such as "foehns," "chinooks," "siroccos," and "Santa Anas" (Chavanier, 1967; Kozlowski, 1976a,b).

Tree Form and Size

Trees exposed to steady winds tend to be small and have one-sided or "flagged" crowns (Fig. 12.3). Often the branches of such trees are swept to the leeward side,

Figure 12.3 A cluster of Engelmann spruce trees at the timberline (10,800 ft) in the Medicine Bow Mountains of Wyoming. A group of "flag" trees rise from a clump of krummholz that never grows above the level of winter snow cover. This clump probably is a clone produced by layering. (From Billings, 1978.)

or the trees lack branches on the windward sides where the buds were killed. Wind that carries particles of ice or snow is a strong abrasive force that often erodes away the buds on the windward sides of trees.

In England, trees often become deformed when the average annual wind speed is 24 km hr^{-1} or higher (varying from about 19 km hr^{-1} in summer to 32 km hr^{-1} in winter) (Gloyne, 1954). Toward the upper limit of intact subalpine forest in North America, the trees are progressively stunted and their lateral branches tend to proliferate near the ground where snow cover is present, forming "mat krummholz." At the highest elevations near timberline the trees grow as mat krummholz only and erect stems are absent (Hadley and Smith, 1986). Shrubs as well as trees are dwarfed by wind at high altitudes and on coastal sand dunes. At the foot of mountains in Costa Rica, canopy trees of *Didymopanax pittieri* average 15 to 23 m high. With increasing altitudes the trees become progressively shorter and on ridge crests are only 5 to 10 m tall where the wind speeds are 1.5 to 1.8 times those at lower elevations (Lawton, 1982).

Daubenmire (1943) and Wardle (1968) described the effects of wind on tree form at high elevations in the Rocky Mountains. Altitudinal differences in wind velocity are relatively small during the summer, but in the winter the force of the wind at the upper timberline increases sharply. As a result, Engelmann spruce trees are both dwarfed and misshapen. Branches develop only on the leeward sides of trees, giving the upper part of the tree a flag shape. The lower branches of such trees form a low mat that is protected from desiccation and blowing snow by snow cover that accumulates early in the autumn. As a result of layering of the lower branches that comprise these mats, isolated trees in the timberline zone expand during periods of favorable climates and form islandlike clumps that tend to retain the general form of the parent trees (Fig. 12.4). Later, death of the oldest trees in such islands creates openings.

Branch tips over the entire surface of the mass of shoots of wind-molded clumps of trees (called mat or cushion krummholz) are dead in the spring. Subsequently the lateral buds produce vigorous shoots, so each dead terminal is replaced by several shoots that expand from lateral buds. Fig. 12.4 shows the change in growth form of Engelmann spruce trees in the Rocky Mountains, from tall and spire-shaped trees below the timberline to cushions of contorted stems and needles (cushion krummholz) on the most severe sites at the extreme upper limits.

The importance of wind in shaping trees at the timberline is emphasized by the confinement of individual trees or clumps of trees to valleys and small areas on the lee side of protecting rocks or ridges. The importance of wind is also emphasized by lowering of the timberline where wind action is unusually severe. Furthermore, individual trees may survive as much as 1000 ft above the average timberline in unusually well protected sites such as deep ravines.

Even gentle, persistent winds may appreciably influence the form of woody plants. In the tundra of the Venezuelan Andes, for example, gentle, prevailing northeast winds were correlated with distinct growth forms of shrubs (Smith, 1972b). The brittle, stiff-branched shrub *Hesperomeles pernettyoides* has a wind-

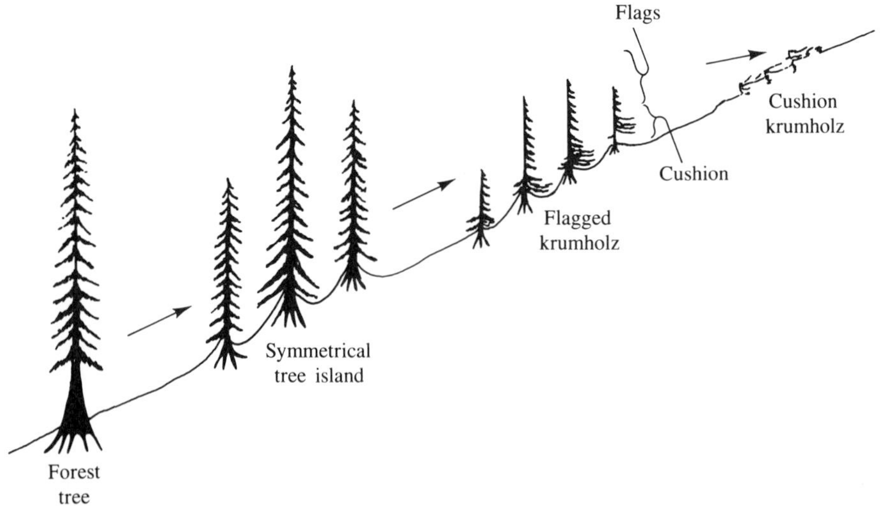

Figure 12.4 Effect of wind on growth forms of Engelmann spruce trees. Increasingly deformed trees are shown from left to right. (From Wardle, 1968.)

sculptured appearance with extensive canopy development on the leeward side but little foliage and many bare branches on the windward side, which is caused by leaf shedding on the windward side. By comparison, *Hypericum laricifolium*, which has a flexible stem and leaves, usually is not injured by the wind.

Growth

Wind greatly alters the rate of vegetative and reproductive growth of free-swaying trees. Whereas wind generally decreases height growth, it increases diameter growth and redistributes growth around the stem and at different stem heights.

Shoot Growth

A reduction in height growth of trees by wind is well documented by measurements in the field and in wind tunnel experiments. Height growth and elongation of lateral shoots of young lodgepole pine trees in a wind tunnel were reduced by about 20% by a wind speed of 8.5 m sec^{-1}. Similar results were obtained by continuous shaking of trees out-of-doors or by shaking them for only 24 min a day (Rees and Grace, 1980a,b). The data were consistent with results obtained with seedlings of several species of broad-leaved trees, including sweet gum (Neel and Harris, 1971), black walnut, and silver maple (Phares *et al.*, 1974). Wind velocities of 6 m sec^{-1} inhibited leaf growth of potted trembling aspen trees by as much as 50%, with the reduction associated with leaf dehydration, which eventually led to necrosis (Flückiger *et al.*, 1978). Exposure to wind also reduced needle elongation of loblolly pine (Telewski and Jaffe, 1981) and Fraser fir (Telewski and Jaffe, 1986).

Cambial Growth

Stems of trees allowed to sway in the wind generally are thicker and more tapered but shorter than those of trees prevented from normal swaying. Acceleration of radial growth by wind is maximal at the point of greatest stress (e.g., at the base of free-swaying trees).

Monterey pine trees that had been prevented from swaying for a few years made less than 70% as much diameter growth, and were less tapered, than trees allowed to sway normally. The trees that were guyed to prevent swaying were very unstable and often toppled when exposed to wind (Jacobs, 1954). When loblolly pine trees were guyed at two stem heights, wood production was decreased in the lower stem and increased in the upper stem. Hence, stem taper was reduced relative to that in free-swaying trees (Burton and Smith, 1972). Larson (1965) exposed some 4-year-old tamarack trees to wind and prevented others from swaying. Wood production was greater in the free-swaying trees, but height growth was greater in trees that were prevented from swaying. The swaying trees also produced more compression wood than those held in rigid positions.

Conventional nursery practices of staking and severely pruning lateral branches of container-grown trees tend to produce thin-stemmed, untapered plants that often are unstable when planted in the field. More tapered, stable trees can be developed by eliminating staking in the nursery, leaving lateral branches on the stem, and spacing containerized plants widely so their tops are free to move in the wind (Leiser *et al.*, 1972). Shaking young sweet gum trees for 30 sec daily for 23 days reduced height growth 20 to 30% and caused formation of terminal buds in six of eight trees (Neel and Harris, 1971).

Cross sections of tree stems exposed to wind from one direction often are elliptical because most of the cambial activity occurs on the leeward side. For example, almost all radial growth of an Arizona cypress tree exposed to strong wind was confined to the leeward side of the stem (Daubenmire, 1984). Annual growth rings were wider on the leeward side than on the windward side of white spruce, eastern white pine, and lodgepole pine trees. In accord with the general inverse relation between ring width and cell length, the tracheids were longest on the windward side, where the annual rings of wood were narrowest (Bannan and Bindra, 1970).

Reaction Wood

Tilting of trees by wind results in redistribution of the amount and nature of cambial growth on the leeward and windward sides of the stem and in formation of abnormal "reaction wood." The reaction wood that forms in conifers is called "compression wood" because it occurs on the lower (leeward) sides of tilted stems. Reaction wood of broad-leaved trees is called "tension wood" because it occurs on the upper sides of tilted stems. Reaction wood has several commercially undesirable characteristics, such as a tendency toward increased shrinking, warping, and weakness.

The structure of compression wood differs from that of normal wood, having shorter and thicker-walled tracheids that are more nearly rounded in cross section and large intercellular spaces (Fig. 12.5). The inner layer (S_3) of the secondary cell wall of compression wood tracheids usually does not form or develops poorly. The inner layer (the middle S_2 layer in normal wood) of cell walls of compression wood is very thick (Kramer and Kozlowski, 1979).

Tension wood of leaning broad-leaved trees often, but not always, is associated with eccentric cambial growth. Tension wood usually has a preponderance of gelatinous fibers that can be identified by the gelatinous appearance of their secondary walls (Fig. 12.6). Tension wood has fewer and smaller vessels and proportionally more thick-walled fibers than normal wood. The cell wall layer designated as S_2 or S_3 in normal wood is replaced in tension wood by an unlignified, often convoluted layer, designated as the G layer. When the S_2 layer is replaced, the new designation is S_2 (G). Sometimes the G layer is produced in addition to the S_1, S_2, and S_3 layers of normal xylem (Fig. 12.7).

Production of compression wood in conifers blown over by severe storms, hurricanes, and cyclones is well documented (Savill, 1983). Examples are eastern white pine in New England, longleaf pine in Mississippi, Monterey pine in Australia, *Abies sachalinensis*, *Picea glehnii*, and *P. jezoensis* in Japan, and slash pine

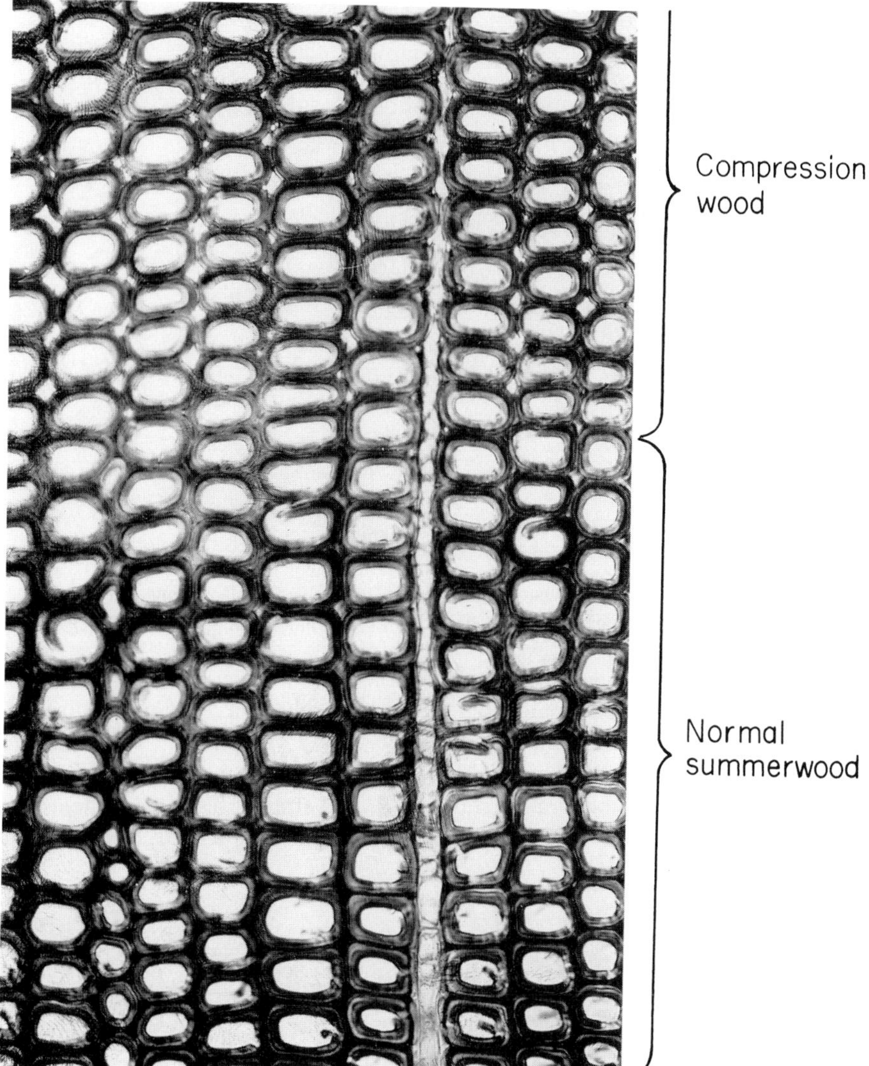

Figure 12.5 Compression wood and normal summerwood of longleaf pine. (Photo courtesy of U.S. Forest Products Laboratory.)

in Rhodesia. Bands of compression wood formed in residual loblolly pine trees shortly after a stand was thinned, reflecting tilting of those trees by wind (Zahner and Whitmore, 1960). Sometimes trees exposed to winds from all directions produce compression wood on all sides of the stem (Hale, 1951).

Stems of trees exposed to wind not only lean in a direction away from the wind but also may develop curved stems. For example, when exposed to wind many

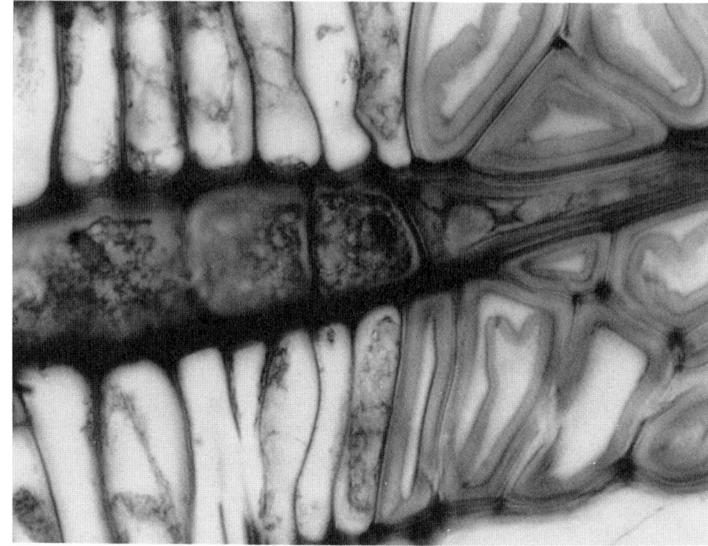

Figure 12.6 Transverse section of the cambium showing subjacent gelatinous fibers of cottonwood. (After Berlyn, 1961; from Kramer and Kozlowski, 1979, by permission of Academic Press.)

species of tropical and subtropical pines tend to develop curved stems and associated compression wood. Examples are *Pinus patula*, *P. pinaster*, *P. kesiya*, and *P. caribaea* var. *hondurensis* (Timell, 1986). Of 770 trees in a study of 23-year-old Monterey pines in Australian plantations, 44% had straight but leaning stems and 25% had curved stems (Nicholls, 1982). Larch trees are particularly susceptible to

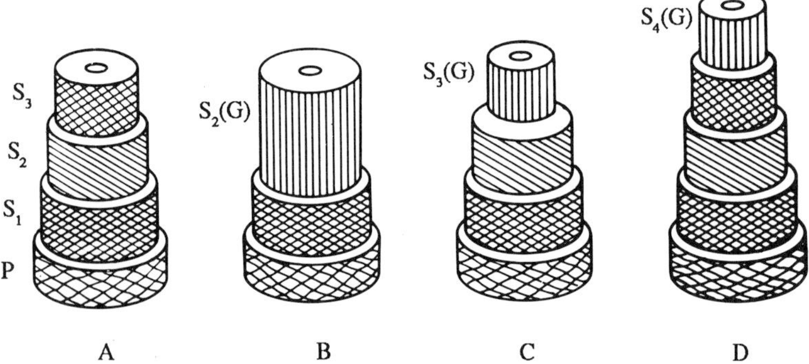

Figure 12.7 Organization of cell walls in normal wood fibers. (A) Normal wood fiber of structure P + S_1 + S_2 + S_3; (B) tension wood fiber of structure P + S_1 + S_2 (G); (C) tension wood fiber of structure P + S_1 + S_2 + S_3 (G); (D) tension wood fiber of structure P + S_1 + S_2 + S_3 + S_4 (G). P, Primary wall; S_1, outer layer of secondary wall; S_2, middle layer of secondary wall; S_3, inner layer of secondary wall; S_4, gelatinous layer. (After Wardrop, 1964; from Kramer and Kozlowski, 1979, by permission of Academic Press.)

bending by wind. Young trees grow rapidly and develop tall, slender stems that are easily bent. They also have deep, well-developed root systems and are more likely to be bent than uprooted by wind. Such trees tend to have curved stems but there is considerable variation among races of larch in development of stem curvature in response to wind (Timell, 1986).

Reproductive Growth

Wind often reduces the yield and quality of fruits and seeds of a wide variety of woody plants. The most obvious effects are caused by mechanical damage to stems and branches. By injuring leaves or inducing their shedding, wind may decrease the photosynthetic surface and the amount of available carbohydrates and hormonal growth regulators. Yield often is reduced following a decrease in physiological efficiency of the foliage. Yields of oranges in Japan were reduced for several years by typhoons that defoliated the trees (Tani, 1967). Reproductive growth may be inhibited by adverse effects on any of several sequential phases of reproduction, including pollen production, pollination, fruit and seed set, and fruit development (Waister, 1972a). Winds also reduce fruit quality by causing undersized and mis-shapen or blemished fruits. On the other hand, wind may benefit reproduction by dispersing pollen and seeds.

In Scotland the yield of raspberries was associated with reduced cane production rather than a direct effect on flowering or fruiting (Waister, 1970). There often is a delayed response of reproductive growth to wind. For example, variations in straw-berry yields were related to wind conditions during the preceding year when the flowers were initiated (Waister, 1972b). The decreased yields were attributed more to bruising of leaves than to effects of wind on plant temperature or soil water balance.

Physiological Processes

The effects of wind on growth are very complex and are mediated by changes in water, food, and hormone relations.

Water Relations

The influence of wind on leaf hydration varies with wind speed, plant species, and plant and atmosphere interactions. At very low speeds, wind often increases transpiration by reducing the thickness of the boundary layer that surrounds leaves in quiet air (Fig. 12.8). However, the effects of wind are complicated by a cooling effect on leaves, which lowers the leaf to air vapor pressure difference. Dixon and Grace (1984) regard the effect of wind on the the vapor pressure gradient as more important than the effect on the boundary layer. A typical pattern is for an initial increase in transpiration as wind speed increases above 1 m sec^{-1}, followed by an

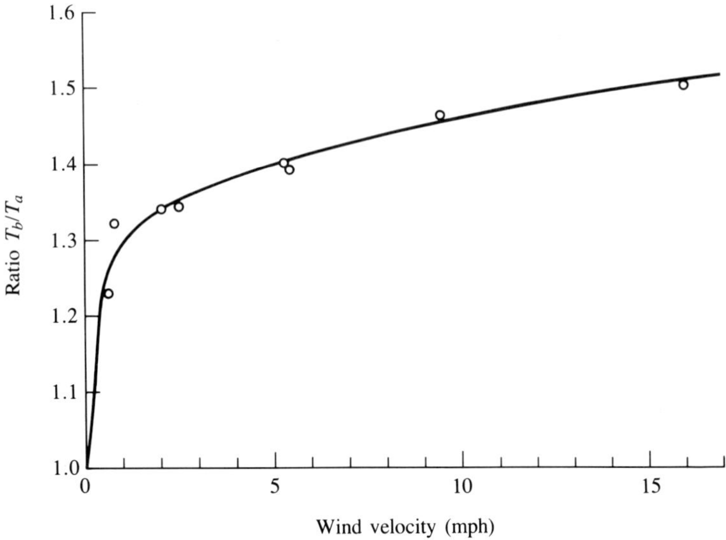

Figure 12.8 Effect of increasing velocity of air movement on transpiration rate of potted sunflowers growing in a sunny greenhouse. The ordinate is the ratio of the rate of plants exposed to wind, T_b, to control plants in quiet air, T_a. Most of the effect occurs at a very low velocity. A velocity of 1 mph equals 44.7 cm sec^{-1}. (After Martin and Clements, 1935; from Kramer and Kozlowski, 1979, by permission of Academic Press.)

eventual decrease as the vapor pressure difference is decreased and stomata begin to close. Hot, dry winds have a particularly strong desiccating effect (see Chapter 5).

Changes in stomatal aperture of plants exposed to wind may involve several mechanisms. For example, stomata often close in response to shaking. Gyratory shaking of soybean plants induced stomatal closure within minutes. Transpiration decreased 17% and leaf water potential increased by 39% within minutes after treatment (Pappas and Mitchell, 1985). Stomatal closure by wind may also result from leaf dehydration as well as a direct response of the guard cells to lowered air humidity as shown for sugar maple (Davies *et al.*, 1974) and cacao (Sena Gomes *et al.*, 1987). Changes in stomatal aperture in response to wind may also involve a CO_2-sensing mechanism of the guard cells. The CO_2 concentration close to the leaf surface is higher when the wind speed is high than when it is low (Mansfield and Davies, 1985). Hence, stomata may open or close through a feedback response depending on the partial CO_2 concentration in the intercellular spaces.

Actual changes in transpiration of different species of woody plants exposed to wind are variable and often related to differences in stomatal responses. Whereas transpiration rates of alder and larch were higher in wind than in still air up to a wind speed of 20 m sec^{-1}, the rates of Norway spruce, Swiss stone pine, mountain ash, and *Rhododendron ferrugineum* were lower (Fig. 12.9). The faster reduction in transpiration of *R. ferrugineum* exposed to wind over that of Swiss stone pine was related to the more rapid stomatal closure in the former (Caldwell, 1970). In cacao the rate of transpiration decreased as wind speed was increased up to 6 m sec^{-1}, apparently in response to a lower leaf to air vapor pressure gradient associated with the cooling effect of wind on the leaves (Sena Gomes and Kozlowski, 1989). Wind at speeds of 5.8 to 26.0 m sec^{-1} increased transpiration of white ash, decreased it in sugar maple, and did not affect the rate of red pine seedlings (Davies *et al.*, 1974). The high rate of transpiration of white ash caused by the increased vapor pressure gradient in wind was not reduced by stomatal closure. In contrast, the stomata of sugar maple closed rapidly, thereby reducing transpiration. The large stomata of white ash are covered by a cuticular ledge; those of sugar maple are not. Apparently the exposed guard cells of sugar maple were dehydrated faster than the covered guard cells of white ash, thus inducing more rapid closure in the former.

Food Relations

Wind has complex effects on photosynthesis. Winds of high velocity often reduce total photosynthesis by causing leaf injury and shedding. At low velocities, wind influences the rate of photosynthesis of leaves by affecting boundary layer thickness

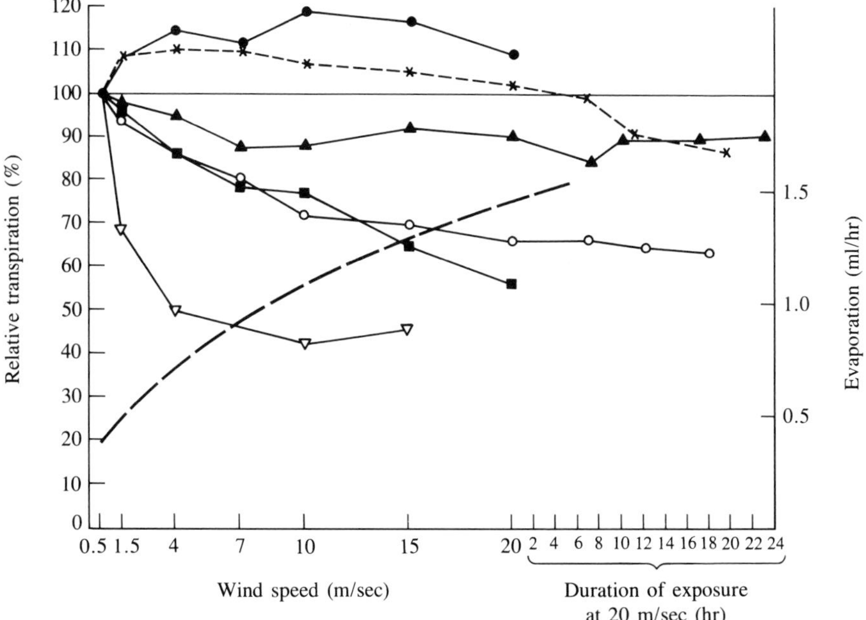

Figure 12.9 Effect of increasing wind speed on transpiration of young potted subalpine plants in a wind tunnel at an air temperature of 20°C, soil temperature of 15°C, and light intensity of 30,000 lux. Evaporation (——) is from a green Piche atmometer under the same conditions. Evaporation increases steadily with increasing wind speed, but transpiration of most species decreases. Alder, *Alnus viridis* (●); larch, *Larix decidua* (x); spruce, *Picea abies* (▲); stone pine, *Pinus cembra* (○); mountain ash, *Sorbus aucuparia* (■); rhododendron, *Rhododendron ferrugineum* (▽). (After Tranquillini, 1969; from Kramer and Kozlowski, 1979, by permission of Academic Press.)

and leaf temperature. Avery (1966) showed that the increased rate of photosynthesis of apple leaves in wind was not simply due to an increased supply of CO_2 to a leaf chamber but resulted partly from reduction in thickness of the boundary layer. Wind-induced changes in leaf temperature influence not only stomatal conductance but CO_2 mesophyll conductance as well. For example, reduction of photosynthesis of *Festuca arundinacea* plants by exposure to wind was associated with a decrease in mesophyll conductance (Grace and Thompson, 1973).

Some other effects of wind on photosynthesis are controversial. Shive and Brown (1978) reported that quaking of eastern cottonwood leaves increased bulk air flow through them. However, Day and Parkinson (1979) doubted the importance of such effects except at very high wind speeds, and Rushin and Anderson (1981) found that quaking of trembling aspen leaves in wind had little effect on stomatal conductance.

High wind velocities may also decrease available carbohydrates by accelerating respiration. For example, increases in respiration of shoots of several species were detected at wind speeds of 3.6 m sec^{-1} or higher. All species responded similarly with increases in respiration of 20 to 40% at wind speeds of 7.2 m sec^{-1}, values commonly measured in the field (Todd *et al.*, 1972). Such increases in respiration may be caused by the mechanical effects of wind (Jaffe, 1980).

Hormone Relations

The effects of wind on cambial growth of trees are also mediated by hormonal changes. Much attention has been given to the effects of changes in auxins and ethylene.

Increase in diameter growth and induction of reaction wood in trees exposed to wind appear to be regulated by redistribution of hormonal growth regulators. Several investigators emphasized that increased cambial growth on the leeward sides of tilted stems occurs because of a high auxin gradient that causes mobilization of food. The role of auxin in initiation and stimulation of cambial growth to produce xylem has been well documented (Kramer and Kozlowski, 1979). When IAA was applied to Monterey pine stems, radial growth near the site of application was greatly stimulated whereas height growth was suppressed (Fraser, 1952). There is some evidence that auxin may stimulate cambial growth by increasing ethylene production. Application of ethrel (which releases ethylene) to eastern white pine stems increased cambial growth (Leopold *et al.*, 1972).

Many investigators attributed formation of compression wood to high auxin levels (for review see Timell, 1986, Vol. 2). Induction of compression wood often follows application of IAA to buds as well as to stems. Compression wood that was induced by applied IAA could not be distinguished from naturally occurring compression wood by appearance or physical properties (Larson, 1969).

Formation of tension wood is associated with auxin deficiency. This is shown by lower concentrations of auxin on the upper sides of tilted stems (Leach and Wareing, 1967), inhibition of tension wood formation by applying auxins to the upper sides of tilted stems, and induction of tension wood by applying auxin antagonists such as TIBA (2,3,5-triiodobenzoic acid) or DNP (2,4-dinitrophenol) (Morey, 1973). The few and small vessels in tension wood also are associated with auxin deficiency. Decreases in vessel frequency from the leaves to the roots are correlated with a gradient of decreasing auxin concentration (Zimmermann and Potter, 1982).

Some investigators have suggested that ethylene plays an important regulatory role in formation of reaction wood. For example, after applying ethrel to Monterey pine stems, Barker (1979) noted the formation of rounded, thick-walled tracheids resembling those of compression wood. Brown and Leopold (1973) stimulated ethylene production in eastern white pine stems by bending them in arcs and concluded that the increased ethylene induced formation of compression wood. Savidge *et al.* (1983) applied the ethylene precursor ACC together with IAA to defoliated lodgepole pine cuttings and noted more differentiation of wood resembling compression wood than was the case after application of IAA alone. However, application of ethrel at concentrations up to 1% to Aleppo pine stems induced formation of xylem with slightly abnormal tracheids but without the essential features of tracheids of well-developed compression wood (Yamamoto and Kozlowski, 1987b). Ethrel-treated seedlings had short but normal-shaped (rectangular) tracheids with only slightly thickened walls and with S_1, S_2, and S_3 wall layers present. By comparison, the rounded, thick-walled tracheids of well-developed compression wood lack an S_3 layer (Coté and Day, 1965). Tilting of Japanese red pine stems induced formation of typical compression wood with rounded tracheids, thick cell walls, and high lignin content. However, treatment of vertically oriented stems with ethrel did not induce formation of typical compression wood. Furthermore, application of ethrel to tilted seedlings counteracted the formation of well-developed compression wood. Yet ethylene contents of stems were increased following treatment with ethrel (Yamamoto and Kozlowski, 1987c). Such observations depreciate the role of ethylene *alone* in causing formation of compression wood.

Some investigators suggested that stress ethylene plays a role in inducing formation of tension wood. When Nelson and Hillis (1978a,b) grew *Eucalyptus gomphocephala* seedlings horizontally, tension wood formed on the upper sides of stems, where ethylene concentrations were higher. They suggested that formation of tension wood may be a response to a high concentration of ethylene or a low auxin and high ethylene level. By comparison, ethylene concentrations were higher on the lower side than the upper side of horizontally oriented apple shoots (Robitaille and Leopold, 1974). Increase in ethylene content of tilted Norway maple seedlings was as great or greater on the lower sides than the upper sides of the stem. Application of ethrel to upright stems did not induce formation of tension wood and applications of ethrel to tilted stems blocked formation of tension wood (Yamamoto and Kozlowski, 1987d). Although ethrel applications altered the wood anatomy of vertically oriented American elm seedlings, it did not induce formation of tension wood (Yamamoto and Kozlowski, 1987a). The weight of evidence indicates that formation of reaction wood is regulated by hormonal interactions, with auxin playing a primary role. However, because ethylene regulates auxin levels by influencing auxin biosynthesis (Abeles, 1973), it may play an indirect role in formation of reaction wood. The emphasis here on auxin and ethylene should not obscure the complexity of control of cambial growth and participation of still other compounds in formation of reaction wood. More research is needed on the mechanism controlling formation of reaction wood.

Dispersal of Air Pollutants

The rate of air movement influences the uptake of air pollutants by affecting the thickness of boundary layers over leaves and by transporting air pollutants. An increase in wind speed in tree canopies often accelerates absorption of gaseous air pollutants by decreasing the thickness of the leaf boundary layer. At high wind speeds the boundary layer is thinner, the path length shorter, and the rate of diffusion higher than when wind speed is low (Grace, 1977).

Air pollutant damage to plants may be caused by local or distant sources of pollution or both. Studies around point sources of pollution such as smelters and power plant smoke stacks illustrate the effects of prevailing winds on dispersion of pollutants. For example, severe SO_2 injury to plants was restricted largely to a narrow strip northeast from an iron-sintering plant in Ontario, Canada, where southwest winds predominated (Fig. 12.10). Near Trail, British Columbia, Canada, the area of severe pollution damage to trees was an ellipse extending north of the nickel and copper smelters for about 25 miles. In Anaconda, Montana, Douglas-fir trees were visibly injured for 20 miles to leeward from a copper smelter.

Air pollutants often move with air masses for long distances. Sulfur dioxide from burning of fossil fuels and nitrogen oxides from vehicles and high-temperature combustion are chemically converted in the atmosphere to acid compounds that may fall in rain hundreds of kilometers from the pollution source. Ozone was transported by air for 275 km from a pollution source in Minneapolis, Minnesota (Pratt *et al.*, 1983). Mortality of pine trees in the San Bernardino Mountains of California was attributed to oxidants formed in the Los Angeles Basin, 95 to 110 km away, and moved in polluted air masses (Miller, 1973). Photochemical oxidants are transported from urban centers along the California coast to inland valleys and across forested mountains to warm deserts. Several air pollutants are transported with prevailing winds from the Ohio Valley and Pennsylvania to New England and Canada. Miller and McBride (1975) reviewed several other studies that demonstrated long-distance transport of air pollutants. Although building of tall stacks on smelters and other point sources of pollution decreases the concentration of pollutants locally, it tends to create regional air pollution problems. For example, almost all the sulfur and 40% of the heavy metals emitted from a tall smoke stack were dispersed for more than 60 km from the smelter (Postel, 1984).

Because of prevailing wind directions, some countries receive the bulk of their pollutants from abroad. For example, most of the air pollutants deposited in the Scandinavian countries originate in the industrialized regions of England and Central Europe. Regional-scale studies indicated that approximately 92 and 82% of the

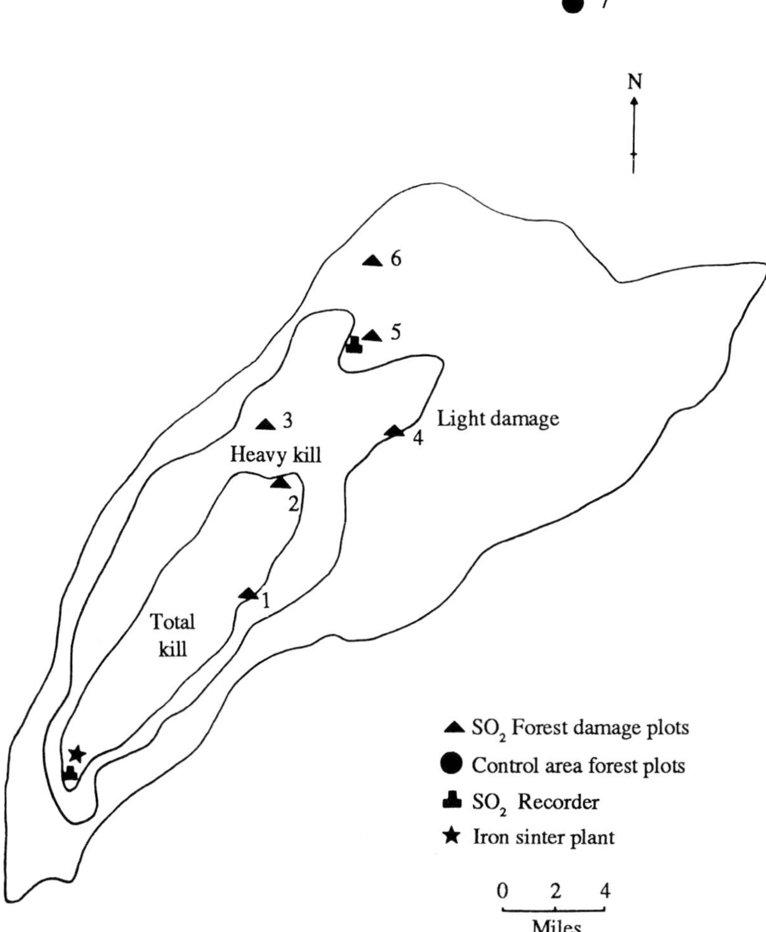

Figure 12.10 Effect of prevailing winds on SO_2 injury to a forest ecosystem at various distances from an iron-sintering plant near Wawa, Ontario, Canada. (Reproduced from Linzon, 1978. Copyright © 1978 by John Wiley & Sons.)

sulfur deposited in Norway and Sweden originated outside their borders (Postel, 1984). About 90% of the sulfur pollutants in Switzerland, 85% in Austria, 77% in the Netherlands, 74% in Finland, 64% in Denmark, 64% in Rumania, and 63% in Czechoslovakia were imported in air masses. By comparison, in England only 20% of the sulfur pollutants originated in other countries, while about two-thirds of those generated moved with prevailing winds to other countries. More sulfur was deposited in Austria than in England, even though Austria produced only about 10% as much.

Dispersal of Spores of Pathogens

Aerial dispersal of fungus spores through space involves liberation, transport, and deposition. In dry spore forms the number of spores set free increases as the wind speed rises. However, spores are not liberated continuously with increasing wind speed, but most are removed above a critical speed characteristic for a given species. For *Helminthosporium maydis* (the cause of corn leaf blight) this speed is about 5 m sec^{-1}, and air of this speed must impinge directly on the conidium. Inasmuch as spores on leaves are well within the air boundary layer, an average wind speed considerably higher than 5 m sec^{-1} is necessary to ensure the required minimal speed (Aylor, 1978).

Many fungus pathogens are spread by wind because they produce vast amounts of small, light spores that are readily ejected into the air. It has been estimated that a large fruiting body of *Fomes applanatus* may produce over 5 trillion spores, discharged at a rate of about 30 billion a day for 6 months, and a single cedar gall of *Gymnosporangium juniperi-virginianae* produced nearly 7.5 trillion basidiospores. The spores of pathogenic fungi usually vary in size from 5 to 125 μm (rarely up to 300 μm) and fall at a rate measured in millimeters per minute in still air, but they can be carried for enormous distances by wind currents (Stakman and Christensen, 1946).

Air turbulence often extends for several thousand feet upward as a result of air flow over tall plants, buildings, and irregular topography. Such turbulence influences spore dissemination by shaking plants, by sweeping eddies into spore-bearing layers of vegetation, and by increasing impaction, boundary layer exchange, and turbulent deposition. Land and sea breezes as well as valley and slope air circulation often are important locally in spore dispersal (Hirst, 1959).

Transport of fungus spores by prevailing winds has been reported over both long and short distances. Spore concentration varies with the origin of air masses, with higher spore concentrations found in tropical air that drifted over landmasses with high spore populations than in polar air that passed over land areas that were relatively free of spores (Pady and Kapica, 1955). As spore-free air masses from the Atlantic Ocean pass over the British Isles, vast numbers of spores are picked up primarily during the day by upward currents due to thermal turbulence. Within a few hours the air may contain more than 160 spores per m^3 at a height of 1000 m (Hirst *et al.*, 1967). In California, spores of the apricot pathogen *Eutypa armeniacae* were transported with the prevailing westerly winds from the San Francisco Bay area and infected trees in the apricot-producing San Joaquin valley (Ramos *et al.*, 1975).

Dissemination of disease-causing fungi over short distances also has been demonstrated. Examples are spores of *Endothia parasitica*, *Gymnosporangium juniperi-virginianae*, and *Cronartium ribicola*. The pattern of short-distance dispersal of spores from a point source has been likened to that of a plume of smoke from a chimney (Ingold, 1965, 1971). The spore cloud takes the form of a horizontal cone with its apex at the point of spore release. As the cloud moves downwind in turbulent air it is diluted and widened by mixing with eddies. At any cross section of the cloud the mean concentration of spores is inversely proportional to the distance from the source, but the concentration is greater around the horizontal axis of the cone than near its surface. This pattern tends to break down with increasing instability of the air and increased distance from the source.

Gymnosporangium juniperi-virginianae, which causes cedar apple rust, overwinters on eastern red cedar. During the spring or early summer, basidiospores are ejected from fungal galls on cedar trees and transported by wind to apple trees. After infection and development of the aecial stage on apple leaves, aeciospores are ejected and distributed by wind, but they can infect only red cedar trees, not apple trees. A single cedar gall may produce as many as 7 million sporidia but these are relatively short-lived and cause severe infection only on apple trees within 1.6 km of the source (MacLachlan, 1935).

Dissemination of blister rust of eastern white pine trees caused by *Cronartium ribicola* also is restricted, even though very large numbers of aeciospores are produced on pine trees because these spores can infect only *Ribes* bushes but not pines. Infected *Ribes* bushes produce urediospores, which can again infect *Ribes* bushes, whereupon teliospores form. The teliospores germinate and produce a promycelium that produces short-lived sporidia that generally are spread to pines for distances of only about 1.6 km. Because of such local spread, white pine trees can be protected by destroying infected *Ribes* bushes within approximately 300 m. Occasionally, however, aeciospores produced on tall pine trees are carried for longer distances by wind and extend the range of the rust for several hundred kilometers by spreading from pines to *Ribes* bushes (Stakman and Christensen, 1946).

The microclimate can be very important in modifying local patterns of spore dissemination by wind, as shown by the spread of spores of *Cronartium ribicola*. In one case, trees just above *Ribes* bushes on a slope were not readily infected, whereas those below them were diseased, because the spores were transported down the slope. The airstream then rose and returned above a temperature inversion to infect pines at higher altitudes (Van Arsdel, 1965).

Some diseases are spread by a combination of rain and wind. For example, sporangia of *Phytophthora palmivora*, which causes shoot blight and seedling root rot of papaya, are readily released in splash droplets and dispersed by wind-borne rain. The sporangia on papaya fruit are not released in moving air under drying conditions (Hunter and Kunimoto, 1974). Wet leaf surfaces and favorable temperatures may be required for spore germination in order for infection to occur, as with apple scab.

Dispersal of Pollen

Transfer of pollen from the anther to the stigma is accomplished chiefly by wind and insects. Wind pollination increases with both latitude and elevation and is common in temperate, deciduous, and boreal forests but extremely uncommon in tropical rain forests, where insect pollination is more common. The tendency for wind pollination is high on remote islands, and higher in early than late successional ecosystems (Whitehead, 1983). Nearly all conifers rely on wind pollination, as do some broad-leaved trees, including *Populus*, *Quercus*, *Fraxinus*, *Ulmus*, *Carya*, and *Platanus*. Insect pollination is characteristic of *Tilia*, *Acer*, and *Salix* and several genera of fruit trees, including *Malus*, *Pyrus*, *Ficus*, and *Persea*.

The effectiveness of wind pollination depends on the number of pollen grains reaching receptive plants. Most airborne pollen comes to rest rather close to the tree that produced it. Therefore, most pollination usually occurs in trees close to a seed tree. For example, Wright (1952, 1953) showed that pollen of some forest trees traveled only a few hundred feet (Table 12.1). Very little pollen was distributed beyond 30 m from an isolated slash pine tree (Table 12.2). Silen (1962) confirmed that pollen counts from a Douglas-fir tree showed a maximum in distribution near the tree, a sharp decline with distance up to 90 m, and then a leveling off at greater distances. The large amount of pollen far from the tree was explained as originating from distant sources.

In addition to direct observation of pollen dispersal, Wright (1962) outlined three other lines of evidence showing that most pollen is distributed for limited distances only: (1) very little gymnosperm pollen remains in the air overnight; (2) uncommon exotic trees that are not completely isolated set very little sound seed to open pollination; and (3) most natural hybrids are reported to occur near their parents.

Some long-distance migration of pollen occurs. For example, Sarvas (1955) found a greater rain of pollen on a ship 20 km from the coast of Finland than in the forest along the shore. In Sweden, Andersson (1963) reported that during a year of heavy pollen production some rocky islands located 4 to 8 km from the nearest forest had a layer of spruce pollen about a centimeter deep. Lanner (1966) suggested that pollen can be lifted upward and carried for great distances by independently

Table 12.1
Variations among Species in Pollen Dispersion[a]

Species	Pollen dispersion distance (ft)
Fraxinus americana	55–150
Pseudotsuga menziesii	60
Populus deltoides	1000 or more
Ulmus americana	1000 or more
Picea abies	130
Cedrus atlantica	240
Cedrus libani	140
Pinus cembroides var. *edulis*	55

[a]After Wright (1953); from Kozlowski (1971b), by permission of Academic Press.

Table 12.2
Pollen Collected at Varying Distances from an Isolated Slash Pine Tree[a]

Distance from source (ft)	Grains per 2.71 cm²		
	Jan. 28	Jan. 29	Jan. 31, Feb. 2, Feb. 3[b]
0	78	204	45
25	28	79	23
50	30	53	7
75	17	38	20
100	7	28	0
150	9	16	0
200	1	13	2
300	0	8	0
400	4	4	0
500	0	6	0
1000	0	6	0
1500	3	1	0
2000	0	2	0
2500	0	2	0
2850	0	0	0
Total	177	460	97

[a]After Wang *et al.* (1960); from Kozlowski (1971b), by permission of Academic Press.

[b]Total grain counts of 3 days' pollen samples.

moving air masses. Masses of pollen may subsequently be washed out of the air and deposited by raindrops.

Conifers are well adapted for wind pollination as shown by accumulation of vast quantities of pollen hundreds of kilometers from conifer forests (Niklas, 1985). Pollen grains of conifers are equipped with air sacs that appreciably increase their buoyancy and the seed cones are adapted for pollen capture by forming pollination droplets and producing stigmalike micropylar lobes bearing long hairs. In addition, conifers have very high pollen grain to ovule ratios and, because they grow in dense stands, produce high densities of airborne pollen. Pine strobili can also deflect unidirectional airflow into eddies around the strobilus axis or over individual ovule-bearing structures, leading to high probability of airborne collision of pollen with ovules.

Because pollen dispersion distances generally are short, open-pollinated seeds usually result from pollination of nearby trees. Hence, most trees growing within a few hundred meters of each other are genetically more alike than trees growing several kilometers apart (Wright, 1976).

Pollen is dispersed by both air turbulence and horizontal movement. Some air movement is necessary for shaking of anthers to release pollen. Dispersal of pollen depends on interactions between the terminal velocity of pollen grains (controlled largely by the density and size) and wind velocity. For the majority of wind-pollinated species the terminal velocities of pollen grains vary from 2 to 6 cm sec^{-1}. By comparison, wind velocities in forests range from 1 to 10 m sec^{-1}. Hence, small, light grains are dispersed more efficiently than large, heavy grains, which settle much faster. Pollen dispersal generally increases as wind velocity increases and settling velocity decreases. The terminal velocity of pollen grains is reduced by partial dehydration. Pollen grains within sporangia usually are completely hydrated but may lose up to 10% of their moisture within seconds after they are released. There is, of course, considerable variation among species in the distance of their pollen dispersal. Pollens of *Picea*, *Pseudotsuga*, *Pinus*, and *Fraxinus* have among the most rapid rates of fall and the shortest dispersion distances. By comparison, the pollens of *Populus*, *Ulmus*, *Juglans*, and *Corylus* exhibit much longer dispersion distances.

Dispersal of Seeds and Fruits

Woody species are dispersed by means of seeds, fruits, fragments of plants, or whole plants. The agents of transport of disseminules include wind, animals, water, and the plant itself. Wind, however, is the most active agent in dispersing seeds (Fahn and Werker, 1972). Wind-dispersed species tend to dominate in dry habitats; animal-dispersed species dominate in wet habitats (Howe and Smallwood, 1982). In forests, most wind-disseminated plants are canopy trees or vines (Keay, 1957).

Seeds are dispersed to various distances by wind. Some are very small and light and carried by wind for great distances (e.g., species of the Ericaceae). The "dust seeds," which have small terminal velocities in still air, are very sensitive to small fluctuations in wind and are affected by small eddies in apparently tranquil environments (Burrows, 1975). Some of the heavier wind-dispersed disseminules have structural modifications such as air-filled sacs (some legumes), hairs (*Salix*, *Populus*, *Platanus*), or wings, which result in gliding or spinning (many conifers, *Pterocarpus*, *Dalbergia*, *Erythrina*, *Acer*, *Liriodendron*, *Tilia*, *Alnus*, *Liquidambar*, and *Betula*). Wind dispersal of some plants consists of movement of entire plants or parts of plants to which the seeds are attached, as in the tumbleweeds and "rollers" of steppes, deserts, and prairies (Ridley, 1930).

Dispersal distances of propagules are determined partly by their rate of fall. A slow rate of descent increases the chances that a seed or fruit will be caught by the wind and increases dispersal distance. The spinning of samaras produces both lift and drag, thereby opposing gravitational forces and slowing the rate of descent. The initial fall distance of samaras, which increases with blade length and seed weight (Table 12.3), comprises a small part of the total fall distance of samaras released from tall trees, but a larger portion of those released from short trees.

The dispersal distances of wind-disseminated seeds are greater for seed released

Table 12.3
Variations in Weight, Size, Disk Loading, Vertical Pull, and Terminal Velocity of Samaras of Seven Species of *Acer*[a]

Species	Samara weight (mg)	Blade length (mm)	Blade width (mm)	Disk loading (mg cm^{-2})	Initial fall (m)	Terminal velocity (m sec^{-1})
Acer rubrum	15.5	14.9	7.7	2.3	0.37	0.79
A. spicatum	30.4	16.7	7.9	3.6	0.43	0.88
A. pensylvanicum	50.9	18.8	9.0	5.5	0.53	1.10
A. negundo	74.4	25.4	12.0	3.7	0.70	0.99
A. saccharum	89.7	22.3	8.5	6.0	0.73	1.17
A. circinatum	133.0	26.0	11.8	6.5	0.63	1.29
A. saccharinum	148.2	33.3	11.7	4.4	0.80	1.12

[a]From Guries and Nordheim (1984). Reprinted from *Forest Science*, published by the Society of American Foresters, 5400 Grosvenor Lane, Bethesda, MD 20814-2198.

from isolated trees than from trees in closed stands. Samaras falling from trees in dense stands may stop and resume spinning several times and travel for short distances. The distance to which seeds travel also varies greatly for different species. For example, the relatively heavy, small-winged seeds of Italian stone pine drop almost straight to the ground, whereas the light, large-winged seeds of cluster pine and Monterey pine travel long distances (Wilgen and von Siegfried, 1986). It has been estimated that a sugar maple tree 15 m high and exposed to a maximum wind speed of 15 mph (7 m sec^{-1}) can disperse samaras over an area 100 m in radius (Green, 1980). Other estimates of variations in potential dispersal distances of seven species of *Acer* are shown in Table 12.4. These distances are overestimates because allowances were not made for interference by surrounding vegetation or effects of turbulence (Guries and Nordheim, 1984).

Approximately two-thirds of the winged nutlets of Ashe's roundleaf birch fell within 28 m, and 5% were recovered 100 m from the source (Ford *et al.*, 1983). In the rain forests of Tasmania, most seeds of Tasmanian false beech and eucryphia fell within 20 m of the source but some were blown up to 150 m (Hickey *et al.*, 1983). It has been estimated that out of a million seeds released by paperbark trees (*Melaleuca quinquenervia*), none is likely to travel more than 1 km. Only 1% will travel beyond 170 m and only 5% of those will be viable. Although seeds can be carried by hurricane-force (100 knot) winds up to 7.1 km, only a very small percentage of these will be viable (Browder and Schroeder, 1981). Under normal wind conditions in a closed forest, regeneration of *Shorea contorta* trees cannot be expected beyond 30 m from seed trees. About 30% of the winged fruits of *Shorea contorta* fell within 10 m of the mother tree, almost 90% within 20 m, and only 5 to 9% at distances greater than 30 m (Tamari and Jacalne, 1984).

Table 12.4
Potential Dispersal Distances of Samaras of Seven Species of *Acer*[a]

Species	Height of release (m)	Potential dispersal distance (m) when wind velocity (m sec^{-1}) is: 2	10	20
Acer rubrum	26	66	330	660
A. spicatum	6	14	70	140
A. pensylvanicum	8	12	60	120
A. negundo	13	26	130	260
A. saccharum	26	40	200	400
A. circinatum	8	12	60	120
A. saccharinum	26	44	220	440

[a]New data based on release of samaras from two-thirds of tree height. From Guries and Nordheim (1984). Reprinted from *Forest Science,* published by the Society of American Foresters, 5400 Grosvenor Lane, Bethesda, MD 20814-2198.

Windbreaks and Shelterbelts

The use of belts of trees as windbreaks is an ancient custom in Europe and a more recent one in North America. A windbreak is a porous barrier that decreases the wind velocity in sheltered areas on the leeward side, and to a lesser extent on the windward side, of the barrier by deflecting the wind (Baer, 1989). The benefits on the leeward side of shelterbelts include a decrease in soil erosion and dust storms, reduction in mechanical damage to plants, increase in crop yields, control of snowdrifting, improved cover and increased food supply for wildlife, and protection of livestock (Stoeckeler, 1962; Baer, 1989). Windbreaks often are useful in protecting young trees in plantations and forest nurseries (Gloyne, 1976). Several tropical and subtropical trees, including cacao, banana, and citrus, are sensitive to wind and often are protected by windbreaks (Alvim and Kozlowski, 1977). Heisler and Herrington (1976) emphasized the benefits of using belts of trees to modify microclimates in metropolitan areas so as to improve human comfort and reduce energy consumption in homes. Tree barriers also are useful in reducing vehicular noise in cities (Reethof and Heisler, 1976).

Wind speed is decreased for a distance to the leeward of a shelterbelt that depends on species, height of the shelterbelt, and the extent of its penetration by wind. A thinly planted belt of trees that allows wind to go through it at a reduced velocity provides a lower degree of shelter beyond the belt than a densely planted belt, but the effect of a penetrable belt extends over a greater distance. Usually a shelterbelt that is moderately permeable to wind provides the best shelter. Frequent severing of the roots of shelterbelt trees along the outer edges may reduce root competition to adjacent crops, thereby increasing the beneficial effects of shelterbelts (Zohar, 1985).

Soil Erosion

When the wind velocity exceeds the threshold necessary to initiate movement of soil particles, the rate of soil erosion is proportional to the wind speed cubed (Hagen, 1976). Hence, even a moderate reduction in wind speed by a barrier of trees appreciably reduces wind erosion. Windbreaks are used to great advantage to decrease soil erosion in cropland areas of North America (Baer, 1989), Australia (Bird, 1981), China (Xiang, 1986), and the USSR (Novitskii, 1980).

Effects on Microclimate

Because wind velocity and turbulence are reduced by shelterbelts, the microclimate is greatly modified in the protected zone. Evaporation on the leeward side is appreciably decreased, often as much as 20 to 30%. The decrease in evaporation varies seasonally and is greatest in the summer and autumn, intermediate in spring, and least in winter. Radiation balance is not altered greatly by shelterbelts, except very close to the belt. Air temperatures usually are slightly higher on the leeward side during the day because transport of sensible heat is reduced; night temperatures are lower because temperature inversions are less likely to be disrupted by air turbulence. The danger of night frosts is somewhat higher in sheltered than in unsheltered regions. This danger is decreased if shelterbelts are partially penetrable by wind. Both the absolute and relative humidity near the ground are lowered on the lee side of shelterbelts. However, these reductions in humidity are local and relatively small (Caborn, 1957).

Shelterbelts significantly alter the local distribution of rain and snow. Inasmuch as rain is usually accompanied by wind, a shelterbelt in a relatively exposed area intercepts rain so that more of it falls over the trees than at some distance to the leeward. Hence, a "rain shadow" develops on the leeward side for a distance that varies with wind velocity. Despite increased rainfall near the tree belt, less soil moisture is available to crops near the belt than far beyond it because of rapid absorption of water by the extensive root systems of the closely spaced trees.

Shelterbelts trap drifting snow and the trapped snow adds to soil moisture recharge as the snowpack melts. Dense, wide shelterbelts tend to accumulate snow very close to the belt; narrow and more permeable belts distribute snow much more uniformly on the leeward side. For example, shelterbelts with one or more rows of densely growing shrubs about 3 m high trapped almost all the snow in drifts 1.5 to 3 m deep within 10 to 25 m on the leeward side. By comparison, narrow belts of trees that had been pruned below and were penetrable by wind distributed snow uniformly in a layer about one-third to two-thirds of a meter deep for a distance of 200 to 400 m (Stoeckeler and Dortignac, 1941). In North America the trapping of snow by shelterbelts is important for crop growth in the Great Plains from Nebraska northward.

Crop Yields

Shelterbelts usually increase crop yields on the leeward side, largely because of increased soil moisture and lowered plant water deficits due to reduced evapotranspiration and the added water from snowmelt. The large areas of trapped snow result in more extensive areas of unfrozen soil than in unprotected areas. This soil condition provides for better infiltration and decreases water loss by runoff in the spring.

The amount by which shelterbelts increase average crop yields varies greatly with the crop, region, the age and composition of the shelterbelt, and other factors. Caborn (1957) summarized the literature showing that shelterbelts increased the yield of a wide variety of crops in Europe and America from a few to several hundred percent. In New Zealand, shelterbelts increased dry matter production of pasture plants (*Lolium perenne*, *Dactylis glomerata*, *Trifolium repens*, and *T. pratense*) by as much as 60% (Radcliffe, 1985). In India, yields of annual field crops (*Arachis hypogea*, *Cajanus americanum*) were increased by up to 60% by shelterbelts of mixed species of trees (Reddi *et al.*, 1981).

The increase in crop yield varies with distance from the belt. Crop yields are reduced in a narrow zone bordering the shelterbelt because of root competition and shading of crops by the shelterbelt trees. This unproductive strip usually is no wider than approximately half the height of the shelterbelt. The harmful effects of competition for soil moisture adjacent to a shelterbelt can be counteracted by using tree species that root deeply, by root pruning of established trees, by cultivating deeply near the shelterbelt to prevent growth of lateral roots, and by using only species that do not develop root suckers (Daubenmire, 1984).

The beneficial effects of shelterbelts on crop yields decrease greatly during droughts. For example, in the Great Plains of the United States, water deficits were lower in plants protected by windbreaks only when soil moisture supplies were relatively high. During droughts, windbreaks increased water deficits in crops (Frank *et al.*, 1977). The effects of shelterbelts on microclimate and yield of a variety of crops are discussed further by Caborn (1957, 1965), Stoeckeler (1962), and Grace (1977). Read (1964) has an excellent discussion of factors influencing selection of tree species for shelterbelts as well as planting, spacing, and management of shelterbelt trees.

Windbreaks do not necessarily consist of trees or shrubs. In some areas, rows of tall-growing plants such as corn, sorghum, or sunflower are used to protect low-growing plants. Windbreaks made of snow fence, matting, or even of earth also are sometimes used.

Summary

Wind topples trees, breaks stems and branches, uproots trees, malforms stems, injures leaves, erodes soil, inhibits plant growth, and increases risks of air pollution, insect attack, disease, and fire. Wind also provides benefits by dispersing pollen and plant propagules.

Although wind decreases the height growth of trees, it often increases the diameter growth. Stems of swaying trees are thicker and more tapered than those of trees prevented from swaying. Wind induces the formation of compression wood on the leeward sides of conifer stems and tension wood on the windward sides of stems of broad-leaved trees.

Wind often affects the yield and quality of fruits and seeds by influencing pollen production, pollination, fruit set, and fruit development.

The effects of wind on tree growth are mediated by changes in water status and in food and hormone relations. Low wind speeds increase transpiration by removing the boundary layer of water vapor surrounding the leaves.

The effects of wind on transpiration are modified by cooling, which lowers the leaf to air vapor pressure difference. Wind may lower transpiration by closing stomata as a result of shaking of leaves, dehydration of leaves, and direct responses of guard cells to lowered air humidity, and in response to high CO_2 concentration near the leaf surface.

High-velocity winds reduce total photosynthesis by inducing leaf abscission and injury. High wind velocities may decrease available carbohydrates by accelerating respiration of leaves. Low-velocity winds often increase photosynthesis by affecting boundary layer thickness and air temperature.

Increase in stem diameter and induction of reaction wood in stems tilted by wind are mediated by hormonal changes. Formation of compression wood is associated with high auxin levels, and formation of tension wood with auxin deficiency. However, the effects of auxin are modified by other hormones.

Wind influences the uptake of gaseous air pollutants by affecting the thickness of the boundary layer and by transporting pollutants. Wind also transports spores of fungus pathogens, pollen, and seeds.

Benefits of belts of trees (windbreaks, shelterbelts) include decreases in soil erosion, dust storms, and mechanical damage to plants; increase in crop yield; control of snowdrift; improvement of cover for wildlife; and protection of livestock. Other benefits include protection of trees in plantations and forest nurseries, modification of microclimates, and reduction of vehicular noise in urban areas.

General References

Caborn, J. M. (1965). "Shelterbelts and Windbreaks." Faber and Faber, London.
Carlquist, S. (1974). "Island Biology." Columbia University Press, New York.

Daubenmire, R. F. (1984). "Plants and Environment." Wiley, New York.
Edmonds, R. L., ed. (1979). "Aerobiology: The Ecological Systems Approach." Dowden, Hutchinson, and Ross, Stroudsburg, PA.
Faegri, K., and van der Pijl, J. (1979). "The Principles of Pollen Ecology." Pergamon Press, Oxford.
Geiger, R. (1966). "The Climate Near the Ground." Harvard University Press, Cambridge, MA.
Grace, J. (1977). "Plant Responses to Wind." Academic Press, London.
Grace, J. (1983). "Plant–Atmosphere Relationships." Chapman and Hall, London.
Grace, J., Ford, E. D., and Jarvis, P. G., eds. (1980). "Plants and Their Atmospheric Environment." Blackwell, London.
Gregory, P. H. (1973). "The Microbiology of the Atmosphere." Wiley, New York.
Harper, J. L. (1977). "Population Biology of Plants." Academic Press, New York.
Ingold, C. T. (1984). "The Biology of Fungi." Hutchinson, London.
Mudd, J. B., and Kozlowski, T. T., eds. (1975). "Responses of Plants to Air Pollution." Academic Press, New York.
Nobel, P. S. (1981). Wind as an Ecological Factor. *Physiol. Plant Ecol.* **12A,** 475–500.
Savill, P. S. (1983). Silviculture in windy climates. *For. Abstr.* **44,** 473–488.
Smith, W. H. (1981). "Air Pollution and Forests." Springer-Verlag, New York.
Stanley, R. G., and Linskens, H. F. (1974). "Pollen: Biology, Biochemistry, Management." Springer-Verlag, New York.
Timell, T. E. (1986). "Compression Wood," Vol. 3. Springer-Verlag, New York.
van der Pijl, J. (1982). "Principles of Dispersal in Higher Plants." Springer-Verlag, New York.
Waister, P. D. (1972). Wind damage in horticultural crops. *Hort. Abstr.* **42,** 609–615.

Chapter 13

Cultural Practices

Introduction

In the first chapter of this book it was stated that the quantity and quality of plant growth depend on the hereditary potential of plants and the environment in which a plant is growing, operating through its physiological processes. In later chapters we discussed some important physiological processes and environmental factors. In this final chapter we will discuss important cultural practices and why they are useful. It also was stated in Chapter 1 that the full potential for growth of woody plants is seldom attained because various environmental stresses continually inhibit essential physiological processes. These include abiotic stresses such as drought, extreme temperatures, shading, wind, air pollution, fire, infertile soil, poor soil structure, and inadequate soil aeration. Important biotic stresses include plant competition, human activities, attacks by insects and plant pathogens, and occasionally feeding by higher animals such as deer and beavers.

Most of the increases in yield of forest and horticultural crops have resulted from improvements in management (Jones, in Cannell and Jackson, 1985). However, cultural practices are helpful only if they protect plants from injury or increase the overall efficiency of the physiological processes that control growth and economic yield. Foresters, arborists, farmers, and gardeners have quite different objectives in terms of products, but they have one common objective: to manage their crops in ways that will minimize the effects of the omnipresent stresses on growth and yield. Important cultural practices include site preparation for planting, fertilization, irrigation, stand thinning, pruning, and application of chemicals, including pesticides and growth regulators. These practices will now be discussed, together with the reasons why they are beneficial for production of planting stock and growth of outplanted trees.

Production of Planting Stock

Maximum productivity of forest and fruit trees depends on combined use of genetically superior planting stock and intensive management of plantations and orchards. Intensive cultural practices alone do not assure maximum productivity.

Genetic variation in most species of forest trees is very high and can be readily exploited, but it is low in a few species such as red pine (Fowler and Morris, 1978). Genetic variation in growth involves multiple gene effects that are expressed through physiological and developmental processes (Namkoong *et al.*, 1988). Maximum gains in genetic improvement of forest trees are possible by use of relatively few of the best genetic parents to supply planting stock. Superior trees are obtained by both selection and breeding. Selection based entirely on the visible characteristics of trees (phenotypes) usually is followed by progeny testing to ascertain if the selected trees are genetically superior.

Selection and hybridization of superior offspring have been used to develop forest planting stock with such attributes as high growth rate, good form, drought tolerance, cold tolerance, pest resistance, and tolerance of low soil fertility (Wright, 1976; Zobel and Talbert, 1984). Genetic improvements often result from new gene combinations. The greatly increased growth rates of hybrid poplars, for example, reflect hybrid vigor or "heterosis."

Early selection techniques also have been developed for desirable characteristics of fruit trees, including pest and disease resistance, frost resistance, adaptation to warm winters, precocity, high quality of fruits, and semidwarfing rootstocks. Other improvements include introduction of mutants such as nectarines, the Washington navel orange, and red mutants of the apple variety Delicious (Alston and Spiegel-Roy, 1985). Methods of breeding fruit trees are discussed by Moore and Janick (1983).

Breeding of forest trees involves different problems from breeding of fruit trees. Forestry encompasses many useful species growing in a wide variety of environments, which often cannot be controlled. Furthermore, several different products commonly are required from a single species and sometimes from the same tree. Forest stands also are managed for multiple objectives. Hence it may be necessary to breed forest trees for widely different environmental regimes as well as for different products (Namkoong *et al.*, 1988). In general, the forest tree breeder favors sparsely flowering, fast-growing, tall trees that usually are propagated by seeds.The fruit tree breeder aims to select compact, early-flowering varieties with a high ratio of fruits to shoots. Constraints on progress in forest tree breeding are discussed by Bridgewater and Franklin (1985).

Seedling Production

Millions of tree seedlings are produced in forest nurseries scattered all over the world. The first requirement for producing forest planting stock is an adequate supply of high-quality seed. Immediate needs can be met by collecting seeds from outstanding phenotypes (plus trees) in natural forest stands. Unless better sources of seeds are known, seeds for conventional plantings should be obtained from trees of native stock growing as near as possible to the site to be planted. This is because natural selection has tended to produce populations of trees that are adapted to the environmental conditions in which they have evolved (Kozlowski, 1979). The U.S. Forest Service suggests that, if possible, planting stock should be produced from seed collected within 160 km geographically and 300 m vertically of the site where it will be used.

For species with extensive north–south ranges, the southern seed sources tend to grow faster, continue shoot growth later into the autumn, and are less cold-hardy. Plants obtained from seeds of trees from a wet region generally grow faster and are less deeply rooted than plants grown from seeds from a dry region. Elevational trends also are important and selection may produce differences, in response to stress, between trees of high and low elevations (Wright, 1976).

To meet long-term seed needs, seed production areas (seed stands) commonly are used, in which poor phenotypes are removed from the stands. The remaining good phenotypes intermate and seeds eventually are collected. Although seed production areas rarely are progeny-tested, they are reliable sources of well-adapted, relatively inexpensive seeds (Zobel and Talbert, 1984).

Genetically improved seeds in appreciable quantities can be produced in vegetative or seedling seed orchards. Vegetative seed orchards are obtained through grafts, cuttings, or tissue culture plantlets. Seedling seed orchards are established by planting seedlings and subsequent roguing of the stand to retain the best trees for seed production (Zobel and Talbert, 1984).

Vegetative Propagation

To ensure that the offspring resemble their parents, gardeners and horticulturists have vegetatively propagated plants desired for their beauty or the quality of their fruit. When woody plants are reproduced from seed they are likely to show considerable variation from the parent because they usually are highly heterozygous. Gardeners and horticulturists have known for centuries that plants propagated from seed do not always resemble the parents. However, they learned that they could obtain offspring resembling the parents through vegetative propagation by grafting, budding, layering, or rooting of cuttings. Now that forest geneticists are expending considerable effort on selecting superior trees, the ability to produce numerous identical offspring from a single superior tree (cloning) has become very important in forestry (see the section on Biotechnology).

Cuttings

Twigs from a few kinds of forest trees and many ornamental shrubs root readily when their bases are placed in sand or soil and kept moist. Sometimes bottom heat beneath the cutting bed is beneficial, as is maintenance of high humidity above the bed by a plastic enclosure or a misting system. The condition of the plants from which cuttings are taken is very important. For example, young, half-hardened twigs of azaleas root well, whereas mature wood of roses is best, and cuttings from mature trees are less likely to root than cuttings from young trees. This is troublesome because by the time many kinds of trees are old enough to show their desirable characteristics, they are difficult to propagate vegetatively. Harmer (1988) reported that epicormic shoots formed on mature English oak after girdling, or even on sections of logs, root well and suggested that this might be a good way to mass propagate superior trees.

Layering

Some trees and shrubs that are difficult to reproduce by cuttings can be propagated readily by layering, that is, by burying a part of an attached branch in the soil until roots form, then cutting it off and planting it. In air layering, a branch is wounded, wrapped in moist sphagnum, and enclosed in waterproof plastic until roots form. The ball of sphagnum containing roots is then cut off and planted. Sometimes a synthetic plant hormone is applied to the wound to stimulate root formation.

Grafting and Budding

Grafting and budding are very old methods of propagating desirable trees and shrubs. For example, most apple and citrus trees are produced by grafting twigs from desirable varieties onto sturdy rootstocks, while most peaches are propagated by budding. In grafting, the cambium of a detached branch (the scion) is placed in contact with the cambium of a rooted plant (the stock) and bound in place (Fig. 13.1). In budding, a sliver of bark and wood bearing a healthy bud is sliced from the parent and bound to a wound on the stock plant in such a manner that the cambia of the two units are in contact. If all goes well, the tissues of stock and scion grow together, forming a new composite plant. Grafting and budding provide opportunities to combine a root system tolerant of local soil conditions with a top producing the desired type of flowers, fruit, or foliage. For example, apples sometimes are grafted on Malling and pear on quince rootstocks to produce dwarf trees and citrus on trifoliate orange rootstocks to improve fruit quality. Three-story rubber trees sometimes are produced by grafting a stem segment with high latex yield between a sturdy root system and a top with leaves resistant to a troublesome leaf disease.

Figure 13.1 Wrapping a side veneer graft of loblolly pine with grafting rubber. (Photo courtesy of S. E. McKeand.)

There is considerable variability in the success of grafting (McKeand *et al.*, 1987). Usually grafts of closely related plants, such as within a species, are most successful, but occasionally incompatibility is found among individual plants within a species. Sometimes the union is mechanically weak because of poor growth, sometimes translocation across the union is poor, and sometimes failure does not occur for a decade or more. The causes of dwarfing and draft incompatibility are not fully understood (Kramer and Kozlowski, 1979, pp. 157–159). More information concerning classic methods of vegetative propagation can be obtained from books such as Hartmann and Kester (1982).

Biotechnology

In recent years, interest has accelerated greatly in improving the genetic quality of forest, fruit, and ornamental plants by the techniques of genetic engineering, gene splicing and recombinant DNA, cloning, and plant tissue culture.

Tissue culture, the most important component of many biotechnology programs, shows potential for (1) micropropagation for biomass energy production; (2) production of disease-free and disease-resistant plants; (3) induction and selection of mutants resistant to insects and diseases, droughts, temperature, herbicides, and so on; (4) production of hybrids through anther culture; (5) hybridization through embryo rescue; (6) somatic hybrids through fusion of protoplasts; (7) transformation through uptake of foreign genetic material; (8) introduction of capacity for nitrogen fixation; and (9) cryopreservaton of germplasm (Bonga and Durzan, 1982; Bajaj, 1986; Hanover and Keathley, 1988).

One advantage of cell or tissue culture propagation over sexual propagation is the opportunity to produce numerous clones from a desirable parent. Another advantage is the reduction in time between selection of a desirable parent and production of new planting stock. Eight to ten years are required from the time scions from a selected loblolly pine tree are grafted on trees in a seed orchard until they produce seed, but plantlets can be produced in one or two years (Amerson *et al.*, in Henke *et al.*, 1985). Tissue cultures can be obtained from various embryonic tissues such as needle fascicles, winter buds, and cotyledons. The rather complex procedure required to produce rooted plants is discussed in papers by Amerson *et al.* and other papers in Henke *et al.* (1985).

Many genera of plants propagated commercially by tissue culture are listed in Table 13.1, but various economic and technical problems limit its usefulness (Zimmerman and Lutz *et al.*, in Henke *et al.*, 1985). One problem is the genotypic and phenotypic variation (sometimes termed somaclonal variation) found in some tissue cultures. There also is uncertainty concerning the quality of the plants produced. For example, first-year plant size is said to be smaller than for seedlings of loblolly pine and the root systems are said to be less efficient in absorbing minerals (Amerson *et al.*, in Kramer and Rose, 1986). However, it seems likely that tissue culture will become increasingly useful as the problems are solved.

Cultural Practices

The success of outplanted seedlings depends on their morphological and physiological condition. The quality of seedlings in terms of survival depends first of all on their capacity to regenerate new roots rapidly after transplanting, that is, their root growth potential or RGP. This depends on the climate, soil, and cultural situations to which they are exposed prior to outplanting. Length of photoperiod and hours of low temperature during the dormant season are among the climatic factors likely to influence seedling physiology. For example, in some years certain low-altitude nurseries in California are said to produce seedlings of some species of conifers with

Table 13.1
Genera of Woody Plants Commercially Produced in Tissue Culture[a]

Acer	*Elaeis*	*Potentilla*
Actinidia	*Escallonia*	*Prunus*
Amelanchier	*Eucalyptus*	*Pyrus*
Arctostaphylos	*Ficus*	*Rhododendron*
Betula	*Forsythia*	*Ribes*
Buddleia	*Garrya*	*Rosa*
Camellia	*Hydrangea*	*Rubus*
Campsis	*Hypericum*	*Salix*
Castanea	*Kalmia*	*Sequoia*
Celtis	*Lagerstroemia*	*Sequoiadendron*
Clematis	*Lapageria*	*Simmondsia*
Corylopsis	*Leucothoe*	*Spiraea*
Corylus	*Magnolia*	*Syringa*
Cotinus	*Malus*	*Vaccinium*
Crataegus	*Nandina*	*Viburnum*
Daphne	*Pinus*	*Vitis*
Deutzia	*Populus*	*Weigela*

[a]From Zimmerman (1985).

low growth potential because the nights are too warm to break dormancy (Stone *et al.*, 1963; Stone and Norberg, 1979). This is seldom a problem with seedlings of species native to mild climates, although some families of the pines native to the southeastern United States require exposure to several hundred hours of low temperature to break bud dormancy and increase root growth potential.

Nursery practices such as density of planting of seedlings, fertilization, irrigation, and wrenching or undercutting the root system have important effects on seedling quality (Duryea and Brown, 1984; Cannell, 1986; Kramer and Rose, 1986). It has been demonstrated frequently that too high a density of planting results in small seedlings with small stem diameters and less root and leaf surface that are less likely to succeed than larger seedlings.

Seedling Morphology

Wakeley (1954) reported that morphology is an unreliable predictor of seedling success, and Feret *et al.* (1986) found that survival of outplanted loblolly pine was only weakly correlated with seedling size, but consistently correlated with RGP. However, Lopushinsky and Beebe (1976) reported that seedlings of Douglas-fir and ponderosa pine with large root systems grew better than those with small root systems. Several experiments indicate that southern pine seedlings with large root collars usually survive and grow better than those with small root collars (South *et al.*, 1985; South, 1987). Some of the discrepancies in the literature may result from differences in cultural practices or site conditions. Carlson (1985b) suggested that large root systems are beneficial not only because they provide more absorbing surfaces, but also because they allow more sites for development of new roots. In general, a good loblolly pine seedling is 20 to 25 cm tall with a firm, woody stem 4 to 5 mm in diameter, fibrous root system, well-developed buds, and an abundance of needles (Kramer and Rose, 1986), in other words, a sturdy seedling. Although a large leaf area may increase water loss, it is needed to carry on photosynthesis and supply the carbohydrates required for growth. Occasionally, seedlings grow so tall that some needles are clipped off, but severe top pruning and clipping may reduce growth after outplanting (Barnett, 1984).

Seedling Physiology

The most important physiological requirement for seedling survival is rapid development of new roots and occupation of a large volume of soil to improve absorption of water and minerals. Some reserve food also is essential because it may be several weeks after transplanting before photosynthesis returns to normal. However, food supply is seldom an important limitation on survival of properly managed seedlings. Most deaths of transplanted seedlings result from dehydration because the root systems are too limited in extent to supply enough water to maintain turgor and growth. Thus the RGP at the time of planting is very important. This often is determined by growing a sample of seedlings under favorable conditions for a few weeks and then determining the number of new white roots (Feret *et al.*, 1986, and others, in South, 1986). The RGP varies widely among lots of seedlings, but usually increases from a low value in the autumn to a high value in late winter and early spring, and then decreases.

Fertilization

The addition of fertilizers to nursery soil is essential to maintain production of high-quality seedlings because they require an abundant supply of minerals to make good growth, and minerals are leached from the soil by rain and irrigation and thus removed for seedling use. Today most nurseries are so well fertilized that consistent correlations between seedling quality and mineral nutrition are difficult to establish (Duryea and Landis, 1984). However, better correlation of fertilizer supply with seedling needs would be desirable, both economically and physiologically. Perhaps foliar analysis can be adapted to use in nurseries, as van den Driessche (1988) reported that survival of Douglas-fir seedlings is correlated with the N content of their needles. Fertilization usually is stopped near the end of the growing season, because if it is continued too late, it delays cessation of growth and development of dormancy, thereby increasing the possibility of frost injury. However, in some instances, fertilization at the end of the growing season was beneficial, as shown in Table 13.2.

Irrigation and Soil

In most climates and on most soils, irrigation is necessary to avoid injury during droughts. Deep, coarse-textured soils are preferable for nurseries because they are easily worked and usually well drained, are a good medium for root growth, and

Table 13.2
Growth and Assimilation on Three Harvest Dates of Douglas-Fir Seedlings Fertilized or Unfertilized in the Nursery in October and Outplanted the Following February[a]

Characteristic	Unfertilized	Fertilized	Δ (%)
New shoot growth (g)			
May 15	0.8 (0.1)	1.3 (0.1)[b]	+63
June 15	3.1 (0.5)	5.0 (0.6)[c]	+61
Sept. 5	4.5 (0.3)	6.5 (0.4)[b]	+44
Relative growth ($g\ g^{-1}$)			
May 15	0.08 (0.01)	0.12 (0.01)[b]	+50
June 15	0.25 (0.02)	0.39 (0.03)[b]	+56
Sept. 5	0.42 (0.01)	0.55 (0.02)[b]	+31
Relative growth rate ($mg\ g^{-1}\ day^{-1}$)			
May 15	3.3 (0.2)	3.2 (0.4)[d]	
June 15	3.9 (0.3)	5.1 (0.3)[c]	+31
Sept. 5	2.5 (0.1)	3.3 (0.1)[b]	+32
Net assimilation rate ($mg\ cm^{-2}\ day^{-1}$)			
May 15	0.12 (0.02)	0.13 (0.01)[d]	
June 15	0.12 (0.01)	0.16 (0.01)[e]	+24
Sept. 5	0.16 (0.05)	0.23 (0.05)[b]	+44
Leaf area ratio ($cm^2\ g^{-1}$)			
May 15	29.8 (0.9)	31.4 (0.7)[d]	
June 15	39.3 (1.1)	45.1 (1.2)[b]	+15
Sept. 5	36.7 (0.8)	43.0 (1.2)[b]	+17

[a]From Margolis and Waring (1986).
[b]$p < 0.01$.
[c]$p < 0.05$.
[d]Not significant.
[e]$p < 0.10$.

seedlings are more easily lifted from sandy than from clay soils. However, such soils become deficient in available water sooner than heavy soils. Generally it is desirable to restrict watering toward the end of the growing season to encourage cessation of growth and development of dormancy. Application of too much water can cause injury to root systems from deficient aeration. Both water and fertilizer must be applied judiciously, preferably by experienced workers.

Storage

Seedlings should be physiologically dormant when lifted for storage. Care must be taken to preserve the RGP of seedlings during the period between lifting and planting. Seedlings lifted early in the winter are best kept in cold storage to reduce water loss and excessive use of food in respiration. Many seedlings are injured by exposure to sun and dry air during transport or at the planting site. In some circumstances it has been found beneficial to dip or puddle root systems in clay slurry, which provides a protective coating and decreases water loss during exposure prior to planting (Slocum and Maki, 1956; Bacon and Hawkins, 1977; Johnston and Ward, 1986). This is an adaptation of a procedure used long ago by gardeners when transplanting bare-rooted trees and shrubs.

Container-Grown Seedlings

Many forest tree seedlings are grown in containers and ornamental trees and shrubs often are grown in large containers for several years before outplanting. The containers are filled with various combinations of sand, peat, vermiculite, and shredded pine bark, which are low in minerals and have a low capacity to retain ions. Sometimes a slow-release fertilizer is incorporated in the mix before the containers are filled, but this tends to be leached out by the frequent irrigations that are required. Thus the initial fertilization must be supplemented by frequent applications of fertilizer, preferably in the irrigation water if trickle irrigation is used.

As plants increase in size, the supply of mineral nutrients should be increased either by increasing the concentration of the nutrient solution or by applying it more often. Recognizing that the amounts of nutrients required per unit of time increase with increasing biomass, Ingestad and Lund (1979, 1986) introduced the concept of relative addition rate (RA) of mineral nutrients. The RA is analogous to the relative growth rate of plants and is expressed as the amount of nutrient to be added per unit of time in relation to the amount of nutrients present in the plant. Ingestad (1981, 1982) emphasized that during the exponential period of seedling growth, mineral nutrients ideally should be added in exponentially increasing amounts. Timmer and Armstrong (1987) showed that both dry weight increase and root development of red pine seedlings in containers were greater when fertilizer was applied at an exponential rather than constantly increasing rate. Furthermore, superior seedlings could be grown by this method with only a fourth of the fertilizer dose conventionally used for production of seedlings in containers. Dixon *et al.* (1983) reported that inoculation of container-grown white oak seedlings with a mycorrhiza-forming fungus improved root and shoot growth the first year after outplanting. Wright and Niemiera (1987) discuss the fertilization of container-grown seedlings in more detail.

Freezing Injury

In some situations, dormant container-grown seedlings and even large plants can be injured by low root temperatures. Even if the roots are not permanently injured, cold and frozen soil may reduce water uptake so much that shoots of plants exposed to sun and wind suffer from desiccation. Lindstrom (1986) reported that the root growth potential of Scotch pine and to a lesser extent of Norway spruce seedlings was reduced by subfreezing temperatures. This could be a problem even in mild climates because the soil in small containers may freeze even when soil in a nursery bed is frozen only on the surface.

Site Preparation

Various cultural treatments, alone and in combination, are used to improve regeneration of harvested forests, plantations, and orchards by optimizing the availability of water, mineral nutrients, light, soil oxygen, and growing space. Such practices include disposal of logging slash; reducing competition from undesirable plants by prescribed burning, mechanical means, or use of herbicides; and soil manipulation to increase seed germination and growth of seedlings. Although most site preparation treatments are applied during seedling establishment, some are initiated prior to harvesting of crop trees or at various other times during the rotation (Smith, 1986). Productivity is most likely to be increased by cultural practices that reduce the time to canopy closure. Thereafter site preparation has much less influence on tree growth (Miller, 1981).

Prescribed Burning

In addition to its use to decrease fuel accumulation, prescribed burning often is practiced to prepare seedbeds prior to planting. Slash usually is burned where trees were felled, piled by machine, or windrowed. Prescribed burning is discussed in more detail in Chapter 11.

Mechanical Site Preparation

Planting sites often are prepared by disposing of logging slash, windrowing, chopping of vegetation, harrowing, disking, bedding, and various combinations of these. Elimination of slash usually is essential to decrease fire hazards and to stimulate regeneration by eliminating heavy shade and facilitating hand or machine planting of seedlings. Windrowing, the heaping of logging waste into rows by blading or root raking, is preferably done along the contour on sloping land. Chopping concentrates vegetation on the ground and facilitates burning of plant debris. Harrowing and plowing generally improve compacted sites and those with a dense mat of roots just below the soil surface. Bedding, the mounding up of surface soil, litter, and logging debris into low ridges or beds, improves drainage on flat, poorly drained soil; increases availability of soil moisture, mineral nutrients, and soil oxygen; and reduces plant competition. On wet sites, bedding provides ridges of well-drained, well-aerated soil in which roots can rapidly become established. Bedding also accelerates mineralization of organic matter. If the A_1 soil horizon is thin and the B horizon is mechanically resistant to root growth, the presence of topsoil, humus, and litter greatly improves conditions for root growth, which eventually is reflected in stimulation of top growth (Haines *et al.*, 1975).

Herbicides

Because of their effectiveness in control of herbaceous and woody weed plants, together with their low costs and ease of application, herbicides often are used to decrease competition from undesirable vegetation during regeneration of forests. Elimination of undesirable plants increases the availability of water, light, and mineral nutrients to crop trees (Chapter 3). Ebert and Dumford (1976) noted that acceleration of tree growth following application of triazine herbicides was associated with increased absorption of N by seedlings, presumably because of decreased plant competition. Certain herbicides may affect mineral nutrition by influencing development of mycorrhizae. For example, roots of Douglas-fir seedlings treated with napropamide were infected with a greater diversity of mycorrhizal fungi than seedlings treated with DCPA or bifenox (Trappe, 1983).

There is a voluminous literature on both short- and long-term beneficial effects of herbicides on establishment and growth of forest trees, and only a few examples will be cited. Following the removal of competing weeds with herbicides and mechanical preparation of land prior to planting on heavy clay soils along the southern shore of Lake Superior in Wisconsin, reasonably acceptable establishment of trees was obtained (Kuntz and Kozlowski, 1963). In Australia, stem volume of Monterey pine trees was increased by 10 to 140%, and mortality reduced by 40%, 1 year after elimination of herbaceous weeds (Nambiar and Zed, 1980). In North Carolina, Mississippi, Arkansas, and Oklahoma, an increase in biomass of up to 1200% was recorded 1 year after loblolly pine was planted in weeded plots (Nelson *et al.*, 1981). Twenty years after young longleaf pine trees were released with herbicides from the competition of broad-leaved trees, volume increased by 40% over controls, the equivalent of 8 years of growth (Michael, 1980).

Drainage

The injurious effects of flooding and inadequate soil aeration were discussed in Chapter 8. Orchards and nurseries often are drained and in northern Europe some land is drained for forests. However, in general, drainage of land for growth of forest trees is uncommon because of the high cost.

In some regions, stimulation of tree growth on drained peat-lands is associated with lowering of the groundwater table, increased substrate aeration (Campbell, 1980) , and consequent greater rooting depth of some species (Boggie, 1972). The decreased water content of drained peat also reduces substrate heat capacity and thermal conductivity. Hence, in the spring the surface layers of drained peatlands warm faster than those of undrained peatlands (Pessi, 1958). In Alberta, Canada, bud opening and flowering of tamarack and dwarf birch *(Betula pumila)* occurred 2 to 6 days earlier on drained than on undrained peatlands (Lieffers and Rothwell, 1987).

An example of profitable drainage for tree production is the wetlands of the coastal plain of the southeastern United States, where the water table varies from as much as 30 cm above the soil surface in the winter to 60 to 200 cm or more below during dry summer weather. Because of the large areas available at low initial cost, this land was very attractive for commercial forestry, but it was soon found that drainage was necessary to obtain satisfactory growth of loblolly pine. In some areas drainage canals were installed with a density of 8 to 16 lineal miles per square mile (5 to 10 km km^{-2}) and seedlings were planted on raised beds to provide better aeration. Fertilization with P also was necessary. Under such intensive management, loblolly pine grew rapidly and attained a height of 80 ft (24 m) at 25 years compared to 45 ft (13.5 m) on unmanaged sites (Campbell and Hughes, 1981). Terry and Hughes (1975) cited several studies that showed increases in volume growth of southern pines of 80 to 1300% as a result of drainage of North Carolina wetlands.

It should be remembered that wetlands are among the most fragile and threatened ecosystems on earth. Although they are one of our greatest natural assets, wetlands have unfortunately had a poor image. Too often they have been considered valuable only if drained and converted to agriculture or other uses. As a result, approximately half of the world's wetlands have been lost since 1900. The United States alone has lost 87 million ha (approximately 54%) of its original wetlands, with agricultural development accounting for about 87% of recent losses (Maltby, 1986).

The possible benefits of drained wetlands should be very carefully weighed against the potential benefits they provide if undrained. As emphasized by Maltby (1986), wetlands are among the world's most productive ecosystems. They can produce up to eight times as much plant growth as the average wheat field. Ugandan papyrus marshes can be twice as productive as lush tropical forests. In addition to supplying such economically important species as rice, sago, oil palm, mangroves, crayfish, shrimps, oysters, waterfowl, fish, and fur-bearing animals, wetlands support grazing land, control floods, maintain water quality, absorb toxic chemicals, and clean up polluted waters. The many benefits of wetlands often extend for great distances from their location. Whereas drainage of wetlands may increase local yields in the short term, it may lead to reduction in yield in other parts of the ecosystem or reduce the capacity of an entire ecosystem for sustained yield of useful products. Maltby (1986) cited several examples of errors that have been made in draining wetlands.

Richardson (1983) emphasized that although pocosins of the North Carolina coastal plain are ecologically very valuable, they are being drained and rapidly lost. Richardson suggested that a "multiobjective programming" approach (Cohen, 1983) be used in dealing with wetlands. This would permit evaluation of several effects of drainage and provide specific and continuous trade-offs among competing objectives. Accordingly, decision making would include concepts of sustained yields, irreversibility (e.g., prevention of permanent loss of ecosystem processing), values of undeveloped wetlands to future generations, and mitigation and reclamation procedures for each developmental activity.

Drainage of soils often is essential for growth of shade and ornamental trees and shrubs. Hardpans at various soil depths and bedrock beneath shallow soils occur commonly in urban soils and restrict root growth and drainage. Accumulation of water in low areas or collection from pavements and building runoff often necessitate installation of drains. Methods of providing drainage in urban areas by penetrating semi-impervious layers and hardpans and installing internal drainage systems are discussed by Harris (1983).

Responses of Trees to Site Preparation

There are many examples of beneficial effects of site preparation treatments and only a few will be cited here. For example, establishment and growth of forest trees have been improved by chopping of vegetation (Burns and Hebb, 1972; Mason and Cullen, 1986), harrowing or plowing (Haines *et al.*, 1975), bedding (Haines and Pritchett, 1965; Baker, 1973,) and elimination of competing plants with herbicides (Smith, 1986).

In Tanzania, survival at 2 years of *Eucalyptus tereticornis* was higher on disked or bedded sites than on unworked soils (Chamsama and Hall, 1987). In the western United States, site preparation suppressed nonconiferous vegetation and increased the amount of soil resources available to ponderosa pine, sugar pine, and white fir (Lanini and Radosevich, 1986). Methods such as brush raking, which reduced shrub canopy development, or other forms of site preparation in combination with herbicides, were most effective in increasing the availability of water and light to the conifers. Height, stem diameter, and increase in canopy volume of conifer species were greatest when competing shrubs were small and least when they were large. Predawn and midday water potentials of conifers were highest when shrub growth was low.

Potential Hazards of Site Preparation

The effectiveness of site preparation on tree growth depends greatly on the skill and experience of forest managers and varies with management practices, soils, and species characteristics, topography, and economic considerations. In the United States not only is site preparation very costly but it may even be harmful. Hence forest managers generally should "do as much site preparation as necessary, but as little as possible" (Haines *et al.*, 1975).

Slash disposal can be beneficial as well as harmful. Whereas thick and dense layers of slash impede establishment of young trees, thin layers of slash benefit reproduction by improving the microclimate and protecting seedlings from desiccation, temperature extremes, and competition from intolerant vegetation (Smith, 1986). Sometimes, particularly on sloping land, intensive site preparation results in soil erosion and nutrient loss and displacement. Chopping of vegetation and bedding generally have little effect on nutrient reserves, but burning (Chapter 11) and windrowing can lead to significant nutrient losses. Judicious burning temporarily increases soil fertility on certain sites but decreases it on others. For example, on fertile soils in Tasmania the nutritional status of the soil was improved by burning

(Ellis *et al.*, 1982) whereas the extensive oxidation of organic matter by burning on sandy soils of southeastern Australia decreased soil fertility (Farrell *et al.*, 1981).

As emphasized by Smith (1986), no silvicultural treatment can impair site productivity more than scraping off the surface soil. This is particularly true if mineral nutrients, which are concentrated in the litter and upper layers of the mineral soil, are moved laterally beyond the future extent of tree roots. Windrowing sometimes is particularly dangerous because it moves topsoil and humus, hence mineral nutrients, along with other plant debris. Windrowing displaced 373, 18, and 27 kg ha^{-1} of N,P, and K, respectively, from the interwindrow area (Morris *et al.*, 1983). These amounts were equivalent to those removed in six stemwood harvests and comprised 10% of the site's nutrient reserves. In some cases, nutrient displacement can be minimized by shortening the distance between windrows and decreasing their size. Unfortunately windrowing often has been used on sites that could be adequately prepared for planting by chopping, burning, or application of herbicides, practices that have less harmful effects on nutrient displacement.

Skillful use of herbicides requires knowledge of responses of both the plants to be favored and those to be eliminated. The influence of herbicides on seed germination and subsequent plant growth varies greatly with the specific chemical; the rate, method, time, and number of applications; plant species; soil type; weather; and other factors (Sutton, 1958; Kozlowski, 1960; Kozlowski and Kuntz, 1963; Kozlowski *et al.*, 1967).

When improperly used, herbicides can be toxic to desirable as well as undesirable vegetation. Sutton (1970) emphasized that unless herbicides are used with regard for ecological considerations, they can be quite harmful. For example, when grass is killed in cold regions, the tendency of soil to be heaved by frost is accentuated, and planted trees are more likely to be "left behind" as the soil settles back after heaving. Sutton pointed out that white spruce trees that appeared to be well established after 4 years in Ontario were heaved from the soil and thrown flat after weeds were controlled so the soil around the trees was bare.

Contact herbicides such as sodium arsenite, paraquat, and cacodylic acid sometimes cause necrotic spots on leaves as a result of spray drift. Applying herbicides that might contribute to injury of nontarget plants by spray drift should be avoided on windy days. If possible, sprayers and spreaders used for applying herbicides (especially hormone-type herbicides) should not be used for applying other chemicals.

Certain herbicides control weeds effectively when applied to the soil surface. Some of these compounds do not leach readily into the deeper soil layers and hence are not absorbed by trees (Kozlowski and Torrie, 1965) However, if persistent herbicides are applied to the soil surface and later mixed into the soil, the toxic residues may injure trees (Winget *et al.*, 1963). Ideally, for new seedbeds, nursery managers prefer selective herbicides that can be applied immediately after seeding to control weeds. However, tree seedlings in the cotyledon stage of development are extremely susceptible to certain herbicides (Sasaki and Kozlowski, 1968a; Kozlowski, 1976b). DCPA (dimethyl-2,3,5,6,-tetrachloroterephthalate) applied at rates up to 4 lb $acre^{-1}$ as a preemergence or postemergence spray to red pine seedbeds was not toxic to young pine seedlings and also controlled several weeds, but other herbicides were toxic (Kuntz *et al.*, 1964; Kozlowski and Torrie, 1965). Direct contact of red pine seeds with various herbicides was followed by abnormal growth of subsequently emerged seedlings (Kozlowski and Sasaki, 1968a,b; Wu *et al.*, 1971; Wu and Kozlowski, 1972; Kozlowski, 1986b).

Certain herbicides inhibit photosynthesis, the amount of reduction varying with specific chemicals and dosage, species and genetic materials, age of plants, and method of application (Sasaki and Kozlowski, 1967; Kozlowski and Keller, 1966). Photosynthetic reduction by herbicides may be associated with injury to mesophyll tissues. Structurally related herbicides may have different effects on photosynthesis. For example, the rate of photosynthesis of tree seedlings was reduced by several triazine herbicides, but was reduced more by simazine and atrazine than by propazine or ipazine (Sasaki and Kozlowski, 1967)

Effects on photosynthesis of several herbicides applied to the soil surface and then mixed into the soil varied appreciably. DCPA did not inhibit photosynthesis of young seedlings; EPTC reduced it slightly; and atrazine and 2,4-D decreased photosynthesis appreciably. Monuron, however, had the greatest inhibitory effect (Sasaki and Kozlowski, 1967). Total photosynthesis is influenced by the effects of herbicides on respiration, with some herbicides increasing respiration and others decreasing it (Sasaki and Kozlowski, 1968b,c).

Fertilization

In addition to their use in stimulating growth of seedlings, fertilizers often are used to improve growth and yield of older forest, fruit, and ornamental trees and shrubs. The physiological basis for fertilization was discussed in Chapter 6. Because the objectives and problems connected with the use of fertilizers vary with the types of plants, we will discuss the various kinds of uses separately.

Nurseries

The use of fertilizers is standard practice in nurseries. As resources become more limiting, greater emphasis undoubtedly will be placed on use of fertilizers to accelerate production of high-quality planting stock. Response to nursery fertilization is influenced by soil and plant nutritional status, seedling density, length of growing season, soil organic matter, previous cropping practices, and seedling genotype (Fisher and Mexal, 1984).

Plantation Establishment

Judicious application of "starter" fertilizer at or soon after planting often increases tree growth and decreases mortality. Application of P or N + P to newly planted slash and loblolly pine seedlings in the southeastern United States significantly increased growth (Bengston, 1979; Pritchett and Smith, 1974). In Malaysia, fertilization of newly planted Tenasserim pine seedlings increased height growth by 26 to 43% during the first 2 years, the amount depending on the soil type (Sandralingam and Carmean, 1974). Starter fertilization also increased growth of Monterey pine in Australia (Flinn, 1985). In the United States, controlled-release fertilizer applied to 1-year-old Douglas-fir seedlings increased shoot and root growth during the following two seasons. Starter fertilization increased internode length of the first flush and the number of needle primordia that formed late in the first growing season (Carlson and Preisig, 1980).

The amount of increase in growth accompanying fertilization depends on genotype, site preparation, and water supply (Stone, 1973). Good site preparation and control of competing vegetation with herbicides plus fertilization at planting resulted in a large increase in volume of loblolly pine seedlings at 5 years (Wilhite and McKee, 1985). Response to fertilization usually increases with increasing soil moisture until aeration becomes limiting (Schomaker, 1969).

Established Forest Stands

The growth response of forest stands to fertilizers varies widely depending on such factors as species and genotype, site quality, age of trees or stand, amount and type of fertilizer, time of fertilizer application, and specific response measured. Lack of response to fertilizers often results from adding deficient elements in inadequate amounts or from adding mineral nutrients that are not deficient in soils. In Germany the addition of P fertilizers to P-deficient sites increased height and volume growth of young Norway spruce plantations for approximately 10 years. Nitrogen deficiency did not develop until 10 to 15 years after planting, and applying N fertilizers prior to that time did not stimulate growth (Fiedler *et al.*, 1983). Correct diagnosis in deficiency of specific mineral elements is important before fertilizers are applied. Visual symptoms of deficiencies, soil analyses, and foliar analysis are helpful (Kramer and Kozlowski, 1979, pp. 343–346; and Chapter 6 of this volume).

Responses of established forest trees to fertilizers may vary with the stage of development of forest stands. Accelerating the time to canopy closure and full site occupation is important in increasing site productivity in any short- or medium-rotation silvicultural system, and fertilization can accomplish this. Growth increases from fertilization in plantations are most likely before canopy closure occurs and near the end of a rotation (Fig. 13.2). When trees are not fully utilizing a site (stage I) their growth is highly dependent on soil supplies of mineral nutrients, and almost any mineral element may be deficient. At this stage the efficiency of mineral uptake is very low but the effect of fertilizers on subsequent stand development is maximal. Availability of soil nutrients is critical for crown development and responses to several applied mineral elements may be expected. By the time the canopy is closed (stage II), the rate of uptake of soil nutrients has increased greatly but the absorbed minerals are being recycled rapidly. Furthermore, both retention and absorption of

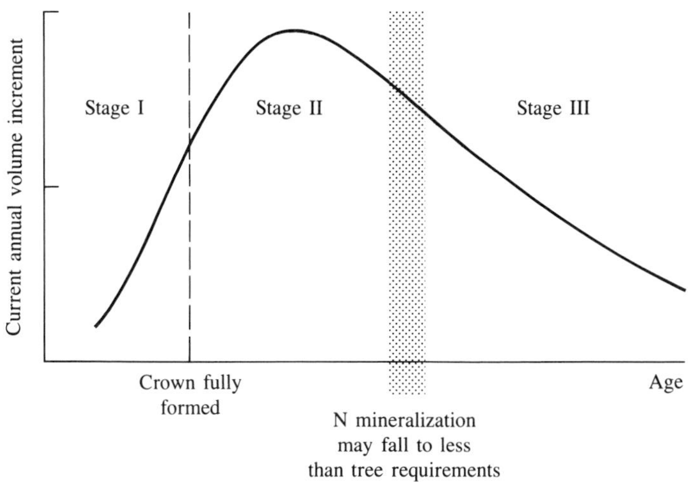

Figure 13.2 Three nutritional stages in the life of a forest stand. (From Miller, 1981.)

nutrients deposited from the atmosphere are high. Because of increased input and recycling of minerals, the requirement for additional soil nutrients is low at stage II and responses to fertilizers are small. By the time stage III is reached, immobilization of N in the biomass and humus can lead to N deficiency and site deterioration, and significant responses to fertilizer again are likely to occur (Miller, 1981).

A voluminous literature shows that growth of established forest trees often is reduced by mineral deficiencies that frequently can be corrected by applying fertilizers. For example, fertilizers are used routinely to stimulate growth of conifer plantations in Australia (Attiwill, 1982; Flinn, 1985), New Zealand (Will, 1985), Europe (Tamm, 1968); McIntosh, 1984), and North America (Bengston, 1979; Pritchett, 1979). In Europe, where most commercial forests grow on poor sites, fertilizers are commonly used to stimulate growth of trees and increase their tolerance to various abiotic and biotic stresses. In thinned plantations of Monterey pine in Australia, "later age" fertilization sometimes is used but may be ineffective because drought often inhibits growth (Flinn, 1985).

In the United States, fertilizers are particularly beneficial in the northwestern and southeastern states. In the Pacific Northwest, volcanic soils often are deficient in N and N fertilizers stimulate tree growth over a broad range of sites. Nitrogen fertilizers such as granular urea are commonly applied by helicopters to 15- to 60-year-old Douglas-fir stands. Tree growth often is stimulated for 5 to 10 years and sometimes even longer. Over a 15-year period a single application of ammonium nitrate fertilizer (157, 314, and 471 kg N ha^{-1}) to 35-year-old Douglas-fir trees increased volume growth by 51, 88, and 110%, respectively. Average height growth of surviving fertilized trees was 30 to 90% greater than that of unfertilized trees (Miller and Tarrant, 1983). The effects of fertilizers last longer on thinned than on unthinned stands. For example, stimulation of growth of a 53-year-old Douglas-fir plantation by combined fertilizer and thinning was evident for 20 to 30 years (Binkley and Reid, 1984). Although lodgepole pine can grow on very infertile sites, it often responds dramatically to fertilizers. Weetman *et al.* (1985) reviewed several Canadian studies that showed increases in volume increment of forest trees up to 50% within 10 years following applications of N of at least 150 kg ha^{-1}.

Fertilizer application is a standard management practice in pine forests of the southeastern United States. Significant responses to N or N + P fertilizers have been shown for 10- to 25-year-old loblolly and slash pine stands in the southern coastal plain (Table 13.3). Urea and ammonium nitrate are commonly used as N sources.

As might be expected, the magnitude of response to fertilizers varies considerably over the southeastern states, where four species of pine (loblolly, shortleaf, slash, and longleaf) dominate a variety of sites. These sites vary from dry coarse sands of Florida, clay soils of the piedmont plateau, poorly drained flatwoods, rocky soils of the Ozark Mountains, to pocosin swamps of the eastern coastal plain.

As mentioned earlier, stimulation of growth of fertilized trees reflects both increased photosynthetic efficiency of the foliage and increase in amount of foliage. Over a 7-year period, 37% of the increase in stem growth of fertilized, 24-year-old Douglas-fir trees was attributed to increased photosynthetic efficiency and the remainder to a greater leaf area (Brix, 1983). The increased photosynthetic efficiency was effective only during the first 3 to 4 years, after which increase in photosynthetic surface became more important (Fig. 13.3). Water use efficiency also was improved by fertilization (Brix and Mitchell, 1986).

There has been some concern lest the stimulation of growth by fertilizers be accompanied by undesirable changes in wood quality. However, Schmidtling, (1973) found that fertilization over a 9-year period increased production of longleaf, slash, and loblolly pines without a decrease in fiber length or wood density. Mineral deficiency sometimes increases the percentage of latewood in older trees, but observations are somewhat contradictory. Likewise, the effects of fertilization on wood density are variable, although reduction in density is reported in rapidly growing

Table 13.3
Radial Growth of Slash Pine 5 Years after Fertilizers Were Applied to 21-Year-Old Trees[a]

Fertilizer	Amount (lb)	Nitrogen per tree (lb)	Radial growth (mm)	Percentage of control
Control	0	0	15.7	100
3-18-6	20	0.6	17.8	113
3-18-6	40	1.2	19.0	121
7-7-7	20	1.4	19.8	126
7-7-7	40	2.8	20.1	128

[a]After Hoekstra and Asher (1962). Reprinted from the *Journal of Forestry*, published by the Society of American Foresters, 5400 Grosvenor Lane, Bethesda, MD 20814-2198.

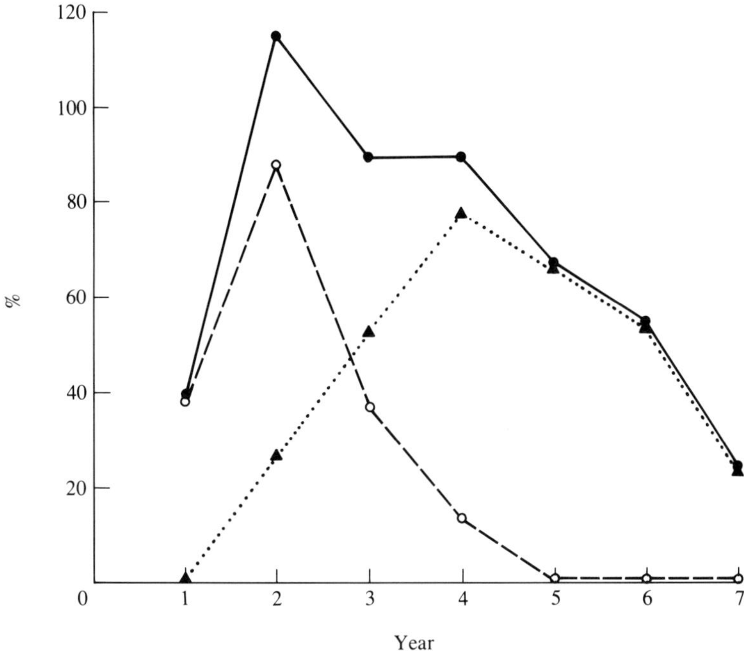

Figure 13.3 Contributions of photosynthetic efficiency (E; ○) and foliage mass (▲) to growth increase (●) of fertilized Douglas-fir trees. (From Brix, 1983.)

trees. In general, the increase in wood production induced by fertilizers more than compensates for any decrease in desirable wood properties (Goddard *et al.*, 1976). The somewhat equivocal literature on the topic was surveyed by Bevege (1984, pp. 312–318) and Zobel and van Buijtenen (1989, pp. 220–231). Much useful material on fertilization of forests can be found in Bowen and Nambiar (1984).

Shade and Ornamental Trees

Not only are landscape trees often grown with restricted root and crown space, but they also are subjected to the harmful effects of poor soil aeration, herbicides, deicing salts, mechanical injury and unusually severe air pollution (Kozlowski, 1985b). Regular application of fertilizers can do much to stimulate growth of such trees, counteract the influences of insects and diseases, and prevent or at least postpone their death.

Growth of shade and ornamental trees is limited more often by N deficiency than by deficiency of any other element. The amount of soil N varies throughout the year and often is low in the spring, hence annual additions of N fertilizers often are necessary for maximum growth. Although complete fertilizers (NPK) are widely used for improving growth of ornamentals, in most cases the amounts of P and K in soil are adequate. According to Harris (1983), responses of shade and ornamental trees to additions of P and K are the exception rather than the rule. Nevertheless, some investigators recommended additions of P to stimulate root growth under very adverse urban conditions (e.g., on acid or sandy soils, and on soils low in organic matter or with low cation exchange capacity) (Pirone, 1978; Tattar, 1978). Because P does not move readily in soil it should be placed in holes close to the absorbing roots rather than applied to the soil surface.

Neely *et al.* (1970) studied the effects of fertilizers applied for 3 consecutive years on shade trees in Illinois. The soils at five different test sites included sandy soils, fertile deep topsoils of prairie origin, fertile but shallow topsoils of forest origin, and infertile shallow topsoils with heavy clay subsoils. The combinations of nutrient elements applied included N alone, PK, NPK, and NPK plus micronutrients. Different N sources were evaluated, including ammonium nitrate, ammonium sulfate, and urea forms. The fertilizers were applied by four methods: (1) broadcast on the soil around the trees, (2) placed as dry fertilizer into holes around the trees, (3) injected as fertilizer solution into the soil, and (4) applied to the foliage as water-soluble sprays.

Nitrogen was the only nutrient that significantly stimulated increase in stem diameter, the response to N plus P or K being similar to the response to N alone. Increased growth of the fertilized trees continued for 1 to 2 years after fertilization was discontinued, the amount of increase depending on the amount of N applied.

Nitrogen applied dry in holes at a rate of 6 lb per 1000 ft^2 of soil area (29.3 g/sq m) is commonly used by arborists for established trees. Higher rates of N applied to the soil surface during the summer were toxic to grass. There was little difference in growth response to the four N sources used, but the time of fertilizer application had an important effect on growth. Trees receiving all or a portion of the N in June or October grew less in diameter than those receiving all the N in April. However, trees fertilized in October grew more than the unfertilized trees (Himelick *et al.*, 1965).

Reproductive Growth

Mineral deficiencies reduce flower, fruit, and seed production of all kinds of plants. Foresters, horticulturists, and growers of ornamental plants have learned that fertilization increases flower and seed production.

Forest Trees

Fertilization is widely used on seed orchards and mature stands selected for seed production, often in combination with thinning, to increase flowering and seed production. In temperate zones, fertilizer usually is applied early in the spring. With pines and other trees that require 2 years to mature seed, the effect of fertilization is seen on flowering the spring after fertilization and on seed production the second autumn after fertilization. Because the greatest effect is on the next crop, annual fertilization is necessary to maintain high yields. Of course, seed production is related to other factors in the environment, especially water supply and unseasonable freezes. Wenger (1957) found that seed production of loblolly pine was positively correlated with rainfall of the second preceding year and negatively correlated with the preceding seed crop. He also suggested that the increase in seed production following stand thinning might result from greater water supply rather than from increased light intensity. Griffin *et al.* (1984) reported that in dry years in Australia, seed production of Monterey pine was increased by irrigation. Because of the usual variations in weather, several years of data are needed to evaluate the effects of various cultural practices on flowering and seed production.

Fruit Trees

Fertilization of fruit trees presents somewhat more specific targets than does fertilization of forest trees. Fruit growers must consider flower initiation, fruit set and development, fruit quality, and storage life.

Flower initiation In apple and several other kinds of fruit trees, application of N stimulates the production of flower buds and also decreases the tendency of apple trees to flower heavily in alternate years. Nitrogen fertilizers stimulated floral induction in lime trees (Arora and Yamadigni, 1986).

Fruit set Deficiencies of macronutrients, especially N, and occasionally deficiencies of micronutrients reduce the percentage of flowers that set fruit. Nitrogen fertilizers increase fruit set in species as different as apple (Oland, 1963) and lime (Govind and Prasad, 1982). Although N fertilizers traditionally are applied in the spring, later applications often are useful. Hill-Cottingham and Williams (1967) found that summer application of N to apple trees was superior to spring application for increasing fruit set the next spring, but autumn application was better yet (Fig. 13.4). Oland (1963) reported that fruit set and yield of apples given a postharvest urea spray were 50% greater than in controls or trees fertilized by application to the soil. Delap (1967) found that summer and autumn applications of N to apple trees were superior to spring applications because they increased the longevity of ovules and the period during which stigmas were receptive to pollen. Of course, autumn fertilization is most effective on trees with low N reserves (Hennerty *et al.*, 1980).

On some soils, correction of deficiencies of micronutrients resulted in significant increases in fruit set. For example, application of a Zn spray increased fruit set of sweet lime (Arora and Yamadigni, 1986) and Mn plus Zn improved fruit set of cranberries (De Moranville and Deubert, 1987). Obviously, application of micronutrients increases fruit set only where deficiencies exist, and the timing of application also is important. Prebloom application of B to prune trees was ineffective whereas later applications were useful (Davison, 1971). Boron sprays applied in the autumn to prune and cherry trees increased fruit set the next year (Chaplin *et al.*, 1977; Westwood and Stevens, 1979). Application of B to Stayman apple trees, which usually have poor fruit set, resulted in increased set, but it was ineffective on Jonathan and Golden Delicious, which usually set heavy crops (Bramlage and Thompson, 1962).

Although application of N fertilizer usually is beneficial, it must be performed

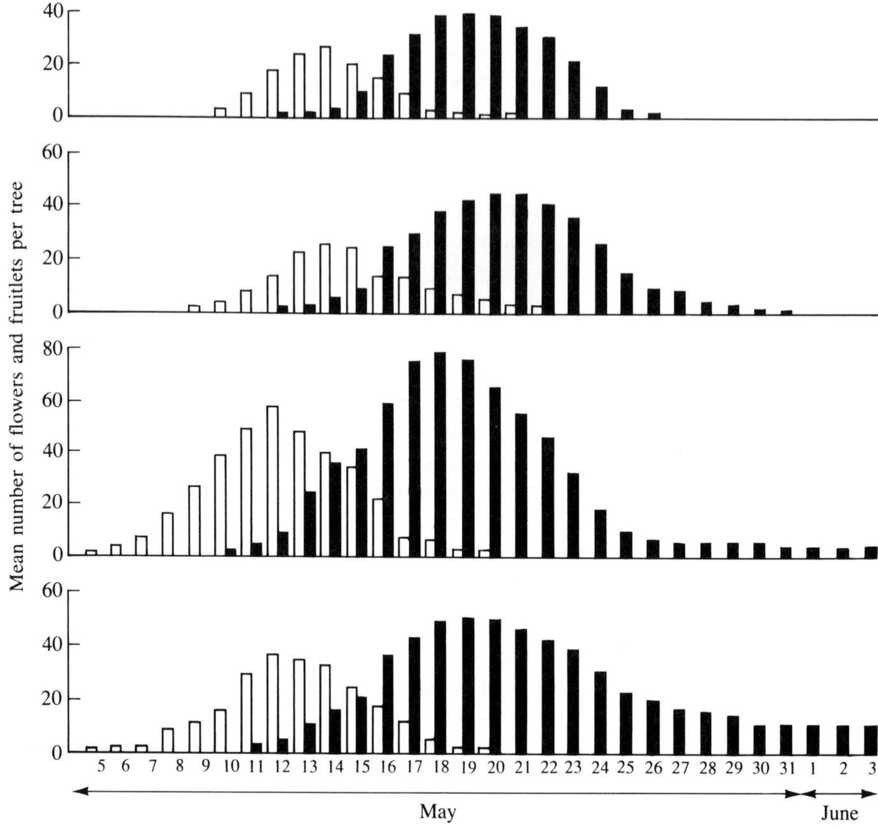

Figure 13.4 Effect of nitrogen fertilizers applied at various times on flowering (open bars) and fruitlet abscission (filled bars) of Lord Lambourne apple trees. *Top to bottom*: without nitrogen; spring application; summer application; autumn application. (From Kramer and Kozlowski, 1979; after Hill-Cottingham and Williams, 1967, by permission of Academic Press.)

with caution. If too many fruits are set, the demand for carbohydrates by the growing fruits may exceed the photosynthetic capacity of the leaves and so result in small fruits. There is an extensive literature dealing with the fertilization of temperate zone fruits and readers are referred to books by Atkinson *et al.* (1980), Childers (1983), Shoemaker (1978a), Westwood, (1978), and Woodroof (1970) for more extensive discussions. Alvim and Kozlowski (1977) and Samson (1980) discuss fertilization of tropical fruit trees.

Fruit quality When fertilizing fruit trees, horticulturists must consider effects on fruit quality, both at harvest and during storage, as well as fruit yield. Heavy applications of N to increase tree growth and fruit yield may result in soft fruits subject to preharvest drop, cork spot, and bitter pit. Apples high in N may show a high incidence of scald, cork spot, bitter pit, and internal breakdown in storage

(Swietlik and Faust, 1984). Both the quantity of N and the form in which it is applied are important. Ammonium N is said to reduce Ca uptake and can contribute to Ca deficiency, so $CaNO_3$ fertilization is recommended for apples. However, Bramlage *et al.* (1980) concluded that the amount of N affects fruit quality more than the form in which it is applied.

Localized deficiency of Ca seems to be associated with a wide variety of fruit disorders, including bitter pit, cork spot, Jonathan spot, internal breakdown, low-temperature breakdown, senescent breakdown, water core, and cracking of apples; cork spot of pears; cracking of prunes and cherries; soft nose of mango; and soft spot of avocado (Shear, 1975). Calcium appears to have an important role in maintaining membrane integrity, lowering the rate of respiration, and postponing changes in ultrastructure in apple cells (Sharples, 1980). Fruit quality seems to be related to the ratio of Ca to other elements, the ratios of N, K, and Mg to Ca being related to the occurrence of bitter pit of apple. According to Bramlage *et al.* (1980), the effects of N, P, K, Mg, and B are exerted primarily through interactions with Ca. Johnson *et al.* (1983) stated that although spraying apple fruits with Ca before harvest improves storage quality, it is ineffective if the background concentration in the fruit is too low.

The supply of P seldom affects fruit quality at picking but P deficiency is associated with rapid deterioration of fruit in storage in areas where it is deficient. Injury from K deficiency is rare, but high levels sometimes are associated with bitter pit, scald, and storage breakdown of apples. Although B deficiency is common, an excess of B also can cause physiological disorders in fruits.

Overfertilization

For a long time it was assumed that forest soils rarely contained an excess of mineral nutrients and tree growth usually would benefit from added nitrogen. However, it appears that many forests in the north temperate zone are receiving an excess of nitrogen from polluted air, which is modifying soil chemistry and mineral uptake and possibly contributing to the forest decline usually attributed to acid rain (Aber *et al.*, 1989; Schulze, 1989). If air pollution continues to increase, problems with several elements may be expected.

Applications of excess fertilizers to forest nursery stock, fruit trees, and shade trees may cause injury. The amount of injury varies with the species and genotype, type of fertilizer, and time of application. A high salt concentration in the soil solution reduces its osmotic potential, thereby decreasing water absorption and leading to leaf dehydration, stomatal closure, reduction in photosynthesis, leaf injury, and plasmolysis of root cells. Sometimes excessive fertilization, especially with N, late in the summer prolongs growth so that trees fail to develop adequate coldhardiness and are injured by early freezes. Overfertilization sometimes stimulates production of large numbers of branches, flowers, and fruits. Pines sometimes produce clusters of cones. Other responses include fasciation or flattening of stems and internal bark necrosis (Boyce, 1948).

Injury to woody plants from overapplication of minor elements also has been reported. For example, spraying apple trees too often with B may cause chlorosis, dieback of shoots, fruit drop, and fruit breakdown in storage (Yogaratnam and Johnson, 1982). Excessive use of Ca sprays may cause symptoms resembling Fe chlorosis, which develop as a result of high accumulation of this element in the soil (Beyers and Terblanche, 1971). Overapplication of Zn fertilizers may severely injure shoots and fruits. However, the extent of such injury can be reduced by use of Zn chelates and organic complexes (Swietlik and Faust, 1984).

Managers of forest nurseries often are tempted to apply high rates of fertilizers to produce very large seedlings. Unfortunately, overfertilization has been reported to result in seedlings of low physiological quality with reduced capacity for survival when outplanted. Conifer seedlings receiving very high applications of mineral nutrients, particularly N, often have low root–shoot ratios and thin tracheid walls. Such seedlings may be very succulent and susceptible to drought and frost injury (Wilde, 1958). For example, high N applications stimulated shoot growth more than root growth of loblolly pine seedlings and reduced their drought tolerance. Seedlings grown at N concentrations that were optimal with normal water supply survived drought best (Pharis and Kramer, 1964). Heavy applications of P fertilizer produced healthy-appearing loblolly and shortleaf pine seedlings but reduced their subsequent survival in plantations (Lynch *et al.*, 1943).

An excess of fertilizer can reduce seedling growth, especially on dry and sandy soils. Application of 300 kg ha^{-1} of N or K reduced growth of pine seedlings in the southeastern United States (Pritchett and Robertson, 1960). In another case an excess of $NaNO_3$ inhibited root growth and produced P deficiency in seedlings growing in soil containing adequate P (Maftoun and Pritchett, 1970).

Irrigation

The injurious effects of water stress caused by drought have been mentioned repeatedly in this book because it is the most common cause of reduction in growth of plants. In some areas such as those with a mediterranean climate, there are well-defined wet and dry seasons, but in most of the temperate zone, droughts are common and usually random in occurrence and severity (Tables 7.1 and 7.4). From earliest recorded history (Kramer, 1983a, pp. 107–108), irrigation has been used to decrease the damage from droughts to crops and orchards, and in recent decades it has been used in forest nurseries. We will briefly review methods of irrigation and the use of irrigation in nurseries, orchards, and forests and for ornamental trees and shrubs.

Methods of Irrigation

The four principal methods of applying water are basin and furrow irrigation, sprinkler irrigation, trickle or drip irrigation and subsurface irrigation. Application of water in furrows or basins formed by low dikes around trees is the oldest method. It can be used only on level land, requires much labor, wastes water, and often results in flooding low areas of orchards. Improved sprinkler systems have largely supplanted furrow irrigation because they can be used on rolling land, permit better and more uniform control of the amount of water applied, and use less labor than furrow irrigation does. Sometimes fertilizers are applied in the irrigation water (fertigation) although the possibility of salt injury to leaves exists. Occasionally leaching from leaves results in chlorosis and wet leaf surfaces favor the spread of leaf pathogens. The cooling effect and high humidity caused by sprinkling are beneficial because they tend to keep the stomata open. Sprinkler irrigation is widely used on orchards (Zekri and Parsons, 1988) and occasionally on forests (Axelsson and Axelsson, 1986). A secondary benefit from the use of sprinkler irrigation is the possibility of using it to prevent frost injury to orchards (Parsons *et al.*, 1985).

Trickle or drip irrigation involves distributing water through tubing and allowing it to trickle out on the soil through nozzles of adjustable capacity (Bresler, 1977). It permits placement of measured volumes of water in the immediate vicinity of root systems, thus reducing losses from evaporation. Sometimes the system is installed belowground, with outlets that allow water to escape in the immediate vicinity of the root system. Subsurface irrigation is suitable for fertigation, can be used on sloping land, conserves water, and decreases salt accumulation, but it is expensive to install. Christersson (1986) obtained equally good results with all three systems on willow cuttings growing in a sandy soil in southern Sweden. However, Zekri and Parsons (1988, 1989) reported less water stress and better tree and fruit growth of Florida grapefruit irrigated with overhead sprinklers than with drip systems, while microsprinklers gave intermediate results. Apparently drip systems did not wet a sufficiently large volume of soil, but this problem might be solved with more outlets per tree.

Irrigation Timing

An important and troublesome problem in irrigation is proper timing, because too frequent irrigation wastes water and may waterlog the soil and cause salt accumulation if nutrients or other minerals are abundant in irrigation water, whereas too infrequent irrigation results in reduced growth of plants. The required frequency depends on the rate of evapotranspiration, the water-holding capacity of soil, and the kind of plant. Plants should be the best indicators of water stress because they integrate both soil and atmospheric factors that control plant water status. However, direct measurement of plant water status usually is impractical for foresters and orchardists, and wilting is unreliable and only occurs after plant water stress has developed. Other methods include change in leaf color or orientation, and measurement of premature stomatal closure by use of porometers or the infiltration method (Alvim and Havis, 1954). Fiscus (1984) computerized a viscous flow porometer that controlled irrigation of corn very efficiently. Idso *et al.* (1980), Jackson (1982), and others have used remote sensing of leaf temperature as an indicator of the need to irrigate herbaceous crops. This method was used on pecans by Sammis *et al.* (1986) and should be applicable to orchards and forests in dry climates, where freely transpiring leaves are cooler than the surrounding air. Irrigation timing should also take into account the stage of development of fruit trees, because reproductive processes are sensitive to water stress, as noted in Chapter 7. Irrigation timing is discussed briefly in Kramer (1988a). We will now discuss a few special uses of irrigation.

Nurseries

The desirability of irrigating forest tree and ornamental tree and shrub nurseries seems self-evident. However, efficient irrigation requires considerable experience and good judgment. Excessive irrigation of tree seedlings late in the growing season can result in succulent seedlings that fail to become dormant before frost and do not store or transplant well. On the other hand, inadequate irrigation may stop growth too early in the season and result in stunted seedlings. Burger *et al.* (1987) recommended that irrigation of container-grown shrubs in California nurseries be based on the rate of evapotranspiration from a grass plot and the measured rate of evapotranspiration from the various species of plants growing in containers in the nursery. They found that *Pyracantha* and *Buddleia* required about four times as much water as *Photinia* and Scotch broom; other species were intermediate in water requirements. It was found convenient to group plants with high and low water use in different parts of the nursery.

Forests and Orchards

Irrigation of orchards has been practiced so long that its benefits scarcely require discussion. Irrigation of apple trees was discussed by Landsberg and Jones, citrus by Kriedemann and Barrs, and tea by Squire and Callander in Kozlowski (1981). Irrigation of orchards has been discussed in numerous papers in *California Agriculture* and in *Horticultural Reviews* and *Forestry Abstracts* cites scattered papers on irrigation of forest stands. Irrigation of tropical trees is discussed by various authors in "Ecophysioloqy of Tropical Crops" (1977), edited by Alvim and Kozlowski. Armitage (1985) provided information on irrigation in arid regions.

Lucier and Hinckley (1982) reported that irrigation of black walnut, white oak, and sugar maple growing in an oak–hickory forest increased the leaf water potential of all three species. Hillerdal-Hagstromer *et al.* (1982) irrigated a 20-year-old Scotch pine stand in central Sweden for three summers and found an increase in leaf water potential in only one of the three summers. A combination of irrigation and fertilization gave the best growth. Axelsson and Axelsson (1986) irrigated and fertilized stands of Scotch pine and Norway spruce for 12 years and found that the effect of fertilization was much greater than the effect of irrigation. They claimed that the decrease in production of fine roots was an important factor in the increase in aboveground biomass of the fertilized trees.

It seems probable that the cost of irrigation may be a limiting factor except for nurseries, seed orchards, and perhaps for plantings on infertile soil that can be irrigated with sewage plant effluent. Griffin *et al.* (1984) reported that in two of four seasons, irrigation of Monterey pine in New Zealand significantly increased seed yield. They also found a significant interaction between irrigation and fertilization with nitrogen in dry years. Stewart *et al.* (1986) reported that irrigation of forest stands in Australia with recycled water increased growth of trees, but an economic analysis (Stewart and Salmon, 1986) indicated that sprinkler irrigation would not be profitable on Monterey pine or eucalyptus, and flooding irrigation of Monterey pine would only be marginally profitable. Obviously the economic returns are affected by a number of factors, and Jarvis (1986) concluded that too little information was available concerning the effects of irrigation on forest stands to permit generalizations about its value.

In recent decades there has been increasing interest in the use of wastewater and wet or dry sludge from sewage disposal plants to irrigate and fertilize established forests and to aid in reclamation of difficult sites such as landfills and strip-mined lands (Wilson *et al.*, 1985). Such usage serves several purposes in addition to disposing of troublesome wastes. It is effective in expediting reclamation of landfills, strip-mined areas, and mine dumps with vegetation coverage. It improves the growth of established forests on infertile soils and, at the same time, improves the quality of the runoff water by removing phosphates, nitrates, and other ions from the wastes. Finally, application of wastewater reduces the inhibitory effects of drought on growth and sometimes adds to groundwater supplies.

Human waste has been used as fertilizer for centuries in China and some other countries, but it was seldom polluted by industrial waste. One problem in industrial countries is that undesirable concentrations of heavy metals such as Cd, Zn, and Pb occur in some sewage plant effluent. However, apparently these elements seldom accumulate to injurious levels, either in the plants growing on treated soil or in the birds and animals living on such sites. There also is a small possibility of transmitting pathogenic organisms to animals and humans. However, this is very unlikely in forests.

In the Orlando, Florida, area, disposal of urban wastewater has become a troublesome problem and its use to irrigate citrus orchards was welcomed (Wood, 1988). The sandy soil of the Orlando area minimizes the danger of soil waterlogging and much of the water percolates down into the groundwater.

Wastewater usually is applied by sprinkler systems similar to those used on orchards and farms (Zekri and Parsons, 1989). Sludge usually is applied from trucks equipped with a spreader system, and it sometimes is disked or plowed into bare soil. The numerous problems encountered in the use of sludge and wastewater were reviewed in the book edited by Sopper and Kerr (1979).

Ornamental Trees and Shrubs

Following transplanting it often is necessary to water trees and shrubs for 2 to 3 years while the root systems are becoming established. Most shade trees will not have a fully redeveloped root system until 3 to 6 years after transplanting (Neely and Himelick, 1987).

A newly planted tree is most easily watered if a circular mound of soil 3 to 4 in. high is prepared at the edge of the planting hole. Such a mound serves as the dike of a reservoir that should be filled with water at 7- to 10-day intervals during dry periods of the growing season. The water in the reservoir will adequately soak the backfill and soil ball. Established trees should also be watered regularly. Daily watering that wets only the surface soil to a depth of an inch or so is of little value because most absorbing roots are at a greater depth and will be left in dry soil by superficial watering. Wetting the soil thoroughly at longer intervals with the equivalent of 2 in. of rainfall is more useful. However, allowing the hose to run in the reservoir for prolonged periods can result in waterlogging. Probably more ornamental trees and shrubs are killed or injured by overwatering than by lack of supplemental watering (Neely and Himelick, 1987).

Thinning of Forest Stands

Naturally regenerated stands often are greatly overstocked, resulting in severe competition for water and minerals and in slow growth early in the life of the stand. If left alone, natural thinning is inefficient because it results from severe competition that reduces the rate of growth. On the other hand, if stand density is too low, growing space is wasted and self-pruning is delayed, resulting in wood with many knots, stems with much taper, and a reduced economic return over time per unit of land. When a stand of trees is thinned the amount of growing space for the crowns and roots of residual trees is increased. Usually the rates of vegetative and reproductive growth of the trees left standing are accelerated because of greater availability of light, water, and mineral nutrients. Also, in many thinning systems growth is directed to trees of superior form. Good management of thinning requires consideration of the species, the soil on which it is growing, and especially the use that is to be made of the trees. If the trees are to be harvested at an early age for fiber, a much higher stand density can be allowed than if they are to be grown to maturity for saw timber.

Several studies from different regions show increased availability of resources following thinning of forest stands. For example, thinning results in greater availability of light, soil water, and mineral nutrients for the remaining trees. Under closely spaced loblolly pine trees, soil moisture was nearly depleted by midsummer, whereas water was not exhausted under widely spaced trees (Fig. 13.5A). Thinning of loblolly pine stands in Arkansas alleviated midsummer water deficits (Bassett, 1964a). Both soil moisture content and leaf hydration were higher in trees of a thinned red pine stand than in trees of an unthinned stand (Sucoff and Hong, 1974). It was estimated that seasonal photosynthesis might be as much as 21% higher in residual trees of a thinned stand of lodgepole pine as a result of increased leaf hydration and additional light (Donner and Running, 1986). Thinning also increases availability of mineral nutrients to the remaining trees. For example, both foliar N and P concentrations were lower in closely spaced than in widely spaced Douglas-fir trees (Cole and Newton, 1986).

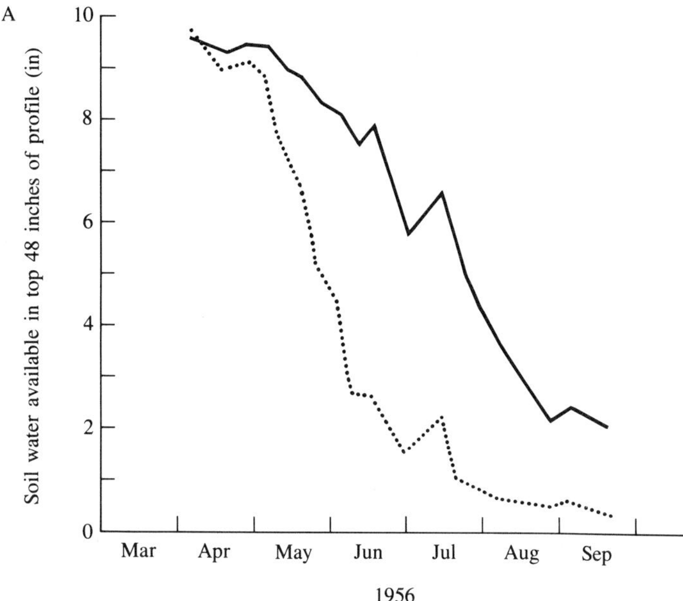

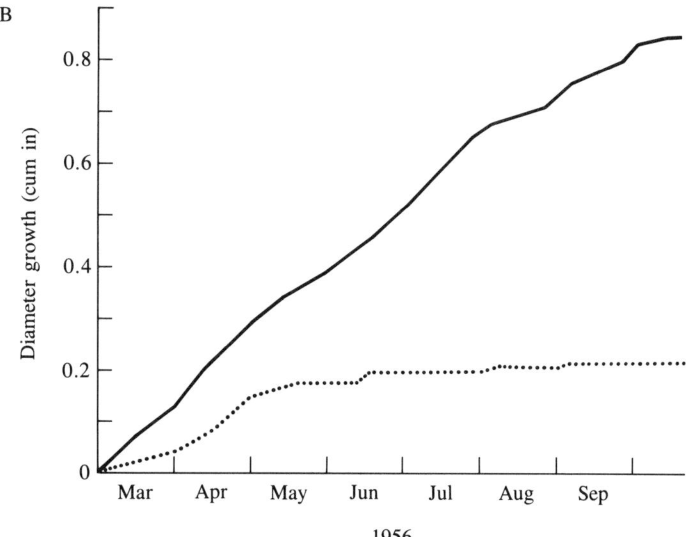

Figure 13.5 Effects of stand thinning on (A) soil water depletion and (B) diameter growth of loblolly pine trees during one growing season. Thinned (——); unthinned (····). (From Zahner and Whitmore, 1960. Reprinted from *Journal of Forestry*, published by the Society of American Foresters, 5400 Grosvenor Lane, Bethesda, MD 20814-2198.)

Vegetative Growth

In addition to an increase in photosynthesis of residual trees in thinned stands, both crown size and leaf area increase, thus adding to the photosynthetically active surface. These changes are followed by increased downward transport in the stem of carbohydrates and hormonal growth regulators. As a result there is an eventual increase in cambial growth and redistribution of wood production along the tree stem.

There are many examples of accelerated cambial growth of the residual trees in thinned forest stands and only a few will be given here. For example, diameter growth in unthinned stands of loblolly pine was only 30% of that of trees in thinned

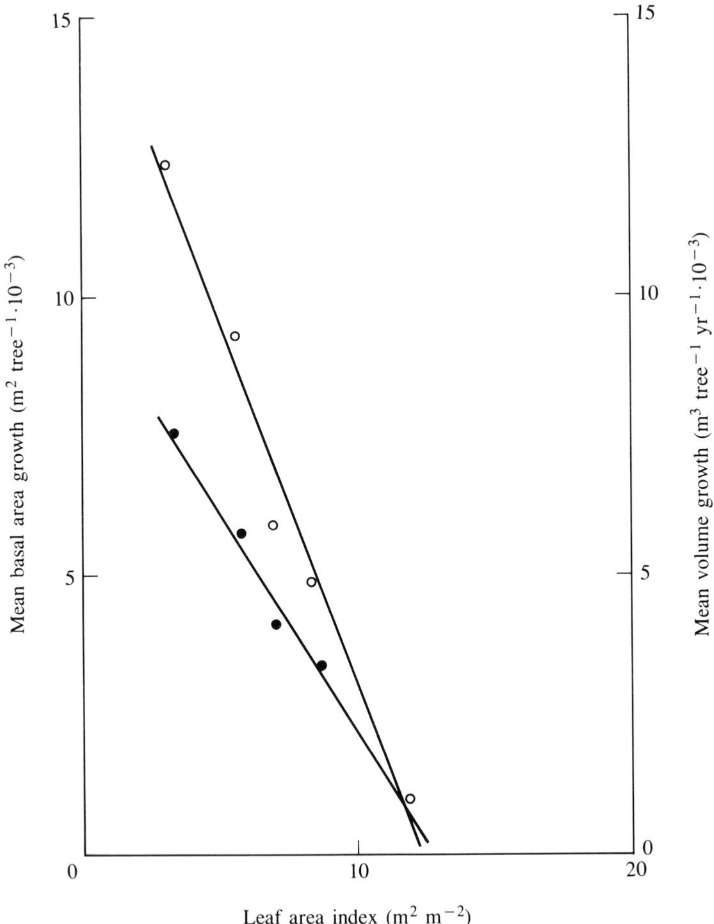

Figure 13.6 Effects of stand thinning on growth of Douglas-fir trees. Both volume (○) and basal area (●) growth were increased as the leaf area index was decreased by thinning. (From Waring *et al.*, 1981.)

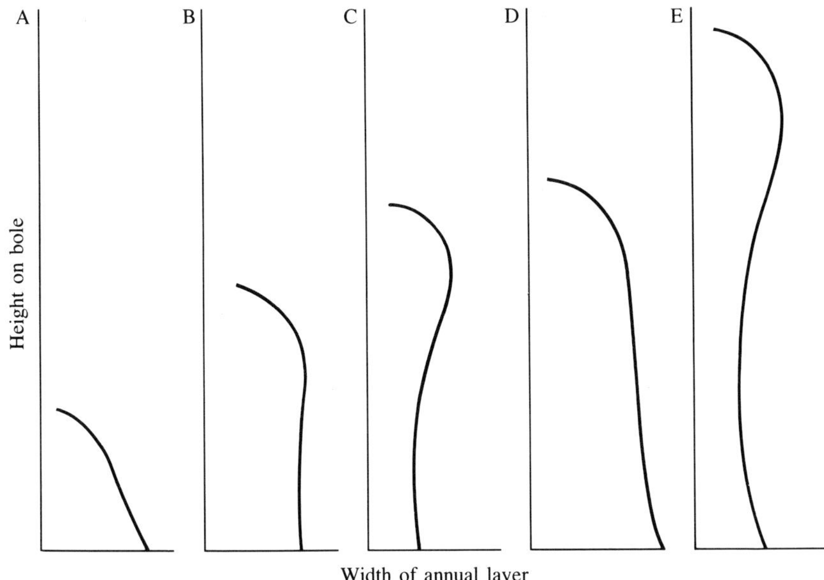

Figure 13.7 Variations in thickness of the annual ring at various stem heights in plantation-grown conifers of varying age. (A) 8 years, when crowns extend to the base of the tree; (B) 16 years, crowns closing; (C) 20 years, lower branches are dead; (D) 21 years, shortly after thinning when crowns have been exposed to full light; (E) 30 years, competition is again severe and crowns have closed. The horizontal scale is greatly exaggerated. (After Farrar, 1961; from Kramer and Kozlowski, 1979, by permission of Academic Press.)

stands (Fig. 13.5B). The effects of thinning on growth of 36-year-old Douglas-fir trees are shown in Fig. 13.6. Stem diameter growth was greatly stimulated by thinning. In the most heavily thinned stands, basal area growth after 15 years was more than 8 times that of the controls, whereas volume growth was 12 times and net assimilation rate almost 4 times greater. Growth efficiency increased almost linearly as the amount of canopy was reduced (Waring *et al.*, 1981). Similar relationships were found for lodgepole pine (Mitchell *et al.*, 1983) and ponderosa pine (Larsson *et al.*, 1983).

The release of a tree by removal of adjacent trees not only increases total wood production but also produces a more tapered stem by stimulating wood production most in the lower stem. Farrar (1961) showed that before competition was severe in young red pine plantations, most wood was laid down in the lower stem. However, when tree crowns began to close, more wood was added in the upper stem than in the lower stem (Fig. 13.7). During plantation establishment, trees must be planted relatively close together to develop nearly cylindrical stems. After this has been accomplished the stand can be thinned to concentrate wood production in selected trees. Original wide spacing of plantation trees may produce trees with long crowns, many knots, and stems with too much taper.

Variations in Response to Thinning

Responses to thinning vary greatly with crown class, species and genotype, age of trees, and duration of suppression. Dominant trees with large and physiologically efficient crowns often show little response to thinning. Trees of intermediate and suppressed crown classes show proportionally greater response.

The effect of thinning on diameter growth of trees may not be apparent for several years (Fig. 13.8). Sometimes the residual trees of very heavily thinned

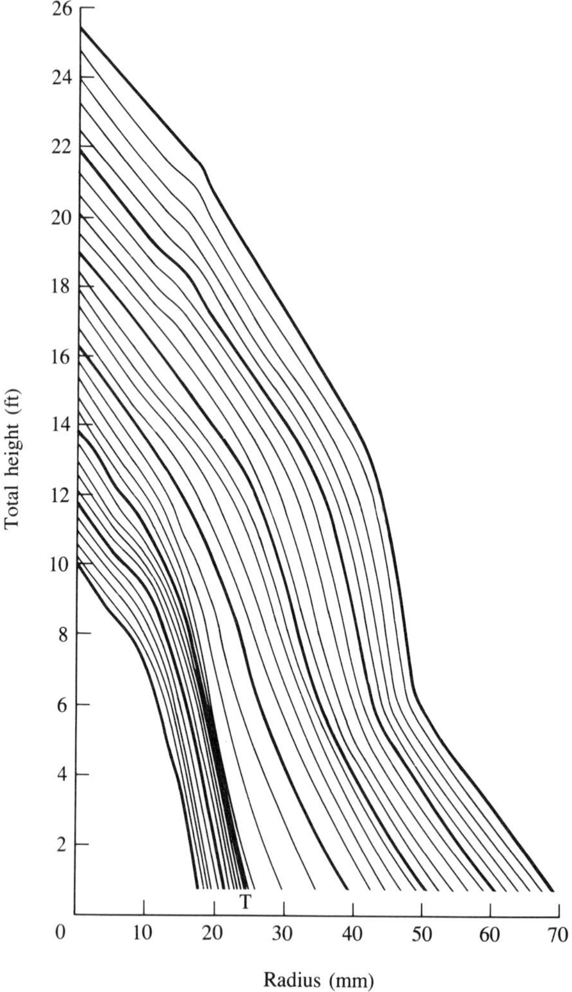

Figure 13.8 Effect of thinning closely grown ponderosa pine trees on cambial growth of residual trees. Before thinning the annual increment was greatest in the upper stem. After thinning the annual increment became greater in the lower stem. (After Myers, 1963. Reprinted from *Forest Science*, published by the Society of American Foresters, 5400 Grosvenor Lane, Bethesda, MD 20814-2198.)

stands exhibit "thinning shock" characterized by leaf chlorosis, growth reduction, windthrow, sunscald, production of epicormic shoots, and death of trees. Usually it is assumed that within the range of stem densities that occur in managed forest stands, thinning has little effect on height growth of residual trees. However, sometimes, especially on poor sites, thinning may decrease height growth. For example, the first year after a 27-year-old Douglas-fir stand was thinned the rate of height growth of the remaining trees was appreciably inhibited, and growth was reduced even more in subsequent years. After 20 years, however, height growth had accelerated and the prospects were for the thinned stand to have taller trees than those in adjacent unthinned stands (Harrington and Reukema, 1983).

Most species of forest trees that initially grow in very dense stands, such as those self-seeded after fire, undergo rapid self-thinning so that vigorous stands are maintained. However, some species do not thin naturally but develop a stagnated physiological state, especially on poor sites. This is characteristic of several conifers, including lodgepole pine, Scotch pine, jack pine, Virginia pine, and sand pine. Such stagnated stands usually have limited capacity to respond to thinning. Stagnated 20-year-old lodgepole pine trees had short internodes as well as short, narrow, and few needles. The rate of height growth of such trees was only about one-fourth of that of vigorous trees. Worrall *et al.* (1985) attributed the low capacity of stagnated lodgepole trees to respond to thinning to their much higher than normal use of photosynthate in root growth and/or respiration. As pointed out by Decker (1944), four trees 5 in. in diameter have the same basal area as one tree 10 in. in diameter, but twice as much circumference and probably twice as much respiration in the cambial region. Hence a dense stand of small stems consumes more carbohydrates in respiration than the same basal area in fewer large stems. It is also possible that the slow response to thinning is associated with photoinhibition of photosynthesis in the suddenly exposed foliage (Chapter 4). The sudden change in microclimate may also have other unidentified effects.

Epicormic Branching

Sudden exposure of trees by heavy thinning may stimulate dormant buds on stems or branches of certain species to produce epicormic shoots, sometimes called "water sprouts." White oak trees in heavily thinned stands produced an average of over 35 epicormic shoots per tree within 2 years. Trees of moderately thinned stands averaged 21 epicormic shoots, and trees of unthinned stands produced less than 7 such shoots (Table 13.4). Epicormic shoots often produce knots and degrade lumber. For example, epicormic branching of Louisiana hardwoods reduced the grade of lumber in 40% of the logs sampled (Hedlund, 1964).

The effect of stand thinning on production of epicormic branches differs among species. For example, the number of epicormic shoots produced varied as follows: very many—white oak, northern red oak; many—American basswood, black cherry, and chestnut oak; few—American beech, hickory, tulip, poplar, red maple, sugar maple, and sweet birch; and very few—white ash (H. C. Smith, 1966; Trimble and Seegrist, 1973).

Table 13.4
Effect of Thinning on Production within Two Growing Seasons of Epicormic Shoots in White Oak and Black Oak Stands[a]

	Residual basal area per acre (ft^2)					
	White oak			Black oak		
	115	70	30	115	70	30
			Tree size and vigor			
Diameter (in.)	11.7	12.5	11.7	14.1	13.6	13.8
Height (ft)	75.7	79.2	76.4	83.4	85.4	80.4
			New epicormic shoots			
North	1.1	4.9	10.7	0.7	1.8	6.3
South	2.0	5.3	6.4	1.3	4.5	3.5
East	1.8	5.1	9.4	2.1	5.2	5.8
West	1.7	6.1	8.0	1.3	5.0	4.2
Total	6.6	21.4	34.5	5.4	16.5	19.8

[a]After Ward (1966). Reprinted from *Forest Science*, published by the Society of American Foresters, 5400 Grosvenor Lane, Bethesda, MD 20814-2198.

Wood Quality

Increasing stem diameter by thinning of forest stands generally improves wood quality. This is largely because the outer stemwood is stronger, has fewer knots, and warps less than the inner wood core. Thinning usually prolongs seasonal diameter growth and hence increases the proportion of the dense latewood (which forms late in the growing season), thereby increasing the strength of the wood.

Insects and Diseases

Thinning of stands often alters the susceptibility of the residual trees to insects and diseases. Thinning increased the abundance of insect vectors of black stain root disease, including *Hylastes nigrinus, Pissodes fasciatus,* and *Steremnius carinata* in a Douglas-fir plantation. The increase was attributed to attraction of the insects to resin and the dead inner bark of slash (Harrington and Reukema, 1983; Witkosky *et al.*, 1986). The number of insect vectors in stands thinned in May was lower than in stands thinned in September, hence the time of thinning was important in controlling the activity of the insect vectors in thinned plantations.

Thinning of lodgepole pine stands reduced the incidence of *Atropellis* canker, which causes cambial necrosis, stem deformity, and growth inhibition (Stanek *et al.*, 1986). Thinning was recommended before the stands were 15 years old, by which time they become susceptible to *Atropellis* canker. Such early thinning before cankers can develop within stems and produce inoculum reduces infection.

When the density of 120-year-old Douglas-fir stands was reduced by thinning or by attacks of *Dendroctonus ponderosae* beetles, the surviving trees showed increased resistance to beetle attack over a 3-year period (Waring and Pitman, 1985).

Susceptibility of white spruce trees to *Dendroctonus rufipennis* beetles at the beginning of an outbreak in Alaska was more closely related to the rate of recent radial growth than to stem diameter. Trees that were killed first had larger than average diameters but had grown slowly during the last 5 years. Trees with larger than average stem diameters with faster than average radial growth were less frequently killed by beetles (Hard *et al.*, 1983). Greater than average stem diameters and faster growth of trees were characteristics of stands with fewer stems per acre. The data indicated that resistance to *Dendroctonus* beetles could be increased by thinning to reduce competition and increase the vigor of residual trees (Hard, 1985). Amman *et al.* (1988) cited evidence that change in microclimate caused by thinning reduced infestation of lodgepole pine by mountain pine beetle. This also might be true in other species.

Wind Damage

As emphasized in Chapter 12, thinning of forest stands in windy areas often increases windthrow, with the greatest damage being likely during the 2- to 5-year period during which the crowns are reclosing. In unthinned Monterey pine stands in Australia that were more than 30 m high, only 0.2% of the trees were windthrown during a gale, whereas 22% of the trees in a stand that had been thinned 5 years earlier were windthrown (Cremer *et al.*, 1977). In England and Ireland, because of the dangers of damage by strong winds, conventional thinning is not recommended for many plantations. However, timber quality can be maintained by early thinning, when the trees are only 5 to 6 m high and before they are exposed to substantial risk of wind damage (Savill, 1983). The effects of wind on trees are discussed further in Chapter 12.

Reproductive Growth

In forest stands, most fruits and seeds are produced on vigorous, dominant trees and very suppressed trees may produce no seeds. Thinning of stands not only increases vigor of the residual trees but also stimulates reproductive growth. For example, Wenger (1954) found that thinning a loblolly pine stand during the winter of 1946–1947 increased the number of flower buds produced during the next growing season. Although the first year that thinning could influence production of mature cones was a very poor seed year, each residual tree produced an average of 51 cones compared with only 5 cones per tree in an unthinned stand. The effect of thinning continued into 1950 and 1951, when the released trees produced 107 and 102 cones per tree, while the unreleased trees bore only 16 and 48 cones during the same 2 years.

Pruning

Removal of portions of plants, usually whole branches or parts of branches but sometimes buds, roots, flowers, and fruits, has been practiced for centuries and was even recorded in the Bible. Fruit growers and arborists prune trees routinely; foresters do so much less often. The objectives of pruning of trees by these three groups are somewhat different.

Forest Trees

Sometimes it is desirable to artificially prune the lower branches of forest trees to produce knot-free lumber. However, this should be done selectively only on trees likely to survive because the economic benefits generally do not compensate for the costs of pruning, especially as most young forest trees will die of suppression before they reach harvest size. Furthermore, many forest trees shed branches naturally.

Natural Pruning

Several species of conifers and broad-leaved trees shed lateral branches either by physiological abscission, a process called cladoptosis, or through the action of biotic and mechanical agents.

Cladoptosis, which is largely restricted to small branches within the crown, is common in such conifers as bald cypress, redwood, sugi or Japanese cedar, eastern hemlock, eastern arborvitae, incense cedar, Port Orford cedar, Alaska cedar, *Podocarpus vitiensis,* and species of *Agathis, Araucaria,* and *Pinus.* Among the many broad-leaved trees of the temperate zone that abscise lateral branches are poplars, willows, elms, maples, walnuts, ashes, cherries, and oaks. Cladoptosis also occurs in tropical trees such as species of *Albizia, Sonneratia, Persea, Xylopia, Canangia, Castilloa,* and *Antiaris* (Millington and Chaney, 1973).

Natural pruning of lower branches without development of an abscission layer is common in competing forest trees. The first step in natural pruning involves the sequential senescence and death of branches beginning from the ground upward. The dead branches are attacked by saprophytic fungi and insects that cause their decay, weakening, and breakage, often by wind or snow. When a branch dies it usually is sealed off from the stem by deposits of resins in conifers and by gums or tyloses in broad-leaved trees. The activity of fungi, which are responsive to temperature and moisture, often determines the rate of branch pruning. The amount of natural pruning in dense stands varies greatly among species. For example, longleaf pine exhibits good natural pruning and Virginia pine does not (Fig. 13.9). Although most broad-leaved trees prune well naturally, there are differences among species. Although scarlet oak shows poor natural pruning of large branches, southern red oak and white oak exhibit good natural pruning (Zahner *et al.*, 1985). Millington and Chaney (1973) provide a good review of natural shedding of branches of forest trees.

Artificial Pruning

When pruning of forest trees is considered advisable it usually is postponed until the lower branches are dead or dying because pruning of live branches may wound the stem. It is also preferable to prune during the dormant season when the bark is tight

Figure 13.9 Variations in natural pruning of forest trees. Good natural pruning of longleaf pine (*top*) and poor natural pruning of Virginia pine (*bottom*).(After Bond, 1964, and U.S. Forest Service; from Millington and Chaney, 1973, by permission of Academic Press.)

and there is less likelihood of stimulating growth of epicormic branches from dormant buds.

The removal of live branches of forest trees decreases the amount of photosynthetic surface, but it also decreases the amount of respiring tissue. Many suppressed basal branches with only a few leaves consume in respiration all the carbohydrates produced in photosynthesis and so do not contribute carbohydrates for stem growth. Removal of such unproductive or parasitic branches is obviously desirable because it does not lessen stem growth but instead reduces the number of knots and decreases attacks by fungi. Height growth depends on carbohydrates and hormones produced in the upper crown, and a large part of the crown can be removed by pruning from below without appreciably inhibiting height growth. As may be seen in Fig. 13.10, removal of 30 to 70% of the live crown had little effect on height growth of red pine.

The influence of removal of live branches on wood production and stem form is the reverse of the effect of thinning. Thinning a stand stimulates cambial growth at the base of residual trees, but pruning tends to inhibit growth at the stem base, because the accretion of xylem following branch removal is concentrated in the upper stem. Pruning, therefore, tends to reduce stem taper, but the degree of reduction depends on pruning severity and timing as well as on the crown characteristics of the tree prior to pruning, the effect being greater on open-grown trees than on those in closed stands. Above 30 or 40% of tree height, the more branches removed, the greater is the decrease in wood production and the upward displace-

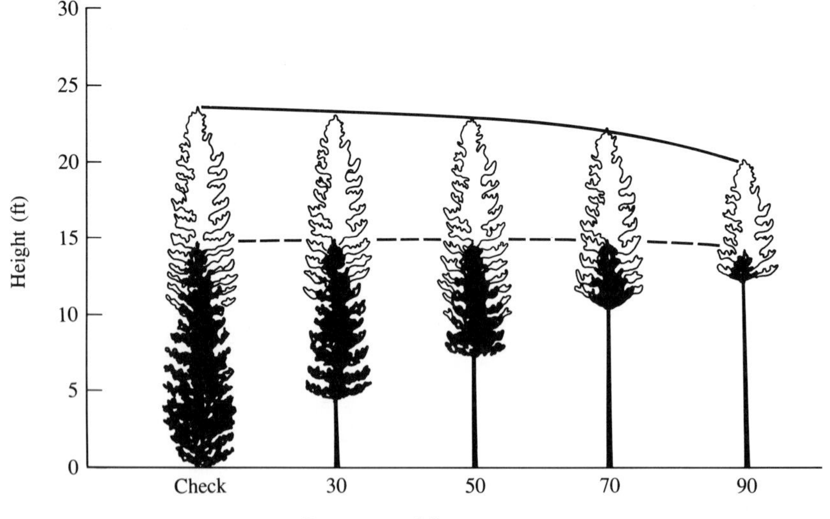

Figure 13.10 The effect of removal of various percentages of live crown on height growth of red pine trees. 1955 (——); 1950 (----). (After Slabaugh, 1957; from Kramer and Kozlowski, 1979, by permission of Academic Press.)

ment of growth increment leading to a decrease in stem taper (Larson, 1963a). The degree of correction of stem form by branch pruning of strongly tapered trees also is modified by the age of trees, with stems of old trees being less responsive to change in taper than those of young trees. The failure of many pruning experiments to alter stem form has been traced to the removal of too few branches or to excessive delay before pruning. As emphasized by Larson (1963a,b), tree stems become more cylindrical with increasing age and with greater stand density. Thus, either a delay in pruning or removal of only a few branches from trees in closed stands may not be followed by noticeable changes in stem form.

Pruning experiments on loblolly pine in which the remaining crown was 50, 35, or 20% of the tree height demonstrated that the crown size of pruned trees had a marked effect on the amount of diameter growth below the crown but a relatively minor effect on growth within the crown (Fig. 13.11). Experiments by Labyak and Schumacher (1954) also indicated that a crown occupying one-third of the tree height supported normal stem growth in loblolly pine. In addition to altering the amount of wood produced, pruning also tends to alter wood quality. In heavily pruned trees with small crowns there is a more abrupt transition to latewood down the stem, and an increase in the percentage of latewood in the lower stem (Larson, 1969; Zobel and van Buijtenen, 1989, pp. 241–243).

Fruit Trees

Fruit trees are pruned regularly, primarily to ensure fruit of high quality and for convenience in spraying and picking fruit. One of the major objectives of pruning is to increase exposure of fruits and adjacent leaves in the bearing part of the canopy to light.

Within the major fruiting zone of apple trees, the light intensities may vary from 30 to 95% of full sunlight (Jackson, 1970). Large fruits of good color develop only under conditions of high light intensity (Jackson, 1970). Fruit size, red color, soluble solids content, sugar content, and acidity of apples are increased by high light intensity on the fruit, adjacent leaves, or both (Jackson, 1985). In addition to improving fruit quality, pruning reduces the size of the canopy, thereby facilitating spraying and fruit harvesting.

The effects of pruning on vegetative and reproductive growth of fruit trees vary with species, cultivar, tree vigor, and time and severity of pruning (Mika, 1986). Dormant pruning of leafless trees stimulates shoot growth while reducing the number of fruits, but increases fruit size. The new shoots grow faster and continue growing for a longer time in the summer than they do on unpruned trees. Early in the growing season the leaf area of pruned trees is decreased but it increases rapidly during the growing season, and often is higher later in the season than it is on unpruned trees. Sometimes overpruning results in excessive growth of new shoots in the crown periphery, thereby inhibiting penetration of light to the fruit-bearing zone. Because carbohydrates from sprouting shoots are rarely translocated to fruits, such sprouting may inhibit fruit growth. By removing growing shoots, summer

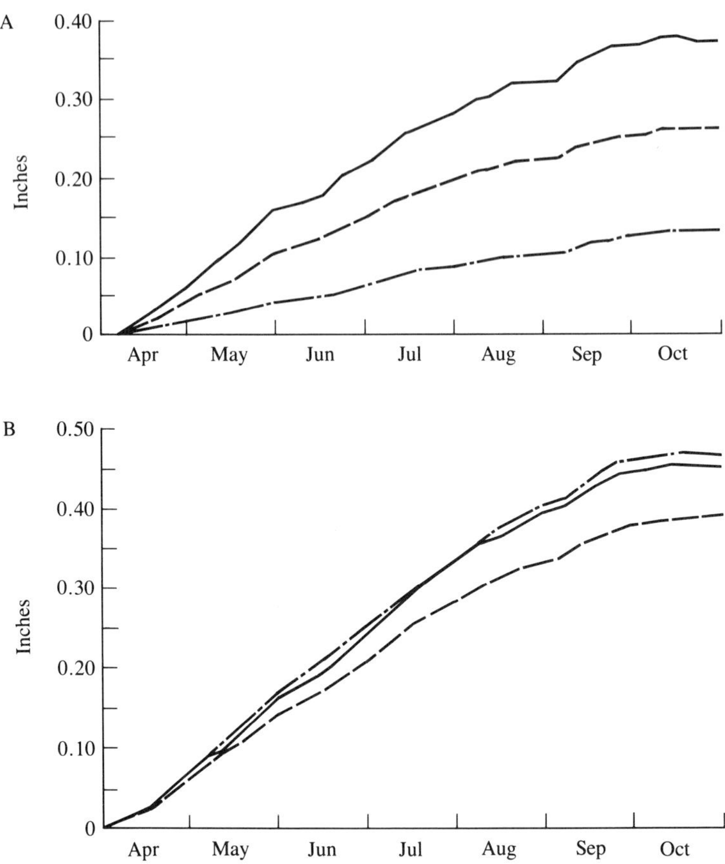

Figure 13.11 Effect on diameter growth of pruning loblolly pine trees to various percentages of their height. (A) Diameter growth at breast height (135 cm). (B) Diameter growth at 80% of tree height. Class 50 trees (——) were pruned to 50% of their height, class 35 trees (– – –) to 35% of their height, and class 20 trees (–·–) to 20% of their height. Reducing crown size decreased diameter growth much more at breast height than at 80% of height. (After Young and Kramer, 1952; from Kramer and Kozlowski, 1979, by permission of Academic Press.)

pruning increases light intensity in the fruit-bearing portion of the canopy, improves fruit color, and stimulates formation of flower buds. In contrast to dormant pruning, summer pruning decreases shoot growth and promotes development of spur shoots. Early summer pruning of apple trees is followed by some regrowth of shoots and restoration of leaf area. However, regrowth of shoots usually does not follow late-summer pruning, hence the amount of leaf surface is reduced. Dormant pruning increases the N,P, and K contents of fruits. However, Ca and Mg contents are decreased as they readily move into the rapidly growing new shoots. Summer pruning often lowers the N,P, and K content of fruits (Mika, 1986). For detailed information on methods of pruning fruit trees, the reader is referred to Childers (1983).

Ornamental Trees and Shrubs

The objectives of pruning ornamental trees usually are different from those for forest or fruit trees. Shade trees and shrubs are pruned for many reasons, including elimination of dead, diseased, injured, broken, and crowded branches; training young plants to create topiary, espalier, or bonsai forms; controlling tree size to prevent crowding and avoid street wires, to facilitate spraying, and to prevent the obstruction of views and traffic; influencing flowering and fruiting; invigorating stagnating trees; and preventing excessive water loss from the crowns of transplanted trees (Harris, 1983).

When transplanting shade trees it often is necessary to compensate for the decreased water-absorbing capacity of the reduced root system by lowering transpirational water loss through decreasing the number of leafy or bud-bearing branches. The extent of such pruning depends on the condition of the tree and the care it will receive after it is moved. Large trees require more pruning than small ones, and bare-rooted trees more than those in containers or balled and burlapped trees. In fact, properly managed containerized or balled and burlapped trees, even evergreens, often require no pruning (Kozlowski, 1976a).

The time to prune depends on the species and results desired. Light prunings can be made at any time during the year. Pruning of broken, dead, or weak branches has a negligible dwarfing effect irrespective of the time when done. The best time for heavy pruning of shade trees usually is late in the dormant season. However, for maximum dwarfing the trees should be pruned when seasonal shoot growth is near completion, from late spring to mid summer. Summer pruning should be moderate to prevent stimulating growth of new shoots. Bonsai (Fig. 1.2) are a good example of the cumulative effect of frequent, severe root and shoot pruning. Harris (1983) has a good discussion of methods of pruning shade trees.

Christmas Trees

The quality of Christmas trees can be greatly improved by pruning so as to foster the loss of apical dominance. Young unpruned conifers often have long internodes in the main stem and branches, which give the tree a spindly appearance that lowers its desirability as a Christmas tree. Hence growers routinely "shear" young trees by cutting back the terminal leader and current-year lateral shoots or by debudding shoots. This practice inhibits shoot elongation and stimulates expansion of dormant buds as well as bud formation and expansion of new buds into branches. Thus many new lateral shoots form along branches and contribute to development of bushy, high-quality Christmas trees (Fig. 13.12). Species with dormant buds along each

Figure 13.12 Shaping of Scotch pine trees by cutting back of shoots. *Left*, Spindly control tree. *Right*, Well-shaped and bushy tree that had been sheared. (U.S. Forest Service photo.)

internode, such as firs and spruces can be sheared at any season except early summer. However, shearing of pines, which generally do not bear dormant buds along internodes, should be done while the shoots are elongating (Smith, 1986).

Pruning and Plant Diseases

Diseases such as fire blight may be increased by the transmission of pathogens on pruning tools through pruning wounds, but the major effect of pruning is stimulation of succulent shoots that may be more susceptible than normal shoots to invasion by pathogens (Mika, 1986). An example is susceptibility of shoots of recently pruned apple trees to fire blight.

Healing of Pruning Wounds

Pruning wounds are potential invasion sites for pathogenic fungi that cause wood decay. Trees respond to wounding by altering their metabolism and producing chemicals that form a protective zone behind a wound. In spite of these barriers, pruning wounds often are invaded by pioneer microorganisms, including bacteria and some fungi that overcome the chemical protection barriers. These pioneer organisms do not induce wood decay directly but are followed by other microorganisms that then cause decay.

When the chemical protective barriers are overcome, additional changes occur in the host tree that tend to compartmentalize the wounded tissues. Barriers to invasion develop, including plugging of vessels with gums in some species, formation of tyloses in others, and production of thick-walled xylem and ray cells by the cambium. The barriers separate the wounded tissues from those formed after wounding. In healthy trees, spread of decay is prevented by such compartmentalization. The extent to which the decay-causing organisms invade the tree varies with species, tree vigor, and environmental conditions (Kramer and Kozlowski, 1979, p. 95).

The rate of wound closure also varies with species, genotype, and tree vigor (Neely, 1988a). Wounds of green ash and sweet gum closed faster than those of Bradford pear, honey locust, or river birch (Martin and Sydnor, 1987). Pruning wounds heal most rapidly in vigorous trees. Because cambial growth of trees in the north temperate zone occurs primarily during May, June, and July, wounds made prior to May heal rapidly; those made after July heal slowly. In suppressed trees, wounds in the upper stem, where the rate of cambial growth is most rapid, heal faster than those in the lower stem. In contrast, in vigorous, open-grown trees, wounds in the lower stem (where the rate of cambial growth is highest) may be expected to heal faster. The rate of wound closure and capacity to compartmentalize wounds, which varied appreciably among red maple cultivars, were correlated with the previous year's rate of shoot elongation (Gallagher and Sydnor, 1983).

Shigo (1984a) recommended that in severing dead branches the cut should be made as close as possible to the living callus collar, which should not be injured or removed (Fig. 13.13). He suggested that when living branches are pruned, the cut

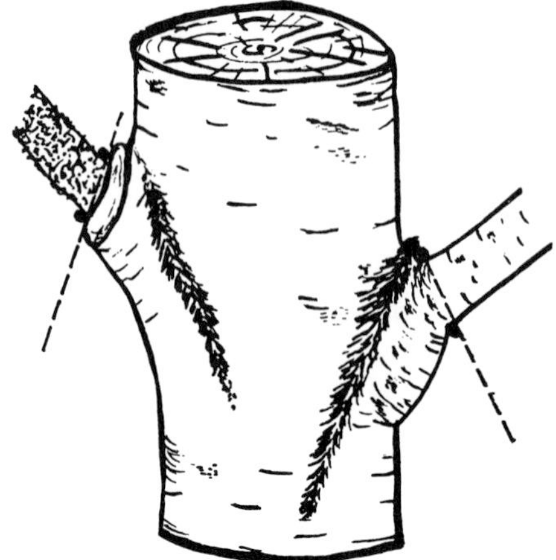

Figure 13.13 Recommendations of Shigo for branch pruning. When living branches are pruned, the cut should be made as close as possible to the branch collar without injuring or removing the collar. Dead branches should be cut very close to the living callus collar without injuring or removing it. (Drawing by A. L. Shigo.)

should be made as close as possible to the branch collar, without leaving a protruding stub. Because the decay associated with the pruning wound is compartmentalized in the stem, Shigo considered it inadvisable to make flush cuts into the healthy callus collar. He reasoned that such cuts may cause extension of decay into tissues that otherwise will remain free of it. Furthermore, flush cuts increase the size of the wound.

Neely (1988a,b) compared the rate of closure of branch wounds following pruning by two methods: (1) branch removal outside the branch collar (Shigo method) and (2) a cut through the branch collar but not necessarily flush with the stem (conventional method). The size of the wound compared to branch size was much greater in the conventional method. However, wounds left by the Shigo method closed less in the first year than those made through the branch collar. By the end of the first growing season, the amount of wood exposed by the conventional method was approximately equal to that exposed by the Shigo method. After the second growing season, more of the conventional than the Shigo cuts were fully closed.

A variety of wound dressings have been applied to pruning wounds. These include asphalt-type materials, shellac, house paints, petrolatum, and fungicides. Although some systemic fungicides applied to pruning wounds were effective in controlling fungi that induced decay of wood, no one chemical acted as an ideal wound treatment. Products with a latex base were useful in the short term but not in the long term (Mercer *et al.*, 1983). Neely (1970) concluded that wound dressings that did not contain fungicides had a negligible influence on the rate of wound healing. In fact a petrolatum dressing reduced the rate of wound closure. These paints are usually dark in color and thus they lessen the visual impact of freshly cut surfaces of pruning wounds. Shigo and Wilson (1977) noted that wound dressings did not prevent infection by decay fungi. They acknowledged that wound dressings have a strong psychological appeal but questioned their usefulness in accelerating wound closure. Nevertheless, demonstrations of effectiveness of mixtures of some systemic fungicides in controlling pathogenic fungi that invade trees through pruning woods are encouraging and should be investigated further (Gendle *et al.*, 1983). A practical way of accelerating wound closure is to concentrate on stimulating cambial growth by fertilizing and irrigating trees, and by removing less valuable trees that may be crowding injured trees.

Root Pruning

Evergreens that are transplanted with bare roots, or even with a root ball of soil, undergo a particularly severe physiological shock because their capacity for absorbing water is suddenly greatly reduced although water loss by transpiration continues. During lifting and handling of planting stock many of the small absorbing roots are lost, disrupting the previous close contact of the root system with a large volume of water-supplying soil. Such loss of roots often is associated with dehydration of transplanted trees (Kozlowski and Davies, 1975a,b). Repeated root pruning sometimes is used in nurseries to develop compact root systems and to condition seedlings to withstand transplanting shock.

In the short term, root pruning decreases the root–shoot ratio, leads to dehydration of shoots and stomatal closure, lowers the rate of transpiration, and decreases tree growth. Root pruning of peach trees reduced absorption of water for about 10 days (Richards and Rowe, 1977) and in transplanted Norway spruce trees for 7 weeks (Parviainen, 1979). In both species the rate of water absorption recovered when new roots had regenerated and the absorbing root surface had increased to prepruning levels.

As mentioned earlier, root pruning and "wrenching," which breaks many roots, are widely used in forest nurseries to increase drought tolerance as discussed by Kramer and Rose (1986). The amount of reduction of transpiration and photosynthesis may be expected to vary with the severity of root pruning. For example, removal of 28 or 59% of the roots of apple trees decreased transpiration by 29% and 45%, respectively, and photosynthesis by 35 and 47%, within a day (Geisler and Ferree, 1984b) Eventually as new roots form, water absorption increases, the leaves rehydrate, stomata reopen, and photosynthesis increases. Recovery of photosynthesis after root pruning of young apple trees required 3 to 7 weeks (Maggs, 1964), and 4 weeks in tropical pines (Abod *et al.*, 1979).

Root pruning is followed by increased allocation of photosynthates to the roots, resulting in decreased shoot growth, and increased root growth, and a progressive increase in the root–shoot ratio. Translocation of photosynthetic products to the roots of Monterey pine seedlings approximately tripled within a month after the roots were pruned (Rook, 1971). Stimulation of root formation on pruned root systems involves subtle relations among carbohydrates, N compounds, and growth hormones (Haissig, 1974, 1982).

Several investigators emphasized the beneficial effects of root pruning for conditioning seedlings for transplanting. For example, undercutting of roots of Monterey pine seedlings in nursery beds produced plants with a compact mass of fibrous roots and a high root–shoot ratio. The transplants that had received the undercutting treatment survived drought better than control seedlings did because the root systems of the former were more efficient in absorbing water (Rook, 1969, 1971; also see Chapter 7). In many arid zone nurseries the roots of containerized plants are pruned routinely (Goor and Barney, 1968).

Shade trees can also be conditioned in the nursery to withstand transplanting by root pruning every few years. This stimulates regeneration of roots from each severed root and confines most of the absorbing roots within the size of the root ball (Fig. 13.14), with the replacement roots originating from the callus formed near the severed surface. At first many small roots form but subsequently one root usually becomes dominant (Fig. 13.15). Pruning of the root system of landscape-size Engelmann spruce trees in the nursery, 5 years before transplanting, led to a quadrupling of the root surface area and a doubling of the percentage of the whole root system in the root ball (Watson and Sydnor, 1987).

Responses to root pruning vary with the time of treatment. Root pruning in the spring is much more effective than pruning in the autumn in increasing root growth. Because the rate of photosynthesis is low in the autumn carbohydrates are less available for transport to the roots (Geisler and Ferree, 1984a). Root pruning of 4-year-old apple trees at either the dormant or full bloom growth stages reduced

Figure 13.14 Effect of root pruning before transplanting on production of roots. (A) An unpruned tree with very few roots in the root ball; (B) a root-pruned tree with many fine roots in the root ball. (From Watson, 1986.)

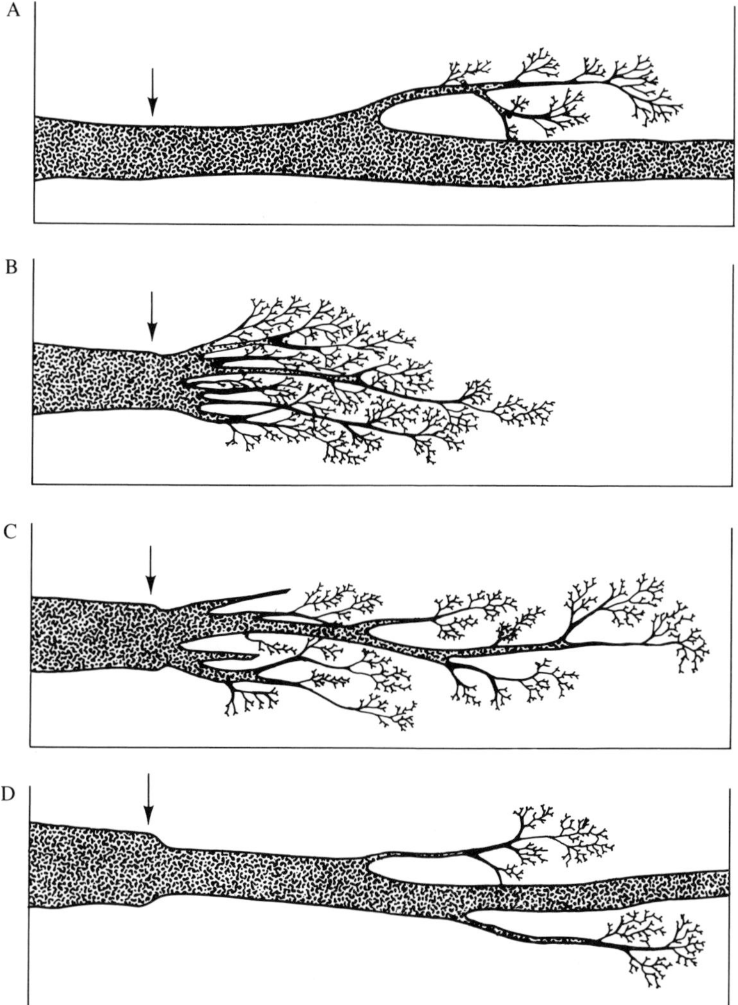

Figure 13.15 Stages in the development of adventitious roots that form at the end of a severed root. (From Watson, 1986.)

both shoot growth and cambial growth (Table 13.5). Regeneration of roots in November was high in trees root pruned at the dormant or full bloom stage. Root regeneration was less in trees pruned at the time of June fruit drop, and least in trees that were pruned just before the apple crop was harvested.

Flowering sometimes can be stimulated by root pruning, alone or in combination with other treatments. For example, root pruning of apple trees before June drop of fruits greatly stimulated flowering in the subsequent year. Root pruning later in the season was less effective (Hoad and Abbott, 1983).

Table 13.5

Effect of Time of Root Pruning on Trunk Cross-Sectional Area (TCSA) Increment, Shoot Length, Root Suckering, Leaf Area, Light Penetration of the Canopy, Fruit Set, and Cropping Efficiency of 4-Year-Old Apple Trees[a]

Growth stage at pruning	TCSA increment (cm^2)	Shoot length (cm)	Root suckers (no. per tree)	Dormant pruning time (min. per tree)	Leaf area (cm^2)	Light penetration, bottom one-third of canopy (% full sun)	Fruit set, no. fruit per no. flower clusters	Crop efficiency (kg yield cm^{-2} TCSA increment)
Control	12.3	35.9	8.0	11.0	27.9	19	0.29	4.5
Dormant	5.8	19.9	1.6	4.1	15.6	39	0.25	6.8
Full bloom	5.8	21.1	4.4	4.8	16.9	28	0.35	8.8
June drop	10.6	35.4	3.1	9.7	27.4	19	0.30	4.6
Preharvest	9.1	37.5	5.3	9.5	27.1	17	0.39	5.2
LSD 5%	3.1	6.0	NS[b]	2.8	1.6	11	NS	2.1

[a]From Schupp and Ferree (1987).

[b]NS, not significant.

Growth Retardants and Pesticides

Application of various chemicals such as fungicides, insecticides, and growth retardants is important in the management of fruit and shade trees. Unfortunately, large amounts of such chemicals often have been applied with little regard to their physiological effects on plants. Growers are reminded that certain applied chemicals sometimes adversely influence physiological processes and their misuse may result in phytotoxicity, growth inhibition, and even death of woody plants. Much injury by chemicals is the result of overdosage, improper application, and application to the wrong species or cultivar.

As emphasized by Tattar (1978), weather conditions may increase or decrease the toxicity of some chemicals. When properly applied they usually are effective and safe over a considerable temperature range but they may be phytotoxic at very high or very low temperatures. For example, when certain fungicides are applied at temperatures above 33°C (90°F), the leaves may be injured. At temperatures below freezing, some fungicides break down and the decomposition products may then injure trees. When certain pesticides are maintained in solution on leaves when the air humidity is very high, they may break down and cause injury. For these reasons, Tattar (1978) recommended that trees should be sprayed with pesticides when temperature extremes are not anticipated and when the pesticides in solution will dry rapidly.

Growth-Retarding Chemicals

Several growth-inhibiting chemicals have been used in the management of forest, fruit, and shade trees. They have been applied as foliage sprays, wound dressings, bark bands, and stem injections to control tree height, avoid interference with power lines, better manage fruit trees, reduce the cost of storing and shipping nursery stock, and increase flowering through suppression of vegetative growth. On the basis of their activity, the growth-retarding chemicals are classified into the following two groups (Table 13.6).

1. *Subapical inhibitors*, which reduce elongation of internodes without arresting cell division in apical meristems.
2. *General growth inhibitors*, which arrest activity of apical meristems or kill terminal buds. They also decrease apical dominance and, by stimulating growth of lateral buds, often inhibit terminal growth.

Vegetative Growth

There are many examples of suppression of vegetative growth by growth-retarding chemicals. Maleic hydrazide (MH) and chlorflurenol applied as sprays or painted on the bark are used to control top growth of trees (Harris, 1983). Naphthaleneacetic acid (NAA) and chlorflurenol applied to pruning wounds inhibit sprouting (Fuller *et al.*, 1965; Domir, 1978). Of 12 chemicals injected into 25 species of detopped

Table 13.6
Some Examples of Growth-Inhibiting Chemicals

Common name	Structure	Trade name
	Subapical Inhibitors	
Daminozide (SADH)	Succinic acid -2, 2-dimethylhydrazide	Alar, B-9
Chlormequat (CCC)	(2-Chloroethyl)trimethyl-ammonium chloride	Cycocel
CBBP (Phosphon)	2,4-Dichlorobenzyl-tributyl-phosphonium chloride	Phosphon-D
Ancymidol (El-531)	*a*-Cyclopropyl-2,4-methoxy-propyl-2,5-pyrimidine	A-Rest
Paclobutrazol	*B*-(4-Chlorophenyl)methyl-(1,1-dimethyl)-1*H*-1,2,4-triazol-1-ethanol	PP333, Cultar
Uniconazole	(*E*)-1-(*p*-Chlorophenyl)-4,4-dimethyl-2-(1,2,4-triazol-l-yl)-1-penten-3-ol	S-3307
Flurprimidol	*a*-[l-Methylethyl)-*a*-[4-trifluoro-methoxy)phenyl]-5-pyrimidinemethanol	EL-500
	General Inhibitors	
Maleic hydrazide (MH)	1,2-Dihydro-3,6-pyridazinedione	Royal Slo-Gro
Chlorflurenols (morphactins)	Methyl-2-chloro-9-hydroxyfluorene-9-carboxylate	Atrinal
Fluoridamide	*N*-4-Methyl-3,3-(1,1,1,-trifluoromethyl) sulfonyl amino phenyl acetamide	Sustar

woody plants, MH and dikegulac were the most effective sprout- inhibiting compounds. Both chemicals controlled sprouting for 1 year and, in several instances, for 2 years. The results were more consistent with dikegulac than with MH (Domir and Roberts, 1983). The sterol paclobutrazol reduces internode length and leaf size of many deciduous fruit crops (Elfving *et al.*, 1987; Curry, 1988). The objectives of applying growth-suppressing chemicals to young conifers range from controlling growth of bonsai trees to producing manageable trees for use as ornamentals. Such uses are feasible since concomitant treatment of Arizona cypress seedlings with gibberellic acid (GA) and daminozide reduced vegetative growth by as much as 70% (Pharis *et al.*, 1965). Use of growth-suppressing chemicals in nurseries to keep transplant stock to a size for easy handling also has considerable merit, especially since the growth-retarding effects can be counteracted by applying gibberellic acid prior to outplanting trees (Kuo and Pharis, 1975). However, growth-retarding chemicals may be toxic to very young seedlings in the cotyledon stage of development (Kozlowski, 1985b).

Differences in plant responses to growth-retarding chemicals may be expected because of variations in their mechanisms of action (Cathey, 1964). Subapical inhibitors often act by blocking gibberellin synthesis. Shoot elongation of conifers often is arrested in proportion to the amount by which endogenous GA production is decreased by these compounds (Pharis *et al.*, 1967). The actions of daminozide and chlormequat often are the reverse of those of the GAs, with endogenous levels of GA-like compounds positively correlated with the amount of shoot elongation (Pharis, 1977). Daminozide may also act by increasing conversion of biologically active GAs to other GAs (or conjugates), which are less active than their precursors (Kuo and Pharis, 1975). Synthesis of GA also is inhibited by triazoles such as paclobutrazol, uniconazole, flurprimidol, and triapenthanol. However, not all effects of subapical inhibitors can be explained by suppression of GA synthesis. For example, inhibition of vegetative growth of some plants by chlormequat cannot be overcome by applying gibberellins (Audus, 1972). In contrast to subapical inhibitors, the general growth inhibitors do not appear to interact directly with any of the known classes of plant growth hormones. Maleic hydrazide appears to act on a process involved in cell division but just where is not clear (Kefford, 1976).

Reproductive Growth

As the season progresses the balance between gibberellins and cytokinins in trees changes in the direction of the latter, thus favoring flowering. This balance also can be modified by applying chemicals that suppress GA biosynthesis. For example, trees treated with daminozide or chlormequat exhibit many characteristics of those grafted on dwarfing rootstocks, that is, short internodes, early cessation of shoot growth, and precocious and prolific formation of flower buds. The triazoles also variously stimulate flowering. Dilute foliar applications of paclobutrazol increased flowering of several apple cultivars in the year after treatment. Applying paclobutrazol to the roots was as effective as targeting the foliage but the effect was evident a year later (Tukey, 1983, 1986). Uniconazole enhanced flowering by stimulating production of spurs and short shoots and by accelerating the maturity of buds (Tukey, 1989).

The mechanisms of action of various triazoles on flowering are not identical and apparently are mediated, at least in part, by effects on hormones other than GA. For example, unlike paclobutrazol, uniconazole appears to lower auxin levels (Lürrsen, 1987).

Potential Adverse Effects of Growth-Retarding Chemicals

Several investigators have pointed out some undesirable effects of growth retardants on trees, such as reduction in the rate of photosynthesis. Although a single injection (3 g liter^{-1}) of daminozide or MH reduced photosynthesis in silver maple and American sycamore seedlings, the daminozide-treated plants recovered after 2 to 3 weeks whereas the MH-treated seedlings did not (Roberts and Domir, 1983).

Although inhibitors of vegetative growth ideally should not reduce the size or shape of fruits (Quinlan, 1981), some adverse effects on fruit quality have been reported. For example, the high rates of daminozide needed to control vegetative growth of apple trees may reduce fruit size and length, (Williams and Edgerton, 1983; Stinchcombe *et al.*, 1984). Such negative effects can be reduced by using moderate rates of paclobutrazol and treatment with GA_3 or a combination of GA_{4-7} plus benzyladenine at or near anthesis (Curry, 1988). Much more research is needed to explore the full potential of the use of growth retardants.

Fungicides

As a general rule, fungicides have been found to be less toxic than other chemicals when applied to trees. Nevertheless, phytotoxicity following application of certain fungicides to trees has been reported (Daines *et al.*, 1957; Cayford and Waldron, 1967; Ilenne and Atkinson, 1973; Andersen *et al.*, 1985; Kozlowski, 1986a).

Some fungicides lower the rate of photosynthesis by reducing the light intensity reaching the leaf, plugging stomatal pores, or affecting metabolism. For example, sulfur fungicides and copper oxychloride, which form an opaque layer on the leaf surface, reduced photosynthesis (Barner, 1961).

Insecticides

Some investigators reported that certain insecticides reduce photosynthetic efficiency and growth of trees. The effects of insecticides on photosynthesis were discussed by Wedding *et al.* (1952), Westwood *et al.* (1960), and Ayers and Barden (1975).

Many insects can be controlled by systemic chemicals without causing appreciable injury. However, this usually requires closer regulation of dosage per tree than does use of conventional insecticides. Norris (1967) emphasized that when systemic chemicals are applied to the soil around tree roots, adsorption of the chemical on soil particles, degradation of the chemical, and capacity of roots for selective adsorption of the chemical reduce the possibility of phytoxicity from errors in dosage. However, when chemicals are injected directly into trees the chances that high dosages will cause injury are greater.

Integrated Pest Management

Synthetic pesticides introduced after World War II brought great benefits to society but gradually they also created some problems. For example, many insects, some rodents, and plant pathogens developed resistance to pesticides, and some weed ecotypes began to show resistance to herbicides (Muir, 1978). In some areas, pesticides no longer controlled target pests effectively but interfered with control of other pests and released some species from natural controls so that they became pests. Pesticides also modified the physiology of certain crop plants unfavorably, killed pollinators and other wildlife, and created health hazards for humans (Smith, 1978).

In recent decades, attention has been directed to the development of systems of integrated pest management (IPM). Truly integrated pest management relies primarily on natural control of pests, together with a combination of pest-suppressing techniques such as cultural methods, pest-specific diseases, resistant crop varieties, sterile insects, insect attractants, release of parasites of predators, or application of pesticides as needed. IPM uses control methods that consider the role of all the pests in a given environment, the relationships among these pests, and other factors that affect growth and yield of crop plants. To be effective IPM requires multidisciplinary participation (Cutler, 1978).

Several investigators have successfully used IPM in fruit orchards. In apple orchards, a predator–prey system uses a predaceous mite that attacks spider mites, together with chemical control of other pests. A system for control of the codling moth involves capture of male moths at pheromone-baited traps. This information, which is coupled with a time model for insect development, is then used to time the application of pesticides to control hatching larvae (Croft, 1978). A citrus parasite of the California red scale has been used in southern California. In Florida, a parasite was introduced to control snow scale on citrus (Huffaker *et al.*, 1978). IPM also has been applied to control insect pests of urban trees, for example, in the San Francisco Bay area, where each of the more than 100 species of street trees has its own complex of insect pests (Olkowski *et al.*, 1978). Principles of IPM are discussed in more detail by Coppel (1977), Pimental and Perkins (1980), Huffaker (1980) and Sill (1982).

Planting for High Yield

As the demand for yield of wood and food increased and extensive acreages of forest lands were lost to agriculture, urban, and industrial use, attention was directed to cultural practices that might increase yield per unit of land. This objective can be achieved by planting of genetically improved trees at very close spacing followed by intensive cultural practices.

Short-Rotation Forestry

There is much interest in the productivity of fast-growing forest trees planted at close spacings, together with intensive culture. The trees may be harvested, usually at 3 to 10 years, for a variety of uses, including pulp, lumber, energy, chemicals, and food for animals. Short-rotation systems have been called "silage silviculture" (Herrick and Brown, 1967) and "intensive energy forestry" (Siren *et al.*, 1987). Early yields in short-rotation systems may double those in conventional plantations (Smith and Dowd, 1981).

A compelling argument for short-rotation forestry has been an increasing need for renewable energy, particularly in developing countries. Hence short-rotation forests commonly are grown for biomass (widely considered to be any plant mass available for harvesting). Probably "phytomass" would be a better term (Green,

1980) because biomass is somewhat imprecise and, as now widely used, conflicts with the well established ecological definition of biomass, namely, the dry matter of living organisms (autotrophs or phytomass and heterotrophs or zoomass), including humanity, present at a specific time (Whittaker *et al.*, 1975).

Both conifers and broad-leaved plants are used in short-rotation forestry, with important broad-leaved species (e.g., willows and poplars) capable of reproducing by coppicing. In the tropics, introduction of *Pinus caribaea* var. *hondurensis* was encouraged by fast early height growth of up to 275 cm annually, volume production of about 14 m^3 $year^{-1}$, and absence of serious insect and fungus pests (Kozlowski and Greathouse, 1970). In Sweden, suitable species for short-rotation forestry include willows, poplars, and native alders (Siren *et al.*, 1987). In Brazil, closely spaced, high-yielding *Eucalyptus* species are grown for charcoal to operate steel mills, for pulp, and for fuelwood. Under ideal conditions, yields are 40 to 50 tons ha^{-1} $year^{-1}$ of dry matter, but the average is closer to 25 tons (Hall and de Groot, 1987).

Some idea of the potential yield of *Eucalyptus* biomass on good agricultural land was obtained by Sachs *et al.* (1988). In Davis, California, the biomass growth rate of selected river red gum *(Eucalyptus camaldulensis)* in an intensively managed plantation (1100 trees ha^{-1}) was 25 to 27 U.S. tons $acre^{-1}$ for the third and fourth year after planting. After 3 years the trees were more than 11 m high and had a DBH of 9.4 cm. Because the rate of growth did not increase after the third year, a 3-year rotation was indicated. About 80% of the total biomass was available for chipping or cordwood. More than 20 dry tons $acre^{-1}$ $year^{-1}$ (7.3 metric tons ha^{-1}) were marketable as pulp-quality wood chips.

In the temperate zone, much attention has been given to short rotation plantation culture of poplars for several reasons.

1. Poplars produce more biomass in short-rotation intensive culture than other deciduous or evergreen species (Dickmann and Stuart, 1983). On silt loam soils in the Mississippi Delta, dominant cottonwoods in thinned plantations grew 3.6 m in height annually for the first 4 years and increased in diameter up to 3 cm annually during the first 15 years (Krinard and Johnson, 1980). When 5 deciduous species were separately grown in closely spaced plantations (0.9 × 0.6 m) in Kentucky, total biomass production ranged from 92 metric tons ha^{-1} for hybrid poplar to 28 metric tons ha^{-1} for green ash. The relative ranking was hybrid poplar > American sycamore > European alder > river birch > green ash. Furthermore, the hybrid poplar contained the highest proportion of bolewood (Wittwer and Immel, 1978).

The rapid early growth of poplars in intensive culture is associated with their unusually high leaf area index, which may vary between 16 and 45, depending on spacing of trees (Isebrands *et al.*, 1977). However, these high values may have been exaggerated by the effects of lateral illumination on small plots. This difficulty could have been avoided by discarding trees from the margins of the plots. The high potential of poplars for short-rotation forestry also is related to their capacity to reproduce readily by vegetative propagation. Because poplars sprout abundantly from stumps or roots, harvested plantations are readily replaced by coppicing.

It should not be assumed that the rate of early growth of minirotation hardwood stands will be greater than that of later growth of older hardwood plantations on good sites. Cannell and Smith (1980) reviewed the literature on productivity of 1- to 5-year-old closely planted poplars, American sycamore, chinese elm, and red alder growing in temperate regions. Yields of vigorous clones growing on fertile sites, with weed control but without irrigation, were about 10 to 12 tons ha^{-1} $year^{-1}$ at 4 to 5 years after planting and about 1 to 30% greater after coppicing. Values of 7 to 15 tons ha^{-1} $year^{-1}$ were not uncommon for more productive 11- to 26-year-old hardwoods with only 250 to 300 trees ha^{-1} at the end of the rotation (Ek and Dawson, 1976a,b).

2. There is a strong demand for conventional use of poplar (pulp, composition board, lumber, etc.) and nontraditional uses (biomass for feedstock for animals and for energy through direct combustion or conversion to alcohol or other fuels) (Dickmann and Stuart, 1983).

3. Wide genetic variability among poplars provides excellent opportunities for genetic improvement and higher yield. For example, poplars vary in stem and crown form (Cooper and Ferguson, 1981; Nelson *et al.*, 1981), growth rate (Pallardy and Kozlowski, 1979b; Phelps *et al.*, 1982); stomatal characteristics (Pallardy and Kozlowski, 1979a); rooting capacity (Ying and Bagley, 1977); responses to environmental stress such as drought (Pallardy and Kozlowski, 1979c,d; Mazzoleni and Dickmann, 1988), and air pollution (Kimmerer and Kozlowski, 1981); and wood properties such as specific gravity, fiber length, energy content and chemical composition (Sastry and Anderson, 1980; Reddy and Jokela, 1982).

The ultimate goal of short-rotation forestry is to maximize sustained utilization of a site without increasing environmental hazards. Usually, however, water and mineral nutrients are depleted more by short-rotation plantations that by conventional rotation plantations (see Chapter 6). Nevertheless, assuming similar soil and site conditions, the environmental impact of short-rotation forestry should be less than that of most agricultural practices (Riekerk, 1983).

Fertilization to replace depleted mineral nutrients often is necessary in short rotation plantations. However, fertilizers should be applied only after soils and plants are analyzed for mineral deficiencies because trees on very fertile sites do not always respond to fertilizers (Blackmon, 1977). Mineral nutrients should be replaced in poplar stands in accordance with the variable responses of specific clones (Dickmann and Stuart, 1983). Where possible, species that fix N such as alders, black locust, or autumn olive, may be planted as the biomass crop or to supply N to some other biomass crop (Hansen and Dawson, 1982; Dickmann and Stuart, 1983).

Irrigation often increases the yield of short rotation plantations. The available energy (after deducting energy inputs for site preparation, fertilization, weed control, irrigation, and harvesting) in 10-year-old *Populus* × *tristis* plantations in Wisconsin was 43% higher than for highly productive nonirrigated, intensively cultured stands in the eastern United States. Hence, the energy invested in irrigation was used very efficiently in biomass production (Zavitkovski, 1979).

Agroforesty

There is much interest, particularly in the tropics, for increasing sustainable productivity of land by planting various combinations of herbaceous plants and woody perennials, including pasture, with or without grazing animals present (Fig. 13.16). Agroforestry (AF) systems are not new, having originated in Burma more than a century ago using teak as the tree crop. Agroforestry then spread throughout Asia to Africa and Latin America. However, worldwide sustained attention to the potential benefits, problems, and needed AF research was not evident until after establishment of the International Council for Research in Agroforestry (ICRAF) in 1977 (von Maydell, 1985).

Important objectives of AF are to provide benefits greater than those from agriculture or forestry alone. The potential benefits include increased overall yields, maintenance of soil fertility, lowered risks of crop failure, easier management, control of pests, and, especially, meeting the needs of the local population. In AF systems the woody perennials have both a production role (material output such as fuel and fodder) and a service role (in providing shelter or in nutrient cycling) (Torres, 1983).

One of the most important AF products in developing countries is wood for fuel because the intensive use of wood for energy is accentuated by rapidly increasing population rates, decreasing areas of forest land, and depletion of wood supplies. In many of the least developed countries, fuelwood is the single largest source of energy. In Malawi, Tanzania, Ethiopia, Somalia, Chad, and Nepal, fuelwood provides more than 90% of all energy needs. In addition, fuelwood is important to varying degrees to industries in developing countries. In India, coal is the most important fuel; fuelwood provides only 6% of the total need. By comparison, fuelwood provides 57% of industry needs in Sri Lanka, 64% in Kenya, 69% in Mozambique, and 88% in Tanzania. Even in Brazil, with its modern industrial base, fuelwood provides 21% of the energy for industry (Barnard, 1987), and in many Latin American countires it is the only fuel for domestic use.

Agroforestry has considerable potential for easing the energy crisis by providing a variety of products, including (1) wood for direct combustion (firewood, sawmill wastes, etc.), (2) pyrolytic conversion products (charcoal, oil, gas), (3) producer gas from wood or charcoal feedstocks, (4) ethanol from fermentation of high-carbohydrate fruits (e.g., babassu palm, bread fruit, carob) or other tree parts, (5) methanol from destructive distillation or catalytic synthesis processes using woody feedstocks, and (6) other potential fuels from experimental biomass energy technologies using tree products (Raintree and Lundgren, 1985). Another benefit of using biomass for fuel is that it replaces the use of an equivalent amount of fossil fuel and thus represents no net contribution to the "greenhouse effect."

An essential feature of AF systems is close interaction, which may be competitive or complementary, between woody perennials and nonwoody plants. Because of wide variations in size, form, growth rate, and physiological responses of these two plant groups, AF associations often involve complexities greater than those in many forests or fruit orchards.

Figure 13.16 Agroforestry systems in Kenya. *Top*, *Grevillea robusta* intercropped with maize. *Bottom*, *Leucaena leucocephala* intercropped with cowpea. (Photo courtesy of International Council for Research in Agroforestry.)

Although AF systems are best known in developing nations, they also show some promise in industrialized countries. Production of livestock on forage growing under forests has been practiced in the southern United States for a long time. Potentially useful AF systems in other parts of the United states include (1) multi-cropping of farm crops under intensively managed, high-value hardwood plantations, and (2) animal grazing and intercropping under managed conifer forests or plantations (Gold and Hanover, 1987). Grazing by sheep during the first 12 years of a Douglas-fir–Oregon white oak plantation increased both height and diameter growth of Douglas-fir trees. However, after grazing was discontinued, its influence diminished on subsequent tree growth (Jaindl and Sharrow, 1988).

In Brazil, AF systems are dominated by extensive cultivation of perennial crops such as cashew, coconut, or palms. Often cattle, sheep, and donkeys are grazed under these plants. Annual subsistence crops (e.g., maize, beans, and cassava) commonly are cultivated in stands of perennial crops. On very large mechanized holdings, grain, sorghum, peanuts, sesame, or cotton are planted (Johnson and Nair, 1985). In Venezuela the major AF systems are of two types: multispecies associations in integrated coffee production systems and silvopastoral systems. The multilayered coffee systems are used in premontane moist forests of the Andes. Typically a 5- to 10-m high layer is formed over coffee trees by fruit trees such as orange and banana. Species forming a third layer 10 to 20 m high include avocado, *Ingo*, and *Erythrina*. Silvopastoral systems are found in the semiarid zones (Escalante, 1985). In the South Pacific region the major Af systems include combinations of tree crops such as coffee, coconut, and cacao together with N-fixing trees (e.g., *Casuarina*, *Gliricidia*, and *Leucaena*) and food crops such as cassava, taro, sweet potato, and yams (Vergara and Nair, 1985).

Different forms of agroforestry often are used in the same region. In Java, for example, tree gardening (cultivation of a wide variety of crops in a multistoried AF system) is practiced on private lands. On state forest lands, the taungya system is practiced in which trees are planted at some time during the slash and burn cycle (Wiersum, 1982). In Burmese the word, "taungya" means hill (taung) cultivation (ya). Originally taungya was the local term for shifting cultivation, which subsequently was used to describe the afforestation method (Nair, 1985b). In the humid lowlands of Nigeria, multispecies mixes are used for production of food, fodder, and wood products for home consumption and for cash crops. Woody species include *Daniellia oliveri, Gliricidia sepium, Parkia clappertoniana,* and *Pterocarpus africana.* Agrosilvopastoral systems with tree–crop–livestock mixes are common around homesteads (Nair, 1985a). In arid and semiarid regions of India, agrosilvicultural systems employ multipurpose trees and shrubs on farmlands. Woody species include *Cajanus cajun, Derris indica, Prosopis cineraria,* and *Tamarindus indica.* In the tropical highlands, common woody plants include *Albizia* spp. and *Grevillea robusta.* In silvopastoral systems the woody species include *Derris indica, Emblica officinalis Psidium guajava,* and *Tamarindus indica.* Nair (1985a) compiled a detailed list of agroforestry systems, including herbaceous crop plants and woody species used in tropical countries.

Research Needs

Certain agroforestry systems involve as many as 30 species of plants in a complex space–time continuum, with a multitude of interactions that have not been adequately studied. As many as 2000 woody perennials have been identified as potentially promising in AF systems (von Carlowitz, 1985). Only recently has serious attention been given to study and quantify the many interactions among the components of AF systems. Much more is known about growth characteristics and useful cultural practices for the herbaceous crops than for the woody perennials, hence there is an urgent need for research on the physiological ecology of the actual and potential woody components of these systems. Such information can provide a rationale for use of species mixtures and cultural practices that will maximize use of radiant energy, optimize water use efficiency, and minimize losses of mineral nutrients, runoff, and loss of soil. When used in conjunction with field, greenhouse, and laboratory experiments, controlled environments can effectively provide the needed information (Kozlowski and Huxley, 1983).

Research is particularly needed on morphological and physiological characteristics of various plants that govern their use in AF systems. Data are needed on ways in which temporal and spatial arrangements of various plants exploit resource pools as a basis for species selection so as to maximize resource sharing and sustainable yield of harvested products and services (Huxley, 1987). As emphasized by Huxley (1983), attention should be directed toward studies of growth and yield of the whole agricultural crop, effects of tree–crop interface on the herbaceous crop and on trees, and growth of trees as a single crop. Specific research needs in AF are addressed in more detail in the book edited by Huxley (1983).

High-Density Fruit and Nut Orchards

Early yields of fruits or nuts per unit of land area often have been increased by very close spacing of dwarf trees. Examples include apple, peach, apricot, cherry, plum, pecan, and macadamia. In addition to their potential for high yield, the use of dwarf trees in high-density systems facilitates fruit harvesting; reduces pruning requirements; allows for better use of soil, water, and light in a limited area; and facilitates application of pesticides.

Dwarf trees generally are obtained by budding or grafting desirable types on dwarfing rootstocks. The roots of dwarfing rootstocks possess a genetically determined capacity for slow growth, which in turn limits the rate of top growth. Most apple rootstocks now used were developed at the East Malling Research Station in England and are designated Ml, M2, and so on. Those developed jointly by the East Malling and Merton Stations are designated MMlOl, MMlO2, and so on. The relative sizes of apple trees grafted on different rootstocks are shown in Fig. 13.17. The actual sizes of trees on a given rootstock are modified by climate, cultural practices, and the scion cultivar. Vigor of a specific cultivar is expressed regardless of the rootstock. Hence, a vigorous variety of apple such as Gravenstein may be twice as large on a M9 dwarfing rootstock as a less vigorous variety such as Golden Delicious (Westwood, 1978). Whereas all apple rootstocks are *Malus pumila,* pear rootstocks may represent several different species of *Pyrus* or a different genus. Prunes are propagated primarily on peach seedling rootstocks. Peach cultivars generally are propagated on peach rootstocks, but sometimes on apricot or almond seedlings.

Several different forms of high-density systems have been used in orchards (Ryugo, 1988). In hedgerow plantings the trees are closely spaced in rows with alleys wide enough to accommodate machinery (Fig. 13.18). The hedgerow system is adaptable to compact trees, which may be freestanding, staked, or trellised (Childers, 1983). In bedding or ultradense systems, the trees are planted so close together that they are not readily identified as individual trees within or between rows. An example is the meadow orchard (Fig. 13.19) used with apple, peach, and some other fruit trees (Luckwill, 1978). Apple trees differentiate flowers on 1-year-old wood. The following year fruit is set and harvested in the autumn. The trees are then cut back to short stumps and a new biennial cycle begins the following year. Meadow orchards have been adapted for annual cropping of peaches in Israel (Erez, 1978). The optimal planting density for meadow orchards is one that produces a full field cover with foliage, without creating too heavily shaded areas so maximum production of high-quality fruit is obtained (Erez, 1982). As many as 70,000 to 100,000 trees ha^{-1} have been used in meadow orchards of apple (Hudson, 1971; Jackson, 1980).

The Tatura system uses a rigid trellis on which the major tree branches are supported for optimal exposure to light and mechanized harvesting (Chalmers and van den Ende, 1975b). In spindle systems the trees support themselves and are pruned often to balance vegetative development (Bargioni *et al.*, 1985, 1986).

The chief advantage of high-density systems is increased yield at an early age. For example, genetically dwarfed peaches planted at 2500 to 3750 trees ha^{-1} in the

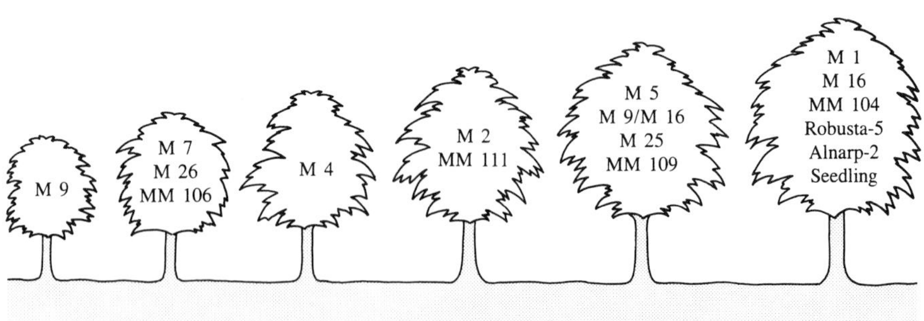

Figure 13.17 Approximate relative sizes of apple trees on different rootstocks. The clonal stocks used with apple originated in Canada (Robusta–5), Sweden (Alanarp–2), and England (M and MM series). (From "Temperate-Zone Pomology," by Melvin N. Westwood. Copyright © 1978 by W. H. Freeman and Company. Reprinted with permission.)

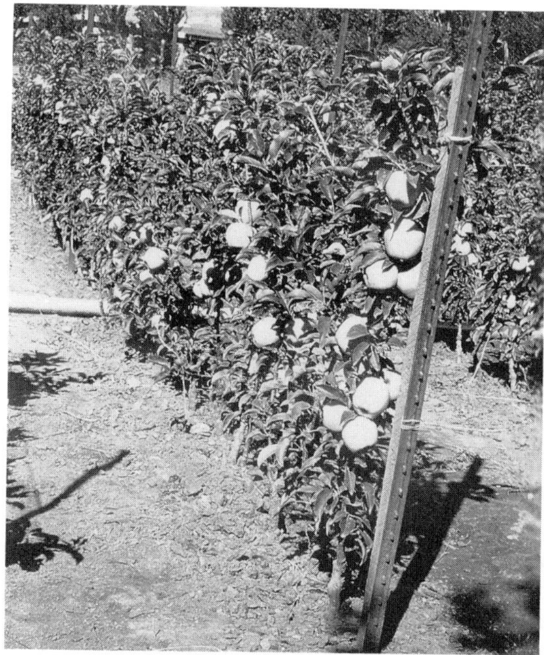

Figure 13.18 Hedgerows of apple trees showing second year crop of Spur Golden trees on M9 rootstocks. The trees were planted at 1.2 × 0.3-m spacings to give 4364 trees per hectare. (Photo courtesy of M. N. Westwood.)

third growing season produced 15 to 22 tons of fruit per acre whereas standard peaches planted at about one-tenth of that density produced only 3 tons acre^{-1}. After 11 years the accumulated yield of a high-density peach orchard was more than double that of a standard orchard (Phillips and Weaver, 1975). However, the difference in yield usually begins to decrease during the sixth growing season (Leuty and Pree, 1980). High-density orchards sometimes become uneconomical sooner than standard orchards because crowding reduces light penetration into the canopy (Chalmers *et al.*, 1981). Yield and quality of pecans were high in the early years of a high-density planting, but declined as the tree canopy closed and nuts were produced only in the upper portions of tree crowns (Witt *et al.*, 1989). There obviously is need for more research on the management and economics of high-density planting.

Summary

Increasing productivity of trees depends on the use of genetically improved planting stock and intensive management of plantations and orchards. Superior planting stock can be obtained by selection and breeding.

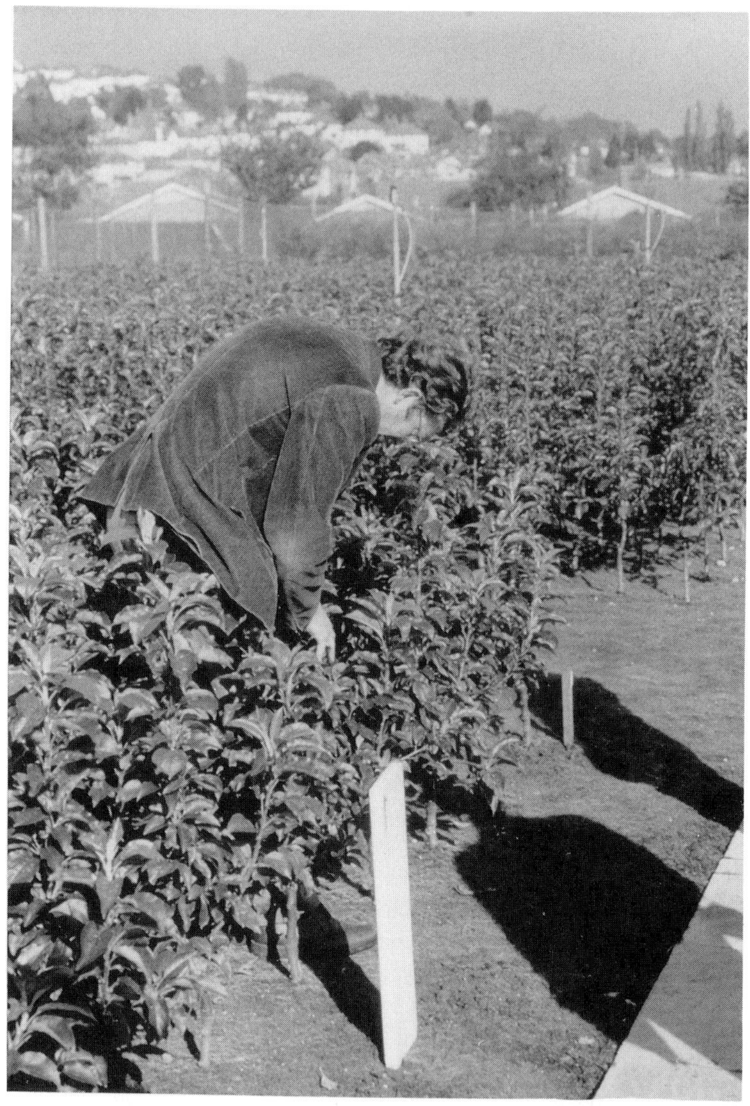

Figure 13.19 Apple meadow orchard. (Photo courtesy of G. C. Martin.)

Production of high-quality planting stock requires seeds from outstanding phenotypes in natural forest stands, vegetatively propagated or seedling seed orchards, or plants obtained by vegetative propagation, or cell and tissue culture. Management of tree nurseries requires close attention to fertilizing and watering practices.

Important cultural practices include site preparation, fertilization, irrigation, thinning of forest stands, pruning of branches, and use of chemicals such as herbicides, growth retardants, and pesticides. These practices are effective only if they protect plants from injury or promote the activity of the physiological processes that control growth.

Site preparation treatments include disposal of logging slash, and reducing plant competition by prescribed burning, mechanical means, and use of herbicides. Planting sites often are prepared by windrowing, chopping of vegetation, bedding, harrowing, disking, drainage, and combinations of these. Most treatments are applied during seedling establishment. Appropriate treatments and intensity of site preparation depend greatly on experience and vary with management practices, soil and species characteristics, topography, and costs.

Fertilizers often are used judiciously to improve growth and yield of forest nursery plants, container-grown plants, recently planted trees, and trees in established stands. Fruit trees are fertilized to increase flower initiation and fruit set and improve fruit quality.

Irrigation is important for seed orchards, nurseries, fruit trees, and ornamentals, which are watered by basin and furrow irrigation, sprinklers, or subsurface irrigation. Fertilizers sometimes are applied in the irrigation water (fertigation). Proper timing of irrigation is important because too-frequent irrigation wastes water, waterlogs the soil, and sometimes causes salt accumulation, whereas too-infrequent irrigation inhibits growth.

Thinning of forest stands increases both vegetative and reproductive growth of the trees left standing. Growth is accelerated because of greater availability of light, soil water, and mineral nutrients. Thinning also leads to more tapered trees because wood production is stimulated most in the lower stem. Responses to thinning vary with crown class, with dominant trees showing less response than more suppressed trees. Thinning may induce production of epicormic shoots, alter susceptibility to insects and diseases, and increase windthrow of trees.

Pruning of branches is routinely practiced by horticulturists but much less often by foresters. Pruning decreases the number of fruits but increases fruit size and improves their quality. Ornamental trees are pruned largely to enhance aesthetic effects. Nursery trees sometimes are root pruned to develop compact root systems and condition trees to withstand transplanting shock.

Application of chemicals such as herbicides, growth retardants, and pesticides often is indispensable. Injury by chemicals may result from overdosage, improper application, and application to the wrong species or cultivar.

Unusually high yields of forest and fruit trees have been obtained by planting genetically improved trees at very close spacings, followed by intensive culture. Examples are short-rotation forests, agroforestry systems, and closely spaced dwarf fruit and nut trees.

General References

Atkinson, D., Jackson, J. E., Sharples, P. O., and Waller, W. M. eds. (1980). "Mineral Nutrition of Fruit Trees." Butterworth, London.

Bajaj, Y. P. S., ed. (1986). "Biotechnology in Agriculture and Forestry. Vol. I. Trees." Springer-Verlag, New York.
Binkley, D. (1986). "Forest Nutrition Management." Wiley, New York.
Bowen, G. D., and Nambiar, E. K. S., eds. (1984). "Nutrition of Forest Plantations," Academic Press, London.
Cannell, M. G. R., and Jackson, J. E. (1985). "Attributes of Trees as Crop Plants". Institute of Terrestrial Ecology, Huntingdon, England.
Childers, N. F. (1983) "Modern Fruit Science". Horticultural Publications. Gainesville, FL.
Duryea, M. L., and Brown, G. N., eds. (1984). "Seedling Physiology and Reforestation Success" Junk, The Hague.
Duryea, M. L., and Landis, T. D. (1984). "Forest Nursery Manual: Production of Bareroot Seedlings." Junk, The Hague.
Goor, A. Y., and Barney, C. W. (1976). "Forest Tree Planting in Arid Zones". Ronald Press, New York.
Hanover, J. W., and Keathley, D. E., eds (1988). "Genetic Manipulation of Woody Plants." Plenum, New York.
Harris, R. W. (1983). "Arboriculture." Prentice-Hall, Englewood Cliffs, NJ.
Hummel, F. C., Palz, W., and Grassi, G., eds. (1988). "Biomass Forestry in Europe: A Strategy for the Future." Elsevier, London and New York.
Kozlowski, T. T. (1971a). "Growth and Development of Trees." Vol. I. Seed Germination, Ontogeny, and Shoot Growth." Academic Press, New York.
Kozlowski, T. T. (1971b). "Growth and Development of Trees. Vol. II. Cambial Growth, Root Growth, and Reproductive Growth." Academic Press, New York.
Kramer, P. J., and Kozlowski, T. T. (1979). "Physiology of Woody Plants." Academic Press, New York.
Maltby, E. (1986). "Waterlogged Wealth". International Institute for Environment and Development. London and Washington, D.C.
Mika, A. (1986). Physiological responses of fruit trees to pruning. *Hort. Rev.*, **8,** 337–378.
Neely, D., and Himelick, E. B. (1987). Fertilizing and watering trees. Illinois Natural History Survey Circular 56, Champaign, IL.
Pritchett, W. L. (1979). "Properties and Management of Forest Soils". Wiley, New York.
Quinlan, J. D. (1981). New chemical approaches to the control of tree form and size. *Acta Hortic.* **120,** 95–106.
Ryugo, K. (1988). "Fruit Culture: Its Science and Art." Wiley, New York.
Samson, J. A. (1984). "Tropical Fruits". Longman, London and New York.
Shepherd, K. R. (1986). "Plantation Silviculture." Martinus Nijhoff, Dordrecht.
Shoemaker, J. S. (1978). "Tree Fruit Production." Avi, Westport, CT.
Smith, D. M. (1986). "The Practice of Silviculture," 8th ed. Wiley, New York.
South, D. B. (1986). Proceedings of the International Symposium on Nursery Management Practices for the Southern Pines. School of Forestry, Auburn University, Auburn, AL.
Steppler, H. A., and Nair, P. K. R. (1987). "Agroforestry: A Decade of Development." International Council for Research in Agroforestry, Nairobi, Kenya.
Swietlik, D., and Faust, M. (1984). Foliar nutrition of fruit crops. *Hort. Rev.* **6,** 287–355.
Tinus, R. W., and McDonald, S. E. (1979). How to grow tree seedlings in containers in greenhouses. USDA Forest Service, Gen. Tech. Rept. RM–6O.
Van der Walk, A., ed. (1989). "Northern Prairie Wetlands." Iowa State University Press, Ames, IA.
Zobel, B. J., and Talbert, J. (1984). "Applied Forest Tree Improvement." Wiley, New York.
Zobel, B. J., and van Buijtenen, J. P. (1989). "Wood Variation: Its Causes and Control." Springer-Verlag, Berlin.
Zobel, B. J., van Wyck, G., and Stahl, P. (1987). "Growing Exotic Trees." Wiley, New York.

Bibliography

Abeles, F. B. (1973). "Ethylene in Plant Biology." Academic Press, New York.

Aber, J. D., Nadelhoffer, K. J., Steudler, P., and Melillo, J. M. (1989). Nitrogen saturation in northern forest ecosystems. *BioScience* **39,** 378–386.

Abod, S. T., Shepherd, V. R., and Bachelard, E. P. (1979). Effects of light intensity, air and soil temperatures on root regenerating potential of *Pinus caribaea* var. *hondurensis* and *Pinus kesiya* seedlings. *Aust. For. Res.* **9,** 173–184.

Abrams, M., and Knapp, A. K. (1986). Seasonal water relations of three gallery forest hardwood species in northeast Kansas. *For. Sci.* **32,** 687–696.

Acevedo, E., Hsiao, T. C., and Henderson, D. W. (1971). Immediate and subsequent growth responses of maize leaves to changes in wter status. *Plant Physiol.* **48,** 631–636.

Ackermann, R. F., and Farrar, J. L. (1965). The effect of light and temperature on the germination of jack pine and lodgepole pine seeds. *Univ. Toronto Fac. For. Tech. Rep.* 5.

Adams, C. M., Dengler, N. G., and Hutchinson, T. C. (1984). Acid rain effects on foliar histology of *Artemisia tilesii. Can. J. Bot.* **62,** 463–474.

Adams, M. B., and Allen, H. L. (1985). Nutrients in foliage of semi-mature loblolly pine. *Plant Soil* **86,** 27–34.

Adams, M. B., and Strain, B. R. (1969). Seasonal photosynthetic rates in stems of *Cercidium floridum* Benth. *Photosynthetica* **3,** 55–62.

Adams, M. B., Allen, H. L., and Davey, C. B. (1986). Accumulation of starch in roots and foliage of loblolly pine (*Pinus taeda* L.): Effects of season, site and fertilization. *Tree Physiol.* **2,** 35–46.

Addicott, F. T. (1982). "Abscission." Univ. of California Press, Berkeley.

Addicott, F. T., and Lyon, J. L. (1973). Physiological ecology of abscission. *In* "Shedding of Plant Parts" (T. T. Kozlowski, ed.), pp. 85–124. Academic Press, New York.

Ahlgren, C. E. (1974a). Introduction. *In* "Fire and Ecosystems" (T. T. Kozlowski and C. E. Ahlgren, eds.), pp. 1–5. Academic Press, New York.

Ahlgren, C. E. (1974b). Effects of fires in temperate forests: North Central United States. *In* "Fire and Ecosystems" (T. T. Kozlowski and C. E. Ahlgren, eds.), pp. 195–223. Academic Press, New York.

Ahlgren, C. E., and Ahlgren, I. (1984). "Lob Trees in the Wilderness." Univ. of Minnesota Press, Minneapolis.

Ahlgren, C. E., and Hansen, H. L. (1957). Some effects of temporary flooding on coniferous trees. *J. For.* **55,** 647–650.

Ahlgren, I.F. (1974). The effects of fire on soil organisms. *In* "Fire and Ecosystems" (T.T. Kozlowski and C.E. Ahlgren, eds.), pp. 47–72. Academic Press, New York.

Alben, A. O. (1958). Waterlogging of subsoil associated with scorching and defoliation of Stuart pecan trees. *Proc. Am. Soc. Hortic. Sci.* **72,** 219–223.

Alberte, R. S., Fiscus, E. L., and Naylor, A. W. (1975). The effects of water stress on the development of the photosynthetic apparatus in greening leaves. *Plant Physiol.* **55,** 317–321.

Alberte, R. S., Thornber, J. P., and Fiscus, E. L. (1977). Water stress effects on the content and organization of chlorophyll in mesophyll and bundle sheath chloroplasts of maize. *Plant Physiol.* **59,** 351–353.

Albertson, F. W., and Weaver, F. E. (1945). Injury and death and recovery of trees in prairie climate. *Ecol. Monogr.* **15,** 393–443.

Alberty, C.A., Pellett, H. M., and Taylor, D. H. (1984). Characterization of soil compaction at construction sites and woody plant response. *J. Environ. Hortic.* **2,** 48–53.

Albrektson, A. (1980). Relations between tree biomass fractions and conventional silvicultural measurements. *Ecol. Bull.* **32,** 315–327.

Alexandre, D. Y. (1980). Caractère saisonnier de la fructification dans une foret hygrophile de Côte d'Ivóire. *Rev. Ecol. (Terre Vie)* **34,** 335–350.

Al-Khatib, K., and Paulsen, G. M. (1989). Enhancement of thermal injury to photosynthesis in wheat plants and thylakoids by high light intensity. *Plant Physiol.* **90,** 1041–1048.

Allaway, W. G., and Milthorpe, F. L. (1976). Structure and functioning of stomata. *In* "Water Deficits and Plant Growth" (T. T. Kozlowski, ed.), Vol. 4, pp. 57–102. Academic Press, New York.

Allen, H. L. (1990). Plant responses to rising carbon dioxide and potential interactions with air pollutants. *J. Environ. Qual.* **19,** 15–34.

Alston, F. H., and Spiegel-Roy, P. (1985). Fruit tree breeding: strategies, achievements and constraints. *In* "Attributes of Trees as Crop Plants" (M. G. R. Cannell and J. E. Jackson, eds.), pp. 49–67. Institute of Terrestrial Ecology, Huntingdon, England.

Alston, R. E., and Turner, B. G. (1963). "Biochemical Systematics." Prentice-Hall, Englewood Cliffs, NJ.

Alvim, P. de T. (1973). Factors affecting flowering of coffee. *In* "Genes, Enzymes and Populations" (A. M. Srb, ed.), Vol. 2, pp. 193–202. Plenum, New York.

Alvim, P. de T. (1977). Cacao. *In* "Ecophysiology of Tropical Crops" (P. de T. Alvim and T. T. Kozlowski, eds.), pp. 279–313. Academic Press, New York.

Alvim, P. de T., and Havis, J. R. (1954). An improved filtration series for studying stomatal opening as illustrated with coffee. *Plant Physiol.* **29,** 97–98.

Alvim, P. de T., and Kozlowski, T. T., eds. (1977). "Ecophysiology of Tropical Crops." Academic Press, New York.

Amman, G. D., McGregor, M. D., Schmitz, R. F., and Oakes, R. D. (1988). Susceptibility of lodgepole pine to infestation by mountain pine beetles following partial cuttings of stands. *Can J. For. Res.* **18,** 688–695.

Amthor, J. S. (1984). Does acid rain directly influence plant growth. *Environ. Pollut., Ser. A* **36,** 1–6.

Andersen, J. L., Campana, R. J., Shigo, A. L., and Shortle, W. C. (1985). Wound response of *Ulmus americana.* 1. Results of chemical injection in attempts to control Dutch elm disease. *J. Arboric.* **11,** 137–142.

Andersen, P. C., Lombard, P. B., and Westwood, M. N. (1984).Leaf conductance, growth, and survival of willow and deciduous fruit tree species under flooded soil conditions. *J. Am. Soc. Hortic. Sci.* **109,** 132–138.

Anderson, J. M., and Osmond, C. B. (1987). Shade–sun responses: Compromises between acclimation and photoinhibition. *In* "Photoinhibition" (D. J. Kyle, C. B. Osmond, and C. J. Arntzen, eds.), pp. 1–38. Elsevier, Amsterdam.

Anderson, J. M., Chow, W. S., and Goodchild, D. J. (1988). Thylakoid membrane organization in sun/shade acclimation. *Aust. J. Plant Physiol.* **15,** 11–26.

Anderson, M. C. (1981). The geometry of leaf distribution in some south-eastern Australian forests. *Agric. Meteorol.* **25,** 195–205.

Andersson, E. (1963). Seed stands and seed orchards in the breeding of conifers. *Proc. World Consult. For. Genet. For. Tree Improve.,* Vol. II, FAO/FORGEN 63C8/1/, pp. 1–18.

Angeles, G., Evert, R. F., and Kozlowski, T. T. (1986). Development of lenticels and adventitious roots in flooded *Ulmus americana* seedlings. *Can. J. For. Res.* **16,** 585–590.

Antonovics, J., Bradshaw, A. D., and Turner, R. G. (1971). Heavy metal tolerance in plants. *Adv. Ecol. Res.* **7,** 1–85.

Appel, D. N., and Stipes, R. J., (1986). A description of declining and blighted pin oaks in eastern Virginia. *J. Arboric.* **12,** 155–158.

Arend, J. L. (1941). Infiltration rates of forest soils in the Missouri Ozarks as affected by woods burning and litter removal. *J. For.* **39,** 726–728.

Armentano, T. V., and Ralston, C. W. (1980). The role of temperate zone forests in the global carbon cycle. *Can. J. For. Res.* **10,** 53–60.

Armitage, F. B. (1985). "Irrigated Forestry in Arid and Semiarid Lands." Int. Dev. Res. Cent., Ottawa, Canada.

Armstrong, W. (1968). Oxygen diffusion from the roots of woody species. *Physiol. Plant.* **21,** 539.

Armstrong, W. (1975). Waterlogged soils. *In* "Environment and Plant Ecology" (J. R. Etherington, ed.), pp. 181–218. Wiley, New York.

Arora, R. K., and Yamadigni, R. (1986). Effect of different doses of nitrogen and zinc sprays on flowering, fruit set and final fruit retention in sweet lime (*C. limettioides* Tanaka). *Haryana Agric. Univ. J. Res.* **16,** 233–239.

Asakawa, S., and Inokuma, T. (1961). Light sensitivity in germination of *Pinus thunbergii* and *Picea glehnii* seeds. *J. Jpn. For. Soc.* **43,** 331–335.

Asen, S. (1977). Flavonoid chemical markers as an adjunct for the identification of cultivars. *Hort. Science* **12,** 410.

Ashby, W. C., and Fritts, H. C. (1972). Tree growth, air pollution, and climate near Laporte, Indiana. *Bull. Am. Meteorol. Soc.* **53,** 246–251.

Ashton, P. S. (1958). Light intensity measurements in rain forest near Santarem, Brazil. *J. Ecol.* **46,** 65–70.

Atkinson, C. J. J., Winner, W. E., and Mooney, H. A. (1988). Gas exchange and SO_2 fumigation studies with irrigated and unirrigated field grown *Diplacus aurantiacus* and *Heteromeles arbutifolia. Oecologia* **75,** 386–393.

Atkinson, D., and Johnson, M. G. (1979). The effect of orchard soil management on the uptake of nitrogen by established apple trees. *Sci. Food Agric.* **30,** 129–135.

Atkinson, D., and White, G. C. (1980). Some effects of orchard soil management on the mineral nutrition of apple trees. *In* "Mineral Nutrition of Fruit Trees" (D. Atkinson, J. E. Jackson, R. O Sharples, and W. M. Waller, eds.), pp. 241–254. Butterworth, London.

Atkinson, D., Jackson, J. E., Sharples, R. O., and Waller, W. M., eds. (1980). "Mineral Nutrition of Fruit Trees." Butterworth, London.

Attiwill, P. M. (1982). Phosphorous in Australian forests. *In* "Phosphorous in Australia" (A. B. Costin and C. H. Williams, eds.), pp. 113–134. School of Botany, University of Melbourne, Australia.

Auclair, D. (1976). Effets des poussières sur la photosynthèse. I. Effets des poussières de ciment et de chârbon sur la photosynthèse de l'epiceá. *Ann. Sci. For.* **33,** 247–255.

Auclair, D. (1977). Effets des poussières sur la photosynthèse. II. Influence des polluants particulaires sur la photosynthèse du pin sylvestre et du peuplier. *Ann. Sci. For.* **34,** 47–57.

Audus, L. J. (1972). "Plant Growth Substances. Vol. I. Chemistry and Physiology." Barnes and Noble, New York.

Augé, R. M., Schekel, K. A., and Wample, R. L. (1986). Osmotic adjustment in leaves of VA mycorrhizal and nonmycorrhizal rose plants in response to drought stress. *Plant Physiol.* **82,** 765–770.

Augsberger, C. K. (1984). Light requirement of neotropical tree seedlings: A comparative study of growth and survival. *J. Ecol.* **72,** 777–795.

Aussenac, G., and Granier, A. (1988). Effects of thinning on water stress and growth in Douglas-fir. *Can. J. For. Res.* **18,** 100–105.

Avery, D. J. (1966). The supply of air to leaves in assimilation chambers. *J. Exp. Bot.* **17,** 655–677.

Avery, D. J. (1977). Maximum photosynthetic rate—A case study in apple. *New Phytol.* **78,** 55–63.

Avery, T. E., and Burkhart, H. E. (1983). "Forest Measurements." McGraw-Hill, New York.

Axelrod, M. C., Coyne, P. I., Bingham, G. E., Kircher, J. R., Miller, P. R., and Hung, R. C. (1980). Canopy analysis of pollutant injured ponderosa pine in the San Bernardino National Forest. *Gen. Tech. Rep. PSW (Pac. Southwest For. Range Exp. Stn.)* **PSW-43,** 227.

Axelsson, E., and Axelsson, B. (1986). Changes in carbon allocation patterns in spruce and pine trees following irrigation and fertilization. *Tree Physiol.* **2,** 189–204.

Ayers, J. C., and Barden, J. A. (1975). Net photosynthesis and dark respiration of apple leaves as affected by pesticides. *J. Am. Soc. Hortic. Sci.* **100,** 24–28.

Aylor, D. (1978). Dispersal in time and space: Aerial pathogens. *In* "Plant Disease" (J. G. Horsfall and E. B. Cowling, eds.), Vol. 2, pp. 159–180. Academic Press, New York.

Ayres, P. G. (1978). Water relations of diseased plants. *In* "Water Deficits and Plant Growth" (T. T. Kozlowski, ed.), Vol 5, pp. 1–60. Academic Press, New York.

Ayres, P. G. (1984). The interactions between environmental stress injury and biotic disease physiology. *Annu. Rev. Phytopathol.* **22,** 53–75.

Azevedo, J., and Morgan, D. L. (1974). Fog precipitation in coastal California forests. *Ecology* **55,** 1135–1141.

Babalola, O., Boersman, L., and Youngberg, C. T. (1968). Photosynthesis and transpiration of Monterey pine seedlings as a function of soil water suction and soil temperature. *Plant Physiol.* **43,** 515–521.

Bacon, G. J., and Hawkins, P. J. (1977). Studies on the establishment of open root Caribbean pine planting stock in southern Queensland. *Aust. For.* **40,** 173–181.

Bacone, J., Bazzaz, F. A., and Boggess, W. R. (1976). Correlated photosynthetic responses and habitat factors of two successional tree species. *Oecologia* **23,** 63–74.

Baer, N. W. (1989). Shelterbelts and windbreaks in the Great Plains. *J. For.* **87,** 32–36.

Bahari, Z. A., Pallardy, S. G., and Parker, W. C. (1985). Photosynthesis, water relations, and drought adaptation in six woody species of oak–hickory forests in central Missouri. *For. Sci.* **31,** 557–569.

Bajaj, Y. P. S., ed. (1986). "Biotechnology in Agriculture and Forestry. I. Trees." Springer-Verlag, Berlin.

Baker, F. S. (1950). "Principles of Silviculture." McGraw-Hill, New York.

Baker, J. B. (1973). Bedding and fertilization influence on slash pine development in the Florida sandhills. *For. Sci.* **19,** 135–138.

Baker, J. M., and van Bavel, C. H. M. (1986). Resistance of plant roots to water loss. *Agron. J.* **78,** 641–644.

Baker, N. R., and Long, S. P., eds. (1986). "Photosynthesis in Contrasting Environments." Elsevier, New York.

Ballard, R. (1984). Fertilization of plantations. *In* "Nutrition of Plantation Forests" (G. D. Bowen and E. K. S. Nambiar, eds.), pp. 327–360. Academic Press, Toronto.

Bannan, M. W., and Bindra, M. (1970). The influence of wind on ring width and cell length in conifer stems. *Can. J. Bot.* **48,** 255–259.

Barber, D. A., and Martin, J. K. (1976). The release of organic substances by cereal roots. *New Phytol.* **76,** 69–80.

Bargioni, G., Loreti, F., and Pisani, P. L. (1985). Ten years of research on peach and nectarine in a high density system in the Verona area. *Acta Hortic.* **173,** 299–309.

Bargioni, G., Loreti, F., and Pisani, P. L. (1986). Performance of Suncrest and Stark Redgold established at different densities. *Acta Hortic.* **160,** 319–320.

Barker, J. E. (1979). Growth and wood properties of *Pinus radiata* in relation to applied ethylene. *N. Z. J. For. Sci.* **9,** 15–19.

Barley, K. P. (1963). Influence of soil strength on growth of roots. *Soil Sci.* **96,** 175–180.

Barley, K. P. (1970). The configuration of the root system in relation to nutrient uptake. *Adv. Agron.* **22,** 159–201.

Barlow, P. W. (1982). "The plant forms cells, not cells the plant": The origin of de Bary's aphorism. *Ann. Bot. (London)* [N.S.] **49,** 269–271.

Barnard, G. W. (1987). Woodfuel in developing countries. *In* "Biomass" (D. O. Hall and R. P. Overend, eds.), pp. 349–366. Wiley, New York.

Barner, J. (1961). Wirkungen von organischen und anorganischen Fungiziden auf die innere Blattstruktur und Stoffproduktion der Pflanzen. *Mitt. Biol. Bundesanst. Land.- Forstwirtsch., Berlin-Dahlem* **104,** 178–183.

Barnes, R. L. (1963). Organic nitrogen compounds in tree xylem sap. *For. Sci.* **9,** 98–102.

Barnes, R. L. (1972). Effects of chronic exposure to ozone on photosynthesis and respiration of pines. *Environ. Pollut.* **3,** 133–138.

Barnett, J. P. (1984). "Top Pruning and Needle Clipping of Container-grown Southern Pine Seedlings," Proc. Southern Nursery Conf., pp. 39–45. U.S.D.A. For. Serv., New Orleans, Louisiana.

Barney, C. W. (1951). Effects of soil temperature and light intensity on root growth of loblolly pine seedlings. *Plant Physiol.* **26,** 146–163.

Baskerville, G. L. (1965a). Dry matter production in immature stands. *For. Sci. Monogr.* **9,** 362–478.

Baskerville, G. L. (1965b). Estimation of dry weight of tree components and total standing crop in conifer stands. *Ecology* **46,** 867–869.

Baskerville, G. L. (1966). Dry-matter production in immature balsam fir (*Abies balsamea*) stands: Roots, lesser vegetation, and total stand. *For. Sci.* **12,** 49–53.

Bassett, J. R. (1964a). Diameter growth of loblolly pine trees as affected by soil moisture availability. *U.S., For. Serv. Res. Pap. SO* **SO–59.**

Bassett, J. R. (1964b). Tree growth as affected by soil moisture availability. *Soil Sci. Soc. Am. Proc.* **28,** 436–438.

Bassett, J. R. (1966). Seasonal diameter growth of loblolly pines. *J. For.* **64,** 674–676.

Bate, G. C., and Canvin, D. T. (1971). A gas-exchange system for measuring the productivity of plant populations in controlled environments. *Can. J. Bot.* **49,** 601–608.

Bates, L. M., and Hall, A. E. (1981). Stomatal closure with soil water depletion not associated with changes in bulk leaf water status. *Oecologia* **50,** 62–65.

Bauer, H., Wiener, R., Hatheway, W. H., and Larcher, W. (1985). Photosynthesis of *Coffea arabica* after chilling. *Physiol. Plant.* **64,** 449–454.

Bazzaz, F. A. (1979). The physiological ecology of plant succession. *Annu. Rev. Ecol. Syst.* **10,** 351–371.

Bazzaz, F. A. (1983). Characteristics of populations in relation to disturbance in natural and non-modified ecosystems. *In* "Disturbance and Ecosystems: Components of Response" (H. A. Mooney and M. Godron, eds.), pp. 259–275. Springer-Verlag, New York.

Bazzaz, F. A. (1984). Dynamics of wet tropical forests and their species strategies. *In* "Physiological Ecology of Plants of the Wet Tropics" (E. Medina, H. A. Mooney, and C. Vazquez-Yanes, eds.), pp. 233–243. Junk, The Hague.

Bazzaz, F. A., and Carlson, R. W. (1984). The response of plants to elevated CO_2. I. Competition among an assembly of annuals at different levels of soil moisture. *Oecologia* **62,** 196–198.

Bazzaz, F. A., and Garbutt, K. (1988). The response of annuals in competitive neighborhoods: Effects of elevated CO_2. *Ecology* **69,** 937–946.

Bazzaz, F. A., and Peterson, D. L. (1984). Photosynthetic and growth responses of silver maple (*Acer saccharinum* L.) seedlings to flooding. *Am. Midl. Natur.* **112,** 261–272.

Bazzaz, F. A., and Pickett, S. T. A. (1980). The physiological ecology of tropical succession: A comparative review. *Annu. Rev. Ecol. Syst.* **11,** 287–310.

Bazzaz, F. A., and Sipe, T. W. (1987). Physiological ecology, disturbance, and ecosystem recovery. *In* "Potentials and Limitations of Ecosystem Analysis" (E.-D. Schulze and H. Zwölfer, eds.), pp. 203–227. Springer-Verlag, New York.

Bazzaz, F. A., Carlson, R. W., and Harper, J. L. (1979). Contribution to reproductive effort by photosynthesis of flowers and fruits. *Nature (London)* **279,** 554–555.

Beadle, C. L., and Jarvis, P. G. (1977). The effects of shoot water status on some photosynthetic partial processes in Sitka spruce. *Physiol. Plant.* **41,** 7–13.

Beadle, C. L., Jarvis, P. G., and Neilson, R. E. (1979). Leaf conductance as related to xylem water potential and carbon dioxide concentration in Sitka spruce. *Physiol Plant.* **45,** 158–166.

Beard, J. S. (1946). The natural vegetation of Trinidad. *Oxford For. Mem.* **20.**

Beardsell, M. F., and Cohen, D. (1974). Endogenous abscisic acid–plant water stress relationships under controlled environmental conditions. *In* "Mechanisms of Regulation of Plant Growth" (R. L. Bieleski, A. R. Ferguson, and M. M. Creswell, eds.). Bull. 12, pp. 411–415. Royal Society of New Zealand, Wellington.

Beaton, J. D. (1959). The influence of burning on the soil in the timber area of Lac le Jeune, British Columbia. I. Physical properties. II. Chemical properties. *Can. J. Soil Sci.* **39,** 1–5, 6–11.

Beaufait, W. R. (1960). Some effects of high temperatures on the cones and seeds of jack pine. *For. Sci.* **6,** 194–199.

Bedinger, M. S. (1981). Hydrology of bottomland hardwood forests of the Mississippi embayment. *In* "Wetlands of Bottomland Hardwood Forests" (J. R. Clark and J. Benforado, eds.), pp. 161–176. Elsevier, New York.

Begg, J. E. (1980). Morphological adaptations of leaves to water stress. *In* "Adaptation of Plants to Water and High Temperature Stress" (N. C. Turner and P. J. Kramer, eds.), pp. 33–42. Wiley, New York.

Bell, A. A. (1981). Biochemical mechanisms of disease resistance. *Annu. Rev. Plant Physiol.* **32,** 21–81.

Bell, A. A. (1982). Plant–pest interaction with environmental stress and breeding for pest resistance: Plant diseases. *In* "Breeding Plants for Less Favorable Environments" (M. N. Christiansen and C. F. Lewis, eds.), pp. 335–363. Wiley, New York.

Bell, T. I. W. (1968). Effect of fertilizer and density pretreatment on spruce seedling survival and growth. *For. Rec.* **67,** 1–67.

Bella, I. E. (1971). A new competition model for individual trees. *For. Sci.* **17,** 364–372.

Benecke, U. (1980). Photosynthesis and transpiration of *Pinus radiata* under natural conditions in a forest stand. *Oecologia* **44,** 192–198.

Bengtson, G. W. (1979). Forest fertilization in the United States: Progress and outlook. *J. For.* **78,** 222–229.

Bennett, C. W. (1956). Biological relations of plant viruses. *Annu. Rev. Plant. Physiol.* **7,** 143–170.

Bennett, J. H. (1981). Photosynthesis and gas diffusion in leaves of selected crop plants exposed to ultraviolet-B radiation. *J. Environ. Qual.* **10,** 271–275.

Bennett, J. H., Lee, E. H., and Heggestad, H. E. (1984). Biochemical aspects of plant tolerance to ozone and oxyradicals: Superoxide dismutase. *In* "Gaseous Air Pollutants and Plant Metabolism" (M. J. Koziol and F. R. Whatley, eds.), pp. 413–424. Butterworth, London.

Bennett, J. P., Resh, H. R., and Runeckles, V. C. (1974). Apparent stimulation of plant growth by air pollutants. *Can. J. Bot.* **52,** 35–41.

Berlyn, G. P. (1961). Recent advances in woody anatomy. The cell wall in secondary xylem. *For. Prod. J.* **14,** 467–476.

Berry, F. H. (1969). Decay in the upland oak stands of Kentucky. *USDA For. Serv. Res. Pap. NE* **NE–126.**

Berry, J., and Björkman, O. (1980). Photosynthetic response and adaptation to temperature in higher plants. *Annu. Rev. Plant Physiol.* **31,** 491–543.

Berry, J. A. (1975). Adaptation of photosynthetic processes to stress. *Science* **188,** 644–650.

Berryman, A. A. (1986). "Forest Insects, Principles and Practice of Population Management." Plenum, New York.

Bervaes, J. C. A. M., Ketchie, D. O., and Kuiper, P. J. C. (1978). Cold hardiness of pine needles and apple bark as affected by alteration of day length and temperature. *Physiol. Plant.* **44,** 365–368.

Bevege, D. I. (1984). Wood yield and quality in relation to tree nutrition. *In* "Nutrition of Plantation Forests" (G. D. Bowen and E. K. S. Nambiar, eds.), pp. 293–326. Academic Press, London.

Bewley, J. D. (1979). Physiological aspects of desiccation tolerance. *Annu. Rev. Plant Physiol.* **30,** 195–238.

Bewley, J. D., Larsen, K. M., and Papp, J. E. T. (1983). Water-stress-induced changes in the pattern of protein synthesis in maize seedling mesocotyls. A comparison with the effects of heat shock. *J. Expt. Bot.* **34,** 1126–1133.

Beyers, E., and Terblanche, J. H. (1971). Identification and control of trace element deficiencies. *Decid. Fruit Grower* **21,** 132–137.

Bhattacharya, N. C., Biswas, P. K., Bhattacharya, S., Sionit, N., and Strain, B. R. (1985). Growth and yield response of sweet potato to atmospheric CO_2 enrichment. *Crop Sci.* **25,** 975–981.

Bidwell, R. G. S. (1983). Carbon nutrition of plants: Photosynthesis and respiration. *In* "Plant Physiology" (F. C. Steward and R. G. S. Bidwell, eds.), Vol. 7, pp. 287–457. Academic Press, New York.

Bier, J. E. (1959). The relation of bark moisture to the development of canker diseases caused by native, facultative parasites. I. Cryptodiaporthe canker on willow. *Can J. Bot.* **37,** 229–238.

Billings, W. D. (1978). "Plants and the Ecosystem." Wadsworth, Belmont, California.

Billings, W. D., Peterson, K. M., Loken, J. O., and Mortensen, D. A. (1984). Interaction of increasing carbon dioxide and soil nitrogen on the carbon balance of tundra microcosms. *Oecologia* **65,** 26–29.

Bingham, R. T., and Squillace, A. E. (1957). Phenology and other features of flowering of pines, with special reference to *Pinus monticola,* Dougl. North Region, *U.S For. Serv., Intermt. For. Range Exp. Stn., Res. Pap.* **53.**

Binkley, D. (1986). "Forest Nutrition Management." Wiley, New York.

Binkley, D., and Reid, P. (1984). Long-term responses of stem growth and leaf area to thinning and fertilization in a Douglas-fir plantation. *Can. J. For. Res.* **14,** 656–660.

Binkley, D., Cromack, K., Jr., and Fredriksen, R. L. (1982). Nitrogen accretion and availability in some snowbush ecosystems. *For. Sci.* **28,** 720–724.

Bird, R. (1981). The benefits of tree planting and factors which influence the effectiveness of shelterbelts. *Trees, Victoria's Resour.* **23,** 4–6.

Birkholz-Lambrecht, A. F., Lester, D. T., and Smalley, E. B. (1977). Temperature, host genotype, and fungus genotype in early testing for Dutch elm disease resistance. *Plant Dis. Rep.* **51,** 238–242.

Biswell, H. H. (1974). Effects of fire on chaparral. *In* "Fire and Ecosystems" (T. T. Kozlowski and C. E. Ahlgren, eds.), pp. 321–364. Academic Press, New York.

Biswell, H. H. (1989). "Prescribed Burning in California Wildlands Vegetation Management." Univ. of California Press, Berkeley.

Biswell, H. H., and Schultz, A. M. (1957). Surface runoff and erosion as related to prescribed burning. *J. For.* **55,** 372–375.

Björkman, E. (1970). Forest tree mycorrhiza: The conditions for its formation and the significance for tree growth and afforestation. *Plant Soil* **32,** 589–610.

Björkman, O. (1971). Interaction between the effects of oxygen and CO_2 concentration on quantum yield and light-saturated rate of photosynthesis in leaves of *Atriplex patula* ssp. *spicata. Year Book—Carnegie Inst. Washington* **70,** 520.

Björkman, O., and Powles, S. B. (1984). Inhibition of photosynthetic reactions under water stress: Interaction with light level. *Planta* **161,** 490–504.

Björkman, O., Badger, M. R., and Armond, P. A. (1980). Response and adaptation of photosynthesis to high temperatures. *In* "Adaptation of Plants to Water and High Temperature Stress" (N. C. Turner and P. J. Kramer, eds.), pp. 233–249. Wiley (Interscience), New York.

Black, M., and Wareing, P. F. (1955). Growth studies in woody species. VII. Photoperiodic control of germination in *Betula pubescens* Ehrh. *Physiol. Plant.* **8,** 300–316.

Blackmon, B. G. (1977). Cottonwood response to nitrogen related to plantation age and site. *Res. Note SO (U.S. For. Serv.)* **SO-229.**

Blaker, N.S., and MacDonald, J. D. (1981). Predisposing effects of soil moisture extremes on the susceptibility of rhododendron to *Phytophthora* root and crown rot. *Phytopathology* **71,** 831–834.

Blanche, C. A., Hodges, J. D., and Nebeker, T. E. (1985). A leaf area–sapwood area ratio developed to rate loblolly pine tree vigour. *Can. J. For. Res.* **15,** 1181–1184.

Blanke, M. M., and Lenz, F. (1989). Fruit photosynthesis. *Plant, Cell Environ.* **12,** 31–46.

Boardman, K. (1977). Comparative photosynthesis of sun and shade plants. *Annu. Rev. Plant Physiol.* **28,** 355–377.

Boerner, R. E. J. (1982). Fire and nutrient cycling in temperate ecosystems. *BioScience* **32,** 187–192.

Boggie, R. (1972). Effect of water table height on root development of *Pinus contorta* on deep peat in Scotland. *Oikos* **23,** 304–312.

Bollard, E. G. (1958). Nitrogenous compounds in tree xylem sap. *In* "The Physiology of Forest Trees" (K. V. Thimann, ed.), pp. 83–93. Ronald Press, New York.

Bond, G. (1976). The results of the IBP survey of root nodule formation in non-leguminous angiosperms. *In* "Symbiotic Nitrogen Fixation in Plants" (P. S. Nutman, ed.), pp. 443–474. Cambridge Univ. Press, London and New York.

Bonga, J. M., and Durzan, D. J., eds. (1982). "Tissue Culture in Forestry". Martinus Nijhoff Dr. W. Junk, The Hague, Netherlands.

Bongi, G., and Long, S. P. (1987). Light-dependent damage to photosynthesis in olive leaves during chilling and high temperature stress. *Plant, Cell Environ.* **10,** 241–249.

Bonte, J. (1982). Effects of air pollutants on flowering and fruiting. *In* "Effects of Gaseous Air Pollution in Agriculture and Horticulture" (M. H. Unsworth and D. P. Ormrod, eds.), pp. 207–223. Butterworth, London.

Borchert, R. (1973). Simulation of rhythmic tree growth under constant conditions. *Physiol. Plant.* **29,** 173–180.

Borchert, R. (1980). Phenology and ecophysiology of a tropical tree, *Erythrina poeppigiana* D. F. Cook. *Ecology* **61,** 1065-1071.

Borger, G. A. (1974). Development and shedding of bark. *In* "Shedding of Plant Parts" (T. T. Kozlowski, ed.), pp. 205–236. Academic Press, New York.

Bormann, B. T. (1983). Ecological implications of phytochrome-mediated seed germination in red alder. *For. Sci.* **4,** 734-738.

Bormann, F. H. (1953). Factors determining the role of loblolly pine and sweetgum in early old-field succession in the Piedmont of North Carolina. *Ecol. Monogr.* **23,** 339–358.

Bormann, F. H. (1956). Ecological implications of changes in the photosynthetic response of *Pinus taeda* seedlings during ontogeny. *Ecology* **37,** 70–75.

Bormann, F. H., and Kozlowski, T. T. (1962). Measurement of tree ring growth with dial-gauge dendrometers and vernier tree ring bands. *Ecology* **43,** 289–294.

Bormann, F. H., and Likens, G. E. (1979). "Pattern and Process in a Forested Ecosystem." Springer-Verlag, New York.

Bormann, F. H., Likens, G. E., and Melillo, J. M. (1977). Nitrogen budget for an aggrading northern hardwood forest ecosystem. *Science* **196,** 981–983.

Boss, W. F. and Morré, D. J., eds. (1989). "Second Messengers in Plant Growth and Development." Liss, New York.

Botkin, D. B., Smith, W. H., Carlson, R. W., and Smith, T. L. (1972). Effects of ozone on white pine saplings. Variation in inhibition and recovery of net photosynthesis. *Environ. Pollut.* **3,** 273–289.

Boubals, D., Vergnes, A., and Bobo, H. (1955). Essais de fongicides organique dans la lutte contre le mildious de la Vigne effectuées en 1954. *Prog. Agric. Vitic.* **143,** 64–74.

Bourdeau, P. F. (1954). Oak seedling ecology determining segregation of species in Piedmont oak-hickory forests. *Ecol. Monogr.* **24,** 297–320.

Bourdeau, P. F., and Laverick, M. L. (1958). Tolerance and photosynthetic adaptability to light intensity in white pine, red pine, hemlock, and ailanthus seedlings. *For. Sci.* **4,** 196–207.

Bowen, G. D. (1984). Tree roots and the use of soil nutrients. *In* "Nutrition of Plantation Forests" (G. D. Bowen and E. K. S. Nambiar, eds.), pp. 147–179. Academic Press, London.

Bowen, G. D. (1985). Roots as a component of tree productivity. *In* "Attributes of Trees as Crop Plants" (M. G. R. Cannell and J. E. Jackson, eds.), pp. 303–315. Institute of Terrestrial Ecology, Huntingdon, England.

Bowen, G. D., and Nambiar, E. K. S., eds. (1984). "Nutrition of Forest Plantations." Academic Press, Orlando, Florida.

Bowman, W. D., and Roberts, S. W. (1985). Seasonal changes in tissue elasticity in chaparral shrubs. *Physiol. Plant.* **65,** 233–236.

Bowman, W. D., and Strain, B. R. (1987) Interaction between CO_2 enrichment and salinity stress in the C_4 non-halophyte. *Andropogon glomeratus* (Walter) BSP. *Plant, Cell Environ.* **10,** 267–270.

Boyce, J. S. (1948) "Forest Pathology." McGraw-Hill, New York.

Boyce, S. G. (1954). The salt spray community. *Ecol. Monogr.* **24,** 29–67.

Boyce, S. G., and Cost, N. D. (1974). Timber potentials in the wetland hardwoods. *In* "Water Resources, Utilization, and Conservation in the Environment" (M. C. Blount, ed.), pp. 130–151. Taylor Printing Co., Reynolds, Georgia.

Boyer, J. S. (1968). Relationships of water potential to growth of leaves. *Plant Physiol.* **43,** 1056–1062.

Boyer, J. S. (1970). Leaf enlargement and metabolic rates in corn, soybean, and sunflower at various leaf water potentials. *Plant Physiol.* **46,** 236–239.

Boyer, J. S. (1971a). Nonstomatal inhibition of photosynthesis in sunflower at low leaf water potentials and high light intensities. *Plant Physiol.* **48,** 532–536.

Boyer, J. S. (1971b). Recovery of photosynthesis in sunflower after a period of low leaf water potential. *Plant Physiol.* **47,** 816–820.

Boyer, J. S. (1985a). Water transport. *Annu. Rev. Plant Physiol.* **36,** 473–516.

Boyer, J. S. (1985b). Molecular mechanisms of photosynthetic response to low water potentials. *Plant Physiol.* **77,** Suppl., 18.

Boyer, J. S., Armond, P. A., and Sharp, R. E. (1987). Light stress and leaf water relations. *In* "Photoinhibition" (D. J. Kyle, C. B. Osmond, and C. J. Arntzen, eds.), pp. 111–122. Elsevier, Amsterdam.

Boyle J. R. (1975). Nutrients in relation to intensive culture of forest crops. *Iowa State J. Res.* **49,** 297–303.

Boynton, D. (1940). Soil atmosphere and the production of new rootlets by apple tree root systems. *Proc. Am. Soc. Hortic. Sci.* **37,** 19–26.

Boynton, D., DeVilliers, J. I., and Reuther, W. (1938). Are there different critical oxygen levels for the different phases of root activity? *Science* **88,** 569–570.

Boysen-Jensen, P. (1929). Studier over Skovtraeernes Forhold til Lyset. *Dan. Skovforen. Tidsskr.* **14,** 5–31.

Bozarth, C. S., Mullet, J. E., and Boyer, J. S. (1987). Cell wall proteins at low water potentials. *Plant Physiol.* **85,** 261–267.

Bradbury, I. K., and Malcolm, D. C. (1978). Dry matter accumulation by *Picea sitchensis* seedlings during winter. *Can. J. For. Res.* **8,** 208–213.

Bradford, K. J. (1983a). Effects of soil flooding on leaf gas exchange of tomato. *Plant Physiol.* **73,** 475–479.

Bradford, K. J. (1983b). Involvement of plant growth substances in the alteration of leaf gas exchange of flooded tomato plants. *Plant Physiol.* **73,** 480–483.

Bradshaw, A. D., Goode, D. A., and Thorpe, E. H. P., eds. (1986). "Ecology and Design in Landscape." Blackwell, Oxford.

Braekke, F. H., and Kozlowski, T. T. (1975). Shrinking and swelling of stems of *Pinus resinosa* and *Betula papyrifera* in northern Wisconsin. *Plant Soil* **43,** 387–410.

Bramlage, W. J., and Thompson, A. H. (1962). The effects of early-season sprays of boron on fruit set, color, finish, and storage life of apples. *Proc. Am. Soc. Hortic. Sci.* **80,** 64–72.

Bramlage, W. J., Drake, M., and Lord, W. J. (1980). The influence of mineral nutrition on the quality and storage performance of pome fruits grown in North America. *In* "Mineral Nutrition of Fruit Trees" (D. Atkinson, J. E. Jackson, R. O. Sharples, and W. M. Waller, eds.), pp. 29–39. Butterworth, London.

Bramlett, D. L. (1972). Cone crop development records for six years in shortleaf pine. *For.* Sci. **18,** 31–33.

Brand, D. G., Weetman, G. F., and Rehsler, P. (1987). Growth analysis of perennial plants: The relative production rate and its yield components. *Ann. Bot.* (*London*) [N.S.] **59,** 45–53.

Brandle, J. R., Campbell, W. F., Sisson, W. B., and Caldwell, M.M. (1977). Net photosynthesis, electron transport capacity, and ultrastructure of *Pisum sativum* L. exposed to ultraviolet-B radiation. *Plant Physiol.* **60,** 165–169.

Brasseur, G. P., Farman, J. C., Isaksen, I. S. A., Kruger, B. G., Labitzke, J. D., Mahlman, J. D., McCormick, M. P., Solomon, P., Stolarski, R. S., Turco, R., and Watson, R. T. (1988). Group report: Changes in Antarctic ozone. *In* "The Changing Atmosphere" (F. S. Rowland and I. S. A. Isaksen, eds.), pp. 235–256. Wiley, Chichester.

Braun, G. (1977a). Über die Ursachen und Kriterien der Immissionsresistenz bei Fichte, *Picea abies* (L.) Karst. I. Morphologischanatomische Immissionsresistenz. *Eur. J. For. Pathol.* **7,** 23–43.

Braun, G. (1977b). Über die Ursachen und Kriterien der Immissionsresistenz bei Fichte, *Picea abies* (L.) Karst. II. Reflecktorische Immissionsresistenz. *Eur. J. For. Pathol.* **7,** 129–152.

Bray, J. R., and Gorham, E. (1964). Litter production in forests of the world. *Adv. Ecol. Res.* **2,** 101–157.

Bresler, E. (1977). Trickle-drip irrigation: Principles and application to soil-water management. *Adv. Agron.* **29,** 344–393.

Brewer, R. F., Garber, M. J., Guillemet, F. B., and Sutherland, F. H. (1967). The effect of accumulated fluoride on yields and fruit quality of Washington navel oranges. *Proc. Am. Soc. Hortic. Sci.* **91,** 150–156.

Bridgewater, F. E., and Franklin, E. C. (1985). Forest tree breeding: Strategies, achievements and constraints. *In* "Attributes of Trees as Crop Plants" (M. G. R. Cannell and J. E. Jackson, eds.), pp. 26–48. Institute of Terrestrial Ecology, Huntingdon, England.

Briggs, G. M., Jurik, T. W., and Gates, D. M. (1986). A comparison of rates of above ground growth and carbon dioxide assimilation by aspen on sites of high and low quality. *Tree Physiol.* **2,** 29–34.

Brink, V. C. (1954). Survival of plants under flood in the lower Fraser River Valley, B. C. *Ecology* **35,** 94–95.

Brix, H. (1960). Determination of viability of loblolly pine seedlings after wilting. *Bot Gaz. (Chicago)* **121,** 220–223.

Brix, H. (1962). The effect of water stress on the rates of photosynthesis and respiration in tomato plants and loblolly pine seedlings. *Physiol. Plant.* **15,** 10–20.

Brix, H. (1979). Effects of plant water stress on photosynthesis and survival of four conifers. *Can. J. For. Res.* **9,** 160–165.

Brix, H. (1983). Effects of thinning and nitrogen fertilization on growth of Douglas-fir: Relative contributions of foliage quality and efficiency. *Can. J. For. Res.* **13,** 167–175.

Brix, H., and Mitchell, A. K. (1986). Thinning and nitrogen fertilization effects on soil and tree water stress in a Douglas-fir stand. *Can. J. For. Res.* **16,** 1334–1338.

Broadfoot, W. M., and Williston, H. L. (1973). Flooding effects on southern forests. *J. For.* **71,** 484–487.

Brokaw, N. V. L. (1985). Treefalls, regrowth, and community structure in tropical forests. *In* "The Ecology of Natural Disturbance and Patch Dynamics" (S. T. A. Pickett and P. S. White, eds.), pp. 53–69. Academic Press, Orlando, Florida.

Brooks, M G. (1951). Effect of black walnut trees and their products on other vegetation. *Bull.—W. Va., Agric. Exp. Stn.* **347.**

Brough, D. H., Jones, H. G., and Grace, J. (1986). Diurnal changes in water content of the stems of apple trees as influenced by irrigation. *Plant, Cell Environ.* **9,** 1–7.

Browder, J. A., and Schroeder, P. B. (1981). "Melaleuca Seed Dispersal and Perspectives on Control," Proc. Melaleuca Symp., 1980, pp. 17–21. South. Fla. Environ. Res. Found., Miami.

Brown, A. A., and Davis, K. P. (1973). "Forest Fire: Control and Use." McGraw-Hill, New York.

Brown, C. L., and Kormanik, P. P. (1967). Suppressed buds on lateral roots of *Liquidambar styraciflua*. *Bot. Gaz. (Chicago)* **128,** 208–211.

Brown, G. N., and Bixby, J. A. (1975). Soluble and insoluble protein patterns during induction of freezing tolerance in black locust seedlings. *Physiol. Plant.* **34,** 187–191.

Brown, K., and Higginbotham, K. O. (1986). Effects of carbon dioxide enrichment and nitrogen supply on growth of boreal tree seedlings. *Tree Physiol.* **2,** 223–232.

Brown, K. M., and Leopold, A. C. (1973). Ethylene and the regulation of growth in pine. *Can. J. For. Res.* **3,** 143–145.

Brown, R. T. (1967). Influence of naturally occurring compounds on germination and growth of jack pine. *Ecology* **48,** 542-546.

Brown, S., and Lugo, A. E. (1984). Biomass of tropical forests: A new estimate based on forest volume. *Science* **223,** 1290–1293.

Bryson, R. A., and Murray J. T. (1977). "Climates of Hunger." Univ. of Wisconsin Press, Madison.

Bucher, J. B., and Keller, T. (1978). Einwirkungen niedriger SO_2-Konzentrationen im mehrwochigen Begasungsversuch auf Waldbäume. *Ber. Ver. Dtsch. Ing.* **314,** 237–242.

Bunce, J. A. (1977). Leaf elongation in relation to leaf water potential in soybean. *J. Exp. Bot.* **28,** 156–161.

Bunce, J. A., Miller, L. N., and Chabot, B. F. (1977). Competitive exploitation of soil water by five eastern North American tree species. *Bot. Gaz. (Chicago)* **138,** 168–173.

Burger, D. W., Hartin, J. S., Hodel, D. R., Lukazewski, T. A., Tjosvold, S. A., and Wagner, S. A. (1987). Water use in California ornamental nurseries. *Calif. Agric.* **41**(9 and 10), 7–8.

Burke, J. J., and Orzech, K. A. (1988). The heat-shock response in higher plants: A biochemical model. *Plant, Cell Environ.* **11,** 441–444.

Burns, G. P. (1923). Studies on the tolerance of New England forest trees. IV. Minimum light referred to a standard. *Bull.—Vt., Agric. Exp. Stn.* **193.**

Burns, R. M., and Hebb, E. A. (1972). Site preparation and reforestation of droughty, acid sands. *U.S., Dep. Agric., Agric. Handb.* **426.**

Burrows, F. M. (1975). Wind-borne seed and fruit movement. *New Phytol.* **75,** 405–418.

Burton, J. D., and Smith, D. M. (1972). Guying to prevent windsway influences loblolly pine growth and wood properties. *U.S., For. Serv., Res. Pap. SO* **S0–80.**

Butin, H., and Shigo, A. L. (1981). Radial shakes and "frost cracks" in living oak trees. *USDA For. Serv. Res. Pap. NE* **NE–478.**

Butler, J. D. (1986). Grass interplanting in horticultural cropping systems. *HortScience* **21,** 394–397.

Byram, G. M., and Doolittle, W. T. (1950). A year of growth for a shortleaf pine. *Ecology* **31,** 27–35.

Caborn, J.M. (1957). Shelterbelts and microclimates. *For. Comm. Bull.* **29.**

Caborn, J. M. (1965). "Shelterbelts and Windbreaks." Faber & Faber, London.

Caldwell, M. M. (1970). The effect of wind on stomatal aperture, photosynthesis, and transpiration of *Rhododendron ferrugineum* L. and *Pinus cembra* L. *Centralbl. Gesamte Forstwes,* **87,** 193–201.

Caldwell, M. M. (1976). Root extension and water absorption. *In* "Water and Plant Life" (O. L. Lange, L. Kappen, and E.-D. Schulze, eds.), pp. 63–85. Springer-Verlag, Berlin.

Caldwell, M. M. (1981). Plant response to solar ultraviolet radiation. *Encycl. Plant Physiol., New Ser.* **12A,** 169–198.

Caldwell, M. M., and Richards, J. H. (1989). Hydraulic lift: Water efflux from upper roots improves effectiveness of water uptake by deep roots. *Oecologia* **79,** 1–5.

Caldwell, W. W., Robberecht, R., and Flint, S. D. (1983). Internal filters: Prospects for UV-acclimation in higher plants. *Physiol. Plant.* **58,** 445–450.

Callander, B. A. (1978). Eddy correlation measurements of sensible heat flux density. Ph.D. Thesis, University of Strathclyde.

Campbell, G. S. (1977). "An Introduction to Environmental Biophysics." Springer-Verlag, Berlin and New York.

Campbell, J. A. (1980). Oxygen flux measurements in organic soils. *Can. J. Soil Sci.* **60,** 641–650.

Campbell, R. G., and Hughes, J. H. (1981). Forest management systems in North Carolina pocosins: Weyerhaeuser. *In* "Pocosin Wetlands" (C. Richardson, ed.), pp. 199–213. Hutchinson Ross, Stroudsburg, Pennsylvania.

Campbell, R. K., and Sorenson, F. C. (1973). Cold-acclimation in seedling Douglas-fir related to phenology and provenance. *Ecology* **54,** 1148–1151.

Campbell, W. A., and Copeland, O. L. (1954). Little leaf disease of shortleaf and loblolly pines. *U.S., Dep. Agric., Circ.* **940.**

Cannell, M. G. R. (1975). Crop physiological aspects of coffee bean yield: A review. *J. Coffee Res.* **5,** 7–20.

Cannell, M. G. R. (1984). Spring frost damage on young *Picea sitchensis*. I. Occurrence of damaging frosts in Scotland compared with western North America. *Forestry* **57,** 159–175.

Cannell, M. G. R. (1985). Dry matter partitioning in tree crops. *In* "Attributes of Trees as Crop Plants" (M. G. R. Cannell and J. E. Jackson, eds.), pp. 160–193. Institute of Terrestrial Ecology, Huntingdon, England.

Cannell, M. G. R. (1986). Physiology of southern pine seedlings. *In* "Nursery Management Practices for the Southern Pines" (D. B. South, ed.), pp. 251–274. Auburn University, Auburn, Alabama.

Cannell, M. G. R., and Jackson, J. E., eds. (1985). "Attributes of Trees as Crop Plants." Institute of Terrestrial Ecology, Huntingdon, England.

Cannell, M. G. R., and Sheppard, L. J. (1982). Seasonal changes in the frost hardiness of provenances of *Picea sitchensis* in Scotland. *Forestry* **55,** 137–153.

Cannell, M. G. R., and Smith, R. I. (1980). Yields of minirotation closely spaced hardwoods in temperate regions: Review and appraisal. *For. Sci.* **26,** 415–428.

Cannell, M. G. R., Rothery, P., and Ford, E. D. (1984). Competition within stands of *Picea sitchensis* and *Pinus contorta*. *Ann. Bot.* (*London*) [N.S.] **53,** 349–362.

Cannell, M. G. R., Murray, M. B., and Sheppard, L. J. (1985). Frost avoidance by selection for late budburst in *Picea sitchensis*. *J. Appl. Ecol.* **22,** 931–941.

Cannell, M. G. R., Milne, R., Sheppard, L. J., and Unsworth, M. H. (1987). Radiation interception and productivity of willow. *J. Appl. Ecol.* **24,** 261–278.

Cannell, R. Q. (1977). Soil aeration and compaction in relation to root growth and soil management. *Appl. Biol.* **2,** 1–86.

Cannon, H. L. (1960). Botanical prospecting for ore deposits. *Science* **132,** 591–598.

Carlisle, D., and Cleveland, G. B. (1958). Plants as a guide to mineralization. *Calif., Div. Mines, Spec. Rep.* **50.**

Carlson, C. (1980). Kraft mill gases damage Douglas-fir in western Montana. *Eur. J. For. Pathol.* **10,** 145–151.

Carlson, R. W. (1979). Reduction in the photosynthetic rate of *Acer*, *Quercus*, and *Fraxinus* species caused by sulphur dioxide and ozone. *Environ. Pollut.* **18,** 159–170.

Carlson, R. W., and Bazzaz, F. A. (1977). Growth reduction in American sycamore (*Platanus occidentalis* L.) caused by Pb–Cd interaction. *Environ. Pollut.* **12,** 243–253.

Carlson, R. W., and Bazzaz, F. A. (1982). Photosynthetic and growth response to fumigation with SO_2 at elevated CO_2 for C_3 and C_4 plants. *Oecologia* **54,** 50–54.

Carlson, W. C. (1985a). Effects of natural chilling and cold storage on budbreak and root growth potential of loblolly pine (*Pinus taeda* L.). *Can. J. For. Res.* **15,** 651–656.

Carlson, W. C. (1985b). Root system considerations in the quality of loblolly pine seedlings. *South. J. Appl. For.* **10,** 87–92.

Carlson, W. C., and Harrington, C. A. (1987). Cross-sectional area relationships in root systems of loblolly and shortleaf pine. *Can. J. For. Res.* **17,** 556–558.

Carlson, W. C., and Preisig, C. L. (1980). Effects of controlled-release fertilizers on the shoot and root development of Douglas-fir seedlings. *Can. J. For. Res.* **11,** 230–242.

Carlson, W. C., Harrington, C. A., Farnum, P., and Hallgren, S. W. (1988). Effects of root severing treatments on loblolly pine. *Can. J. For. Res.* **18,** 1376–1385.

Carmean, W. H. (1975). Forest site quality evaluation in the United States. *Adv. Agron.* **27,** 209–269.

Carmi, A., Hesketh, J. D., Enos, W. T., and Peters, D. B. (1983). Interrelationships between shoot growth and photosynthesis, as affected by root growth restriction. *Photosynthetica* **17,** 240–245.

Carter, G. A., Miller, J. H., Davis, D. E. and Patterson, R. M. (1984). Effect of vegetative competition on the moisture and nutrient status of loblolly pine. *Can. J. For. Res.* **14,** 1–9.

Carter, M. C. (1972). Net photosynthesis in trees. *In* "Net Carbon Dioxide Assimilation in Higher Plants" (C. Black, ed.), pp. 55–74. Am. Soc. Plant Physiol., Atlanta, Georgia.

Cassel, D. K. (1983). Effects of soil characteristics and tillage practices on water storage and its availability to plant roots. *In* "Crop Reactions to Water and Temperature Stresses in Humid, Temperate Climates" (C. D. Raper, Jr. and P. J. Kramer, eds.), pp. 167–186. Westview Press, Boulder, Colorado.

Cathey, H. M. (1964). Physiology of growth retarding chemicals. *Annu. Rev. Plant Physiol.* **15,** 271–302.

Cayford, J. H., and Waldron, R. M. (1967). Effect of captan on the germination of white spruce, jack, and red pine seed. *For. Chron.* **43,** 381–384.

Ceulemans, R., and Impens, I. (1983). Net CO_2 exchange rate and shoot growth of young poplar (*Populus*) clones. *J. Exp. Bot.* **34,** 866–870.

Chabot, B. F., and Bunce, J. A. (1979). Drought stress effects on leaf carbon balance. *In* "Topics in Plant Population Biology" (O. T. Solbrig, S. K. Jain, G. B. Johnson, and P. H. Raven, eds.), pp. 338–355. Columbia Univ. Press, New York.

Chabot, B. F., and Mooney, H. A., eds. (1985). "Physiological Ecology of North American Plant Communities." Chapman & Hall, New York.

Chabot, J. F., and Leopold, A.C. (1982). Ultrastructural changes of membranes with hydration in soybean seeds. *Am. J. Bot.* **69,** 623–633.

Chalmers, D. J. (1983). Water relations of peach trees and orchards. *In* "Water Deficits and Plant Growth" (T. T. Kozlowski, ed.), Vol. 7, pp. 197–232. Academic Press, New York.

Chalmers, D. J., Mitchell, P. D., and van Heek, L. (1981). Control of peach tree growth and productivity by regulated water supply, tree density, and summer pruning. *J. Am. Soc. Hortic. Sci.* **106,** 307–312.

Chalmers, D. J., and van den Ende, B. (1975a). Productivity of peach trees: Factors affecting dry-weight distribution during tree growth. *Ann. Bot.* (*London*) [N.S.] **39:**423–432.

Chalmers, D. J., and van den Ende, B. (1975b). The tatura trellis—A new design for high yielding orchards. *J. Agric.* **73,** 473–476.

Chamshama, S. A. O., and Hall, J. B. (1987). Effects of site preparation and fertilizer application at planting on *Eucalyptus tereticornis* at Morogoro, Tanzania. *For. Ecol. Manage.* **18,** 103–112.

Chandler, C., Cheyney, P., Thomas, P., Trabaud, L. and Williams, D. (1983). "Fire in Forestry. Vol. II. Forest Fire Management and Organization." Wiley, New York.

Chandler, R. F., Jr. (1941). The amount and mineral nutrient content of freshly fallen needle litter of some northeastern conifers. *Soil Sci. Soc. Am. Proc.* **8,** 409–411.

Chandler, R. F. (1944). The amount and mineral nutrient content of freshly fallen needle litter of some northeastern conifers. *Soil Sci. Soc. Am., Proc.* **8,** 409–411.

Chaney, W. R. (1981). Sources of water. *In* "Water Deficits and Plant Growth" (T. T. Kozlowski, ed.), Vol. 6, pp. 1–47. Academic Press, New York.

Chaney, W. R., and Kozlowski, T. T. (1969). Diurnal expansion and contraction of leaves and fruits of English Morello Cherry. *Ann. Bot.* (*London*) [N.S.] **33,** 991–999.

Chaney, W. R., and Strickland, R. C. (1974). Effect of cadmium and sulfur dioxide on pollen germination. *Proc. North Am. For. Biol. Workshop, 3rd, 1974,* pp. 372–373.

Chaney, W. R., and Strickland, R. C. (1984). Relative toxicity of heavy metals to red pine pollen germination and germ tube elongation. *J. Environ. Qual.* **13,** 391–393.

Chang, C. W. (1975). Fluorides. *In* "Responses of Plants to Air Pollution" (J. B. Mudd and T. T. Kozlowski, eds.), pp. 57–95. Academic Press, New York.

Chapela, I. H., and Boddy, L. (1988). Fungal colonization of attached beech branches. *New Phytol.* **110,** 47–57.

Chapin, F. S., and Shaver, G. R. (1985). Arctic. *In* "Physiological Ecology of North American Plant Communities" (B. F. Chabot and H. A. Mooney, eds.), pp. 16–40. Chapman & Hall, New York.

Chaplin, M. H., Stebbins, R. L., and Westwood, M. N. (1977). Effect of fall applied boron sprays on fruit set and yield of 'Halian' prune (*Prunus domestica* L.). *HortScience* **12,** 500–501.

Charles-Edwards, D. A. (1982). "Physiological Determinants of Crop Growth." Academic Press, Sydney.

Chazdon, R. L., and Pearcy, R. W. (1986a). Photosynthetic responses to light variation in rain forest species. I. Induction under constant and fluctuating light conditions. *Oecologia* **69,** 517–523.

Chazdon, R. L., and Pearcy, R. W. (1986b). Photosynthetic responses to light variation in rain forest species. II. Carbon gain and light utilization during sunflecks. *Oecologia* **69,** 524–531.

Chavanier, G. (1967). New observations on avocado growing in Morocco. *Calif. Avocado Soc. Yearb.* **51,** 111–113.

Cherry, J. H., ed. (1989). "Environmental Stress in Plants." Springer-Verlag, Berlin.

Chiarello, N. R., Mooney, H. A., and Williams, K. (1989). Growth, carbon allocation and cost of plant tissues. *In* "Plant Physiological Ecology" (R. W. Pearcy, J. Ehleringer, H. A. Mooney, and P. W. Rundel, eds.), pp. 327–365. Chapman & Hall, London and New York.

Chiba, O., and Tanaka, T. (1968). The effect of sulphur dioxide on the development of pine needle blight caused by *Rhizosphaera kalkhoffii* Bubak (I). *J. Jpn. For. Soc.* **50,** 135–139.

Childers, N. F. (1983). "Modern Fruit Science." Horticultural Publications, Gainesville, Florida.

Childers, N. F., and White, D. G. (1942). Influence of submersion of the roots on transpiration, apparent photosynthesis, and respiration of young apple trees. *Plant Physiol.* **17,** 603–618.

Chirkova, T. V. (1968). Features of the O_2 supply of roots of certain woody plants in anaerobic conditions. *Fiziol. Rast. (Moscow)* **15,** 565–568.

Chirkova, T. V., and Gutman, T. S. (1972). Physiological role of branch lenticels in willow and poplar under conditions of root anaerobiosis. *Sov. Plant Physiol. (Engl. Transl.)* **19,** 289–295.

Chow, W. S., Qian, L., Goodchild, D. J., and Anderson, J. M. (1988). Photosynthetic acclimation of *Alocasia macrorrhiza* (L.) G. Don to growth irradiance: Structure, function and composition of chloroplasts. *Aust. J. Plant Physiol.* **15,** 107–122.

Christensen, N. L. (1973). Fire and the nitrogen cycle in California chaparral. *Science* **181,** 66–68.

Christensen, N. L. (1985). Shrubland fire regimes and their evolutionary consequences. *In* "The Ecology of Natural Disturbance and Patch Dynamics" (S. T. A. Pickett and P. S. White, eds.), pp. 85–100. Academic Press, Orlando, Florida.

Christensen, N. L. (1987). The biogeochemical consequences of fire and their effects on the vegetation of the coastal plain of the southeastern United States. *In* "The Role of Fire in Ecological Systems" (L. Trabaud, ed.), pp. 1–21. SPB Academic Publishing, The Hague, Netherlands.

Christensen, N. L., and Muller, C. H. (1975a). Effects of fire on factors controlling plant growth in *Adenostoma* chaparral. *Ecol. Mongr.* **45,** 29–55.

Christensen, N. L., and Muller, C. H. (1975b). Relative importance of factors controlling germination and seedling survival in *Adenostoma* chaparral. *Am. Midl. Nat.* **93,** 71–78.

Christersson, L. (1986). High technology biomass production by *Salix* clones on a sandy soil in southern Sweden. *Tree Physiol.* **2,** 261–272.

Christiansen, E., Waring, R. H., and Berryman, A. A. (1987). Resistance of conifers to bark beetle attack: Searching for general relationships. *For. Ecol. Manage.* **22,** 89–106.

Christiansen, M. N. (1964). Influence of chilling upon seedling development of cotton. *Plant Physiol.* **38,** 520–522.

Christiansen, M. N., and Lewis, C. F., eds. (1982). "Breeding Plants for Less Favorable Environments." Wiley, New York.

Chrosciewicz, A. (1976). Burning for black spruce regeneration on a lowland cutover site in southeast Manitoba. *Can. J. For. Res.* **6,** 179–186.

Chung, H. H., and Barnes, R. L. (1977). Photosynthate allocation in *Pinus taeda*. I. Substrate requirements for synthesis of shoot biomass. *Can. J. For. Res.* **7,** 106–111.

Chung, H. H., and Barnes, R. L. (1980a). Photosynthate allocation in *Pinus taeda*. II. Seasonal aspects of photosynthate allocation to different biochemical fractions in shoots. *Can. J. For. Res.* **10,** 338–347.

Chung, H. H., and Barnes, R. L. (1980b). Photosynthate allocation in *Pinus taeda*. III. Photosynthate economy: Its production, consumption, and balance in shoots during the growing season. *Can. J. For. Res.* **10,** 348–356.

Chung, H. H., and Kramer, P. J. (1975). Absorption of water and ^{32}P through suberized and unsuberized roots of loblolly pine. *Can. J. For. Res.* **5,** 229–235.

Clark, F. B., and Liming, F. G. (1953). Sprouting of blackjack oak in the Missouri Ozarks. *U.S., For. Serv., Cent. States For. Exp. Stn., Tech. Pap.* **137.**

Clark, F. E., and Kemper, W. D. (1967). Microbial activity in relation to soil water and soil aeration. *Agron. Monogr.* **11,** 472–480.

Clarkson, D. T. (1985). Factors affecting mineral nutrient acquisition by plants. *Annu. Rev. Plant Physiol.* **36,** 77–115.

Clarkson, D. T., and Hanson, J. B. (1980). The mineral nutrition of higher plants. *Annu. Rev. Plant Physiol.* **31,** 239–311.

Clayton-Greene, K. A. (1983). The tissue water relationships of *Callitris columellaris*, *Eucalyptus melliodora*, and *Eucalyptus microcarpa* investigated using the pressure-volume technique. *Oecologia* **57,** 368–373.

Cleary, B. D., and Waring, R. H. (1969). Temperature: Collection of data and its analysis for the interpretation of plant growth and distribution. *Can. J. Bot.* **47,** 167–173.

Cleland, R. (1967). A dual role of turgor pressure in auxin-induced cell elongation in *Avena* coleoptiles. *Planta* **77,** 182–191.

Cleland, R. (1971). Cell wall extension. *Annu. Rev. Plant Physiol.* **22,** 197–223.

Cleland, R. (1988). Molecular events of photoinhibitory inactivation in the reaction centre of photosystem II. *Aust. J. Plant Physiol.* **15,** 135–150.

Clemens, J., and Jones, P. G. (1978). Modification of drought resistance by water stress conditioning in *Acacia* and *Eucalyptus*. *J. Exp. Bot.* **29,** 895–904.

Clemens, J., Kirk, A. M., and Mills, P. D. (1978). The resistance to water-logging of three *Eucalyptus* species, effect of flooding and of ethylene-releasing growth substances on *E. robusta*, *E. grandis*, and *E. saligna*. *Oecologia* **34,** 125–131.

Clements, F. E. (1936). Nature and structure of the climax. *J. Ecol.* **24,** 252–284.

Clough, J. M., Peet, M. M., and Kramer, P. J. (1981). Effects of high atmospheric CO_2 and sink size on rates of photosynthesis of a soybean cultivar. *Plant Physiol.* **67,** 1007–1010.

Coe, J. M., and McLaughlin, S. B. (1980). Winter season corticular photosynthesis in *Cornus florida*, *Acer rubrum*, *Quercus alba*, and *Liriodendron tulipifera*. *For. Sci.* **26,** 561–566.

Cohon, J. L. (1978). "Multiobjective Programming and Planning." Academic Press, New York.

Coile, T. S. (1937). Distribution of forest tree roots in North Carolina Piedmont soils. *J. For.* **35,** 247–257.

Coile, T. S. (1940). Soil changes associated with loblolly pine succession on abandoned agricultural land of the Piedmont Plateau. *Duke Univ. Sch. For. Bull.* **5.**

Coile, T. S., and Schumacher, F. X. (1953). Relation of soil properties to site index of loblolly and shortleaf pines in the Piedmont region of the Carolinas, Georgia, and Alabama. *J. For.* **51,** 739–744.

Cole, E. C., and Newton, M. (1986). Nutrient, moisture, and light relations in 5-year-old Douglas-fir plantations under variable competition. *Can J. For. Res.* **16,** 727–732.

Cole, D. W. & Rapp, M. (1981). Elemental cycling in forest ecosystems. *In* "Dynamic Principles of Forest Ecosystems" (D. E. Reichle, ed.), pp. 341–409. Cambridge Univ. Press, London and New York.

Cole, D. W., Henry, C. L., and Nutter, W. L. (1986). "The Forest Alternative for Treatment and Utilization of Municipal Wastes." Univ. of Washington Press, Seattle.

Coley, P. D. (1987). Interspecific variations in plant antiherbivory properties: The role of habitat quality and rate of disturbance. *New Phytol.* **106,** Suppl., 251–263.

Coley, P. D., Bryant, J. P,. and Chapin, F. S., III (1985). Resource availability and plant antiherbivore defense. *Science* **230,** 895–899.

Collins, W. S., Chang, S. H., Raines, G., Canney, F., and Ashley, R. (1983). Airborne biogeochemical mapping of hidden mineral deposits. *Econ. Geol.* **78,** 737–749.

Colombo, S. J., and Asseltine, M. F. (1989). Root hydraulic conductivity and root growth potential of black spruce (*Picea mariana*) seedlings. *Tree Physiol.* **5,** 73–81.

Colombo, S. J., Glerum, C., and Webb, D. P. (1989). Winter hardening in first-year black spruce (*Picea mariana*) seedlings. *Physiol. Plant.* **76,** 1–9.

Comstock, J., and Ehleringer, J. (1984). Photosynthetic response to slowly decreasing leaf water potentials in *Encelia frutescens*. *Oecologia* **61,** 241–248.

Conard, S. G., and Radosevich, S. R. (1982). Growth responses of white fir to decreased shading and root competition by montane chaparral shrubs. *For. Sci.* **28,** 309–320.

Connell, J. H. (1975). Some mechanisms producing structure in natural communities: A model and evidence from field experiments. *In* "Ecology and Evolution of Communities" (M. Cody and J. Diamond, eds.), pp. 460–469. Harvard Univ. Press, Cambridge, Massachusetts.

Connell, J. H., and Slatyer, R. O. (1977). Mechanisms of succession in natural communities and their role in community stability and organization. *Am. Nat.* **111,** 1119–1144.

Connor, D. J. (1983). Plant stress factors and their influence on production of agroforestry plant associations. *In* "Plant Research and Agroforestry" (P. A. Huxley, ed.), pp. 401–426. Pillans & Wilson, Edinburgh.

Connor, D. J., Begge, N. J., and Turner, N. C. (1977). Water relations of mountain ash (*Eucalyptus regnans* F. Muell.) forests. *Aust. J. Plant Physiol.* **4,** 753–762.

Conroy, J. P., Smillie, R. M., Küppers, M., Bevege, D. I., and Barlow, E. W. (1986). Chlorophyll a fluorescence and photosynthetic and growth responses of *Pinus radiata* to phosphorus deficiency, drought stress, and high CO_2. *Plant Physiol.* **81,** 423–429.

Constantinidou, H. A., and Kozlowski, T. T. (1979a). Effects of sulfur dioxide and ozone on *Ulmus americana* seedlings. I. Visible injury and growth. *Can. J. Bot.* **57,** 170–175.

Constantinidou, H. A., and Kozlowski, T. T. (1979b). Effects of sulfur dioxide and ozone on *Ulmus americana* seedlings. II. Carbohydrates, proteins, and lipids. *Can. J. Bot.* **57,** 176–184.

Constantinidou, H. A., Kozlowski, T. T., and Jensen, K. (1976). Effects of sulfur dioxide on *Pinus resinosa* seedlings in the cotyledon stage. *J. Environ. Qual.* **5,** 141–144.

Cook, J. M., Mark, A. F., and Shore, B. F. (1980). Responses of *Leptospermum scoparium* and *L. ericoides* (Myrtaceae) to waterlogging. *N.Z. J. Bot.* **18,** 233–246.

Cooke, J. R., and Rand, R. H. (1980). Diffusion resistance models. *In* "Predicting Photosynthesis for Ecosystem Models" (J. B. Hesketh and J. W. Jones, eds.), Vol. 1, pp. 94–121. CRC Press, Boca Raton, Florida.

Cooper, D. T., and Ferguson, R. B. (1981). Evaluation of bole straightness in cottonwood usual visual scores. *Res. Note SO* (*U.S. For. Ser.*) **S0–277.**

Coppel, H. C. (1977). "Biological Insect Pest Suppression." Springer-Verlag, New York.

Corak, S. J., Blevins, D. G., and Pallardy, S. G. (1987). Water transfer in an alfalfa–maize association. *Plant Physiol.* **84,** 582–586.

Corns, G. W. (1988). Compaction by forestry equipment and effects on coniferous seedling growth on four soils in the Alberta foothills. *Can. J. For. Res.* **18,** 75–84.

Cosgrove, D. (1986). Biophysical control of plant cell growth. *Annu. Rev. Plant Physiol.* **37,** 377–405.

Costonis, A. C. (1973). Injury to eastern white pine by sulfur dioxide and ozone alone and in mixtures. *Eur. J. For. Path.* **3,** 50–55.

Costonis, A. C., and Sinclair, W. A. (1972). Susceptibility of healthy and ozone-injured needles of *Pinus strobus* to invasion by *Lophodermium pinastri* and *Aureobasidium pullulans*. *Eur. J. For. Pathol.* **2,** 65–73.

Coté, W. A., Jr., and Day, A. C. (1965). Anatomy and ultrastructure of reaction wood. *In* "Cellular Ultrastructure of Woody Plants" (W. A. Coté, ed.), pp. 391–418. Syracuse Univ. Press, Syracuse, New York.

Couey, H. M. (1982). Chilling injury of crops of tropical and subtropical origin. *HortScience* **17,** 162–165.

Coutts, M. P. (1981). Effects of waterlogging on water relations of actively growing and dormant Sitka spruce seedlings. *Ann. Bot.* (*London*) [N.S.] **47,** 747–753.

Coutts, M. P. (1982). The tolerance of tree roots to waterlogging. V. Growth of woody roots of Sitka spruce and lodgepole pine in waterlogged soil. *New Phytol.* **90,** 467–476.

Coutts, M. P. (1986). Components of tree stability in Sitka spruce on peaty gley soil. *Forestry* **59,** 173–197.

Coutts, M. P., and Armstrong, W. (1976). Role of oxygen transport in the tolerance of trees to waterlogging. *In* "Tree Physiology and Yield Improvement" (M. G. R. Cannell and F. T. Last, eds.), pp. 361–385. Academic Press, New York and London.

Coutts, M. P., and Philipson, J. J. (1978a). The tolerance of tree roots to waterlogging. I. Survival of Sitka spruce and lodgepole pine. *New Phytol.* **80,** 63–69.

Coutts, M. P., and Philipson, J. J. (1978b). The tolerance of tree roots to waterlogging. II. Adaptation of Sitka spruce and lodgepole pine to waterlogged soil. *New Phytol.* **80,** 71–77.

Covington, W. W., and Sackett, S. S. (1984). The effect of a prescribed fire in southwestern ponderosa pine on organic matter and nutrients in woody debris and forest floor. *For. Sci.* **30,** 183–192.

Covington, W. W., and Sackett, S. S. (1986). Effect of periodic burning on soil nitrogen concentrations in ponderosa pine. *Soil Sci. Soc. Am. J.* **50,** 452–457.

Cowan, I. R. (1982). Regulation of water use in relation to carbon gain in higher plants. *Encycl. Plant Physiol., New Ser.* **12B,** 535–562.

Cowling, D. W., and Lockyer, D. R. (1976). Growth of perennial ryegrass (*Lolium perenne* L.) exposed to a low concentration of sulphur dioxide. *J. Exp. Bot.* **27,** 411–417.

Crane, A. J. (1985). Possible effects of rising CO_2 on climate. *Plant, Cell Environ.* **8,** 371–379.

Crane, W. J. B., and Raison, R. J. (1980). Removal of phosphorous in logs when harvesting *Eucalyptus delegatensis* and *Pinus radiata* on short and long rotations. *Aust. For.* **43,** 253–260.

Craul, P. J. (1985). A description of urban soils and their desired characteristics. *J. Arboric.* **11,** 330–339.

Crawford, R. M. M., and Finegan, D. M. (1989). Removal of ethanol from lodgepole pine roots. *Tree Physiol.* **5,** 53–61.

Cremer, K. W., Meyers, B. J., van der Duys, F., and Craig, I. E. (1977). Silvicultural lessons from the 1974 windthrow in radiata pine plantations near Canberra. *Aust. For.* **40,** 274–292.

Cremer, K. W., Borough, C. J., McKinnell, F. H., and Carter, P. R.(1982). Effects of stocking and thinning on wind damage in plantations. *N. Z. J. For. Sci.* **12,** 245–268.

Crist, C. R., and Schoeneweiss, D. F. (1975). The influence of controlled stresses on susceptibility of European white birch stems to attack by *Botryosphaeria dothidea*. *Phytopathology* **65,** 369–373.

Critchley, C. (1988). The molecular mechanism of photoinhibition—-Facts and fiction. *Aust. J. Plant Physiol.* **15,** 27–41.

Croft, B. A. (1978). Potentials for research and implementation of integrated pest management on deciduous tree fruits. *In* "Plant Control Strategies" (E.H. Smith and D. Pimentel, eds.), pp. 101–115. Academic Press, New York.

Cromack, K., Jr., and Monk, C. D. (1975). Litter production, decomposition, and nutrient cycling in a mixed hardwood watershed and a white pine watershed. *In* "Mineral Cycling in Southeastern Ecosystems" (F.G. Howell *et al.*, eds.), ERDA Symp. Ser. (CONF-740513), pp. 609-624. Natl. Tech. Inf. Serv., Springfield, Virginia.

Crossley, A., and Fowler, D. (1986). The weathering of Scots pine epicuticular wax in polluted and clean air. *New Phytol.* **103,** 207–218.

Curran, M. (1985). Gas movements in the roots of *Avicennia marina* (Forsk.) Vierh. *Aust. J. Plant Physiol.* **12,** 97–108.

Curran, M., Cole, M., and Allaway, W. G. (1986). Root aeration and respiration in young mangrove plants (*Avicennia marina* (Forsk.) Vierh.). *J. Exp. Bot.* **37,** 1225–1233.

Currie, D. J., and Paquin, U. (1987). Large-scale geographical patterns of species richness of trees. *Nature (London)* **329,** 326–327.

Curry, E. A. (1988). Chemical control of vegetative growth of deciduous fruit trees with paclobutrazol and RSW0411. *HortScience* **23,** 470–473.

Curry, J. R., and Church, T. W. (1952). Observations on winter drying of conifers in the Adirondacks. *J. For.* **50,** 114–116.

Cutler, M. R. (1978). The role of USDA in integrated pest management. *In* "Pest Control Strategies" (E. H. Smith and D. Pimentel, eds.), pp. 9–20. Academic Press, New York.

Daines, R. H., Lukens, R. J., Brennan, E., and Leone, I. A.(1957). Phytotoxicity of captan as influenced by formulation, environment, and plant factors. *Phytopathology* **47,** 567–572.

Dale, J. E., and Sutcliffe, J. F. (1986). Phloem transport. *In* "Plant Physiology" (F. C. Steward, J. F. Sutcliffe, and J. E. Dale, eds.), Vol. 9, pp. 455–549. Academic Press, Orlando, Florida.

Daley, P. F., Raschke, K., Ball, J. T., and Berry, J. A. (1989). Topography of photosynthetic activity of leaves obtained from video images of chlorophyll fluorescence. *Plant Physiol.* **90,** 1233–1238.

Daniel, T. W., Helms, J. A., and Baker, F. S. (1979). "Principles of Silviculture." McGraw-Hill, New York.

Darley, E. F. (1966). Studies on the effect of cement-kiln dust on vegetation. *J. Air Pollut. Control Assoc.* **16,** 145–150.

Darrall, N. M. (1989). The effect of air pollutants on physiological processes in plants. *Plant, Cell Environ.* **12,** 1–30.

Daubenmire, R. F. (1943). Vegetational zonation in the Rocky Mountains. *Bot. Rev.* **9,** 325–393.

Daubenmire, R. F. (1984). "Plants and Environment." Wiley, New York.

Daubenmire, R. F., and Deters, M. E. (1947). Comparative studies of growth in deciduous and evergreen trees. *Bot. Gaz. (Chicago)* **109,** 1–12.

Davey, C. B., and Wollum, A. E. (1984). Nitrogen fixation systems in forest plantations. *In* "Nutrition of Plantation Forests" (G. D. Bowen and E. K. S. Nambiar, eds.), pp. 361–377. Academic Press, London.

Davies, D. D. (1980). Anaerobic metabolism and the production of organic acids. *In* "The Biochemistry of Plants. Vol. 2. Metabolism and Respiration" (D. D. Davies, ed.), pp. 581–611. Academic Press, New York.

Davies, D. D., Grego, S., and Kenworthy, P. (1974). The control of the production of lactate and ethanol by higher plants. *Planta* **118,** 297–310.

Davies, F. S., and Flore, J. A. (1986). Gas exchange and flooding stress of highbush and rabbiteye blueberries. *J. Am. Soc. Hortic. Sci.* **111,** 565–571.

Davies, W. J. (1977). Stomatal responses to water stress and light in plants grown in controlled environments and in the field. *Crop Sci.* **17,** 735–740.

Davies, W. J., and Kozlowski, T. T. (1974). Stomatal responses of five woody angiosperm species to light intensity and humidity. *Can. J. Bot.* **52,** 1525–1534.

Davies, W. J., and Kozlowski, T. T. (1975a). Effects of applied abscisic acid and plant water stress on transpiration of woody angiosperms. *For. Sci.* **22,** 191–195.

Davies, W. J., and Kozlowski, T. T. (1975b). Effect of applied abscisic acid and silicone on water relations and photosynthesis of woody plants. *Can. J. For. Res.* **5,** 90–96.

Davies, W. J., Kozlowski, T. T., and Pereira, J. (1974). Effect of wind on transpiration and stomatal aperture of woody plants. *Bull.—R. Soc. N. Z.* **12,** 433–438.

Davies, W. J., Mansfield, T. A., and Wellburn, A. R. (1979). A role for abscisic acid in drought endurance and drought avoidance. *In* "Plant Growth Substances 1979" (F. Skoog, ed.), pp. 242–253. Springer-Verlag, Berlin.

Davies, W. J., Metcalfe, J., Lodge, T. A. and DaCosta, A. R. (1986). Plant growth substances and the regulation of growth under drought. *Aust. J. Plant Physiol.* **13,** 105–125.

Davis, D. D. (1977). Response of ponderosa pine primary needles to separate and simultaneous ozone and PAN exposures. *Plant Dis. Rep.* **61,** 640–644.

Davis, D. D., and Gerhold, H. D. (1976). Selection of trees for tolerance of air pollutants. *USDA For. Serv. Gen. Tech. Rep. NE* **NE–22,** 61–66.

Davis, E. A., and Pase, C. P. (1977). Root system of shrub live oak: Implications for water yield in Arizona chaparral. *J. Soil Water Conserv.* **32(4),** 174–180.

Davis, J. T., and Sparks, D. (1974). Assimilation and translocation patterns of carbon–14 in the shoots of fruiting pecan trees. *J. Am. Soc. Hortic. Sci.* **99,** 468–480.

Davison, R. M. (1971). Effect of early season sprays of trace elements on fruit setting of apples. *N. Z. J. Agric. Res.* **14,** 931- 935.

Day, W., and Parkinson, K. J. (1979). Importance to gas exchange of mass flow of air through leaves. *Plant Physiol.* **64,** 345–346.

Dean, T. J., Pallardy, S. G., and Cox, G. S. (1982). Photosynthetic responses of black walnut (*Juglans nigra*) to shading. *Can. J. For. Res.* **12,** 725–730.

De Bano, L. F. (1981). Water repellent soils: A state of the art. *USDA For. Serv. Gen. Tech. Rep. PSW* **PSW-46.**

De Bano, L. F., Eberlein, G. E., and Dunn, P. H. (1979). Effects of burning on chaparral soil. I. Soil nitrogen. *Soil Sci. Soc. Am. J.* **43,** 504–509.

Decker, J. P. (1944). Effect of temperature on photosynthesis and respiration in red and loblolly pines. *Plant Physiol.* **19,** 679–688.

Decker, W. L. (1983). Probability of drought for humid and subhumid regions. *In* "Crop Reactions to Water and Temperature Stresses in Humid, Temperate Climates" (C. D. Raper, Jr. and P. J. Kramer, eds.), pp. 11–19. Westview Press, Boulder, Colorado.

Delap, A. V. (1967). The effect of supplying nitrate at different seasons on the growth, blossoming and nitrogen content of young apple trees in sand culture. *J. Hortic. Sci.* **42,** 149–167.

Dellinger, G. P., Brink, E. L., and Allmon, A. D. (1976). Tree mortality caused by flooding at two midwestern reservoirs. *Proc. 3rd Annu. Conf. Southeast. Fish Game Comm.* pp. 645–648.

De Lucia, E. H. (1986). Effect of low root temperature on net photosynthesis, stomatal conductance and carbohydrate concentration in Engelmann spruce (*Picea engelmannii* Parry ex Engelm.) seedlings. *Tree Physiol.* **2,** 143–154.

Demaree, J. B., and Wilcox, M. S. (1947). Fungi pathogenic to blueberries in the eastern United States. *Phytopathology* **37,** 487- 506.

Demmig, B., Winter, K., Krüger, A., and Czygan, F.-C. (1987). Photoinhibition and zeaxanthin formation in intact leaves. *Plant Physiol.* **84,** 218–224.

Demmig-Adams, B., Winter, K., Krüger, A., and Czygan, F. (1989). Zeaxanthin synthesis, energy dissipation and photoprotection of photosystem II at chilling temperatures. *Plant Physiol.* **90,** 894–898.

De Moranville, C. J., and Deubert, K. H. (1987). Effect of commercial calcium–boron and manganese–zinc formulations on fruit set of cranberries. *J. Hortic. Sci.* **62,** 163–169.

Denne, M. P., and Atkinson, L. D. (1973). A phytotoxic effect of captan on the growth of conifer seedlings. *Forestry* **46,** 49–53.

de Silva, C. A. (1961). Wind damage in rubber plantations. *Q. J.—Rubber Res. Inst. Ceylon* **37,** 91–92.

Devine, T. E. (1982). Genetic fitting of crops to problem soils. *In* "Breeding Plants for Less Favorable Environments" (M. N. Christiansen and C. F. Lewis, eds.), pp. 143–173. Wiley, New York.

De Wald, L. E., and Feret, P. P. (1987). Changes in loblolly pine root growth potential from September to April. *Can. J. For. Res.* **17,** 635–643.

Dewers, R. S., and Moehring, D. M. (1970). Effects of soil water stress on initiation of ovulate primordia in loblolly pine. *For. Sci.* **16,** 219–221.

de Yoe, D. R., and Brown, G. N. (1979). Glycerolipid and fatty acid changes in eastern white pine chloroplast lamellae during the onset of winter. *Plant Physiol.* **64,** 924–929.

Dickerson, B. P. (1976). Soil compaction after tree-length skidding in northern Mississippi. *Soil Sci. Soc. Am. J.* **40,** 965–966.

Dickmann, D. I., and Kozlowski, T. T. (1968). Mobilization by *Pinus resinosa* cones and shoots of C^{14}-photosynthate from needles of different ages. *Am. J. Bot.* **55,** 900–906.

Dickmann, D. I., and Kozlowski, T. T. (1970). Mobilization and incorporation of photoassimilated ^{14}C by growing vegetative and reproductive tissues of adult *Pinus resinosa* Ait. trees. *Plant Physiol.* **45,** 284–288.

Dickmann, D. I., and Kozlowski, T. T. (1971). Cone size and seed yield in red pine. *Am. Midl. Nat.* **85,** 431–436.

Dickmann, D. I., and Stuart, K. W. (1983). "The Culture of Poplars." Dept. of Forestry, Michigan State University, East Lansing.

Dickson, R. E., Hosner, J. F., and Hosley, N. W. (1965). The effects of four water regimes upon the growth of four bottomland tree species. *For. Sci.* **11,** 299–305.

Dixon, M., and Grace, J. (1984). Effect of wind on the transpiration of young trees. *Ann. Bot.* (*London*) [N.S.] **53,** 811–819.

Dixon, M., Webb, E. C., Thorne, C. J., and Tipton, K. F. (1979). "Enzymes." Longman, London.

Dixon, R. K., Wright, G. M., Behrns, G. T., Teskey, R. O., and Hinckley, T. M. (1980). Water deficit and root growth of ectomycorrhizal white oak seedlings. *Can. J. For. Res.* **10,** 545–548.

Dixon, R. K., Pallardy, S. G., Garrett, H. E., and Cox, G. S. (1983). Comparative water relations of container-grown and bare-root ectomycorrhizal and non-ectomycorrhizal *Quercus velutina* seedlings. *Can. J. Bot.* **61,** 1559–1565.

Dochinger, L. S., Bender, F. W., Fox, F. O., and Heck, W. W. (1970). Chlorotic dwarf of eastern white pine caused by ozone and sulphur dioxide interaction. *Nature (London)* **225,** 476.

Doley, D., and Leyton, L. (1968). Effects of growth regulating substances and water potential on the development of secondary xylem in *Fraxinus*. *New Phytol.* **67,** 579–594.

Domanski, R., and Kozlowski, T. T. (1968). Variations in kinetin-like activity in buds of *Betula* and *Populus* during release from dormancy. *Can. J. Bot.* **46,** 397–403.

Domir, S. C. (1978). Chemical control of tree height. *J. Arboric.* **4,** 145–153.

Domir, S. C., and Roberts, B. R. (1983). Tree growth retardation by injection of chemicals. *J. Arboric.* **9,** 217–224.

Donnelly, J. R., and Shane, J. B. (1986). Forest ecosystem responses to artificially induced soil compaction. I. Soil physical properties and tree diameter growth. *Can. J. For. Res.* **16,** 750–754.

Donner, S. L., and Running, S. W. (1986). Water stress response after thinning *Pinus contorta* stands in Montana. *For. Sci.* **32,** 614–625.

Dormling, I. (1982). Frost resistance during bud flushing and shoot elongation in *Picea abies*. *Silvae Fenn.* **16,** 167–177.

Dougherty, P. M., Teskey, R. O., Phelps, J. E., and Hinckley, T. M. (1979). Net photosynthesis and early growth trends of a dominant white oak (*Quercus alba* L.). *Plant Physiol.* **64,** 930–935.

Downs, R. J. (1962). Photocontrol of growth and dormancy in woody plants. *In* "Tree Growth" (T. T. Kozlowski, ed.), pp. 133–148. Ronald Press, New York.

Downs, R. J., and Hellmers, H. (1975). "Environment and the Control of Plant Growth." Academic Press, New York.

Downton, W. J. S., Grant, W. J. R., and Loveys, B. R. (1987). Carbon dioxide enrichment increases yield of Valencia orange. *Aust. J. Plant Physiol.* **14,** 493–501.

Downton, W. J. S., Loveys, B. R., and Grant, W. J. R. (1988). Stomatal closure fully accounts for the inhibition of photosynthesis by abscisic acid. *New Phytol.* **108,** 263–266.

Drew, A. P., and Ferrell, W. K. (1977). Morphological acclimation to light intensity in Douglas-fir seedlings. *Can J. Bot.* **55,** 666–674.

Drew, M. C. (1983). Plant injury and adaptation to oxygen deficiency in the root environment: A review. *Plant Soil* **75,** 179–199.

Duchelle, S. F., Skelly, J. M., and Chevone, B. I. (1982). Oxidant effects on forest tree seedling growth in the Appalachian Mountains. *Water, Air, Soil Pollut.* **18,** 363–373.

Duchesne, L. C., and Larson, D. W. (1989). Cellulose and the evolution of plant life. *BioScience* **39,** 238–241.

Duddridge, J. A., Malibari, A., and Read, D. J. (1980). Structure and function of mycorrhizal rhizomorphs with special reference to their role in water transport. *Nature (London)* **287,** 834–836.

Duncan, D. P., and Hodson, A. C. (1958). Influence of the tent caterpillar upon the aspen forests of Minnesota. *For. Sci.* **4,** 71–93.

Dunford, E. G., and Fletcher, P. W. (1947). Effect of removal of stream-bank vegetation upon water yield. *Am. Geophys. Union Trans.* **28,** 105–110.

Dunford, E. G., and Niederhoff, C. H. (1944). Influence of aspen, young lodgepole pine and open grassland types upon factors affecting water yield. *J. For.* **42,** 673–677.

Durbin, R. D. (1978). Abiotic diseases induced by unfavorable water relations. *In* "Water Deficits and Plant Growth" (T. T. Kozlowski, ed.), Vol. 5, pp. 101–117. Academic Press, New York.

Duryea, M. L., and Brown, G. N., eds. (1984). "Seedling Physiology and Reforestation Success." Martinus Nijhoff/Dr. W. Junk, The Hague, Netherlands and Boston, Massachusetts.

Duryea, M. L., and Landis, T. D., eds. (1984). "Forest Nursery Manual: Production of Bareroot Seedlings." Martinus Nijhoff/Dr. W. Junk, The Hague, Netherlands.

Duryea, M. L., and McClain, K. M. (1984). Altering seedling physiology to improve reforestation success. *In* "Seedling Physiology and Reforestation Success" (M. L. Duryea, and G. N. Brown, eds.), pp. 77–114. Martinus Nijhoff/Dr. W. Junk, The Hague, Netherlands and Boston, Massachusetts.

Eagles, C. F., and Wareing, P. F. (1964). The role of growth substances in the regulation of bud dormancy. *Physiol. Plant.* **17,** 697–709.

Eames, A. J., and MacDaniels, L. H. (1947). "An Introduction to Plant Anatomy." McGraw-Hill, New York

Eaton, F. M., and Ergle, D. R. (1948). Carbohydrate accumulation in the cotton plant at low moisture levels. *Plant Physiol.* **23,** 169–187.

Ebert, E., and Dumford, S. W. (1976). Effects of triazine herbicides on the physiology of plants. *Residue Rev.* **65,** 1–103.

Eckersten, H. (1986). Simulated willow growth and transpiration: The effect of high and low resolution weather data. *Agric. For. Meteorol.* **38,** 289–306.

Eckert, R. T., and Houston, D. B. (1980). Photosynthesis and needle elongation response of *Pinus strobus* clones to low level sulfur dioxide exposures. *Can. J. For. Res.* **10,** 357–361.

Edwards, P. J. (1977). Studies of mineral cycling in a montane rain forest in New Guinea. II. The production and disappearance of litter. *J. Ecol.* **65,** 971–992.

Edwards, P. J. (1982). Studies of mineral cycling in a montane rain forest in New Guinea. V. Rates of cycling in throughfall and litterfall. *J. Ecol.* **70,** 807–827.

Edwards, P. J., and Grubb, P. J. (1982) Studies of mineral cycling in a montane rain forest in New Guinea. IV. Soil characteristics and the division of mineral elements between the vegetation and soil. *J. Ecol.* **70,** 649–666.

Egler, F. E. (1954). Vegetation science concepts. I. Initial floristic composition, a factor in old field vegetation development. *Vegetatio* **4,** 412–417.

Eglinton, G., and Hamilton, R. J. (1967). Leaf epicuticular waxes. *Science* **156,** 1322–1335.

Ehleringer, J. (1980). Leaf morphology and reflectance in relation to water and temperature stress. *In* "Adaptations of Plants to Water and High Temperature Stress" (N. C. Turner and P. J. Kramer, eds.), pp. 295–308. Wiley, New York.

Einspahr, D., and McComb, A. L. (1951). Site index of oaks in relation to soil and topography in southeastern Iowa. *J. For.* **49,** 719–723.

Ek, A. R., and Dawson, D. H. (1976a). Actual and projected growth and yields of *Populus tristis* number 1 under intensive culture. *Can. J. For. Res.* **6,** 132–144.

Ek, A.R., and Dawson, D. H. (1976b). Yields of intensively grown *Populus*: Actual and projected. *USDA For. Serv. Gen. Tech. Rep. NC* **NC–21,** 5–9.

Elfving, D. C., Chu, C. L., Lougheed, E. C., and Cline, R. A. (1987). Effects of daminozide and paclobutrazol treatments on fruit ripening and storage behavior of 'McIntosh' apple. *J. Am. Soc. Hortic. Sci.* **112,** 910–915.

Elliott, K. J., and White, A. S. (1987). Competitive effects of various grasses and forbs on ponderosa pine seedlings. *For. Sci.* **33,** 356–366.

Ellis, R. C., Lowry, R. K., and Davies, S. K. (1982). The effect of regeneration burning upon the nutrient status of soil in two forest types in southern Tasmania. *Plant Soil* **65,** 171–186.

Ellmore, G. S., and Ewers, F. W. (1985). Hydraulic conductivity in trunk xylem of elm, *Ulmus americana*. *IAWA Bull.* [N.S.] **6,** 303–307.

Emmingham, W. H., and Waring, R. H. (1977). An index of photosynthesis for comparing forest sites in western Oregon. *Can. J. For. Res.* **7,** 165–174.

Enoch, H. Z., and Kimball, B. A. eds. (1986). "Carbon Dioxide Enrichment of Greenhouse Crops," Vol. 2. CRC Press, Boca Raton, Florida.

Environmental Protection Agency (1977). "National Air Quality and Emissions Trends Report, 1976," EPA-450/1-77-002. U.S.E.P.A., Research Triangle Park, North Carolina.

Environmental Protection Agency (1978a). "Diagnosing Vegetation Injury Caused by Air Pollution." EPA-450/3-78-005. U.S.E.P.A., Research Triangle Park, North Carolina.

Environmental Protection Agency (1978b). "Research Outlook. 1978," EPA-600/9-78-001. U.S.E.P.A., Washington, D.C.

Environmental Protection Agency (1984a). "National Air Pollutant Emission Estimates, 1940–1982." EPA-450/4-83-024. U.S.E.P.A., Research Triangle Park, North Carolina.

Environmental Protection Agency (1984b). "Air Quality Criteria for Ozone and Other Photochemical Oxidants," Vol. III, EPA-600/8-84-020A. U.S.E.P.A., Corvallis, Oregon.

Epstein, E. (1972). "Mineral Nutrition of Plants: Principles and Perspectives." Wiley, New York.

Epstein, E., Norlyn, J. D., and Cabot, C. (1988). Silicon and plant growth. *Plant Physiol.* **86,** Suppl., 134.

Erez, A. (1978). Adaptation of the peach to the meadow orchard system. *Acta Hortic.* **65,** 245–250.

Erez, A. (1982). Peach meadow orchard: Two feasible systems. *HortScience* **17,** 138–142.

Esau, K. (1965). "Plant Anatomy." Wiley, New York.

Esau, K. (1977). "Anatomy of Seed Plants," 2nd ed. Wiley, New York.

Escalante, E. E. (1985). Promising agroforestry systems in Venezuela. *Agrofor. Syst.* **3,** 209–221.

Eschrich, W., Burchardt, R., and Essiamah, S. (1989). The induction of sun and shade leaves of the European beech (*Fagus sylvatica* L.): Anatomical studies. *Trees* **3,** 1–10.

Evans, G. C. (1972). "The Quantitative Analysis of Plant Growth." Univ. of California Press, Berkeley.

Evans, J. R. (1988). Acclimation by the thylakoid membranes to growth irradiance and the partitioning of nitrogen between soluble and thylakoid proteins. *Aust. J. Plant Physiol.* **15,** 93–106.

Evans, L. S., and Miller, P. R. (1972). Ozone damage to ponderosa pine: A histological and histochemical appraisal. *Am. J. Bot.* **59,** 299–304.

Evans, L. T. (1975). "Crop Physiology." Cambridge Univ. Press, Cambridge.

Evans, L. T. (1980). The natural history of crop yield. *Am. Sci.* **68,** 388–397.

Evans, P. T., and Malmberg, R. L. (1989). Do polyamines have roles in plant development? *Annu. Rev. Plant Physiol. Plant Mol. Biol.* **40,** 235–269.

Evans, W. G. (1972). The attraction of insects to forest fires. *Proc. Tall Timbers Conf. Ecol. Anim. Control Habitat Manage.,* Vol. 3, pp. 115–127.

Evelyn, J. (1670). "Sylva." J. Martin and J. Allestry, London.

Evenari, M. (1960). Plant physiology and arid zone research. *Arid Zone Res.* **18,** 175–195.

Evert, R. F., and Kozlowski, T. T. (1967). Effect of isolation of bark on cambial activity and development of xylem and phloem in trembling aspen. *Am. J. Bot.* **55,** 860–874.

Evert, R. F., Kozlowski, T. T., and Davis, J. D. (1972). Influence of phloem blockage on cambial growth of *Acer saccharum. Am. J. Bot.* **49,** 632–641.

Facteau, T. J., and Rowe, R. E. (1977). Effect of hydrogen fluoride and hydrogen chloride on pollen tube growth and sodium fluoride on pollen germination in "Tilton" apricot. *J. Am. Soc. Hortic. Sci.* **102,** 95–96.

Facteau, T. J., and Rowe, R. E. (1981). Response of sweet cherry and apricot pollen tube growth to high levels of sulfur dioxide. *J. Am. Soc. Hortic. Sci.* **106,** 77–79.

Facteau, T. J., Wang, S. Y., and Rowe, R. E. (1973). The effect of hydrogen fluoride on pollen germination and pollen tube growth in *Prunus avium* L. cv. 'Royal Ann.' *J. Am. Soc. Hortic. Sci.* **98,** 234–236.

Fahn, A., and Werker, E. (1972). Anatomical mechanisms of seed dispersal. *In* "Seed Biology" (T. T. Kozlowski, ed.), Vol. I, pp. 151–221. Academic Press, New York.

Fajer, E. D., Bowers, M. D., and Bazzaz, F. A. (1989). The effects of enriched carbon dioxide atmospheres on plant herbivore interactions. *Science* **243,** 1198–1200.

Faller, N., Herwig, K., and Kuhn, H. (1970). Die Aufnahme von Schwefeldioxyd ($S^{35}O_2$) aus der Luft. I. Einfluss auf den pflanzlichen Ertrag. *Plant Soil* **33,** 177–191.

Farmer, R. E., Jr. (1968). Sweet gum dormancy release: Effects of chilling, photoperiod, and genotype. *Physiol. Plant.* **21,** 1241–1248.

Farmer, R. E., Jr., and Bonner, F. T. (1967). Germination and initial growth of eastern cottonweed as influenced by moisture stress, temperature, and storage. *Bot. Gaz. (Chicago)* **128,** 211–215.

Farnum, P., Timmis, R., and Kulp, J. G. (1983). Biotechnology of forest yield. *Science* **219,** 694–702.

Farquhar, G. D., and Sharkey, T. D. (1982). Stomatal conductance and photosynthesis. *Annu. Rev. Plant Physiol.* **33,** 317–345.

Farrar, J. F., Relton, J., and Rutter, A. J. (1977). Sulphur dioxide and the scarcity of *Pinus sylvestris* in the industrial Pennines. *Environ. Pollut.* **14,** 63–68.

Farrar, J. L. (1961). Longitudinal variations in the thickness of the annual ring. *For. Chron.* **37,** 323–331.

Farrell, P. W., Flinn, D. W., Squire, R. O., and Craig, F. G. (1981). On the maintenance of productivity of radiata pine monocultures on sandy soils in southeast of Australia. *Proc. IUFRO Congr., 17th,* Div. I, pp. 117–128.

Fawcett, H. S. (1936). "Citrus Diseases and Their Control." McGraw-Hill, New York.

Federer, C. A., and Tanner, C. B. (1966). Spectral distribution of light in the forest. *Ecology* **47,** 555–560.

Fege, A. S., and Brown, G. N. (1984). Carbohydrate distribution in dormant *Populus* shoots and hardwood cuttings. *For. Sci.* **30,** 999–1010.

Feld, S. J. (1982). Studies on the role of irrigation and soil water matrix potential on *Phytophthora parasitica* root rot of citrus. Ph.D. Dissertation, University of California, Riverside.

Fellows, R. J., and Boyer, J. S. (1978). Altered ultrastructure of cells of sunflower leaves having low water potentials. *Protoplasma* **93,** 381–395.

Fenner, M. (1987). Seedlings. *New Phytol.* **106,** Suppl., 35–47.

Fenton, R. H., and Bond, A. R. (1964). The silvics and silviculture of Virginia pine in southern Maryland. *USDA For. Serv. Res. Pap. NE* **NE–27.**

Feret, P. P., Freyman, R. C., and Kreh, R. E. (1986). Variation in root growth potential of loblolly pine from seven nurseries. *In* "Nursery Management Practices for the Southern Pines" (D. B. South, ed.), pp. 317–328. Auburn University, Auburn, Alabama.

Ferlin, P., Flühler, H., and Palomski, J. (1982). Immissionsbedingte Fluorbelastung eines Föhrenstandortes im unteren Pfynwald. *Schweiz. Z. Forstwes.* **133,** 139–157.

Ferrell, G. T. (1978). Moisture stress threshold of susceptibility to fir engraver beetles in pole-size white fir. *For. Sci.* **24,** 85–94.

Fetcher, N., Jaeger, C. H., Strain, B. R., and Sionit, N. (1988). Long-term elevation of atmospheric CO_2 concentration and the carbon exchange rates of saplings of *Pinus taeda* L. and *Liquidambar styraciflua* L. *Tree Physiol.* **4,** 255–262.

Fidler, J. C., Wilkinson, B. G., Edney, K. L., and Sharples, R. O. (1973). "The Biology of Apple and Pear Storage." Commonwealth Agricultural Bureau, Slough, England.

Fiedler, H. J., Nebe, W., and Lerch, J. (1983). Fertilization with phosphorous and nitrogen in young spruce stands (*Picea abies* Karst.). *Fert. Res.* **4,** 155–164.

Field, C. B. (1988). On the role of photosynthetic responses in constraining the habitat distribution of rainforest plants. *Aust. J. Plant Physiol.* **15,** 343–348.

Field, C. B., and Goulden, M. L. (1988). Hydraulic lift: Broadening the sphere of plant–environment interactions. *Trees* **3,** 189–190.

Field, C. B., Merino, J., and Mooney, H. A. (1983). Compromises between water-use efficiency and nitrogen-use efficiency in five species of California evergreens. *Oecologia* **60,** 384–389.

Fielding, J. M. (1955). The seasonal and daily elongation of shoots of Monterey pine and the daily elongation of roots. *For. Timb. Bur., Aust. Leaflet* **75.**

Filer, T. H. (1975). Mycorrhizae and soil microflora in a green tree reservoir. *For. Sci.* **24,** 36–39.

Fiscus, E. L. (1984). Integrated stomatal opening as an indicator of water stress in *Zea. Crop Sci.* **24,** 245–249.

Fisher, J. T., and Mexal, J. G. (1984). Nutrition management: A physiological basis for yield improvement. *In* "Seedling Physiology and Reforestation Success." (M. L. Duryea and G. N. Brown, eds.), pp. 271–299. Martinus Nijhoff/Dr. W. Junk, The Hague, Netherlands and Boston, Massachusetts.

Fisher, R. F., Woods, R. A., and Glavicik, M. R. (1979). Allelopathic effects of goldenrod and aster on young sugar maple. *Can. J. For. Res.* **8,** 1–9.

Fitter, A. H. (1987) An architectural approach to the comparative ecology of plant root systems. *New Phytol.* **106,** Suppl., 61–77.

Flake, R. H., von Rudloff, E., and Turner, B. L. (1969). Quantitative study of clinal variation in *Juniperus virginiana* using terpenoid data. *Proc. Natl. Acad. Sci. U.S.A.* **64,** 487–494.

Flinn, D. W. (1985). Practical aspects of the nutrition of exotic conifer plantations and native eucalypt forests in Australia. *In* "Research for Forest Management" (J. J. Landsberg and W. Parsons, eds.), pp. 73–93. CSIRO, Melbourne, Australia.

Flint, H. L. (1972). Cold hardiness of twigs of *Quercus rubra* L. as a function of geographic origin. *Ecology* **53,** 1163–1170.

Flückiger, W., Oertli, J. J., and Flückiger-Keller, H. (1978). The effect of wind gusts on leaf growth and foliar water relations of aspen. *Oecologia* **34,** 101–106.

Foil, P. R., and Ralston, C. W. (1967). The establishment and growth of loblolly pine seedlings on compacted soil. *Soil Sci. Soc. Am. Proc.* **331,** 565–568.

Foote, K. C., and Schaedle, M. (1976). Diurnal and seasonal patterns of photosynthesis and respiration by stems of *Populus tremuloides* Michx. *Plant Physiol.* **58,** 651–655.

Foott, J. H., Ough, C. S., and Wolpert, J. A. (1989) Rootstock effects on wine grapes. *Calif. Agric.* **43** (4), 27–29.

Ford, E. D.(1982). High productivity in a pole stage Sitka spruce stand and its relation to canopy structure. *Forestry* **55,** 1–17.

Ford, E. D. (1984). The dynamics of plantation growth. *In* "Nutrition of Plantation Forests" (G. D. Bowen and E. K. S. Nambiar, eds.), pp. 17–52. Academic Press, Toronto.

Ford, R. H., Sharik, T. L., and Feret, P. P. (1983). Seed dispersal of the endangered Virginia round-leaf birch (*Betula uber*). *For. Ecol. Manage.* **6,** 115–128.

Fordham, R. (1972). Observations on the growth of roots and shoots of tea (*Camellia sinensis* L.) in southern Malawi. *J. Hortic. Sci.* **47,** 221–229.

Foster, D. R. (1988). Species and stand response to catastrophic winds in central New England U.S.A. *J. Ecol.* **76,** 135–151.

Foster, R. S., and Blaine, J. (1978). Urban tree survival: Trees in the sidewalk. *J. Arboric.* **4,** 14–17.

Fowells, H. A. (1965). Silvics of forest trees of the United States. *U.S., Dep. Agric., Agric. Handb.* **271.**

Fowells, H. A., and Schubert, G. H. (1956). Seed crops of forest trees in the pine region of California. *U.S., Dep. Agric., Tech. Bull.* **1150.**

Fowler, D. P., and Morris, R. W. (1977). Genetic diversity in red pine: Evidence for low genetic heterozygosity. *Can. J. For. Res.* **77,** 343–347.

Foyer, C. H. (1984). "Photosynthesis." Wiley, New York.

Fralish, J. S., and Loucks, O. L. (1975). Site quality evaluation models for aspen (*Populus tremuloides*) in Wisconsin. *Can. J. For. Res.* **5,** 523–528.

Frank, A. B., Harris, D. G., and Willis, W. O. (1977). Plant water relationships of spring wheat as influenced by shelter and soil water. *Agron. J.* **69,** 906–910.

Franklin, J. F., and Dyrness, C. T. (1973). Natural vegetation of Oregon and Washington. *U.S., For. Serv., Pac. Northwest For. Range Exp. Stn., Gen. Tech. Rep. PNW* **PNW–8.**

Franklin, J. F., Moir, W. H., Douglas, G. W., and Wiberg, C. (1971). Invasion of subalpine meadows by trees in the Cascade Range, Washington and Oregon. *Arct. Alp. Res.* **3,** 215–224.

Fraser, A. I., and Gardiner, J. B. H. (1967). Rooting and stability in Sitka spruce. *For. Comm. Bull.* (*U.K.*) **40.**

Fraser, D. A. (1952). Initiation of cambial activity in some forest trees in Ontario. *Ecology* **33,** 259–273.

Fraser, D. A. (1962). Apical and radial growth of white spruce (*Picea glauca*) (Moench Voss) at Chalk River, Ontario, Canada. *Can. J. Bot.* **40,** 659–668.

Freeland, R. O. (1952). Effect of age of leaves upon the rate of photosynthesis in some conifers. *Plant Physiol.* **27,** 685–690.

Freer-Smith, P. H. (1984). Response of six broad-leaved trees during long-term exposure to SO_2 and NO_2. *New Phytol.* **97,** 49–61.

Fretz, T. A. (1977). Identification of *Juniperus horizontalis* Moench. cultivars by foliage monoterpenes. *Sci. Hortic. (Amsterdam)* **6,** 142–148.

Friedrich, G., and Schmidt, G. (1959). Untersuchungen über das assimilatorische Verhalten von Apfel, Birne, Kirsche und Pflaume unter Verwendung einer neu entwickelten Apparatur. *Arch. Gartenbau* **7,** 321–346.

Friend, D. J. C. (1984). Shade adaptation of photosynthesis in *Coffea arabica*. *Photosynth. Res.* **5,** 325–334.

Fritschen, L., Bovee, K., Buettner, K., Charlson, R., Monteith, L., Pickford, S., Murphy, J., and Darley, E. (1970). Slash fire atmospheric pollution. *U.S., For. Serv., Res. Pap. PNW* **PNW–97.**

Fritts, H. C. (1959). The relations of radial growth to maximum and minimum temperatures in three tree species. *Ecology* **40,** 261–265.

Fritts, H. C. (1976). "Tree Rings and Climate." Academic Press, New York.

Froehlich, H. A., Miles, D. W. R., and Robbins, R. W. (1985). Soil bulk density recovery on compacted skid trails in central Idaho. *Soil Sci. Soc. Am. J.* **49,** 1015–1017.

Froehlich, H. A., Miles, D. W. R., and Robbins, R. W. (1986). Growth of young *Pinus ponderosa* and *Pinus contorta* on compacted soil in central Washington. *For. Ecol. Manage.* **15,** 285–291.

Fryer, J. H., and Ledig, F. T. (1972). Microevolution of the photosynthetic temperature optimum in relation to the elevational complex gradient. *Can. J. Bot.* **50,** 1231–1235.

Fuchinoue, H. (1982). The winter desiccation damage of tea plant in Japan. *In* "Plant Cold Hardiness and Freezing Stress" (P. H. Li and A. Sakai, eds.), Vol. 2, pp. 499–510. Academic Press, New York.

Fuchs, Y., and Lieberman, M. (1968). Effects of kinetin, IAA, and gibberellin on ethylene production, and their interactions in growth of seedlings. *Plant Physiol.* **43,** 2029–2036.

Fujii, J. A., and Kennedy, R. A. (1985). Seasonal changes in the photosynthetic rate in apple trees. A comparison between fruiting and nonfruiting trees. *Plant Physiol.* **78,** 519–524.

Fuller, R. G., Bell, D. E., and Kazmaier, H. E. (1965). Tree wound dressings for sprout control in pole line clearing. *Edison Electr. Inst. Bull.* **33,** 290–294.

Gadgil, P. D. (1972). Effect of waterlogging on mycorrhizas of radiata pine and Douglas-fir. *N. Z. J. For. Sci.* **2,** 222–226.

Gaff, D. F. (1980). Protoplasmic tolerance of extreme water stress. *In* "Adaptation of Plants to Water and High Temperature Stress" (N. C. Turner and P. J. Kramer, eds.), pp. 207–230. Wiley, New York.

Gaff, D. F., and Carr, D. J. (1961). The quantity of water in the cell wall and its significance. *Austr. J. Biol. Sci.* **14,** 299–311.

Gallagher, P. W., and Sydnor, T. D. (1983). Variation in wound response among cultivars of red maple. *J. Am. Soc. Hortic. Sci.* **198,** 744–746.

Galloway, J. N., Likens, G. E., and Hawley, M. E. (1984). Acid precipitation: Natural versus anthropogenic components. *Science* **226,** 829–831.

Gara, R. I., Geiszler, D. R., and Littke, W. R. (1984). Primary attraction of the mountain pine beetle to lodgepole pine in Oregon. *Ann. Entomol. Soc. Am.* **77,** 333–334.

Garner, W. W., and Allard, H. A. (1923). Further studies in photoperiodism, the response of the plant to the relative length of day and night. *J. Agric. Res.* **28,** 871–920.

Garrec, J. P., Plebin, R., and Audin, M. (1981). Effets du fluor sur les teneurs en sucres et en acides amines d'aiguilles de sapin *Abies alba*. Mill. *Environ. Pollut., Ser. A* **26,** 281–295.

Garsed, S. G., Farrar, J. F., and Rutter, A. J. (1979). The effects of low concentrations of sulphur dioxide on the growth of four broadleaved tree species. *J. Appl. Ecol.* **16,** 217–226.

Gashwiler, J. S. (1967). Conifer seed survival in a western Oregon clearcut. *Ecology* **48,** 431–438.

Gates, D. M. (1973). Plant temperatures and energy budget. *In* "Temperature and Life" (H. Precht, J. Christophersen, H. Hensel, and W. Larcher, eds.), pp. 87–101. Springer-Verlag, New York.

Gates, D. M. (1976). Energy exchange and transpiration. *In* "Water and Plant Life" (O. L. Lange, L. Kappen, and E. -D. Schulze, eds.), pp. 137–147. Springer-Verlag, Berlin and New York.

Gates, D. M. (1980). "Biophysical Ecology." Springer-Verlag, New York and Berlin.

Gates, D. M., Keegan, H. J., Schleter, J. C., and Weidner, V. R. (1965). Spectral properties of plants. *Appl. Opt.* **4,** 11–20.

Gäumann, E. A. (1950). "Principles of Plant Infection" (English translation by W. B. Brierley). Crosby, Lockwood, London.

Geiger, D. R., and Savonick, S. A. (1975). Effect of temperature, anoxia, and other metabolic inhibitors on translocation. *Encycl. Plant Physiol., New Ser.* **1,** 256–286.

Geisler, D., and Ferree, D. C. (1984a). Response of plants to root pruning. *Hortic. Rev.* **6,** 155–188.

Geisler, D., and Ferree, D. C. (1984b). The influence of root pruning on water relations, net photosynthesis and growth of young 'Golden Delicious' apple trees. *J. Am. Soc. Hortic. Sci.* **109,** 827–831.

Geiszler, D. R., Gara, R. I., Driver, C. H., Gallucci, V. F., and Martin, R. E. (1980). Fire, fungi, and beetle influences on a lodgepole pine ecosystem in South-Central Oregon. *Oecologia* **46,** 239–243.

Gendle, P., Clifford, D. R., Mercer, P. C., and Kirk, S. A. (1983). Movement, persistence, and performances of fungitoxicants applied as pruning wound treatments on apple trees. *Ann. Appl. Biol.* **102,** 281–291.

Genys, J. B., and Heggestad, H. E. (1978). Susceptibility of different species, clones, and strains of pines to acute injury caused by ozone and sulfur dioxide. *Plant Dis. Rep.* **62,** 687–691.

Geurten, I. (1950). Untersuchungen über den Gaswechsel von Baumrinden. *Forstwiss. Centralbl.* **69,** 704–743.

Gholz, H. L. (1982). Environmental limits on above ground net primary production, leaf, area, and biomass in vegetation zones of the Pacific Northwest. *Ecology* **54,** 152–159.

Gholz, H. L. (1986). Canopy development and dynamics in relation to primary production. *In* "Crown and Canopy Structure in Relation to Productivity" (T. Fujimori and D. Whitehead, eds.), pp. 224–242. Forestry and Forest Products Research Institute, Ibaraki, Japan.

Gholz, H. L., and Fisher, R. F. (1982). Organic matter production and distribution in slash pine (*Pinus elliottii*) plantations. *Ecology* **63,** 1827–1839.

Gibbs, R. D. (1935) Studies of wood. II. The water content of certain Canadian trees, and changes in the water-gas system during seasoning and flotation. *Can. J. Res.* **12,** 727–760.

Gibbs, R. D. (1974). "Chemotaxonomy of Flowering Plants." McGill-Queen's Univ. Press, Montreal.

Gibson, A., and Bachelard, E. P. (1986). Germination of *Eucalyptus sieberi* L. Johnson seeds. II. Internal water relations. *Tree Physiol.* **1,** 67–77.

Giddings, J. L. (1962). Development of tree-ring dating as an archeology aid. *In* "Tree Growth" (T. T. Kozlowski, ed.), pp. 119–132. Ronald Press, New York.

Gifford, R. M., Thorne, J. H., Hitz, W. D., and Giaquinta, R. T. (1984). Crop productivity and photoassimilate partitioning. *Science* **225,** 801–808.

Gilbert, O. L. (1983). The growth of planted trees subject to fumes from brickworks. *Environ. Pollut., Ser. A* **31,** 301–310.

Gill, A. M. (1981). Fire adaptive traits of vascular plants. *Gen. Tech. Rep. WO (U.S., For. Serv.)* **WO-26,** 208–230.

Gill, A. M., and Groves, R. H. (1981) Fire regimes in heathlands and their plant ecologic effects. *In* "Ecosystems of the World. Vol. 9b. Heathland and Related Shrublands" (R. L. Specht, ed.), pp. 61–84. Elsevier, Amsterdam.

Gill, A. M., and Tomlinson, P. B. (1975). Aerial roots: An array of forms and functions. *In* "The Development and Function of Roots" (J. G. Torrey and D. T. Clarkson, eds.), pp. 237–260. Academic Press, London.

Gill, C. J. (1970). The flooding tolerance of woody species—A review. *For. Abstr.* **31,** 671–688.

Gill, C. J. (1975). The ecological significance of adventitious rooting as a response to flooding in woody species, with special reference to *Alnus glutinosa* L. Gaertn. *Flora (Jena)* **164,** 85–97.

Gimmingham, C. H. (1972). "Ecology of Heathlands." Chapman & Hall, London.

Ginter-Whitehouse, D. L., Hinckley, T. M., and Pallardy, S. G. (1983). Spatial and temporal aspects of water relations of three tree species with different vascular anatomy. *For. Sci.* **29,** 317–329.

Givnish, T. J. (1986a). Optimal stomatal conductance, allocation of energy between leaves and roots, and the marginal cost of transpiration. *In* "On the Economy of Plant Form and Function" (T. J. Givnish, ed.), pp. 171–213. Cambridge Univ. Press, Cambridge, England.

Givnish, T. J., ed. (1986b). "On the Economy of Plant Form and Function." Cambridge Univ. Press, Cambridge.

Givnish, T. J. (1988). Adaptation to sun and shade: A whole-plant perspective. *Aust. J. Plant Physiol.* **15,** 63–92.

Glerum, C. (1976). Frost hardiness of forest trees. *In* "Tree Physiology and Yield Improvement" (M. G. R. Cannell and F. T. Last, eds.), pp. 403–420. Academic Press, New York.

Glerum, C., and Farrar, J. L. (1965). A note on internal frost damage in white spruce needles. *Can. J. Bot.* **43,** 1590–1591.

Glerum, C., and Farrar, J. L. (1966). Frost ring formation in the stems of some coniferous species. *Can. J. Bot.* **44,** 879–886.

Glock, W. S. (1955). Tree growth. II. Growth rings and climate. *Bot. Rev.* **21,** 73–188.

Gloyne, R. W. (1954). Some effects of shelterbelts on local microclimate. *Forestry* **27,** 85–95.

Gloyne, R. W. (1976). Shelter in agriculture, forestry, and horticulture—A review. *A.D.A.S. Q. Rev.* **21,** 197–207.

Goddard, R. E., Zobel, B. J., and Hollis, C. A. (1976). Response of *Pinus taeda* and *Pinus elliottii* to varied nutrition. *In* "Tree Physiology and Yield Improvement" (M. G. R. Cannell and F. T. Last, eds.), pp. 449–462. Academic Press, London.

Gold, M. A., and Hanover, J. W. (1987). Agroforestry systems for the temperate zone. *Agrofor. Syst.* **5,** 109–121.

Gold, W. G., and Caldwell, M. M. (1983). The effects of ultraviolet-B radiation on plant competition in terrestrial ecosystems. *Physiol. Plant.* **58,** 435–444.

Goldstein, G., Meinzer, F., and Monastero, B. (1984). The role of capacitance in the water balance of Andean giant rosette species. *Plant, Cell Environ.* **7,** 179–186.

Gollan, T., Turner, N. C., and Schulze, E.-D. (1985). The responses of stomata and leaf gas exchange to vapour pressure deficits and soil water content. III. In the sclerophyllous woody species *Nerium oleander*. *Oecologia* **65,** 356–362.

Good, N. F., and Good, R. E. (1972). Population dynamics of tree seedlings and saplings in a mature eastern hardwood forest. *Bull. Torrey Bot. Club* **99,** 172–178.

Goode, J. E., and Higgs, K. H. (1973). Water, osmotic and pressure potential relationships in apple leaves. *J. Hortic. Sci.* **48,** 203–215.

Goode, J. E., and Hyrycz, K. J. (1964). The response of Laxton's Superb apple trees to different soil moisture conditions. *J. Hortic. Sci.* **39,** 254–276.

Goor, A. Y., and Barney, C. W. (1968). "Forest Tree Planting in Arid Zones." Ronald Press, New York.

Gordon, A. G., and Gorham, E. (1963). Ecological aspects of air pollution from an iron sintering plant at Wawa, Ontario. *Can. J. Bot.* **41,** 1063–1078.

Gordon, G. T. (1973). Damage from wind and other causes in mixed white fir–red fir stands adjacent to clearcuttings. *USDA For. Serv. Res. Pap. PSW* **PSW–90.**

Gosz, J. R. (1980). Biomass distribution and production budget for a nonaggrading forest ecosystem. *Ecology* **61,** 507–514.

Govind, S., and Prasad, A. (1982). Effect of nitrogen nutrition on fruit-set, fruit drop and yield in sweet orange. *Punjab Hortic. J.* **22**(1/2), 15–20.

Gower, S. T., Grier, C. C., and Vogt, K. A. (1989). Aboveground production and N and P use by *Larix occidentalis* and *Pinus contorta* in the Washington Cascades, USA. *Tree Physiol.* **5,** 1–11.

Grace, J. (1978). "Plant Response to Wind." Academic Press, London.

Grace, J. (1987). Climatic tolerance and distribution of plants. *New Phytol.* **106,** Suppl., 113–130.

Grace, J., and Thompson, J. R. (1973). The after-effect of wind on photosynthesis and transpiration of *Festuca arundinacea*. *Physiol. Plant.* **28,** 541–547.

Graham, D., and Patterson, B. D., (1982). Responses of plants to low, non freezing temperatures: Proteins, metabolism, and acclimation. *Annu. Rev. Plant Physiol.* **33,** 347–372.

Gratkowski, H. (1961a). "Brush Problems in Southwestern Oregon." USDA, For. Serv. Pac. Northwest For. Range Exp. Stn., Portland, Oregon.

Gratkowski, H. (1961b). Brush seedlings after controlled burning of brushlands in southwestern Oregon. *J. For.* **45,** 118–120.

Greacen, E. L., and Sands, R. (1980). Compaction of forest soils; A review. *Aust. J. Soil Res.* **18,** 163–189.

Green, D. S. (1980). The terminal velocity and dispersal of spinning samaras. *Am. J. Bot.* **67,** 1218–1224.

Green, K., and Wright, R. (1977). Field response of photosynthesis to CO_2 enhancement in ponderosa pine. *Ecology* **58,** 687–692.

Green, V. E. (1980). Use phytomass and epiphytotic when referring to plants. *Agron. J.* **72,** 1068.

Green, W. E. (1947). Effect of water impoundment on tree mortality and growth. *J. For.* **45,** 118–120.

Greenham, D. W. P. (1976). The fertilizer requirements of fruit trees. *Proc. Fert. Soc.* **157,** 1–32.

Greenland, D. J., and Kowal, J. M. L. (1960). Nutrient content of moist tropical forest of Ghana. *Plant Soil* **12,** 154–174.

Gregory, J. D., Guiness, W. M., and Davey, C. B. (1982). Fertilization and irrigation stimulate flowering and cone production in a loblolly pine seed orchard. *South. J. Appl. For.* **6,** 44–48.

Greszta, J., Braniewski, S., and Nosek, A. (1982). The effect of dusts from different emitters on the height increment of the seedlings of selected tree species. *Fragm. Florist. Geobot.* **28,** 67–75.

Grier, C. C. (1975). Wildfire effects on nutrient distribution and leaching in a coniferous ecosystem. *Can. J. For. Res.* **5,** 599–607.

Grieve, B. J., and Hellmuth, E. O. (1970). Ecophysiology of western Australian plants. *Oecol. Plant.* **5,** 33–68.

Griffin, A. R., Crane, W. J. B., and Cromer, R. N. (1984). Irrigation and fertilizer effects on productivity of a *Pinus radiata* seed orchard: Response to treatment of an established orchard. *N. Z. J. For. Sci.* **14,** 289–302.

Griffin, D. H., Quinn, K., and McMillen, B. (1986). Regulation of hyphal growth rate of *Hypoxylon mammatum* by amino acids: Stimulation by proline. *Exp. Mycol.* **10,** 307–314.

Grill, D. (1973). A study by scanning electron microscope of spruce needles exposed to SO_2. *Phytopathol. Z.* **78,** 75–80.

Grill, D., Liegl, E., and Windisch, E. (1979). E. (1979). Holzanatomische Untersuchungen an abgasbelasteten Bäumen. *Phytopathol. Z.* **94,** 335–342.

Grime, J. P. (1965). Shade tolerance in flowering plants. *Nature* (*London*) **208,** 161–163.

Grime, J. P. (1966). Shade avoidance and shade tolerance in flowering plants. *In* "Light as an Ecological Factor" (R. Bainbridge, G. C. Evans, and O. Rackham, eds.), pp. 187–207. Blackwell, Oxford.

Grochowska, M. J. (1973). Comparative studies on physiological and morphological features of bearing and non-bearing spurs of the apple tree. I. Changes in starch content during growth. *J. Hortic. Sci.* **48,** 347–356.

Grodzinski, W., Weiner, J., and Maycock, P. F. (1984). "Forest Ecosystems in Industrial Regions." Springer-Verlag, New York.

Gross, H. L. (1972). Crown deterioration and reduced growth associated with excessive seed production by birch. *Can. J. Bot.* **50,** 2431–2437.

Guderian, R., and Kueppers, H. (1980). Responses of plant communities to air pollution. *Gen. Tech. Rep. PSW Pac. Southwest For. Range Exp. Stn.* **PSW-43,** 187–199.

Guderian, R., and Stratmann, H. (1968). Freilandversuche zur Ermittlung von Schwefeldioxidwirkungen auf die Vegetation. *Forschungsber. Landes Nordrhein-Westfalen* **1920,** 114.

Guries, R. P., and Nordheim, E. V. (1984). Flight characteristics and dispersal potential of maple samaras. *For. Sci.* **30,** 434–440.

Hacskaylo, J., Finn, R. F., and Vimmerstedt, J. P. (1969). Deficiency symptoms of some forest trees. *Ohio Agric. Res. Dev. Cent., Res. Bull.* **1015.**

Hader, D., and Tevini, M. (1987). "General Photobiology." Pergamon, Oxford.

Hadley, J. L., and Smith, W. K. (1986). Wind effects on needles of timberline conifers: Seasonal influences on mortality. *Ecology* **67,** 12–19.

Hagen, L. J. (1976). Windbreak design for optimum wind erosion control. *In* "Great Plains Symposium on Shelterbelts" (R. W. Tinus, ed.), Great Plains Agric. Counc., Lincoln, NE, Publ. No. 78, pp. 31–36.

Hagglund, B. (1981). Evaluation of forest site productivity. *For. Abstr.* **42,** 515–527.

Hahn, G. G., Hartley, C., and Rhoads, A. S. (1920). Hypertrophied lenticels on roots of conifers and their relation to moisture and aeration. *J. Agric. Res.* **20,** 253–265.

Haines, L. H., Maki, T. E., and Sanderford, S. G. (1975). The effect of mechanical site preparation treatments on soil productivity and tree (*Pinus taeda* L. and *P. elliottii* var. *elliottii)* growth. *In* "Forest Soils and Forest Land Management" (B. Bernier and C. H. Winget, eds.), pp. 379–395. Laval Univ. Press, Quebec.

Haines, L. W. and Pritchett, W. L. (1965). The effect of site preparation on the availability of soil nutrients and on slash pine growth. *Proc.—Soil Crop Sci. Soc. Fla.* **25,** 356–374.

Haines, S. G., Haines L. W., and White, G. (1978). Leguminous plants increase sycamore growth in northern Alabama. *Soil Sci. Soc. Am. J.* **42,** 130–132.

Haissig, B. E. (1974). Metabolism during adventitious root primordium initiation and development. *N. Z. J. For. Sci.* **4,** 324–337.

Haissig, B. E. (1982). Carbohydrate and amino acid concentrations during adventitious root primordium development in *Pinus banksiana* Lamb. cuttings. *For. Sci.* **28,** 813–821.

Halbwachs, G. (1984). Organismal responses of higher plants to atmospheric pollutants: Sulphur dioxide and fluoride. *In* "Air Pollution and Plant Life" (M. Treshow, ed.), pp. 175–214. Wiley, Chichester.

Hale, C. R., and Weaver, R. J. (1962). The effect of developmental stage on direction of translocation of photosynthate in *Vitis vinifera. Hilgardia* **33,** 89–131.

Hale, J. D. (1951). Variations in properties of wood caused by structural differences. *In* "Canadian Woods, Their Properties and Uses" (T. A. McElhanney, ed.), pp. 57–104. For. Branch, For. Prod. Lab., Ottawa, Canada.

Hall, D.O. and de Groot, P.J. (1987). Introduction: The biomass framework. *In* "Biomass" (D.O. Holland and R.P. Overend, eds.), pp. 3–24. Wiley, New York.

Hall, J. B., and Swaine, M. D. (1980). Seed stocks in Ghanaian forest soils. *Biotropica* **12,** 256–263.

Hall, M. A., Kapuya, J. A., Sivakumaran, S., and John, A. (1977). The role of ethylene in the responses of plants to stress. *Pestic. Sci.* **8,** 217–223.

Hall, T. F., Penfound, W. T., and Hess, A. D. (1946). Water level relationships of plants in the Tennessee Valley with particular reference to malaria control. *J. Tenn. Acad. Sci.* **21,** 18–59.

Hall, T. F., and Smith, G. E. (1955). Effects of flooding on woody plants. West Sandy dewatering project, Kentucky Reservoir. *J. For.* **53,** 281–285.

Hanes, T. L. (1988). California chaparral. *In* "Terrestrial Vegetation of California" (M.G. Barbour and J. Major, eds.), pp. 417–469. Wiley, New York.

Hanke, D. E. (1989). Second messengers in plant growth and development. *Science* **246,** 511–512.

Hanover, J. W., and Keathley, D. E., eds. (1988). "Genetic Manipulation of Woody Plants." Plenum, New York.

Hansen, E. A., and Dawson, J. O. (1982). Effect of *Alnus glutinosa* on hybrid *Populus* height growth in a short-rotation intensively cultured plantation. *For. Sci.* **28,** 49–59.

Hansen, H. L. (1980). The Lake States region. *In* "Regional Silviculture of the United States" (J. W. Barrett, ed.), 2nd ed. pp. 67–105. Wiley, New York.

Hansen, P. (1970). ^{14}C-Studies on apple trees. IV. Photosynthetic consumption in fruits in relation to the leaf-fruit ratio and to the leaf-fruit position. *Physiol. Plant.* **23,** 805–810.

Hanson, P. J., McLaughlin, S. B., and Edwards, N. T. (1988a). Net CO_2 exchange of *Pinus taeda* shoots exposed to variable ozone levels and rain chemistries in field and laboratory settings. *Physiol. Plant.* **74,** 635–642.

Hanson, P. J., Isebrands, J. G., Dickson, R. E., and Dixon, R. K. (1988b). Ontogenetic patterns of CO_2 exchange of *Quercus rubra* L. leaves during three flushes of shoot growth. I. Median flush leaves. *For. Sci.* **34,** 55–68.

Harbinson, J., and Woodward, F. I. (1984). Field measurements of the gas exchange of woody plants species in simulated sunflecks. *Ann. Bot.* (*London*) [N.S.] **53,** 841–851.

Harcombe, P. A. (1987). Tree life tables. *BioScience* **37,** 557–568.

Harcombe, P. A., and Marks, P. L. (1983). Five years of tree death in a *Fagus–Magnolia* forest, southeast Texas (USA). *Oecologia* **57,** 49–54.

Hard, J. S. (1985). Spruce beetles attack slowly growing spruce. *For. Sci.* **31,** 839–850.

Hard, J. S., Werner, R. A., and Holsten, E. H. (1983). Susceptibility of white spruce to attack by spruce beetles during the early years of an outbreak in Alaska. *Can. J. For. Res.* **13,** 678–684.

Hardy, R. W. F., and Havelka, U. D. (1976). Photosynthesis as a major factor limiting N_2 fixation by field grown legumes with emphasis on soybeans. *In* "Symbiotic Nitrogen Fixation in Plants" (P.S. Nutman, ed.), pp. 421–439. Cambridge Univ. Press, London.

Harley, J. L., and Smith, S. E. (1983). "Mycorrhizal Symbiosis." Academic Press, New York.

Harlow, W. M., Harrar, E. S., and White, F. M. (1979). "Textbook of Dendrology." McGraw-Hill, New York.

Harmer, R. (1988) Production and use of epicormic shoots for the vegetative propagation of mature oak. *Forestry* **61,** 305–316.

Harmon, M. E., Ferrell, W. K., and Franklin, J. E. (1990). Effects on carbon storage of conversion of old-growth forests to young forests. *Science* **247,** 699–702.

Harrington, C. A., and Reukema, D. L. (1983). Initial shock and long-term stand development following thinning in a Douglas-fir plantation. *For. Sci.* **29,** 33–46.

Harris, M. D. (1975). Effects of initial flooding on forest vegetation at two Oklahoma lakes. *J. Soil Water Conserv.* **30,** 294–295.

Harris, R. W. (1983). "Arboriculture." Prentice-Hall, Englewood Cliffs, New Jersey.

Harris, W. F., Kinerson, R. S. Jr., and Edwards, N. T. (1977). Comparison of below-ground biomass of natural deciduous forests and loblolly pine plantations. *Range Sci. Dep. Sci. Ser.* (*Colo. State Univ.*) **26,** 29–37.

Hart, J. W. (1988). "Light and Plant Growth." Unwin Hyman, London.

Hartage, R. H. (1968). Permeability to air of different pore-volume fractions. *Z. Pflanzenernaehr. Bodenkd.* **120,** 31–45.

Hartmann, H. T., and Kester, D. E. (1982). "Plant Propagation: Principles and Practices." Prentice-Hall, Englewood Cliffs, New Jersey.

Hatano, K., and Asakawa, S. (1964). Physiological processes in forest tree seeds during maturation, storage, and germination. *Int. Rev. For. Res.* **1,** 279–323.

Havas, P., and Huttunen, S. (1972). The effects of air pollution on the radial growth of Scots pine (*Pinus sylvestris*). *Biol. Conserv.* **4,** 361–368.

Havis, J. R. (1971). Water movement in stems during freezing. *Cryobiology* **8,** 581–585.

Havis, J. R. (1976). Root hardiness of woody ornamentals. *HortScience* **11,** 385–386.

Haynes, R. J. (1981). Some observations on the effect of grassing-down, nitrogen fertilization and irrigation on the growth, leaf nutrient content and fruit quality of young Golden Delicious apple trees. *J. Sci. Food Agric.* **32,** 1005–1013.

Haynes, R. J., and Goh, K. M. (1980). Seasonal levels of available nutrients under grassed-down, cultivated, and zero-tilled orchard soil management practices. *Aust. J. Soil Res.* **18,** 363–373.

Head, G. C. (1967). Effects of seasonal changes in shoot growth on the amount of unsuberized root in apple and plum trees. *J. Hortic. Sci.* **42,** 169–180.

Heagle, A. S., and Johnston, J. W. (1979). Variable responses of soybeans to mixtures of ozone and sulfur dioxide. *J. Air Pollut. Control. Assoc.* **29,** 729–732.

Heath, O. V. S., and Meidner, H. (1967). Compensation points and carbon dioxide enrichment for lettuce grown under glass in winter. *J. Exp. Bot.* **18,** 746–751.

Heath, R. L. (1980). Initial events in injury to plants by air pollutants. *Annu. Rev. Plant Physiol.* **31,** 395–431.

Heck, W. W. (1973). Air pollution and the future of agricultural production. *Adv. Chem. Ser.* **122,** 118–129.

Hedlund, A. (1964). Epicormic branching in north Louisiana delta. *Res. Note SO* (*U.S. For. Serv.*) **S0–8.**

Hegarty, T.W. (1975). Effects of fluctuating temperature on germination and emergence of seeds in different moisture environments. *J. Exp. Bot.* **26,** 203–211.

Heiberg, S. O., and White, D. P. (1951). Potassium deficiency of reforested pine and spruce stands in northern New York. *Soil Sci. Soc. Am. Proc.* **15,** 369–376.

Heide, O. M. (1974). Growth and dormancy in Norway spruce ecotypes (*Picea abies*). I. Interaction of photoperiod and temperature. *Physiol. Plant.* **30,** 1–12.

Heikilla, J. J., Papp, E. T., Schultz, G. A., and Bewley, J. D. (1984). Induction of heat shock protein messenger RNA in maize mesocotyls by water stress, abscisic acid, and wounding. *Plant Physiol.* **76,** 270–274.

Heilman, P. (1981). Root penetration of Douglas-fir seedlings into compacted soil. *For. Sci.* **27,** 660–666.

Heinicke, A. J. (1932). The effect of submerging the roots of apple trees at different seasons of the year. *Proc. Am. Soc. Hortic. Sci.* **29,** 205–207.

Heinicke, D. R., and Childers, N.F. (1937). The daily rate of photosynthesis during the growing season of 1935, of a young apple tree of bearing age. *Mem.—N.Y., Agric. Exp. Stn.* (*Ithaca*) **201.**

Heinrichs, E. A., ed. (1988). "Plant Stress—Insect Interactions." Wiley, Interscience, New York.

Heinselman, M. L. (1970). Landscape evolution, peatland types and the environment in the Agassiz Peatlands Natural Area, Minnesota. *Ecol. Monogr.* **46,** 59–84.

Heisler, G. M., and Herrington, L. P. (1976). Selection of trees for modifying metropolitan climates. *USDA For. Serv. Gen. Tech. Rep. NE* **NE-22,** 31–37.

Hellkvist, J., Richards, G. P., and Jarvis, P. G. (1974). Vertical gradients of water potential and tissue water relations in Sitka spruce trees measured with the pressure chamber. *J. Appl. Ecol.* **11,** 637–667.

Hellmers, H. (1962). Temperature effect upon optimum tree growth. *In* "Tree Growth" (T. T. Kozlowski, ed.), pp. 275–287. Ronald Press, New York.

Hellmers, H. (1963). Effect of soil and air temperatures on growth of redwood seedlings. *Bot. Gaz.* (*Chicago*) **124,** 172–177.

Hellmers, H. (1966). Growth response of redwood seedlings to thermoperiodism. *For. Sci.* **12,** 276–283.

Hellmers, H., and Bonner, J. (1960). Photosynthetic limits of forest tree yields. *Proc. Soc. Am. For., 1959,* pp. 32–35.

Hellmers, H., and Rook, D. A. (1973). Air temperature and growth of radiata pine seedlings. *N. Z. J. For. Sci.* **3,** 217–285.

Hellmers, H., Horton, J. S., Juhren, G., and O'Keefe, J. (1955). Root systems of some chaparral plants in southern California. *Ecology* **36,** 667–678.

Hellmers, H., Genthe, M. K., and Ronco, F. (1970). Temperature affects growth and development of Engelmann spruce. *For. Sci.* **16,** 447–452.

Helms, J. A. (1964). Apparent photosynthesis of Douglas-fir in relation to silvicultural treatment. *For. Sci.* **10,** 432–442.

Helms, J. A. (1976). Factors affecting net photosynthesis in trees: An ecological viewpoint. *In* "Tree Physiology and Yield Improvement" (M. G. R. Cannell and F. T. Last, eds.), pp. 55–109. Academic Press, London.

Henke, R. R., Hughes, K. W., Constantin, M. J., and Hollaender, A., eds. (1985). "Tissue Culture in Forestry and Agriculture." Plenum, New York.

Hennerty, M. J., O'Kennedy, B. T., and Titus, J. S. (1980). Conservation and reutilization of bark proteins in apple trees. *In* "Mineral Nutrition of Fruit Trees" (D. Atkinson, J. E. Jackson, R. O. Sharples, and W. M. Waller, eds.), pp. 369–377. Butterworth, London.

Henwood, K. (1973). A structural model of forces in buttressed tropical rain forest trees. *Biotropica* **5,** 83–93.

Herrick, A.M., and Brown, C.L. (1967). A new concept in cellulose production: Silage sycamore. *Agric. Sci. Rev.*, pp. 8–13.

Hett, J. M., and Loucks, O. L. (1971). Sugar maple (*Acer saccharum* Marsh.) seedling mortality. *J. Ecol.* **59,** 507–520.

Heyward, F. (1938). Soil temperatures during forest fires in the longleaf pine region. *J. For.* **36,** 478–491.

Hickey, J. E., Blakesley, A. J., and Turner, B. (1983). Seedfall and germination of *Nothofagus cun-*

ninghamii (Hook.) Oerst., *Ecryphia lucida* (Labill.) Baill and *Atherosperma moschatum* Labill. Implications for regeneration practice. *Aust. For. Res.* **13,** 21–28.

Hill, A. G. C., and Campbell, C. K. G. (1949). Prolonged dormancy of deciduous fruit trees in warm climates. *Emp. J. Exp. Agric.* **17,** 259–264.

Hill-Cottingham, D. G., and Williams, R. R. (1967). Effect of time of application of fertilizer nitrogen in the growth, flower development, and fruit set of maiden apple trees, var. Lord Lambourne, and on the distribution of total nitrogen within the trees. *J. Hortic. Sci.* **42,** 319–338.

Hillerdal-Hagstrommer, K., Mattson-Djos, E., and Hellkvist, J. (1982). Field studies of water relations and photosynthesis in Scots pine. II. Influence of irrigation and fertilization on needle water potentials of young pine trees. *Physiol. Plant.* **54,** 295–301.

Himelick, E. B., Neely, D., and Crowley, W. R., Jr. (1965). Experimental field studies on shade tree fertilization. *Biol. Notes (Ill. Nat. Hist. Surv.)* **53,** 1–12.

Hincha, D. K., Höfner, R., Schwab, K. B., Heber, U., and Schmitt, J. M. (1987). Membrane rupture is the common cause of damage to chloroplast membranes in leaves injured by freezing or excessive wilting. *Plant Physiol.* **83,** 251–253.

Hinckley, T. M., Chambers, J. L., Bruckerhoff, D. N., Roberts, J. E., and Turner, J. (1974). Effect of mid-day shading on stem diameter, xylem pressure potential, leaf surface resistance, and net assimilation rate in a white oak sapling. *Can. J. For. Res.* **4,** 296–300.

Hinckley, T. M., Duhme, F., Hinckley, A. R., and Richter, H. (1980). Water relations of drought hardy shrubs: Osmotic potential and stomatal resistivity. *Plant, Cell Environ.* **3,** 131–140.

Hirst, J. M. (1959). Spore liberation and dispersal. *In* "Plant Pathology: Problems and Progress, 1908–1958" (C. S. Holton *et al.*, eds.), pp. 529–538. Univ of Wisconsin Press, Madison.

Hirst, J. M., Stedman, O. J., and Hogg, W. H. (1967). Long-distance spore transport: Methods of measurement, vertical spore profiles and the detection of immigrant spores. *J. Gen. Microbiol.* **48,** 329–355.

Hoad, G. V., and Abbott, D. L. (1983). Hormonal control of growth and reproductive development in apple. *In* "Regulation of Photosynthesis in Fruit Trees" (A. N. Lakso and F. Long, eds.), Spec. Rep. N. Y. State Agric. Exp. Stn., Geneva.

Hodges, J. D. (1967). Patterns of photosynthesis under natural environmental conditions. *Ecology* **48,** 234–242.

Hoekstra, P. E., and Asher, W. C. (1962). Diameter growth of pole-size slash pine after fertilization. *J. For.* **60,** 341–342.

Hoffman, J. S. (1984). Carbon dioxide and future forests. *J. For.* **82,** 164–167.

Hoffman, J. S., and Wells, J. B. (1987). Forests: Past and projected changes in greenhouse gases. *In* "The Greenhouse Effect, Climate Change, and U.S. Forests" (W. E. Shands and J. S. Hoffman, eds.), pp. 19–41. Conservation Foundation, Washington, D.C.

Hogue, E. (1982). Biochemical aspects of stress physiology of plants and some consideration of defense mechanisms in conifers. *Eur. J. For. Pathol.* **12,** 280–296.

Holbrook, S.H. (1944). "Burning an Empire. The Story of American Forest Fires." Macmillan, New York.

Holch, A. E. (1931). Development of roots and shoots of certain deciduous tree seedlings in different forest sites. *Ecology* **12,** 259–298.

Holt, M. A., and Pellett, N. E. (1981). Cold hardiness of leaf and stem organs of *Rhododendron* cultivars. *J. Am. Soc. Hortic. Sci.* **106,** 608–612.

Holzer, K. (1973). Die Vererbung von physiologischen und morphologischen Eigenschaften der Fichte. II. Mutterbaummerkmall Unveröffentliches Manuskript.

Hook, D. D. (1984). Adaptations to flooding with fresh water. *In* "Flooding and Plant Growth" (T. T. Kozlowski, ed.), pp. 265–294. Academic Press, Orlando, Florida.

Hook, D. D., and Brown, C. L. (1973). Root adaptations and relative flood tolerance of five hardwood species. *For. Sci.* **19,** 225–229.

Hook, D. D., Brown, C. L., and Kormanik, P. O. (1970a). Lenticels and water root development of swamp tupelo under various flooding conditions. *Bot. Gaz. (Chicago)* **131,** 217–224.

Hook, D. D., Langdon, O. G., Stubbs, J., and Brown, C. L. (1970b). Effects of water regimes on the survival, growth, and morphology of tupelo seedlings. *For. Sci.* **16,** 304–311.

Hook, D.D., Brown, C.L., and Kormanik, P.P. (1971). Inductive flood tolerance in swamp tupelo [*Nyssa sylvatica* var. *biflora* (Walt.) Sarg.]. *J. Exp. Bot.* **22,** 78–89.

Hoover, M. D. (1944). Effect of removal of forest vegetation upon water yields. *Trans. Am. Geophys. Union* **25,** 969–977.

Hoover, M. D., Olson, D. R., and Greene, G. E. (1953). Soil moisture under a young loblolly pine plantation. *Soil Sci. Soc. Am. Proc.* **17,** 147–150.

Hopen, H. J., and Ries, S. K. (1962). The mutually compensating effect of carbon dioxide concentrations and light intensities on the growth of *Cucumis sativus* L. *Proc. Am. Soc. Hortic. Sci.* **81,** 358–364.

Horler, D. N. H., Barber, J., and Barringer, A. R. (1980). Effects of heavy metals on the absorbance and reflectance of plants. *Int. J. Remote Sens.* **1,** 121–136.

Horler, D. N. H., Dockray, M., Barber, J., and Barringer, A. R. (1983). Red edge measurements for remotely sensing plant chlorophyll content. *Adv. Space Res.* **3,** 273–277.

Horsley, S. B. (1977a). Allelopathic inhibition of black cherry by fern, grass, goldenrod, and aster. *Can. J. For. Res.* **7,** 205–216.

Horsley, S. B. (1977b). Allelopathic inhibition of black cherry. II. Inhibition by woodland grass, ferns, and club moss. *Can. J. For. Res.* **7,** 515–519.

Horsley, S. B. (1986). Evaluation of hayscented fern interference with black cherry. *Am. J. Bot.* **73,** 668–669.

Hosner, J. F. (1957). Effects of water upon the seed germination of bottomland trees. *For. Sci.* **3,** 67–71.

Hosner, J. F. (1958). The effects of complete inundation upon seedlings of six bottomland tree species. *Ecology* **39,** 371–373.

Hosner, J. F. (1960). Relative tolerance to complete inundation of fourteen bottomland tree species. *For. Sci.* **6,** 246–251.

Hosner, J. F. (1962). The southern bottomland region. *In* "Regional Silviculture of the United States" (J. W. Barrett, ed.), pp. 296–333. Wiley, New York.

Hosner, J. F., and Leaf, A. L. (1962). The effect of soil saturation upon the dry weight, ash content and nutrient absorption of various bottomland species. *Soil Sci. Soc. Am. Proc.* **26,** 401–404.

Hosner, J. F., Leaf, A. L., Dickson, R., and Hart, J. B. (1965). Effects of varying soil moisture upon the nutrient uptake of four bottomland species. *Soil Sci. Soc. Am. Proc.* **29,** 313–316.

Houston, D. B., and Dochinger, L. S. (1977). Effects of ambient air pollution on cone, seed, and pollen characteristics in eastern white and red pine. *Environ. Pollut.* **12,** 1–5.

Howard, A. (1925). The effect of grass on trees. *Proc. R. Soc. London, Ser. B* **97,** 284–321.

Howe, H. F., and Smallwood, J. (1982). Ecology of seed dispersal. *Annu. Rev. Ecol. Syst.* **13,** 201–228.

Hsiao, T. C. (1970). Rapid changes in levels of polyribosomes in *Zea mays* in response to water stress. *Plant Physiol.* **46,** 281–285.

Hsiao, T. C. (1973). Plant responses to water stress. *Annu. Rev. Plant Physiol.* **24,** 519–570.

Hsiao, T. C., and Bradford, K. J. (1983). Physiological consequences of cellular water deficits. *In* "Limitations to Efficient Water Use in Crop Production" (H. M. Taylor, W. R. Jordan, and T. R. Sinclair, eds.), pp. 227–265. Am. Soc. Agron., Crops Sci. Soc. Am., Soil Sci. Soc. Am., Madison, Wisconsin.

Hsiao, T. C., Acevedo, E., Ferreres, E., and Henderson, D. W. (1976). Water stress, growth, and osmotic adjustment. *Philos. Trans. R. Soc. London, Ser. B* **273,** 479–500.

Huber, B. (1923). Transpiration in verschiedener Stammhöhe. *Z. Bot.* **15,** 465–501.

Huber, B. (1956). Die Gefässleitung. *Encycl. Plant Physiol.* **3,** 541–583.

Huck, M. G. (1983). Root distribution, growth, and activity with reference to agroforestry. *In* "Plant Research and Agroforestry" (P. A. Huxley, ed.), pp. 527–542. International Council for Research in Agroforestry, Nairobi, Kenya.

Huck, M. G,. Klepper, B., and Taylor, H. M. (1970). Diurnal variations in root diameter. *Plant Physiol.* **45,** 529–530.

Hudson, J. P. (1971). Meadow orchards. *Agriculture (London)* **78,** 157–160.

Hudson, J.P. (1977). Plants and the weather. *In* "Environmental Effects on Crop Physiology" (J. J. Landsberg and C. V. Cutting, eds.), pp. 1–20. Academic Press, London.

Huffaker, C.B., ed. (1980). "New Technology of Pest Control." Wiley, New York.

Huffaker, R. C., Radin, R. L., Kleinkopf, C. J., and Cox, G. E. (1970). Effect of mild water stress on enzymes of nitrate assimilation and of the carboxylative phase of photosynthesis in barley. *Crop Sci.* **10,** 471–474.

Huffaker, C.B., Shoemaker, C.A., and Gutierrez, A.P. (1978). Current status, urgent needs, and future prospects of integrated pest management. *In* "Pest Control Strategies" (E.H. Smith and D. Pimentel, eds.), pp. 237–259. Academic Press, New York.

Hunt, F. M. (1951). Effect of flooded soil on growth of pine seedlings. *Plant Physiol.* **26,** 363–368.

Hunt, R. (1982). "Plant Growth Curves: The Functional Approach to Plant Growth." Arnold, London.

Hunter, J. E., and Kunimoto, R. K. (1974). Dispersal of *Phytophthora palmivora* sporangia by wind-blown rain. *Phytopathology* **64,** 202–206.

Hutchinson, B. A., and Matt, D. R. (1977). The distribution of solar radiation within a deciduous forest. *Ecol. Monogr.* **47,** 185- 207.

Hutchinson, T. C., and Whitby, L. M. (1977). The effects of acid rainfall and heavy metal particulates on a boreal forest ecosystem near the Sudbury smelting region of Canada. *Water, Air, Soil Pollut.* **7,** 421–428.

Hütte, P. (1968). Experiments on wind blow and wind damage in Germany: Site and susceptibility of spruce forest to storm damage. *Forestry, Suppl.* pp. 20–27.

Huttunen, S. (1984). Interactions of disease and other stress factors with atmospheric pollution. *In* "Air Pollution and Plant Life" (M. Treshow, ed.), pp. 321–356. Wiley, Chichester.

Huxley, P. A. (1983). "Plant Research and Agroforestry." Pillans & Wilson, Edinburgh.

Huxley, P. A. (1985). The tree/crop interface—Or simplifying the biological/environmental study of mixed cropping agroforestry systems. *Agrofor. Syst.* **3,** 251–266.

Huxley, P. A., ed. (1987). Agroforestry experimentation: Separating the wood from the trees? *Agrofor. Syst.* **5,** 251–275.

Idso, S.B.(1984). The case for carbon dioxide. *J. Environ. Sci.* **27,** 19–22.

Idso, S. B., Reginato, R. J., Hatfield, J. L., Walker, G. K., Jackson, R. D., and Pinter, P. G., Jr. (1980). A generalization of the stress-degree-day concept of yield prediction to accommodate a diversity of crops. *Agric. Meteorol.* **21,** 205–211.

Idso, S. B., Kimball, B. A., Anderson, M. G. and Mauney, J. R. (1987). Effects of atmospheric CO_2 enrichment on plant growth: The interactive role of air temperature. *Agric., Ecosyst. Environ.* **20,** 1–10.

Ingestad, T. (1979). Mineral nutrient requirements of *Pinus silvestris* and *Picea abies* seed. *Physiol. Plant.* **45,** 373–380.

Ingestad, T. (1981). Nutrition and growth of birch and grey alder seedlings in low conductivity solutions and at a varied relative rate of nutrient addition. *Physiol. Plant.* **52,** 454–466.

Ingestad, T. (1982). Relative addition rate and external concentration: Driving variables used in plant nutrition research. *Plant, Cell Environ.* **5,** 433–453.

Ingestad, T., and Lund, A.-B. (1979). Nitrogen stress in birch seedlings. I. Growth techniques and growth. *Physiol. Plant.* **45,** 137–148.

Ingestad, T., and Lund, A.B. (1986). Theory and techniques for steady state mineral nutrition and growth of plants. *Scand. J. For. Res.* **1,** 439–453.

Ingold, C. T. (1965). "Spore Liberation." Oxford Univ. Press (Clarendon), Oxford.

Ingold, C. T. (1971). "Fungal Spores: Their Liberation and Dispersal." Oxford Univ. Press, London and New York.

Ingram, D. L. (1986). Root cell membrane heat tolerance of two dwarf hollies. *J. Am. Soc. Hortic. Sci.* **111,** 270–272.

Irving, R. M., and Lanphear, F. O. (1967). Environmental control of cold hardiness in plants. *Plant Physiol.* **42,** 1191–1196.

Isaac, L. A., and Hopkins, L. G. (1937). The forest soil of the Douglas-fir region, and changes wrought upon it by logging and slash burning. *Ecology* **18,** 264–279.

Isebrands, J. G., and Nelson, N. D. (1982). Crown architecture of short-rotation, intensively cultured *Populus.* II. Branch morphology and distribution of leaves within the crown of *Populus* 'Tristis' as related to biomass production. *Can. J. For. Res.* **12,** 853–864.

Isebrands, J. G., Promnitz, L. C., and Dawson, D. H. (1977). Leaf area development in short rotation intensive cultured *Populus* plots. *TAPPI For. Biol. Wood Chem. Conf.* [*Conf. Pap.*], pp. 201–209.

Jackson, D. S. (1965). Species siting, climate, soil and productivity. *N. Z. J. For.* **10,** 90–102.

Jackson, J. E. (1970). Aspects of light climate within apple orchards. *J. Appl. Ecol.* **7,** 207–216.

Jackson, J. E. (1980). Light interception and utilization by orchard systems. *Hortic. Rev.* **2,** 208–267.

Jackson, J. E. (1981). Theory of light interception by orchards and a modelling approach to optimizing orchard design. *Acta Hortic.* **114,** 69–79.

Jackson, J. E. (1985). Future fruit orchard design: Economics and biology. *In* "Attributes of Trees as Crop Plants" (M. G. R. Cannell and J. E. Jackson, eds.), pp. 441–459. Institute of Terrestrial Ecology, Huntingdon, England.

Jackson, J. E., and Palmer, J. W. (1977). Effects of shade on the growth and cropping of apple trees. II. Effects on components of yield. *J. Hortic. Sci.* **52,** 253–266.

Jackson, L. W. R. (1967). Effect of shade on leaf structure of deciduous tree species. *Ecology* **48,** 498–499.

Jackson, M. B., and Drew, M. C. (1984). Effects of flooding on growth and metabolism of herbaceous plants. *In* "Flooding and Plant Growth" (T. T. Kozlowski, ed.), pp. 47–128. Academic Press, New York.

Jackson, M. B., Herman, B., and Goodenough, A. (1982). An examination of the importance of ethanol in causing injury to flooded plants. *Plant, Cell Environ.* **5,** 163–172.

Jackson, R. D. (1982). Canopy temperature and plant water stress. *Adv. Irrig.* **7,** 43–85.

Jackson, W. T. (1942). The role of adventitious roots in recovery of shoots following flooding of the original root systems. *Am. J. Bot.* **42,** 816–819.

Jackson, W. T. (1943). The relative importance of factors causing injury to shoots of flooded tomato plants. *Am. J. Bot.* **43,** 637–639.

Jacobi, J. D., Gerrish, G., and Mueller-Dombois, D. (1983). *Metrosideros* dieback in Hawaii: Vegetation changes in permanent plots. *Pac. Sci.* **37,** 327–337.

Jacobs, M. R. (1954). The effect of wind sway on the form and development of *Pinus radiata* D. Don. *Aust. J. Bot.* **2,** 35–51.

Jacobs, M. R. (1955). Growth habits of the eucalypts. *For. Timber Bur. Aust.*, pp. 1–262.

Jacobson, J. S. (1982). Ozone and the growth and productivity of agricultural crops. *In* "Effects of Gaseous Air Pollution in Agriculture and Horticulture" (M. H. Unsworth and D. P. Ormrod, eds.), pp. 293–304. Butterworth, London.

Jacoby, G. C., Jr., Sheppard, P. R., and Sieh, K. E. (1988). Irregular occurrence of large earthquakes along the San Andreas fault: Evidence from trees. *Science* **241,** 196–199.

Jaffe, M. J. (1980). Morphogenetic responses of plants to mechanical stimuli or stress. *BioScience* **30,** 239–243.

Jaindl, R. G., and Sharrow, S. H. (1988). Oak/Douglas-fir/sheep: A three-crop silvopastoral system. *Agrofor. Syst.* **6,** 147–152.

James, R. L., Cobb, F. W., Jr., Miller, P. R., and Parmeter, J. R., Jr. (1980). Effects of oxidant air pollution on susceptibility of pine roots to *Fomes annosus. Phytopathology* **70,** 560–563.

Jancarik, V. (1961). Vyskyt drevokaznych hub v kourem poskozovani oblasti krusnych hor. *Lesnictvi* **7** 667–692.

Janos, D. P. (1980). Vesicular-arbuscular mycorrhizae affect lowland tropical rain forest plant growth. *Ecology* **61,** 151–162.

Jarvis, P. G. (1981a). Production efficiency of coniferous forest in the U.K. *In* "Physiological Processes Limiting Plant Productivity" (C. B. Johnson, ed.), pp. 81–107. Butterworth, London.

Jarvis, P. G. (1981b). Plant water relations in models of tree growth. *Stud. For. Suec.* **160,** 51–60.

Jarvis, P. G. (1985). Transpiration and assimilation of tree and agricultural crops: The omega factor. *In* "Attributes of Trees as Crop Plants" (M. G. R. Cannell and J. E. Jackson, eds.), pp. 460–480. Institute of Terrestrial Ecology, Huntingdon, England.

Jarvis, P. G. (1986). Coupling of carbon and water interactions in forest stands. *Tree Physiol.* **2,** 347–368.

Jarvis, P. G. (1989). Atmospheric carbon dioxide and forests. *Philos. Trans. R. Soc. London, Ser. B* **324,** 369–392.

Jarvis, P. G., and Leverenz, J. W. (1983). Productivity of temperate, deciduous and evergreen forest. *Encycl. Plant Physiol., New Ser.* **12D,** 233–280.

Jarvis, P. G., and Mansfield, T. A., eds. (1981). "Stomatal Physiology." Cambridge Univ. Press, Cambridge.

Jarvis, P. G., James, G. B., and Landsberg, J. J. (1975). Coniferous forest. *In* "Vegetation and the Atmosphere. Vol. 2. Case Studies" (J. L. Monteith, ed.), pp. 171–240. Academic Press, London.

Jeffrey, P. W. (1967). Phosphate nutrition of Australian heath plants. I. The importance of proteoid roots in *Banksia* (Proteaceae). *Austr. J. Bot.* **15,** 403–411.

Jenny, H. (1941). "Factors of Soil Formation; A System of Quantitative Pedology." McGraw-Hill, New York.

Jensen, K. F. (1973). Response of nine forest tree species to chronic ozone fumigation. *Plant Dis. Rep.* **57,** 914–917.

Jensen, K. F. (1977). Sulfur dioxide affects growth of forest tree species. *Proc. Am. Phytopathol. Soc.* **4,** 89.

Jensen, K. F. (1981). Ozone fumigation decreased the root carbohydrate content and dry weight of green ash seedlings. *Environ. Pollut., Ser. A* **26,** 147–152.

Jensen, K. F. (1982). An analysis of the growth of silver maple and eastern cottonwood seedlings exposed to ozone. *Can. J. For. Res.* **12,** 420–424.

Jensen, K. F. (1983). Growth relationships in silver maple seedlings fumigated with O_3 and SO_2. *Can. J. For. Res.* **13,** 298–302.

Jensen, K. F., and Kozlowski, T. T. (1974). Effect of SO_2 on photosynthesis of quaking aspen and white ash seedlings. *Proc. North. Am. For. Biol. Workshop, 3rd, 1974,* p. 359.

Jensen, K. F., and Kozlowski, T. T. (1975). Absorption and translocation of sulfur dioxide by seedlings of four forest tree species. *J. Environ. Qual.* **4,** 379–381.

Jester, J. R., and Kramer, P. J. (1939). The effect of length of day on the height growth of certain forest tree seedlings. *J. For.* **37,** 796–803.

Johnson, A. F., Abrahamson, W. G., and McCrea, K. D. (1986). Comparison of biomass recovery after fire of a seeder (*Ceratiola ericoides*) and a sprouter (*Quercus inopina*) species from south-central Florida. *Am. Midl. Nat.* **116,** 423–428.

Johnson, D. S., Stinchcombe, G. R., and Stott, K. G. (1983). Effect of soil management on mineral composition and storage quality of Cox's Orange Pippin apples. *J. Hortic. Sci.* **58,** 317–326.

Johnson, D. V., and Nair, P. K. R. (1985). Perennial crop-based agroforestry systems in northeast Brazil. *Agrofor. Syst.* **2,** 281–292.

Johnson, D. W., West, D. C., Todd, D. E., and Mann, L. K. (1982). Effects of sawlog vs. wholetree harvesting on the nitrogen, phosphorus, potassium and calcium budgets of an upland mixed oak forest. *Soil Sci. Soc. Am. J.* **46,** 1304–1309.

Johnson, D. W., Van Miegroet, H., Cole, D. W., and Richter, D. D. (1983). Contributions of acid deposition and natural processes to cation leaching from forest soils: A review. *J. Air Pollut. Contr. Assoc.* **33,** 1036–1041.

Johnson, E. A., and Kovner, J. L. (1956). Effect on streamflow of cutting a forest understory. *For. Sci.* **2,** 82–91.

Johnson, G. A., Brown, J., and Kramer, P. J. (1987). Magnetic resonance microscopy of changes in water content in stems of transpiring plants. *Proc. Natl. Acad. Sci. U.S.A.* **84,** 2752–2755.

Johnson, J. D., Seiler, J. R., and McNabb, K. L. (1986). Manipulation of pine seedling physiology by water stress conditioning. *In* "Nursery Management Practices for the Southern Pines" (D. B. South, ed.), pp. 290–302. Auburn University, Auburn, Alabama.

Johnson, R. L. (1978). Timber harvests from wetlands. *In* "Wetland Functions and Values: The State of Our Understanding" (P. E. Greeson, J. R. Clark, and J. E. Clark, eds.), pp. 598–605. Am. Water Resour. Assoc., Minneapolis, Minnesota.

Johnson, S. W. (1868). "How Crops Grow." Orange Judd Co., New York.

Johnston, T. N., and Ward, D. (1986). Impact of clay clipping and nursery handling procedures on field growth and survival of Honduran Caribbean pine. *In* "Nursery Management Practices for the Southern Pines" (D. B. South, ed.), pp. 515–523. Auburn University, Auburn, Alabama.

Joly, R. J., and Zaerr, J. B. (1987). Alteration of cell-wall water content and elasticity in Douglas-fir during periods of water deficit. *Plant Physiol.* **83,** 418–422.

Jones, H. G. (1987). Repeat flowering in apple caused by water stress or defoliation. *Trees* **1,** 135–138.

Jones, H. G. (1989). Water stress and stem conductivity. *In* "Environmental Stress in Plants" (J. H. Cherry, ed.), pp. 17–24. Springer-Verlag, Berlin.

Jones, H. G., Flowers, T. J., and Jones, M. B., eds. (1989). "Plants under Stress." Cambridge Univ. Press, Cambridge.

Jones, L. (1961). Effect of light on germination of forest tree seeds. *Proc. Int. Seed Test. Assoc.* **26,** 437–452.

Jones, R. H., and Raynal, D. J. (1987). Root sprouting in American beech: Production, survival, and the effect of parent tree vigor. *Can. J. For. Res.* **17,** 539–544.

Jonsson, B., and Sundberg, R. (1980). Has the acidification by atmospheric pollution caused a growth reduction in Swedish forests? A comparison of growth between regions with different soil properties. *Res. Note, Inst. Skogsprod. (Stockholm)* No. 20.

Jorgensen, J. R., and Wells, C. G. (1971). Apparent nitrogen fixation in soil influenced by prescribed burning. *Soil Sci. Soc. Am. Proc.* **35,** 806–810.

Junttila, O. (1986). Effects of temperature on shoot growth in northern provenances of *Pinus sylvestris* L. *Tree Physiol.* **1,** 185–192.

Junttila, O., and Heide, O. M. (1981). Shoot and needle growth in *Pinus sylvestris* as related to temperature in northern Fennoscandia. *For. Sci.* **27,** 423–430.

Jurik, T. W. (1986). Temporal and spatial patterns of specific leaf weight in successional northern hardwood tree species. *Am. J. Bot.* **73,** 1083–1092.

Jurik, T. W., Briggs, G. M., and Gates, D. M. (1985). "Carbon Dynamics of Northern Hardwood Forests: Gas Exchange Characteristics," DOE/EV/ 10091 - 1 TR019. NII Service, U.S. Dept. of Energy, Washington, D.C.

Kable, P. F., Fliegel, P., and Parker, K. G. (1967). *Cytospora* canker on sweet cherry in New York State: Association with winter injury and pathogenicity to other species. *Plant Dis. Rep.* **51,** 155–157.

Kahdr, A. H., Wallace, A., and Romney, E. M. (1965). Mineral nutritional problems of trifoliate orange rootstock. *Calif. Agric.* **6**(9), 12–13.

Kaiser, W. M. (1982). Correlation between changes in photosynthetic activity and changes in total protoplast volume in leaf tissue from hygro-, meso- and xerophytes under osmotic stress. *Planta* **154,** 538–545.

Kalma, J. D., and Stanhill, G. (1969). The radiation climate of an irrigated orange plantation. *Sol. Energy* **12,** 491–508.

Kan, M., Saito, H., and Shidei, T. (1965). Studies of the productivity of evergreen broad leaved forests. *Bull. Kyoto Univ. For.* **37,** 55–75 (in Japanese).

Kappen, L. (1981). Ecological significance of resistance to high temperature. *Encycl. Plant Physiol, New Sec.* **12A,** 439–474.

Karnosky, D. F. (1976). Threshold levels for foliar injury to *Populus tremuloides* by sulphur dioxide and ozone. *Can. J. For. Res.* **6,** 166–169.

Karnosky, D. F. (1977). Evidence for genetic control of response to sulfur dioxide and ozone in *Populus tremuloides*. *Can. J. For. Res.* **7,** 437–440.

Kasana, M. S., and Mansfield, T. A. (1986). Effects of air pollutants on the growth and functioning of roots. *Proc.—Indian Acad. Sci.* [*Ser.*] *Plant Sci.* **96,** 429–441.

Kaufmann, M. R. (1968). Water relations of pine seedlings in relation to root and shoot growth. *Plant Physiol.* **43,** 281–288.

Kaufmann, M. R. (1969). Effects of water potential on germination of lettuce, sunflower, and citrus seeds. *Can. J. Bot.* **47,** 1761–1764.

Kaufmann, M. R. (1972). Water deficits and reproductive growth. *In* "Water Deficits and Plant Growth" (T. T. Kozlowski, ed.), Vol. 3, pp. 91–124. Academic Press, New York.

Kaufmann, M. R. (1976). Stomatal response of Engelmann spruce to humidity, light, and water stress. *Plant Physiol.* **57,** 898–901.

Kaufmann, M. R. (1982). Evaluation of season, temperature and water stress effects on stomata using a leaf conductance model. *Plant Physiol.* **69,** 1023–1026.

Kaufmann, M. R., and Troendle, C. A. (1981). The relationship of leaf area and foliage biomass to sapwood conducting area in four subalpine forest tree species. *For. Sci.* **27,** 477–482.

Kaurin, A., Stushnoff, C., and Junttila, O. (1982). Vegetative growth and frost hardiness of cloudberry (*Rubus chamaemorus*) as affected by temperature and photoperiod. *Physiol. Plant.* **55,** 76–81.

Kawada, K., Grierson, W., and Soule, J. (1979). Seasonal resistance to chilling injury of 'Marsh' grapefruit as related to winter field temperature. *Citrus Ind.* **60**(10), 5–9.

Kawase, M. (1978). Anaerobic elevation of ethylene concentration in waterlogged plants. *Am. J. Bot.* **65,** 736–740.

Kawase, M. (1979). Cellulase activity in waterlogged herbaceous horticultural crops. *HortScience* **16,** 30–34.

Kawase, M. (1981). Anatomical and morphological adaptations of plants to water-logging. *HortScience* **16,** 8–12.

Kayll, A. J. (1974). Use of fire in land management. *In* "Fire and Ecosystems" (T. T. Kozlowski and C. E. Ahlgren, eds.), pp. 483–511. Academic Press, New York.

Keay, R. W. J. (1957). Wind-dispersed species in a Nigerian forest. *J. Ecol.* **45,** 471–478.

Keeley, J. E., and Zedler, P. H. (1978). Reproduction of chaparral shrubs after fire: A comparison of sprouting and seeding strategies. *Am. Midl. Nat.* **99,** 142–161.

Keeling, C. D. (1986). "Atmospheric CO_2 Concentrations—Mauna Loa Observatory, Hawaii 1958–1986, NDP-001/R1. Carbon Dioxide Inf. Cent., Oak Ridge Natl. Lab., Oak Ridge, Tennessee.

Kefford, N. P. (1976). Dislocation of developmental processes. *In* "Herbicides: Physiology, Biochemistry, Ecology" (L. J. Audus, ed.), 2nd ed., Vol. 1, pp. 427–442. Academic Press, London.

Keller, T. (1966). Über den Einfluss von transpirationshemmenden Chemikalien (Antitranspirantien) auf Transpiration, CO_2-Aufnahme und Wurzelwachstum von Jungfichten. *Forstwiss. Centralbl.* **85,** 65–79.

Keller, T. (1973a). On the phytotoxicity of dust-like fluoride compounds. *Staub—Reinhalt. Luft* **33,** 379–381.

Keller, T. (1973b). Über die schädigende Wirkung des Fluors. *Schweiz. Z. Forstwes.* **124,** 700–706.

Keller, T. (1977a). Der Einfluss von Fluorimmissionen auf die Nettoassimilation von Waldbaumarten. *Mitt. Eidg. Anst. Forstl. Versuchswes.* **53,** 161–198.

Keller, T. (1977b). Begriff und Bedeutung der 'latenten Immissionsschädigung.' *Allg. Forst- Jagdzg.* **148,** 115–120.

Keller, T. (1977c). The effect of long term low SO_2 concentrations upon photosynthesis of conifers. *Proc. Int. Clean Air Congr., 4th, 1977,* pp. 81–83.

Keller, T. (1978). Der Einfluss einer SO_2-Belastung zu verschiedenen Jahreszeiten auf CO_2-Aufnahme und Jahrringbau der Fichte. *Schweiz. Z. Forstwes.* **129,** 381–393.

Keller, T. (1980a). The effect of a continuous springtime fumigation with SO_2 on CO_2 uptake and structure of the annual ring in spruce. *Can. J. For. Res.* **10,** 1–6.

Keller, T. (1980b). The simultaneous effect of soil-borne NaF and air pollutant SO_2 on CO_2-uptake and pollutant accumulation. *Oecologia* **44,** 283–285.

Keller, T. (1983). Air pollutant deposition and effects on plants. *In* "Effects of Accumulation of Air

Pollutants in Forest Ecosystems" (B. Ulrich and J. Pankrath, eds.), pp. 285–294. Reidel Publ., Dordrecht, The Netherlands.

Keller, T., and Beda, H. (1984). Effects of SO_2 on the germination of conifer pollen. *Environ. Pollut., Ser. A* **33,** 237–243.

Keller, T., and Schwager, A. (1971). Der Nachweis unsichtbarer ('physiologischer') Fluor-Immissionsschädigungen an Waldbäumen durch eine einfache kolorimetrische Bestimmung der Peroxidase-Aktivität. *Eur. J. For. Pathol.* **1,** 6–18.

Kelly, J. M., Parker, G. R., and McFee, W. W. (1979). Heavy metal accumulation and growth of seedlings of five forest species as influenced by soil cadmium level. *J. Environ. Qual.* **8,** 361–364.

Kender, W. J., and Spierings, F. H. F. G. (1975). Effects of sulfur dioxide, ozone, and their interactions on 'Golden Delicious' apple trees. *Neth. J. Plant Pathol.* **81,** 149–151.

Kendrick, R. E., and Frankland, B. (1983). "Phytochrome and Plant Growth," 2nd ed. Arnold, London.

Kenk, G., and Fischer, G. (1988). Evidence from nitrogen fertilization in the forests of Germany. *Environ. Pollut.* **54,** 199–218.

Kennedy, H. E., Jr., and Krinard, R. M. (1974). 1973 Mississippi river floods impact on natural hardwood forests and plantations. *Res. Note SO (U.S. For. Serv.)* **SO–177.**

Kenrick, J. R., and Bishop. D. G. (1986). The fatty acid composition of phosphatidylglycerol and sulfoquinosyldiacylglycerol of higher plants in relation to chilling sensitivity. *Plant Physiol.* **81,** 946–949.

Kent, B. M., and Dress, P. (1979). On the convergence of forest stand spatial pattern over time: The case of random initial spatial pattern. *For. Sci.* **25,** 445–451.

Kercher, J. R., Axelrod, M. C., and Bingham, G. E. (1980). Forecasting effects of SO_2 pollution on growth and succession in a western conifer forest. *USDA For. Serv. Gen. Tech. Rep. PSW* **PSW–43,** pp. 200–202.

Kerr, R. A. (1989). Greenhouse skeptic out in the cold. *Science* **246,** 1118–1119.

Ketchie, D. O., and Ballard, A. L. (1968). Environments which cause heat injury to Valencia oranges. *Proc. Am. Soc. Hortic. Sci.* **93,** 166–172.

Ketchie, D. O., and Burts, W. D. (1973). The relation of lipids to cold acclimation in "Red Delicious" apple trees. *Cryobiology* **10,** 529.

Key, J. L., Lin, C. Y., and Chen, Y. M. (1981). Heat shock proteins of higher plants. *Proc. Natl. Acad. Sci. U.S.A.* **78,** 3526–3530.

Keyes, M. R., and Grier, C. C. (1981). Above- and below-ground net production in 40-year-old Douglas-fir stands on low and high productivity sites. *Can. J. For. Res.* **11,** 599–605.

Khan, A. A., and Malhotra, S. S. (1977). Effects of aqueous sulphur dioxide on pine needle glycolipids. *Phytochemistry* **16,** 539–543.

Kienholz, R. (1934). Leader, needle, cambial, and root growth of certain conifers and their relationships. *Bot. Gaz. (Chicago)* **96,** 73–92.

Kimball, B. A. (1983). Carbon dioxide and agricultural yield: An assemblage and analysis of 430 prior observations. *Agron. J.* **75,** 779–788.

Kimball, B. A. (1986). Influence of elevated CO_2 on crop yield. *In* "Carbon Dioxide Enrichment of Greenhouse Crops" (H. Z. Enoch and B. A. Kimball, eds.), Vol. 2, pp. 105–115. CRC Press, Boca Raton, Florida.

Kimmerer, T. W., and Kozlowski, T. T. (1981). Stomatal conductance and sulfur uptake of five clones of *Populus tremuloides* exposed to sulfur dioxide. *Plant Physiol.* **67,** 990–995.

Kimmerer, T. W., and Kozlowski, T. T. (1982). Ethylene, ethane, acetaldehyde, and ethanol production by plants under stress. *Plant Physiol.* **69,** 840–847.

Kimmerer, T. W., and Stringer, M. A. (1988). Alcohol dehydrogenase and ethanol in the stems of trees. *Plant Physiol.* **87,** 693–697.

Kimmins, J. P. (1987). "Forest Ecology." Macmillan, New York.

Kinerson, R. S. (1975). Relationships between plant surface area and respiration in loblolly pine. *J. Appl. Ecol.* **12,** 965–971.

Kinerson, R. J., Higginbotham, K. O., and Chapman, R. C. (1974). Dynamics of foliage distribution within a forest canopy. *J. Appl. Ecol.* **11,** 347–353.

Kinerson, R. S., Ralston, C., and Wells, C. (1977). Carbon cycling in a loblolly pine plantation. *Oecologia* **29,** 1–10.

King, D. A. (1986). Tree form, height growth and susceptibility to wind damage in *Acer saccharum*. *Ecology* **67,** 980–990.

Kira, T. (1975). Primary productivity of forests. *In* "Photosynthesis and Productivity in Different Environments" (J. P. Cooper, ed.), IBP, Vol. 3, pp. 5–40. Cambridge Univ. Press, Cambridge.

Kira, T. (1976). "Terrestrial Ecosystems—A General View," Handbook of Ecology, Vol. 2. Kyoritsu Shuppan, Tokyo, Japan (in Japanese).

Kira, T., and Kumura, A. (1983). Dry matter production and efficiency in various types of plant canopies. *In* "Plant Research and Agroforestry" (P. A. Huxley, ed.), pp. 347–364. Pillans & Wilson, Edinburgh.

Kira, T., and Shidei, T. (1967). Primary production and turnover of organic matter in different forest ecosystems of the western Pacific. *Jpn. J. Ecol.* **17,** 80–87.

Klein, R. M. (1978). Plants and near-ultraviolet radiation. *Bot. Rev.* **44,** 1–127.

Klein, R. M., and Perkins, T. D. (1988). Primary and secondary causes and consequences of contemporary forest decline. *Bot. Rev.* **54,** 1–43.

Kleinig, H. (1989). The role of plastids in isoprenoid biosynthesis. *Annu. Rev. Plant Physiol. Plant Mol. Biol.* **40,** 39–59.

Klemmedson, J. O. (1976). Effect of thinning and slash burning on nitrogen and carbon in ecosystems of young dense ponderosa pine. *For. Sci.* **22,** 45–53.

Knight, H. (1966). Loss of nitrogen from the forest floor by burning. *For. Chron.* **42,** 149–152.

Knoerr, K. R. (1967). Contrasts in energy balances between individual leaves and vegetated surfaces. *In* "International Symposium on Forest Hydrology" (W. E. Sopper and H. W. Lull, eds.), pp. 391–401. Pergamon, Oxford.

Koch, R. (1882). "Über die Midzbrandimpfung: Eine Entgegnung auf den von Pasteur in Genf gehaltenen Vortrag." Theodore Fischer, Kassel and Berlin.

Kohut, R. J., Davis, D. D., and Merrill, W. (1976). Response of hybrid poplar to simultaneous exposure to ozone and PAN. *Plant Dis. Rep.* **60,** 777–780.

Kolattukudy, P. E. (1975). Biochemistry of cutin, suberin and waxes, the lipid barriers on plants. *In* "Recent Advances in the Chemistry and Biochemistry of Plant Lipids" (T. Galliard and E. I. Mercer, eds.), pp. 203–246. Academic Press, London.

Koller, D. and Hadas, A. (1982). Water relations in the germination of seeds. *Encycl. Plant Physiol., New Ser.* **12B,** 402–431.

Komarek, E. V. (1974). Effects of fire on temperate forests and related ecosystems: Southeastern United States. *In* "Fire and Ecosystems" (T. T. Kozlowski and C. E. Ahlgren, eds.), pp. 251–277. Academic Press, New York.

Korstian, C. F., and Coile, T. S. (1938). Plant competition in forest stands. *Duke Univ. Sch. For. Bull.* **3.**

Kosiyachinda, S., and Young, R. E. (1976). Chilling sensitivity of avocado fruit at different stages of respiratory climacteric. *J. Am. Soc. Hortic. Sci.* **101,** 665–667.

Kossuth, S. V., and Ross, S. D., eds. (1987). Hormonal control of tree growth. *Plant Growth Regul.* **6,** 1–215.

Kozlowski, T. T. (1943). Transpiration rates of some forest tree species during the dormant season. *Plant Physiol.* **18,** 252- 260.

Kozlowski, T. T. (1949). Light and water in relation to growth and competition of Piedmont forest tree species. *Ecol. Monogr.* **19,** 207–231.

Kozlowski, T. T. (1960). Some problems in the use of herbicides in forestry. *Proc.—North Cent. Weed Control Conf.* **17,** 1–10.

Kozlowski, T. T. (1962). Photosynthesis, climate, and tree growth. *In* "Tree Growth" (T. T. Kozlowski, ed.), pp. 149–170. Ronald Press, New York.

Kozlowski, T. T. (1963). Growth characteristics of forest trees. *J. For.* **61,** 655–662.
Kozlowski, T. T. (1965). Expansion and contraction of plants. *Adv. Front. Plant Sci.* **10,** 63–77.
Kozlowski, T. T. (1967a). Growth and development of *Pinus resinosa* seedlings under controlled temperatures. *Adv. Front. Plant Sci.* **19,** 17–27.
Kozlowski, T. T. (1967b). Diurnal variations in stem diameters of small trees. *Bot. Gaz. (Chicago)* **128,** 60–68.
Kozlowski, T. T. (1968a). Soil water and tree growth. *In* "The Ecology of Southern Forests" (N. E. Linnartz, ed.), pp. 30–57. Louisiana State Univ. Press, Baton Rouge.
Kozlowski, T. T. (1968b). Diurnal changes in diameters of fruits and tree stems of Montmorency cherry. *J. Hortic. Sci.* **43,** 1–15.
Kozlowski, T. T. (1968c). Water balance in shade trees. *Proc. Int. Shade Tree Conf., 44, 1968,* pp. 29–42.
Kozlowski, T. T. (1968d). "Water Deficits and Plant Growth," Vol. 1. Academic Press, New York.
Kozlowski, T. T. (1969). Tree physiology and forest pests. *J. For.* **69,** 118–122.
Kozlowski, T. T. (1971a). "Growth and Development of Trees," Vol. 1. Academic Press, New York.
Kozlowski, T. T. (1971b). "Growth and Development of Trees," Vol. 2. Academic Press, New York.
Kozlowski, T. T., ed. (1972a). "Water Deficits and Plant Growth." Vol. 3. Academic Press, New York.
Kozlowski, T. T. (1972b). Shrinking and swelling of plant tissues. *In* "Water Deficits and Plant Growth" (T. T. Kozlowski, ed.), Vol. 3, pp. 1–64. Academic Press, New York.
Kozlowski, T. T. (1972c). Physiology of water stress. *USDA For. Serv. Gen. Tech. Rep. INT* **INT–1,** 229–244.
Kozlowski, T. T. (1973). Extent and significance of shedding of plant parts. *In* "Shedding of Plant Parts" (T. T. Kozlowski, ed.), pp. 1–44. Academic Press, New York.
Kozlowski, T. T., ed. (1973). "Shedding of Plant Parts." Academic Press, New York.
Kozlowski, T. T. (1976a). Drought resistance and transplantability of shade trees. *USDA For. Serv. Gen. Tech. Rep. NE.* **NE-22,** 77–90.
Kozlowski, T. T. (1976b). Susceptibility of young tree seedlings to environmental stresses. *Am. Nurseryman* **144** 12–13, 55–59.
Kozlowski, T. T. (1976c). Water supply and leaf shedding. *In* "Water Deficits and Plant Growth" (T. T. Kozlowski, ed.), Vol. 4, pp. 191–231. Academic Press, New York.
Kozlowski, T. T., ed. (1978a). "Water Deficits and Plant Growth," Vol. 5. Academic Press, New York.
Kozlowski, T. T. (1978b). How healthy plants grow. *In* "Plant Pathology: An Advanced Treatise" (J. G. Horsfall and E. B. Cowling, eds.), Vol. 3, pp. 19–51. Academic Press, New York.
Kozlowski, T. T. (1979). "Tree Growth and Environmental Stresses." Univ. of Washington Press, Seattle.
Kozlowski, T. T. (1980a). Impacts of air pollution on forest ecosystems. *BioScience* **30,** 88–93.
Kozlowski, T. T. (1980b). Responses of shade trees to pollution. *J. Arboric.* **6,** 29–41.
Kozlowski, T. T., ed. (1981). "Water Deficits and Plant Growth," Vol. 6. Academic Press, New York.
Kozlowski, T. T. (1982a). Water supply and tree growth. Part I. Water deficits. *For. Abstr.* **43,** 57–95.
Kozlowski, T. T. (1982b). Water supply and tree growth. Part II. Flooding. *For. Abstr.* **43,** 145–161.
Kozlowski, T. T. (1982c). Physiology of tree growth. *In* "Introduction to Forest Science" (R. A. Young, ed.), pp. 71–91. Wiley, New York.
Kozlowski, T. T., ed. (1983a). "Water Deficits and Plant Growth," Vol. 7. Academic Press, New York.
Kozlowski, T. T. (1983b). Reduction in yield of forest and fruit trees by water and temperature stress. *In* "Crop Reactions to Water and Temperature Stresses in Humid, Temperate Climates" (C. D. Raper and P.J. Kramer, eds.), pp. 67–88. Westview Press, Boulder, Colorado.
Kozlowski, T. T. (1984a). Extent, causes, and impacts of flooding. *In* "Flooding and Plant Growth" (T. T. Kozlowski, ed.), pp. 1–7. Academic Press, New York.
Kozlowski, T. T. (1984b). Responses of woody plants to flooding. *In* "Flooding and Plant Growth" (T. T. Kozlowski, ed.), pp. 129–163. Academic Press, New York.

Kozlowski, T. T. (1984c). Plant responses to flooding of soil. *BioScience* **34,** 162–167.

Kozlowski, T. T. (1985a). Soil aeration, flooding, and tree growth. *J. Arboric.* **11,** 85–96.

Kozlowski, T. T. (1985b). Tree growth in response to environmental stresses. *J. Arboric.* **11,** 97–111.

Kozlowski, T. T. (1985c). Effects of SO_2 on plant community structure. *In* "Sulfur Dioxide and Vegetation: Physiology, Ecology, and Policy Issues" (W. E. Winner, H. A. Mooney, and R. Goldstein, eds.), pp. 431–451. Stanford Univ. Press, Stanford, California.

Kozlowski, T. T. (1985d). Measurement of effects of environmental and industrial chemicals on terrestrial plants. *In* "Methods for Estimating Risk of Chemical Injury: Human and Non-human Biota and Ecosystems" (V. B. Vouk, G. C. Butler, D. G. Hoel, and D. B. Peakall, eds.), pp. 573–609. Wiley, London.

Kozlowski, T. T. (1985e). Effect of direct contact of *Pinus resinosa* seeds and young seedlings with *N*-diethylamino succinamic acid, (2-chlorethyl)trimethylammonium chloride, or maleic hydrazide. *Can. J. For. Res.* **15,** 1000–1004.

Kozlowski, T. T. (1986a). Effects on seedling development of direct contact of *Pinus resinosa* seeds or young seedlings with captan. *Eur. J. For. Pathol.* **16,** 87–90.

Kozlowski, T. T. (1986b). Effects of 2,3,6-TBA on seed germination, early development, and mortality of *Pinus resinosa* seedlings. *Eur. J. For. Pathol.* **16,** 385–390.

Kozlowski, T. T. (1986c). The impact of environmental pollution on shade trees. *J. Arboric.* **12,** 29–37.

Kozlowski, T. T. (1986d). Soil aeration and growth of forest trees. *Scand. J. For. Res.* **1,** 113–123.

Kozlowski, T. T. (1991). Effects of environmental stresses on deciduous trees. *In* "The Integrated Response of Plants to Stress" (H. A. Mooney, W. E. Winner, and E. J. Pell, eds.). Academic Press, San Diego, California.

Kozlowski, T. T., and Ahlgren, C. E., eds. (1974). "Fire and Ecosystems." Academic Press, New York.

Kozlowski, T. T., and Borger, G. A. (1971). Effect of temperature and light intensity early in ontogeny on growth of *Pinus resinosa* seedlings. *Can. J. For. Res.* **1,** 57–65.

Kozlowski, T. T., and Clausen, J. J. (1966). Shoot growth characteristics of heterophyllous woody plants. *Can. J. Bot.* **44,** 827–843.

Kozlowski, T. T., and Constantinidou, H. A. (1986a). Responses of woody plants to environmental pollution. Part I. Sources, types of pollutants, and plant responses. *For. Abstr.* **47,** 5–51.

Kozlowski, T. T., and Constantinidou, H. A. (1986b). Responses of woody plants to environmental pollution. Part II. Factors affecting responses to pollution. *For. Abstr.* **47,** 105–132.

Kozlowski, T. T., and Davies, W. J. (1975a). Control of water balance in transplanted trees. *J. Arboric.* **1,** 1–10.

Kozlowski, T. T., and Davies W. J. (1975b). Control of water loss in shade trees. *J. Arboric.* **1,** 81–90.

Kozlowski, T. T., and Gentile, A. C. (1958). Respiration of white pine buds in relation to oxygen availability and moisture content. *For. Sci.* **4,** 147–152.

Kozlowski, T. T., and Gentile, A. C. (1959). Influence of the seed coat on germination, water absorption and oxygen uptake of eastern white pine seed. *For. Sci.* **5,** 389–395.

Kozlowski, T. T., and Greathouse, T. E. (1970). Shoot growth characteristics of tropical pines. *Unasylva* **24,** 1–10.

Kozlowski, T. T., and Huxley, P. A. (1983). The role of controlled environments in agroforestry research. *In* "Plant Research and Agroforestry" (P. A. Huxley, ed.). pp. 551–567. Pillans & Wilson, Edinburgh.

Kozlowski, T. T., and Keller, T. (1966). Food relations of woody plants. *Bot. Rev.* **32,** 293–382.

Kozlowski, T. T., and Kuntz, J. E. (1963). Effect of simazine, atrazine, propazine, and eptam on growth of pine seedlings. *Soil Sci.* **95,** 164–174.

Kozlowski, T. T., and Pallardy, S. G. (1979). Stomatal responses of *Fraxinus pennsylvanica* seedlings during and after flooding. *Physiol. Plant.* **46,** 155–158.

Kozlowski, T. T., and Pallardy, S. G. (1984). Effects of flooding on water, carbohydrate, and mineral relations. *In* "Flooding and Plant Growth" (T. T. Kozlowski, ed.), pp. 165–193. Academic Press, Orlando, Florida.

Kozlowski, T. T., and Sasaki, S. (1968a). Effects of direct contact of pine seeds or young seedlings with commercial formulations, active ingredients, or inert ingredients of triazine herbicides. *Can. J. Plant Sci.* **48,** 1–7.

Kozlowski, T. T., and Sasaki, S. (1968b). Germination and morphology of red pine seeds and seedlings in contact with EPTC, CDEC, CDAA, 2,4-D and picloram. *Proc. Am. Soc. Hortic. Sci.* **93,** 655–662.

Kozlowski, T. T., and Sasaki, S. (1970). Effects of herbicides on seed germination and development of young pine seedlings. *Proc. Int. Symp. Seed Physiol. Woody Plants, 1968,* pp. 19–24.

Kozlowski, T. T., and Torrie, J. H. (1965). Effect of soil incorporation of herbicides on seed germination and growth of pine seedlings. *Soil Sci.* **100,** 139–146.

Kozlowski, T. T., and Winget, C. H. (1964). The role of reserves in leaves, branches, stems, and roots on shoot growth of red pine. *Am. J. Bot.* **51,** 522–529.

Kozlowski, T. T., Winget, C. H., and Torrie, J. H. (1962). Daily radial growth of oak in relation to maximum and minimum temperature. *Bot. Gaz. (Chicago)* **124,** 9–17.

Kozlowski, T. T., Sasaki, S., and Torrie, J. H. (1967). Influence of temperature on phytotoxicity of triazine herbicides to pine seedlings. *Am. J. Bot.* **54,** 790–796.

Kozlowski, T. T., Torrie, J. H., and Marshall, P. E. (1973). Predictability of shoot length from bud size in *Pinus resinosa* Ait. *Can. J. For. Res.* **3,** 34–38.

Kramer, P. J. (1936). The effect of variation in day length on the growth and dormancy of trees. *Plant Physiol.* **12,** 881–883.

Kramer, P. J. (1937). Photoperiodic stimulation of growth by artificial light as a cause of winter killing. *Plant Physiol.* **12,** 881–883.

Kramer, P. J. (1943). Amount and duration of growth of various species of tree seedlings. *Plant Physiol.* **18,** 239–251.

Kramer, P. J. (1946). Absorption of water through suberized roots of trees. *Plant Physiol.* **21,** 37–41.

Kramer, P. J. (1951). Causes of injury to plants resulting from flooding of the soil. *Plant Physiol.* **26,** 722–736.

Kramer, P. J. (1957). Some effects of various combinations of day and night temperatures and photoperiod on the height growth of loblolly pine seedlings. *For. Sci.* **3,** 45–55.

Kramer, P. J. (1958). Thermoperiodism in trees. *In* "The Physiology of Forest Trees" (K. V. Thimann, ed.), pp. 573–580. Ronald Press, New York.

Kramer, P. J. (1978). The use of controlled environments in research. *HortScience* **13,** 447–451.

Kramer, P. J. (1981). Carbon dioxide concentration, photosynthesis, and dry matter production. *BioScience* **31,** 29–33.

Kramer, P. J. (1982). Water and plant productivity or yield. *In* "Handbook of Agricultural Productivity" (M. Rehcigl, Jr., ed.), Vol. 1, pp. 41–47. CRC Press, Boca Raton, Florida.

Kramer, P. J. (1983a). "Water Relations of Plants." Academic Press, New York.

Kramer, P. J. (1983b). Problems in water relations of plants and cells. *Int. Rev. Cytol.* **85,** 253–285.

Kramer, P. J. (1986). The role of physiology in forestry. *Tree Physiol.* **2,** 1–16.

Kramer, P. J. (1988a). Measurement of plant water status: Historical perspectives and current concerns. *Irrig. Sci.* **9,** 275–287.

Kramer, P. J. (1988b). Changing concepts regarding plant water relations. *Plant, Cell Environ.* **11,** 565–568.

Kramer, P. J., and Bullock, H. C. (1966). Seasonal variations in the proportions of suberized and unsuberized roots of trees in relation to the absorption of water. *Am. J. Bot.* **53,** 200–204.

Kramer, P. J., and Clark, W. S. (1947). A comparison of photosynthesis in individual pine needles and entire seedlings at various light intensities. *Plant Physiol.* **22,** 51–57.

Kramer, P. J., and Decker, J. P. (1944). Relation between light intensity and rate of photosynthesis of loblolly pine and certain hardwoods. *Plant Physiol.* **19,** 350–358.

Kramer, P. J., and Kozlowski, T. T. (1960). "Physiology of Trees." McGraw-Hill, New York.

Kramer, P. J., and Kozlowski, T. T. (1979). "Physiology of Woody Plants." Academic Press, New York.

Kramer, P. J., and Rose, R. W., Jr. (1986). Physiological characteristics of loblolly pine in relation to field performance. *In* "Nursery Management Practices for the Southern Pines" (D. B. South, ed.), pp. 416–440. Auburn University, Auburn, Alabama.

Kramer, P. J., and Sionit, N. (1987). Effects of increasing CO_2 concentration on the physiology and growth of forest trees. *In* "The Greenhouse Effect, Climate Change, and U.S. Forests" (W. E. Shands and J. S. Hoffman, eds.), pp. 219–246. Conservation Foundation, Washington, D. C.

Kramer, P. J., and Wetmore, T. H. (1943). Effects of defoliation on cold resistance and diameter growth of broad leaved evergreens. *Am. J. Bot.* **30,** 428–431.

Kramer, P. J., Riley, W. S., and Bannister, T. T. (1952). Gas exchange of cypress (*Taxodium distichum*) knees. *Ecology* **33,** 117–121.

Kramer, P. J., Hellmers, H. and Downs, R. J. (1970). SEPEL: New phytotrons for environmental research. *BioScience* **20,** 1201–1208.

Krasny, M. E., Zasada, J. C., and Vogt, K. A. (1988). Adventitious rooting of four Salicaceae species in response to a flooding event. *Can. J. Bot.* **66,** 2597–2598.

Kress, L. W., Skelly, J. M., and Hinkelmann, K. H. (1982). Growth impact of O_3, NO_2, and/or SO_2 on *Pinus taeda. Environ. Monit. Assess.* **1,** 229–239.

Kriebel, H. B. (1957). Patterns of genetic variation in sugar maple. *Ohio, Agric. Exp. Stn., Res. Bull.* **791,** 1–56.

Kriebel, H. B., and Wang, C. (1962). The interaction between provenance and degree of chilling in bud-break of sugar maple. *Silvae Genet.* **11,** 125–130.

Kriedemann, P. E. (1971). Photosynthesis and transpiration as a function of gaseous diffusive resistances in orange leaves. *Physiol. Plant.* **24,** 218–225.

Kriedemann, P. E., and Buttrose, M. S. (1971). Chlorophyll content and photosynthetic activity within woody shoots of *Vitis vinifera* (L.). *Photosynthetica* **5,** 22–27.

Kriedemann, P. E., Sward, R. J., and Downton, W. J. S. (1976). Vine response to carbon dioxide enrichment during heat therapy. *Aust. J. Plant Physiol.* **3,** 605–618.

Krinard, R. M. and Johnson, R. L. (1980). Fifteen years of cottonwood plantation growth and yield. *South. J. Appl. For.* **4,** 180–185.

Krizek, D. T., and Dubik, S. P. (1987). Influence of water stress and restricted root volume on growth and development of urban trees. *J. Arboric.* **13,** 47–55.

Krueger, K. W., and Ferrell, W. K. (1965). Comparative photosynthetic and respiratory responses to temperature and light by *Pseudotsuga menziesii* var. *menziesii* and var. *glauca* seedlings. *Ecology* **46,** 794–801.

Kubler, H. (1983). Mechanism of frost crack formation in trees—A review and synthesis. *For. Sci.* **29,** 559–568.

Kubler, H. (1987). Origin of frost cracks in stems of trees. *J. Arboric.* **13,** 93–97.

Kubler, H. (1988). Frost cracks in stems of trees. *Arboric. J.* **12,** 163–175.

Kuhns, M. R., Garrett, H. E., Teskey, R. O., and Hinckley, T. M. (1985). Root growth of black walnut trees related to soil temperature, soil water potential, and leaf water potential. *For. Sci.* **31,** 617–629.

Kummerow, J. (1980). Adaptation of roots in water-stressed native vegetation. *In* "Adaptation of Plants to Water and High Temperature Stress" (N. C. Turner and P. J. Kramer, eds.), pp. 57–73. Wiley, New York.

Kuntz, J. E., and Kozlowski, T. T. (1963). Effect of herbicides and land preparation on replanting of heavy soil in northern Wisconsin. *Univ. Wisc., For. Res. Note* **89.**

Kuntz, J. E., Kozlowski, T. T., Wojahn, K. E., and Brener, W. H. (1964). Nursery weed control with Dacthal. *Tree Plant. Notes* **61,** 8–10.

Kuo, C. G., and Pharis, R. P. (1975). Effects of AMO-1618 and B-995 on growth and endogenous gibberellin content of *Cupressus arizonica* seedlings. *Physiol. Plant.* **34,** 288–292.

Kwesiga, F. R., and Grace, J. R. (1986). The role of the red far-red ratio in the response of tropical tree seedlings to shade. *Ann. Bot.* (*London*) [N.S.] **57,** 283–290.

Kwesiga, F. R., Grace, J. R., and Sandford, A. P. (1986). Some photosynthetic characteristics of tropical timber trees as affected by the light regime during growth. *Ann. Bot.* (*London*) [N.S.] **58,** 23–32.

Kyle, D J. (1987). The biochemical basis for photoinhibition of photosystem II. *In* "Photoinhibition" (D. J. Kyle, C. B. Osmond, and C. J. Arntzen, eds.), pp. 197–226. Elsevier, Amsterdam.

Labanauskas, C. K., and Handy, M. F. (1972). Nutrient removal by Valencia orange fruit from citrus orchards in California. *Calif. Agric.* **26,** 3–4.

Labanauskas, C. K., Stolzy, L. H., Klotz, L. J., and DeWolfe, T. A. (1965). Effects of soil temperature and oxygen on the amounts of macronutrients and micronutrients in citrus seedlings (*Citrus sinensis* var. 'Bessie'). *Soil Sci. Soc. Am. Proc.* **29,** 60–64.

Labanauskas, C. K., Stolzy, L. H., and Handy, M. F. (1972). Concentrations and total amounts of nutrients in citrus seedlings (*Citrus sinensis* Osbeck) and in soil as influenced by differential soil oxygen treatments. *Soil Sci. Soc. Am. Proc.* **36,** 454–457.

Labyak, L. F., and Schumacher, F. X. (1954). The contribution of its branches to the main stem growth of loblolly pine. *J. For.* **52,** 333–337.

Laessle, A. M. (1965). Spacing and competition in natural stands of sand pine. *Ecology* **46,** 65–72.

Lakso, A. N. (1979). Seasonal changes in stomatal response to leaf water potential in apple. *J. Am. Soc. Hortic. Sci.* **104,** 58–60.

La Marche, V. C., Jr., and Mooney, H. A. (1972). Recent climatic change and development of the bristlecone pine (*Pinus longaeva* Bailey) krummholz zone. Mt. Washington, Nevada. *Arct. Alp. Res.* **4,** 61–72.

La Marche, V. C., Jr., Graybill, D. A., Fritts, H. C., and Rose, M. R. (1985). Increasing atmospheric carbon dioxide: Tree ring evidence for growth enhancement in natural vegetation. *Science* **225,** 1019–1021.

Lambers, H. (1985). Respiration in intact plants and tissues: Its regulation and dependence on environmental factors, metabolism, and invaded organisms. *Encycl. Plant Physiol., New Ser.* **18,** 418–473.

Lamoreaux, R. J., and Chaney, W. R. (1977). Growth and water movement in silver maple seedlings affected by cadmium. *J. Environ. Qual.* **6,** 201–205.

Lamoreaux, R. J., and Chaney, W. R. (1978). Photosynthesis and transpiration of excised silver maple leaves exposed to cadmium and sulphur dioxide. *Environ. Pollut.* **17,** 259–268.

Landis, T. D., and Evans, A. K. (1974). A relationship between *Fomes applanatus* and aspen windthrow. *Plant Dis. Rep.* **58,** 110–113.

Landsberg, J. J. (1986). "Physiological Ecology of Forest Production." Academic Press, London.

Landsberg, J. J., Blanchard, T. W., and Warrill, B. (1976). Studies on the movement of water through apple trees. *J. Exp. Bot.* **27,** 579–596.

Lange, O. L., Lösch, R., Schulze, E.-D., and Kappen, L. (1971). Responses of stomata to changes in humidity. *Planta* **100,** 76–86.

Lange, O. L., Kappen, L., and Schulze, E.-D., eds. (1976). "Water and Plant Life." Springer-Verlag, Berlin.

Langenfeld-Heyser, R. (1989). CO_2 fixation in stem slices of *Picea abies* (L.) Karst microautoradiographic studies. *Trees* **3,** 24–32.

Langenheim, J. H., Osmond, C. B., Brooks, A., and Farrar, P. J. (1984). Photosynthetic responses to light in seedlings of selected Amazonian and Australian rainforest tree species. *Oecologia* **63,** 215–224.

Lanini, W. T., and Radosevich, S. R. (1986). Response of three conifer species to site preparation and shrub control. *For. Sci.* **9,** 497–506.

Lanner, R. M. (1966). Needed: A new approach to the study of pollen dispersion. *Silvae Genet.* **15,** 50–52.

Lanner, R. M. (1985). On the insensitivity of height growth to spacing. *For. Ecol. Manage.* **13,** 143–148.

Larcher, W. (1961). Zur Assimilationsökologie der immergrünen *Quercus pubescens* im nördlichen Gardaseegebiet, *Planta* **56,** 607–617.

Larcher, W. (1969). The effect of environmental and physiological variables on the carbon dioxide exchange of trees. *Photosynthetica* **3,** 167–198.

Larcher, W. (1980). "Physiological Plant Ecology." Springer-Verlag, Berlin.

Larcher, W. (1983). "Physiological Plant Ecology." Springer-Verlag, Berlin.

Larcher, W. (1985a). Kälte und Frost. *In* "Handbuch der Pflanzenkrankheiten" (P. Sorauer, ed.), Vol. I, pp. 107–326. Parey, Berlin.

Larcher, W. (1985b). Water stress in high mountains. *Eidg. Anst. Forstl. Versuchswes.* **270,** 11–19.

Larcher, W., and Bauer, H. (1981). Ecological significance of resistance to low temperature. *Encycl. Plant Physiol., New Ser.* **12A,** 404–437.

Larson, P. R. (1962). The indirect effect of photoperiod on tracheid diameter in red pine. *Am. J. Bot.* **49,** 132–137.

Larson, P. R. (1963a). Stem form and silviculture. *Proc. Soc. Am. For.*, pp. 103–107.

Larson, P. R. (1963b). Stem form development in forest trees. *For. Sci. Monogr.* **5.**

Larson, P. R. (1964). Some indirect effects of environment on wood formation. *In* "The Formation of Wood in Forest Trees" (M. H. Zimmermann, ed.), pp. 345–365. Academic Press, New York.

Larson, P. R. (1965). Stem form of young *Larix* as influenced by wind and pruning. *For. Sci.* **11,** 412–424.

Larson, P. R. (1969). Wood formation and the concept of wood quality. *Yale Sch. For. Bull.* **74.**

Larsson, S., Oren, R., Waring, R. H., and Barnett, J. W. (1983). Attacks of mountain pine beetle as related to tree vigor of ponderosa pine. *For. Sci.* **29,** 395–402.

Lassoie, J. P. (1973). Diurnal dimensional fluctuations in a Douglas-fir stem in response to tree water status. *For. Sci.* **19,** 251–255.

Lassoie, J. P., and Hinckley, T. M., eds. (1990). "Techniques and Approaches in Forest Tree Ecophysiology." CRC Press, Boca Raton, Florida.

Lassoie, J. P., Hinckley, T. M., and Grier, C. C. (1985). Coniferous forests of the Pacific Northwest. *In* "Physiological Ecology of North American Plant Communities" (B. F. Chabot and H. A. Mooney, eds.), pp. 127–161. Chapman & Hall, New York and London.

Last, F. T. (1982). Effects of atmospheric sulphur compounds on natural and man-made terrestrial and aquatic ecosystems. *Agric. Environ.* **7,** 299–387.

Lawrence, W. T., and Oechel, W. C. (1983). Effects of soil temperature on the carbon exchange of taiga seedlings. II. Photosynthesis, respiration, and conductance. *Can. J. For. Res.* **13,** 850–859.

Lawton, R. O. (1982). Wind stress and elfin stature in a montane rain forest tree: An adaptive explanation. *Am. J. Bot.* **69,** 1224–1230.

Leach, R. W. A., and Wareing, P.F. (1967). Distribution of auxin in horizontal woody stems in relation to gravimorphism. *Nature (London)* **214,** 1025–1027.

Lechowicz, M. J. (1984). Why do temperate deciduous trees leaf out at different times? Adaptations and ecology of forest communities. *Am. Nat.* **124,** 821–842.

Ledig, F. T. (1974). Concepts of growth analysis. *Proc. North Am. For. Biol. Workshop, 3rd, 1974,* pp. 166–182.

Ledig, F. T. (1976). Physiological genetics, photosynthesis and growth models. *In* "Tree Physiology and Yield Improvement" (M. G. R. Cannell and F. T. Last, eds.), pp. 21–54. Academic Press, London.

Ledig, F. T., and Perry, T. O. (1967). Variation in photosynthesis and respiration among loblolly pine progenies. *South. Conf. For. Tree Improve., 9th,* pp. 120–128.

Lee, J. C., and Kramer, P. J. (1987). Forestry research needs and strategies. *In* "The Greenhouse Effect, Climate Change, and U.S. Forests" (W. E. Shands and J. S. Hoffman, eds.), pp. 295–302. Conservation Foundation, Washington, DC.

Lees, J. C. (1972). Soil aeration response to drainage intensity in basin peat. *Forestry* **45,** 135–143.

Lee-Stadelmann, O. Y., and Stadelmann, E. J. (1976). Cell permeability and water stress. *In* "Water and Plant Life" (O. L. Lange, L. Kappen, and E.-D. Schulze, eds.), pp. 268–280. Springer-Verlag, Berlin and New York.

Legg, M. H., and Schneider, G. (1977). Soil deterioration in campsites: Northern forest types. *Soil Sci. Soc. Am. Proc.* **41,** 437–441.

Leiser, A. T., Harris, R. W., Neel, P. L., Long, D., Slice, N. W., and Maire, R. G. (1972). Staking and pruning influence trunk development of young trees. *J. Am. Soc. Hortic. Sci.* **97,** 498- 503.

Leite, R. M. de O., and Alvim, P. de T. (1978). Efeito do vento e da radiacao solar na ruptura do pulvinulo foliar do cacaueiro (*Thebroma cacao* L.). Commissao Executiva da Recuperaco da Lavoura Cacaueira. *Inf. Tec.—Centro Pesqui. Cacao, 1977/1978,* pp. 65–66.

Lemon, E. R., ed. (1983). "CO_2 and Plants." Westview Press, Boulder, Colorado.

Leopold, A. C. (1980). Temperature effects on soybean imbibition and leakage. *Plant Physiol.* **65,** 1096–1098.

Leopold, A. C. (1983). Volumetric components of seed imbibition. *Plant Physiol.* **73,** 677–680.

Leopold, A. C., Brown, K. M., and Emerson, F. H. (1972). Ethylene in the wood of stressed trees. *HortScience* **7,** 175.

Leopold, A. C., Musgrave, M. E., and Williams, K. M. (1981). Solute leakage resulting from leaf desiccation. *Plant Physiol.* **68,** 1222–1225.

Lerner, R. H., and Evenari, M. (1961). The nature of the germination inhibitor present in leaves of *Eucalyptus rostrata. Physiol. Plant.* **14,** 221–229.

Leshem, B. (1965). The annual activity of intermediary roots of the Aleppo pine. *For. Sci.* **11,** 291–298.

Lester, D. T. (1970). Variation in seedling development of balsam fir associated with seed origin. *Can. J. Bot.* **48,** 1093–1097.

Leuty, S. J., and Pree, D., J. (1980). The influence of tree population and summer pruning on productivity, growth and quality of peaches. *J. Am. Soc. Hortic. Sci.* **105,** 702–705.

Leverenz, J. W., and Lev, D. J. (1987). Effects of carbon dioxide- induced climate changes on the natural ranges of six major commercial tree species in the western United States. *In* "The Greenhouse Effect, Climate Change, and U.S. Forests" (W. E. Shands and J. S. Hoffman, eds.), pp. 123–155. Conservation Foundation, Washington, D. C.

Leverenz, J. W., and Öquist, G. (1987). Quantum yields of photosynthesis at temperatures between −2°C and 35°C in a cold-tolerant C_3 plant (*Pinus sylvestris*) during the course of one year. *Plant, Cell Environ.* **10,** 287–295.

Levitt, J. (1980a). "Responses of Plants to Environmental Stresses," 2nd ed., Vol. 1. Academic Press, New York.

Levitt, J. (1980b). "Responses of Plants to Environmental Stresses," 2nd ed., Vol. 2. Academic Press, New York.

Lewandowska, M., and Öquist, G. (1980). Structural and functional relationships in developing *Pinus silvestris* chloroplasts. *Physiol. Plant.* **48,** 39–46.

Lewandowska, M., Hart, J. W., and Jarvis, P. G. (1976). Photosynthetic electron transport in plants of Sitka spruce subjected to differing light environments during growth. *Physiol. Plant.* **37,** 269–274.

Lewis, A. R. (1988). Buttress arrangement in *Pterocarpus officinalis* (Fabaceae): Effects of crown asymmetry and wind. *Biotropica* **20,** 280–285.

Lewis, L. N., Coggins, C. W., Jr., and Hield, H. Z. (1964). The effect of biennial bearing and NAA on the carbohydrate and nitrogen composition of Wilking Mandarin leaves. *Proc. Am. Soc. Hortic. Sci.* **84,** 147–151.

Li, P. H., and Weiser, C. J. (1967). Evaluation of extraction and assay methods for nucleic acids from red osier dogwood and RNA, DNA, and protein changes during cold acclimation. *Proc. Am. Soc. Hortic. Sci.* **91,** 716–727.

Lichenthaler, H. K., Buschmann, C., Doll, M., Fietz, H.-J., Bach, T., Kozel, U., Meier, D., and Rahmsdorf, U. (1981). Photosynthetic activity, chloroplast ultrastructure, and leaf characteristics of high-light and low-light plants and of sun and shade leaves. *Photosynth. Res.* **2,** 115–141.

Lieberman, D., Lieberman, M., Peralta, R., and Hartshorn, G. S. (1985). Mortality patterns and stand turnover rates in a wet tropical forest in Costa Rica. *J. Ecol.* **73,** 915–924.

Lieffers, V. J., and Rothwell, R. L. (1986a). Effects of water table and substrate temperature on root and top growth of *Picea mariana* and *Larix laricina* seedlings. *Can. J. For. Res.* **16,** 1201–1206.

Lieffers, V. J., and Rothwell, R. L. (1986b). Rooting of peatland black spruce and tamarack in relation to depth of water table. *Can. J. Bot.* **65,** 817–821.

Lieffers, V. J., and Rothwell, R. L. (1987). Effects of drainage on substrate temperature and phenology of some trees and shrubs in an Alberta peatland. *Can. J. For. Res.* **17,** 97–104.

Lieth, H. (1975). Primary productivity of the major vegetation units of the world. *In* "Primary Productivity of the Biosphere" (H. Lieth and R. H. Whittaker, eds.), pp. 203–215. Springer-Verlag, Berlin and New York.

Lieth, H., and Whittaker, R. H., eds. (1975). "Primary Productivity of the Biosphere." Springer-Verlag, Berlin and New York.

Likens, G. E., and Bormann, F. H. (1974). Linkages between terrestrial and aquatic ecosystems. *BioScience* **24,** 447–456.

Likens, G. E., Bormann, F. H., Johnson, N. M., Fisher, D. W., and Pierce, R. S. (1970). Effects of forest cutting and herbicide treatment on nutrient budgets in the Hubbard Brook watershed ecosystem. *Ecol. Monogr.* **40,** 23–47.

Likens, G. E., Bormann, F. H., Pierce, R. S., Eaton, J. S., and Johnson, N. M. (1979). "Biogeochemistry of a Forested Ecosystem." Springer-Verlag, New York.

Lincoln, D. E., Sionit, N., and Strain, B. R. (1984). Growth and feeding responses of *Pseudoplusia includens* (Lepidoptera: Noctuidae) to host plants grown in controlled and carbon dioxide atmospheres. *Environ. Entomol.* **13,** 1527–1530.

Lindberg, S. E., Lovett, G. M., Richter, D. D., and Johnson, D. W. (1986). Atmospheric deposition and canopy interactions of major ions in a forest. *Science* **231,** 141–145.

Linder, S., and Axelsson, B. (1982). Changes in carbon uptake and allocation as a result of irrigation and fertilization in a young *Pinus sylvestris* stand. *In* "Carbon Uptake and Allocation in Subalpine Ecosystems as a Key to Management" (R. H. Waring, ed.), pp. 38–44. Oregon State University, Corvallis.

Linder, S., and Rook, D. A. (1984). Effects of mineral nutrition on carbon dioxide exchange and partitioning of carbon in trees. *In* "Nutrition of Plantation Forests" (G. D. Bowen and E. K. S. Nambiar, eds.), pp. 211–236. Academic Press, London.

Linder, S., and Troeng, E. (1981). The seasonal course of respiration and photosynthesis in strobili of Scots pine. *For. Sci.* **27,** 267–276.

Linder, S., McMurtrie, R. E., and Landsberg, J. J. (1985). Growth of *Eucalyptus*: A mathematical model applied to *Eucalyptus globulus*. *In* "Crop Physiology of Forest Trees" (P. M. A. Tigerstedt, P. Puttonen, and V. Koski, eds.), pp. 117–126. Dept. of Plant Breeding, University of Helsinki, Finland.

Lindstrom, A. (1986). Freezing temperatures in the root zone—Effects on growth of containerized *Pinus sylvestris* and *Picea abies* seedlings. *Scand. J. For. Res.* **1,** 371–377.

Lindstrom, A., and Nystrom, C. (1987). Seasonal variation in root hardiness of container-grown Scots pine, Norway spruce, and lodgepole pine seedlings. *Can. J. For. Res.* **17,** 787–793.

Linzon, S. N. (1958). The influence of smelter fumes on the growth of white pine in the Sudbury region. *Ont., Dep. Lands For., Can., Dep. Agric., Publ.,* pp. 1–45.

Linzon, S. N. (1966). Damage to eastern white pine by sulfur dioxide, semi-mature-tissue needle blight and ozone. *J. Air Pollut. Control. Assoc.* **16,** 140–144.

Linzon, S. N. (1978). Effects of airborne sulfur pollutants on plants. *In* "Sulfur in the Environment. Part II. Ecological Impacts" (J. O. Nriagu, ed.), pp. 109–162. Wiley, New York.

Linzon, S. N. (1986). Effects of gaseous pollutants on forests in eastern North America. *Water, Air, Soil Pollut.* **31,** 537–550.

Little, C. H. A. (1975). Inhibition of cambial activity in *Abies balsamea* by internal water stress: Role of abscisic acid. *Can. J. Bot.* **53,** 3041–3050.

Little, S. (1974). Effects of fire on temperate forests: Northeastern United States. *In* "Fire and Ecosystems" (T. T. Kozlowski and C. E. Ahlgren, eds.), pp. 225–250. Academic Press, New York.

Lloyd, J., Syvertsen, J. P., and Kriedemann, P. E. (1987). Salinity effects on leaf water relations and gas exchange of Valencia orange, *Citrus sinensis* (L.) Osbeck, on rootstocks with different salt exclusion characteristics. *Aust. J. Plant Physiol.* **14,** 605–617.

Loach, K. (1967). Shade tolerance in tree seedlings. I. Leaf photosynthesis and respiration in plants raised under artificial shade. *New Phytol.* **66,** 607–621.

Loehle, C. (1988a). Tree life history strategies: The role of defenses. *Can. J. For. Res.* **18,** 209–222.

Loehle, C. (1988b). Forest decline: Endogenous dynamics, tree defenses, and the elimination of spurious correlation. *Vegetatio* **77,** 65–78.

Logan, K. T., and Krotkov, G. (1968). Adaptations of the photosynthetic mechanism of sugar maple (*Acer saccharum*) seedlings grown in various light intensities. *Physiol. Plant.* **22,** 104–116.

LoGullo, M. A., and Salleo, S. (1988). Different strategies of drought resistance in three Mediterranean sclerophyllous trees growing in the same environmental conditions. *New Phytol.* **108,** 267–276.

Lopushinsky, W., and Beebe, T. (1976). Relationship of shoot–root ratio to survival and growth of outplanted Douglas-fir and ponderosa pine seedlings. *USDA For. Serv. Res. Note PNW* **PNW-274.**

Lorenc-Plucinska, G. (1982). Effects of sulphur dioxide on CO_2 exchange in SO_2-tolerant and SO_2-susceptible Scots pine seedlings. *Photosynthetica* **16,** 140–144.

Lorimer, C. G. (1983). Tests of age-independent competition indices for individual trees in natural hardwood stands. *For. Ecol. Manage.* **6,** 343–360.

Lorimer, C. G. (1989). Relative effect of small and large disturbances on temperate hardwood forest structure. *Ecology* **70,** 565–567.

Lorio, P. L., Jr. (1986). Growth-differentiation balance: A basis for understanding southern pine beetle–tree interactions. *For. Ecol. Maange.* **14,** 159–273.

Lorio, P. L., Jr., and Sommers, R. A. (1986). Evidence of competition for photosynthates between growth processes and oleoresin synthesis in *Pinus taeda* L. *Tree Physiol.* **2,** 301–306.

Lorio, P. L., Jr., Sommers, R. A., Blanche, C. A., Hodges, J. D., and Nebeker, T. E. (1990). Modeling the resistance to bark beetles based on growth and differentiation balance principles. *In* "Process Modeling of Forest Growth Responses to Environmental Stress" (R. K. Dixon, R. S. Meldahl, G. A. Ruark, and W. E. Warren, eds.). Timber Press, Portland, Oregon.

Lovett, G. M., Reiners, W. A., and Olson, R. K., (1982). Cloud droplet deposition in subalpine fir forests: Hydrologic and chemical inputs. *Science* **218,** 1303–1304.

Lowe, W. J., Hocker, H. W., and McCormack, M. L., Jr. (1977). Variation in balsam fir provenances planted in New England. *Can. J. For. Res.* **7,** 63–67.

Lucier, A. A., and Hinckley, T. M. (1982). Phenology, growth and water relations of irrigated and non-irrigated black walnut. *For. Ecol. Manage.* **4,** 127–142.

Luckwill, L. C. (1978). Meadow orchards and fruit walls. *Acta Hortic.* **65,** 237–243.

Ludlow, M. M., and Björkman, O. (1984). Paraheliotropic leaf movement in siratro as a protective mechanism against drought-induced damage to primary photosynthetic reactions: Damage by excessive light and heat. *Planta* **161,** 505–518.

Lugo, A. E., Applefield, M., Pool, D. J., and McDonald, R. B (1983). The impact of Hurricane David on the forest of Dominica. *Can. J. For. Res.* **13,** 201–211.

Lugo, A. E., Brown, S., and Chapman, J. (1988). An analytical review of production rates and stemwood biomass of tropical forest plantations. *For. Ecol. Manage.* **23,** 179–200.

Lundegärdh, H. (1931). "Environment and Plant Development." Arnold, London.

Lürssen, K. (1987). The use of inhibitors for gibberellin and sterol biosynthesis to probe hormone action. *In* "Hormone Action in Plant Development—A Critical Appraisal" (G. V. Hoad, J. R. Lenton, M. B. Jackson, and R. K. Atkin, eds.), pp. 133–144. Butterworth, London.

Luukkanen, O. (1978). Investigations on factors affecting net photosynthesis in trees: Gas exchange in clones of *Picea abies* (L.) Karst. *Acta For. Fenn.* **162,** 1–63.

Luukkanen, O., and Kozlowski, T. T. (1972). Gas exchange in six *Populus* clones. *Silvae Genet.* **21,** 220–229.

Luxmoore, R. J. (1981). CO_2 and phytomass. *BioScience* **31,** 626.

Lynch, D. W., Davis, W. C., Roof, L. R., and Korstian, C. F. (1943). Influence of nursery fungicide–fertilizer treatments on survival and growth in a southern pine plantation. *J. For.* **41,** 411–413.

Lyon, T. L., and Buckman, H. O. (1943). "The Nature and Properties of Soil," 4th ed. Macmillan, New York.

McCauley, K. J., and Cook, S. A. (1980). *Phellinus weirii* infestation of two mountain hemlock forests in the Oregon Cascades. *For. Sci.* **25,** 23–29.

McClaugherty, C. A., Aber, J. D., and Melillo, J. M. (1982). The role of fine roots in the organic matter and nitrogen budgets of two forested ecosystems. *Ecology* **63,** 1481–1490.
McClenahen, J. R. (1979). Effects of ethylene diurea and ozone on the growth of tree seedlings. *Plant Dis. Rep.* 63, 320–323.
McClenahen, J. R. (1983). The impact of an urban-industrial area on deciduous forest tree growth. *J. Environ. Qual.* **12,** 64–69.
McColl, J. G., and Powers, R. F. (1984). Consequences of forest management on soil–tree relationships. *In* "Nutrition of Plantation Forests" (G. D. Bowen and E. K. S. Nambiar, eds.), pp. 379–412. Academic Press, London.
McComb, A. O., and Loomis, W. E. (1944). Subclimax prairie. *Bull. Torrey Bot. Club* **71,** 46–76.
McCool, P. M., Mange, J. A., and Taylor, O. C. (1979). Effects of O_3 and HCl gas on development of mycorrhizal fungus (*Glomus fasciculatus*) and growth of 'Troyer' citrange. *J. Am. Soc. Hortic. Sci.* **104,** 151–154.
McCracken, I. W., and Kozlowski, T. T. (1965). Thermal contraction in twigs. *Nature* (*London*) **208,** 910–912.
McCree, K. J., and Davis, S. D. (1974). Effect of water stress and temperature on leaf size and number of epidermal cells in grain sorghum. *Crop Sci.* **14,** 751–755.
McDermott, R. E. (1954). Effects of saturated soil on seedling growth of some bottomland hardwood species. *Ecology* **35,** 36–41.
MacFall, J. S., Johnson, G. A., and Kramer, P. J. (1990). Observation of a water-depletion region surrounding loblolly pine roots by magnetic resonance imaging. *Proc. Natl. Acad. Sci. USA* **87,** 1203–1207.
McGee, A. B., Schmierbach, M. R., and Bazzaz, F. A. (1981). Photosynthesis and growth of populations of *Populus deltoides* from contrasting habitats. *Am. Midl. Nat.* **105,** 305–311.
McGregor, W. H. D., and Kramer, P. J. (1963). Seasonal trends in rates of photosynthesis and respiration of loblolly pine. *Am. J. Bot.* **50,** 760–765.
MacHattie, L. B. (1963). Winter injury of lodgepole pine foliage. *Weather* **19,** 301–307.
McIntosh, R. (1984). "Fertiliser Experiments in Established Conifer Stands," Forest Record. Forestry Commission, United Kingdom.
McKeand, S. E., Jett, J. B., Sprague, J. B., and Todhunter, M. N. (1987). Summer wax grafting of loblolly pine. *South. J. Appl. For.* **11,** 96–99
McKee, W. H. Jr., and Wilhite, L. P. (1986). Loblolly pine response to bedding and fertilization varies by drainage class on low Atlantic Coastal Plain sites. *South. J. Appl. For.* **10,** 16–21.
MacLachlan, J. D. (1935). The dispersal of viable basidiospores of the *Gymnosporangium* rusts. *J. Arnold Arbor., Harv. Univ.* **16,** 411–422.
McLaughlin, S. B., McConathy, R. K., Duvick, D. N., and Mann, L. K. (1982). Effects of chronic air pollution stress on photosynthesis, carbon allocation, and growth of white pine trees. *For. Sci.* **28,** 60–70.
McLaughlin, S. B., Blasing, I. J., Mann, L. K., and Duvick, D. N. (1983). Effects of acid rain and gaseous pollutants on forest productivity: A regional approach. *J. Air Pollut. Control Assoc.* **33,** 1042–1049.
MacLean, D. A., and Wein, R. W. (1977). Nutrient accumulation for postfire jack pine and hardwood succession patterns in New Brunswick forest stands. *Can. J. For. Res.* **7,** 562–578.
McMinn, R. G. (1963). Characteristics of Douglas-fir root systems. *Can. J. Bot.* **41,** 105–122.
McMinn, J. W., and McNab, W. H. (1971). Early growth and development of slash pine under drought and flooding. *USDA For. Serv. Res. Pap. SE* **SE–89.**
McNeil, R. C., Lea, R., Ballard, R., and Allen, H. L. (1988). Predicting fertilizer response of loblolly pine using foliar and needle-fall nutrients sampled in different seasons. *For. Sci.* **34,** 698–707.
McWilliam, J. R. (1983). Physiological basis for chilling stress and the consequences for crop production. *In* "Crop Reactions to Water and Temperature Stresses in Humid, Temperate Climates" (C. D. Raper, Jr. and P. J. Kramer, eds.), pp. 113–132. Westview Press, Boulder, Colorado.
McWilliam, J. R. (1986). The national and international importance of drought and salinity effects on agricultural production. *Aust. J. Plant Physiol.* **13,** 1–13.

McWilliam, J. R., Kramer, P. J., and Musser, R. L. (1982). Temperature-induced water stress in chilling-sensitive plants. *Aust. J. Plant Physiol.* **9,** 343–352.

Mader, D. R. (1976). Soil-site productivity of natural stands of white pine in Massachusetts. *Soil Sci. Soc. Am. J.* **40,** 112–115.

Maftoun, M., and Pritchett, W. L. (1970). Effects of added nitrogen on the availability of phosphorus to slash pine on two lower coastal plain soils. *Soil Sci. Soc. Am. Proc.* **34,** 685–690.

Maggs, D. H. (1964). Growth rates in relation to assimilate supply and demand. I. Leaves and roots as limiting regions. *J. Exp. Bot.* **15,** 574–583.

Maggs, D. H. (1965). Growth rates in relation to assimilate supply and demand. II. The effect of particular leaves and growing regions in determining dry matter distribution in young apple leaves. *J. Exp. Bot.* **16,** 387–404.

Maguire, D. A., and Hann, D. W. (1987). Equations for predicting sapwood area at crown base in southwestern Oregon. *Can. J. For. Res.* **17,** 236–241.

Majdi, H., and Persson, H. (1989). Effects of road-traffic pollutants (lead and cadmium) on tree fine-roots along a motor road. *Plant Soil,* **119,** 1–5.

Malhotra, S. S., and Khan, A. A. (1978). Effects of sulphur dioxide fumigation on lipid biosynthesis in pine needles. *Phytochemistry* **17,** 241–244.

Malhotra, S. S., and Khan, A. A. (1980). Effects of sulphur dioxide and other air pollutants on acid phosphatase activity in pine seedlings. *Biochem. Physiol. Pflanz.* **175,** 228–236.

Malhotra, S. S., and Sarkar, S. K. (1979). Effects of sulphur dioxide on sugar and free amino acid content of pine seedlings. *Physiol Plant.* **47,** 223–228.

Maltby, E. (1986). "Waterlogged Wealth." International Institute for Environment and Development, London and Washington, DC.

Mansfield, T. A. (1986). The physiology of stomata: New insights into old problems. *In* "Plant Physiology" (F. C. Steward, J. F. Sutcliff, and J. E. Dale, eds.), Vol. 9, pp. 155–224. Academic Press, Orlando, Florida.

Mansfield, T. A., and Davies, W. J. (1981). Stomata and stomatal mechanisms. *In* "The Physiology and Biochemistry of Drought Resistance in Plants" (L. G. Paleg and D. Aspinall, eds.), pp. 315–346. Academic Press, Sydney.

Mansfield, T. A., and Davies, W. J. (1985). Mechanisms for leaf control of gas exchange. *BioScience* **46,** 158–164.

Mansfield, T. A., and Freer-Smith, P. H. (1984). The role of stomata in resistance mechanisms. *In* "Gaseous Air Pollutants and Plant Metabolism" (M. J. Koziol and F. R. Whatley, eds.), pp. 131–146. Butterworth, London.

Mantai, K. E., and Bishop, N. I. (1967). Studies on the effects of ultraviolet irradiation on photosynthesis and on the 520 nm light–dark difference spectra in green algae and isolated chloroplasts. *Biochim. Biophys. Acta* **131,** 350–356.

Marbut, C. F., and Manifold, C. B. (1926). The soils of the Amazon basin in relation to agricultural possibilities. *Geogr. Rev.* **16,** 414–442.

Margolis, H. A., and Waring, R. H. (1986). Carbon and nitrogen allocation patterns of Douglas-fir seedlings fertilized with nitrogen in autumn. II. Field performance. *Can. J. For. Res.* **16,** 903–909.

Markhart, A. H., Peet, M. M., Sionit, N., and Kramer, P. J. (1980). Low temperature acclimation of root fatty acid composition, leaf water potential, gas exchange and growth of soybean seedlings. *Plant, Cell Environ.* **3,** 435–441.

Marks, G. C., and Kozlowski, T. T., eds. (1973). "Ectomycorrhizae." Academic Press, New York.

Marks, P. L. (1975). On the relation between extension growth and successional status of deciduous trees of the northeastern United States. *Bull. Torrey Bot. Club* **102,** 172–177.

Marquard, R. D., and Hanover, J. W. (1984). The effect of shade on flowering of *Picea glauca*. *Can. J. For. Res.* **14,** 830–832.

Marr, J. W. (1948). Ecology of the forest–tundra ecotone on the east coast of Hudson Bay. *Ecol. Monogr.* **18,** 117–144.

Marschner, H. (1986). "The Mineral Nutrition of Higher Plants." Academic Press, London.

Marshall, J. D., and Waring, R. H. (1984). Conifers and broadleaf species: Stomatal sensitivity differs in western Oregon. *Can. J. For. Res.* **14,** 905–908.

Marshall, J. D., and Waring, R. H. (1985). Predicting fine root production and turnover by monitoring root starch and soil temperature. *Can. J. For. Res.* **15,** 791–800.

Marshall, P. E., and Furnier, G. R. (1981). Growth responses of *Ailanthus altissima* seedlings to SO_2. *Environ. Pollut., Ser. A* **25,** 149–153.

Marshall, P. E., and Kozlowski, T. T. (1974a). The role of cotyledons in growth and development of woody angiosperms. *Can. J. Bot.* **52,** 239–245.

Marshall, P. E., and Kozlowski, T. T. (1974b). Photosynthetic activity of cotyledons and foliage leaves of young angiosperm seedlings. *Can. J. Bot.* **52,** 2023–2032.

Marshall, P. E., and Kozlowski, T. T. (1976). Importance of photosynthetic cotyledons for early growth of woody angiosperms. *Physiol. Plant.* **37,** 336–340.

Marth, P. C., and Gardner, T. E. (1939). Evaluation of a variety of peach seedling stocks with respect to "wet feet" tolerance. *Proc. Am. Soc. Hortic. Sci.* **37,** 335–337.

Martin, E. V., and Clements, F. E. (1935). Studies of the effect of artificial wind on growth and transpiration in *Helianthus annuus. Plant Physiol.* **10,** 613–636.

Martin, J. M., and Sydnor, T. D. (1987). Differences in wound closure rates in 12 tree species. *HortScience* **22,** 442–444.

Martin, U., Pallardy, S. G., and Bahari, Z. A. (1987). Dehydration tolerance of leaf tissues of six woody angiosperm species. *Physiol. Plant.* **69,** 182–186.

Martinez-Ramos, M., Alvarez-Buylla, E., and Sarukhan, J. (1989). Tree demography and gap dynamics in a tropical rain forest. *Ecology* **70,** 555–558.

Martsolf,. J. D., Ritter, C. M., and Hatch, A. H. (1975). Effect of white latex paint on temperature of stone fruit tree trunks in winter. *J. Am. Soc. Hortic. Sci.* **100,** 122–129.

Marx, D. H. (1980). Growth of loblolly and shortleaf pine seedlings after two years on a strip-mined coal spoil in Kentucky is stimulated by *Pisolithus* ectomycorrhizae and "starter" fertilizer pellets. *Abstr., North Am. Conf. Mycorrhizae, 3rd, 1977.*

Marx, D. H., and Artman, J. D. (1979). *Pisolithus tinctorius* ectomycorrhizae improve survival and growth of pine seedlings on acid coal spoils in Kentucky and Virginia. *Reclam. Rev.* **2,** 23–31.

Marx, D. H., and Cordell, C. H. (1986). Bayleton (triadmefon) affects ectomycorrhizal development on slash and loblolly pine seedlings in nurseries. *In* "Nursery Management Practices for the Southern Pines" (D. B. South, ed.), pp. 460–475. Auburn University, Auburn, Alabama.

Marx, D. H., Hatch, A. B., and Mendocino, J. F. (1977). High soil fertility decreases sucrose content and susceptibility of loblolly pine roots to ectomycorrhizal infection by *Pisolithus tinctorius. Can. J. Bot.* **55,** 1569–1574.

Marx, D. H., Cordell, C. E., Kenney, J. G., Mexal, J. G., Artman, J. D., Riffle, J. W. and Molina, R. J. (1984). Commercial vegetative inoculum of *Pisolithus tinctorius* and inoculum techniques for development of ectomycorrhizae on bare-root tree seedlings. *For. Sci. Monogr.* **25.**

Masle, J., and Farquhar, G. D. (1988). Effects of soil strength on the relation of water-use efficiency and growth to carbon isotope discrimination in wheat seedlings. *Plant Physiol.* **86,** 32–38.

Masle, J., and Passioura, J. B. (1987). The effect of soil strength on the growth of young wheat plants. *Aust. J. Plant Physiol.* **14,** 643–656.

Mason, E. G., and Cullen, A. W. J. (1986). Growth of *Pinus radiata* on ripped and unripped Taupo pumice soil. *N. Z. J. For. Sci.* **16,** 3–18.

Mathew, M., and George, C. M. (1967). Wind damage in rubber plantations. *Rubber Board. Bull. (India)* **9**(3), 39–44.

Matson, P. A., and Waring, R.H. (1984). Effects of nutrient and light limitation on mountain hemlock: Susceptibility to laminated root rot. *Ecology* **65,** 1517–1524.

Matsushima, J., and Brewer, R. F. (1972). Influence of sulfur dioxide and hydrogen fluoride as a mix on reciprocal exposure on citrus growth and development. *J. Air Pollut. Control. Assoc.* **22,** 710–713.

Mattoon, W. R. (1908). The sprouting of shortleaf pine in the Arkansas National Forest. *For. Q.* **6,** 158–159.

Mattson, W. J., and Haack, R. A. (1987a). The role of drought in outbreaks of leaf-eating insects. *In* "Insect Outbreaks: Ecological and Evolutionary Perspectives" (P. Barbosa and J. C. Schultz, eds.), pp. 365–407. Academic Press, Orlando, Florida.

Mattson, W. J., and Haack, R. A. (1987b). Effects of drought on host plants, phytophagous insects, and their natural enemies to induce insect outbreaks. *BioScience* **37,** 110–118.

Mattson, W. J., Levieux, J., and Bernard-Dagan, E., eds. (1988). "Mechanisms of Woody Plant Defenses against Insects. Search for Pattern." Springer-Verlag, New York,

Matyssek, R. (1986). Carbon, water and nitrogen relations in evergreen and deciduous conifers. *Tree Physiol.* **2,** 177–187.

Matziris, D. I., and Nakos, G. (1978). Effect of simulated 'acid rain' on juvenile characteristics of Aleppo pine (*Pinus halepensis* Mill.). *For. Ecol. Manage.* **1,** 267–272.

Maxie, E. C. (1957). Heat injury in prunes. *Proc. Am. Soc. Hortic. Sci.* **69,** 116–121.

May, L. H., and Milthorpe, F. L. (1962). Drought resistance of crop plants. *Field Crop Abstr.* **15,** 171–179.

Mazzoleni, S., and Dickman, D. I. (1988). Differential physiological and morphological responses of two hybrid *Populus* clones to water stress. *Tree Physiol.* **4,** 61–70.

Mead, D. J. (1984). Diagnosis of nutrient deficiencies in plantations. *In* "Nutrition of Plantation Forests" (G. D. Bowen and E. K. S. Nambiar, eds.), pp. 259–291. Academic Press, London.

Meiners, T. M., Smith, D. W., Sharik, T. L., and Beck, D. E. (1984). Soil and plant water stress in an Appalachian oak forest in relation to topography and stand age. *Plant Soil* **80,** 171–179.

Mejnartowicz, L. E. (1984). Enzymatic investigations on tolerance in forest trees. *In* "Gaseous Air Pollutants and Plant Metabolism" (M. J. Koziol and F. R. Whatley, eds.), pp. 381–398. Butterworth, London.

Mellenthin, W. M., and Wang, C. Y. (1976). Preharvest temperatures in relation to postharvest quality of 'Anjou' pears. *J. Am. Soc. Hortic. Sci.* **101,** 302–305.

Mercer, P. C., Kirk, S. A., Gendle, P., and Clifford, D. R. (1983). Chemical treatments for control of decay in pruning wounds. *Ann. Appl. Biol.* **102,** 435–439.

Mergen, F. (1954). Mechanical aspects of wind-breakage and windfirmness. *J. For.* **52,** 119–125.

Metz, L. J. (1952). Weight and nitrogen and calcium content of the annual litter fall of forests in the South Carolina Piedmont. *Soil Sci. Soc. Am. Proc.* **16,** 38–41.

Metz, L. J., Wells, C. G., and Swindell, B. F. (1966). Sampling soil and foliage in a pine plantation. *Soil Sci. Soc. Am. Proc.* **30,** 397–399.

Meyer, B. S. (1956). The hydrodynamic system. *Encycl. Plant Physiol.* **3,** 596–614.

Meyer, B. S., Anderson, D. B., Bohning, R. H., and Fratianne, D. G. (1973). "Introduction to Plant Physiology." Van Nostrand-Reinhold, Princeton, New Jersey.

Meyer, J., Schneider, B. U., Werk, K. S., Oren, T., and Schulze, E. -D. (1988). Performance of two *Picea abies* (L.) Karst. stands at different stages of decline. V. Root tip and ectomycorrhiza development and their relation to aboveground and soil nutrients. *Oecologia* **77,** 7–13.

Michael, J. L. (1980). Long-term impact of aerial application of 2,4,5-T to longleaf pine (*Pinus palustris*). *Weed Sci.* **28,** 255–257.

Michelana, V. A., and Boyer, J. S. (1982). Complete turgor maintenance at low water potential in the elongating region of maize leaves. *Plant Physiol.* **69,** 1145–1149.

Mika, A. (1986). Physiological responses of fruit trees to pruning. *Hortic. Rev.* **8,** 337–378.

Mikola, P. (1950). Puiden kasvun vaihteluista ja niden merkityksesta kasvututkimuksissa. *Comm. Inst. For. Fenn.* **385,** 1–131.

Mikola, P. (1962). Temperature and tree growth near the northern timber line. *In* "Tree Growth" (T. T. Kozlowski, ed.), pp. 265–274. Ronald Press, New York.

Milburn, J. A. (1966). The conduction of sap. I. Water conduction and cavitation in water stressed leaves. *Planta* **69,** 34–42.

Miller, D. H. (1959). Transmission of insolation through pine forest canopy, as it affects the melting of snow. *Mitt. Schweiz. Anst. Forstl. Versuchswes.* **35,** 57–79.

Miller, E. C. (1938). "Plant Physiology," 2nd ed. McGraw-Hill, New York.

Miller, H. G. (1981). Forest fertilization: Some guiding concepts. *Forestry* **54,** 157–167.

Miller, H. G. (1984). Dynamics of nutrient cycling in plantation ecosystems. *In* "Nutrition of Plantation Forests" (G. D. Bowen and E. K. S. Nambiar, eds.), pp. 53–78. Academic Press, London.

Miller, H. G. (1986). Carbon × nutrient interactions—The limitations to productivity. *Tree Physiol.* **2,** 373–385.

Miller, P. R. (1973). Oxidant-induced community change in a mixed conifer forest. *Adv. Chem. Ser.* **122,** 101–117.

Miller, P. R., and McBride, J. R. (1975). Effects of air pollutants on forests. *In* "Responses of Plants to Air Pollution" (J. B. Mudd and T. T. Kozlowski, eds.), pp. 195–235. Academic Press, New York.

Miller, P. R., Parmeter, J. R., Flick, B. H., and Martinez, C. W. (1969). Ozone damage response of ponderosa pine seedlings. *J. Air Pollut. Control Assoc.* **19,** 435–438.

Miller, R. E., and Tarrant, R. F. (1983). Long-term growth response of Douglas-fir to ammonium nitrate fertilizer. *For. Sci.* **29,** 127–137.

Miller, W. F., Dougherty, P. M., and Switzer, G. L. (1987). Effect of rising carbon dioxide and potential climate change on loblolly pine distribution, growth, survival and productivity. *In* "The Greenhouse Effect, Climate Change, and U.S. Forests" (W. E. Shands and J. S. Hoffman, eds.), pp. 157–187. Conservation Foundation, Washington, D. C.

Millington, W. F., and Chaney, W. R. (1973). Shedding of shoots and branches. *In* "Shedding of Plant Parts" (T. T. Kozlowski, ed.), pp. 149–204. Academic Press, New York.

Minckler, L. S., Woerheide, J. D., and Schlesinger, R. C. (1973). Light, soil moisture, and tree reproduction in hardwood forest openings. *USDA For. Serv. Res. Pap. NC* **NC–89.**

Mirov, N. T. (1961). Composition of gum turpentines of pines. *U.S., Dep. Agric., Tech Bull.* **1239.**

Misaghi, I. J. (1982). "Physiology and Biochemistry of Plant–Pathogen Interactions." Plenum, New York.

Mitcham-Butler, E. J., Hinesley, L. E., and Pharr, D. M. (1987). Soluble carbohydrate concentration of Fraser fir foliage and its relationship to post-harvest needle retention. *J. Am. Soc. Hortic. Sci.* **112,** 672–676.

Mitchell, J. E., Waide, J. B., and Todd, R. L. (1975). A preliminary compartment model of the nitrogen cycle in a deciduous forest ecosystem. *In* "Mineral Cycling in Southeastern Ecosystems" (F. G. Howell *et al.*, eds.), ERDA Symp. Ser. (CONF-7Y0513), pp. 41–57. Natl. Tech. Inf. Serv., Springfield, Virginia.

Mitchell, R. G., Waring, R. H., and Pitman, G. B. (1983). Thinning lodgepole pine increases tree vigor and resistance to mountain pine beetle. *For. Sci.* **29,** 204–211.

Miyanishi, K., and Kellman, M. (1986). The role of fire in recruitment of two neotropical savanna shrubs, *Miconia albicans* and *Clidemia sericea*. *Biotropica* **18,** 224–230.

Mizutani, F., Yamada, M., Sugiura, A., and Tomana, T. (1979). Differential water tolerance among *Prunus* species and the effect of waterlogging on the growth of peach scions on various root stocks. *Engeigaku Kenkyu Shuroku* (Stud. Inst. Hortic., Kyoto Univ.) **9,** 28–35.

Mohr, H., and Shropshire, W. (1983). An introduction to photomorphogenesis for the general reader. *Encycl. Plant Physiol., New Ser.* **15A,** 24–38.

Möller, C. M., Müller, D., and Nielsen, J. (1954). Graphic presentation of dry matter production of European beech. *Forstl. Forsoegsvaes. Dan.* **21,** 327–335.

Molyneaux, D. E., and Davies, W. J. (1983). Rooting patterns and water relations of three pasture grasses growing in drying soil. *Oecologia* **58,** 220–224.

Monsi, M., and Saeki, K. (1953). Über den Lichtfaktor in den Pflanzengesellschaften und seine Bedeutung für die Stoffproduktion. *Jpn. J. Bot.* **14,** 22–52.

Monson, R. K., and Fall, R. (1989). Isoprene emission from aspen leaves. *Plant Physiol.* **90,** 267–274.

Monteith, J. L. (1973). "Principles of Environmental Physics." Arnold, London.

Monteith, J. L. (1977). Climate and efficiency of crop production in Britain. *Philos. Trans. R. Soc. London, Ser. B* **281,** 277–294.

Mooney, H. A. (1969). Dark respiration of related evergreen and deciduous mediterranean plants during induced drought. *Bull. Torrey Bot. Club* **96,** 550–555.

Mooney, H. A., and Dunn, E. L. (1970). Photosynthetic systems of Mediterranean climate shrubs and trees of California and Chile. *Am. Midl. Nat.* **104,** 447–453.

Mooney, H. A., and Hays, R. I. (1973). Carbohydrate storage cycles in two Californian Mediterranean-climate trees. *Flora (Jena)* **162,** 295–304.

Mooney, H. A., and Shropshire, F. (1967). Population variability in temperature related photosynthetic acclimation. *Oecol. Plant.* **2,** 1–13.

Mooney, H. A., Björkman, O., and Collatz, G. J. (1977). Photosynthetic acclimation to temperature and water stress in the desert shrub *Larrea divaricata. Year Book—Carnegie Inst. Washington* **76,** 328–335.

Mooney, H. A., Winner, W. E., and Pell, E. J., eds. (1991). "The Integrated Response of Plants to Stress." Academic Press, San Diego, California.

Moore, J. A., Budelsky, C. A., and Schlesinger, R. C. (1973). A new index representing individual tree competitive status. *Can. J. For. Res.* **3,** 495–500.

Moore, J. N., and Janick, J., eds. (1983). "Methods in Fruit Breeding." Purdue Univ. Press, West Lafayette, Indiana.

Moore, P. D. (1988). Blow, blow thou winter winds. *Nature* **336,** 313.

Morey, P. R. (1973). "How Trees Grow." Arnold, London.

Morgan, D. C., Rook, D. A., Warrington, I. J., and Turnbull, H. L. (1983). Growth and development of *Pinus radiata* D. Don.: The effect of light quality. *Plant, Cell Environ.* **6,** 691–701.

Morgan, J. M. (1984). Osmoregulation and water stress in higher plants. *Annu. Rev. Plant Physiol.* **35,** 299–319.

Morison, J. I. L. (1985). Sensitivity of stomata and water use efficiency to high CO_2. *Plant, Cell Environ.* **8,** 467–474.

Morison, J. I. L., and Gifford, R. M. (1984). Plant growth and water supply in high CO_2 concentrations. I. Leaf area, water use and transpiration. *Aust. J. Plant Physiol.* **11,** 361–374.

Morris, L. A., Pritchett, W. L., and Swindel, B. F. (1983). Displacement of nutrients into windrows during site preparation of a flatwood forest. *Soil Sci. Soc. Am. J.* **47,** 591–594.

Morrison, I. K. (1984). Acid rain. *For. Abstr.* **45,** 483–506.

Morrow, P. A., and Mooney, H. A. (1974). Drought adaptations in two California evergreen sclerophylls. *Oecologia* **15,** 205–222.

Moshkov, B. S. (1935). Photoperiodismus und Frostharte ausdauernder Gewachse. *Planta* **23,** 774–803.

Moss, A. E. (1940). Effect of wind-driven salt water. *J. For.* **38,** 421–425.

Moulds, F. R. (1957). Exotics can succeed in forestry as in agriculture. *J. For.* **55,** 563–566.

Mroz, G. D., Jurgensen, M. F., Harvey, A. E., and Larsen, M. J. (1980). Effects of fire on nitrogen in forest floor horizons. *Soil Sci. Soc. Am. J.* **44,** 395–400.

Mroz, G. D., Frederick, D. J., and Jurgensen, M. F. (1985a). Site and fertilizer effects on northern hardwood stump sprouting. *Can. J. For. Res.* **15,** 535–543.

Mroz, G. D., Gale, M. R., Jurgensen, M. F., Frederick, D. J., and Clark, A. (1985b). Composition, structure, and aboveground biomass of two old-growth northern hardwood stands in upper Michigan. *Can. J. For. Res.* **15,** 78–82.

Mudd, J. B., Banerjee, S. K., Dooley, M. M., and Knight, K. L. (1984). Pollutants and plant cells: Effects on membranes. *In* "Gaseous Air Pollutants and Plant Metabolism" (M. J. Koziol and F. R. Whatley, eds.), pp. 105–116. Butterworth, London.

Mueller-Dombois, D. (1986). Perspectives for an etiology of stand-level dieback. *Annu. Rev. Ecol. Syst.* **17,** 221–243.

Mueller-Dombois, D. (1987). Natural dieback in forests. *BioScience* **37,** 575–583.

Muir, W. (1978). Pest control—A perspective. *In* "Pest Control Strategies" (E. H. Smith and D. Pimentel, eds.), pp. 3–7. Academic Press, New York.

Murphy, T. M. (1983). Membranes as targets of ultraviolet radiation. *Physiol. Plant.* **58,** 381–388.

Myers, B. A., and Neales, T. F. (1986). Osmotic adjustment, induced by drought, in seedlings of three *Eucalyptus* species. *Aust. J. Plant Physiol.* **13,** 597–603.

Myers, B. J., Robichaux, R. H., Unwin, G. L., and Craig, I. E. (1987). Leaf water relations and anatomy of a tropical rainforest tree species vary with crown position. *Oecologia* **74,** 81–85.

Myers, C. A. (1963). Vertical distribution of annual increment in thinned ponderosa pine. *For. Sci.* **9,** 394–404.

Nadelhoffer, K. J., Aber, J. D., and Melillo, J. M. (1985). Fine root production in relation to net primary production along a nitrogen availability gradient in temperate forests: A new hypothesis. *Ecology* **66,** 1377–1390.

Nagao, A., and Asakawa, S. (1963). Light-sensitivity in the germination of *Abies* seeds. *J. Jpn. For. Soc.* **45,** 375–377.

Nair, P. K. R. (1985a). Classification of agroforestry systems. *Agrofor. Syst.* **3,** 97–128.

Nair, P, K. R. (1985b). Fruit trees in tropical agroforestry systems. Working Paper No. 32. International Council for Research in Agroforestry, Nairobi, Kenya.

Nair, P. K. R. (1989). "Agroforestry Systems in the Tropics." Kluwer, Dordrecht, The Netherlands.

Nakashizuka, T. (1989). Role of uprooting in composition and dynamics of an old-growth forest in Japan. *Ecology* **70,** 1273–1278.

Nambiar, E. K. S., and Zed, P. G. (1980). Influence of weeds on the water potential, nutrient content and growth of young radiata pine. *Aust. For. Res.* **10,** 279–288.

Namkoong, G., Kang, H. C., and Brouard, V. S. (1988). "Tree Breeding: Principles and Strategies." Springer-Verlag, New York.

Nash, T. H., Fritts, H. C., and Stokes, M. A. (1975). A technique for examining non-climatic variation in widths of annual tree rings with special references to air pollution. *Tree-Ring Bull.* **35,** 15–24.

Nasrulhaq-Boyce, A., and Mohamed, M. A. H. (1987). Photosynthetic and respiratory characteristics of Malayan sun and shade ferns. *New Phytol.* **105,** 81–88.

National Academy of Sciences (1977a). Effects of nitrogen oxides on vegetation. *In* "Nitrogen Oxides." Natl. Acad. Sci., Washington, D. C.

National Academy of Sciences (1977b.) "Ozone and Other Photochemical Oxidants." Natl. Acad. Sci., Washington, D. C.

National Academy of Sciences (1977c). "Nitrogen Oxides." Natl. Acad. Sci., Washington, D. C.

National Research Council (1979a). "Airborne Particles." University Park Press, Baltimore, Maryland.

National Research Council (1979b). "Carbon Dioxide and Climate: A Scientific Assessment." National Academy Press, Washington, D. C.

Naveh, Z. (1973). The ecology of fire in Israel. *Proc. 13th Annu. Tall Timbers Fire Ecol. Conf.*, pp. 131–170.

Naveh, Z. (1974). Effects of fire in the Mediterranean region. In "Fire and Ecosystems" (T. T. Kozlowski and C. E. Ahlgren, eds.), pp. 401–434. Academic Press, New York.

Navratil, S., and McLaughlin, M. C. (1979). Field survey techniques can detect SO_2 pollution effects on white pine up to 120 km. *Phytopathology* **69,** 918.

Neales, T. F., and Incoll, L. D. (1968). The control of leaf photosynthesis rate by the level of assimilate concentration in the leaf: A review of the hypothesis. *Bot. Rev.* **34,** 107–125.

Neel, P. L., and Harris, R. W. (1971). Motion-induced inhibition of elogation and induction of dormancy in *Liquidambar. Science* **173,** 58–59.

Neely, D. (1968). Bleeding necrosis of sweetgum in Illinois and Indiana. *Plant Dis. Rep.* **52,** 223–225.

Neely, D. (1970). Healing of wounds on trees. *J. Am. Soc. Hortic. Sci.* **95,** 536–540.

Neely, D. (1988a). Tree wound closure. *J. Arboric.* **14,** 148–152.

Neely, D. (1988b). Closure of branch pruning wounds with conventional and 'Shigo' cuts. *J. Arboric.* **14,** 261–264.

Neely, D., and Himelick, E. B. (1987). Fertilizing and watering trees. *Circ.—Ill. Nat. Hist. Surv.* **56.**

Neely, D., Himelick, E. B., and Crowley, W. R., Jr. (1970). Fertilization of established trees. *Bull.—Ill. Nat. Hist. Surv.* **30,** 235–266.

Neilson, R. E., Ludlow, M. M., and Jarvis, P. G. (1972). Photosynthesis in Sitka spruce (*Picea sitchensis*) (Bong. Carr). II. Response to temperature. *J. Appl. Ecol.* **9,** 721–745.

Nelson, N. D. (1984). Woody plants are not inherently low in photosynthetic capacity. *Photosynthetica* **18,** 600–605.

Nelson, N. D., and Hillis, W. E. (1978a). Association between altitude and xylem ethylene levels in *Eucalyptus pauciflora*. *Aust. For. Res.* **8,** 69–73.

Nelson, N. D., and Hillis, W. E. (1978b). Ethylene and tension wood formation in *Eucalyptus gomphocephala*. *Wood Sci. Technol.* **12,** 309–315.

Nelson, N. D., and Isebrands, J. G. (1983). Late-season photosynthesis and photosynthate distribution in an intensively-cultured *Populus nigra* × *laurifolia* clone. *Photosynthetica* **17,** 537–549.

Nelson, N. D., Burk, T., and Isebrands, J. G. (1981). Crown architecture of short-rotation intensively cultured *Populus*. I. Effects of clone characteristics. *Can. J. For. Res.* **11,** 73–81.

Nelson, L. R., Pederson, R. C., Autry, J. L., Dudley, S., and Walstad, J. D. (1981). Impacts of herbaceous weeds in young loblolly pine plantations. *South. J. Appl. For.* **5,** 153–158.

Newman, E. I. (1969). Resistance to water flow in soil and plant. I. Soil resistance in relation to amounts of root: Theoretical estimates. *J. Appl. Ecol.* **6,** 1–12.

Newsome, R. D., Kozlowski, T. T., and Tang, Z. C. (1982). Responses of *Ulmus americana* seedlings to flooding of soil. *Can. J. Bot.* **60,** 1688–1695.

Ng, F. S. P. (1980). Germination ecology of Malaysian woody plants. *Malay. For.* **43,** 406–437.

Ni, B., and Pallardy, S. G. (1987). Stomatal and nonstomatal inhibition of photosynthesis in water-stressed oak seedlings. *Plant Physiol., Suppl.* **83,** 49.

Nicholls, J. W. P. (1982). Wind action, leaning trees and compression wood in *Pinus radiata* D. Don. *Aust. For. Res.* **12,** 75–91.

Nielsen, D. G., Terrell, L. E., and Weidensaul, T. C. (1977). Phytotoxicity of ozone and sulfer dioxide to laboratory fumigated Scotch pine. *Plant Dis. Rep.* **61,** 699–703.

Niklas, K. J. (1985). The aerodynamics of wind pollution. *Bot. Rev.* **51,** 328–386.

Nitsch, J. P. (1962). Photoperiodic regulation of growth in woody plants. *Adv. Hortic. Sci. Their Applic., Proc. Int. Hortic. Congr., 16th, 1962,* pp. 14–23.

Nobel, P. S. (1974). "An Introduction to Biophysical Plant Physiology." Freeman, San Francisco, California.

Nobel, P. S. (1976). Photosynthetic rates of sun versus shade leaves of *Hyptis emoryi* Torr. *Plant Physiol.* **58,** 218–223.

Nobel, P. S. (1980). Leaf anatomy and water use efficiency. *In* "Adaptations of Plants to Water and High Temperature Stress" (N. C. Turner and P. J. Kramer, eds.), pp. 43–55. Wiley, New York.

Nobel, P. S. (1983). "Biophysical Plant Physiology and Ecology." Freeman, San Francisco, California.

Noble, I. R. (1981). Predicting successional change. *Gen. Tech. Rep. WO (U.S. For. Serv.)* **WO-26,** 278–300.

Noble, I. R., and Slatyer, R. O. (1980). The use of vital attributes to predict successional changes in plant communities subject to recurrent disturbances. *Vegetatio* **43,** 1–21.

Noble, L. (1981). Predicting successional change. *Gen. Tech. Rep. WO (U.S. For. Serv.)* **WO-26,** 278–300.

Noble, R. D., and Jensen, K. F. (1980). Effects of sulfur dioxide and ozone on growth of hybrid poplar leaves. *Am. J. Bot.* **67,** 1005–1009.

Noland, T. L., and Kozlowski, T. T. (1979). Effect of SO_2 on stomatal aperture and sulfur uptake of woody angiosperm seedlings. *Can. J. For. Res.* **9,** 57–62.

Norby, R. J. (1987). Nodulation and nitrogenase activity in nitrogen-fixing woody plants stimulated by CO_2 enrichment of the atmosphere. *Physiol. Plant.* **41,** 77–82.

Norby, R. J., and Kozlowski, T. T. (1980). Allelopathic potential of ground cover species on *Pinus resinosa* seedlings. *Plant Soil* **57,** 363–374.

Norby, R. J., and Kozlowski, T. T. (1981a). Response of SO_2-fumigated *Pinus resinosa* seedlings to post fumigation temperature. *Can. J. Bot.* **59,** 470–475.

Norby, R. J., and Kozlowski, T. T. (1981b). Relative sensitivity of three species of woody plants to SO_2 at high or low exposure temperature. *Oecologia* **51,** 33–36.

Norby, R. J., and Kozlowski, T. T. (1981c). Interactions of SO_2-concentration and post-fumigation temperature on growth of five species of woody plants. *Environ. Pollut. Ser. A* **25,** 27–39.

Norby, R. J., and Kozlowski, T. T. (1982). The role of stomata in sensitivity of *Betula papyrifera* Marsh. seedlings to SO_2 at different humidities. *Oecologia* **53,** 34–39.

Norby, R. J., and Kozlowski, T. T. (1983). Flooding and SO_2 stress interaction in *Betula papyrifera* and *B. nigra* seedlings. *For. Sci.* **29,** 739–750.

Norby, R. J., O'Neill, E. G., and Luxmoore, R. J. (1986). Effects of atmospheric CO_2 enrichment on the growth and mineral nutrition of *Quercus alba* seedlings in nutrient-poor soil. *Plant Physiol.* **82,** 83–89.

Norby, R. J., Weerasuriya, Y., and Hanson, P. J. (1989). Induction of nitrate reductase activity in red spruce needles by NO_2 and HNO_3 vapor. *Can. J. For. Res.* **19,** 889–896.

Norris, D. M. (1967). Systemic insecticides in trees. *Annu. Rev. Entomol.* **12,** 127–148.

Norris, R. F., and Bukovac, M. J. (1968). Structure of the pear leaf cuticle with special reference to cuticular penetration. *Am. J. Bot.* **55,** 975–983.

Novitskii, Z. B. (1980). Effectiveness of shelterbelts against wind erosion. *Lesn. Khoz.* **8,** 39–40.

Nutman, F. J. (1933). The root system of *Coffea arabica,* Part II. The effect of some soil conditions in modifying the "normal" root system. *Empire J. Exp. Agric.* **1,** 285–286.

Nygren, M., and Kellomaki, S. (1983). Effect of shading on leaf structure and photosynthesis in young birches, *Betula pendula* Roth. and *B. pubescens* Ehrh. *For. Ecol. Manage.* **7,** 119–132.

Nyman, B. (1961). Effect of red and far-red irradiation on the germination process in seeds of *Pinus sylvestris* L. *Nature (London)* **191,** 1219–1220.

Oberarzbacher, P. (1977). Beiträge zur physiologischen Analyse des Hohenzuwachses von verschiedenen Fichtenklonen entlang eines Hohenprofils im Wipptal (Tirol) und in Klimakammern. Dissertation, Univ. Innsbruck.

Oberbauer, S. F., and Strain, B. R., (1986). Effects of canopy position and irradiance on the leaf physiology and morphology of *Pentaclethra macroloba* (Mimosaceae). *Am. J. Bot.* **73,** 409–416.

Oberbauer, S. F., Strain, B. R., and Fetcher, N. (1985). Effect of CO_2-enrichment on seedling physiology and growth of two tropical tree species. *Physiol. Plant.* **65,** 352–356.

Odening, W. R., Strain, B. R., and Oechel, W. C. (1974). The effect of decreasing water potential on net CO_2 exchange of intact desert shrubs. *Ecology 55,* 1086–1095.

Ohmart, C. P., and Williams, C. B. (1979). The effects of photochemical oxidants on radial growth increment for five species of conifers in San Bernardino National Forest. *Plant Dis. Rep.* **63,** 1038–1042.

Oland, K. (1963). Changes in the content of dry matter and major nutrient elements of apple foliage during senescence and abscission. *Physiol. Plant.* **16,** 682–694.

Oleksyn, J., and Bialobok, S. (1986). Net photosynthesis, dark respiration and susceptibility to air pollution of 20 European provenances of Scots pine *Pinus sylvestris* L. *Environ. Pollut., Ser. A* **40,** 287–302.

Oliver, C. D. (1981). Forest development in North America following major disturbances. *For. Ecol. Manage.* **3,** 153–168.

Oliver, H. R., and Mayhead, G. J. (1974). Wind measurements in a pine forest during a destructive gale. *Forestry* **47,** 185–194.

Olkowski, W., Olkowski, H., Drlik, T., Heidler, N., Minter, R., Zuparko, R., Laub, L., and Orthel, L. (1978). Pest control strategies: Urban integrated pest management. *In* "Pest Control Strategies" (E. H. Smith and D. Pimentel, eds.), pp. 215–234. Academic Press, New York.

Olofinboba, M. O., and Kozlowski, T. T. (1973). Accumulation and utilization of carbohydrate reserves in shoot growth of *Pinus resinosa. Can. J. For. Res.* **3,** 346–353.

Olsen, C. (1961). Competition between trees and herbs for nutrient elements in calcareous soil. *Symp. Soc. Exp. Biol.* **15,** 145–155.

Olson, J. S., Stearns, F., and Nienstaedt, H. (1959). Eastern hemlock seeds and seedlings. Response to photoperiod and temperature. *Bull.—Conn. Agric. Expt. Stn., New Haven* **620.**

Omasa, K., Hashimoto, Y., Kramer, P. J., Strain, B. R., Aiga, I., and Kondo, J. (1985). Direct observation of reversible and irreversible stomatal response of attached sunflower leaves to SO_2. *Plant Physiol.* **79,** 153–158.

O'Neill, E. G., Luxmoore, R. J., and Norby, R. J. (1987). Increases in mycorrhizal colonization and seedling growth in *Pinus echinata* and *Quercus alba* in an enriched CO_2 atmosphere. *Can. J. For. Res.* **17,** 878–883.

O'Neill, S. D., and Leopold, A. C. (1982). An assessment of phase transitions in soybean membranes. *Plant Physiol.* **70,** 1405–1409.

Oosting, H. J. (1956). "The Study of Plant Communities." Freeman, San Francisco, California.

Oosting, H. J., and Billings, W. D. (1942). Factors affecting vegetational zonation on coastal dunes. *Ecology* **23,** 131–142.

Opik, H. (1980). "The Respiration of Higher Plants." Arnold, London.

Oppenheimer, H. R. (1932). Zur Kenntnis des hochsomerlichen Wasserbilanz mediterranean Gehölze. *Ber. Dtsch. Bot. Ges.* **50,** 185–243.

Oppenheimer, H. R. (1951). Summer drought and water balance of plants growing in the Near East. *J. Ecol.* **39,** 356–362.

Öquist, G. (1983). Effects of low temperature on photosynthesis. *Plant, Cell Environ.* **6,** 281–300.

Öquist, G., Greer, D. H., and Ogren, E. (1987). Light stress at low temperature. *In* "Photoinhibition" (D. J. Kyle, C. B. Osmond, and C. J. Arntzen, eds.), pp. 67–88. Elsevier, Amsterdam.

O'Reilly, C., and Parker, W. H. (1982). Vegetative phenology in a clonal seed orchard of *Picea glauca* and *Picea mariana* in northwestern Ontario. *Can. J. For. Res.* **12,** 408–413.

Oren, R., Schulze, E.-D., Werk, K. S., Meyer, J., Schneider, B. U., and Heilmeier, H. (1988a). Performance of two *Picea abies* (L.) Karst. stands at different stages of decline. I. Carbon relations and stand growth. *Oecologia* **75,** 25–37.

Oren, R., Werk, K. S., Schulze, E.-D., Meyer, J. Schneider, B. U., and Schramel, P. (1988b). Performance of two *Picea abies* (L.) Karst. stands at different stages of decline. VI. Nutrient concentration. *Oecologia* **77,** 151–162.

Ormrod, D. P. (1978). "Pollution in Horticulture." Elsevier, Amsterdam.

Osborne, D. J. (1974). Internal factors regulating abscission. *In* "Shedding of Plant Parts" (T. T. Kozlowski, ed.), pp. 125–147. Academic Press, New York.

Osmond, C. B. (1987). Photosynthesis and carbon economy of plants. *New Phytol.* **106,** Suppl., 161–175.

Osmond, C. B., Björkman, O., and Anderson, D. J. (1980). "Physiological Processes in Plant Ecology." Springer-Verlag, Berlin.

Osmond, C. B., Austin, M. P., Berry, J. A., Billings, W. D., Boyer, J. S., Dacey, W. H., Nobel, P. S., Smith, S. D., and Winner, W. E. (1987). Stress physiology and the distribution of plants. *BioScience* **37,** 49–57.

Osonubi, O., and Davies, W. J. (1978). Solute accumulation in leaves and roots of woody plants subjected to water stress. *Oecologia* **32,** 323–332.

Osonubi, O., and Osundina, M. A. (1987). Comparisons of the responses to flooding of seedlings and cuttings of *Gmelina*. *Tree Physiol.* **3,** 147–156.

Osonubi, O., Oren, R., Werk, K. S., Schulze, E.-D., and Heilmeier, H. (1988). Performance of two *Picea abies* (L.) Karst. stands at different stages of decline. IV. Xylem sap concentrations of magnesium, calcium, potassium and nitrogen. *Oecologia* **77,** 1–6.

Othieno, C. A. (1978). Supplementary irrigation of young clonal tea in Kenya. II. Internal water status. *Exp. Agric.* **14,** 309–316.

Ovington, J. D. (1956). The form, weights, and productivity of tree species grown in close stands. *New Phytol.* **55,** 289–388.

Ovington, J. D. (1957). Dry matter production in *Pinus sylvestris* L. *Ann. Bot.* (*London*) [N.S.] **21,** 287–314.

Ovington, J. D. (1962). Quantitative ecology and the woodland ecosystem concept. *Adv. Ecol. Res.* **1,** 103–192.

Ovington, J. D. (1965). Organic production, turnover, and mineral cycling in woodlands. *Biol. Rev. Cambridge Philos. Soc.* **40,** 295–366.

Padila, G. P., Dickinson, M. B., and Kolattukudy, P. E. (1988). Transcriptional activation of a cutinase gene in isolated fungal nuclei by plant cutin monomers. *Science* **242,** 922–925.

Pady, S. M., and Kapica, L. (1955). Fungi in air over the Atlantic Ocean. *Mycologia* **47,** 34–50.

Pallardy, S. G. (1981). Comparative water relations of closely related woody plants. *In* "Water Deficits and Plant Growth" (T. T. Kozlowski, ed.), Vol. 6, pp 511–548. Academic Press, New York.

Pallardy, S. G., and Kozlowski, T. T. (1979a). Frequency and length of stomata of 21 *Populus* clones. *Can. J. Bot.* **57,** 2519–2523.

Pallardy, S. G., and Kozlowski, T. T. (1979b). Early root and shoot growth of *Populus* clones. *Silvae Genet.* **28,** 153–156.

Pallardy, S. G., and Kozlowski, T. T. (1979c). Stomatal response of *Populus* clones to light intensity and vapor pressure deficit. *Plant Physiol.* **64,** 112–114.

Pallardy, S. G., and Kozlowski, T. T. (1979d). Relationship of leaf diffusion resistance of *Populus* clones to leaf water potential and environment. *Oecologia* **40,** 371–380.

Pallardy, S. G., and Kozlowski, T. T. (1980). Cuticle development in the stomatal region of *Populus* clones. *New Phytol.* **85,** 363–368.

Pallardy, S. G., and Kozlowski, T. T. (1981). Water relations of *Populus* clones. *Ecology* **57,** 367–373.

Pan, E., and Bassuk, N. (1985). Effects of soil type and compaction on the growth of *Ailanthus altissima* seedlings. *J. Environ. Hortic.* **3,** 158–162.

Pappas, T., and Mitchell, C. A. (1985). Influence of seismic stress on photosynthetic productivity, gas exchange, and leaf diffusive resistance of *Glycine max* (L.) Merrill cv. Wells, II. *Plant Physiol.* **79,** 285–289.

Parke, J. L., Lindeman, R. G., and Black, C. H. (1983). The role of ectomycorrhizae in the drought tolerance of Douglas fir seedlings. *New Phytol.* **95,** 83–95.

Parker, A. F. (1961). Bark moisture relations in disease development: Present status and future needs. *Recent Adv. Bot.* **2,** 1535–1537.

Parker, J. (1950). The effect of flooding on the transpiration and survival of some southeastern forest tree species. *Plant Physiol.* **25,** 453–460.

Parker, J. (1952). Desiccation in conifer leaves: Anatomical changes and determination of the lethal level. *Bot. Gaz.* (*Chicago*) **114,** 189–198.

Parker, J. (1968). Drought-resistance mechanisms. *In* "Water Deficits and Plant Growth" (T. T. Kozlowski, ed.), Vol. 1, pp. 195–234. Academic Press, New York.

Parker, W. C., and Pallardy, S. G. (1985a). Genotypic variation in tissue water relations of leaves and roots of black walnut (*Juglans nigra*) seedlings. *Physiol. Plant.* **64,** 105–110.

Parker, W. C., and Pallardy, S. G. (1985b). Drought-induced leaf abscission and whole-plant drought tolerance of seven black walnut families. *Can. J. For. Res.* **15,** 818–821.

Parmeter, J. R., and Cobb, F. W., Jr. (1972). "Long-term Impingement of Aerobiology Systems on Plant Production Systems," US/IBB Aerobiol. Program Handb. No. 2, pp. 61–68. Univ. of Michigan Press, Ann Arbor.

Parsons, L. R., Combs, B. S., and Tucker, D. P. H. (1985). Citrus freeze protection with micro-sprinkler irrigation during advective freeze. *HortScience* **20,** 1078–1080.

Parsons, R. F. (1969). Physiological and ecological tolerances of *Eucalyptus incrassata* and *E. socialis* to edaphic factors. *Ecology* **50,** 386–390.

Parviainen, J. (1979). Einfluss des Verpflanzens und des Wurvelschnittes auf den Tagesverlauf des Xylem Wasserpotentials bei Fichtenpflanzen. *Forst archiv* **50,** 148–153.

Passioura, V. B. (1988). Water transport in and to roots. *Annu. Rev. Plant Physiol. Plant Mol. Biol.* **39,** 245–265.

Paton, D. M. (1982). A mechanism for frost resistance in *Eucalyptus*. *In* "Cold Hardiness and Freezing Stress" (P. H. Li and A. Sakai, eds.), Vol 2. pp. 77–92. Academic Press, London.

Patt, J., Caimeli, D., and Zafrir, T. (1966). Influence of soil physical condition on root development and on production of citrus trees. *Soil Sci.* **102,** 82–84.

Patterson, D. T. (1975). Nutrient return in the stemflow and throughfall of individual trees in the piedmont deciduous forest. *In* "Mineral Cycling in Southeastern Ecosystems" (F. G. Howell, J. B. Gentry, and M. H. Smith, eds.), ERDA Symp. Ser. (CONF-740513), pp. 800–812. Natl. Tech. Inf. Serv., Springfield, Virginia.

Patterson, D. T. (1986). Responses of soybean (*Glycine max*) and three C_4 grass weeds to CO_2 enrichment during drought. *Weed Sci.* **34,** 203–210.

Patterson, J. C. (1976). Soil compaction and its effects upon urban vegetation. *USDA For. Serv. Gen. Tech. Rep. NE* **NE-22,** 91–101.

Pauley, S. S., and Perry T. O. (1954). Ecotypic variation of the photoperiodic response in *Populus. J. Arnold Arbor., Harv. Univ.* **35,** 167–188.

Pearcy, R. W. (1977). Acclimation of photosynthetic and respiratory carbon dioxide exchange to growth temperature in *Atriplex lentiformis* (Torr.) Wats. *Plant Physiol.* **59,** 795–799.

Pearcy, R. W. (1988). Photosynthetic utilisation of lightflecks by understory plants. *Aust. J. Plant. Physiol.* **15,** 223–238.

Pearcy, R. W., and Björkman, O. (1983). Physiological effects. *In* "CO_2 and Plants" (E. R. Lemon, ed.), pp. 65–105. Westview Press, Boulder, Colorado.

Pearcy, R. W., and Calkin, H. W. (1983). Carbon dioxide exchange of C_3 and C_4 tree species in the understory of a Hawaiian forest. *Oecologia* **58,** 26–31.

Pearcy, R. W., and Franceschi, V. R. (1986). Photosynthetic characteristics and chloroplast ultrastructure of C_3 and C_4 tree species grown in high- and low-light environments. *Photosynth. Res.* **9,** 317–331.

Pearcy R. W., and Troughton, J. H. (1975). C_4 photosynthesis in tree form *Euphorbia* species from Hawaiian rainforest sites. *Plant. Physiol.* **55,** 1054–1056.

Pearcy, R. W., Ehleringer, J., Mooney, H. A., and Rundel, P. W., eds. (1989). "Plant Physiological Ecology." Chapman and Hall, London and New York.

Peet, M. M., and Kramer, P. J. (1980). Effects of decreasing source–sink ratio in soybeans on photosynthesis, photorespiration, transpiration and yield. *Plant, Cell Environ.* **3,** 201–206.

Peet, R. K., and Christensen, N. L. (1980). Succession: A population process. *Vegetatio* **43,** 131–140.

Peet, R. K., and Christensen, N. L. (1987). Competition and tree death. *BioScience* **37,** 586–594.

Pellett, H. (1971). Comparison of cold hardiness levels of root and stem tissue. *Can. J. Plant Sci.* **51,** 193–195.

Perchorowicz, J. T., Raynes, D. A., and Jensen, R. G. (1981). Light limitation of photosynthesis and regulation of ribulose bisphosphate carboxylase in wheat seedlings. *Proc. Natl. Acad. Sci. U.S.A.* **78,** 2985–2989.

Percy, K. E., and Riding, R. T. (1978). The epicuticular waxes of *Pinus strobus* subjected to air pollutants. *Can. J. For. Res.* **8,** 474–477.

Percy, R. E., and Riding, R. T. (1981). Histology and histochemistry of elongating needles of *Pinus strobus* subjected to a long-duration, low concentration exposure of sulfur dioxide. *Can. J. Bot.* **59,** 2558–2567.

Pereira, J. S., and Kozlowski, T. T. (1976). Influence of light intensity, temperature, and leaf area on stomatal aperture and water potential of woody plants. *Can. J. For. Res.* **7,** 145–153.

Pereira, J. S., and Kozlowski, T. T. (1977). Variations among woody angiosperms in response to flooding. *Physiol. Plant.* **41,** 184–192.

Pereira, J. S., Tenhunen, J. D., Lange, O. L., Beyschlag, W., Meyer, A., and David, M. M. (1986). Seasonal and diurnal patterns in leaf gas exchange of *Eucalyptus globulus* trees growing in Portugal. *Can. J. For. Res.* **16,** 177–184.

Perry, T. O. (1962). Racial variation in the day and night requirements of red maple and loblolly pine. *For. Sci.* **8,** 336–344.

Perry, T. O., and Baldwin, G.W. (1966). Winter breakdown of the photosynthetic apparatus of evergreen species. *For. Sci.* **12,** 298–300.

Perry, T. O., and Wang, C. W. (1960). Genetic variation in the winter chilling requirement for date of dormancy break for *Acer rubrum*. *Ecology* **41,** 790–794.

Pessi, Y. (1958). On the influence of bog draining upon thermal conditions in the soil and air near the ground. *Acta. Agric. Scand.* **24,** 359–374.

Petersen, T. D., Newton, M., and Zedaker, S. M. (1988). Influence of *Ceanothus velutinus* and associated forbs on the water stress and stemwood production of Douglas-fir. *For. Sci.* **34,** 333–343.

Petty, J. A., and Worrell, R. (1981). Stability of coniferous tree stems in relation to damage by snow. *Forestry* **54,** 115–128.

Pezeshki, S. R., and Chambers, J. L. (1985a). Stomatal and photosynthetic response of sweet gum (*Liquidambar styraciflua*) to flooding. *Can. J. For. Res.* **15,** 371–375.

Pezeshki, S. R., and Chambers, J. L. (1985b). Responses of cherrybark oak (*Quercus falcata* var. *pagodaefolia*) seedlings to short-term flooding. *For. Sci.* **31,** 760–771.

Pezeshki, S. R., and Chambers, J. L. (1986). Variation in flood-induced stomatal and photosynthetic responses of three bottomland tree species. *For. Sci.* **32,** 914–923.

Phares, R. E., Kolar, C. M., Hendricks, T. R., and Ashby, W. C. (1974). Motion induced effects on growth of black walnut, silver maple, and sweetgum seedlings under two light regimes. *Proc. North Am. For. Biol. Workshop, 3rd, 1974.*

Pharis, R. P. (1976). Probable roles of plant hormones in regulating shoot elongation, diameter growth, and crown form of forest trees. *In* "Tree Physiology and Yield Improvement" (M. G. R. Cannell and F. T. Last, eds.), pp 291–306. Academic Press, London.

Pharis, R. P., and Ferrell, W. K. (1966). Differences in drought resistance between coastal and inland sources of Douglas-fir. *Can. J. Bot.* **44,** 1651–1659.

Pharis, R. P., and King, R. W. (1985). Gibberellins and reproductive development in seed plants. *Annu. Rev. Plant Physiol.* **36,** 517–568.

Pharis, R. P., and Kramer, P. J. (1964). The effects of nitrogen and drought on lobolly pine seedlings. I. Growth and composition. *For. Sci.* **10,** 143–150.

Pharis, R. P., and Kuo, C. G. (1977). Physiology of gibberellins in conifers. *Can. J. For. Res.* **7,** 299–325.

Pharis, R. P., Ruddat, M., Phillips, C. C., and Heftmann, E. (1965). Gibberellin, growth retardants and apical dominance in Arizona cypress. *Naturwissenschaften* **52,** 88–89.

Pharis, R. P., Ruddat M., and Phillips, C. (1967). Response of conifers to growth retardants. *Bot. Gaz. (Chicago)* **128,** 105–109.

Pharis, R. P., Hellmers, H., and Schuurmans, E. (1970). Effects of subfreezing temperatures on photosynthesis of evergreen conifers under controlled conditions. *Photosynthetica* **4,** 273–279.

Phelps, J. E., Isebrands, J. G., and Jowett, J. D. (1982). Raw material quality of short-rotation intensively cultured *Populus* clones. I. A comparison of stem and branch properties at three spacings. *IAWA Bull.* [N.S.] **3,** 193–200.

Philipson, J. J., and Coutts, M. P. (1978). The tolerance of tree roots to waterlogging. III. Oxygen transport in lodgepole pine and Sitka spruce roots of primary structure. *New Phytol.* **80,** 341–349.

Philipson, J. J., and Coutts, M. P. (1980). The tolerance of tree roots to waterlogging. *New Phytol.* **85,** 489–494.

Phillips, D. R., and Van Loon, D. H. (1984). Biomass removal and nutrient drain as affected by total tree harvest in southern pine and hardwood stands. *J. For.* **82,** 547–550.

Phillips, I. D. J., and Wareing, P. F. (1958). Studies in the dormancy of sycamore. I. Seasonal changes in growth substance content of the shoot. *J. Exp. Bot.* **9,** 350–364.

Phillips, J. (1974). Effects of fire in forest and savanna ecosystems of sub-sahara Africa. *In* "Fire and Ecosystems" (T. T. Kozlowski and C. E. Ahlgren, eds.), pp. 435–481. Academic Press, New York.

Phillips, J. H. H., and Weaver, G. M. (1975). A high density peach orchard. *HortScience* **10,** 580–582.

Phung, H. T., and Knipling, E. B. (1976). Photosynthesis and transpiration of citrus seedlings under flooded conditions. *HortScience* **11,** 131–133.

Pickett, S. T. A., Collins, S. L., and Arnesto, J. J. (1987). Models, mechanisms and pathways of succession. *Bot. Rev.* **53,** 335–371.

Pielou, E. C. (1959). The use of point-to-point distances in the study of the pattern of plant populations. *J. Ecol.* **47,** 607–613.

Pielou, E. C. (1960). A single mechanism to account for regular, random, and aggregate populations. *J. Ecol.* **48,** 575–584.

Pigott, C. D., and Huntley, J. P. (1981). Factors controlling the distribution of *Tilia cordata* at the northern limits of its geographical range. III. Nature and causes of seed sterility. *New Phytol.* **87,** 817–839.

Pimental, D., and Perkins, J. H., eds. (1980). "Pest Control: Cultural and Environmental Aspects." Westview Press, Boulder, CO.

Pirone, P. P. (1978). "Diseases and Pests of Ornamental Plants." Wiley, New York.

Pisek, A., and Larcher, W. (1954). Zusammenhang zwischen Austrocknungsresistenz und Frosthärte bei Immergrünen. *Protoplasma* **44,** 30–46.

Piskornik, Z. (1969). Effect of industrial air pollutants on photosynthesis in hardwoods. *Bull. Inst. Pap. Chem.* **43,** 8509.

Pitcher, R. S., and Webb, P. C. R. (1949). A fungus disease of raspberries induced by insect attack. *Nature (London)* **163,** 574–575.

Pollard, D. F. W., and Logan, K. T. (1977). The effects of light intensity, photoperiod, soil moisture potential, and temperature on bud morphogenesis in *Picea* species. *Can. J. For. Res.* **7,** 415–421.

Pomeroy, K. B. (1949). The germination and initial establishment of lobolly pine under various surface soil conditions. *J. For.* **47,** 541–543.

Ponnamperuma, F. N. (1972). The chemistry of submerged soils. *Adv. Agron.* **24,** 29–96.

Ponnamperuma, F. N. (1984). Effects of flooding on soils. *In* "Flooding and Plant Growth" (T. T. Kozlowski, ed.), pp. 9–45. Academic Press, New York.

Pook, E. W. (1986). Canopy dynamics of *Eucalyptus maculata* Hook. IV. Contrasting responses to two severe droughts. *Aust. J. Bot.* **34,** 1–14.

Popescu, I., and Necsulescu, H. (1967). The harmful effect of prolonged inundation on plantations of black poplars in the Braila Marshes. *Rev. Padurilor* **82,** 20–23.

Popp, M., Kramer, D., Lee, H., Diaz, M., Ziegler, H., and Lüttge, U. L. (1987). Crassulacean acid metabolism in tropical dicotyledonous trees. *Trees* **1,** 238–247.

Portis, A. R., Salvucci, M. E., and Ogren, W. L. (1986). Activation of ribulosebisphosphate carboxylase/oxygenase at physiological CO_2 and ribulosebisphosphate concentrations by Rubisco activase. *Plant Physiol.* **82,** 967–971.

Possingham, J. V. (1970). Aspects of the physiology of grape vines. *In* "Physiology of Tree Crops" (L. C. Luckwill and C. V. Cutting, eds.), pp. 335–349. Academic Press, London.

Postel, S. (1984). "Air Pollution, Acid Rain, and the Future of Forests," Worldwatch Paper No. 58. Worldwatch Institute, Washington, D. C.

Potvin, C. (1985). Amelioration of chilling effects by CO_2 enrichment. *Physiol. Veg.* **4,** 345–352.

Powell, D. B. B. (1974). Some effects of water stress in late spring on apple trees. *J. Hortic. Sci.* **49,** 257–272.

Powell, D. B. B., and Thorpe, M. R. (1977). Dynamic aspects of plant–water relations. *In* "Environmental Effects on Crop Physiology" (J. J. Landsberg and C. V. Cutting, eds.), pp. 259- 270. Academic Press, New York.

Powles, S. B. (1984). Inhibition of photosynthesis induced by visible light. *Annu. Rev. Plant Physiol.* **35,** 15–44.

Powles, S. B., Berry, J. A., and Björkman, O. (1983). Interaction between light and chilling temperature on the inhibition of photosynthesis in chilling-sensitive plants. *Plant, Cell Environ.* **6,** 117–123.

Pratt, G. C., Hendrickson, R. H., Chevone, B. I., Christopherson, D. A., and Krupa, S. V. (1983). Ozone and oxides of nitrogen in the rural upper-midwestern U.S.A. *Atmos. Environ.* **17,** 2013–2023.

Price, D. T., Black, T. A., and Kelliher, F. M. (1986). Effects of salal understory removal on photosynthetic rate and stomatal conductance of young Douglas-fir trees. *Can. J. For. Res.* **16,** 90–97.

Pritchett, W. L. (1979). "Properties and Management of Forest Soils." Wiley, New York.

Pritchett, W. L., and Robertson, W. L. (1960). Problems relating to research in forest fertilization with southern pines. *Soil Sci. Soc. Am. Proc.* **24,** 510–512.

Pritchett, W. L., and Smith, W. H. (1974). Management of wet savanna forest soils for pine production. *Bull.—Fla., Agric. Exp. Stn.* **762.**

Proctor, J., and Woodell, S. R. J. (1975). The ecology of serpentine soils. *Adv. Ecol. Res.* **9,** 255–366.

Proebsting, E. L. (1937). "Kelsey spot" of plums in California. *Proc. Am. Soc. Hortic. Sci.* **34,** 272–274.

Proebsting, E. L. Jr., and Middleton, J. E. (1980). The behavior of peach and pear trees under extreme drought stress. *J. Am. Soc. Hortic. Sci.* **105,** 380–385.

Puckett, L. J. (1982). Acid rain, air pollution, and tree growth in southeastern New York. *J. Environ. Qual.* **11,** 376–381.

Pukacki, P., and Pukacka, S. (1987). Freezing stress and membrane injury of Norway spruce (*Picea abies*) tissues. *Physiol. Plant.* **69,** 156–160.

Pukkala, T., and Kolström, T. (1987). Competition indices and the prediction of radial growth in Scots pine. *Silva Fenn.* **21,** 55–67.

Pulkkinen, P., Poykko, T., Tigerstedt, P. M. A., and Velling, P. (1989). Harvest index in northern temperate cultivated conifers. *Tree Physiol.* **5,** 83–98.

Putnam, J. A., Furnival, G. M., and McKnight, J. S. (1960). Management and inventory of southern hardwoods. *U.S., Dep. Agric., Agric. Handb.* **181.**

Putz, F. E., Coley, P. D., Lu, K., Montalvo, A., and Aiello, A. (1983). Uprooting and snapping of trees: Structural determinants and ecological consequences. *Can. J. For. Res.* **13,** 1011–1020.

Pyne, S. J. (1984). "Introduction to Wildland Fire Management in the United States." Wiley, New York.

Quamme, H. A. (1985). Avoidance of freezing injury in woody plants by deep supercooling. *Acta Hortic.* **168,** 11–30.

Queen, W. H. (1967). Radial movement of water and ^{32}P through suberized and unsuberized roots of grape. Ph.D. Dissertation, Duke University, Durham, North Carolina.

Quinlan, J. D. (1981). New chemical approaches to the control of tree form and size. *Acta Hortic.* **120,** 95–106.

Radcliffe, J. E. (1985). Shelterbelt increases dryland pasture growth in Canterbury. *Proc. N. Z. Grassl. Assoc.* **46,** 51–56.

Radin, J. W. (1983). Physiological consequences of cellular water deficits: Osmotic adjustment. *In* "Limitations to Efficient Water Use in Crop Production" (H. M. Taylor, W. R. Jordan, and T. R. Sinclair, eds.), pp. 267–276. Am. Soc. Agron., Crops Sci. Soc. Am., Soil Sci. Soc. Am., Madison, Wisconsin.

Radin, J. W., and Matthews, M. A. (1989). Water transport properties of cortical cells in roots of nitrogen—and phosphorus—deficient cotton seedlings. *Plant Physiol.* **89,** 264, 268.

Radwanski, S. A., and Wickens, G. E. (1967). The ecology of *Acacia albida* on mantle soils in Zalingeri, Jebel Marsa, Sudan. *J. Appl. Ecol.* **4,** 569–579.

Raintree, J. B., and Lundgren, B. O. (1985). Agroforestry potentials for biomass production in integrated land use systems. "*In* "Biomass Energy Systems: Building Blocks for Sustainable Agriculture." World Resources Institute, Washington, D. C.

Raison, R. J. (1979). Modification of the soil environment by vegetation fires, with particular reference to nitrogen transformations. A review. *Plant Soil* **51,** 73–108.

Raison, R. J., Khanna, P. K., and Woods, P. V. (1985). Mechanisms of element transfer to the atmosphere during vegetation fires. *Can. J. For. Res.* **15,** 132–140.

Ramos, D. E., Moller, W. J., and English, H. (1975). Production and dispersal of ascospores of *Eutypa armeniacae* in California. *Phytopathology* **65,** 1364–1371.

Ranney, J. W., and Cushman, J. H. (1982). "Short Rotation Woody Crops Program," Publ. No. 2000. Environ. Sci. Div., Oak Ridge Natl. Lab., Oak Ridge, Tennessee.

Raschke, K. (1986). The influence of the CO_2 content of the ambient air on stomatal conductance and the CO_2 concentration in leaves. *In* "Carbon Dioxide Enrichment of Greenhouse Crops" (H. Z. Enoch and B. A. Kimball, eds.), Vol. 2, pp. 87–102. CRC Press, Boca Raton, Florida.

Rashid, G. H. (1987). Effects of fire on soil carbon and nitrogen in a Mediterranean oak forest of Algeria. *Plant Soil* **103,** 89–93.

Rauner, H. L. (1975). Deciduous forests. *In* "Vegetation and the Atmosphere. Vol. II. Case Studies" (J. L. Monteith, ed.), pp. 241–264. Academic Press, London.

Raven, J. A. (1985). Trees as producers of exudates and extractives. *In* "Trees as Crop Plants" (M. G. R. Cannell and J. R. Jackson, eds.), pp. 253–270. Institute of Terrestrial Ecology, Huntingdon, England.

Read, D. J., and Armstrong, W. (1972). A relationship between oxygen transport and the formation of the ectotrophic mycorrhizal sheath in conifer seedlings. *New Phytol.* **71,** 49–53.

Read, R. A. (1964). Tree windbreaks for the central Great Plains. *U.S., Dep. Agric., Agric. Handb.* **250.**

Reddi, G. H. S., Rao, Y. Y., and Rao, M. S. (1981). The effect of shelterbelts on the productivity of annual field crops. *Indian For.* **107,** 624–629.

Reddy, D. V., and Jokela, J. J. (1982). Variation in specific gravity within trees and between clones in the cottonwood evaluation test in southern Illinois. *Proc. North Am. Poplar Counc. Annu. Meet.,* Rhinelander, Wisconsin, pp. 127–133.

Redfern, D. B., and Cannell, M. G. R. (1982). Needle damage in Sitka spruce caused by early autumn frosts. *Forestry* **54,** 39–45.

Reed, H. S., and Bartholomew, E. T. (1930). The effects of desiccating winds on citrus trees. *Calif., Agric. Exp. Stn., Bull.* **484,** 1–59.

Reed, J. F. (1939). Root and shoot growth of shortleaf and loblolly pines in relation to certain environmental conditions. *Duke Univ. Sch. For. Bull.* **4.**

Rees, D. J., and Grace, J. (1980a). The effects of wind on the extension growth of *Pinus contorta* Douglas. *Forestry* **53,** 143–153.

Rees, D. J., and Grace, J. (1980b). The effects of shaking on extension growth of *Pinus contorta. Forestry* **53,** 155–166.

Reethof, G., and Heisler, G. M. (1976). Trees and forests for noise abatement and visual screening. *U.S. For. Serv. Gen. Tech. Rep. NE* **NE–22,** 39–48.

Regehr, D. L., Bazzaz, F. A., and Boggess, W. R. (1975). Photosynthesis, transpiration, and leaf conductance of *Populus deltoides* in relation to flooding and drought. *Photosynthetica* **9,** 52–61.

Rehfeldt, G. E. (1986). Adaptive variation in *Pinus ponderosa* from intermountain regions. I. Snake and Salmon River basins. *For. Sci.* **32,** 79–92.

Reich, P. B. (1983). Effects of low concentrations of O_3 on net photosynthesis, dark respiration, and chlorophyll contents in aging hybrid poplar leaves. *Plant Physiol.* **73,** 291–296.

Reich, P. B. (1987). Quantifying plant response to ozone: A unifying theory. *Tree Physiol.* **3,** 63–91.

Reich, P. B., and Amundson, R. G. (1985). Ambient levels of ozone reduce net photosynthesis in tree crop species. *Science* **230,** 566–570.

Reich, P. B., and Borchert, R. (1982). Phenology and ecophysiology of the tropical tree, *Tabeonia neochrysantha* (Bignoniaceae). *Ecology* **63,** 294–299.

Reich, P. B., and Lassoie, J. P. (1985). Influence of low concentrations of ozone on growth, biomass partitioning and leaf senescence in young hybrid poplar plants. *Environ. Pollut., Ser. A* **39,** 39–51.

Reich, P. B., Teskey, R. O., Johnson, P. S., and Hinckley, T. M. (1980). Periodic root and shoot growth in oak. *For. Sci.* **26,** 590–598.

Reich, P. B., Lassoie, J. P., and Amundson, R. G. (1984). Reduction in growth of hybrid poplar following field exposure to low levels of O_3 and/or SO_2. *Can. J. Bot.* **62,** 2835–2841.

Reich, P. B., Schoettle, A. W., Stroo, H. F., Troiano, J., and Amundson, R. G. (1985). Effects of O_3, SO_2 and acidic rain on mycorrhizal infection in northern red oak seedlings. *Can. J. Bot.* **63,** 2049–2055.

Reich, P. B., Stroo, H. F., Schoettle, A. W., and Amundson, R. G. (1986). Acid rain and ozone influence mycorrhizal infection in tree seedlings. *J. Air Pollut. Control Assoc.* **36,** 724–726.

Reich, P. B., Schoettle, A. W., Stroo, H. F., Troiano, J., and Amundson, R. G. (1987). Effects of O_3 and acid rain on white pine seedlings grown in five soils. I. Net photosynthesis and growth. *Can. J. Bot.* **65,** 977–987.

Reich, P. B., Schoettle, A. W., Stroo, H. F., and Amundson, R. G. (1988). Effects of ozone and acid rain on white pine (*Pinus strobus*) seedlings grown in five soils. III. Nutrient relations. *Can. J. Bot.* **66,** 1517–1531.

Reid, D. M., and Bradford, K. J. (1984). Effects of flooding on hormone relations. *In* "Flooding and Plant Growth" (T. T. Kozlowski, ed.), pp. 195–219. Academic Press, New York.

Reifsnyder, W. E., and Lull, H. W. (1965). Radiant energy in relation to forests. *U.S., Dep. Agric., Tech. Bull.* **1344.**

Reiners, N. M., and Reiners, W. A. (1965). Natural harvesting of trees. *William L. Hutcheson Mem. For. Bull.* **2,** 9–17.

Reinert, R. A., Heagle, A. S., and Heck, W. W. (1975). Plant responses to pollutant combinations. *In*

"Responses of Plants to Air Pollution" (J. B. Mudd and T. T. Kozlowski, eds.), pp. 159–177. Academic Press, New York.

Remphrey, W. R., Davidson, C. G., and Blouw, M. J. (1987). A classification and analysis of crown forms in green ash (*Fraxinus pennsylvanica*). *Can. J. Bot.* **65,** 2185–2195.

Renwick, J. A. A., and Potter, J. (1981). Effects of sulfur dioxide on volatile terpene emission from balsam fir. *J. Air. Pollut. Control Assoc.* **31,** 65–66.

Reyes, D. M., Stolzy, L. H., and Labanauskas, C. K. (1977). Temperature and oxygen effects in soil on nutrient uptake in jojoba seedlings. *Agron. J.* **69,** 647–650.

Reynolds, E. R. C. (1975). Tree rootlets and their distribution. *In* "The Development and Function of Roots" (J. G. Torrey and D. T. Clarkson, eds.), pp. 163–177. Academic Press, London.

Rice, E. L. (1984). "Allelopathy," 2nd ed. Academic Press, Orlando, Florida.

Richards, D., and Cockcroft, B. (1974). Soil physical properties and root concentrations in an irrigated apple orchard. *Aust. J. Exp. Agric. Anim. Husb.* **14,** 103–107.

Richards, D., and Rowe, D. N. (1977). Root–shoot interactions in peach: The function of the root. *Ann. Bot.* (*London*) [N.S.] **41,** 1211–1216.

Richards, J. H., and Caldwell, M. M. (1987). Hydraulic lift: Substantial nocturnal transport of water between soil layers by *Artemisia tridentata* roots. *Oecologia* **73,** 486–489.

Richards, L. A., and Wadleigh, C. H. (1952). Soil water and plant growth. *In* "Soil Physical Conditions and Plant Growth" (B. T. Shaw, ed.), pp. 73–251. Academic Press, New York.

Richards, P. W. (1966). "Tropical Rain Forest." Cambridge Univ. Press, London and New York.

Richardson, C. J. (1983). Pocosins: Vanishing wastelands or valuable wetlands? *BioScience* **33,** 626–633.

Richardson, S. D. (1953). Root growth of *Acer pseudoplatanus* L. in relation to grass cover and nitrogen deficiency. *Meded. Landbouwhogesch. Wageningen* **53,** 75–97.

Richter, D. D., Ralston, C. W., and Harms, W. R. (1982). Prescribed fire: Effects on water quality and forest nutrient cycling. *Science* **215,** 661–663.

Richter, D. D., King, K. S., and Witter, J. A. (1989). Moisture and nutrient status of extremely acid umbrepts in the Black Mountains of North Carolina. *J. Soil Sci. Soc. Am.* **53,** 1222–1228.

Richter, H. (1974). Erhöhte Saugspannungswerte und morphologische Veränderungen durch transversale Einschnitte in einem Taxus—Stamm. *Flora* **163,** 291–309.

Richter, W. (1984). A structural approach to the function of buttresses of *Quararibea asterolepis*. *Ecology* **65,** 1429–1435.

Ridge, C. R., Hinckley, T. M., Stettler, R. F., and Van Volkenburgh, E. (1986). Leaf growth characteristics of fast-growing poplar hybrids. *Tree Physiol.* **1,** 209–216.

Riding, R. T., and Boyer, K. A. (1983). Germination and growth of pine seedlings exposed to 10–20 pphm sulphur dioxide. *Environ. Pollut., Ser. A* **30,** 245–253.

Ridley, H. N. (1930). "The Dispersal of Plants throughout the World." Reeve, Ashford.

Riekerk, H. (1983). Environmental impact of intensive silviculture in Florida. *USDA For. Serv. Gen. Tech Rep. PNW* **PNW-163,** 264–271.

Robberecht, R., Caldwell, M. M., and Billings, W. D. (1980). Leaf ultraviolet optical properties along a leaf latitudinal gradient in the arctic–alpine life zone. *Ecology* **61,** 612–619.

Roberts, B. R. (1976). The response of field grown white pine seedlings to different sulphur dioxide environments. *Environ. Pollut.* **11,** 175–180.

Roberts, B. R., and Domir, S. C. (1983). The influence of daminozide and maleic hydrazide on growth and net photosynthesis of silver maple and American sycamore seedlings. *Sci. Hortic.* **19,** 367–372.

Roberts, B. R., Townsend, A. M., and Dochinger, L. S. (1971). Photosynthetic response to SO_2 fumigation in red maple. *Plant Physiol.* **47,** Suppl., 30.

Roberts, J. K. M. (1984). Study of plant metabolism *in vivo* using NMR spectroscopy. *Annu. Rev. Plant Physiol.* **35,** 375–386.

Roberts, J. K. M., Andrade, F. H., and Anderson, I. C. (1985). Further evidence that cytoplasmic acidosis is a determinant of flooding intolerance in plants. *Plant Physiol.* **77,** 492–494.

Roberts, L. (1989). Global warming: Blaming the sun. *Science* **246,** 992–993.

Roberts, W. O. (1987). Time to prepare for global climatic change. *In* "The Greenhouse Effect, Climate Change, and U.S. Forests" (W. E. Shands and J. S. Hoffman, eds.), pp. 9–17. Conservation Foundation, Washington, D.C.

Robinson, T. W. (1952). Phreatophytes and their relation to water in the western United States. *Trans. Am. Geophysical Union* **33,** 57–61.

Robitaille, G. (1981). Heavy-metal accumulation in the annual rings of balsam fir *Abies balsamea* (L.). *Environ. Pollut., Ser. B* **2,** 193–202.

Robitaille, H. A., and Leopold, A. C. (1974). Ethylene and the regulation of apple stem growth under stress. *Physiol. Plant.* **32,** 301–304.

Rock, B. N., Vogelmann, J E., Williams, D. L., Vogelmann, A. F., and Hoshizaki, T. (1986). Remote detection of forest damage. *BioScience* **36,** 439–445.

Rodin, L. E., and Bazilevich, N. I. (1967). "Production and Mineral Cycling in Terrestrial Vegetation." Oliver & Boyd, Edinburgh.

Rogers, H. H., Bingham, G. E., Cure, J. D., Smith, J. M., and Surano, K. A. (1983). Responses of selected plant species to elevated CO_2 in the field. *J. Environ. Qual.* **12,** 569–574.

Rogers, H. H., Sionit, N., Cure, J. D., Smith, J. M., and Bingham, G. E. (1984). Influence of carbon dioxide on water relations of soybeans. *Plant Physiol.* **74,** 233–238.

Rom, C., and Brown, S. A. (1979). Water tolerance of apples on clonal rootstocks and peaches on seedling rootstocks. *Compact Fruit Tree* **12,** 30–33.

Rook, D. A. (1969). Water relations of wrenched and unwrenched *Pinus radiata* seedlings on being transplanted into conditions of water stress. *N. Z. J. For.* **14,** 50–58.

Rook, D. A. (1971). Effect of undercutting and wrenching on growth of *Pinus radiata* D. Don seedlings. *J. Appl. Ecol.* **8,** 477–490.

Rook, D. A. (1973). Conditioning radiata pine seedlings to transplanting by restricted watering. *N. Z. J. For. Sci.* **3,** 54–59.

Roques, A., Kerjean, M., and Auclair, D. (1980). Effets de la pollution atmosphérique par le fluor et le dioxyde de soufre sur l'appareil reproducteur femelle de *Pinus silvestris* en Forêt de Roumare (Seine-Maritime, France). *Environ. Pollut., Ser. A* **21,** 191–201.

Rosen, C. J., and Carlson, R. M. (1984). Influence of root zone oxygen stress on potassium and ammonium absorption by Myrobalan plum rootstock. *Plant Soil* **80,** 345–353.

Rosenberg, C. R., Hutnik, R. J., and Davis, D. D. (1979). Forest composition at varying distances from a coal-burning power plant. *Environ. Pollut.* **19,** 307–317.

Rosenthal, G. A., and Janzen, D. H., eds. (1979). "Herbivores: Their Interaction with Secondary Plant Metabolites." Academic Press, New York.

Rosenquist, E. A. (1961). Manuring of rubber in relation to wind damage. *Proc. Natl. Rubber Res. Conf.,* Kuala Lumpur, Malaysia, *1960,* pp. 81–87.

Rosenzweig, M. L. (1968). Net primary productivity of terrestrial communities: Prediction from climatological data. *Am. Nat.* **102,** 67–74.

Ross, S. D., Pharis, R. P., and Binder, W. D. (1983). Growth regulators and conifers: Their physiology and potential uses in forestry. *In* "Plant Growth Regulating Chemicals. Vol. II" (L. G. Nickell, ed.), pp. 35–78. CRC Press, Boca Raton, Florida.

Ross, S. D., Webber, J. E., Pharis, R. P., and Owens, J. N. (1985). Interaction between gibberellin $A_{4/7}$ and root pruning on the reproductive and vegetative process in Douglas-fir. I. Effects on flowering. *Can. J. For. Res.* **15,** 341–347.

Rowe, R. N., and Beardsell, D. V. (1973). Waterlogging of fruit trees. *Hortic. Abstr.* **43,** 533–548.

Ruark, G. A., and Bockheim, J. G. (1987). Below-ground biomass of 10-, 20-, and 32-year-old *Populus tremuloides* in Wisconsin. *Pedobiologia* **30,** 207–217.

Ruark, G. A., and Bockheim, J. G. (1988). Biomass, net primary production, and nutrient distribution for an age sequence of *Populus tremuloides* ecosystems. *Can. J. For. Res.* **18,** 435–443.

Ruark, G. A., Mader, D. L., and Tattar, T. A. (1982). The influence of soil compaction and aeration on the root growth and vigor of trees—A literature review. Part 1. *Arboric. J.* **6,** 251–265.

Rubin, S. (1981). Air pollution constraints on increased coal use by industry. *J. Air. Pollut. Control Assoc.* **31,** 349–360.

Ruhland, W., ed. (1956). Encyclopedia of Plant Physiology. Springer-Verlag, Berlin.

Rundel, P. W. (1981). Fire as an ecological factor. *Encycl. Plant Physiol., New Ser.* **12A,** 501–538.

Rundel, P. W. (1982). Water uptake by organs other than roots. *Encycl. Plant Physiol., New Ser.* **12B,** 111–134.

Rundel, P. W., and Stecker, R. E. (1977). Morphological adaptations of tracheid structure to water stress gradients in the crown of *Sequoiadendron giganteum. Oecologia* **27,** 135–139.

Runkle, J. R. (1981). Gap regeneration in some old-growth forests of the eastern United States. *Ecology* **62,** 1041–1051.

Runkle, J. R. (1982). Patterns of disturbance in some old-growth mesic forests of eastern North America. *Ecology* **63,** 1533–1546.

Runkle, J. R. (1985). Disturbance regimes in temperate forests. *In* "The Ecology of Natural Disturbance and Patch Dynamics" (S. T. A. Pickett and P. S. White, eds.), pp. 17–33. Academic Press, New York.

Runkle, J. R., and Yetter, T. C. (1987). Treefalls revisited: Gap dynamics in the southern Appalachains. *Ecology* **68,** 417–424.

Rushin, J. W., and Anderson, J. E. (1981). An examination of the leaf quaking adaptation and stomatal distribution in *Populus tremuloides* Michx. *Plant Physiol.* **67,** 1264–1266.

Rushton, B. S., and Toner, A. E. (1989). Wind damage to leaves of sycamore (*Acer pseudoplatanus* L.) in coastal and noncoastal stands. *Forestry* **62,** 67–88.

Russell, E. W. (1973). "Soil Conditions and Plant Growth." Longman, London.

Rutter, N., and Edwards, R. S. (1968). Deposition of air-borne marine salt at different sites over the college farm, Aberystwyth (Wales), in relation to wind and weather. *Agric. Meteorol.* **5,** 235–254.

Runyon, E. H. (1936). Ratio of water content to dry weight in leaves of the creosote bush. *Bot. Gaz. (Chicago)* **97,** 518–553.

Ryugo, K. (1988). "Fruit Culture: Its Science and Art." Wiley, New York.

Sachs, P. M., Ripperda, J., Forister, G., Miller, G., Kasemsap, P., Murphy, M,. and Beyl, G. (1988). Maximum biomass yields on prime agricultural land. *Calif. Agric.* November–December, pp. 23–24.

Sakai, A. (1970). Mechanism of desiccation damage of conifers wintering in soil-frozen areas. *Ecology* **51,** 657–664.

Sakai, A., and Larcher, W. (1987). "Frost Survival of Plants." Springer-Verlag, Berlin and New York.

Sakai, A., and Weiser, C. J. (1973). Freezing resistance of trees in North America with reference to tree regions. *Ecology* **54,** 118–126.

Salisbury, F. B., and Ross, C. W. (1978). "Plant Physiology," 2nd ed. Wadsworth Publ. Co., Belmont, California.

Salisbury, F. B., and Ross, C. W. (1985). "Plant Physiology," 3rd ed. Wadsworth Publ. Co., Belmont, California.

Salvucci, M. E., Portis, A. R., and Ogren, W. L. (1986). Light and CO_2 response of ribulose–1,5-bisphosphate carboxylase/oxygenase activation in *Arabidopsis* leaves. *Plant Physiol.* **80,** 655–659.

Sammis, T. W., Riley, W. R., and Lugg, D. G. (1986). Scheduling irrigation on pecans using the crop water stress index. *Proc. West. Pecan Conf., 21st,* Las Cruses, New Mexico, pp. 13–34.

Samson, J. A. (1980). "Tropical Fruits." Longman, London and New York.

Sanchez, P. A. (1973). A review of soils research in tropical Latin America. *N.C., Agric. Exp. Stn., Tech. Bull.* **219.**

Sanchez, P. A., Bandy, D. E., Villachica, J. H., and Nicholaides, J. J. (1982). Amazon basin soils: Management for continuous crop production. *Science* **216,** 821–827.

Sanchez-Diaz, M. F., and Kramer, P. J. (1971). Behavior of corn and sorghum under water stress and during recovery. *Plant Physiol.* **48,** 613–616.

Sandenburgh, R., Taylor, C., and Hoffman, J. S. (1987). How forest product companies can respond to rising carbon dioxide and climate change. *In* "The Greenhouse Effect, Climate Change and U.S. Forestry" (W. E. Shands and J. S. Hoffman, eds.), pp. 247–257. Conservation Foundation, Washington, D.C.

Sanderson, P. L., and Armstrong, W. A. (1980a). The responses of conifers to some of the adverse factors associated with waterlogged soils. *New Phytol.* **85,** 351–362.

Sanderson, P. L., and Armstrong, W. A. (1980b). Phytotoxins in periodically waterlogged forest soils. *J. Soil Sci.* **31,** 643–653.

Sandford, A. P., and Jarvis, P. G. (1986). Stomatal response to humidity in selected conifers. *Tree Physiol.* **2,** 89–103.

Sandralingam, P., and Carmean, W. H. (1974). The effect of fertilization on the growth of newly planted *Pinus merkusii* in Bahau forest reserve. *Malays. For.* **37,** 161–166.

Sands, R., and Bowen, G. D. (1978). Compaction of sandy soils in radiata pine forests. II. Effects of compaction on root configuration and growth of radiata pine seedlings. *Aust. For. Res.* **8,** 163–170.

Sands, R., and Nambiar, E. K. S. (1984). Water relations of *Pinus radiata* in competition with weeds. *Can. J. For. Res.* **14,** 233–237.

Sands, R., Fiscus, E. L., and Reid, C. P. P. (1982). Hydraulic properties of pine and bean roots with varying degrees of suberization, vascular differentiation and mycorrhizal infection. *Aust. J. Plant Physiol.* **9,** 559–569.

Santantonio, D., Hermann, R. K., and Overton, W. S. (1977). Root biomass studies in forest ecosystems. *Pedobiologia* **17,** 1–31.

Sarkar, S. K., and Malhotra, S. S. (1979). Effects of SO_2 on organic acid content and malate dehydrogenase activity in jack pine seedlings. *Biochem. Physiol. Pflanz.* **174,** 438–445.

Sarvas, R. (1955). Ein Beitrag Zur Fenverbreitung des Blütenstaubes einger Waldbäume. 2. *Forstgenet. Forstpflanz.* **4,** 137–142.

Sasaki, S. (1980a). Storage and germination of some Malaysian legume seeds. *Malays. For.* **43,** 161–165.

Sasaki, S. (1980b). Storage and germination of dipterocarp seed. *Malays. For.* **43,** 290–308.

Sasaki, S., and Kozlowski, T. T. (1967). Effects of herbicides on carbon dioxide uptake of pine seedlings. *Can. J. Bot.* **45,** 961–971.

Sasaki, S. and Kozlowski, T. T. (1968a). Effects of herbicides on seed germination and early seedling development of *Pinus resinosa*. *Bot. Gaz.* (*Chicago*) **129,** 238–246.

Sasaki, S., and Kozlowski, T. T. (1968b). Effects of herbicides on respiration of red pine (*Pinus resinosa*) Ait. seedlings. I. *S*-Triazine and chlorophenoxy acid herbicides. *Adv. Front. Plant Sci.* **22,** 187–202.

Sasaki, S., and Kozlowski, T. T. (1968c). Effects of herbicides on respiration of red pine (*Pinus resinosa* Ait.) seedlings. II. Monuron, diuron, DCPA, dalapon, CDEC, CDAA, EPTC, and NPS. *Bot. Gaz.* (*Chicago*) **129,** 286–293.

Sasaki, S., and Kozlowski, T. T. (1968d). The role of cotyledons in early development of pine seedlings. *Can. J. Bot.* **46,** 1173–1183.

Sasaki, S., and Kozlowski, T. T. (1970). Effects of cotyledon and hypocotyl photosynthesis on growth of young pine seedlings. *New Phytol.* **69,** 493–500.

Sasaki, S., Tan, C. H., and Zolfatah, bin H. (1979). Some observations on the unusual flowering and fruiting of dipterocarps. *Malays. For.* **42,** 38–45.

Sasek, T. W., and Strain, B. R. (1990). Implications of atmospheric CO_2 enrichment and climatic change for the geographical distribution of two introduced vines in the U.S.A. *Clima. Change,* **16,** 31–51.

Sasek, T. W., DeLucia, E. H., and Strain, B. R. (1985). Reversibility of photosynthetic inhibition in cotton after long-term exposure to elevated CO_2 concentrations. *Plant Physiol.* **78,** 619–622.

Sastry, C. B. R., and Anderson, H. W. (1980). Clonal variation in gross heat combustion of juvenile *Populus* hybrids. *Can. J. For. Res.* **10,** 245–249.

Satoo, T. (1956). Drought resistance of some conifers at the first summer after their emergence. *Bull. Tokyo Univ. For.* **51,** 1–108.

Satoo, T. (1966a). Production and distribution of dry matter in forest ecosystems. *Tokyo Univ. For., Misc. Inf.* **16,** 1–15.

Satoo, T. (1966b). Variation in response of conifer seed germination to soil moisture conditions. *Tokyo Univ. For., Misc. Inf.* **16,** 17–20.

Saunders, P. J. W. (1971). Modification of the leaf surface and its environment by pollution. *In* "Ecology of Leaf Surface Micro-Organisms" (T. F. Preece and C. H. Dickinson, eds.), pp. 81–89. Academic Press, New York.

Saunders, P. J. W. (1973). Effects of atmospheric pollution on leaf surface microflora. *Pestic. Sci.* **4,** 589–595.

Savidge, R. A., Mutumba, G. M. C., Heald, J. K., and Wareing, P. F. (1983). Gas chromatography–mass spectroscopy identification of l-amino-cyclopropane-l-carboxylic acid in compression wood vascular cambium of *Pinus contorta* Dougl. *Plant Physiol.* **71,** 434–436.

Savill, P. S. (1976). The effects of drainage and ploughing of surface water gleys on rooting and windthrow of Sitka spruce in Northern Ireland. *Forestry* **49,** 133–141.

Savill, P. S. (1983). Silviculture in windy climates. *For. Abstr.* **44,** 473–488.

Scalabrelli, G., and Couvillon, G. A. (1986). The effect of temperature and bud type on rest completion and the GDH °C requirement for budbreak in 'Redhaven' peach. *J. Am. Soc. Hortic. Sci.* **111,** 537–540.

Scarascia-Mugnozza, G., Hinckley, T. M., and Stettler, R. F. (1986). Evidence for nonstomatal inhibition of net photosynthesis in rapidly dehydrated shoots of *Populus*. *Can. J. For. Res.* **16,** 1371–1375.

Schaedle, M. (1975). Tree photosynthesis. *Annu. Rev. Plant Physiol.* **26,** 101–115.

Schaedle, M., and Brayman, A. A. (1986). Ribulose-1,5-bisphosphate carboxylase activity of *Populus tremuloides* Michx. bark tissues. *Tree Physiol.* **1,** 53–56.

Scheffer, T. C., and Hedgcock, G. C. (1955). Injury to northwestern forest trees by sulfur dioxide from smelters. *U.S., Dep. Agric., Tech. Bull.* **117.**

Schindlbeck, W. E. (1977). Biochemische Beiträge zur Immissionforschung. *Forstwiss. Centralbl.* **96,** 67–71.

Schlesinger, W. H., and Gill, D. S. (1978). Demographic studies of the chaparral shrub, *Ceanothus megacarpus* in the Santa Ynez Mountains, California. *Ecology* **59,** 1256–1263.

Schlesinger, W. H., Gray, J. T., Gill, D. S., and Mahall, B. E. (1982). *Ceanothus megacarpus* chaparral: A synthesis of ecosystem processes during development and annual growth. *Bot. Rev.* **48,** 71–117.

Schmidtling, R. C. (1973). Intensive culture increases growth without affecting wood quality of young southern pines. *Can. J. For. Res.* **3,** 565–573.

Schneider, S. H., and Dennett, R. P. (1975). Climatic barriers to long-term energy growth. *Ambio* **4,** 66–71.

Schneider, S. H., and Londer, R. (1984). "The Coevolution of Climate and Life." Sierra Club Books, San Francisco.

Schoch, P., and Binkley, D. (1986). Prescribed burning increased nitrogen availability in a mature loblolly pine stand. *For. Ecol. Manage.* **14,** 13–22.

Schoeneweiss, D. F. (1978a). Water stress as a predisposing factor in plant disease. *In* "Water Deficits and Plant Growth" (T. T. Kozlowski, ed.), Vol. 5., pp. 61–99. Academic Press, New York.

Schoeneweiss, D. F. (1978b). The influence of stress on diseases of nursery and landscape plants. *J. Arboric.* **4,** 217–225.

Scholander, P. F. (1968). How mangroves desalinate seawater. *Physiol. Plant.* **21,** 251–261.

Scholander, P. F., van Dam, L., and Scholander, S. I. (1955). Gas exchange in roots of mangroves. *Am. J. Bot.* **42,** 92–98.

Scholz, F., and Stephan, B. R. (1982). Growth and reaction to drought of 43 *Abies grandis* provenances in a greenhouse study. *Silvae Genet.* **31,** 27–35.

Scholz, F., Timmann, T., and Krusche, D. (1980). Genotypic and environmental variance in the response of Norway spruce families to HF-fumigation. *In* "Papers Presented to the Symposium on the Effects of Airborne Pollution on Vegetables," p. 277. United Nations Economic Commission for Europe, Warsaw, Poland.

Schomaker, C. E. (1969). Growth and foliar nutrition of white pine seedlings as influenced by simultaneous changes in moisture and nutrient supply. *Soil Sci. Soc. Am. Proc.* **33,** 614–618.

Schönherr, J. (1976). Water permeability of isolated cuticular membranes: The effect of cuticular waxes on diffusion of water. *Planta* **131,** 159–164.

Schönherr, J., and Ziegler, H. (1980). Water permeability of *Betula* periderm. *Planta* **147,** 345–354.

Schreiber, J. D., Duffy, P. D., and McClurkin, D. C. (1976). Dissolved nutrient losses in storm runoff from five southern pine watersheds. *J. Environ. Qual.* **5,** 201–205.

Schroeder, C. A., and Wieland, P. A. (1956). Diurnal fluctuation in size in various parts of the avocado tree and fruit. *Proc. Am. Soc. Hortic. Sci.* **68,** 253–258.

Schulte, P. J., and Hinckley, T. M. (1987). Abscisic acid relations and the response of *Populus trichocarpa* stomata to leaf water potential. *Tree Physiol.* **3,** 103–113.

Schulte, P. J., Hinckley, T. M., and Stettler, R. F. (1987). Stomatal response of *Populus* to leaf water potential. *Can. J. Bot.* **65,** 255–260.

Schultz, H. R., and Matthews, M. A. (1988). Resistance to water transport in shoots of *Vitis vinifera* L. *Plant Physiol.* **88,** 718–724.

Schulze, E.-D. (1989). Air pollution and forest decline in a spruce (*Picea abies*) forest. *Science* **244,** 776–783.

Schulze, E.-D., and Kuppers, M. (1979). Short-term and long-term effects of plant water deficits on stomatal response to humidity in *Corylus avellana* L. *Planta* **146,** 319–326.

Schulze, E.-D., Lange, O. L., Buschbom, U., Kappen, L., and Evenari, M. (1972). Stomatal responses to changes in humidity in plants growing in the desert. *Planta* **108,** 259–270.

Schulze, E.-D., Cermak, J., Matyssek, R., Penska, M. Zimmermann, M., Vasicek, F., Gries, W., and Kucera, J. (1985). Canopy transpiration and water fluxes in the xylem of the trunk of *Larix* and *Picea* trees—A comparison of xylem flow, porometer and cuvette measurements. *Oecologia* **66,** 475–483.

Schupp, E. W., Howe, H. F., Augspurger, C. K., and Levey, D. J. (1989). Arrival and survival in tropical treefall gaps. *Ecology* **20,** 562–564.

Schupp, J. R., and Ferree, D. C. (1987). Effect of root pruning at different stages of growth on growth and fruiting of apple trees. *HortScience* **22,** 307–390.

Schütt, P. (1985). Vernetzte Problemstellung—vernetzte—Forschung?—Betrachtungen zur Waldsterbenforschung. *Forstarchiv* **56,** 179–181.

Schütt, P., and Cowling, E. P. (1985). Waldsterben, a general decline of forests in central Europe: Symptoms, development, and possible causes. *Plant Dis.* **69,** 548–558.

Schwarz, M., and Gale, J. (1984). Growth response to high salinity at high levels of carbon dioxide. *J. Exp. Bot.* **35,** 193–196.

Scott, F. M. (1964). Lipid deposition in intercellular space. *Nature* (*London*) **203,** 164–165.

Scriber, J. M. (1977). Limiting effects of low leaf-water content on the nitrogen utilization, energy budget, and larval growth of *Hylaphora cecropia* (Lepidoptera: Saturniidae). *Oecologia* **28,** 269–287.

Seidel, K. W. (1972). Drought resistance and internal water balance of oak seedlings. *For. Sci.* **18,** 34–40.

Sena Gomes, A. R., and Kozlowski, T. T. (1980a). Growth responses and adaptations of *Fraxinus pennsylvanica* seedlings to flooding. *Plant Physiol.* **66,** 267–271.

Sena Gomes, A. R., and Kozlowski, T. T. (1980b). Responses of *Melaleuca quinquenervia* seedlings to flooding. *Physiol. Plant.* **49,** 373–377.

Sena Gomes, A. R., and Kozlowski, T. T. (1980c). Effects of flooding on growth of *Eucalyptus camaldulensis* and *E. globus* seedlings. *Oecologia* **46,** 139–142.

Sena Gomes, A. R., and Kozlowski, T. T. (1980d). Responses of *Pinus halepensis* seedlings to flooding. *Can. J. For. Res.* **10,** 308–311.

Sena Gomes, A. R., and Kozlowski, T. T. (1986). The effects of flooding on water relations and growth of *Theobroma cacao* var. *catongo* seedlings. *J. Hortic. Sci.* **61,** 265–276.

Sena Gomes, A. R., and Kozlowski, T. T. (1987). Effects of temperature on growth and water relations of cacao (*Theobroma cacao* var. *comum*) seedlings. *Plant Soil* **103,** 3–11.

Sena Gomes, A. R., and Kozlowski, T. T. (1988a). Stomatal characteristics, leaf waxes, and transpiration rates of *Theobroma cacao* and *Hevea brasiliensis* seedlings. *Ann. Bot.* (*London*) [N.S.] **61,** 425–432.

Sena Gomes, A. R., and Kozlowski, T. T. (1988b). Physiological and growth responses to flooding of seedlings of *Hevea brasiliensis*. *Biotropica* **20,** 286–293.

Sena Gomes, A. R., and Kozlowski, T. T. (1989). Responses of seedlings of two varieties of *Theobroma cacao* to wind. *Trop. Agric.* **66,** 137–141.

Sena Gomes, A. R., Kozlowski, T. T., and Reich P. B. (1987). Some physiological responses of *Theobroma cacao* var. *catongo* seedlings to air humidity. *New Phytol.* **107,** 591–602.

Sen Gupta, A., and Berkowitz, G. A. (1988). Chloroplast osmotic adjustment and water stress effects on photosynthesis. *Plant Physiol.* **88,** 200–206.

Senser, M. (1982). Frost resistance in spruce (*Picea abies* (L). Karst.). III. Seasonal changes in the phospho- and galactolipids of spruce needles. *Z. Pflanzenphysiol.* **105,** 229–239.

Shackel, K., and Matthews, M. A. (1986). Dynamic relation between expansion and cellular turgor in growing leaves of grape (*Vitis vinifera* L.). *Plant Physiol.* **80,** Suppl., 105.

Shackel, K. A., Matthews, M. A., and Morrison, J. C. (1987). Dynamic relationship between expansion and turgor in growing grape (*Vitis vinifera* L.) leaves. *Plant Physiol.* **84,** 1166–1171.

Shainsky, L. D., and Radosevich, S. R. (1986). Growth and water relations of *Pinus ponderosa* seedlings in competitive regimes with *Arctostaphylos patula* seedlings. *J. Appl. Ecol.* **23,** 957–966.

Shands, W. E., and Hoffman, J. S., eds. (1987). "The Greenhouse Effect, Climate Change, and U.S. Forests." Conservation Foundation, Washington, D.C.

Shanklin, J., and Kozlowski, T. T. (1984). Effect of temperature preconditioning on responses of *Fraxinus pennsylvanica* seedlings to SO_2. *Environ. Pollut., Ser. A* **36,** 311–326.

Shanklin, J., and Kozlowski, T. T. (1985a). Effect of flooding of soil on growth and subsequent responses of *Taxodium distichum* seedlings to SO_2. *Environ. Pollut., Ser. A* **38,** 199–212.

Shanklin, J., and Kozlowski, T. T. (1985b). Effect of temperature regime on growth and subsequent responses of *Sophora japonica* seedlings to SO_2. *Plant Soil* **88,** 399–405.

Sharkey, T. D., and Ogawa, T. (1987). Stomatal response to light. *In* "Stomatal Function" (E. Zeiger, G. D. Farquhar, and I. R. Cowan, eds.), pp. 195–208. Stanford Univ. Press, Stanford, California.

Sharp, R. E., and Boyer, J. S. (1986). Photosynthesis at low water potentials in sunflower: Lack of photoinhibitory effects. *Plant Physiol.* **82,** 90–95.

Sharp, R. E., and Davies, W. J. (1979). Solute regulation and growth by roots and shoots of water-stressed maize plants. *Planta* **147,** 43–49.

Sharples, R. O. (1980). The influence of orchard nutrition on the storage quality of apples and pears grown in the United Kingdom. *In* "Mineral Nutrition of Fruit Trees" (D. Atkinson, J. E. Jackson, R. O. Sharples, and W. M. Waller, eds.), pp. 17–28. Butterworth, London.

Shaw, C. G., III, and Taes, E. H. A. (1977). Impact of *Dothistroma* needle blight and *Armillaria* root rot on diameter growth of *Pinus radiata*. *Phytopathology* **66,** 1319–1323.

Shaybany, B., and Martin, G. C. (1977). Abscisic acid identification and its quantification in leaves of *Juglans* seedlings during waterlogging. *J. Am. Soc. Hortic. Sci.* **102,** 300–302.

Shear, C. B. (1975). Calcium-related disorders of fruit and vegetables. *HortScience* **10,** 361–365.

Sheriff, D. W., and Whitehead, D. (1984). Photosynthesis and wood structure in *Pinus radiata* D. Don during dehydration and immediately after rewatering. *Plant, Cell Environ.* **7,** 53–62.

Shigo, A. L. (1984a). Tree decay and pruning. *Arboric. J.* **8,** 1–12.

Shigo, A. L. (1984b). Compartmentalization: A conceptual framework for understanding how trees grow and defend themselves. *Annu. Rev. Phytopathol.* **22,** 189–214.

Shigo, A. L., and Wilson, C. L. (1977). Wound dressings on red maple and American elm: Effectiveness after five years. *J. Arboric.* **3,** 81–87.

Shive, J. B., Jr., and Brown, K. W. (1978). Quaking and gas exchange in lines of cottonwood (*Populus deltoides, Marsh.*). *Plant Physiol.* **61,** 331–333.

Shoemaker, J. S. (1978a). "Tree Fruit Production." Avi Publ. Co., Westport, Connecticut.

Shoemaker, J. S. (1978b). "Small Fruit Culture." Avi Publ. Co., Westport, Connecticut.

Shoulders, E. (1967). Fertilizer application, inherent fruitfulness, and rainfall affect flowering of longleaf pine. *For. Sci.* **13,** 376–383.

Shoulders, E., and Ralston, C. W. (1975). Temperature, root aeration and light influence slash pine nutrient uptake rates. *For. Sci.* **21,** 401–410.

Silen, R. R. (1962). Pollen dispersal considerations for Douglas-fir. *J. For.* **60,** 790–795.

Silen, R. R. (1973). First and second season effect on Douglas-fir cone initiation from a single shade period. *Can. J. For. Res.* **3,** 428–435.

Sill, W. H., Jr. (1982). "Plant Protection: An Integrated Interdisciplinary Approach." Iowa State Univ. Press, Ames.

Silverborg, S. B., and Ross, E. W. (1968). Ash dieback disease development in New York State. *Plant Dis. Rep.* **52,** 105–107.

Siminovitch, D. (1981). Common and disparate elements in the processes of adaptation of herbaceous and woody plants to freezing—A perspective. *Cryobiology* **18,** 166–185.

Siminovitch, D., Rheaume, B., Pomeroy, K., and Lepage, M. (1968). Phospholipid, protein, and nucleic acid increases in protoplasm and membrane structures associated with development of extreme freezing resistance in black locust tree cells. *Cryobiology* **5,** 202–225.

Simmonds, N. W. (1982). "Bananas." Longman, London.

Simons, R. K. (1970). Phloem tissue development response to freeze injury to trunks of apple trees. *J. Am. Soc. Hortic. Sci.* **95,** 182–190.

Simons, R. K., and Lott, R. V. (1963). The morphological and anatomical development of apple injured by late spring frost. *Proc. Am. Soc. Hortic. Sci.* **83,** 88–100.

Sinclair, T. R., and Ludlow, M. M. (1985). Who taught plants thermodynamics? The unfulfilled potential of water potential. *Aust. J. Plant Physiol.* **12,** 213–217.

Sinclair, T. R., Spaeth, S. C., and Vendeland, J. S. (1981). Microclimatic limitations to crop yield. *In* "Breaking the Climate/Soil Barriers to Crop Yield" (M. H. Miller, D. M. Brown, and E. G. Beauchamps, eds.), pp. 3–27. University of Guelph, Guelph, Ontario, Canada.

Sinclair, W. A. (1969). Polluted air: Potent new selective force in forests. *J. For.* **67,** 305–309.

Singh, N. K., Nelson, D. E., La Rosa, P. C., Bracker, C. E., Handa, A. K., Hasegawa, P. M., and Bressan, R. A. (1989). Osmotin: A protein associated with osmotic stress adaptation in plant cells. *In* "Environmental Stress in Plants" (J. H. Cherry, ed.), pp. 67–87. Springer-Verlag, Berlin.

Sionit, N., and Kramer, P. J. (1986). Woody plant reactions to CO_2 enrichment. *In* "Carbon Dioxide Enrichment of Greenhouse Crops" (H.Z. Enoch and B. A. Kimball, eds.), pp. 69–84. CRC Press, Boca Raton, Florida.

Sionit, N., Hellmers, H., and Strain, B. R. (1980). Growth and yield of wheat under CO_2 enrichment and water stress. *Crop Sci.* **20,** 687–690.

Sionit, N., Strain, B. R., and Beckford, H. A. (1981). Environmental controls on the growth and yield of okra. 1. Effects of temperature and of CO_2 enrichment at low temperature. *Crop Sci.* **21,** 885–888.

Sionit, N., Strain, B. R., Hellmers, H., Riechers, G. H., and Jaeger, C. H. (1985). Long-term atmospheric CO_2 enrichment affects the growth and development of *Liquidambar styraciflua* and *Pinus taeda* L. seedlings. *Can. J. For. Res.* **15,** 468–471.

Siren, G. (1963). Tree rings and climatic forecasts. *New Sci.* **346,** 18–20.

Siren, G., Sennerby-Forsse, L., and Ledin, S. (1987). Energy plantations—Short rotation forestry in Sweden. *In* "Biomass" (D. O. Hall and R. P. Overend, eds.) pp. 119–143. Wiley, New York.

Siwecki, R., and Kozlowski, T. T. (1973). Leaf anatomy and water relations of excised leaves of six *Populus* clones. *Arbor. Kornickie* **8,** 83–105.

Slabaugh, P. E. (1957). Effects of live crown removal on the growth of red pine. *J. For.* **55,** 904–906.

Slatyer, R. O. (1957). The significance of the permanent wilting percentage in studies of plant and soil water relations. *Bot. Rev.* **23,** 585–636.

Slatyer, R. O. (1960). Aspects of the tissue water relationships of an important arid zone species (*Acacia aneura* F. Muell.) in comparison with two mesophytes. *Bull. Res. Counc. Isr., Sect. D* **8,** 159–168.

Slatyer, R. O. (1961). Internal water balance of *Acacia aneura* F. Muell. in relation to environmental conditions. *Arid Zone Res.* **16,** 137–146.

Slatyer, R. O. (1967). "Plant–Water Relationships." Academic Press, London.

Slatyer, R. O., and Bierhuizen, J. F. (1964a). Transpiration from cotton leaves under a range of environmental conditions in relation to internal and external diffusive resistances. *Aust. J. Biol. Sci.* **17,** 115–130.

Slatyer, R. O., and Bierhuizen, J. F. (1964b). The influence of several transpiration suppressants on transpiration, photosynthesis, and water-use efficiency of cotton leaves. *Aust. J. Biol. Sci.* **17,** 131–146.

Slatyer, R. O., and Ferrar, P. J. (1977). Altitudinal variation in the photosynthetic characteristics of snow gum, *Eucalyptus pauciflora* Sieb. ex Spreng. V. Rate of acclimation to an altered growth environment. *Aust. J. Plant Physiol.* **4,** 595–609.

Slocum, G. K., and Maki, T. E. (1956). Exposure of loblolly pine planting stock. *J. For.* **54,** 313–315.

Slowik, K., Labanauskas, C. K., Stolzy, L. H., and Zentmyer, G. A. (1979). Influence of rootstocks, soil oxygen, and soil moisture on the uptake and translocation of nutrients in young avocado plants. *J. Am. Soc. Hortic. Sci.* **104,** 172–175.

Small, J. A., and Monk, C. D. (1959). Winter changes in tree radii and temperature. *For. Sci.* **5,** 229–233.

Smirnoff, N., and Stewart, G. R. (1985). Nitrate assimilation and translocation by higher plants: Comparative physiology and ecological consequences. *Physiol. Plant.* **64,** 133–140.

Smith, A. P. (1972a). Buttressing of tropical trees: A descriptive model and a new hypothesis. *Am. Nat.* **106,** 32–46.

Smith, A. P. (1972b). Notes on wind-related growth patterns of Paramo plants in Venezuela. *Biotropica* **4,** 10–16.

Smith, A. P. (1973). Stratification of temperate and tropical forests. *Am. Nat.* **107,** 671–683.

Smith, D. M. (1986). "The Practice of Silviculture." Wiley, New York.

Smith, D. W., and Linnartz, N. E. (1980). The southern hardwood region. *In* "Regional Silviculture of the United States" (J. W. Barrett, ed.), 2nd ed., pp. 145–230. Wiley, New York.

Smith, H., and Kefford, N. P. (1964). The chemical regulation of the dormancy phases of bud development. *Am. J. Bot.* **51,** 1002–1012.

Smith, H., and Morgan, D. C. (1982). The spectral characteristics of the visible radiation incident upon the surface of the earth. *In* "Plants and the Daylight Spectrum" (H. Smith, ed.), pp. 1–20. Academic Press, London.

Smith, H. C. (1966). Epicormic branching on eight species of Appalachian hardwoods. *USDA For. Serv. Res. Note NE* **NE–53.**

Smith, K. A., and Restall, S. W. F. (1971). The occurrence of ethylene in anaerobic soil. *J. Soil Sci.* **22,** 430–443.

Smith, R. F. (1978). History and complexity of integrated pest management. *In* "Pest Control Strategies" (E. H. Smith and D. Pimentel, eds.), pp. 41–53. Academic Press, New York.

Smith, R. H. (1966a). Resin quality as a factor in the resistance of pines to bark beetles. *In* "Breeding Pest Resistant Trees" (H. D. Gerhold, E. J. Schreiner, R. E. McDermott, and J. A. Wisnieski, eds.), pp. 189–196. Pergamon, New York.

Smith, R. H. (1966b). The monoterpene composition of *Pinus ponderosa* xylem resin and of *Dendroctonus brevicomis* pitch tubes. *For. Sci.* **12,** 63–68.

Smith, R. L. (1966). "Ecology and Field Biology." Harper & Row, New York.

Smith, T. A. (1985). Polyamines. *Annu. Rev. Plant Physiol.* **36,** 117–143.

Smith, W. H. (1974). Air pollution. Effects on the structure and function of the temperate forest ecosystem. *Environ. Pollut.* **6,** 111–129.

Smith, W. H. (1976). Character and significance of forest tree root exudates. *Ecology* **57,** 324–331.

Smith, W. H. (1977). Tree root exudates and the forest soil ecosystem: Exudate chemistry, biological significance, and alteration by stress. *Range Sci. Dep. Sci. Ser.* (*Colo. State Univ.*) **26,** 289–301.

Smith, W. H. (1981). "Air Pollution and Forests." Springer-Verlag, New York.

Smith, W. H., and Dowd, M. L. (1981). Biomass production in Florida. *J. For.* **79,** 508–511.

Smithberg, M. H., and Weiser, C. J. (1968). Patterns of variation among climatic races of red osier dogwood. *Ecology* **49,** 495–505.

Smit-Spinks, B., Swanson, B. T., and Markhart, A. H. (1985). The effect of photoperiod and thermoperiod on cold acclimation and growth of *Pinus sylvestris*. *Can. J. For. Res.* **15,** 453–460.

Sojka, R. E., and Stolzy, L. H. (1981). Stomatal response to soil oxygen. *Calif. Agric.* **35,** 18–19.

Solarova, J., Pospisilova, J., and Slavik, B. (1981). Gas exchange regulation by changing of epidermal conductance with antitranspirants. *Photosynthetica* **15,** 365–400.

Solomon, A. M., and West, D. C. (1987). Simulating forest ecosystem responses to expected climate change in eastern North America: Applications to decision making in the forest industry. *In* "The Greenhouse Effect, Climate Change, and U.S. Forests" (W. E. Shands and J. S. Hoffman, eds.), pp. 189–217. Conservation Foundation, Washington, D.C.

Sommer, N. F. (1955). Sunburn predisposes walnut trees to branch wilt. *Phytopathology* **45,** 607–613.

Sopper, W. E., and Kerr, S. N., eds. (1979). "Utilization of Municipal Sewage Effluent on Forest and Disturbed Land." Penn. State Univ. Press, University Park.

South, D. B., ed. (1986). "Nursery Management Practices for the Southern Pines." School of Forestry, Auburn University, Auburn, Alabama.

South, D. B. (1987). A re-evaluation of Wakeley's 'critical tests' of morphological grades of southern pine nursery stock. *South. Afr. For. J.* **142,** 56–59.

South, D. B., Boyer, J. N., and Bosch, L. (1985). Survival and growth of loblolly pine as influenced by seedling grade: 13-year results. *South. J. Appl. For.* **9,** 76–81.

Southwick, S. M., and Davenport, T. L. (1986). Characterization of water stress and low temperature effects on flower induction in citrus. *Plant Physiol.* **81,** 26–29.

Specht, R. L. (1981). Growth indices—Their role in understanding the growth, structure, and distribution of Australian vegetation. *Oecologia* **50,** 347–356.

Sperry, J. S., Holbrook, N. M., Zimmermann, M. H., and Tyree, M. T. (1987). Spring filling of xylem vessels in wild grapevines. *Plant Physiol.* **83,** 414–417.

Sperry, J. S., Donnelly, J. R., and Tyree, M. T. (1988). Seasonal occurrence of xylem embolism in sugar maple. *Am. J. Bot.* **75,** 1212–1218.

Spies, T. A., and Franklin, J. F. (1989). Gap characteristics and vegetation response in coniferous forests of the Pacific Northwest. *Ecology* **7,** 543–545.

Sprugel, D. G. (1976). Dynamic structure of wave-regenerated *Abies balsamea* forests in the northeastern United States. *J. Ecol.* **64,** 889–910.

Spurr, S. H., and Barnes, B. V. (1980). "Forest Ecology," 3rd ed. Wiley, New York.

Squillace, A. E. (1971). Inheritance of monoterpene composition in cortical oleoresin of slash pine. *For. Sci.* **17,** 381–387.

Squillace, A. E., and Silen, R. R. (1962). Racial variation in ponderosa pine. *For. Sci. Monogr.* **2.**

Squire, G. R., and Callander, B. A. (1981). Tea plantations. *In* "Water Deficits and Plant Growth. Vol. 6. Woody Plant Communities" (T. T. Kozlowski, ed.), pp. 471–510. Academic Press, New York.

Stahle, D. W., Cleaveland, M. K., and Duvick, J. G. (1988). North Carolina climate changes reconstructed from tree rings: A. D. 372 to 1985. *Science* **240,** 1517–1520.

Stakman, E. C., and Christensen, C. M. (1946). Aerobiology in relation to plant disease. *Bot. Rev.* **12,** 205–253.

Stålfelt, M. G. (1956). Morphologie und Anatomie des Blattes als Transpirationsorganen. *Encycl. Plant Physiol.* **3,** 324–341.

Stanek, W., Hopkins, J. C., and Simmons, C. S. (1986). Effects of spacing in lodgepole pine stands on incidence of *Atropellis* canker. *For. Chron.* **62,** 91–95.

Stansell, J. R., Klepper, B., Browning, V. D., and Taylor, H. M. (1973). Plant water status in relation to clouds. *Agron. J.* **65,** 677–678.

Stark, N. M. (1977). Fire and nutrient cycling in a Douglas-fir/larch forest. *Ecology* **58,** 16–30.

Stark, N. M., and Jordan, C. F. (1978). Nutrient retention by the root mat of an Amazonian rain forest. *Ecology* **59,** 434–437.

Stark, N. M., and Spitzner, C. (1985). Xylem sap analysis for determining the nutrient status and growth of *Pinus ponderosa*. *Can. J. For. Res.* **15,** 783–790.

Stearns, F., and Olson, J. (1958). Interactions of photoperiod and temperature affecting seed germination in *Tsuga canadensis*. *Am. J. Bot.* **45,** 55–58.

Steiner, K. C., McCormick, L. H., and Canavera, D. S. (1980). Differential responses of paper birch provenances to aluminum in solution culture. *Can. J. For. Res.* **10,** 25–29.

Stephens, G. R., Turner, N. C., and De Roo, H. C. (1972). Some effects of defoliation by gypsy moth (*Porthetria dispar* L.) and elm spanworm (*Ennomos subsignarius* Hbn.) on water balance and growth of deciduous forest trees. *For. Sci.* **18,** 326–330.

Steponkus, P. L. (1984). Role of the plasma membrane in freezing injury and cold acclimation. *Annu. Rev. Plant Physiol.* **35,** 543–584.

Stewart, H. T. L., and Salmon, G. R. (1986). Irrigation of tree plantations with recycled water. 2. Some economic analyses. *Aust. For.* **49,** 89–96.

Stewart, H. T. L., Allender, E., Sandell, P., and Kube, P. (1986). Irrigation of tree plantations with recycled water. 1. Research developments and case studies. *Aust. For.* **49,** 81–88.

Stinchcombe, G. R., Copas, E., Williams, R. R., and Arnold, G. (1984). The effects of paclobutrazol and daminozide on the growth and yield of cider apple trees. *J. Hortic. Sci.* **59,** 323–327.

Stoeckeler, J. H. (1960). Soil factors affecting the growth of quaking aspen forests in the Lake States. *Minn., Agric. Exp. Stn., Tech. Bull.* **233.**

Stoeckeler, J. H. (1962). Shelterbelt influence on the Great Plains field environment and crops. *U.S., Dep. Agric., Prod. Res. Rep.* **62.**

Stoeckeler, J. H., and Dortignac, E. J. (1941). Snowdrifts as a factor in growth and longevity of shelterbelts in the Great Plains. *Ecology* **22,** 117–124.

Stoeckeler, J. H., and Mason, J. W. (1956). Regeneration of aspen cutover areas in northern Wisconsin. *J. For.* **54,** 13–16.

Stolzy, L. H., and Sojka, R. E. (1984). Effects of flooding on plant disease. *In* "Flooding and Plant Growth" (T. T. Kozlowski, ed.), pp. 221–264. Academic Press, New York.

Stolzy, L. H., Letey, J., Klotz, L. J., and DeWolfe, T. A. (1965). Soil aeration and root-rotting fungi as factors in decay of citrus feeder roots. *Soil Sci.* **99,** 403–406.

Stone, E. C., and Juhren, G. (1951). The effect of pine on the germination of *Rhus ovata* Wats. *Am. J. Bot.* **38,** 368–372.

Stone, E. C., and Norberg, E. A. (1979). Root growth capacity: One key to bare-root survival. *Calif. Agric.* **33,** 14–15.

Stone, E. C., and Vasey, R. B. (1968). Preservation of coast redwood on alluvial flats. *Science* **159,** 157–161.

Stone, E. C., Schubert, G. H., Benseler, R. W., Baron, F. J., and Krugman, S. L. (1963). Variation in the root-regenerating potentials of ponderosa pine from four California nurseries. *For. Sci.* **9,** 217–225.

Stone, E. C., Grah, R. F., and Zinke P. J. (1972). Preservation of the primeval redwoods in the Redwood National Park. *Am. For.* **78,** Part I, 50–55; Part II, 48–59.

Stone, E. L. (1968). Microelement nutrition of forest trees: A review. *In* "Forest Fertilization: Theory and Practice," pp. 132–175. Tennessee Valley Authority, Muscle Shoals, Alabama.

Stone, E. L. (1973). Biological objectives in forest fertilization. *USDA For. Serv. Gen. Tech. Rep. NE* **NE–3.**

Stone, E. L., and Stone, M. H. (1954). Root collar sprouts in pine. *J. For.* **52,** 487–491.

Strain, B. R. (1987). Direct effects of increasing atmospheric CO_2 on plants and ecosystems. *Trends Ecol. Evol.* **2,** 18–21.

Strain, B. R., and Bazzaz, F. A. (1983). Terrestrial plant communities. *In* "CO_2 and Plants" (E. R. Lemon, ed.), pp. 177–222. Westview Press, Boulder, Colorado.

Strain, B. R., and Cure, J. D. (1986). "Direct Effects of Atmospheric CO_2 Enrichment on Plants and Ecosystems: A Bibliography with Abstracts, ORNG/CDIC–13. Natl. Tech. Inf. Serv., U.S. Dept. of Commerce, Springfield, Virginia.

Strain, B. R., Higginbotham, K. O., and Mulroy, J. C. (1976). Temperature preconditioning and photosynthetic capacity of *Pinus taeda* L. *Photosynthetica* **10,** 47–52.

Stroo, H. F., Reich, P. B., Schoettle, A. W., and Amundson, R. G. (1988). Effects of ozone and acid rain on white pine (*Pinus strobus*) seedlings grown in five soils. II. Mycorrhizal infection. *Can. J. Bot.* **66,** 1510–1516.

Stumpf, P. K., and Conn, E. E., eds. (1980–). "The Biochemistry of Plants," Vols. 1–12. Academic Press, New York.

Sucoff, E. (1972). Water potential in red pine: Soil moisture, evapotranspiration, crown position. *Ecology* **53,** 681–686.

Sucoff, E., and Hong, S. G. (1974). Effects of thinning on needle water potential in red pine. *For. Sci.* **20,** 25–29.

Sumimoto, M., Shiraga, M., and Kondo, T. (1975). Ethane in pine needles preventing the feeding of the beetle, *Monochamus alternatus. J. Insect Physiol.* **21,** 713–722.

Surano, K. A., Daley, P. F., Houpis, J. L., Shinn, J. H., Helms, J. A., Palassou, R. J., and Costella, M. P. (1986). Growth and physiological responses of *Pinus ponderosa* Dougl. ex P. Laws to long-term elevated CO_2 concentration. *Tree Physiol.* **2,** 243–259.

Sutton, R. F. (1958). Chemical herbicides and their uses in the silviculture of forests of eastern Canada. *For. Res. Div. Tech. Note* **68.**

Sutton, R. F. (1970). Chemical herbicides and forestation. *For. Chron.* **46,** 458–465.

Suwannapinunt, W., and Kozlowski, T. T. (1980). Effect of SO_2 on transpiration, chlorophyll content, growth, and injury in young seedlings of woody angiosperms. *Can. J. For. Res.* **10,** 78–81.

Swaine, M. D., and Whitmore, T. C. (1988). On the definition of ecological species groups in tropical rain forests. *Vegetatio* **75,** 81–86.

Swaine, M. D., Lieberman, D., and Putz, F. E. (1987). The dynamics of tree populations in tropical forest: A review. *J. Trop. Ecol.* **3,** 359–366.

Swank, W. T., and Douglass, J. E. (1974). Streamflow greatly reduced by converting deciduous hardwood stands to pine. *Science* **185,** 857–859.

Swanson, F. J. (1981). Fire and geomorphic processes. *Gen. Tech. Rep. WO (U.S., For. Serv.)* **WO-26,** 421–444.

Sweeney, J. R., and Biswell, H. H. (1961). Quantitative studies on the removal of litter and duff by fire under controlled conditions. *Ecology* **42,** 572–575.

Sweet, G. B. (1975). Flowering and seed production. *In* "Seed Orchards" (R. Faulkner, ed.), pp. 72–82. *British For. Comm. Bull.* **54,** London.

Swietlik, D., and Faust, M. (1984). Foliar nutrition of fruit crops. *Hort. Rev.* **6,** 287–355.

Switzer, G. L., and Nelson, L. E. (1972). Nutrient accumulation and cycling in loblolly pine (*Pinus taeda* L.) plantation ecosystems: The first twenty years. *Soil Sci. Soc. Am. Proc.* **36,** 143–147.

Switzer, G. L., Nelson, L. E., and Smith, W. H. (1968). The mineral cycle in forest stands. *In*"Forest Fertilization: Theory and Practice," pp. 1–9. Tennessee Valley Authority, Muscle Shoals, Alabama.

Syvertsen, J. P. (1984). Light acclimation in citrus leaves. II. CO_2 assimilation and light, water, and nitrogen use efficiency. *J. Amer. Soc. Hort. Sci.* **109,** 812–817.

Tadaki, Y., Sato, A., Sakurai, S., Takeuchi, I., and Kawahara, T. (1977). Studies on the production structure of forest. XVII. Structure and primary production in subalpine "dead tree strips" *Abies* forest near Mt. A. Saki. *Jpn. J. Ecol.* **27,** 83–90.

Tai, E. A. (1977). Banana. *In* "Ecophysiology of Tropical Crops"(P. de T. Alvim and T. T. Kozlowski, eds.), pp. 441–460. Academic Press, New York.

Talboys, P. W. (1968). Water deficits in vascular disease. *In* "Water Deficits and Plant Growth" (T. T. Kozlowski, ed.), Vol. 2, pp. 255–311. Academic Press, New York.

Tamari, C., and Jacalne, D. V. (1984). Fruit dispersal of dipterocarps. *Bull. For. For. Prod. Res. Inst. (Jpn.)* **325,** 127–140.

Tamm, C. O. (1968). The evolution of forest fertilization in European silviculture. *In* "Forest Fertilization: Theory and Practice," pp. 242–247. Tennessee Valley Authority, Muscle Shoals, Alabama.

Tang, Z. C., and Kozlowski, T. T. (1982a). Some physiological and morphological responses of *Quercus macrocarpa* seedlings to flooding. *Can. J. For. Res.* **10,** 308–311.

Tang, Z. C., and Kozlowski, T. T. (1982b). Physiological, morphological, and growth responses of *Platanus occidentalis* seedlings to flooding. *Plant Soil* **66,** 243–255.

Tang, Z. C., and Kozlowski, T. T. (1982c). Some physiological and growth responses of *Betula papyrifera* seedlings to flooding. *Physiol. Plant.* **55,** 415–420.

Tang, Z. C., and Kozlowski, T. T. (1983). Responses of *Pinus banksiana* and *Pinus resinosa* seedlings to flooding. *Can. J. For. Res.* **13,** 633–639.

Tang, Z. C., and Kozlowski, T. T. (1984). Ethylene production and morphological adaptations of woody plants to flooding. *Can J. Bot.* **62,** 1659–1664.

Tani, N. (1967). On the prevention measure of the damage from typhoon in South Kyushu. *Bull. Kyushu Agric. Exp. Stn.* **12**(3/4), 343–387.

Tans, P. P., Fung, I. Y., and Takahashi, T. (1990). Observational constraints on the global atmospheric CO_2 budget. *Science* **247,** 1431–1438.

Tappeiner, J. C., II, and Helms, J. A. (1971). Natural regeneration of Douglas-fir and white fir on exposed sites in the Sierra Nevada of California. *Am. Midl. Nat.* **86,** 358–370.

Tarrant, R. F. (1949). Douglas-fir site quality and soil fertility. *J. For.* **47,** 716–720.

Tattar, T. A. (1978). "Diseases of Shade Trees." Academic Press, New York.

Taylor, A. H., and Zisheng, Quin. (1988). Tree replacement patterns in subalpine *Abies—Betula* forests, Wolong Natural Reserve, China. *Vegetatio* **78,** 141–149.

Taylor, B. K., and May, L. H. (1967). The nitrogen nutrition of the peach tree. II. Storage and mobilization of nitrogen in young trees. *Aust. J. Biol. Sci.* **20,** 389–411.

Taylor, G., and Davies, W. J. (1986). Yield turgor of growing leaves of *Betula* and *Acer*. *New Phytol.* **104,** 347–353.

Taylor, G., and Davies, W. J. (1988). The influence of photosynthetically active radiation and simulated shadelight on the control of leaf growth of *Betula* and *Acer*. *New Phytol.* **108,** 393–398.

Taylor, G. E., Jr., McLaughlin, S. B., and Shriner, D. S. (1982). Effective pollutant dose. *In* "Effects of Gaseous Air Pollution in Agriculture and Horticulture" (M. H. Unsworth and D. P. Ormrod, eds.), pp. 458–460. Butterworth, London.

Taylor, H. M., and Ratliff, L. F. (1969). Root elongation rates of cotton and peanuts as a function of soil strength and soil water content. *Soil Sci.* **108,** 113–119.

Taylor, J. S., and Dumbroff, E. B. (1975). Bud, root, and growth-regulator activity in *Acer saccharum* during the dormant season. *Can. J. Bot.* **53,** 321–331.

Tazaki, T., Ishihara, K., and Usijima, T. (1980). Influence of water stress on the photosynthesis and productivity of plants in humid areas. *In* "Adaptation of Plants to Water and High Temperature Stress" (N. C. Turner and P. J. Kramer, eds.), pp. 309–321. Wiley, New York.

Teigen, O. (1975). Spire-og etableringsforsok med gran og furu i kunstig forsuret mineraljord. *Intern. Rep.—SNSF Proj.* No. 10/75.

Telewski, F. W., and Jaffe, M. J. (1981). Thigmomorphogenesis: Changes in the morphology and chemical composition induced by mechanical perturbation in 6-month-old *Pinus taeda* seedlings. *Can. J. For. Res.* **11,** 380–387.

Telewski, F. W., and Jaffe, M. J. (1986). Thigmomorphogenesis: Field and laboratory studies of *Abies fraseri* in response to wind or mechanical perturbation. *Physiol. Plant.* **66,** 211–218.

Templin, E. (1962). Zur Populationsdynamik einiger Kiefernschadinsekten in rauchgeschädigten Beständen. *Wiss. Z. Tech. Univ., Dresden* **11,** 631–637.

Teoh, T. S., Aylmore, L. A. G., and Quirk, J. P. (1967). Retention of water by plant cell walls and implications for drought resistance. *Aust. J. Biol. Sci.* **20,** 41–50.

Teramura, A. H. (1983). Effects of ultraviolet-B radiation on the growth and yield of crop plants. *Physiol. Plant.* **58,** 415–427.

Terashima, I., Wang, S. C., Osmond, C. B., and Farquhar, G. D. (1988). Characterization of nonuniform photosynthesis induced by abscisic acid in leaves having different mesophyll anatomies. *Plant Cell Physiol.* **29,** 385–394.

Terry, T. A., and Hughes, J. H. (1975). The effects of intensive management on planted loblolly pine (*Pinus taeda* L.) growth on poorly drained soils of the Atlantic coastal plain. *In* "Forest Soils and Forest Land Management" (B. Bernier and C. H. Winget, eds.), pp. 351–377. Laval Univ. Press, Quebec.

Teskey, B. J. E., and Shoemaker, J. S. (1978). "Tree Fruit Production." Avi Publ. Co., Westport, Connecticut.

Teskey, R. O., and Hinckley, T. M. (1981). Influence of temperature and water potential on root growth of white oak. *Physiol. Plant.* **52,** 363–369.

Teskey, R. O., and Shrestha, R. B. (1985). A relationship between carbon dioxide, photosynthetic efficiency and shade tolerance. *Physiol. Plant.* **63,** 126–132.

Teskey, R. O., Hinckley, T. M., and Grier, C. C. (1984). Temperature induced changes in the wter relations of *Abies amabilis* (Dougl.) Forbes. *Plant Physiol.* **74,** 77–80.

Teskey, R. O., Grier, C. C., and Hinckley, T. M. (1985). Relationship between root system size and water inflow capacity of *Abies amabilis* growing in a subalpine forest. *Can. J. For. Res.* **15,** 669–672.

Teskey, R. O., Fites, J. A., Samuelson, L. J., and Bongarten, B. C. (1986). Stomatal and nonstomatal limitations to net photosynthesis in *Pinus taeda* L. under different environmental conditions. *Tree Physiol.* **2,** 131–142.

Thielges, B. A. (1972). Intraspecific variation in foliage polyphenols of *Pinus* (subsection Sylvestres). *Silvae Genet.* **21,** 114–119.

Thimann, K. V. (1980). "Senescence in Plants." CRC Press, Boca Raton, FL.

Thompson, C. R., and Taylor, O. C. (1969). Effects of air pollutants on growth, leaf drop, fruit drop, and yield of citrus trees. *Environ. Sci. Technol.* **3,** 934–940.

Thompson, M. A. (1981). Tree rings and air pollution. A case study of *Pinus monophylla* growing in east-central Nevada. *Environ. Pollut., Ser. A* **26,** 251–266.

Thomson, A. J., and Moncrief, S. M. (1982). Prediction of bud burst in Douglas-fir by degree-day accumulation. *Can. J. For. Res.* **12,** 448–452.

Thor, E., and Barnett, P. E. (1973). Taxonomy of *Abies* in the southern Appalachians: Variations in balsam monoterpenes and wood properties. *For. Sci.* **20,** 32–40.

Thornthwaite, C. W. (1948). An approach toward a national classification of climate. *Geogr. Rev.* **38,** 55–94.

Thornthwaite, C. W., and Mather, J. R. (1957). Instructions and tables for computing potential evapotranspiration and the water balance. *Drexel Inst. Technol. Lab. Climatol., Publ. Climatol.* **10,** 181–311.

Thorud, D. B., and Frisell, S. S. (1979). Time changes in soil density following compaction under an oak forest. *Minn. For. Res. Notes* **257.**

Tibbits, T. W. (1979). Humidity and plants. *BioScience* **29,** 358–368.

Tiedemann, A. R. (1987). Combustion losses of sulfur from forest foliage and litter. *For. Sci.* **33,** 216–223.

Timell, T. E. (1986). "Compression Wood," Vols. 1–3. Springer-Verlag, Berlin and New York.

Timmer, V. R., and Armstrong, G. (1987). Growth and nutrition of containerized *Pinus resinosa* at exponentially increasing nutrient additions. *Can. J. For. Res.* **17,** 644–647.

Ting, I. P. (1982). "Plant Physiology." Addison-Wesley, Reading, Massachusetts.

Ting, I. P. (1985). Crassulacean acid metabolism. *Annu. Rev. Plant Physiol.* **36,** 595–622.

Ting, I. P., and Dugger, W. M., Jr. (1971). Ozone resistance in tobacco plants: Possible relationship to water balance. *Atmos. Environ.* **5,** 147–150.

Ting, I. P., and Mukerji, S. K. (1971). Leaf ontogeny as a factor in susceptibility to ozone: Amino acid and carbohydrate changes during expansion. *Am. J. Bot.* **58,** 497–504.

Tingey, D. T., Wilhour, R. G., and Standley, C. (1976). The effect of chronic ozone exposures on the metabolite content of ponderosa pine seedlings. *For. Sci.* **22,** 234–241.

Tinker, P. B. (1976). Roots and water. Transport of water to plant roots in soil. *Philos. Trans. R. Soc. London, Ser. B* **273,** 445–461.

Tinker, B., and Läuchli, A. (1986). "Advances in Plant Nutrition," Vol. 2. Praeger, New York.

Tinus, R. W. (1972). CO_2 enriched atmosphere speeds growth of ponderosa pine and blue spruce seedlings. *Tree Plant. Notes* **23,** 12–18.

Todd, G. W., Chadwick, D. L., and Tosi, S.-D. (1972). Effect of wind on plant respiration. *Physiol. Plant.* **27,** 342–346.

Tolley, L. C. (1982). The effects of atmospheric carbon dioxide enrichment, irradiance and water stress on seedling growth and physiology of *Liquidambar styraciflua* and *Pinus taeda*. Ph.D. Dissertation, Duke Univerisity, Durham, North Carolina.

Tolley, L. C., and Strain, B. R. (1984a). Effects of CO_2 enrichment on growth of *Liquidambar styraciflua* and *Pinus taeda* seedlings under different irradiance levels. *Can. J. For. Res.* **14,** 343–354.

Tolley, L. C., and Strain, B. R. (1984b). Effects of atmospheric CO_2 enrichment and water stress on growth of *Liquidambar styraciflua* L. and *Pinus taeda* L. seedlings. *Can. J. Bot.* **62,** 2135–2139.

Tomlinson, P. B. (1986). "The Botany of Mangroves." Cambridge Univ. Press, Cambridge.

Toole, V. K., Toole, E. H., Hendricks, S. B., Borthwick, H. S., and Snow, A. B., Jr. (1961). Responses of seeds of *Pinus virginiana* to light. *Plant Physiol.* **36,** 285–290.

Topa, M. A., and McLeod, K. W. (1986a). Responses of *Pinus clausa*, *Pinus serotina* and *Pinus taeda* seedlings to anaerobic solution culture. II. Changes in tissue nutrient concentration and net acquisition. *Physiol. Plant.* **68,** 532–539.

Topa, M. A., and McLeod, K. W. (1986b). Aerenchyma and lenticel formation in pine seedlings: A possible avoidance mechanism to anaerobic growth conditions. *Physiol. Plant.* **68,** 540–550.

Torres, F. (1983). Role of woody perennials in animal agroforestry. *Agrofor. Syst.* **1,** 131–163.

Torrey, J. G., and Clarkson, D. T., eds. (1975). "The Development and Function of Roots." Academic Press, New York.

Toumey, J. W. (1929). Initial root habit in American trees and its bearing on regeneration. *Proc. Int. Bot. Congr., 4th, 1926,* Vol. 1, pp. 713–728.

Toumey, J. W., and Kienholz, R. (1931). Trenched plots under forest canopies. *Bull.—Yale Univ., Sch. For.* **30.**

Tranquillini, W. (1955). Die Bedeutung des Lichtes und der Temperatur für die Kohlensäureassimilation von *Pinus cembra* Jungwachs an einem hochalpinen Standort. *Planta* **46,** 154–178.

Tranquillini, W. (1962). Beitrag zur Kausalanalyse des Wettbewerbs ökologisch verschiedener Holzarten. *Ber. Dtsch. Bot. Ges.* **75,** 353–364.

Tranquillini, W. (1965). Über den Zusammenhang zwischen Entwicklungzustand und Dürreresistenz junger Zirben (*Pinus cembra* L.) im Pflanzengarten. *Mitt. Forstl. Bundesversuchsanst., Mariabrunn.* **66,** 241–271.

Tranquillini, W. (1969). Photosynthese und Transpiration einiger Holzarten bei verschieden starkem Wind. *Centralbl. Gesamte Forstwes.* **86,** 35–48.

Tranquillini, W. (1973). Der Wasserhaushalt junger Forstpflanzen nach dem Versetzen und seine Beeinflussbarkeit. *Centralbl. Gesamte Forstwes.* **90,** 46–52.

Tranquillini, W. (1979). "Physiological Ecology of the Alpine Timberline." Springer-Verlag, New York.

Tranquillini, W., and Unterholzner, R. (1968). Das Wachstum zweijärhriger Lärchen einheitlicher Herkunft in verschiedener Seehöhe. *Centralbl. Gesamte Forstwes.* **85,** 43–49.

Transeau, E. N. (1905). Forest centers of eastern North America. *Am. Nat.* **39,** 875–889.

Trapp, E. (1938). Untersuchung über die Verteilung der Helligkeit in einem Buchenbestand. *Bioklimatologie, Ser. B* **5,** 153–158.

Trappe, J. M. (1983). Effects of herbicides bifenox, DCPA, and napropamide on mycorrhiza development of ponderosa pine and Douglas-fir seedlings in six western nurseries. *For. Sci.* **29,** 464–468.

Treshow, M. (1975). Interaction of air pollutants and plant diseases. *In* "Responses of Plants to Air Pollution" (J. B. Mudd and T. T. Kozlowski, eds.), pp. 307–334. Academic Press, New York.

Treshow, M. (1980). Pollution effects on plant distribution. *Environ. Conserv.* **7,** 279–284.

Treshow, M., and Anderson, F. K. (1982). Ecological assessment of potential fluoride effects on plants. *In* "Fluoride Emissions: Their Monitoring and Effects on Vegetation and Ecosystems" (F. Murray, ed.), pp. 177–189. Academic Press, Sydney.

Treshow, M., and Pack, M. R. (1970). Fluoride. *In* "Recognition of Air Pollution Injury to Vegetation. A Pictorial Atlas" (J. S. Jacobson and A. C. Hill, eds.), pp. D1-D7. Air Pollut. Control Assoc., Pittsburgh, Pennsylvania.

Trewavas, A. (1981). How do plant growth substances work? *Plant, Cell Environ.* **4,** 203–228.

Trewavas, A. (1986). Understanding the control of plant development and the role of growth substances. *Aust. J. Plant Physiol.* **13,** 447–457.

Trimble, G. R., Jr., and Seegrist, D. W. (1973). Epicormic branching on hardwood trees bordering forest openings. *USDA For. Serv. Res. Pap. NE* **NE–261.**

Trimble, G. R., Jr., Reinhart, K. G., and Webster, H. H. (1963). Cutting the forest to increase water yields. *J. For.* **61,** 635–640.

Tripepi, R. P., and Mitchell, C. A. (1984). Stem hypoxia and root respiration of flooded maple and birch seedlings. *Physiol. Plant.* **60,** 567–571.

Trochoulias, T., and Lahav, E. (1983). The effect of temperature on growth and dry matter production of macadamia. *Sci. Hortic.* **19,** 167–176.

Tromp, J. (1970). Storage and mobilization of nitrogenous compounds in apple trees with special reference to arginine. *In* "Physiology of Tree Crops" (L. C. Luckwill and C. V. Cutting, eds.), pp. 143–159. Academic Press, London.

Trought, M. C. T., and Drew, M. C. (1980). The development of waterlogging damage in wheat seedlings (*Triticum aestivum* L.). I. Shoot and root growth in relation to changes in the concentrations of dissolved gases and solutes in the soil solution. *Plant Soil* **54,** 77–94.

Trousdell, K. B., and Hoover, M. D. (1955). A change in groundwater after clearcutting of loblolly pine in the Coastal Plain. *J. For.* **53,** 493-498.

Tsukahara, H., and Kozlowski, T. T. (1984). Effect of flooding on growth of *Larix leptolepis* seedlings. *J. Jpn. For. Soc.* **66,** 33–66.

Tsukahara, H., and Kozlowski, T. T. (1985). Importance of adventitious roots to growth of flooded *Platanus occidentalis* seedlings. *Plant Soil* **88,** 123–132.

Tsukahara, H., Kozlowski, T. T., and Shanklin, J. (1985). Tolerance of *Pinus densiflora*, *Pinus thunbergii*, and *Larix leptolepis* seedlings to SO_2. *Plant Soil* **88,** 385–397.

Tsukahara, H., Kozlowski, T. T., and Shanklin, J. (1986). Effects of SO_2 on two age classes of *Chamaecyparis obtusa* seedlings. *J. Jpn. For. Soc.* **68,** 349–353.

Tsukahara, H., Kozlowski, T. T., and Shanklin, J. (1987). Responses of *Betula platyphylla* var. *japonica* seedlings to SO_2. *J. Yamagata Agric. For. Soc.* **44,** 5–12.

Tukey, H. B., Jr. (1970). The leaching of substances from plants. *Annu. Rev. Plant Pysiol.* **21,** 305–324.

Tukey, L. D. (1983). Vegetative control and fruiting on mature apple trees treated with PP–333. *Acta Hortic.* **137,** 103–109.

Tukey, L. D. (1986). Cropping characteristics of bearing apple trees annually sprayed with paclobutrazol (PP–333). *Acta Hortic.* **179,** 481–488.

Tukey, L. D. (1989). Uniconazole—A new triazole growth regulant for apple. *Acta Hortic.* (in press).

Tumanov, I. I. (1967). The frost hardening process in plants. *In* "The Cell and Environmental Temperature" (A. S. Troshin, ed.), pp. 6–14. Pergamon, Oxford.

Tuomi, J., Niemelä, P., and Mannila, R. (1982). Resource allocation on dwarf shoots of birch (*Betula pendula*): Reproduction and leaf growth. *New Phytol.* **91,** 483–487.

Turgeon, R. (1989). The source–sink transition in leaves. *Annu. Rev. Plant Physiol. Mol. Biol.* **40,** 119–138.

Turner, N. C. (1986). Crop water deficits: A decade of progress. *Adv. Agron.* **39,** 1–51.

Turner, N. C., and Begg, J. E. (1978). Reponses of pasture plants to water deficits. *In* "Plant Relations in Pastures" (J. R. Wilson, ed.), pp. 50–66. CSIRO, Melbourne.

Turner, N. C., and Begg, J. E. (1981). Plant-water relations and adaptations to stress. *Plant Soil* **58,** 97–131.

Turner, N. C., and Jones, M. M. (1980). Turgor maintenance by osmotic adjustment: A review and evaluation: *In* "Adaptation of Plants to Water and High Temperature Stress" (N. C. Turner and P. J. Kramer, eds.), pp. 87–103. Wiley, New York.

Tydeman, R. M. (1964). The relation between time of leaf break and flowering in seedling apples. *Annu. Rep. East Malling Res. Stn.*, Kent, pp. 70–72.

Tyree, M. T., and Dixon, M. A. (1986). Water stress induced cavitation and embolism in some woody plants. *Physiol. Plant.* **66,** 397–405.

Tyree, M. T., and Jarvis, P. G. (1982). Water in tissues and cells. *Encycl. Plant Physiol., New Ser.* **12B,** 35–77.

Tyree, M. T., and Sperry, J. S. (1988). Do woody plants operate near the point of catastrophic xylem-dysfunction caused by dynamic water stress? Answers from a model. *Plant Physiol.* **88,** 574–580.

Tyree, M. T., and Sperry, J. S. (1989). Vulnerability of xylem to cavitation and embolism. *Annu. Rev. Plant Physiol. Mol. Biol.* **40,** 19–38.

Tyree, M. T., Dixon, M. A., Tyree, E. L., and Johnson, R. (1984). Ultrasonic acoustic emissions from the sapwood of cedar and hemlock: An examination of three hypotheses concerning cavitations. *Plant Physiol.* **75,** 988–992.

Uriu, K. (1964). Effect of post-harvest soil moisture depletion on subsequent yield of apricots. *Proc. Am. Soc. Hortic. Sci.* **84,** 93–97.

Uriu, K., and Magness, J. R. (1967). Deciduous tree fruits and nuts. *In* "Irrigation of Agricultural Lands" (R. M. Hagan, H. R. Haise, and T. W. Edminster, eds.), Monogr. 2, pp. 686–703. Am. Soc. Agron., Madison, Wisconsin.

Uriu, K., Davenport, D., and Hagan, R. M. (1975). Preharvest antitranspirant spray on cherries. I. Effect on fruit size. II. Postharvest fruit benefits. *Calif. Agric.* **29**(10), 7–11.

USDA Forest Service (1974). Woody plant seed manual. *U.S., Dep. Agric., Agric. Handb.* **450.**

Usher, R. W., and Williams, W. T. (1982). Air pollution toxicity to eastern white pine in Indiana and Wisconsin. *Plant Dis.* **66,** 199–204.

Vaartaja, O. (1959). Evidence of photoperiodic ecotypes in trees. *Ecol. Monogr.* **29,** 91–111.

Valoras, N., Letey, J., Stolzy, L. H., and Frolich, E. F. (1964). The oxygen requirements for root growth of three avocado varieties. *Proc. Am. Soc. Hortic. Sci.* **85,** 172–178.

Van Arsdel, E. P. (1965). Micrometeorology and plant disease epidemiology. *Phytopathology* **55,** 945–950.

van Bavel, C. H. M., and Verlinden, F. J. (1956). Agricultural drought in North Carolina. *N.C., Agric. Exp. Stn. Tech. Bull.* **122.**

van Buijtenen, J. P., Bilan, M. V., and Zimmerman, R. H. (1976). Morpho-physiological characteristics related to drought resistance in *Pinus taeda*. *In* "Tree Physiology and Yield Improvement" (M. G. R. Cannell and F. T. Last, eds.), pp. 348–359. Academic Press, London.

van den Driessche, R. (1984). Relationship between spacing and nitrogen fertilization of seedlings in the nursery, seedling mineral nutrition, and outplanting performance. *Can. J. For. Res.* **14,** 431–436.

van den Driessche, R. (1988). Nursery growth of conifer seedlings using fertilizers of different solubilities and application time, and their forest growth. *Can. J. For. Res.* **18,** 172–180.

van den Driessche, R,. Connor, D. J., and Tunstall, B. R. (1971). Photosynthetic response of brigalow to irradiance, temperature, and water potential. *Photosynthetica* **5,** 210–217.

Van Dorsser, J. C., and Rook, D. A. (1972). Conditioning of radiata pine seedlings by undercutting and wrenching: Description of methods, equipment, and seedling response. *N. Z. J. For.* **17,** 61–73.

Van Ryn, D. M., Jacobson, J. S., and Lassoie, J. P. (1986). Effects of acidity on *in vitro* pollen germination and tube elongation in four hardwood species. *Can. J. For. Res.* **16,** 397–400.

Van Ryn, D. M., Lassoie, J. P., and Jacobson, J. S. (1988). Effects of acid mist on *in vivo* pollen tube growth in red maple. *Can. J. For. Res.* **18,** 1049–1052.

Van Volkenburgh, E., and Davies, W. J. (1977). Leaf anatomy and water relations of plants grown in controlled environments and in the field. *Crop Sci.* **16,** 353–358.

Vartanian, N., and Berkaloff, A. (1989). Drought adaptability of *Agrobacterium rhizogenes*-induced roots in oilseed rape (*Brassica napus*, var. *oleifera*). *Plant, Cell Environ.* **12,** 197–204.

Vazquez-Yanes, C., and Segovia, A. O. (1984). Ecophysiology of seed germination in the tropical humid forests of the world: A review. *In* "Physiological Ecology of Plants of the Wet Tropics" (E. Medina, H. A. Mooney, and C. Vazquez-Yanes, eds.), pp. 37–50. Junk, The Hague.

Vazquez-Yanes, C., and Smith, H. (1982). Phytochrome control of seed germination in the tropical rain forest pioneer trees (*Cecropia obtusifolia* and *Piper auritum*) and its ecological significance. *New Phytol.* **92,** 477–485.

Veblen, T. T. (1985). Forest development in tree-fall gaps in the temperate rain forests of Chile. *Natl. Geogr. Res.* **1,** 161–184.

Vergara, N. T., and Nair, P. K. R. (1985). Agroforestry in the South Pacific region—An overview. *Agrofor. Syst.* **3,** 363–379.

Vertucci, C. W., and Leopold, A. C. (1984). Bound water in soybean seed and its relation to respiration and imbibitional damage. *Plant Physiol.* **75,** 114–117.

Vertucci, C. W., and Leopold, A. C. (1987a). Oxidative processes in soybean and pea seeds. *Plant Physiol.* **84,** 1038–1043.

Vertucci, C. W., and Leopold, A. C. (1987b). Water binding in legume seeds. *Plant Physiol.* **85,** 224–231.

Vertucci, C. W., and Leopold, A. C. (1987c). The relationship between water binding and desiccation tolerance in tissues. *Plant Physiol.* **85,** 232–238.

Vierstra, R. D., and Quail, P. H. (1983). Purification and initial characterization of 124 kilodalton phytochrome. *Biochemistry* **22,** 2498–2505.

Villiers, T. A. (1972). Seed dormancy. *In* "Seed Biology" (T. T. Kozlowski, ed.), pp. 219–281. Academic Press, New York.

Vince-Prue, D. (1975). "Photoperiodism and Plants." McGraw-Hill, Maidenhead, England.

Virgin, H. I. (1965). Chlorophyll formation and water deficit. *Physiol. Plant* **18,** 994–1000.

Viro, P. J. (1974). Effects of forest fire on soil. *In* "Fire and Ecosystems" (T. T. Kozlowski and C. E. Ahlgren, eds.), pp. 7–45. Academic Press, New York.

Vité, J. P. (1961). The influence of water supply on oleoresin exudation pressure and resistance to bark beetle attack in *Pinus ponderosa*. *Contrib. Boyce Thompson Inst.* **21,** 37–66.

Vitousek, P. (1984). Litterfall, nutrient cycling, and nutrient limitations in tropical forests. *Ecology* **65,** 285–298.

Vogt, K. A., Edmonds, R. L., and Grier, C. C. (1981). Seasonal changes in biomass and vertical distribution of mycorrhizal and fibrous-textured conifer fine roots in 23- and 180-year-old subalpine *Abies amabilis* stands. *Can. J. For. Res.* **11,** 223–229.

Vogt, K. A., Grier, C. C., and Vogt, D. J. (1986). Production, turnover, and nutrient dynamics of above- and below-ground detritus of world forests. *Adv. Ecol. Res.* **15,** 303–377.

von Carlowitz, P. G. (1985). Some considerations regarding principles and practice of information collection on multipurpose trees. *Agrofor. Syst.* **3,** 181–195.

von Maydell, H. J. (1985). The contribution of agroforestry to world forestry development. *Agrofor. Syst.* **3,** 83–90.

Vose, J. M. (1988). Patterns of leaf area distribution within crowns of nitrogen- and phosphorus-fertilized loblolly pine trees. *For. Sci.* **34,** 564–573.

Vose, J. M., and Allen, H. L. (1988). Leaf area, stemwood growth and nutrition relationships in loblolly pine. *For. Sci.* **34,** 547–563.

Vowinckel, T., Oechel, W. C., and Boll, W. G. (1975). The effect of climate on the photosynthesis of *Picea mariana* at the subarctic treeline. I. Field measurements. *Can. J. Bot.* **53,** 604–620.

Wadleigh, C. H. (1946). The integrated soil moisture stress upon a root system in a large container of saline soil. *Soil Sci.* **61,** 225–238.

Wadleigh, C. H., Gauch, H. G., and Magistad, O. C. (1946). Growth and rubber accumulation in guayule as conditioned by soil salinity and irrigation regimes. *U.S., Dep. Agric., Tech. Bull.* **925.**

Wagenbreth, D. (1965). Das Auftreten von zwei Letalstufen bei Hitzeeinwirkung auf Pappelblätter. *Flora (Jena)* **156A,** 116–126.

Waggoner, P. E. (1984). Agriculture and carbon dioxide. *Am. Sci.* **72,** 179–184.

Waggoner, P. E., and Bravdo, B. (1967). Stomata and the hydrologic cycle. *Proc. Natl. Acad. Sci. U.S.A.* **57,** 1096–1102.

Wagle, R. F., and Eakle, T. W. (1979). A controlled burn reduces the impact of a subsequent wildfire in a ponderosa pine vegetation type. *For. Sci.* **25,** 123–129.

Wagner, M. R., and Evans, P. D. (1985). Defoliation increases nutritional quality and allelochemics of pine seedlings. *Oecologia* **67,** 235–237.

Wainwright, S. J. (1984). Adaptations of plants to flooding with salt water. *In* "Flooding and Plant Growth" (T. T. Kozlowski, ed.), pp. 295–343. Academic Press, New York.

Waisel, Y. (1972). "Biology of Halophytes." Academic Press, New York.

Waister, P. D. (1970). Effects of shelter from wind on the growth and yield of raspberries. *J. Hortic. Sci.* **45,** 435–445.

Waister, P. D. (1972a). Wind damage in horticultural crops. *Hortic. Abstr.* **42,** 609–615.

Waister, P. D. (1972b). Wind as a limitation on the growth and yield of strawberries. *J. Hortic. Sci.* **47,** 411–418.

Wakeley, P. C. (1948). Physiological grades of southern pine nursery stock. *Proc. Soc. Am. For.*, pp. 312–322.

Wakeley, P. C. (1954). Planting the southern pines. *U.S. For. Serv. Agric. Monogr.* **18.**

Walker, R. F., West, D. C., McLaughlin, S. B., and Amundsen, C. C. (1989). Growth, xylem pressure potential, and nutrient absorption of loblolly pine on a reclaimed surface mine as affected by an induced *Pisolithus tinctorius* infection. *For. Sci.* **35,** 569–581.

Walker, R. R. (1986). Sodium exclusion and potassium–sodium selectivity in salt-treated trifoliate orange (*Poncirus trifoliata*) and Cleopatra mandarin (*Citrus reticulata*) plants. *Aust. J. Plant Physiol.* **13,** 293–303.

Walter, H. (1973). "Vegetation of the Earth." Springer-Verlag, New York.

Walter, H. (1979). "Vegetation of the Earth," 2nd ed. Springer-Verlag, Berlin and New York.

Walton, D. C. (1980). Biochemistry and physiology of abscisic acid. *Annu. Rev. Plant Physiol.* **7,** 191–214.

Wample, R. L., and Reid, D. M. (1975). Effect of aeration on the flood-induced formation of adventitious roots and other changes in sunflower (*Helianthus annuus* L.). *Planta* **127,** 263–270.

Wample, R. L., and Reid, D. M. (1979). The role of endogenous auxin and ethylene in the formation of adventitious roots and hypocotyl hypertrophy in flooded sunflower plants (*Helianthus annuus*). *Physiol. Plant.* **45,** 219–226.

Wang, C. W., Perry, T. O., and Johnson, A. G. (1960). Pollen dispersion of slash pine with special reference to seed orchard management. *Silvae Genet.* **9,** 65–92.

Wang, C. Y. (1982). Physiological and biochemical responses of plants to chilling stress. *HortScience* **17,** 173–186.

Ward, W. W. (1966). Epicormic branching of black and white oaks. *For. Sci.* **12,** 290–296.

Wardle, J. A., and Allen, R. B. (1983). Dieback in New Zealand *Nothofagus* forests. *Pac. Sci.* **37,** 397–404.

Wardle, P. (1968). Engelmann spruce (*Picea engelmannii* Engel.) at its upper limits on the front range, Colorado. *Ecology* **49,** 483–495.

Wardle, P. (1981). Winter desiccation of conifer needles simulated by artificial freezing. *Arct. Alp. Res.* **13,** 419–423.

Wardrop, A. B. (1964). The reaction anatomy of arborescent angiosperms. *In* "The Formation of Wood in Forest Trees" (M. H. Zimmermann, ed.), pp. 405–456. Academic Press, New York.

Wareing, P. F. (1951). Growth studies in woody species. III. Further photoperiodic effects in *Pinus sylvestris*. *Physiol. Plant.* **4,** 41–56.

Wareing, P. F. (1953). Growth studies in woody species. V. Photoperiodism in dormant buds of *Fagus sylvatica* L. *Physiol. Plant.* **6,** 692–706.

Wareing, P. F. (1954). Growth studies in woody species. VI. The locus of photoperiodic perception in relation to dormancy. *Physiol. Plant.* **7,** 261–277.

Wareing, P. F. (1956). Photoperiodism in woody plants. *Annu. Rev. Plant Physiol.* **7,** 191–214.

Wareing, P. F. (1970). Growth and its coordination in trees. *In* "Physiology of Tree Crops" (L. C. Luckwill and C. V. Cutting, eds.), pp. 1–21. Academic Press, New York.

Wareing, P. F. (1985). Tree growth at cool temperatures and prospects for improvement by breeding. *In* "Attributes of Trees as Crop Plants" (M. G. R. Cannell and J. E. Jackson, eds.), pp. 80–88. Institute of Terrestrial Ecology, Huntingdon, England.

Wareing, P. F., and Longman, K. A. (1960). Studies on the physiology of flowering in forest trees. *G. B. For. Comm. Rep. For. Res., 1959,* pp. 109–110.

Wareing, P. F., Khalifa, M. M., and Treharne, K. J. (1968). Rate-limiting processes in photosynthesis at saturating light intensities. *Nature (London)* **220,** 453–457.

Waring, R. H. (1970). Matching species to site. *In* "Regeneration of Ponderosa Pine" (R. K. Hermann, ed.), pp. 54–61. For. Res. Lab., Oregon State University, Corvallis.

Waring, R. H. (1983). Estimating forest growth and efficiency in relation to canopy leaf area. *Adv. Ecol. Res.* **13,** 327–354.

Waring, R. H. (1987). Characteristics of trees predisposed to die. *BioScience* **37,** 569–574.

Waring, R. H., and Franklin, J. F. (1979). The evergreen coniferous forests of the Pacific Northwest. *Science* **204,** 1248–1254.

Waring, R. H., and Pitman, G. B. (1985). Modifying lodgepole pine stands to change susceptibility to mountain pine beetle attack. *Ecology* **66,** 889–897.

Waring, R. H., and Running, S. W. (1978). Sapwood water storage: Its contribution to transpiration and effect upon water conductance through the stems of old-growth Douglas-fir. *Plant, Cell Environ.* **1,** 131–140.

Waring, R. H., and Schlesinger, W. H. (1985). "Forest Ecosystems: Concepts and Management." Academic Press, Orlando, Florida.

Waring, R. H., Whitehead, D., and Jarvis, P. G. (1979). The contribution of stored water to transpiration in Scots pine. *Can. J. For. Res.* **10,** 555–558.

Waring, R. H., Whitehead, D., and Jarvis, P. G. (1980). Comparison of an isotopic method and the Penman–Monteith equation for estimating transpiration from Scots pine. *Can. J. For. Res.* **10,** 555–558.

Waring, R. H., Newman, K., and Bell, J. (1981). Efficiency of tree crowns and stemwood production at different canopy leaf densities. *Forestry* **54,** 129–137.

Waring, R. H., Aber, J. D., Melillo, J. M., and Moore, B. (1986). Precursors of change in terrestrial ecosystems. *BioScience* **36,** 433–438.

Warren, H. V. (1972). Biogeochemistry in Canada. *Endeavour* **31,** 46–49.

Warren, S. D., Nevill, M. B., Blackburn, W. H., and Garza, N. E. (1986). Soil response to trampling under intensive rotation grazing. *Soil Sci. Soc. Am. J.* **50,** 1336–1341.

Warrington, I. J., Peet, M., Patterson, D. P., Bunce, J., Haslemore, R. M., and Hellmers, H. (1977). Growth and physiological responses of soybean under various thermoperiods. *Aust. J. Plant Physiol.* **4,** 371–380.

Warrington, I. J., Rook, D. A., Morgan, D. C., and Turnbull, H. L. (1988). The influence of simulated shadelight and daylight on growth, development and photosynthesis of *Pinus radiata, Agathis australis* and *Dacrydium cupressinum. Plant, Cell Environ.* **11,** 343–356.

Watson, D. J. (1952). The physiological basis of variation in yield. *Adv. Agron.* **4,** 101–145.

Watson, G. W. (1986). Cultural practices can influence root development for better transplanting success. *J. Environ. Hortic.* **4,** 32–34.

Watson, G. W. (1987). The relationship of root growth and tree vigour following transplanting. *Arboric. J.* **11,** 97–104.

Watson, G. W. (1988). Organic mulch and grass competition influence tree root development. *J. Arboric.* **14,** 200–203.

Watson, G. W., and Sydnor, T. D. (1987). The effect of root pruning on the root system of nursery trees. *J. Arboric.* **13,** 126–130.

Weaver, H. (1974). Effects of fire on temperate forests: Western United States. *In* "Fire and Ecosystems" (T. T. Kozlowski and C. E. Ahlgren, eds.), pp. 279–319. Academic Press, New York.

Wedding, R. T., Riehl, L. A., and Rhoads, W. A. (1952). Effect of petroleum oil sprays on photosynthesis and respiration in citrus leaves. *Plant Physiol.* **27,** 269–278.

Weetman, G. F., Yang, R. C., and Bella, I. E. (1985). Nutrition and fertilization of lodgepole pine. *In* "Lodgepole Pine: The Species and Its Management" (D. M. Baumgartner, R. G. Krebill, J. T. Arnott, and G. F. Weetman, eds.), pp. 225–232. Washington State University, Pullman.

Weiner, J., and Thomas, S. C. (1986). Size variability and competition in plant monocultures. *Oikos* **47,** 211–222.

Weinstein, L. H. (1977). Fluoride and plant life. *J. Occup. Med.* **19,** 49–78.

Weinstein, L. H., and Alscher-Herman, R. (1982). Physiological responses of plants to fluoride. *In* "Effects of Gaseous Air Pollution on Agriculture and Horticulture" (M. H. Unsworth and D. P. Ormrod, eds.), pp. 139–167. Butterworth, London.

Wellburn, A. R. (1988). "Air Pollution and Acid Rain: The Biological Impact." Longman Scientific, New York.

Weller, D. E. (1987). A reevaluation of the −3/2 power rule of self thinning. *Ecol. Monogr.* **57,** 23–43.

Wellner, C. A. (1948). Light intensity related to stand density in mature stands of the western white pine type. *J. For.* **46,** 16–19.

Wells, C. G. (1968). Techniques and standards for foliar diagnosis of N deficiency in loblolly pine. *In* "Forest Fertilization: Theory and Practice," pp. 72–76. Tennessee Valley Authority, Muscle Shoals, Alabama.

Wells, C. G., Jorgensen, J. R., and Burnette, C. E. (1975). Biomass and mineral elements in a thinned loblolly pine plantation at age 16. *USDA For. Serv. Res. Pap. SE* **SE–126.**

Wenger, K. F. (1953). The effect of fertilization and injury on the cone and seed production of loblolly pine seed trees. *J. For.* **51,** 570–573.

Wenger, K. F. (1954). The stimulation of loblolly pine seed trees by preharvest release. *J. For.* **52,** 115–118.

Wenger, K. F. (1955). Light and mycorrhiza development. *Ecology* **36,** 518–520.

Wenger, K. F. (1957). Annual variation in the seed crops of loblolly pine. *J. For.* **55,** 567–569.

Wenkert, W., Lemon, E. R., and Sinclair, T. R. (1978). Leaf elongation and turgor pressure in field-grown soybean. *Agron. J.* **70,** 761–764.

Went, F. W. (1953). The effect of temperature on plant growth. *Annu. Rev. Plant Physiol.* **4,** 347–362.

Went, F. W., and Stark, N. (1968). The biological and mechanical role of soil fungi. *Proc. Natl. Acad. Sci. U.S.A.* **60,** 497–504.

Went, F. W., Juhren, G., and Juhren, M. C. (1952). Fire and biotic factors affecting germination. *Ecology* **33,** 351–364.

Wentzel, K. F., and Ohnesorge, B. (1961). Zum Auftreten von Schadinsekten bei Luftverunreinigung. *Forstarchiv* **32,** 177–186.

Westlake, D. F. (1963). Comparisons of plant productivity. *Biol. Rev. Cambridge Philos. Soc.* **38,** 385–425.

Westman, L. (1974). Air pollution indications and growth of spruce and pine near a sulfite plant. *Ambio* **3,** 189–193.

Westman, W. E., O'Leary, J. F., and Melanson, G. P. (1981). The effects of fire intensity, aspect, and substrate on post-fire growth of California coastal sage scrub. *In* "Components of Productivity of Mediterranean-Climate Regions—Basic and Applied" (N. S. Margaris and H. A. Mooney, eds.), pp. 151–179. Junk, The Hague.

Westwood, M. N. (1978). "Temperate-Zone Pomology." Freeman, San Francisco, California.

Westwood, M. N., and Stevens, G. (1979). Factors influencing cherry and prune set. *Proc. Oreg. Hortic. Soc.* **70,** 175–179.

Westwood, M. N., Batjer, L. P., and Billingsley, H. D. (1960). Effects of several organic spray materials on fruit growth and foliage efficiency of apple and pear. *Proc. Am. Soc. Hortic. Sci.* **76,** 59–67.

Westwood, M. N., Roberts A. N., and Bjornstad, H. O. (1976). Influence of 1976 in-row spacing on yield of 'Golden Delicious' and 'Starking Delicious' apple on M9 rootstock in hedgerows. *J. Am. Soc. Hortic. Sci.* **101,** 309–311.

Whitcomb, C. E. (1981). Response of woody landscape plants to bermudagrass competition and fertility. *J. Arboric.* **7,** 191–194.

White, C. S. (1986). Effects of prescribed fire on rates of decomposition and nitrogen mineralization in a ponderosa pine ecosystem. *Biol. Fertil. Soils* **2,** 87–95.

White, J. (1979). The plant as a metapopulation. *Annu. Rev. Ecol. Syst.* **10,** 109–145.

Whitehead, D. R. (1984). Wind pollination: Some ecological and evolutionary perspectives. *In* "Pollination Biology" (L. Real, ed.), pp. 97–108. Academic Press, Orlando, Florida.

Whitford, L. A. (1956). A theory on the formation of cypress knees. *J. Elisha Mitchell Sci. Soc.* **72,** 80–83.

Whitmore, F. W., and Zahner, R. (1967). Evidence for a direct effect of water stress in the metabolism of cell walls in *Pinus. For Sci.* **13,** 397–400.

Whitmore, T. C. (1983). Secondary succession from seed in tropical rain forests. *For. Abstr.* **44,** 769–779.

Whitmore, T. C. (1984). "Tropical Rain Forests of the Far East." Oxford Univ. Press (Clarendon), London and New York.

Whitmore, T. C. (1988). The influence of tree population dynamics on forest species composition. *In* "Plant Population Ecology" (D. J. Davy, M. J. Hutchings, and A. R. Watkinson, eds.), pp. 271–291. Blackwell, Oxford.

Whitmore, T. C. (1989). Canopy gaps and the two major groups of forest trees. *Ecology* **70,** 536–538.

Whittaker, R. H. (1970). "Communities and Ecosystems." Macmillan, London.

Whittaker, R. H. (1975). "Communities and Ecosystems." Macmillan, New York.

Whittaker, R. H., and Levin, S. A. (1977). The role of mosaic phenomena in natural communities. *Theor. Popul. Biol.* **12,** 117–139.

Whittaker, R. H., and Marks, P. L. (1975). Methods of assessing terrestrial productivity. *In* "Primary Productivity of the Biosphere" (H. Lieth and R. H. Whittaker, eds.), pp. 55–118. Springer-Verlag, Berlin and New York.

Whittaker, R. H., Likens, G. E., and Lieth, H. (1975). "Primary Productivity of the Biosphere" (H. Lieth and R. Whittaker, eds.), Springer-Verlag, Berlin and New York.

Wiebe, H. H. (1975). Photosynthesis in wood. *Physiol. Plant.* **33,** 245–246.

Wiegand, K. M. (1906). Some studies regarding the biology of buds and twigs in winter. *Bot. Gaz. (Chicago)* **41,** 373–424.

Wiersum, K. F. (1982). Tree gardening and taungya in Java. Examples of agroforestry techniques in the humid tropics. *Agrofor. Syst.* **1,** 53–70.

Wilbur, R. B., and Christensen, N. L. (1983). Effects of fire on nutrient availability in a North Carolina coastal plain pocosin. *Am. Midl. Nat.* **110,** 54–63.

Wild, A. (1989). "Russell's Soil Conditions and Plant Growth." Wiley, New York.

Wilde, S. A. (1954). Mycorrhizal fungi: Their distribution and effect on tree growth. *Soil Sci.* **78,** 23–31.

Wilde, S. A. (1958). "Forest Soils." Ronald Press, New York.

Wilgen, B. W., and von Siegfried, W. R. (1986). Seed dispersal properties of three pine species as a determinant of invasive potential. *S. Afr. J. Bot.* **52,** 546–548.

Wilhite, L. P., and McKee, W. H., Jr. (1985). Site preparation and phosphorus application alter early growth of loblolly pine. *South. J. Appl. For.* **9,** 103–109.

Wilkinson, T. G., and Barnes, R. L. (1973). Effects of ozone on $^{14}CO_2$ fixation patterns in pine. *Can. J. Bot.* **51,** 1573–1578.

Will, G. M. (1985). "Nutrient Deficiencies and Fertilizer use in New Zealand Exotic Forests," FRI Bull. Forest Research Institute, Rota Rua, New Zealand.

Willatt, S. T., and Pullar, D. M. (1983). Changes in soil physical properties under grazed pastures. *Aust. J. Soil Res.* **22,** 343–348.

Williams, G. J., and McMillan, C. (1971). Frost tolerance of *Liquidambar styraciflua* native to the United States. *Can. J. Bot.* **49,** 1551–1558.

Williams, M. W., and Edgerton, L. J. (1983). Vegetative growth control of apple and pear trees with ICI PP333 (paclobutrazol) a chemical analog of Boyleton. *Acta Hortic.* **137,** 111–116.

Williams, W. E., Garbutt, K., Bazzaz, F. A., and Vitousek, P. M. (1986). The response of plants to elevated CO_2. IV. Two deciduous forest tree communities. *Oecologia* **69,** 454–459.

Williamson, G. B. (1975). Pattern and seral composition in an old-growth beech–maple forest. *Ecology* **56,** 727–731.

Williamson, M. J. (1966). Premature abscissions and white oak acorn crops. *For. Sci.* **12,** 19–21.

Wilson, C. C. (1948). Fog and atmospheric carbon dioxide as related to apparent photosynthetic activity of some broadleaf evergreens. *Ecology* **29,** 507–508.

Wilson, J. (1980). Macroscopic features of wind damage to leaves of *Acer pseudoplatanus* L. and its relationship with season, leaf age, and windspeed. *Ann. Bot.* (*London*) [N.S.] **46,** 303–311.

Wilson, L. (1984). Microscopic features of wind damage to leaves of *Acer pseudoplatanus* L. *Ann. Bot.* (*London*) [N.S.] **53,** 73–82.

Wilson, S. A., Rahe, T. A., and Webber, W. B., Jr., eds. (1985). Municipal wastewater sludge as a soil amendment for revegetating landfill cover. *J. Soil Water Conserv.* **40,** 296–299.

Winget, C. H., and Kozlowski, T. T. (1964). Winter shrinkage in stems of forest trees. *J. For.* **62,** 335–337.

Winget, C. H., and Kozlowski, T. T. (1965). Seasonal basal area growth as an expression of competition in northern hardwoods. *Ecology* **46,** 786–793.

Winget, C. H., Kozlowski, T. T., and Kuntz, J. E. (1963). Effects of herbicides on red pine nursery stock. *Weeds* **11,** 87–90.

Winner, W. E., and Bewley, J. D. (1978). Contrasts between bryophyte and vascular plant synecological responses in an SO_2-stressed white spruce association in central Alberta. *Oecologia* **33,** 311–325.

Winner, W. E., Koch, G. W., and Mooney, H. A. (1982). Ecology of SO_2 resistance. IV. Predicting metabolic responses of fumigated trees and shrubs. *Oecologia* **52,** 16–21.

Winner, W. E., Mooney, H. A., and Goldstein, R. A. (1985). Introduction. *In* "Sulfur Dioxide and Vegetation: Physiology, Ecology, and Policy Issues" (W. E. Winner, H. A. Mooney, and R. A. Goldstein, eds.), pp. 1–7. Stanford Univ. Press, Stanford, California.

Winter, K. (1981). C_4 plants of high biomass in arid regions of Asia—Occurrence of C_4 photosynthesis in Chenopodiaceae and Polygonaceae from the Middle East. *Oecologia* **48,** 100–106.

Wise, R. R., and Naylor, A. W. (1987). Chilling-enhanced photooxidation. The peroxidative destruction of lipids during chilling injury to photosynthesis and ultrastructure. *Plant Physiol.* **83,** 272–277.

Wise, R. R., McWilliam, J. R., and Naylor, A. W. (1981). The effect of chilling stress on chloroplast ultrastructure and oxygen exchange in cotton and collard. *Plant Physiol.* **67,** S61.

Witkosky, J. J., Schowalter, T. D., and Hansen, E. M. (1986). The influence of time of precommercial thinning on the colonization of Douglas-fir by three species of root-colonizing insects. *Can. J. For. Res.* **16,** 745–749.

Witt, H. J., Albsin, J. R., and Daniell, J. W. (1989). Economic analyses of space management practices in high-density pecan groves. *J. Am. Soc. Hortic. Sci.* **114,** 61–64.

Wittwer, R. F., and Immel, M. J. (1978). A comparison of five tree species for intensive fiber production. *For. Ecol. Manage.* **1,** 249–254.

Wolk, W. D., and Herner, R. C. (1982). Chilling injury of germinating seeds and seedlings. *HortScience* **17,** 169–173.

Wong, S. C. (1979). Elevated atmospheric partial pressure of CO_2 and plant growth. I. Interactions of nitrogen nutrition and photosynthetic capacity in C_3 and C_4 plants. *Oecologia* **44,** 68–74.

Wong, S. C., Cowan, I. R., and Farquhar, G. D. (1979). Stomatal conductance correlates with photosynthetic capacity. *Nature* (*London*) **282,** 424–426.

Wong, T. L., Harris, R. W., and Fissell, R. E. (1971). Influence of high soil temperatures on five woody-plant species. *J. Am. Soc. Hortic. Sci.* **96,** 80–83.

Wood, C. (1988). Urban wastewater irrigates Florida citrus. *Citrus Ind.* **69,** 14–16.

Wood, M. O. (1938). Seedling reproduction of oak in southern New Jersey. *Ecology* **19,** 276–293.

Wood, T., and Bormann, F. H. (1977). Short-term effects of a simulated acid rain upon the growth and nutrient relations of *Pinus strobus* L. *Water, Air, Soil Pollut.* **7,** 479–488.

Woodman, J. N. (1971). Variation of net photosynthesis within the crown of a large forest-grown conifer. *Photosynthetica* **5,** 50–54.

Woodman, J. N. (1987a). Pollution-induced injury in North American forests: Facts and suspicions. *Tree Physiol.* **3,** 1–15.

Woodman, J. N. (1987b). Potential impact of carbon dioxide-induced climate changes on management of Douglas-fir and western hemlock. *In* "The Greenhouse Effect, Climate Change and U.S. Forests" (W. E. Shands and J. S. Hoffman, eds.), pp. 277–283. Conservation Foundation, Washington, D.C.

Woodman, J. N., and Cowling, E. B. (1987). Airborne chemicals and forest health. *Environ. Sci. Technol.* **21,** 120–126.

Woodmansee, R. G., and Wallach, L. S. (1981). Effects of fire regimes on biogeochemical cycles. *Gen. Tech. Rep. WO (USDA For. Serv.)* **WO-26,** 379–400.

Woodroof, J. G. (1970). "Tree Nuts." Avi Publ. Co., Westport, CT.

Woods, D. B., and Turner, N. C. (1971). Stomatal responses to changing light intensity by four tree species of varying shade tolerance. *New Phytol.* **70,** 77–84.

Woods, F. W. (1957). Factors limiting root penetration in deep sands of the southeastern coastal plain. *Ecology* **38,** 357–359.

Woodwell, G. M., Hobbie, J. E., Houghton, R. A., Melillo, J. M., Moore, B., Peterson, B. J., and Shaver, G. R. (1983). Global deforestation: Contribution to atmospheric carbon dioxide. *Science* **222,** 1081–1086.

Worrall, J., Draper, D. A., and Andersen, S. A. (1985). Periphysis in stagnant lodgepole pine: An hypothesis demolished. *In* "Crop Physiology of Forest Trees" (P. M. A. Tigerstedt, P. Puttonen, and V. Koski, eds.), pp. 65–70. Dept. of Plant Breeding, University of Helsinki, Finland.

Wray, S. M., and Strain, B. R. (1987). Competition in old-field perennials under CO_2 enrichment. *Ecology* **68,** 1116–1120.

Wright, E. (1931). The effect of high temperatures on seed germination. *J. For.* **29,** 679–687.

Wright, H. A., and Bailey, A. W. (1982). "Fire Ecology." Wiley, New York.

Wright, H. A., Churchill, F. M., and Stevens, W. C. (1976). Effects of prescribed burning on sediment, water yield, and water quality from dozed juniper lands in central Texas. *J. Range Manage.* **29,** 294–298.

Wright, H. A., Churchill, F. M., and Stevens, W. C. (1982). Soil loss and runoff on seeded vs. non-seeded watersheds following prescribed burning. *J. Range Manage.* **35,** 382–385.

Wright, J. W. (1952). Pollen dispersion of some forest trees. *USDA For. Serv. Res. Pap. NE* **NE-46.**

Wright, J. W. (1953). Pollen dispersion studies; some practical applications. *J. For.* **51,** 114–118.

Wright, J. W. (1962). Genetics of forest tree improvement. *FAO For. For. Prod. Stud.* **16.**

Wright, J. W. (1976). "Introduction to Forest Genetics." Academic Press, New York.

Wright, L. C., Berryman, A. A., and Gurusiddaiah, S. (1979). Host resistance to the fir engraver beetle, *Scolytus ventralis* (Coleoptera: Scolytidae). 4. Effect of defoliation on wound monoterpene and inner bark carbohydrate concentrations. *Can. Entomol.* **111,** 1255–1262.

Wright, L. C., Berryman, A. A., and Wickman, B. E. (1984). Abundance of the fir engraver, *Scolytus ventralis*, and the Douglas-fir beetle, *Dendroctonus pseudotsugae*, following tree defoliation by the Douglas-fir tussock moth, *Orgyia pseudotsugata*. *Can. Entomol.* **116,** 293–305.

Wright, R. D., and Niemiera, A. X. (1987). Nutrition of container-grown woody nursery crops. *Hortic. Rev.* **9,** 75–101.

Wu, C. C., and Kozlowski, T. T. (1972). Some histological effects of direct contact of *Pinus resinosa* seeds and young seedlings with 2,4,5-T. *Weed Res.* **12,** 229–233.

Wu, C. C., Kozlowski, T. T., Evert, R. F., and Sasaki, S. (1971). Effects of direct contact on *Pinus resinosa* seeds and young seedlings with 2,4-D or picloram on seedling development. *Can. J. Bot.* **49,** 1737–1742.

Wuenscher, J. E., and Kozlowski, T. T. (1971). The relationship of gas exchange resistance to tree seedling ecology. *Ecology* **52,** 1016–1023.

Wulff, R. D., and Strain, B. R. (1982). Effects of CO_2 enrichment on growth and photosynthesis in *Desmodium paniculatum*. *Can. J. Bot.* **60,** 1084–1091.

Wyman, D. (1950). Order of bloom. *Arnoldia* **10,** 41–56.

Xiang, K.-F. (1986). Some aspects of establishment and effects of shelterbelt systems in northern China. *In* "Windbreak Technology" (D. L. Hintz and J. R. Brandle, eds.), Great Plains Agric. Counc. Publ. No. 117, pp. 167–170. Rocky Mt. Forest and Range Expt. Sta.

Yadava, D. L., and Doud, S. L. (1980). The short life and replant problems of deciduous fruit trees. *Hortic. Rev.* **2,** 1–116.

Yamamoto, F., and Kozlowski, T. T. (1987a). Effect of ethrel on stem anatomy of *Ulmus americana* seedlings. *IAWA Bull.* [N.S.] **8,** 3–9.

Yamamoto, F., and Kozlowski, T. T. (1987b). Effect of ethrel on growth and stem anatomy of *Pinus halepensis* seedlings. *IAWA Bull.* [N.S.] **8,** 11–19.

Yamamoto, F., and Kozlowski, T. T. (1987c). Effects of flooding, tilting of stems, and ethrel application on growth, stem anatomy and ethylene production of *Pinus densiflora* seedlings. *J. Exp. Bot.* **38,** 293–310.

Yamamoto, F., and Kozlowski, T. T. (1987d). Effect of flooding of soil on growth, stem anatomy, and ethylene production of *Cryptomeria japonica* seedlings. *Scand. J. For. Res.* **2,** 45–50.

Yamamoto, F., and Kozlowski, T. T. (1987e). Effects of flooding of soil and application of NPA and NAA to stems on growth and stem anatomy of *Acer negundo* seedlings. *Environ. Exptl. Bot.* **27,** 329–340.

Yamamoto, F., and Kozlowski, T. T. (1987f). Effects of flooding, tilting of stems, and ethrel application on growth, stem anatomy, and ethylene production of *Acer platanoides* seedlings. *Scand. J. For. Res.* **2,** 141–156.

Yamamoto, F., Kozlowski, T. T., and Wolter, K. E. (1987). Effect of flooding on growth, stem anatomy and ethylene production of *Pinus halepensis* seedlings. *Can. J. For. Res.* **17,** 69–79.

Yarwood, C. E. (1959). Predisposition. *In* "Plant Pathology" (J. G. Horsfall and A. E. Diamond, eds.), Vol. 1, pp. 521–562. Academic Press, New York.

Yarwood, C. E. (1976). Modification of the host response—Predisposition. *Encycl. Plant Physiol., New Ser.* **4,** 703–718.

Yelenosky, G. (1964). Tolerance of trees to deficiencies of soil aeration. *Proc. Int. Shade Tree Conf.* **40,** 127–147.

Yetter, T. C., and Runkle, J. R. (1986). Height growth rates of canopy tree species in southern Appalachian gaps. *Castanea* **5,** 157–167.

Ying, C. C., and Bagley, W. T. (1977). Variation in rooting capability of *Populus deltoides*. *Silvae Genet.* **26,** 204–207.

Yogaratnam, N., and Johnson, D. S. (1982). The application of foliar sprays containing nitrogen, magnesium, zinc and boron to apple trees. II. Effects on the mineral composition and quality of the fruit. *J. Hortic. Sci.* **57,** 159–164.

Yoshida, S. (1974). Studies on lipid changes associated with frost hardiness in cortex in woody plants. *Contrib. Inst. Low Temp. Sci., Hokkaido Univ., Ser. B* **18,** 1–43.

Yoshida, S., and Sakai, A. (1973). Phospholipid changes associated with the cold hardiness of cortical cells from poplar stem. *Plant Cell Physiol.* **14,** 353–359.

Young, H. E., and Kramer, P. J. (1952). The effect of pruning on the height and diameter growth of loblolly pine. *J. For.* **50,** 474–479.

Youngberg, C. T. (1959). The influence of soil conditions following tractor logging on the growth of planted Douglas-fir seedlings. *Soil Sci. Soc. Am. Proc.* **23,** 76–78.

Youngberg, C. T., and Wollum, A. G. (1970). Nonleguminous symbiotic nitrogen fixation. *In* "Tree Growth and Forest Soils" (C. T. Youngberg and C. B. Davey, eds.), pp. 383–395. Oregon State University Press, Corvallis.

Zabadal, T. J. (1974). A water potential threshold for the increase of abscisic acid in leaves. *Plant Physiol.* **53,** 125–127.

Zaerr, J. B. (1971). Moisture stress and stem diameter in young Douglas-fir. *For. Sci.* **17,** 466–469.

Zaerr, J. B. (1983). Short-term flooding and net photosynthesis in seedlings of three conifers. *For. Sci.* **29,** 71–78.

Zahner, R. (1968). Water deficits and growth of trees. *In* "Water Deficits and Plant Growth" (T. T. Kozlowski, ed.), Vol. 2, pp. 191–254. Academic Press, New York.

Zahner, R., and Whitmore, F. W. (1960). Early growth of radically thinned loblolly pine. *J. For.* **58,** 628–634.

Zahner, R., Myers, R. K., and Hutto, C. J. (1985). Crop tree quality in young piedmont oak stands of sprout origin. *South J. Appl. For.* **9,** 15–20.

Zangerl, A. R., and Bazzaz, F. A. (1984). The response of plants to elevated CO_2. II. Competitive interactions among annual plants under varying light and nutrients. *Oecologia* **62,** 412–417.

Zavitkovski, J. (1979). Energy production in irrigated, intensively cultured plantations of *Populus*'Tristis #1' and jack pine. *For. Sci.* **25,** 383–392.

Zeiger, E., Farquhar, G. D., and Cowan, I. R., eds. (1987a). "Stomatal Function." Stanford Univ. Press, Stanford, California.

Zeiger, E., Moritoshi, I., Shimazaki, K., and Ogawa, T. (1987b). The blue-light response of stomata: Mechanism and function. *In* "Stomatal Function" (E. Zeiger, G. D. Farquhar, and I. R. Cowan, eds.), pp. 209–228. Stanford Univ. Press, Stanford, California.

Zekri, M., and Parsons, L. R. (1988). Water relations of grapefruit trees in response to drip, micro-sprinkler, and overhead sprinkler irrigation. *J. Am. Soc. Hortic. Sci.* **113,** 819–823.

Zekri, M. and Parsons, L. R. (1989). Grapefruit leaf and fruit growth in response to drip, microsprinkler, and overhead sprinkler irrigation. *J. Am. Soc. Hortic. Sci.* **114,** 25–29.

Zelawski, W., and Kucharska, J. (1969). Winter depression of photosynthetic activity in seedlings of Scots pine (*Pinus silvestris* L.). *Photosynthetica* **1,** 207–213.

Zhang, J. and Davies, W. J. (1989a). Sequential response of whole plant water relations to prolonged soil drying and the involvement of xylem sap ABA in the regulation of stomatal behaviour of sunflower plants. *New Phytol.* **113,** 167–174.

Zhang, J., and Davies, W. J. (1989b). Abscisic acid produced in dehydrating roots may enable the plant to measure the water status of the soil. *Plant, Cell Environ.* **12,** 73–81.

Zhang, J., Schuur, U., and Davies, W. J. (1987). Control of stomatal behaviour by abscisic acid which apparently originates in the roots. *J. Exp. Bot.* **38,** 1174–1181.

Zimmermann, M. H. (1964). Effect of low temperature on ascent of sap in trees. *Plant Physiol.* **39,** 568–572.

Zimmermann, M. H. (1978). Hydraulic architecture of some diffuse porous trees. *Can. J. Bot.* **56,** 2286–2295.

Zimmermann, M. H. (1983). "Xylem Structure and the Ascent of Sap." Springer-Verlag, Berlin.

Zimmermann, M. H., and Brown, C. L. (1971). "Trees: Structure and Function". Springer-Verlag, Berlin and New York.

Zimmermann, M. H., and Milburn, J. A., eds. (1975). "Transport in Plants." Springer-Verlag, Berlin.

Zimmermann, M. H., and Potter, D. (1982). Vessel-length distribution in branches, stems, and roots of *Acer rubrum* L. *IAWA Bull.* [N.S.] **3,** 103–109.

Zimmermann, R. H. (1985). Application of tissue culture propagation to woody plants. *In* "Tissue Culture in Forestry and Agriculture" (R. R. Henke, K. W. Hughes, M. J. Constantin, A. Hollaender, and C. M. Wilson, eds.). Plenum, New York.

Zisa, R. P., Halverson, H. G., and Stout, B. J. (1980). Establishment and early growth of conifers on compost soils in urban areas. *USDA For. Serv. Res. Pap. NE* **NE–451.**

Zobel, B. J., and Talbert, J. (1984). "Applied Forest Tree Improvement." Wiley, New York.

Zobel, B. J., and van Buijtenen, J. P. (1989) "Wood Variation: Its Causes and Control." Springer-Verlag, Berlin.

Zobel, B. J., van Wyk, G., and Stahl, P. (1987). "Growing Exotic Trees." Wiley, New York.

Zohar, Y. (1985). Root distribution of a eucalypt shelterbelt. *For. Ecol. Manage.* **12,** 305–307.

Zon, R. (1907). A new explanation of the tolerance and intolerance of trees. *Proc. Soc. Am. For.* **2,** 79–94.

Common Name Index

Abelia, glossy	*Abelia* × *grandiflora* (Andre) Rehd.
Abui	*Pouteria ramiflora* Radlk.
Acacia	*Acacia* Mill. spp.
African	*Acacia senegal* (L.) Willd.
apple-ring	*Acacia albida* Delile
Actinidia	*Actinidia* Lindl. spp.
Afara	*Terminalia superba* Engl. and Diels
Ailanthus	*Ailanthus altissima* (Mill.) Swingle
Albizia	*Albizia* Durrazz. spp.
Alder	*Alnus* B. Ehrh. spp.
red	*Alnus rubra* Bong. (*A. oregona* Nutt.)
gray	*Alnus incana* (L.) Moench
speckled	*Alnus rugosa* (Du Roi) K. Spreng.
European black, black, European	*Alnus glutinosa* (L.) Gaertn.
Sitka	*Alnus sinuata* (Regel) Rydb.
Almond	*Prunus dulcis* var. *dulcis* (Mill.) D.A. Webb
Amla	*Emblica officinalis* Gaertn. (*Phyllanthus emblica* L.)
Apesearring	*Pithecelobium racemosum* Ducke
Apple	*Malus* Mill. spp.
Apple	*Malus domestica* Bork. (*M. pumila* Mill.; *M. sylvestris* (L.) Mill.)
Apple, Balsam	*Clusia rosea* Jacq.
Apricot	*Prunus armeniaca* L.
Japanese	*Prunus mume* Siebold and Zucc.
Arborvitae, eastern (see Northern white cedar)	
Arborvitae, giant (see Western red cedar)	
Arborvitae, globe	*Gliricidia* H. B. & K. spp.
Arborvitae, Oriental	*Platycladus orientalis* (L.) Franco
Arrowwood	*Viburnum* L. spp.
Ash, (see also Eucalyptus, Mountain ash)	*Fraxinus* L. spp.
green or red	*Fraxinus pennsylvanica* Marsh.
white or American	*Fraxinus americana* L.
European	*Fraxinus excelsior* L.
Ash, European green	*Alnus viridis* (Chaix) D.C.
Asparagus	*Asparagus aphyllus* L.
Aspen (see also Cottonwood, Poplar)	*Populus* L. spp
bigtooth	*Populus grandidentata* Michx.
trembling or quaking	*Populus tremuloides* Michx.
European	*Populus tremula* L.
Aster	*Aster* L. spp.
Araucaria	*Araucaria* Juss. spp.
Avocado	*Persia americana* Mill.
Azalea	*Rhododendron* L. spp.

Common name	Scientific name
Hiryu	*Rhododendron obtusum* 'Hinodegiri' (Lindl.) Planch.
Balsa	*Ochroma lagopus* Sw.
Bamboo	*Bambuseae* Lindl. tribe
Japanese dwarf	*Sasa borealis* (Hack.) Makino
Banana	*Musa* L. spp.
Baobab tree	*Adansonia digitata* L.
Barbasco	*Magonia pubescens* A. St. Hill
Barberry, Japanese	*Berberis thunbergii* D.C.
Basswood (see also Linden)	*Tilia* L. spp.
American	*Tilia americana* L.
Bayberry (see Wax myrtle)	
Bean	*Phaseolus* L. spp.
Bean, Coral	*Erythrina* L. spp.
Bean, Locust	*Parkia cleppertoniana*
Bear bush	*Garrya fremontii* Torr.
Bears's-breech	*Acanthus* L. spp.
Beech (see also False-beech)	*Fagus* L. spp.
American	*Fagus grandifolia* J.F. Ehrh.
European	*Fagus sylvatica* L.
Japanese	*Fagus crenata* Blume
Beefwood	*Casuarina* L. ex Adans. spp.
Bermuda grass	*Cynodon dactylon* (L.) Pers.
Birch	*Betula* L. spp.
Ashe's roundleaf	*Betula uber* (Ashe) Fern.
dwarf	*Betula pumila* L.
hairy, white, European or European white	*Betula pubescens* J. F. Ehrh.
Himalayan	*Betula utilis* D. Don
Japanese white	*Betula mandschurica* (Regal) Nakai (*B. platyphylla* Sukachev)
paper, white, or canoe	*Betula papyrifera* Marsh.
river	*Betula nigra* L.
silver	*Betula pendula* Roth. (*Betula verrucosa* J.F. Ehrh.)
sweet	*Betula lenta* L.
yellow	*Betula alleghaniensis* Britt.
Blackberry	*Rubus* L. spp.
Blackbryony	*Tamus communis* Link.
Black mangrove	*Avicennia* L. spp.
Bloodwood or dragon blood tree	*Pterocarpus officinalis* Jacq.
Blueberry	*Vaccinium* L. spp.
Bonit kajang or kayu b'linchi	*Xylopia malayana* Hook. fil. and Thom. fil. Ind.
Borneocamphor	*Dryobalanops aromatica* Gaertn.
Bottlebrush	*Metrosideros polymorphus* Hook.f.
Box	*Buxus sempervirens* L.
Broom	*Cytisus* L. spp.
Scotch	*Cytisus scoparius* (L.) Link
warminster	*Cytisus* × *praecox* Bean
Buckeye	*Aesculus* L. spp.
California	*Aesculus californica* (Spach) Nutt.
Ohio	*Aesculus glabra* Willd.
yellow	*Aesculus octandra* Marsh.
Buckthorn	*Rhamnus* L. spp.

Common name	Scientific name
Chestnut, Moreton Bay	*Castanospermum australe* A. Cunn. and C. Fraser
Chile bells	*Lapageria* Ruiz and Pav. spp.
Chinquapin, Japanese evergreen	*Castanopsis cuspidata* (Thunb.) Schottky
Christmas berry	*Heteromeles arbutifolia* (Ait.) M.J. Roem.
Chrysanthemum	*Chrysanthemum* L. spp.
Cinquefoil	*Potentilla* L. spp.
Cinquefoil, shrubby	*Potentilla fruticosa* L.
Claoxylon	*Claoxylon sandwicense* Muell-Arg.
Clematis	*Clematis* L. spp.
vernal	*Clematis cirrhosa* L.
Cloudberry	*Rubus chamaemorus* L.
Clover	*Trifolium* L. spp.
crimson	*Trifolium incarnatum* L.
red	*Trifolium pratense* L.
subterranean	*Trifolium subterraneum* L.
white	*Trifolium repens* L.
Coconut	*Cocos nucifera* L.
Coffee	*Coffea* L. spp.
Arabian	*Coffea arabica* L.
Coffeeberry	*Rhamnus californica* Eschsch.
Copaltree, Venezuelan	*Copaifera venezuelana* Harms & Pittier
Cork tree	*Phellodendron amurense* Rupr.
Corn	*Zea* L. spp.
Cotoneaster	*Cotoneaster* spp.
creeping	*Cotoneaster adpressa praecox* Bois. and Bert.
rock	*Cotoneaster horizontalis* Decne.
Cotton	*Gossypium* L. spp.
Cottonwood (see also Aspen, Poplar)	*Populus* L. spp.
black	*Populus trichocarpa* Torr. and A. Gray
eastern	*Populus deltoides* Bartr. ex Marsh.
Crabapple (see Apple)	
Cranberry	*Vaccinium macrocarpum* Ait.
Cranberry, bush	*Viburnum opulus* L.
Creosote bush	*Larrea* Cav. spp.
Cress, Mouse-ear	*Arabidopsis* Heynh. spp.
Cryptomeria	*Cryptomeria* D. Don spp.
Japanese	*Cryptomeria japonica* (L.f.) D. Don
Cucumber	*Cucumis sativus* L.
Currant	*Ribes* L. spp.
Cypress	*Cupressus* L. spp.; *Taxodium* Rich. spp; *Chamaecyparis* Spach spp.
Arizona	*Cupressus arizonica* Greene
bald	*Taxodium distichum* (L.) L. Rich
Hinoki	*Chamaecyparis obtusa* (Siebold & Zucc.) Endl.
Sargent's	*Cupressus sargentii* Jeps.
Cypress-pine, white	*Callitris columellaris* F.J. Muell.
Dakua	*Podocarpus vitiensis* Seem.
Dammar-pine	*Agathis* Salisb. spp.
Daphne	*Daphne* L. spp.
Daviesia	*Daviesia mimosoides* (R. Br. in) Ait.
Deutzia	*Deutzia* Thunb. spp.
Dogwood	*Cornus* L. spp.

flowering	*Cornus florida* L.
red-osier	*Cornus stolonifera* Michx. (*C. sericea* L.)
Douglas-fir	*Pseudotsuga menziesii* (Mirb.) Franco
bigcone	*Pseudotsuga macrocarpa* (Vasey) Mayr.
Elder, box	*Acer negundo* L.
Elephant's-ear plant	*Alocasia macrorrhiza* (L.) G. Don
Elm	*Ulmus* L. spp.
American	*Ulmus americana* L.
Chinese	*Ulmus parvifolia* Jacq.
Scotch or Wych	*Ulmus glabra* Huds.
winged	*Ulmus alata* Michx.
Encelia	*Encelia* Adans. spp.
California	*Encelia californica* Nutt.
Escallonia	*Escallonia* Mutis ex L.f. spp.
Eucalyptus	*Eucalyptus* L'Her. spp.
alpine ash	*Eucalyptus delegatensis* R.T. Bak.
black mountain ash	*Eucalyptus sieberi* L.A.S. Johnson
broad-leaved peppermint	*Eucalyptus dives* Schau.- *E. dalrympleana* Maid.
cabbage gum	*Eucalyptus pauciflora* Sieber ex A. Spreng.
forest red gum	*Eucalyptus tereticornis* Sm.
jarrah	*Eucalyptus marginata* Donn ex Sm.
manna gum	*Eucalyptus viminalis* Labill.
Murray red gum, red gum, river red gum	*Eucalyptus rostrata* Schlechtend. (*E. camaldulensis* Dehnh.)
rose gum	*Eucalyptus grandis* W. Hill ex Maiden
snow gum	*Eucalyptus coccifera* Hook.f.
spotted gum	*Eucalyptus maculata* Hook.
Swamp mahogany	*Eucalyptus robusta* Sm.
Sydney blue gum	*Eucalyptus saligna* Sm.
Tasmanian blue gum or blue gum	*Eucalyptus globulus* Labill.
tuart	*Eucalyptus gomphocephala* D.C.
Eucryphia	*Eucryphia lucida* (Labill.) Baill.
Euonymus	*Euonymus* L. spp.
Euonymus, wintercreeper	*Euonymus fortunei* 'Colorata' (Rehd.) Rehd.
Euphorbia	*Euphorbia* L. spp.
False-beech, Tasmanian	*Nothofagus cunninghamia* Oerst.
Fescue, tall	*Festuca arundinacea* Schreb.
Fetterbush	*Leucothoe* D. Don spp.
Fig	*Ficus* L. spp.
Filbert	*Corylus* L. spp.
European	*Corylus avellana* L.
Fir (see also Douglas-fir)	*Abies* Mill. spp.
alpine	*Abies lasiocarpa* (Hook.) Nutt.
balsam	*Abies balsamea* (L.) Mill.
Delavay's silver	*Abies delavayi* var. *faxoniana* (Rehd. & Wils.) Jacks. (*Abies faxoniana* Rehd. & Wils.)
Fraser	*Abies fraseri* (Pursh) Poir.
grand	*Abies grandis* (D. Don ex Lamb.) Lindl.
Maries	*Abies mariesii* M.T. Mast
Nikko	*Abies homolepis* Sieb. & Zucc.
noble	*Abies procera* Rehd.
Pacific silver, silver	*Abies amabilis* Dougl. ex J. Forbes
red, California red	*Abies magnifica* A. Murr.

Sakhalin	*Abies sachalinensis* (Friedr. Schmidt) M.T. Mast
European silver	*Abies alba* Mill.
Veitch	*Abies veitchii* Lindl.
white	*Abies concolor* (Gord.) Lindl. ex Hildebr.
Firethorn	*Pyracantha* M.J. Roem. spp.
scarlet	*Pyracantha coccinea* M.J. Roem.
Forsythia	*Forsythia* Vahl. spp.
Fuchsia	*Fuchsia boliviana* Carriere
Germander	*Teucrium creticum* Lam.
Goldenrod	*Solidago* L. spp.
Gooseberry (see Currant)	
Grama	*Bouteloua* Lag. spp.
blue	*Bouteloua gracilis* (H.B.K.) Lag. ex Steud.
Grape	*Vitis* L. spp.
Muscadine	*Vitis rotundifolia* Michx.
Grapefruit	*Citrus paradisi* Macfady.
Greasewood	*Adenostoma fasciculatum* Hook. and Arn.
Greenbrier, Eurasian	*Smilax aspera* L.
Guamo	*Inga* Benth. & Hook. spp.
Guava	*Psidium guajava* L.
Guayule	*Parthenium argentatum* A. Gray
Gum (see also Tupelo and Eucalyptus)	*Nyssa* L. spp.; *Liquidambar* L. spp.
black	*Nyssa sylvatica* Marsh.
sweet or red	*Liquidambar styraciflua* L.
Gum-tree	*Castilla* Sesse. spp.
Hackberry	*Celtis occidentalis* L.
Hawthorn	*Crataegus* L. spp.
Hay-scented fern	*Dennstaedtia punctilobula* (Michx.) T. Moore
Hazel	*Corylus* L. spp.
Hazel, witch	*Hamamelis virginiana* L.
Hazel, winter	*Corylopsis* Siebold and Zucc. spp.
Heath, sand	*Ceratiola* Michx. spp.
Heather	*Calluma* Salisb. spp.
Hemlock	*Tsuga* Carriere spp.
mountain	*Tsuga mertensiana* (Bong.) Carr.
western	*Tsuga heterophylla* (Raf.) Sarg.
eastern or Canadian	*Tsuga canadensis* (L.) Carr.
Hiba, Japanese (see Oriental arborvitae)	
Hickory	*Carya* Nutt. spp.
mockernut	*Carya tomentosa* Nutt.
pecan	*Carya illinoensis* (Wangenh.) K. Koch
shagbark	*Carya ovata* (Mill.) K. Koch
water or swamp	*Carya aquatica* (Michx.f.) Nutt.
Holly	*Ilex* L. spp.
American	*Ilex opaca* Ait.
inkberry	*Ilex glabra* (L.) A. Gray
Japanese	*Ilex crenata* Thunb.
yaupon	*Ilex vomitoria* Ait.
Honeysuckle	*Lonicera* L. spp.
Etruscan	*Lonicera etrusca* Santi.
Horse chestnut (see also Buckeye)	*Aesculus* L. spp.
horse chestnut	*Aesculus hippocastanum* L.
Hydrangea	*Hydrangea* L. spp.

Idigbo	*Terminalia ivorensis* A. Chevalier
Immortelle, mountain	*Erythrina poeppigiana* (Walp.) O.F. Cook
Indian-cucumber-root	*Medeola virginica* L.
Iripilbark tree	*Pentaclethra macroloba* (Willd.) Kuntze
Iris	*Iris* L. spp.
Ivy, English	*Hedera helix* L.
Java-locust	*Parkia javanica* (Lam.) Merrill.
Jhand	*Prosopis cineraria* (L.) McBride
Jojoba	*Simmondsia chinensis* (Link) C.K. Schneid.
Juniper (see also Eastern red cedar)	*Juniperus* L. spp.
Ashe	*Juniperus ashei* Buchh.
common	*Juniperus communis* L.
creeping	*Juniperus horizontalis* Moench
Kalmia	*Kalmia* L. spp.
Katsura tree	*Cercidiphyllum japonicum* Siebold & Zucc. ex J. Hoffm. & H. Schult.
Keranji	*Dialinum maingayi* Baker
Kudzu	*Pueraria lobata* (Willd.) Ohwi
Kuka or Senegal-mahogany	*Khaya senegalensis* (Desr.) A. Juss.
Larch (see also Tamarack)	*Larix* Mill. spp.
eastern or tamarack	*Laris laricina* (DuRoi) K. Koch
European	*Larix decidua* Mill.
Japanese	*Larix kaempferi* (Lamb) Carrl. (*L. leptolepis* (Sieb. and Zucc.) Gord.)
western	*Larix occidentalis* Nutt.
Laurel, California	*Umbellularia californica* (Hook. and Arn.) Nutt.
Laurel	*Laurus* L. spp.
Laurel, Grecian	*Laurus nobilis* L.
Laurel, Indian	*Ficus rotusa* L.
Leadtree (see Leucaena)	
Lemon, rough	*Citrus limon* (L.) Burm. f.
Lettuce	*Lactuca* L. spp.
Lettuce, head	*Lactuca sativa* L.
Leucaena or leadtree	*Leucaena leucocephala* (Lam.) de Wit
Leucothoe, drooping	*Leucothoe fontanesiana* (Steud.) Sleum.
Lilac	*Syringa vulgaris* L.
Lime, sweet or Tahiti	*Citrus aurantifolia* (Christm.) Swingle 'Tahiti'
Linden (see also Basswood)	*Tilia* L. spp.
littleleaf	*Tilia cordata* Mill.
Locust (see also Java-locust, Bean)	*Robinia* L. spp.; *Gleditsia* L. spp.
black	*Robinia pseudoacacia* L.
honey	*Gleditsia triacanthos* L.
Luaun	*Shorea contorta* Vidal
Lupine	*Lupinus* L. spp.
Macadamia nut	*Macadamia integrifolia* Maiden and Betche
Madder	*Rubia tenuifolia* Urv.
Madre	*Gliricidia sepium* H.B. & K.
Madrone	*Arbutus andrachne* L.
Magnolia	*Magnolia* L. spp.
saucer	*Magnolia* × *soulangiana* Soul.-Bod.
southern	*Magnolia grandiflora* L.
star	*Magnolia stellata* (Siebold and Zucc.) Maxim.
Mahaleb	*Prunus mahaleb* L.

Common name	Scientific name
Maize (see Corn)	
Mango	*Mangifera* L. spp.
Mangrove (see also Black mangrove, Burma mangrove, Button mangrove, White mangrove)	*Rhizophora* L. spp.; *Sonneratia* L. f. spp.
Mangrove, American	*Rhizophora mangle* L.
Mangrove, Burma	*Bruguiera gymnorhiza* Lam.
Manzanita	*Arctostaphylos* Adans. spp. Parry
bigberry	*Arctostaphylos glauca* Lindl.
common	*Arctostaphylos manzanita* Parry
eastwood	*Arctostaphylos glandulosa* Eastw.
greenleaf	*Arctostaphylos patula* Greene
littleberry	*Arctostaphylos sensitiva* Jepson
Mexican pointleaf	*Arctostaphylos pungens* H.B.K.
pinkbract pringle	*Arctostaphylos pringlei* Parry
Stanford	*Arctostaphylos stanfordiana* Parry
whiteleaf	*Arctostaphylos viscida* Parry
woollyleaf	*Arctostaphylos tomentosa* (Pursh) Lindl.
Maple	*Acer* L. spp.
bigleaf	*Acer macrophyllum* Pursh.
black	*Acer nigrum* Michx. f.
Japanese	*Acer palmatum* Thunb.
mountain	*Acer spicatum* Lam.
Norway	*Acer platanoides* L.
red	*Acer rubrum* L.
silver	*Acer saccharinum* L.
striped	*Acer pensylvanicum* L.
sugar	*Acer saccharum* Marsh.
sycamore	*Acer pseudoplatanus* L.
vine	*Acer circinatum* Pursh
Matchwood	*Didymopanax* Decne. and Planch. spp.
Merbau or Kwila	*Intsia palembanica* Miq.
Mesquite	*Prosopis* L. spp.
Miconia	*Miconia albicans* (Sw.) Triana
Mimosa (see Albizia)	
Mint	*Mentha* L. spp.
Minyak (see Saputi)	
Monkey flower, bush	*Diplacus aurianticus* (Curtis) Jeps.
Mountain ash	*Sorbus* L. spp.
American	*Sorbus americana* Marsh.
European	*Sorbus aucuparia* L.
white beam	*Sorbus aria* (L.) Crantz.
Mulberry	*Morus* L. spp.
Mulberry, red	*Morus rubra* L.
Mulga	*Acacia aneura* F.J. Muell.
Myrtle, crepe	*Lagerstroemia* L. spp.
Myrtle, Oregon (see California laurel)	
Myrtle, wax	*Myrica* L. spp.
Nectarine (see Peach)	
Nothofagus	*Nothofagus solandri* (Hook.f.) Orst.
Oak	*Quercus* L. spp.
Israelian	*Quercus infectoria* var. *boisseri* (Reuter) Gürke

ballutah, hudzi gerdze pelit, gisyl gerdze pelit — *Quercus aegilops* L. var. *ithaburensis* (Decaisne) Boissier
black — *Quercus velutina* Lam.
blackjack — *Quercus marilandica* Muenchh.
bur — *Quercus macrocarpa* Michx.
California scrub — *Quercus dumosa* Nutt.
California live — *Quercus agrifolia* Nee
canyon live — *Quercus chrysolepis* Liebm.
cherrybark — *Quercus falcata* var. *pagodaefolia* Elliott
chestnut — *Quercus prinus* L.
chinkapin or chinquapin — *Quercus muehlenbergii* Engelm.
cork — *Quercus suber* L.
English — *Quercus robur* L.
interior live — *Quercus wislizenii* A. D.C.
laurel — *Quercus laurifolia* Michx.
leather — *Quercus durata* Jeps.
live — *Quercus virginiana* Mill.
northern pin — *Quercus ellipsoidalis* E.J. Hill
northern red or red — *Quercus rubra* L. (*Q. borealis* Michx. f.)
Nuttall — *Quercus nuttallii* Palmer
Oregon white — *Quercus garryana* Dougl. ex Hook.
overcup — *Quercus lyrata* Walt.
Palestine — *Quercus calliprinos* Webb
pin — *Quercus palustris* Muenchh.
post — *Quercus stellata* Wangenh.
scarlet — *Quercus coccinea* Muenchh.
scrub oak — *Quercus inopina* Ashe
southern red — *Quercus falcata* Michx.
swamp chestnut — *Quercus michauxii* Nutt.
swamp white — *Quercus bicolor* Willd.
turkey — *Quercus laevis* L.
white — *Quercus alba* L.
willow — *Quercus phellos* L.
Oak, silk — *Grevillea robusta* A. Cunn. ex R. Br.
Obeche — *Triplochiton scleroxylon* K. Schum.
Okra — *Abelmoschus esculentus* (L.) Moench
Oleander — *Nerium* L. spp.
Oleaster — *Elaeagnus latifolia* L.
Olive — *Olea* L. spp.
Olive, autumn — *Elaeagnus umbellata* Thunb.
Orange (see also Lemon, Lime, Grapefruit, Osage orange) — *Citrus* L. spp.
Mandarin — *Citrus reticulata* Blanco
trifoliate — *Poncirus trifoliata* (L.) Raf.
Valencia, navel, or sweet — *Citrus sinensis* L.
Osage orange — *Maclura pomifera* (Raf.) Schneid.
Orchard grass — *Dactylis glomerata* L.
Origanum — *Majorana syriaca* L. (*Origanum majorana* L.)
Pachysandra, Japanese — *Pachysandra terminalis* Siebold and Zucc.
Paduk — *Pterocarpus* Jacq. spp.
Palm, oil — *Elaeis guineensis* Jacq.
Palo d'agua — *Heliocarpus* L. spp.

red	*Pinus resinosa* Ait.
Rocky Mountain bristlecone	*Pinus aristata* Engelm.
sand	*Pinus clausa* (Chapm.) Vasey
Scots or Scotch	*Pinus sylvestris* L.
shortleaf	*Pinus echinata* Mill.
singleleaf pinyon	*Pinus monophylla* Torr. & Frem.
slash	*Pinus elliottii* Engelm.
sugar	*Pinus lambertiana* Dougl.
Swiss stone	*Pinus cembra* L.
Table mountain	*Pinus pungens* Lamb.
Virginia	*Pinus virginiana* Mill.
western white	*Pinus monticola* D. Don
Pistache, lentisk	*Pistacia lentiscus* L.
Pistache, Palestine	*Pistacia palaestina* Boiss.
Plane or planetree (see Sycamore)	
Plantain	*Plantago* L. spp.
Plum	*Prunus* L. spp.
cherry	*Prunus cerasifera* J.F. Ehrh.
Japanese	*Prunus salicina* Lindl.
Podocarpus	*Podocarpus* L'Her. ex Pers. spp.
Poplar (see also Aspen, Cottonwood)	*Populus* L. spp.
balsam	*Populus balsamifera* L.
brown twig	*Populus* × *tristis* Fisch. (*P.* 'Tristis')
Carolina or hybrid black	*Populus* × *euramericana* Guinier (*P. deltoides* Bartr. ex Marsh. × *P. nigra* L.)
hybrid	*Populus* × *petrowskiana* (Regel) C.K. Schneid.
hybrid	*Populus deltoides* × *P. trichocarpa* Torr. and A. Gray
hybrid	*Populus deltoides* Bartr. ex Marsh. × *P. caudina* Tenore
hybrid	*Populus nigra* L. × *P. trichocarpa* Torr. and A. Gray
hybrid	*Populus* × *robusta* C.K. Schneid.
Lombardy	*Populus nigra* var. *italica* Moench
Maine	*Populus candicans* Ait. (*P. balsamifera* L. × *P. berolinensis* Dipp.)
yellow (see Tulip poplar)	
Potato	*Solanum tuberosum* L.
Potato, sweet	*Ipomoea batatas* (L.) Lam.
Prasium	*Prasium majus* L.
Prune	*Prunus* L. spp.
Pussyroot	*Aplectrum hyemale* (Muhlenb. ex Willd.) Torr.
Quince	*Cydonia* Mill. spp.
Quince, Japanese	*Chaenomeles speciosa* (Sweet) Nakai
Raspberry (see Blackberry)	
Redberry	*Rhamnus crocea* Nutt.
Redbud	*Cercis* L. spp.
eastern	*Cercis canadensis* L.
Judastree	*Cercis siliquastrum* L.
Redwood, coast	*Sequoia sempervirens* (D. Don) Endl.
Red gum	*Liquidambar styraciflua* L.
Rhododendron (see also Azalea)	*Rhododendron* L. spp.

Alpine-rose	*Rhododendron ferrugineum* L.
Carolina	*Rhododendron carolinianum* Rehd.
Catawba	*Rhododendron catawbiense* Michx.
Korean	*Rhododendron mucronulatum* (Blume) G. Don
Rice	*Oryza* L. spp.
Rose	*Rosa* L. spp.
Rose, rock	*Cistus salviifolius* L., *Cistus villosus* L. (*C. incana* L.)
Rose, daphne	*Daphne cneorum* L.
Rosemary, Florida	*Ceratiola ericoides* Michx.
Rosewood	*Dalbergia* L. f. spp.
Rubber	*Hevea* Aubl. spp.
Ryegrass, perennial	*Lolium perenne* L.
Sage	*Salvia* (Tourn.) L. spp.
black	*Salvia mellifera* Greene
Greek	*Salvia triloba* L.f.
Sagebrush	*Artemisia* L. spp.
Sago	*Cycas circinalis* L.
Salai	*Boswellia serrata* Roxb. ex Colebr.
Salal	*Gaultheria shallon* Pursh
Saltbush	*Atriplex patula* L.
Saputi, minyak or sepetir	*Sindora coriacea* Prain
Sassafras	*Sassafras albidum* (Nutt.) Nees
Savory	*Satureja thymifolia* Scop.
Sequoia, giant	*Sequoiadendron giganteum* Buchh.
Serviceberry or shadbush	*Amelanchier* Medic. spp.
Sesame	*Sesamum* L. spp.
Silk tassel	*Garrya* Lindl. spp.
Silk tree	*Albizia julibrissin* Durrazz.
Smoke tree	*Cotinus* Mill. spp.
Snowbell	*Styrax benzoin* Dryand.
Solomon's-seal, small	*Polygonatum biflorum* (Walt.) Elliott
Sophora	*Sophora japonica* L.
Sorghum	*Sorghum* Moench. spp.
Sourwood	*Oxydendrum arboreum* (L.) DC
Soybean	*Glycine* L. spp.
Spinach	*Spinacia* L. spp.
Spiraea	*Spiraea* L. spp.
Spruce	*Picea* A. Dietr. spp.
black	*Picea mariana* (Mill.) B.S.P.
Colorado blue, blue or Colorado	*Picea pungens* Engelm.
Engelmann	*Picea engelmannii* Parry ex Engelm.
Norway	*Picea abies* (L.) Karst.
red	*Picea rubens* Sarg.
Sakhalin	*Picea glehnii* (Friedr. Schmidt) M.T. Mast
Serbian	*Picea omorika* (Panc.) Purk.
Sitka	*Picea sitchensis* (Bong.) Carr.
white	*Picea glauca* (Moench) Voss
Yeddo	*Picea jezoensis* (Siebold and Zucc.) Carriere
Storax	*Styrax officinalis* L.
Strawberry	*Fragaria* L. spp.

Scientific Name Index

Abelia × *grandiflora* (Andre) Rehd.	glossy abelia
Abelmoschus esculentus (L.) Moench	okra
Abies Mill. spp.	Fir
alba Mill.	European silver fir
amabilis Dougl. ex J. Forbes	Pacific silver fir, silver fir
balsamea (L.) Mill.	balsam fir
concolor (Gord.) Lindl. ex Hildebr.	white fir
delavayi var. *faxoniana* (Rehd. & Wils) Jacks. (*A. faxoniana* Rehd. & Wils.)	Delavay's silver fir
fraseri (Pursh) Poir.	Fraser fir
grandis (D. Don ex Lamb.) Lindl.	grand fir
homolepis Sieb. & Zucc.	Nikko fir
lasiocarpa (Hook.) Nutt.	alpine fir
magnifica A. Murr.	California red fir, red fir
mariesii M.T. Mast	Maries fir
procera Rehd.	noble fir
sachalinensis (Friedr. Schmidt) M.T. Mast	Sakhalin fir
veitchii Lindl.	Veitch fir
Acacia Mill. spp.	Acacia
albida Delile	Apple-ring acacia
aneura F.J. Muell.	mulga
craspedo carpa F. Muell.	
senegal (L.) Willd.	African acacia
Acanthus L. spp.	Bears's-breech
Acer L. spp.	Maple
circinatum Pursh	vine maple
macrophyllum Pursh.	bigleaf maple
negundo L.	box elder
nigrum Michx. f.	black maple
palmatum Thunb.	Japanese maple
pensylvanicum L.	striped maple
platanoides L.	Norway maple
pseudoplatanus L.	sycamore maple
rubrum L.	red maple
saccharinum L.	silver maple
saccharum Marsh.	sugar maple
spicatum Lam.	mountain maple
Actinidia Lindl. spp.	Actinidia
Adansonia digitata L.	baobab tree
Adenostoma fasciculatum Hook and Arn.	greasewood
glandulosa Hook. and Arn.	chamise
Aegialitis Gaertn. spp.	Mangrove
Andropogon virginicus L.	broomsedge
Aesculus L. spp.	Horse chestnut, Buckeye
californica (Spach) Nutt.	California buckeye

glabra Willd.	Ohio buckeye
hippocastanum L.	horse chestnut
octandra Marsh.	yellow buckeye
Agathis Salisb. spp.	Dammar-pine
Agropyron desertorum (Fisch.) Schult.	desert wheatgrass, bunchgrass
Ailanthus altissima (Mill.) Swingle	ailanthus, tree of heaven
Albizia Durrazz. spp.	Albizia, Mimosa
julibrissin Durrazz.	albizia, silk tree
Alnus B. Ehrh. spp.	Alder
glutinosa (L.) Gaertn.	European black alder, European alder, black alder
incana (L.) Moench	gray alder
rubra Bong. (*A. oregona* Nutt.)	red alder
rugosa (Du Roi) K. Spreng.	speckled alder
sinuata (Regel) Rydb.	Sitka alder
viridis (Chaix) D.C.	European green ash
Alocasia macrorrhiza (L.) G. Don	elephant's-ear plant
Amelanchier Medic. spp.	Serviceberry, shadbush
Anacardium occidentale L.	cashew
Antiaris toxicaria Lesch.	upas-tree
Aplectrum hyemale (Muhlenb. ex Willd.) Torr.	pussyroot
Arabidopsis Heynh. spp.	Mouse-ear cress
Arachis hypogaea L.	peanut
Araucaria Juss. spp.	Araucaria
Arbutus andrachne L.	madrone
Arctostaphylos Adans. spp.	manzanita
glandulosa Eastw.	eastwood manzanita
glauca Lindl.	bigberry manzanita
manzanita Parry	common manzanita
patula Greene	greenleaf manzanita
pringlei Parry	pinkbract pringle manzanita
pungens H.B. & K.	Mexican pointleaf manzanita
sensitiva Jepson	littleberry manzanita
stanfordiana Parry	Stanford manzanita
tomentosa (Pursh) Lindl.	woolyleaf manzanita
viscida Parry	whiteleaf manzanita
Asparagus aphyllus L.	asparagus
Aster L. spp.	Aster
Astragalus L. spp.	Milk vetch
Atriplex patula L.	saltbush
Avicennia L. spp.	Black mangrove
Bambuseae Lindl. tribe	Bamboo
Berberis thunbergii D.C.	Japanese barberry
Betula L. spp.	Birch
alleghaniensis Britt.	yellow birch
lenta L.	sweet birch
mandschurica (Regal) Nakai (*B. platyphylla* Sukachev)	Japanese white birch
nigra L.	river birch
papyrifera Marsh.	canoe birch, paper birch, white birch
pendula Roth. (*Betula verrucosa* J.F. Ehrh.)	silver birch
pubescens J.F. Ehrh.	European white birch, European birch, hairy birch, white birch

Scientific name	Common name
pumila L.	dwarf birch
uber (Ashe) Fern.	Virginia round-leaf birch
utilis D. Don	Himalayan birch
Boswellia serrata Roxb. ex Colebr.	salai, andukwood
Bouteloua Lag. spp.	Grama
gracilis (H.B. & K.) Lag. ex Steud.	blue grama
Brassica oleracea L.	Wild cabbage
Bruguiera gymnorhiza Lam.	Burma mangrove
Buddleia L. spp.	Butterfly bush
Buxus sempervirens L.	box
Cactaceae Juss. fam.	Cactus
Cajanus cajan (L.) Millsp.	pigeon pea
Callitris columellaris F.J. Muell.	white cypress-pine
Calocedrus decurrens (Torr.) Florin (*Libocedrus decurrens* Torr.)	incense cedar
Calycotome villosa (Poir.) Lk.	thorny-broom
Camellia japonica L.	camellia
thea (L.) Ktze.	tea
Campsis Lour. spp.	Trumpet creeper
Canangia spp. (D.C.) Hook. f. & T. Thoms., not Aubl.	Ylang ylang
Carica papaya L.	papaya
Carya Nutt. spp.	Hickory
aquatica (Michx. f.) Nutt.	water hickory, swamp hickory
illinoensis (Wangenh.) K. Koch	pecan, pecan hickory
ovata (Mill.) K. Koch	shagbark hickory
tomentosa Nutt.	mockernut hickory
Castanea Mill. spp.	Chestnut
Castanopsis cuspidata (Thung.) Schottky	Japanese evergreen chinquapin
Castanospermum australe A. Cunn. and C. Fraser	Moreton Bay Chestnut
Castilla Sesse. spp.	Gum-tree
Casuarina L. ex Adans. spp.	Beefwood
Catalpa bignonioides Walt.	southern catalpa
Ceanothus L. spp.	Ceanothus
cordulatus Kellogg	mountain whitehorn ceanothus
crassifolius Torr.	hoary leaf ceanothus
cuneatus (Hook.) Nutt.	wedgeleaf ceanothus
foliosus Parry	wavyleaf ceanothus
greggii A. Gray	Gregg ceanothus
integerrimus Hook and Arn.	deerbrush ceanothus
leucodermis Greene	chaparral whitethorn ceanothus
megacarpus Nutt.	big pod ceanothus
velutinus Dougl.	snowbrush ceanothus
Cecropia obtusifolia Bertolini	cecropia
peltata L.	trumpet tree
Cedrus L. spp.	Cedar
atlantica (Endl.) G. Manetti ex Carriere	Atlas cedar
libani A. Rich.	cedar of Lebanon
Celtis laevigata Willd.	sugarberry
occidentalis L.	hackberry
Cephalanthus L. spp.	Buttonbush
Ceratiola Michx. spp.	Sand heath

ericoides Michx.	Florida rosemary
Ceratonia siliqua L.	St. John's bread, Carob
Cercidiphyllum japonicum Sieb. & Zucc. ex J. Hoffm. & H. Schult.	Katsura tree
Cercidium floridum Benth. ex A. Gray	palo verde
Cercis L. spp.	Redbud
canadensis L.	eastern redbud
siliquastrum L.	Judastree redbud
Cereus giganteus Engelm. (*Carnegiea gigantea* Britt. and Rose)	saguaro cactus
Ceriops tagal C. B.	Robinson tangal, tengarl
Chamaecyparis lawsoniana (A. Murr.) Parl.	Port Orford cedar
nootkatensis (D. Don) Spach.	Alaska cedar
obtusa (Siebold & Zucc.) Endl.	Hinoki cypress
Chrysanthemum L. spp.	Chrysanthemum
Cicer arietinum L.	chick-pea
Cistus salviifolius L.	rock rose
villosus L. (*C. incanus* L.)	rock rose
Citrus L. spp.	Citrus, Lemon, Lime, Grapefruit, Orange
sinensis L. 'Valencia'	Valencia orange, sweet orange
aurantifolia (Christm.) Swingle	sweet lime, Tahiti lime
limon (L.) Burm. f.	rough lemon
paradisi Macfady.	grapefruit
reticulata Blanco 'Cleopatra'	tangarine, Cleopatra mandarin orange, spice orange
Claoxylon sandwicense Muell-Arg.	claoxylon
Clematis L. spp.	Clematis
cirrhosa L.	vernal clematis
Clidemia sericea D. Don	
Clusia L. spp.	
Clusia rosea Jaq.	balsam apple
Cocos nucifera L.	coconut
Coffea L. spp.	Coffee
arabica L.	Arabian coffee
Colocasia esculenta (L.) Schott	taro
Conocarpus erectus L.	button mangrove
Copaifera venezuelana Harms & Pittier	Venezuelan copaltree
Copernicia cerifera (Arr. Cam.) Mart.	palm
Cornus L. spp.	Dogwood
florida L.	flowering dogwood
stolonifera Michx. (*C. sericea* L.)	red-osier dogwood, red-stemmed dogwood
Corylopsis Siebold and Zucc. spp.	Winter hazel
Corylus L. spp.	Hazel, Filbert
avellana L.	European hazel
Cotinus Mill. spp.	Smoke tree
Cotoneaster B. Ehrh. spp.	Cotoneaster
adpressa praecox Bois. and Bert.	creeping cotoneaster
horizontalis Decne.	rock cotoneaster
Crataegus L. spp.	Hawthorn
Cryptomeria D. Don spp.	Cryptomeria
japonica (L. f.) D. Don	Japanese cryptomeria
Cucumis sativus L.	cucumber
Cupressus L. spp.	Cypress

Myrica L. spp.	Wax myrtle
Nandina Thunb. spp.	Nandina
Nerium L. spp.	Oleander
Nothofagus cunninghamia Oerst.	Tasmanian false-beech
solandri (Hook. f.) Orst.	nothofagus
Nyssa L. spp.	Gum, Tupelo
aquatica L.	water tupelo
sylvatica Marsh.	black gum, tupelo gum
sylvatica var. *biflora* (Walt.) Sarg.	swamp black tupelo
Ochroma lagopus Sw.	balsa
Olea L. spp.	Olive
Oryza L. spp.	Rice
Oxydendrum arboreum (L.) D.C.	sourwood
Pachysandra terminalis Siebold and Zucc.	Japanese pachysandra
Parkia javanica (Lam.) Merrill.	Java-locust
cleppertoniana	Locust-bean
Parthenium argentatum A. Gray	guayule
Paulownia tomentosa (Thunb.) Sieb. and Zucc.	royal paulownia, princess tree
Pentaclethra macroloba (Willd.) Kuntze	iripilbark tree
Persea americana Mill.	avocado
Phaseolus L. spp.	Bean
Phellodendron amurense Rupr.	Cork tree
Phillyrea media (L.) C.K. Schneid.	phillyrea
Photinia Lindl. spp.	Photinia
Picea A. Dietr. spp.	Spruce
abies (L.) Karst.	Norway spruce
engelmannii Parry ex Engelm.	Engelmann spruce
glauca (Moench) Voss	white spruce
glehnii (Friedr. Schmidt) M.T. Mast	Sakhalin spruce
jezoensis (Siebold and Zucc.) Carriere	Yeddo spruce
mariana (Mill.) B.S.P.	black spruce
omorika (Panc.) Purk.	Serbian spruce
pungens Engelm.	blue spruce, Colorado blue spruce, Colorado spruce
rubens Sarg.	red spruce
sitchensis (Bong.) Carr.	Sitka spruce
Pickeringia montana Nutt. ex Torr. and A. Gray	chaparral pea
Pieris D. Don. spp.	Pieris
floribunda (Pursh ex Sims) Benth. and Hook.	mountain pieris
japonica (Thunb.) D. Don ex G. Don	Japanese pieris
Pinus L. spp.	Pine
aristata Engelm.	Rocky Mountain bristlecone pine
attenuata Lemm.	knobcone pine
banksiana Lamb.	jack pine
caribaea var. *hondurensis* Morelet	Carib pine
cembra L.	Swiss stone pine
cembroides var. *edulis* Zucc.	Mexican stone pine
clausa (Chapm.) Vasey	sand pine
contorta Dougl. ex Loud.	lodgepole pine
coulteri G. Don	Coulter pine
densiflora Sieb. and Zucc.	Japanese red pine
echinata Mill.	shortleaf pine
elliottii Engelm.	slash pine

Triplochiton scleroxylon K. Schum.	obeche
Triticum L. spp.	Wheat
Tsuga Carriere spp.	Hemlock
canadensis (L.) Carr.	eastern hemlock, Canadian hemlock
heterophylla (Ref.) Sarg.	western hemlock
mertensiana (Bong.) Carr.	mountain hemlock
Typha latifolia L.	cattail
Ulmus L. spp.	Elm
alata Michx.	winged elm
americana L.	American elm
glabra Huds.	Scotch elm, Wych elm
parvifolia Jacq.	Chinese elm
Umbellularia californica (Hook. and Arn.) Nutt.	California laurel, Oregon myrtle
Vaccinium L. spp.	Blueberry
macrocarpum Ait.	cranberry
Viburnum carlesii Hemsl.	Koreanspice viburnum
opulus L.	bush cranberry
Vinca minor L.	periwinkle
Vitis L. spp.	Grape
rotundifolia Michx.	Muscadine grape
Weigela florida (Bunge) A. D.C.	weigela
Xylopia malayana Hook. fil. and Thom. fil. Ind.	bonit kajang, kayu b'linchi
Zea L. spp.	Corn

Subject Index